The Geological Society of America
Memoir 162

The Caribbean-South American Plate Boundary and Regional Tectonics

Edited by

William E. Bonini
Robert B. Hargraves
Reginald Shagam

1984

Published by The Geological Society of America, Inc.
3300 Penrose Place, P.O. Box 9140, Boulder, Colorado 80301

Printed in U.S.A.

Library of Congress Cataloging in Publication Data
Main entry under title:

The Caribbean-South American plate boundary and regional
    tectonics.

    (Memoir ; 162)
    Includes bibliographies.
    1. Plate tectonics—Addresses, essays, lectures.
2. Geology—Caribbean Area—Addresses, essays, lectures.
3. Geology—South America—Addresses, essays, lectures.
I. Bonini, William E. (William Emory), 1926-
II. Hargraves, Robert B. III. Shagam, Reginald.
IV. Series: Memoir (Geological Society of America) ;
162.
QE511.4.C36   1984   551.1'36'091821   84-5947
ISBN 0-8137-1162-2

Contents

iii

iv

Contents

Preface

The fact that debate continues on the evolution of the Caribbean plate and its borderlands is a reflection of its complexity. Since the disruption of Pangea, the plates comprising this area have been, to some degree, opened and closed, rotated and translated, together with obduction and accretion of displaced terranes. The geology we see today is the integrated product of all these processes.

The diversity of the subject matter covered by papers assembled in this volume is likewise a reflection of this complexity, for all are considered to bear on the tectonics of the Caribbean-South American plate boundary. Diverse approaches, and sources of new data, are nevertheless essential for the ultimate elucidation of this history. Some of the contributions postulate a distant Pacific source for Caribbean crust, and discuss its age, subsequent geochemical evolution, and its lateral translation until the time of collision with the South American plate. Others consider the geometry, timing, and effects of collision, and the subsequent phases of obduction over the craton. Yet other writers view the process of evolution from the vantage point of the craton, from the time prior to collision, through the ensuing orogeny caused by collision, to the youngest phases of Plio-Pleistocene time.

Not entirely unforeseen yet surprising to the editors is the accumulating evidence for increasingly radical, large-scale rotations and translations in the heretofore sacrosanct cratonic "autochthon." Most evidence is for such occurrence prior to the earliest Cretaceous, but there is also evidence that such movement is continuing today.

Where early workers had no alternative but to search in isolation for clues to the tectonic evolution of the region, there are now sufficient signs of a more precise definition of future research directions. Intensive study of small local, but crucial, problems may yield rewards of regional significance as exemplified by some of the short contributions in this volume. Precise definition and attack of crucial problems will be the motto for the research wave of the future, rather than broad regional studies, in which crucial data is come upon by chance. In all likelihood, definition of absolute and relative age relationships of rocks and tectonic units will continue to be of prime importance whether derived from studies of magnetic anomaly stripes, structural relationships, isotopic dating, or paleontological analyses. The increasing evidence for large-scale movements and rotations of terranes indicates that paleomagnetic studies accompanied by accurate age dating will continue to constitute a prime research tool. Petrologic and geochemical studies should also focus on defining and solving specific problems in terranes composed of widely contrasting igneous and metamorphic rock suites. In many cases, the test of tectonic hypotheses will be provided by the stratigraphy and lithology of associated sedimentary formations. Large-scale integration of detailed local studies into unified hypotheses of tectonic evolution will require, however, the aid of precisely defined and executed large-scale geophysical studies.

W. E. Bonini, Princeton University
R. B. Hargraves, Princeton University
R. Shagam, Ben-Gurion University of the Negev

Acknowledgments

The burden of reviewing and editing a volume such as this falls most heavily on those willing to take the time and effort to provide the meaningful critical reviews necessary. We have been most fortunate, and indeed are most grateful, to have had those listed below share in this volume:

T. A. Anderson, Gulf Oil Exploration and Production Co., Houston, T. H. Anderson, University of Pittsburgh; D. R. Baker, Rice University; M. N. Bass, Chevron Oil Field Research Co., La Habra, California; J. S. Bell, BP Exploration Canada, Ltd., Calgary; C. Benjamini, Ben-Gurion University of the Negev, Israel; A. L. Bloom, Cornell University; H. K. Brueckner, Queens College, C.U.N.Y.; K. C. Burke, SUNY at Albany; J. E. Case, U.S. Geological Survey, Menlo Park; T. W. Donnelly, SUNY at Binghamton; A. G. Fischer, Princeton University; J. A. Grow, U.S. Geological Survey, Woods Hole, Massachusetts; S. A. Hall, University of Houston; R. I. Harker, University of Pennsylvania; H. D. Hedberg, Princeton University; W. J. Hinze, Purdue University; S. Judson, Princeton University; J. N. Kellogg, University of Hawaii; D. V. Kent, Lamont-Doherty Geological Observatory, Columbia University; T. A. Konigsmark, Esso Inter-America, Inc., Coral Gables, Florida; J. W. Ladd, Lamont-Doherty Geological Observatory, Columbia University; C. A. Lawson, NASA, Houston; T. M. Lutz, University of Pennsylvania; W. D. MacDonald, SUNY at Binghamton; J. C. Maxwell, University of Texas, Austin; B. A. Morgan III, U.S. Geological Survey, Reston; W. J. Morgan, Princeton University; K. W. Muessig, Exxon Minerals Co., Houston; C. W. Naeser, U.S. Geological Survey, Denver; F. Nagle, University of Miami; T. C. Onstott, Princeton University; A. Salvador, University of Texas, Austin; E. A. Silver, University of California, Santa Cruz; R. Van der Voo, University of Michigan; F. B. Van Houten, Princeton University; R. C. Vierbuchen, Exxon Production Research, Houston; R. A. Zimmermann, U.S. Geological Survey, Denver.

The Geological Society of America has a long history of interest in publications concerning the Caribbean-South American region. The staff at Boulder has continued in this tradition in supplying much needed help and advice in the planning and publication of this Memoir.

This volume would not have been possible without the active support of the faculty, staff, and resources of the Department of Geological and Geophysical Sciences, Princeton University. We are grateful to them. In particular, however, we would like to cite for special thanks the secretarial staff of the department, who cheerfully bore with us throughout the project: Jayne Bialkowski, Kathy Bilous, Jean Olsen, Sally Shaginaw, and Joan Wyckoff.

W. E. Bonini
R. B. Hargraves
R. Shagam

Geological Society of America
Memoir 162
1984

Map of geologic provinces in the Caribbean region

J. E. Case
Geologic Division, Geophysics Branch
U.S. Geological Survey
345 Middlefield Road, MS 989
Menlo Park, California 94025

T. L. Holcombe
Marine Geology Branch
Sea Floor Geosciences Division
Naval Ocean Research and Development Activity
NSTL, Mississippi 39529

R. G. Martin
Gulf Oil Exploration and Production Company
P.O. Box 36506
Houston, Texas 77236

ABSTRACT

The greater Caribbean region has been divided into more than 100 geologic provinces, some of which are tectonostratigraphic terranes or suspect terranes as defined by Coney, Jones, and Monger (1980). The principal criteria for distinguishing provinces are groups of rocks that differ from their immediate neighbors with respect to: (1) rock lithology, thickness, and age, (2) structural style, (3) presence or absence of outcropping igneous rocks, (4) degree of metamorphism, (5) physiographic expression, (6) nature of crust, and other characteristics. Many of the provinces thus identified are bordered by known major faults, including suture and transform zones; other provinces are bordered by unexposed or cryptic faults; and still other boundaries are drawn on the basis of major changes in rock facies, and so are not boundaries of tectonostratigraphic terranes. Paleomagnetic data, far from complete in the region, indicate that many of the provinces have experienced large tectonic translations and rotations.

Colors and patterns have been used on the map in an attempt to portray provinces that appear to be geologically similar in rock type and age, and style and age of principal deformation. For example, Neogene accretionary prisms of the Lesser Antilles, North Caribbean, South Caribbean, North Panama, and Pacific margin deformed belts are shown by the same color, with slight variations in pattern to indicate apparent structural differences between the deformed belts.

Many provinces defined here can be further subdivided on the basis of existing information, and many changes will be required as new data accumulate.

INTRODUCTION

Since about 1965, various investigators have been compiling maps of tectonostratigraphic terranes of the western United States, Canada, and Mexico. These compilations have been spurred by evidence from paleomagnetic and paleobiogeographic studies that some packets of rocks have been displaced by very large translations or rotations from their sites of formation. Some postulated tranlations are in terms of thousands of kilometers; some postulated rotations are more than 90 degrees. Such displacements are presumably related to general plate-tectonics processes, but the paleogeographic origins of kindred terranes and the mechanisms and rates of displacements are poorly understood.

For this map (in pocket inside back cover) of the Caribbean region, geologic provinces (also termed terranes or blocks), including distinct tectonostratigraphic terranes, were established on the basis of: (1) nature of the internal rock sequence, (2) style and degree of deformation, (3) nature and thickness of crust, where known, (4) presence or absence of Phanerozoic igneous activity, (5) degree and kind of metamorphism, (6) presence or absence and kind of seismicity, (7) paleomagnetic evidence for displacement or rotation, (8) paleobiogeographic evidence for displacement or rotation, (9) isotopic age evidence, especially in shield areas, (10) nature of contacts with adjacent provinces, (11) present or past depositional environments, (12) presence or absence of diagnostic suites of mineral deposits, (13) occurrence of large accumulations of hydrocarbons, (14) geomorphology, and so forth. Most provinces defined on the map are Late Mesozoic and Cenozoic features, although a few Paleozoic and Precambrian crystalline provinces are identified and briefly discussed.

On the map, many of the province boundaries are poorly known and located. Some boundaries are faults, some are unconformities, some are depositional contacts, some are inferred from changes in style of deformation or geomorphic setting, and others are based on seismic, gravity, or magnetic data. Many provinces have distinct subdivisions, of which only a few are shown, and more will undoubtedly be identified as more data accumulate. Some provinces merge with or overlap adjacent ones, and many boundaries shown are speculative. Many of the structural blocks in northern South America are similar to those defined by Campbell (1974).

For western North America, various maps and discussions of tectonostratigraphic terranes have been presented by Hamilton (1969), Jones and others (1981), and Coney, Jones, and Monger (1980), and should be consulted for philosophical background. This map of the Caribbean region includes all geologic provinces, whether they are allochthonous or not with respect to their neighbors; in this report, "allochthonous" refers not only to terranes displaced along thrusts but also to terranes displaced along large strike-slip faults. Provinces greatly allochthonous with respect to one or more neighbors are marked by a star (*) in the descriptions.

For the Caribbean region, some principal modern sources of data are a tectonic map of northern South America (Martín, 1978); a volume on Venezuela by González de Juana (1980); a geologic-tectonic map (Case and Holcombe, 1980); a recent volume on the geology of Central America (Weyl, 1980); another volume on the structural geology of the Caribbean region (Butterlin, 1977); a geological review of the Andes (Zeil, 1979); a volume on the Gulf of Mexico and the Caribbean region (Nairn and Stehli, 1975); a volume on the geology and geophysics edited by Weaver (1977); volumes on the geology of Mexico by López Ramos (1979–1982); and a volume on the tectonics and dynamics of the Caribbean prepared by the Soviet Geophysical Committee, Academy of Sciences, USSR. A major summary of circum-Caribbean magmatism was edited by Alan Smith (1980). The literature on the region, both on land and offshore, is voluminous. Most pertinent recent papers are cited in bibliographies of the above sources. Key references for each province are listed in this report, but the emphasis is on reports published since 1960.

Crustal types: Crustal types are defined as follows: *Continental crust* is of low average density, 2.7-2.9 g/cm^3, has an intermediate to silicic bulk composition in at least its upper part, contains a high proportion of pre-Mesozoic metamorphic and igneous rocks, and is 20 to 45 km thick as determined from limited refraction data and gravity models. *Normal oceanic crust* is mafic, ultramafic, of high density, 2.85-3.0 g/cm^3, and typically is 6-10 km thick (depth to M-discontinuity = 10-15 km below sea level). *Undefined crust* has properties that are intermediate between those of oceanic and continental crust; in some areas it may represent rifted crust; *multiple oceanic crust*, a term used informally in this report, refers to mafic crust causing large positive gravity anomalies and which appears to be not only dense but substantially thicker than normal oceanic crust (thicknesses of 20-30 km). Such thickened crust may have originated by imbricate stacking of slabs of oceanic crust, by volcanic loading, or by downfolding in areas of bidirectional convergence (like a tectogene). Crust under the deep basins of th Caribbean Sea, as identified by seismic refraction, is 10-20 km thick, somewhat thicker than typical oceanic crust. A generalized map of inferred depth to the Mohorovičić discontinuity (depth to M) and crustal character is shown on Figure 1. Alternative, more genetic, classifications of the Caribbean crust have been presented by MacDonald (1972) and others. Many basins have been identified in the Caribbean region; the genesis of some is fairly clear, but others have had a complex history related to changes in tectonic regime with time. Terms such as "fore arc," "back arc," or "pull-apart" are used informally.

In this report, the Caribbean provinces around principal cratonic areas are described as follows; in general, provinces of older rocks and older episodes of deformation will be described first, followed by successively younger belts:

A. Guayana Shield, and adjacent provinces that form broadly arcuate patterns around the northern shield

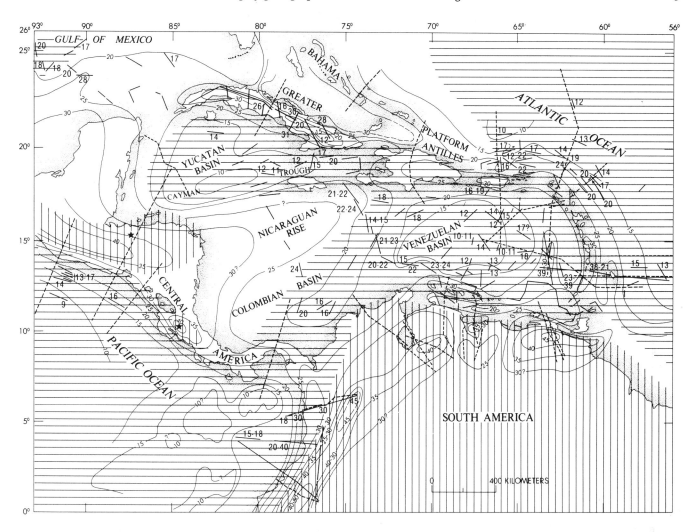

Figure 1. Highly schematic map showing approximate depth to the M-discontinuity in the Caribbean region. Solid straight line segments represent lines of seismic refraction; small numbers indicate depth to layers of seismic velocity of 7.8 km/sec or greater. Lines without numbers are refraction sites where mantle velocities were not detected. Short-dashed lines indicate areas where gravity models were used to control the contours of depth. Stars indicate crustal determinations from local earthquake data. Depths are in kilometers below sea level. Data are from many sources, including: Antoine and Ewing (1963), Arden (1975), Barday (1974), Birch (1970), Bonini (1978), Bonini and others (1977), Bovenko and others (1980), Bowin (1976), Boynton and others (1979), Briceño-Guarupe (1978), Bunce and Fahlquist (1962), Case and Donnelly (1976), Case and others (1971, 1973), Couch (1976), Diebold and others (1981), Dillon and others (1972), Edgar and others (1971), Ewing and others (1957, 1960), Ewing and Worzel (1954), Folinsbee (1972), Fluh and others (1981a), Gettrust and others (1978), Hersey and others (1952), Ibrahim and others (1979, 1981), Keary (1974), Kellogg and Bonini (1982), Kim and others (1979), Ladd and others (1978), Martin and Case (1975), Matumoto and Latham (1977), Matumoto and others (1976), Meissner and others (1976), Meyer and others (1976), Officer and others (1957, 1959), Reblin (1973), Shcherbakova and others (1978, 1980), Shein and others (1978), Shor and Fisher (1961), Silver and others (1975), Soloviev and others (1964a), Talwani and others (1959), Uchupi and others (1971), Victor (1976), Westbrook (1975), Woodcock (1975), Worzel and Ewing (1952). Long vertical pattern indicates areas of continental crust; long horizontal pattern indicates areas of oceanic crust; unpatterned areas are areas where data are scant, and represent undefined or "transitional" crust.

which extend outward to the Caribbean basins on the north and the Cordillera Occidental on the west.

B. Chortis block or southern Nuclear Central America, south of the Motagua-Polochic transform system, and adjacent provinces to the east on the Nicaraguan Rise.

C. Provinces related to convergence between the Cocos plate and Nuclear Central America. Provinces related to convergence between southern Central America, northwest Colombia and the Nazca plate.

D. Motagua transform fault zone.

E. Chiapas Massif and adjacent provinces.

F. Gulf of Mexico provinces.

G. Bahama Platform-Greater Antilles deformed belts.

H. Nicaraguan Rise and vicinity.

I. Interior Caribbean basins and vicinity.

In general, for each province, the brief descriptions include statements about principal rock types and ages, style and age of deformation, postulated nature and thickness of crust, and some recent references. Each province described in the text is keyed numerically to the map.

This report is not intended to be a tectonic synthesis or model, but rather, a brief objective summary of the disposition of major stratigraphic-structural blocks. We believe that understanding the geologic evolution of the region requires lithofacies maps of appropriate time intervals, compiled on maps showing the present geography, and on paleogeographic reconstructions such as those attempted by Burke and others in this volume.

ESCUDO DE GUAYANA PROVINCES

These provinces are described only briefly because criteria for province (or terrane) distinction in Precambrian complexes have not yet been established. The relative significance of emplacement, intrusive, metamorphic, and deformational events as compared to Phanerozoic terranes requires much further study. The most complete descriptions of these Precambrian rocks and events have been provided by Martín Bellizzia (1974), Mendoza (1977), and Martín (1978); a useful review was provided by Gibbs and Barron (1983): much of the material here was prepared by T. C. Onstott (written communication, 1982), and tectonostratigraphic provinces have been proposed by Onstott and Hargraves (1982). Crust for most provinces is continental, but oceanic elements are present in some. No determinations of crustal thickness by refraction surveys have been made. The Guayana Shield had been relatively stable tectonically in terms of compressional tectonics during most of Phanerozoic time, except for its relative westward movement from Africa since Mesozoic opening of the Atlantic Ocean. Vertical uplift, however, was substantial during Cretaceous and Cenozoic time. Precambrian rocks, probably elements of the Guayana Shield, are also exposed in the much younger Cordilleras Central and Oriental of Colombia, in the Sierra de Perijá, Cordillera de Mérida, and possibly in the Cordillera de la Costa of Venezuela. Shield elements underlie sub-Andean basins of Colombia and Venezuela.

*1. **Imataca:** This province includes metasedimentary and metaigneous rocks of amphibolite to granulite facies, and includes widespread iron-formation which has been extensively exploited. The province was uplifted and cooled from peak metamorphism about 2.0 b.y. ago coincident with intrusion of syntectonic granites. It was subjected to a subsequent mild thermal disturbance between 1.1 and 1.2 b.y. Estimates of the age of the protolith range from 2.7 to 3.7 b.y. Foliations in general trend east-northeast. The province is separated from those to the south by a fundamental shear zone, the Guri fault. Paleomagnetic data on Triassic dikes cutting the province indicate a stable South America pole position. In addition to references above: Chase (1965), Dougan (1972), Gaudette and others (1978), Hurley and others (1973), Kalliokoski (1965a, b) Menéndez (1974), Menéndez and others (1974), Montgomery (1979), Montgomery and Hurley (1978), Onstott and Hargraves (1982).

2. **Pastora-Supamo:** This province contains major subdivisions which are not distinguished on the map because of cartographic complexity at this scale (see Martín, 1978). This is a greenstone belt (Pastora) and granite (Supamo) terrane. The Pastora Supergroup is composed of isoclinally folded sequences of basaltic to silicic flows and tuffs, subvolcanic intrusives, turbidites, phyllites, and chemically deposited sediments. The metamorphic facies ranges from lower greenschist near the center of the belt to amphibolite facies at the thermally metamorphosed contacts between the greenstone belts and the Supamo basement. The Supamo is composed of oligoclase, quartz, microcline, biotite ± hornblende ± muscovite gneisses having a wide variety of textural facies which form dome-shaped structures separating the greenstone belts. The terrane formed 2.25 b.y. ago and was subjected to subsequent metamorphic episodes at 2.0 b.y. and 1.5 b.y. Relevant articles are as follows: Benaim (1974), Espejo (1974), Gibbs (1980), Klar (1979), Martín Bellizzia (1974), Menéndez (1968), Onstott and Hargraves (1982).

3. **Cuchivero:** The Cuchivero province consists of sequences of intermediate to silicic ignimbrite sheets, flows, and associated volcanogenic sedimentary rocks forming elongated, open-folded, synformal structures (3A) separated by antiformal structures of comagmatic calc-alkaline plutons. The regional metamorphism was constrained to greenschist facies. The province formed sometime between 1.7 and 1.9 b.y. The principal references are as follows: Gaudette and others (1978), Mendoza (1974), Onstott and Hargraves (1982), and Rios (1972).

4. **Roraima:** A gently deformed sequence of Precambrian sedimentary rocks and great diabase sheets unconformably overlies most older Precambrian provinces. It includes as much as 3,700 meters of sandstone, arkose, conglomerate, siltstone, jasper or chert (probably volcanogenic), and pebbly sandstone. Block faults are relatively common. Isotopic ages of diabase sills are 1.6-1.8 b.y. Bateson (1966), Bosma and de Roever (1976), Gansser (1954, 1974), Hargraves (1968), Keats (1976), Kloos-

terman (1976), Onstott and Hargraves (1982), Reid (1974), Van de Putte (1974), Veldkamp and others (1971).

5. *Parguaza:* This province comprises a large (30,000 km^2), unmetamorphosed, 1.55 b.y. rapakivi granite batholith of homogeneous composition. Relevant references are as follows: Gaudette and others (1978), Mendoza (1972).

6. *Mitu [Vichada]:* This poorly defined province consists mainly of granites and migmatic gneisses with minor amphibolite and sillimanite gneiss intercalations, metamorphosed to upper amphibolite facies. Radiometric data suggest that the province is at least 2.0 b.y. old, but has been subjected to subsequent thermal episodes at 1.5 and 1.2 b.y. Relevant references are: Huguett and others (1979), Kroonenberg (1981), Onstott and Hargraves (1982), Priem (1979).

BASINS MARGINAL TO GUAYANA SHIELD

7. *Guianas Margin Basin:* Phanerozoic strata in this "trailing-edge basin" are as thick as 6,000 meters and include mainly Cretaceous and Tertiary marine, paralic, and continental deposits. The strata are gently warped, but block faults of variable trend are common. Crust is continental (M = 30-40 km). Some references are Bosma and Groeneweg (1970), Krook (1969), Krook and Mulders (1971), Wong and Van Lissa (1978).

8. *Eastern Venezuela Basin:* Folds and faults of this province trend northeast to east. The basin contains as much as 10 km of Mesozoic(?) and Cenozoic marine and continental strata. Late Paleozoic strata have been reported at depth. It is autochthonous on the Guayana Shield to the south, but northward-dipping thrusts occur along the northern boundary. The basin is asymmetric, and the deepest segments are adjacent to the northern mountain front. The basin has prolific hydrocarbon resources. The crust is continental, (M = 25-47 km), but great crustal or mantle diversity is indicated by major changes in the gravity field longitudinally across the basin. Beck (1978), Bonini and others (1977a, b), González Silva (1977), Hay and Armand (1977), Hedberg (1950), Hedberg and others (1947), Lamb and Sulek (1968), Martínez (1970, 1977a), Murany (1972a, 1972b), Rosales (1976).

9. *El Baul:* This northwest-trending, fault-bordered uplift is cored with Paleozoic metamorphic rocks of relatively low grade and Permian or Triassic silicic volcanic and volcanogenic rocks. Paleozoic(?) alkaline plutons are overlain by Cretaceous and Tertiary strata. Paleomagnetic data on Triassic rocks indicate a stable South America pole. The crust is continental (M = 25-30 km). Feo-Codecido (1972b), MacDonald and Opdyke (1974), Martín Bellizzia (1961).

10. *Sub-Andean Basins:* Basins south and southeast of the Venezuelan and Colombian Andes are moderately folded and extensively faulted at depth. The Andes are generally thrust southward or eastward over the basins. Local Paleozoic, thick Cretaceous (as much as 4,500 meters), and thick (as much as 5,000 meters) Cenozoic sequences fill these basins. Most of the Cretaceous deposits are marine clastic, and the Tertiary deposits include older marine clastic and carbonate rocks and younger continental deposits. Sedimentary rocks rest on a basement which is composed of various elements of the Precambrian Guayana Shield. The crust is continental (M = 25-40 km). Deal (1981), Feo-Codecido (1972b), Irving (1975), Martín (1978), Martínez (1972a), Zambrano and others (1971).

SOUTH CARIBBEAN PLATE BOUNDARY ZONE

The boundary between the South American plate and Caribbean plate is not a single, narrow zone, but rather, a broad diffuse zone, having many independently moving blocks, that extends from the sub-Andean thrust belts, on the south and east, to the south margins of the Colombian and Venezuelan Basins. Recent reviews of seismicity for the zone have been provided by Kellogg and Bonini (1982), Pennington (1981), Pérez and Aggarwal (1981), and Vierbuchen (this volume).

UPLIFTED BLOCKS CONTAINING CORES OF PRECAMBRIAN AND PALEOZOIC ROCKS

Several great uplifted mountain ranges of the northern Andes contain cores of Precambrian rocks—probably outliers of the Guayana Shield—metamorphosed Paleozoic rocks, and Paleozoic plutons. They include the Cordillera de Mérida, Cordillera Oriental, Sierra de Perijá, Sierra Nevada de Santa Marta, Cordillera Central, and blocks on the Guajira Peninsula. Paleozoic rocks, in general, display an increasing grade of metamorphism, as well as a deeper-water facies, progressively away from the Guayana Shield and downward in the Paleozoic section.

11. *Cordillera de Merida:* Enormous vertical uplift and a high level of seismicity characterize this complex polydeformed and polycomponent structural block. The basement includes Precambrian(?) and Paleozoic metamorphic and igneous rocks, overlain by Paleozoic sedimentary rocks, Mesozoic redbeds, and Cretaceous marine clastic and carbonate strata, and varying thicknesses of Tertiary marine and continental clastic deposits. The aggregate thickness of Phanerozoic strata may be as much as 10 km. The rocks are complexly deformed during numerous Phanerozoic intervals (summarized by Shagam, 1975). Foliations of metamorphic rocks and folds of younger rocks trend generally northeast. The range is bisected by the active Boconó fault zone which has both vertical and strike-slip components of motion. The flanking faults are thought to dip beneath the range. The crust is continental (M = 35-40 km). Grauch (1971, 1972, 1975), Hargraves and Shagam (1969, several articles this volume), Schubert (1982), Shagam (1972b, 1975, 1977), Vásquez and Dickey (1972).

12. *Cordillera Oriental:* The Cordillera Oriental is a polydeformed uplift having a core of Precambrian and Paleozoic metamorphic and igneous rocks overlain by Paleozoic sedimentary rocks; Mesozoic redbeds, volcanic rocks, evaporites, and marine clastic strata, and local thick Cenozoic deposits. Most folds and faults trend northerly. The thickness of Mesozoic and

younger strata exceeds 10 km in places. The structural block was strongly deformed in several Precambrian and Paleozoic intervals, during Late Cretaceous-Paleogene time, and greatly uplifted in post-Miocene time. The range is cut by numerous Paleozoic and Mesozoic granitoid plutons. The Santander massif, which exposes most of the plutons, is a distinct subprovince. Flanking thrusts dip both easterly and westerly beneath the range. Although the range is slightly allochthonous with respect to its neighbors, paleomagnetic data on Jurassic(?) plutonic rocks yields a stable South American pole (MacDonald and Opdyke, 1974). The crust is continental (M = 35-40 km). Bürgl (1973), Campbell and Bürgl (1965), Etayo-Serna (1976), Irving (1975), Julivert (1970), Shagam (1975).

13. *Sierra de Perija:* The Sierra de Perijá is another post-Miocene uplifted range having many stratigraphic and structural similarities to the Cordillera de Merida and the Cordillera Oriental. Most structural trends are northeast. Paleozoic marine and continental strata occur in isolated localities on both the eastern and western flanks of the Sierra. Paleozoic and Mesozoic volcanic and plutonic rocks, Jurassic calc-alkaline volcanic rocks, and Paleozoic and Mesozoic redbeds occur in the core of the range. The Mesozoic redbeds were deposited in an extensional environment in NNE-SSW-trending graben. Jurassic and Cretaceous marine strata constitute most of the exposures at high elevations. At least five or six episodes of tectonic activity or uplift have been recognized during Paleozoic to post-Miocene time. The range is still seismically active. Paleomagnetic data on Jurassic rocks (Maze and Hargraves, this volume) suggest movement different than in the Cordillera de Mérida. Crust is continental and depth to M is about 30-35 km. Some pertinent references are Bowen (1972), Kellogg (1980, this volume), Kellogg and Bonini (1982), Miller (1962), Shagam (1975).

*14. *Cordillera Central:* This uplifted terrane contains a core of Precambrian and Paleozoic metamorphic and igneous rocks, overlain on the east by Triassic(?) and Jurassic redbeds and calc-alkaline volcanic rocks, and by Cretaceous marine clastic and local carbonate rocks. The west flank includes oceanic and mantle rocks in thrust zones. Large Mesozoic granitoid plutons and a few Tertiary plutons intrude the range. Mesozoic units are as thick as 4,000 meters and are moderately folded and cut by numerous faults, some of which have large strike-slip components. Most structural elements trend north to northeast. Thick Neogene volcanogenic terrace deposits occur on the east flank. Paleomagnetic data just south of the map area suggest very large translations and rotations of Triassic rocks. The crust is continental (M = 35-50 km). Botero (1963), Bürgl (1961, 1973), Feininger (1970), Nelson (1957), Scott (1978), Van Houten (1976).

14A. *Andean Neogene Volcanic Belt:* The most northerly of the active Andean volcanoes are superimposed on the Cordillera Central. Volcanic edifices and associated volcanogenic deposits are mainly calc-alkalic. Volcanism is related presumably to convergence at the Colombia trench. Crust is continental (M = 40-50 km). Ramírez (1968), Hantke and Parodi (1966), Herd (1974).

*15. *Santa Marta:* The Sierra Nevada de Santa Marta is an enormous pyramidal uplift which is a complex polycomponent terrane cored with Precambrian and Paleozoic metamorphic and igneous rocks. Foliations trend about northeast. It is bordered on the north by the great Oca fault (right-slip) and on the west by the Santa Marta fault (left-slip), both of which have large vertical components. Patches of Paleozoic sedimentary rocks and thick sequences of Mesozoic volcanic and volcanogenic sedimentary rocks have been mapped. Mesozoic metamorphic rocks occur in a northerly subterrane (Ruma metamorphic belt). The complex is intruded by Paleozoic, Mesozoic, and Cenozoic granitoid plutons. Crust is continental (M = 15-20 km?), but it is not in isostatic equilibrium. Case and MacDonald (1973), Gansser (1955), Kellogg and Bonini (1982), MacDonald and others (1971), Tschanz and others (1969, 1974).

*16. *Guajira:* This uplift is a complex polycomponent terrane having a core of Precambrian and Paleozoic metamorphic rocks, overlain by a thick sequence of Mesozoic clastic and carbonate deposits and local volcanic rocks which have been highly folded and faulted. Most structural elements trend northeast to east. A northerly subterrane of Mesozoic metamorphic and ultramafic rocks (Ruma metamorphic belt) is in fault contact with the southern subterrane. Tertiary basins are superimposed on the older terranes. Crust is oceanic, transitional, or continental (not in isostatic equilibrium) (M = ±25 km). Paleomagnetic data suggest substantial rotations of Cretaceous rocks in the south. Alvarez (1967), Lockwood (1965), MacDonald (1964), MacDonald and Opdyke (1972), Renz (1960), Rollins (1965), Thomas and MacDonald (1976).

ANDEAN BASINS HAVING A FOUNDATION OF CONTINENTAL OR UNDEFINED CRUST

Five major basins, filled primarily with Tertiary strata, occur between or adjacent to the Andean uplifts. They probably are underlain by continental crust or rifted continental crust. These include the Middle Magdalena, San Jorge-Magdalena, Cesar, Baja Guajira, and Maracaibo Basins.

17. *Middle Magdalena Basin:* The middle segment of the Magdalena Basin is a great intermontane topographic and structural feature between the Cordilleras Central and Oriental. It was probably a "foredeep" during Late Cretaceous and Cenozoic time. It contains 6,000 meters or more of Mesozoic and Cenozoic strata. Mesozoic deposits include older redbeds and Cretaceous marine clastics and local carbonates. Cenozoic strata are continental and locally marine. The province is moderately deformed by folding and thrusting. Structural trends are north to northeast. The crust is continental (M = 30-35 km). Campbell and Bürgl (1965), Irving (1975), Thompson (1966).

18. *San Jorge-Magdalena:* This moderately deformed basin contains 3,000 to 8,000 meters of mainly Tertiary marine and continental strata. Folds and faults trend north and northeast. The basement includes pre-Late Cretaceous elements of Precambrian and Paleozoic crystalline rocks similar to those of the Cor-

dillera Central. Prolific hydrocarbon production occurs in the province. The crust is continental or undefined. (M = ±25 km). Duque-Caro (1979, this volume), Irving (1975).

19. **Cesar Basin:** This asymmetrical Neogene basin, which may be genetically related to both strike-slip and thrust systems, appears to be overthrust by the Sierra de Perijá. It contains Cenozoic marine and continental deposits as much as 1,000 meters thick. Mesozoic strata may be as thick as 1,000 m. It is broadly folded and cut by northeast-trending, high-angle faults. The crust is continental (M = ±30 km). Campbell 1968, Kellogg and Bonini (1982), Polson and Henao (1968).

20. **Baja Guajira Basin:** North of the right-lateral Oca fault, this Neogene basin contains as much as 4,000 meters of Eocene(?)-Pliocene marine and continental clastic deposits. It is (internally) only moderately deformed but may be genetically related to the Oca and Cuiza faults. The crust is undefined but probably includes elements of continental crust. Case and McDonald (1973), Duque-Caro (1976), Irving (1975), Renz (1977), Zambrano and others (1971).

21. **Maracaibo:** One of the world's major petroleum basins contains Jurassic redbeds, Cretaceous marine carbonate and clastic deposits (1,000 meters or more) and 5,000-7,000 meters of Tertiary deltaic, marine, and fluviatile and lacustrine deposits. Metamorphic rocks have been penetrated by some deep drill holes. It is complexly faulted, with complexity increasing with depth, and it is moderately folded, especially near faults. Structures trend north and northeast. Hydrocarbon production is prolific. The crust is continental or undefined (M = 33 to 38 km). Bartok and others (1981), Bonini and others (1977b), Kellogg and Bonini (1982), Renz (1977), Vásquez and Dickey (1972), van Houten and others (this volume), Zambrano and others (1971).

SUTURE ZONES, MESOZOIC AND CENOZOIC DEFORMED BELTS, DISPLACED TERRANES, AND PULL-APART OR SUCCESSOR BASINS OF SOUTH CARIBBEAN PLATE BOUNDARY ZONE

The northern part of the South Caribbean plate boundary zone is a highly complex collage of Mesozoic and Cenozoic deformed belts which have formed principally on oceanic or undefined crust. The arcuate zone extends from the Cordillera Occidental, on the southwest, to at least Trinidad and Tobago on the east. Numerous blocks, especially those containing Cretaceous and Paleogene and older rocks, have probably been transported great distances from their sites of original formation, or have undergone extreme tectonic rotations. Many belts of the Caribbean Coastal Ranges have apparently been thrust southward or southeastward. Many features of this part of the plate boundary zone are described in considerable detail in various chapters of this volume.

*22. **Romeral:** A major tectonic feature of northern South America is the Romeral fault system, a great suture zone of Late Cretaceous and Paleogene age upon which is superimposed a graben-like basin related to right-slip(?) transform motion. The eastward-dipping suture zone separates oceanic crust (M = about 25-30 km) on the west from continental crust (M = 35-45 km) on the east and includes a mega-melange of ultramafic rocks, ophiolite fragments, local blueschists, blocks of continental crust, and Mesozoic pelagic and flyschoid strata. Tertiary plutons intrude the zone. Principal structural features trend north to northeast along the zone. The Tertiary basin (Cauca-Patía) contains as much as 3,000 meters of mainly continental fluviatile and volcanic Tertiary deposits. Paleomagnetic data suggest counterclockwise rotation of Tertiary igneous rocks (MacDonald, 1980). Barrero (1979), Campbell (1968), Case and others (1971), Irving (1975), Mooney (1980).

*23. **Cordillera Occidental:** This complex polycomponent terrane includes local ultramafic bodies and great thicknesses of Cretaceous tholeiitic marine basalts and basaltic andesites, pelagic strata, and turbidites, intruded by Tertiary granitoid plutons. Local metamorphic belts occur. Major fault systems border the east and west flanks of the terrane. Principal deformation, along north-trending fold and fault systems, occurred in pre-middle Eocene time, but enormous vertical uplift continued into Holocene. Crust is oceanic (M = 25-35 km). Barrero (1979), Case and others (1971), Gansser (1973), Irving (1975), Mooney (1980).

*24. **San Jacinto:** Aggregate thickness of Late Cretaceous to Neogene strata in this belt, including marine pelagic, turbidite, clastic, carbonate, fluvial, and lacustrine deposits, is as much as 10 km. The rocks are highly deformed by folds and thrusts that trend north-northeast as a result of Paleogene and Neogene compressional events. Local tonalitic plutons cut Cretaceous strata. The crust is undefined, but probably is oceanic (M = ±25 km). In many respects, the San Jacinto belt is an extension of the Cordillera Occidental. Duque-Caro (1979, this volume), Irving (1975).

*25. **South Caribbean Deformed Belt:** This great Neogene belt contains as much as 10 km of deformed Neogene (and probably older) pelagic and turbidite deposits. Folds and thrusts trending northeast to east are related to oblique convergence of the Colombian Basin, Beata Ridge, and the Venezuelan Basin with northern South America. Numerous mud diapirs occur in the western part of the terrane. This belt includes the Sinu belt described by Duque-Caro (this volume). The crust is oceanic (M = 20-25 km). Case (1974b), Duque-Caro (1979, this volume), Edgar and others (1971), Krause (1971), Ladd and Watkins (1979), Shepard (1973), Silver and others (1975), Talwani and others (1977).

*26. **Leeward Antilles (ABC) Terrane:** A series of uplifted blocks trending easterly off the north coast of Venezuela includes exposures of Middle(?) Cretaceous tholeiites, probable oceanic crust, Late Cretaceous tholeiites and calc-alkaline volcanic rocks, probably a primitive magmatic arc, associated pelagic strata, turbidites, and Paleogene conglomerate, overlain by gently deformed Neogene marine carbonate terraces. Mesozoic rocks metamorphosed up to amphibolite facies occur in the western and parts of the eastern blocks. Most structural trends are easterly. Calc-alkaline plutons range from 88 to 44 m.y. in age. Paleomagnetic data suggest large clockwise rotations since Cretaceous time. The crust is oceanic (M = 25-30 km). Beets and MacGillavry (1977), Hargraves and Skerlec (1982), LaGaay (1969), Santamaría and Schubert (1974), Silver and others (1975), Stearns and others (1982), Skerlec and Hargraves (1980).

27. Paraguaná: This complex uplifted block has a basement of Paleozoic metaigneous rocks and Mesozoic metavolcanic and metasedimentary rocks, partly covered by as much as 3,000 meters of gently deformed Tertiary strata. A large, zoned anorthositic gabbro and a zoned ultramafic complex of probable Mesozoic age are conspicuous elements of the terrane. Paleomagnetic data suggest large displacements and/or rotations. The crust is oceanic or undefined. Feo-Codecido (1971), MacDonald (1968), Martín Bellizzia and de Arozena (1972), Skerlec and Hargraves (1980).

28. *Blanquilla Province:* This poorly defined province includes an exposed trondhjemite pluton (62-64 m.y.), and folded and faulted strata offshore. Age and total thickness of offshore strata are unknown. Crust is oceanic or undefined (M = ±25 km). Peter (1972), Santamaría and Schubert (1974), Silver and others, (1972b, 1975).

BASINS ALONG VENEZUELAN BORDERLAND

Several sedimentary basins or graben occur between the relatively uplifted blocks of the Venezuelan borderland. Some are "pull-apart" basins related to strike-slip fault systems. Others may be passive or successor basins in that the areas remained depressed through time while the adjacent blocks were being tectonically uplifted.

29. *Chichibacoa-Golfo de Venezuela:* These Cenozoic basins, which may in part have a "pull-apart" origin, contain 4,000 to 7,000 meters of Tertiary marine and local paralic deposits. Mesozoic strata in the Golfo de Venezuela include redbeds, carbonates, and minor clastic deposits. The basin has been deformed by many faults of diverse displacement and orientation. The crust is continental or transitional (M = ±25 km). Bonini and others (1977b), Coronel (1967), Renz (1977), Thomas (1972), Zambrano and others (1971).

30. *Los Roques Basin:* This east-trending basin, locally faulted, contains more than 2,000 meters of layered material, presumably comprising Tertiary pelagic and turbidite material. Deeper strata are more highly deformed than shallow strata. The crust is oceanic (M = 20-25 km). Edgar and others (1971), Peter (1972), Silver and others (1975), Talwani and others (1977).

31. *La Orchila Basin:* Structural features east and west of this transverse graben-like basin are interrupted by it. The basin contains up to 2,000 m of moderately deformed strata above rough acoustic basement and is bordered by north-northwest-trending fault systems having both normal and probably strike-slip components of motion. Crust is oceanic or undefined. Peter (1972), Silver and others (1972b, 1975).

32. Bonaire Basin: The Bonaire Basin is a probable offshore extension of the Falcón Basin. It may be a Tertiary "pull-apart" basin subjected to late Neogene compression, especially along the south margin. As much as 4,000 meters of Tertiary strata have been penetrated by drill holes; an igneous-metamorphic basement occurs in the west, overlain by Cretaceous(?) and Paleogene deposits, and by Miocene and younger clastic and carbonate deposits. Numerous unconformities occur, and the degree of deformation increases with depth. The crust is probable transitional to oceanic (M = 20-30 km). Bonini and others (1977a, 1977b), Edgar and others (1971), Galavis and Louder (1970), (1977a, 1977b), Silver and others (1975), Vásquez and Masroua (1973, 1976).

33. Falcón Basin: Folds and faults in this "pull-apart" basin trend east-west to northeast. Deformed Paleogene deposits occur at depth and were involved in southward thrusting during later Paleogene time. Subsidence related to tensional opening between wrench fault systems was accompanied by deposition of as much as 6,000 meters of Neogene clastic deposits and reefal limestones. Tertiary volcanic rocks occur locally. Compressive deformation occurred again in Miocene and subsequent time. The crust is transitional or oceanic (M = 25-30 km). Díaz de Gamaro (1977a, 1977b), Muessig (1978, this volume), Stephan (1977), Wheeler (1963).

34. Cariaco Basin: This complex Neogene basin, probably formed between and over strike-slip faults, is located at the intersection of the east-trending El Pilar and northwest-trending Urica-Orchila Basin right-lateral fault systems. Total sediment thickness is unknown, but 4 km of low density material has been modeled for the Tuy Basin segment south of the main basin, and as much as 1.5 km occur in the Cubagua Basin to the north. The crust is probably transitional (M = 25-30 km). Bader and others (1970), Ball and others (1971), Bonini (1978), Murany (1972), Peter (1972), Schubert (1982), Silver and others (1972b).

DEFORMED TERRANES OF THE CORDILLERA DE LA COSTA AND SERRANIA DEL INTERIOR, VENEZUELA AND TRINIDAD

Mesozoic metamorphic (and locally older) rocks of the Cordillera de la Costa are complex sequences of metasedimentary and metavolcanic rocks, which are probably allochthonous, and which have been displaced from areas to the north or northwest (from the present area of the Falcón -Bonaire Basins, for example). In general, these rocks were emplaced southward into Paleogene flysch units. Some allochthonous blocks are relatively unmetamorphosed. These terranes are only briefly described because they are discussed more fully in other reports of this volume.

35. Siquisique: Paleogene flysch and slope deposition (3,000 m?) of the Falcón Basin culminated in southward thrusting and ophiolite emplacement. This allochthonous terrane contains a melange of ultramafic rocks, gabbros, pillow basalt, pelagic strata (some of Jurassic age), and shelf or slope Cretaceous rocks emplaced in Paleogene flysch. The block is unconformably overlain by various Oligocene-Miocene units. The crust is oceanic or transitional (M = ±30 km). Bellizzia and others (1972), Coronel (1970), Muessig (this volume), Skerlec and Hargraves (1980), Stephan (1977c).

*36. *Aroa-Mision:* Another nearby allochthonous terrane, probably a melange, includes blocks of Mesozoic(?) metasedimentary and metavolcanic rocks, ultramafic rocks, and Precambrian(?) gneissic bodies. The rocks are complexly deformed and locally overlain by thrust sheets of Paleogene strata. Crust is oceanic or undefined (M = ±30 km). Bellizzia and others (1976), Bellizzia and Rodriguez (1968), Rodriguez (1972).

*37. *Cordillera De La Costa:* Extensive sheets of Mesozoic metasedimentary and metavolcanic rocks were emplaced south or southeastward into Paleogene flysch deposits along the southern flank of the Venezuelan coastal ranges, and are bordered on the north by strands of the San Sebastian-El Pilar right-slip transform system in eastern Venezuela and Trinidad. (Displacement on the system is very controversial: estimates of right-slip range from as little as 5 km to more than 400 km. See summary by Schubert, 1982.) Metamorphic grades range from greenschist to amphibolite or higher facies. Foliations, though variable, trend generally east-west. Subterranes include the Barquisimeto (37A), Caracas (37B), Araya-Paria-Trinidad (37C), Margarita (37D), and Tobago (37E). Ultramafic intrusions or bodies along fault slices are common in most of the subterranes, and Late Cretaceous and/or Tertiary granitoid plutons occur in several. Crust is continental or undefined in the south and oceanic or undefined in the north (M = 25 to 35 km). Bellizzia (1972b), Bonini and others (1977a, 1977b), Dengo (1953), Harvey (1972), Maresch (1972b, 1972c, 1975), Menéndez (1965, 1967), Morgan (1969), Taylor (1960), Vierbuchen (this volume).

*38. *El Tinaco-Cacagua-Paracotos Terranes:* At least three terranes occur in area 38: The El Tinaco terrane to the west includes Paleozoic(?) metamorphic rocks of almandine-amphibolite facies; the Cacagua belt includes Cretaceous metasedimentary and metaigneous rocks; and the Paracatos belt includes metasedimentary rocks of zeolite facies containing a foraminiferal fauna of Maestrichtian age. These terranes are highly deformed, and foliations trend generally east-west. Crust is continental (M = 30-35 km). Paleomagnetic data suggest major tectonic rotations. References for terrane 37 plus Menéndez (1966), Piburn (1968), Skerlec and Hargraves (1980).

*39. *Villa de Cura:* An allochthonous sheet of Mesozoic(?) mafic metavolcanic and metasedimentary rocks of blueschist facies was thrust south or southeastward during Paleogene flysch deposition. Foliations trend east-west to north of east. The allochthonous sheet is probably 5-7 km thick. Ultramafic rocks occur in the complex. The underlying crust is continental (M = ±30 km). Paleomagnetic data suggest major tectonic rotations. Bell (1971, 1972), Bellizzia (1972), Bonini and others (1977a), Harvey (1972), Menéndez (1966, 1967), Murray (1972), Oxburgh (1965, 1966), Piburn (1967, 1968), Shagam (1960), Skerlec and Hargraves (1980).

*40. *Guarico-Roblecito:* Along the southern flank of the Venezuelan mountains, the Paleogene flysch belt has been highly

folded and cut by north-dipping thrusts; melange, wildflysch, and olistostromes are common; and exotic blocks of metamorphic and ultramafic rocks from terranes to the north are scattered along the belt. Crust is continental (M = 25-35 km). References for terranes 37-40 plus Beck (1977a, 1978), Bell and Pierson (1971).

41. *Eastern Serrania del Interior:* Cretaceous and Paleogene strata of this province are tightly folded and Neogene rocks are less folded. Structural trends are east-west to north of east. The province is cut by many reverse and strike-slip faults. Cretaceous and Paleogene deposits, as much as 4,000 meters thick, include marine carbonate and clastic deposits. Neogene deposits are mainly clastic. The crust is continental or transitional (M = 35-40 km). Hedberg (1950), Murany (1972a, 1972b), Salvador and Stainforth (1968), Vierbuchen (this volume).

42. *Trinidad Province:* South of the El Pilar right-lateral fault system, superimposed deformed basins contain Cretaceous carbonate and clastic strata, Paleogene flysch and neritic deposits, and Neogene marine and deltaic deposits. Many mud diapirs have deformed the layered sequence. The aggregate thickness of Cretaceous and younger strata may be as much as 10 km. Prolific hydrocarbon occurrences characterize this province. The crust is probably continental to transitional (M = 25-40 km). Barr and Saunders (1968), Kugler (1956), Michelson (1976), Persad (1978), Salvador and Stainforth (1968).

CHORTIS BLOCK AND ADJACENT PROVINCES

A second major cratonic area in the Caribbean region is the Chortis block or superterrane, south of the Motagua-Chamelecon-Jocatan fault zone, in Guatemala, Honduras, and northern Nicaragua. The boundary between this block, which is underlain by continental crust, and southern Central America, which is underlain by oceanic crust, has not been located as yet. An extension of either the Pedro Bank or Hess fracture zones may serve as boundary. Extensive Cenozoic volcanic deposits cover much of the area.

Relations of Precambrian(?) and Paleozoic metamorphic and igneous rocks of the Chortis block to rocks of northern South America and to those in southern Mexico, north of the Motagua-Polochic fault zone are as yet unknown. Some reconstructions for Mesozoic time place the Chortis block to the west, in continuity with basement rocks of southern Mexico; other reconstructions place the Chortis block adjacent to cratonic northern South America; and still other reconstructions, based principally on paleomagnetic data (Gose and others, 1980), show the Chortis block as a completely independent block with respect to the other major cratonic areas.

A similar problem exists for Precambrian and Paleozoic metamorphic and igneous rocks within various fault strands of the Motagua-Polochic transform system.

*43. *Chortis:* This complex polycomponent province in northeastern Honduras and Nicaragua includes Precambrian(?) and Paleozoic metamorphic and igneous rocks; Mesozoic redbeds, carbonates, clastic, and local volcanic rocks; and Cenozoic marine, continental, and volcanic rocks. Aggregate thickness of layered strata probably is greater than 4 km. Deformation of Mesozoic and younger rocks includes moderately complex folding and numerous block, high-angle thrust, and strike-slip faults, mainly formed during the Late Cretaceous-Early Tertiary time. Faults and folds are of diverse trend, and include sets that trend both northwest and northeast. The extensive area covered by Tertiary volcanic rocks constitutes a subprovince. The crust is continental (M = 35-40 km). Dengo (1973), Everitt (1970, 1981), Everitt and Fakundiny (1976), Elvir (1974), Finch (1981), Kim and others (1982), Mills and Hugh (1974), Weyl (1980).

44. *Yolaina:* This poorly known province in northeastern Nicaragua includes Mesozoic continental and marine deposits, and Cenozoic marine, continental, and volcanic rocks having an aggregate thickness of as much as 7,000 meters. The block is moderately folded and faulted. Faults and lineaments trend both northeast and northwest. The crust is probably oceanic in the south but may be transitional or continental in the north (M = 30?-35? km). Dengo (1973), McBirney and Williams (1965), Weyl (1980).

45. *Mosquitia-Nicaraguan Rise Basins:* The southeastern margin of this province has been arbitrarily placed along the northerly-trending San Andres trough or fracture zone. The province overlaps or merges into area 43, 44, 46, 91, and 99. It comprises one or more basins in which Cretaceous and Tertiary strata are 2,000 to 4,000 meters thick. Strata include clastic deposits, evaporites, and carbonate sequences. The province is moderately warped and is cut by many high-angle faults that trend mainly northwest and northeast. These basins have a "trailing-edge" aspect with respect to the Chortis and Yolaina blocks. The crust is undefined but probably is continental in the northwest (M = ±30 km). Arden (1975), Case and Holcombe (1980), Mills and Hugh (1974), Pinet (1971, 1972).

46. *Honduran Borderland Diapiric Province:* Numerous small diapiric structures occur off the north coast of Honduras; they apparently intrude Neogene(?) strata. Their total vertical extent is unknown. They have been interpreted as evaporite diapirs, but some may be clastic and others serpentinite. Crust is probably continental to transitional (M = 25? to 30 km). Above references plus Pinet (1973), Meyerhoff (1973).

NEOGENE IGNEOUS AND DEFORMED BELTS OF CENTRAL AMERICA AND NORTHWEST COLOMBIA

A complex northwest-trending zone of Neogene deformed belts, volcanic rocks, and belts of high seismicity provides a geologic tie between cratonic Central America and northwestern South America. The Cenozoic volcanism, mainly calc-alkalic, and deformation are thought to be genetically related to convergence of the Cocos and Nazca plates with continental blocks on the north and south and with oceanic blocks in southern Central

America. Scant paleontological and drill data indicate that the oceanic rocks along the Pacific margin include Late Jurassic and Cretaceous sequences, including ophiolites, and that some magmatic belts of Late Cretaceous and Paleogene age are superimposed on the older oceanic crust.

Neogene convergence between the Caribbean plate and the Panama block has occurred, as well as between the Nazca plate and Panama block. The contact zone between Neogene deformed belts of eastern Panama-northwest Colombia and those of northern Colombia (South Caribbean deformed belt) is inferred to be the Atrato fault zone.

Elements of these provinces will be described from the Chortis terrane southwestward to the Cocos plate and from the North Panama deformed belt south to the Nazca plate.

47. *Middle America Volcanic Province:* This northwest-trending volcanic belt is composed mainly of Miocene to Holocene calc-alkaline volcanic and related volcanogenic rocks. Volcanism is genetically related to subduction at the Middle America trench. The province is moderately deformed by block-faulting and tilting along northwest and northeast-trending faults; it is segmented along northeast-trending zones. The Nicaraguan depression trends northwest. The crust is oceanic in the south and undefined or continental in the north (M = 25-40 km). Stoiber and Carr (1974), Weyl (1980).

48. *Middle America Forearc Basin:* A moderately deformed forearc basin contains as much as 8 km of strata of Cretaceous and younger marine and local paralic strata. The crust is transitional or oceanic (M = 20-25 km). Couch and Woodcock (1981), Seeley (1979), Seeley and others (1974), von Huene and others (1980).

*49. *Middle America Forearc Ridge (Nicoya):* This terrane comprises exposed blocks of ultramafic rocks, tholeiitic basalts, pelagic, turbiditic, and local reefal materials of Jurassic to Paleogene age in the southeast and an inferred buried extension near the shelf edge to the northwest. The older assemblage is regarded as oceanic crust (M = 17-25 km) overlain by a primitive magmatic arc complex, including pelagic and turbiditic components. Paleogene(?) emplacement of Mesozoic rocks has been inferred. Paleomagnetic data suggest complex northeastward translations and clockwise rotations of Cretaceous and Cenozoic(?) rocks. Dengo (1962b), de Boer (1979), Galli-Olivier (1979), Ladd and others (1978), Schmidt-Effing (1979a), Seely and others (1974), von Huene and others (1980).

50. *Middle America Trench-Slope:* Two to four hundred meters of marine strata of Cretaceous to Neogene age appear to be only moderately deformed, in contrast to accretionary prisms at other convergent margins. Mafic and ultramafic rocks were drilled in DSDP Leg 84. The crust is oceanic (M = 15-20 km). Seely (1979), von Huene and others (1980), Scientific Party Leg 84 (1982).

51. *Cocos Plate:* Oceanic crust (M = 8-12 km) of Eocene(?) to Miocene age is overlain by as much as 400 meters of deep-water deposits. Cocos Ridge (terrane 51B), an oceanic plateau, may represent a locus of volcanism resulting from transla-

tion over a hotspot. The Cocos plate is moderately deformed by faulting and broad warping. Briceño-Guarupe (1978), Lonsdale and Klitgord (1978), van Andel and others (1977).

51A. *Middle America Trench:* More than 400 meters of hemipelagic and turbiditic material, locally block-faulted, comprise trench fill of probable post-Early Miocene age. The crust is oceanic (M = 9-17 km). References for terranes 48-50, plus Ibrahim and others (1979).

52. *Panama Fracture Zone:* This active north-trending fracture zone is the right-slip transform boundary between the Cocos and Nazca plates. The fault zone may have been "leaky." It includes oceanic crust (M = 9-15 km) of probable Miocene age and local, thin, deep-water deposits. Hey (1977), Lowrie and others (1979), Lonsdale and Klitgord (1978), van Andel and others (1971).

*53. *North Panama Deformed Belt:* A major accretionary wedge related to convergence between the Colombian Basin and the Panama block contains as much as 7,000 meters of deformed Cretaceous to Holocene turbidite, pelagic, hemipelagic, and possibly paralic strata. Structural trends are arcuate around northern Panama. The belt emerges on land to the northwest as the Limon Basin. The area is moderately to highly deformed by folding and southward thrusting. The crust is oceanic (M = 20-30 km). Bowin (1976), Dengo (1960, 1962a, 1973), Lu and others (1983), Weyl (1980).

54. *Talamanca-Gatun:* This poorly defined province includes great thicknesses of Tertiary calc-alkaline volcanic rocks and granitoid plutons representing a magmatic arc; it also includes local exposures of moderately deformed Tertiary marine and continental strata. The crust is undefined or oceanic (M = 25-30 km). del Giudice and Recchi (1969), Kesler and others (1977), Terry (1956).

*55. *San Blas-Darien:* Cretaceous tholeiitic basalt, perhaps oceanic, basaltic andesite, probably a magmatic arc, and pelagic strata rest on a foundation of oceanic crust (M = 20-25 km). Paleogene calc-alkalic volcanic rocks and marine strata are several km thick. The terrane is moderately to complexly deformed and cut by Late Cretaceous and Paleogene plutons. Normal faults or north-dipping reverse faults form the southwest margin of the terrane, but faulting on the northeast margin has not been documented. Case (1974), Interoceanic Canal Studies Commission (1968), Terry (1956), Wing and MacDonald (1973).

*56. *Choco Terrane:* Uplifted oceanic crust, and primitive magmatic arc rocks of Late Cretaceous and Paleogene age adjoin deep forearc(?) basins containing as much as 10 km of pelagic, hemipelagic, turbidite, and paralic strata of Cretaceous and Cenozoic age. Faults and broad folds trend north, northeast, and northwest. This terrane is evidently overridden along the eastward-dipping Atrato fault by the South Caribbean deformed belt and Cordillera Occidental. Crust is oceanic (M = 15-30 km): References for terrane 55 plus Case and others (1971), Barlow (1981), Atlantic-Pacific Interoceanic Canal Studies Commission (1969a, 1969b).

57. *Panama-Colombia Trench-Slope:* This poorly known province includes moderately deformed marine deposits of presumed Neogene and possibly older age. The crust is oceanic (M = ±20 km). Briceño-Guarupe (1978), Lowrie (1978), Pennington (1981).

58. *Filled Panama-Colombia Trench:* The fill in this trench presumably comprises Neogene-Holocene pelagic and turbiditic marine strata having little evidence of internal deformation. Rough apparent acoustic basement is probable Miocene oceanic crust (M = 10-15 km). Perhaps this province should be considered part of the Northern Nazca Plate. References for terrane 56 and 57 plus Case (1974).

*59. *Northern Nazca Plate:* Oceanic crust (7-15 km) of Miocene age is overlain by as much as 700 meters of Miocene and younger strata. Numerous high-angle faults have been mapped. References for terrane 57 and 58 plus Lonsdale and Klitgord (1978), Meyer and others (1976), and van Andel and others (1971).

MOTAGUA ZONE

*60. *Motagua:* Within the Motagua-Polochic-Chamelecón fault zone, regarded as a Neogene transform boundary between the North American and Caribbean plates, Precambrian(?) and Paleozoic metamorphic and igneous rocks and Mesozoic metamorphic and igneous rocks have not been well correlated with rocks of the Chiapas Massif to the north or the Chortis block to the south. Episodes of pre-Late Paleozoic metamorphism and deformation took place, and major tectonic convergence involving thrusting of ultramafic rocks, probable ophiolite sequences, and development of greenschist and blueschist metamorphism of Late Mesozoic rocks has occurred, probably in Late Cretaceous and Paleogene time. Relatively unmetamorphosed blocks of Paleozoic, Mesozoic, and Cenozoic marine and continental rocks occur in or adjacent to many fault slices. The crust is mainly continental, based on refraction and gravity data, but a large residual positive anomaly over the El Tambor subterrane (60C) plus abundant pillow basalts and cherts indicate that this is a block of oceanic crust. Burkart (1978), Dengo (1973), Donnelly (1977), Kim and others (1982), Petersen (1980), Roper (1978a, b), Rosenfeld (1981), Schwartz (1977).

CHIAPAS MASSIF AND ADJACENT TERRANES

North of the Motagua-Polochic transform boundary, in southern Mexico, northern Guatemala, and Belize, a third cratonic block having a basement of Precambrian(?) and Paleozoic metamorphic and igneous rocks extends northwest, beyond the map area, to perhaps the Isthmus of Tehuantepec. Exposed basement blocks, especially the Chiapas Massif, are bordered on the northeast by great fold belts, such as the Chiapas-Petén, and these, in turn, are bordered by less deformed provinces, including the Yucatan Platform. These areas have undergone relatively little deformation, other than vertical uplift, during the Neogene.

61. *Chiapas Massif:* A Precambrian(?) and Paleozoic de-

formed igneous and metamorphic complex cores this uplift and is in fault contact with Paleozoic and Mesozoic strata to the east. The complex contains Mesozoic and Tertiary plutons. The crust is continental (M = 35-40 km). López Ramos (1974, 1981).

62. ***Chiapas-Petén:*** As much as 10 km of Late Paleozoic(?), Mesozoic, and Cenozoic strata, including Mesozoic redbeds, carbonates, and evaporites, have been folded and locally thrust in an easterly-trending fold belt mainly in Late Cretaceous-Early Tertiary time. East-trending, left-lateral, strike-slip faults have been mapped. The basement includes Precambrian(?) and Paleozoic metamorphic rocks and granitoid plutons, some of which may be of Mesozoic or Cenozoic age. The crust is continental to transitional(?) (M = 35-40 km). Anderson and others (1973), Bateson and Hall (1977), Bishop (1980), López Ramos (1975, 1981), Viniegra (1971, 1981).

63. ***Yucatán Platform:*** This relatively undeformed to moderately deformed platform is underlain by 3,000-6,000 m of Late Paleozoic to Neogene strata, especially Cretaceous and Tertiary carbonate and caliche sequences. Broad warps and block-faults are principal structures affecting Cretaceous and younger rocks, except in the southwest where salt-pillowing and block, reverse, and left-slip faulting dominate the structural style. The crust is continental to transitional (M = 20-30 km). Prolific hydrocarbon production occurs in the southwest. Bishop (1980), López Ramos (1975, 1981), Viniegra (1981).

64. ***Yucatan Borderland:*** This complex belt may be a continuation of the Chiapas-Petén. At least three different origins have been proposed for this complex horst and graben-basin province: a faulted margin related to sphenochasmic rifting of the Yucatán Peninsula out of the present site of the Yucatán Basin; a left-lateral transform margin (probably Late Cretaceous-Paleogene) between western Cuba and Guatemala-Honduras; or a fault block margin associated with Tertiary subsidence of the Yucatán Basin. Some drill holes have penetrated more than 2,000 meters of Tertiary and Cretaceous carbonate and clastic strata and Mesozoic redbeds above a metamorphic basement, and metamorphic rocks have been dredged off scarps. Some geologists have reported folds with rocks like the San Cayetano Group (Jurassic) strata in the cores of anticlines. Cretaceous deposits are clastic and carbonate, and Tertiary deposits are mainly carbonate with some clastic intercalations. The crust is undefined but probably is transitional (M = 25-30 km). Bishop (1980), Pyle and others (1973), Uchupi (1973), Vedder and others (1973).

GULF OF MEXICO PROVINCES

Two segments of the southern Gulf of Mexico have been defined as different provinces on the basis of some differences in stratigraphy and style and timing of deformation.

65. ***Sigsbee-Campeche:*** A deep ocean basin and submergent continental margin is underlain by as much as 8 km of Triassic(?) to Holocene pelagic, carbonate, turbidite, and terrigenous-clastic deposits, including Jurassic evaporites; Paleozoic strata may be locally present. Crust is undefined but may be

modified or rifted continental crust (M = 16-35 km); thickened oceanic crust of probable Late Jurassic age underlies the extreme northwest part of the province. Pre-Late Jurassic strata are deformed by warping and block-faulting; Late Jurassic and younger strata are deformed only in vicinity of salt anticlines and domes. Buffler and others (1980, 1981), Ibrahim and others (1981), Martin (1980), Martin and Foote (1981).

66. ***Catoche:*** A deep ocean basin in the Catoche area is underlain by 3-7 km of strata of Triassic(?) to Holocene age, principally shallow to deep-marine carbonate rocks and pelagic strata, which rest on a rough basement. Basement rocks include Lower Paleozoic amphibolite and biotite gneiss, and diabase of Triassic age; reported metamorphic ages (Cambrian-Devonian) of basement rocks are similar to some from northern South America and the northern peninsula of Florida. Basement rocks and Mesozoic strata are warped, block-faulted, and tilted. The crust is transitional or undefined (M=20-25 km). Buffler and others (1981), Dallmeyer (1982), Scientific Staff DSDP Leg 77 (1981a).

BAHAMA PLATFORM-GREATER ANTILLES DEFORMED BELTS

In many respects, the northern margin of the Caribbean region is less well understood than the other margins. The Bahama Platform, which is relatively undeformed, contains stratigraphic elements similar to the Florida Platform to the north and the Yucatán Platform to the west. Rocks of the platform are overridden along south-dipping faults by deformed Mesozoic volcanic, volcaniclastic, clastic, and pelagic units in Cuba and possibly off northern Hispaniola. The nature of the crust beneath the Bahama Platform and the Greater Antilles deformed belts has been highly controversial in the past, particularly in Cuba. Recent refraction data obtained by groups from the U.S.S.R. indicate a continental crust in western Cuba, and radiometric dates as old as 950 m.y. have been reported (Somin and Millan, 1977). Relatively recent geologic and tectonic maps of Cuba are contained in an atlas of Cuba, published by the Instituto Cubano de Geodesía y Cartografía in 1978.

67. ***Florida-Bahama Platform:*** A huge, relatively undeformed platform is underlain by as much as 6,000 meters of Jurassic to Neogene carbonates and evaporites. Mesozoic reef facies are prominent. The Cayo Coco subprovince is considered part of the platform. The crustal character is controversial and may be oceanic (in the south), or continental in the north (M=20-30 km). Meyerhoff and Hatten (1974a, 1974b), Shein and others (1978), Sheridan and others (1981), Uchupi and others (1971).

*68. ***Zaza:*** The Zaza belt in northern Cuba perhaps should be considered a superterrane. It includes (1) a northerly belt of mainly allochthonous ultramafic, gabbroic, and Cretaceous tholeiitic volcanic rocks, and associated pelagic sedimentary and volcanogenic rocks that have been strongly deformed and thrust; and (2) a southerly belt comprising mafic Cretaceous volcanic rocks, Late Cretaceous marine sedimentary and volcanogenic strata, and sporadic ultramafic bodies. Paleogene strata include flysch facies, pelagic strata, and volcanogenic deposits. The aggregate thickness of the layered sequence may be as much as 10 km in places. Thrusts, mainly from the south, emplaced these rocks in flysch basins in pre-middle Eocene time. Granitoid plutons of Cretaceous(?) age intrude the terrane. This terrane and other Cuban terranes are thrust northward over the Bahama Platform on high-angle, south-dipping reverse faults. Tertiary basins (68B), as much as 2000 meters thick, are superimposed on the Zaza and other terranes. The crust is probably oceanic in eastern Cuba, based on huge positive gravity anomalies, but is probably continental in western Cuba (M=10-30 km). Bovenko and others (1980), Khudoley and Meyerhoff (1971), Pardo (1975), Shcherbakova and others (1977), Shein and others (1978).

69. Purial: The Purial terrane is a belt where volcanic and volcaniclastic rocks, locally intercalated with carbonates and chert, have been metamorphosed to blueschist, greenschist, and amphibolite facies. The original rocks were probably Mesozoic. The crust is oceanic (M=11-25 km). Above references plus Boiteau and others (1972a), Boiteau and Michard (1976).

70. Sierra Maestra: This calc-alkaline magmatic belt, on trend with Cayman Ridge, includes relatively unmetamorphosed volcanic rocks and associated volcaniclastic strata and marine clastic and carbonate deposits of Late Cretaceous and Paleogene age. The sequence is as thick as 4,000 m and is moderately to strongly deformed. Diorite and granodiorite plutons have been dated as 46 to 58 m.y. Tertiary basins (68B) are superimposed on the northern part of the terrane. The terrane is probably cut off on the south by strands of the Cayman trough transform system. The crust is regarded as oceanic (M=±20 km). Khudoley and Meyerhoff (1971), Lewis and Straczek (1955), Pardo (1975), Shcherbakova and others (1980).

71. Escambray (Sierra de Trinidad): This uplifted province has a core of metasedimentary and metavolcanic(?) rocks of greenschist, amphibolite, and blueschist facies that are probably Mesozoic but which may be Paleozoic in part. Foliations are variable, but northwesterly to westerly trends predominate. These polydeformed rocks have been intruded by diorite for which a controversial isotopic age of 180 m.y. has been reported. The northern contact is a north-dipping fault. The nature of the crust is likewise controversial: it has been regarded as continental; from gravity anomalies the crust is transitional or oceanic (M=±30 km). Boiteau and Michard (1976), Khudoley and Meyerhoff (1971), Pardo (1975), Shcherbakova and others (1980), Somin and Millan (1972).

72. Las Villas: Jurassic, Cretaceous, and Paleogene carbonate, clastic, and chert sequences that contain little volcanic material occur in this polycomponent province. Belts termed Yaguajay, Sagua la Chica, and Jatibonica are included. Facies changes are fairly abrupt from bank carbonates northward to slope and scarp deposits, and pelagic deposits, southward. Thicknesses range from 500 to 4,000 meters in various subareas. Extensive Paleogene flysch deposits occur in some subareas. Structures vary from monoclinal to tightly folded and thrust sequences.

Several Tertiary basins are superimposed and contain 1,000 meters or more of marine and continental strata. The crust is continental or undefined (M=20-25 km). Khudoley and Meyerhoff (1971), Pardo (1975).

73. Cifuentes-Placetas: Early, Middle, and Late Cretaceous deposits of pelagic and shallow carbonate, clastic, and local volcaniclastic deposits form this small province. The thickness is about 500 meters. The strata are tightly folded and cut by imbricate thrusts, and, in places, Eocene flysch deposits are caught up along faults. The crust is regarded as continental or is undefined (M=±25 km). Khudoley and Meyerhoff (1971), Pardo (1975).

74. *Cascarajicara:* This belt contains about 300 meters of massive shallow water limestone of Aptian to Coniaciana ge, identical to those in the Yaguajay belt of terrane 72. The crust is undefined, (M=±25 km). Bryant and others (1969), Pardo (1975).

75. *Organos-Rosario:* Many complex thrust sheets and major strike-slip faults characterize this province. The Rosario belt includes Jurassic clastic and carbonate deposits (700 m); Lower Cretaceous limestone, chert, and sandstone (180 m); about 1,000 meters of Upper Cretaceous carbonate, chert, and conglomerates; and Eocene carbonate and clastic deposits. The Organos belt includes more than 4,000 meters of clastic and local carbonate deposits of Jurassic age; several hundred meters of Middle to Late Cretaceous carbonate, chert, and complex melange (similar to parts of the Zaza terrane). Eocene deposits include fragmental carbonates in melange. The province is overlapped on the west by a Tertiary basin containing more than 500 meters of Tertiary strata. Crust is continental or transitional (M=±25-30 km). Iturralde-Vinent (1977), Khudoley and Meyerhoff (1971), Pardo (1975), Pszczolkowski (1977a, 1977b), Shcherbakova and others (1980).

76. *North Cuban Borderland:* Little information is available for this province. From evidence on a few seismic profiles, as much as one second of layered strata may be present. It is possible that adjacent Tertiary basins may extend into this province. Locally the strata appear to be moderately to highly deformed. The crust is undefined (M=20-25? km). Case and Holcombe (1980), Instituto Cubano de Geodesía y Cartografía (1978).

77. Isla de Pinos (Isla de Juventud): This polycomponent province includes highly deformed metasedimentary rocks (greenschist to amphibolite facies) of controversial Mesozoic(?) or Paleozoic(?) age, local serpentinite masses, and Cretaceous(?) volcanic rocks (probably Zaza terrane) overlapped by Neogene strata. The crust is undefined (M=±20? km). Boiteau and Michard (1976), Khudoley and Meyerhoff (1971), Pardo (1975), Somin and Millan (1972).

78. *South Cuban Borderland:* Little has been published about the south flank of the island of Cuba. From a few widely spaced seismic profiles, the borderland appears to be highly faulted. Basement highs and faults strike northeast off the southwest coast, and faults strike east-west along the southeast coast. The crust is undefined (M=±20 km). Case and Holcombe (1980), Instituto Cubano de Geodesía y Cartografía (1978).

HISPANIOLA PROVINCES

No continental crust has been recognized on Hispaniola, and from the large positive gravity anomalies, most investigators have proposed that the island is underlain by oceanic crust that is substantially thicker than normal oceanic crust. Although strikes of major rock belts are approximately parallel to those of belts in western Cuba and Puerto Rico, many differences occur in the stratigraphy, and attempts at correlations have not been very successful. Neogene deformed belts occur both in the southern and northern parts of Hispaniola. Provinces will be described generally from south to north across the island. Many important references, not all cited here, appear in the *Transactions* of the IX Caribbean Geological Conference (Llinas and others, 1982).

79. *South Haitian Borderland:* Little is known about this province. Geomorphic features strike northwest. Sparse gravity data are not definitive, but the crust is regarded as oceanic to transitional (M=20-25 km). Maurrasse (1980b).

80. *Massif de la Hotte-Bahoruco:* The south peninsula of Haiti and vicinity appears to be cored by uplifted oceanic crust and mafic volcanic rocks that may be equivalent to the Late Cretaceous tholeiitic magmatic rocks of terranes 98, 99, 100, and 101 to the south. Pelagic strata are intercalated with basalts and may be as old as Early Cretaceous. Younger basalt and intercalated pelagic strata carry a Campanian to Maestrichtian fauna. These older rocks are as much as 1,500 m thick. Paleogene deposits and Neogene strata include both carbonate and clastic deposits. The rocks are moderately to strongly folded, and north- and southing-dipping thrusts have been mapped. The crust is oceanic (M=20-25 km). Butterlin (1977), Maurrasse (1980a), Maurrasse and others (1979).

The uplifted block of the Massif de la Selle-Bahoruco area has some stratigraphic differences from the Massif de la Hotte. Older rocks include Early and Late Cretaceous basalts and rudistid-bearing limestones that form the basement. Paleogene strata include carbonate and clastic deposits (more than 600 meters), and Neogene deposits include massive limestone, flysch (1000+ meters), and other clastic deposits. Most thrusts and axial planes dip southerly. The crust is oceanic (M=20-25 km). Above references plus Maurrasse and others (1982), Biju-Duval and others (1983), Máscle and others (1980).

81. *Enriquillo-Cul-de-Sac:* As much as 3,000 meters of Neogene strata have been penetrated by deep wells in this graben-like feature, and Cenozoic strata may be as thick as 6,000 meters. Most of the deposits are marine, but Pliocene evaporites occur in the upper part of the section. Most folds and faults trend about east-west. Both westward and eastward extent of the feature is unknown. Some geologists regard this feature as a continuation of the North Caribbean deformed belt, but others regard it as a separate genetic feature. The crust is probably oceanic (M=20-25 km). Biju-Duval and others (1983), Bowin (1975), Butterlin (1977), Lewis (1980, 1982a).

*82. *North Caribbean Deformed Belt:* A major accretionary wedge of folded and faulted strata occurs along the south-ern margin and slope off Puerto Rico and the Dominican Republic. The belt may extend westward on land, across southern Hispaniola, perhaps to the Cayman trough transform terrane, but this concept has been debated by several geologists. The accretionary terrane is related to convergence at the partially filled Muertos trench (101B) along the north margin of the Venezuelan Basin, and movements may have a component of left slip. Deformed strata offshore exceed 4,000 m in thickness. The belt probably contains Paleogene flyschoid deposits as well as Neogene flysch, shelf, and slope clastic and carbonate deposits; possible local evaporites; and continental deposits. The crust is regarded as oceanic (M=20-25 km). Biju-Duval and others (1983), Ladd and Watkins (1978), Máscle and others (1980).

*83. *Hispaniola Northwestern-South-Central Zone:* This ill-defined terrane includes Cretaceous and Paleogene calc-alkaline volcanic rocks, volcaniclastic deposits, and carbonates. Oligocene to Early Miocene strata are mainly carbonate; younger Miocene deposits include flysch, carbonates, and shelf or slope clastic deposits. Neogene calc-alkaline and, perhaps, alkaline magmatism occurred locally. The Tertiary sequence probably exceeds 4,000 meters in aggregate thickness. The rocks, including Neogene strata, are moderately to strongly deformed. Axial surfaces and faults dip both northeast and southwest, and trend northwest. Gonave Ridge appears to be an anticline cored with Paleogene strata and overlain by Oligocene units. The Artibonite Basin is unknown offshore. One drill hole near the coast penetrated more than 1,200 meters of strata. Wells on Gonave penetrated up to 2,400 meters of strata. The crust is probably oceanic (M=20-25 km). Bowin (1975), Butterlin (1977), Lewis (1980, 1982a).

*84. *Massif du Norde-Cordillera Central:* This terrane in part represents a Late Mesozoic calc-alkaline magmatic arc. Toward the south, it includes Mesozoic(?) basalts and andesites and Cretaceous intermediate to silicic volcanic and plutonic rocks. Local Cretaceous carbonates and volcanogenic strata occur. Paleogene deposits are greatly variable and include carbonate, clastic, and volcanic deposits. Neogene strata include flysch, carbonates, and continental deposits. The sequences are highly folded and faulted. Paleomagnetic data indicate large post-Cretaceous counterclockwise rotations. The northerly part of the terrane includes schists and amphibolites (metamorphosed mafic to intermediate volcanic rocks and associated volcanogenic material), northwest-trending ultramafic bodies in a median belt, keratophyric and andesitic volcanic rocks of unknown age, and blocks of blueschist. The rocks have been intruded by granitoid plutons of Late Cretaceous age. The terrane is locally overlain by Neogene strata of adjacent provinces. The northeasterly margin locally is a southwest-dipping thrust; strong internal deformation characterizes the terrane. The crust is oceanic (M=±25 km). On the island of Tortue (84B), Oligocene(?) or Miocene limestones, perhaps 100 meters thick, are reported to overlie an anticlinal basement. Igneous rocks and Eocene limestones have also been reported. The Tortue block appears to be a horst bound by zones of complex faulting, perhaps strike-slip. Bowin (1975), Butterlin

(1977), Kesler (1971a), Draper and Lewis (1982), Kesler and others (1977), Lewis (1980, 1980b, 1982a, b, c), Vincenz and Dasgupta (1978), Tarasiewicz and others (1980).

85. *Seibo:* This poorly defined province includes Middle Cretaceous carbonates, Upper Cretaceous volcanic rocks, volcanogenic deposits, and marine deposits; and Paleogene marine strata. It is moderately to strongly deformed and cut by Paleogene plutons. A superimposed Neogene basin on the south (San Pedro-Trujillo Basin subprovince) contains Middle Tertiary carbonates. Paleomagnetic data suggest large counterclockwise rotations since Cretaceous. The crust is oceanic (M=25-30? km). Above references plus Bowin (1966), de la Fuente and Ellis (1982), Kesler (1980a), Ladd and others (1981), Lewis (1982a).

86. *Cibao Basin:* This basin formed mainly in Early Tertiary to Miocene time, It is cut by relatively young strike-slip faults, and most faults and folds trend north-northwest. Tertiary strata have an aggregate thickness of as much as 4,000 meters. Neogene deposits include slope, shelf, and paralic sequences of fine to coarse clastic deposits, and carbonates. Oligocene and older units are moderately to strongly deformed but Miocene and younger units are only slightly deformed except near major faults. The crust is oceanic (M=20-25 km). References for terranes 82 and 85, plus Antonini (1979), Palmer (1979), Saunders and others (1982).

*87. *Cordillera Septentrional-Samana:* This polydeformed terrane includes melange and/or olistostrome containing blocks of pre-Late Cretaceous to middle Eocene sedimentary and volcanic rocks, serpentinite and blueschist, and larger belts of marble and other metamorphic rocks (including blueschists) that occur in the east. More coherent, moderately deformed units of Paleogene age include calc-alkaline volcanic rocks, conglomerates, and marine clastic deposits. Much of the terrane is covered by Miocene carbonates (500 m) and younger terrace and paralic carbonates. Older rocks are strongly deformed but Neogene strata are only moderately deformed except near major faults. The crust is regarded as oceanic (M=20-25 km). Above references plus Bowin and Nagle (1982), Eberle and others (1982), Joyce (1982), Joyce and Nagle (1980), Nagle (1974, 1979), Nagle and Redmond (1980).

PUERTO RICO-VIRGIN ISLANDS AND VICINITY

Cretaceous and Paleogene rocks of Puerto Rico and the Virgin Islands and vicinity comprise a magmatic arc having early primitive island arc assemblages and later calc-alkaline assemblages in the classification of Donnelly and Rogers (1980).

*88. *Puerto Rico-Virgin Islands:* This complex polydeformed terrane is a Middle Cretaceous to Paleogene magmatic arc that contains early primitive island-arc elements followed by younger local calc-alkalic, keratophyric, and potassium-rich components. Pre-Neogene volcanic, volcaniclastic, and local carbonate rocks are strongly folded and cut by many faults, especially ones that have strike-slip displacement. Major structures trend west-northwest. Late Cretaceous and Paleogene granitoid

plutons are common. Neogene sedimentary basins on the north, west, and south sides of Puerto Rico are superimposed on the older rocks. The crust is regarded as thickened oceanic (M=25-30 km). Unpublished paleomagnetic data indicate extensive Late Cretaceous-Paleogene rotations (D. P. Elston, personal communication, 1982). Butterlin (1977), Donnelly (1975), Khudoley and Meyerhoff (1971), Krushensky and Elston (1983), Donnelly and Rogers (1978).

*88A. *Bermeja:* This allochthonous(?) complex includes serpentinitized peridotite, Jurassic and Cretaceous pelagic strata, Mesozoic(?) amphibolites, and spilites (midocean ridge basalts). The complex has been cut by thrusts, but the polarity of thrusting is controversial. The crust is regarded as oceanic (M=±30 km). Above references plus Mattson (1973a), Tobisch (1968).

*89. *Anegada:* The Anegada Passage is a complex graben and horst terrane that has long been regarded as a left-slip transform system, but some regard the area as a left-lateral pull-apart system. Exposed strata on St. Croix include 6 to 10 km of Cretaceous turbidites, volcaniclastic deposits, and siliceous carbonates. Neogene strata include Oligocene mudstone (400m) and Miocene marl (22m). Small diorite and gabbro plutons cut older sequences. In offshore areas, layered strata are absent above some basement blocks but range up to 2,000 m in thickness in the graben-basins. Crust is regarded as oceanic or multiple oceanic (M=±25 km). Garrison and others (1972a, 1972b), Holcombe (1979), Marlow and others (1974), Whetten (1966).

89A. *North Saba Slope:* Faults and folds of this subarea trend northwest into the Anegada province. The southern part, Saba Bank, has as much as 2900 meters of Paleogene and Neogene carbonate and clastic strata, moderately deformed, overlying a calc-alkaline volcanic basement (65 m.y.). Crust is regarded as oceanic or transitional. Garrison and others (1972b), Andreieff and others (1979).

90. *Hispaniola-Puerto Rico Borderland:* This poorly defined province is block-faulted, and strata in various blocks are as much as 1,000 m thick above apparent acoustic basement. Reflection data suggest southward underthrusting of the Bahama Platform off northwestern Hispaniola, and patterns of seismicity suggest incorporation of fragments of the Barracuda and Bahama Platform terranes into the borderland. An oceanic crust is probable (M=20-25 km). References for the Puerto Rico trench (118) plus Bunce and others (1974), Garrison and others (1972b), McCann and Sykes (1981), Perfit and others (1980), Uchupi and others (1971).

NICARAGUAN RISE—JAMAICA AND VICINITY

One of the major problems in the Caribbean region is that of the nature and age of the crust beneath the northern Nicaraguan Rise and Jamaica. To the west, the crust is probably continental adjacent to Honduras and Nicaragua; to the east, gravity data suggest oceanic crust beneath Jamaica; to the south, scant drill-hole data and acoustic stratigraphy suggest oceanic crust beneath the southern Nicaraguan Rise.

*91. **Northern Nicaraguan Rise:** This carbonate platform is underlain by as much as 5,000 m of layered strata in some basins. One drill hole penetrated almost 2,000 meters of Tertiary carbonate and clastic strata above an igneous basement. The province is moderately warped and locally block-faulted along northeast- and northwest-trending high-angle faults. The south margin of the province is the northeast trending Pedro Bank fracture zone which may be a fundamental crustal boundary. The crustal nature is not defined by present data (M=25?-30? km). Arden (1975).

*92. **Explorer:** The Explorer terrane is characterized by great fault systems, mainly downthrown to the north, that probably have a large left-slip component although some geologists have argued for very small lateral displacements (see references for terranes 60 and 119). Layered strata are as much as 1,000 m thick in parts of the terrane. Dredge hauls have found Neogene carbonates, of both deep and shallow deposition; Eocene carbonates, fine clastic deposits, and volcanic rocks; and Cretaceous(?) carbonates, plutonic rocks, and volcanic rocks (59-64 m.y.). Amphibolite was dredged at one locality. Deformed Eocene and less-deformed Miocene strata crop out on Swan Island. Older parts of the terrane are regarded as part of a Late Mesozoic-Early Tertiary magmatic arc. The crust is probably transitional or oceanic in the east and transitional or continental in the west (M=15? to 25? km). Banks and Richards (1969), Perfit and Heezen (1978).

93. **Jamaica:** There are as much as 3,000 to 6,500 meters of Cretaceous clastic, volcaniclastic, and carbonate deposits; 5,000 to 6,000 meters of Paleogene clastic, volcaniclastic, volcanic, and carbonate deposits; and 1,500 meters of Neogene carbonates occur in this polycomponent terrane. Fault trends are east-west, northwest, and northeast. Broad warps are common, but tight folds have developed mainly in older clastic strata. Some geologists have proposed five major structural-depositional cycles for Jamaica: (1) an episode of deposition and metamorphism of volcanic and associated sedimentary rocks, probably in the Late Jurassic; (2) local development of metamorphic rocks (volcanic and sedimentary protoliths) and emplacement of serpentinites and deformation in Early Cretaceous followed by episodes of major transgressions and regressions through the Late Cretaceous; (3) sedimentation and volcanism in trough areas in the Paleogene; (4) tectonic quiescence and limestone deposition during the middle Eocene to Late Miocene; and (5) deposition of Late Miocene and younger strata on Pedro Bank and around the periphery of Jamaica. Ambiguous paleomagnetic data suggest counterclockwise rotations of Jamaica since the Cretaceous. Crust is oceanic (M=15-20 km). Burke and others (1980), Testamarta and Gose (1982), Meyerhoff and Krieg (1977a, 1977b), Vincenz and others (1973), Wright (1974).

*93A. **Blue Mountains:** This terrane is separated from the main Jamaica terrane (93) by a major fault zone, possibly a suture, along which amphibolites, greenschists, blueschists, and small serpentinite bodies occur, and by the Wagwater trough, containing Paleogene marine and nonmarine clastic units that

grade to carbonates. Within the Blue Mountains, the rock sequence includes Late Cretaceous island-arc volcanic rocks and volcaniclastic deposits (600-1,500 m), and granitoid plutons; Late Cretaceous to Paleogene clastic, volcaniclastic, and volcanic rocks (1,500–2,500 m); and Paleogene clastic deposits (500-1,500 meters). The province is moderately to strongly deformed. Crust is oceanic (M=±20 km). Krijnen and Lee Chin (1978), Meyerhoff and Kreig (1977a, 1977b), Wadge and Draper (1978).

INTERIOR CARIBBEAN BASINS AND VICINITY

The three main interior basins of the Caribbean Sea appear to be underlain by oceanic crust. The crust beneath the Yucatán Basin is probably Eocene or older and that beneath the Colombian and Venezuelan Basins is Late Cretaceous or older.

94. **Yucatán Basin:** The central Yucatán Basin is remarkably flat and contains as much as 2,000 m of strata in some areas. More than 1,000 m of the upper sequence is composed of turbiditelike reflectors. Beneath is a sequence of variable thickness having the acoustic characteristics of pelagic sediments. Age of the strata is unknown but probably is Eocene (or older) to recent. The topography of the acoustic basement is variable. The northwest flank of the Yucatán Basin is a complex faulted margin. The basin is locally warped and cut by block faults, especially near the margins. The crust is oceanic (M=±15 km). Case (1975), Dillon and others (1972), Dillon and Vedder (1973), Uchupi (1973), Vedder and others (1972).

95. **Belize Fan:** This thick, deep-sea fan overlaps several provinces, including 94, 97, and 119, near their intersection with the Motogua-Polochic transform boundary. Many seismic profiles indicate a thickness of at least 2,000 m of gently warped strata; and one drill hole, just off the Guatemala coast, penetrated almost 4,000 meters of strata and bottomed in Lower Permian limestone (according to A. A. Meyerhoff, personal communication, 1982). The province is moderately warped. The fan deposits appear to be ponded against steep scarps of the Yucatán Borderland, western Cayman Ridge, and the north and south walls of western Cayman trough. The crust is undefined but probably is continental in the west and oceanic in the east (M=20-30 km). References for terrane 94, Case and Holcombe (1980).

96. **Pickle Province:** This poorly defined province, named for Pickle Seamount, is formed of high-standing basement blocks of unknown age and composition bounded by northeast-trending faults. Its history may be similar to that of the Yucatán Basin. Layered strata between basement blocks are 500 m to more than 1,000 m in thickness. Some strata are folded and faulted, but other patches are ponded against basement blocks. The crust is probably oceanic or transitional. (M=10-15 km). Case and Holcombe (1980), Dillon and others (1972), Vedder and others (1972).

97. **Cayman Ridge:** This ridge trends east-northeast, parallel to the Cayman trough transform system; it has extensive areas of apparently exposed acoustic basement on which are perched small patches of strata having thickness of 500 to 1,500 meters.

Dredge hauls have recovered Paleogene carbonate and clastic deposits, volcanic rocks and plutonic rocks (59 to 64 m.y.), and Neogene carbonate and clastic deposits. The island platforms are capped with Neogene carbonate terraces. The province is extensively block-faulted parallel to the ridge. The area has been regarded as part of a Late Cretaceous-Paleogene magmatic arc and may have been continuous with the Explorer terrane prior to separation along the Cayman trough transform system. Crust is oceanic (M=±15 km). Case (1975), Dillon and others (1972), Fahlquist and Davies (1971), Perfit and Heezen (1978).

98. *Colombian Basin:* Deformation of the interior part of this deep-ocean province includes northwest and northeast-trending broad warps and block-faults. The oceanic crust (M=15-20 km) is thought to be pre-Campanian, overlain by Campanian and younger pelagic strata. Tholeiitic basalt sills intrude the older part of the stratigraphic sequence. Cretaceous and Tertiary marine strata include pelagic, hemipelagic, and turbidite sequences, generally 500 to 2,000 meters thick, but locally as much as 6,000 m thick, overlain by the great Neogene Panama-Costa Rica (98A) and Magdalena (98B) fans. The province is being slowly subducted along its southwestern and south margin beneath the North Panama and South Caribbean deformed belts. Biju-Duval and others (1978), Edgar and others (1971, 1973), Houtz and Ludwig (1977a, 1977b), Lu and others (1983), Ludwig and others (1975), Krause (1971), Matthews and Holcombe (1976), Shepard (1973).

99. *Southern Nicaraguan Rise:* This province has many stratigraphic similarities to the Colombian Basin (98), even though it is structurally higher. It is separated from the Colombian Basin by the Hess Escarpment, possibly a northeast-trending fracture zone, and from the Northern Nicaraguan Rise by the Pedro Bank fracture zone. Strata vary in thickness from about 500 to more than 1,000 meters above a generally irregular acoustic basement. One DSDP hole (152), penetrated carbonates, chert, and siliceous limestone (Campanian to Paleogene) above basalts correlative with seismic horizon B" (Late Cretaceous pelagic strata and basalt). Neogene carbonates and Tertiary(?) calc-alkalic and possible alkalic volcanic rocks are exposed on the islands of San Andres and Catalina in the western part of the province. The province is broadly warped and block-faulted, and layered strata are interrupted by Neogene(?) plutons or volcanic rocks in places. The crust is probably oceanic or transitional (M=25-30? km). Arden (1975), Edgar and others (1973), Holcombe (1977), Moore and Fahlquist (1976).

100. *Beata Ridge:* This uplift of oceanic crust (M=15-20 km) is characterized by block-faults that trend northwest and northeast. Deep-marine strata of Santonian and Paleogene age and Oligocene to Pleistocene age have been drilled (367 meters), and local Eocene shallow-water deposits as well as Cretaceous(?) basalt and diabase and Oligocene to Pleistocene strata have been dredged. The province is being subducted beneath the South Caribbean deformed belt. Fox and Heezen (1975), Edgar and others (1971a, 1973), Roemer and others (1976), Rezak and others (1972).

101. *Venezuelan Basin:* This relatively undeformed ocean basin is being slowly subducted beneath both the North Caribbean and South Caribbean deformed belts. Oceanic crust (M=12-25 km) of Cretaceous(?) or older age is overlain by tholeiitic basalt and pelagic strata of Campanian and younger age. Strata above acoustic basement range from 250 meters in the west-central part to more than 4,000 m in the south part of the province. Pelagic and hemipelagic strata dominate lower and distal parts of the province, and turbidites, the younger and proximal parts. Warps and faults trend both northwest and northeast. Crustal character differs from place to place; and a subprovince (101B) of rough basement, like typical oceanic crust, has been identified. Biju-Duval and others (1978), Diebold and others (1981), Edgar and others (1971, 1973), Holcombe (1977), Ladd and Watkins (1978, 1979, 1980), Ladd and others (1977), Matthews and Holcombe (1976), Saunders and others (1973), Silver and others (1975).

LESSER ANTILLES AND VICINITY

102. *Aves Ridge:* Younger strata of this north-trending ridge are mainly autochthonous with respect to neighbors, although the ridge may be cut off by faults at the north and south ends. The ridge consists of a basement of basaltic rocks, calc-alkaline granitoid plutons, and intermediate volcanic rocks overlain by pelagic and shallow-water Tertiary deposits which are as much as 1,500 m thick. This uplift is regarded as a Late Cretaceous-Paleogene magmatic arc. Crust is transitional to oceanic (M=20-25 km). Biju-Duval and others (1978), Bowin (1976), Boynton and others (1979), Fox and Heezen (1975), Kearey (1976).

103. *Grenada Basin:* This backarc basin (with respect to the Lesser Antilles) is moderately to strongly warped and faulted in its northern part but appears to contain a greater thickness of strata and to be less deformed in its southern part. Younger strata of this basin are mainly autochthonous with respect to their neighbors but may be cut off by faults at the southern end. Strata (probably Tertiary) are 1,000 to 2,000 m thick in the north part and as much as 4,000 m in the south. Crust is oceanic or transitional (M=±25 km). Bunce and others (1970), Biju-Duval and others (1978), Bowin (1976), Bouysse and Martin (1979), Boynton and others (1979), Marlow and others (1974).

104. *Lesser Antilles Volcanic Arc:* This Neogene volcanic arc, autochthonous with respect to its immediate neighbors, is mainly calc-alkaline but includes compositional variations toward low SiO_2 in the north and high K_2O in the south. Neogene volcanic rocks are underlain by local Paleogene sedimentary and volcanogenic rocks. Other than block-faulting, deformation of the Neogene sequence is moderate, but older deposits are more highly deformed. Crust is oceanic to undefined and thickened (M=25-35 km). The arc appears to be segmented along transverse lines. Gravity highs along the arc extend at least to Margarita. Andrieff and others (1979), Bouysse and Martin (1979), Boynton and others (1979), Butterlin (1977), Martin-Kaye (1969), Smith (1980), Tomblin (1975), and Westbrook (1975).

105. *Limestone Caribbees:* Gently deformed Neogene marine reef terraces overlie deformed Paleogene pelagic, turbidite, and carbonate strata. Calc-alkaline volcanic and plutonic rocks indicate an early magmatic arc. Orthogonal northwest and northeast-trending block faults are common. Crust is oceanic to transitional and thickened (M=25-30 km). References for province 104, plus Christman (1953), Gunn and Roobol (1976), Lewis and Robinson (1976), Solomiac (1974).

106. *Lesser Antilles Borderland:* This poorly defined province includes lithologic and structural elements of all adjacent terranes. At La Desirade (106A), possible Jurassic plagiogranite, mafic dikes, and pillow basalts are unconformably overlain by several hundred meters of Neogene platform carbonates. Dredge hauls from scarps include greenschist facies metabasalts and metagabbros. Elsewhere in the province, more than 1,000 m of deformed stratified material is present. Dredges include tuffs and graywackes having Late Cretaceous radiolaria, and Eocene shallow-water limestone. Crust is oceanic (M=20-25 km). Andrieff and others (1979), Briden and others (1979), Bunce and others (1970), Chase and Bunce (1969), Fink (1970, 1971, 1974), Fox and Heezen (1975), Marlow and others (1976), Nagle and others (1973), Peter and Westbrook (1976).

107. *Tobago Basin:* This broad, gently deformed forearc basin contains as much as 5-8 km of stratified material, presumably of Paleogene and Neogene age. Ponded turbidites occupy the floor of the deeper parts. Strata interfinger westward with volcanogenic strata of the Lesser Antilles Borderland along a poorly defined boundary. The crust is oceanic (M=±25 to 30 km). Biju-Duval and others (1978), Bowin (1976), Boynton and others (1979), Bunce and others (1970), Chase and Bunce (1969), Westbrook (1975).

108. *Margarita Basin:* This basin contains more than 6,000 meters of Tertiary strata and, in part, is a southwest continuation of the Tobago forearc basin. Strata are only gently deformed except near steep faults. The crustal type is undefined: it may be oceanic, continental, or transitional (M=±25 or more km). Bassinger and others (1971), Bonini (1978), Feo-Codecido (1977), Weeks and others (1971).

109. *Barbados Ridge:* Complexly deformed Paleogene deposits form the exposed core of this forearc ridge, and include turbidites, volcanogenic strata, olistostromes, possible mud volcanoes, and local paralic(?) and deltaic deposits, and conglomerates. Structural trends of Paleogene strata diverge considerably from those of younger rocks. Neogene deposits include fossiliferous marls, tuffs, and clastic deposits. Aggregate thickness of Tertiary strata exceeds 4,000 meters; and 20 km, or more, of relatively low-density material underlies the ridge. It is regarded as part of the Lesser Antilles accretionary prism by many. On the north, the ridge merges with terranes 106 and 111, without a well-defined boundary. Crust is oceanic to transitional (m=25-30 km). Biju-Duval and others (1978), Bunce and others (1970), Poole and Barker (1982), Saunders (1968), Speed (1981), Speed and Larue (1982), Tomblin (1975), Westbrook (1975, 1982).

110. *Barbados Basin:* Moderate faulting and local folding characterize this Neogene forearc basin. One drill hole penetrated more than 1,500 meters of strata, and seismic data suggest thicknesses of 3,000 m or more. Strata of the province merge with those of provinces 109 and 111. Relations to the Trinidad province (42) are poorly known. The crust is oceanic or undefined (M=±25 km). Biju-Duval and others (1978), Mascle and others (1980), Michelson (1976), Persad (1978), Westbrook (1982).

111. *Lesser Antilles Deformed Belt:* This great accretionary wedge contains as much as 10 km, or more, of highly faulted and folded Tertiary strata. Deformation has accompanied westward subduction of the Atlantic Ocean segments of the North and South American plates (provinces 112, 113, and 116). The subducted oceanic plate is probably of Cretaceous age. Evidence from seismicity suggests that the Barracuda Ridge (115), perhaps a leaky transform fault or tilted block of oceanic crust, has also been subducted. The accretionary wedge of sediments is thicker in the south because of proximity to sediment sources from the Rio Orinoco and Rio Amazon. The crust is oceanic (M=15-20 km). Biju-Duval and others (1978), Bowin (1976), Bunce and others (1970), Chase and Bunce (1969), McCann and Sykes (1981, 1982), Peter and Westbrook (1976), Westbrook (1975).

ATLANTIC PROVINCES

112. *South American Plate (Abyssal Hills):* A small segment of the South American plate, informally termed the "Abyssal Hills" province, is characterized by rough acoustic basement, including the Tiburon Rise (112B). As much as 2,500 m of strata overlie the basement which is oceanic crust (M=13-15 km) of possible Jurassic to Late Cretaceous age. Neogene turbidites are ponded against Barracuda Ridge. A DSDP hole penetrated 411 meters of hemipelagic and pelagic strata of Late Cretaceous and Eocene to Holocene age, and 44 meters of pillow basalt. The strata in the terrane are broadly warped and block-faulted. The province is subducting westward beneath the Lesser Antilles deformed belt. Damuth (1973, 1977), Officer and others (1959), Peter and Westbrook (1976), Scientific Staff (1981b).

113. *Orinoco Deep-sea Fan:* This large fan is autochthonous on older and adjacent rocks except along the thrust zone bordering the southern part of the Lesser Antilles deformed belt. Strata are as thick as 4,000 m, or more, and are relatively undeformed except where fan deposits overlie diapiric structures of the Trinidad province (42). Crustal nature is unknown (M=13-25? km). It could be continental (Precambrian Guayana Shield) in the south and oceanic (Cretaceous) in the north. Bowin (1976), Case and Holcombe (1980—data from Marek Truchan), Damuth (1977), Damuth and Kumar (1975a, 1975b).

114. *Demerara:* This marginal plateau province appears to be autochthonous with respect to its neighbors and may be part of the Guianas Margin Basin (terrane 7). A drill hole has penetrated more than 4,900 meters of strata and bottomed in Neocomian rocks. DSDP holes bottomed in Aptian or Albian calcareous clay. Paleogene, Oligocene, and younger strata have been identi-

fied. One dredge haul encountered Late Jurassic or Early Cretaceous fossiliferous sandstone. The older rocks are moderately deformed. The crust is undefined (M=20-25 km). Fox and others (1970), Hayes and others (1970).

115. *Barracuda:* This east-trending ridge is regarded as a leaky transform fault or tilted fault block that separates the Atlantic Ocean segments of the North and South American plates. Basement of the ridge comprises hydrothermally altered and low-grade metamorphosed oceanic basalt (M=±15 km). The ridge is covered by variable thicknesses of layered strata, 500 to 1,500 m thick, and a DSDP hole bottomed in Eocene(?) strata at 475 meters. The ridge is locally block-faulted and warped. It has been postulated that the ridge is subducting beneath the Lesser Antilles deformed belt. Bader and others (1970), Birch (1970), Bonatti (1971), Marlow and others (1974), McCann and Sykes (1981), Officer and others (1959), Peter and Westbrook (1976), Schubert (1974), Westbrook (1982).

116. *Nares:* An enormous province in the northeastern part of the North American plate, comprises oceanic crust (M=10-15 km) of Cretaceous(?) to Jurassic age. Basement outcrops are common and the sedimentary cover is generally thin—varying from less than 200 meters to about 2,000 meters. DSDP hole 28 (424 meters) bottomed in Cenomanian pelagic strata after penetrating Eocene deep-water deposits. DSDP hole 99A (248 meters) bottomed in Oxfordian pelagic carbonates after penetrating a thin Neogene sequence and 191 meters of Lower Cretaceous carbonates. This province is moderately warped and locally block-faulted. Bader and others (1970), Bunce and Hersey (1966), Uchupi and others (1971), Tucholke and Ewing (1974).

117. *Silver Magnetic Province:* This geophysically defined province of oceanic crust (M=±15 km) is represented by a long northwest-trending magnetic high that is nearly perpendicular to the northeast-trending magnetic stripes of the Nares province to the northeast. Its source has been modeled as a northeast-dipping intracrustal mass about 6 km thick having a high susceptibility contrast. Reflection profiles over the anomaly reveal little or no correlation with basement topography. It has been speculated that it is a leaky transform fault or represents volcanism associated with early rifting. Bracey (1968), Griscom and Geddes (1966).

PUERTO RICO TRENCH-CAYMAN THROUGH TRANSFORM SYSTEM

118. *Puerto Rico Trench:* The Puerto Rico trench is a highly controversial segment of the transform boundary between the North American and Caribbean plates. Structurally, this depression displays elements of left-slip and normal faulting, and apparent underthrusting, especially on the south wall. The trench merges on the east into the Lesser Antilles deformed belt. Strata in the trench, including turbidites, are as much as 1,500 meters thick. Dredges from the walls have recovered Late Cretaceous to Pliocene sedimentary rocks, basalt, and serpentinite. Crust is oceanic (M=12-22 km). Ewing and others (1968), Fox and Heezen (1975).

*119. *Cayman Trough Transform:* This terrane is the Eocene(?) and younger left-slip boundary between the North American and Caribbean plates. A small, active north-trending spreading center, the mid-Cayman Rise, occupies the central part of the terrane. Sediments are absent or thin in the axial part of the terrane but progressively thicken to as much as 0.5 km in the east and 0.5 to 2 km in the west. Basement is regarded as ridge-generated oceanic crust (M=10-20 km) of Eocene(?) to Recent age. The terrane is extensively block-faulted, and faults trend north-south in the central portion but trend northeast in the distal eastern portion. Dredge hauls have recovered ultramafic rocks, gabbro, basalt, amphibolite, anorthosite, pelagic carbonate and clastic rocks, and volcanic rocks. Bowin (1976), Holcombe and others (1973), Macdonald and Holcombe (1978), Perfit and Heezen (1978).

CONCLUSIONS

Depending on the frame of reference, all parts of the Caribbean region are allochthonous to variable degrees with respect to their neighbors: during the Neogene, the North American and South American plates have moved relatively westward with respect to the Mid-Atlantic Ridge; the Nazca and Cocos plates moved eastward and northeastward with respect to the East Pacific Rise; the Caribbean plate, as a single entity, has moved eastward with respect to both the North American and South American plates, but in a passive sense, it has moved westward with respect to the Nazca and Cocos plates or with respect to hotspot reference frames. A similar broad pattern probably existed in Paleogene and Late Cretaceous time, but with many variations of relative motion vectors between individual tectono-stratigraphic terranes.

Most recent investigators have recognized a need for at least one, and possibly two, spreading centers, tensional rifts, or great transform faults through the Caribbean region between the Late Paleozoic-Early Mesozoic positions of North and South America. Convergent phenomena (Mesozoic metamorphic belts) are recognized as well as divergent phenomena.

Few published paleomagnetic studies have been made for pre-Aptian rocks of the Caribbean plate (here we include Cuba and the Yucatán Basin with the Caribbean plate, especially in Late Cretaceous-Paleogene time). Many determinations for Aptian and younger Cretaceous-Paleogene rocks of the Colombian and Venezuelan borderland regions and for the Greater Antilles (Jamaica, Hispaniola, and Puerto Rico) suggest formation at equatorial latitudes and rotations of up to 90 degrees or more. Declinations for the Greater Antilles are mostly westerly and for northern South America easterly. If such northerly motions occurred, then the seafloor *south* of the proto-Greater Antilles was probably also moving relatively north and eastward so that the mafic volcanic and primitive island arcs of the western Cordillera of Colombia were being emplaced by oblique convergence.

During the same general time interval, the Chortis block of middle America was undergoing a complex set of translations and rotations prior to final emplacement during the Neogene.

We believe that in order to reconstruct the highly complex evolution of the region a series of lithofacies maps for appropriate time slices must be prepared, one set using the present geographic base and other sets using paleogeographic bases for various proposed past positions of the major plates. Data from private company drill holes and seismic reflection will be critical elements of such reconstructions.

ACKNOWLEDGMENTS

We are grateful for the continuing stimulus provided by hundreds of colleagues who study the Caribbean region. We wish to especially thank those who reviewed all or parts of this map of geologic provinces and the brief descriptions. Most reviewers provided corrections of fact, much new information—some unpublished, and suggestions for reorganization of the original manuscript. Many had pertinent comments on such topics as definitions of basin types, the nature of magmatic arcs, particularly primitive arcs, and on crustal definitions. Many reviewers had adverse reactions about our descriptions of areas of our own inexpertise. We have tried to reconcile descriptions in some areas where two or more reviewers had differing views.

Reviewers included: John Albers, William Bonini, Carl Bowin, Kevin Burke, T. W. Donnelly, Herman Duque-Caro, Anthony Eva, Warren Hamilton, John Lewis, Ernesto López Ramos, James Kellogg, Florentin Maurrasse, William Maze, Arthur Meyerhoff, William Muehlberger, Karl Muessig, Tullis Onstott, Amos Salvador, Eli Silver, John Suppe, R. B. Van Houten, and Graham Westbrook.

Christine Reid provided invaluable editorial assistance.

REFERENCES CITED

Alvarez, Walter, 1967, Geology of the Simarua and Carpintero areas, Guajira Peninsula, Colombia [Ph.D. thesis]: Princeton, N.J., Princeton University, 168 p.

Anderson, T. H., Burkart, B., Clemons, R. E., Bohnenberger, O. H., and Blount, D. N., 1973, Geology of the western Altos Cuchumatanes, northwestern Guatemala: Geological Society of America Bulletin, v. 84, p. 805–826.

Andrieff, Patrick, Bouysse, Philippe, and Westercamp, Denis, 1979, Reconnaissance géologique de l'arc insulaire des Petites Antilles. Résultats d'une campagne à la mer de prélèvements de roches entre Sainte-Lucie et Anguilla (ARCANTE 1): Bulletin du Bureau de Recherches Géologiques et Minières (deuxième série), Section IV, nos. 3–4, p. 227–270.

Antonini, G. S., 1979, Physical geography of northwest Dominican Republic, *in* Lidy, B., and Nagle, F., eds., Hispaniola: Tectonic focal point of the northern Caribbean—Three geologic studies in the Dominican Republic: Miami Geological Society, p. 69–95.

Arden, D. D., Jr., 1975, Geology of Jamaica and the Nicaragua Rise, *in* Nairn, A.E.M., and Stehli, F. G., eds., The Ocean basins and margins, v. 3, The Gulf of Mexico and the Caribbean: New York, Plenum Press, p. 617–661.

Atlantic-Pacific Interoceanic Canal Study Commission (IOCS), 1969a, Engineering feasibility studies, Atlantic-Pacific interoceanic canal, geology, final report-route 25: v. 2, 20 p.

Atlantic-Pacific Interoceanic Canal Study Commission (IOCS), 1969b, Summary of geology and rock physical properties, route 25: IOCS Memorandum NCG-33, NCG-203-25, 43 p.

Bader, R. G., and others, eds., 1970, Initial report of the Deep-Sea Drilling Project, v. 4, Washington, D.C., U.S. Government Printing office, 753 p.

Ball, M. M., Harrison, C.G.A., Supko, P. R., and Bock, W. D., 1971, Normal faulting on the southern boundary of the Caribbean Sea, Unare Bay, Northern Venezuela, *in* Transactions, Caribbean Geological Conference, V, St. Thomas, Virgin Islands, 1968, Queens College Press, p. 17–22.

Banks, N. G., and Richards, M. L., 1969, Structure and bathymetry of western end of Bartlett Trough, Caribbean Sea, *in* McBirney, A. R., ed., Tectonic relations of northern Central America and the western Caribbean—the Bonacca Expedition: American Association of Petroleum Geologists Memoir 11, p. 221–228.

Barlow, C. A., 1981, Radar geology and tectonic implications of the Chocó Basin, Colombia, South America [M.S. thesis]: Fayetteville, Ark., University of Arkansas, 113 p.

Barr, K. W., and Saunders, J. B., 1968, An outline of the geology of Trinidad, *in* Transactions, Caribbean Geological Conference, IV, Port-of-Spain, Trinidad, 1965, p. 1–10.

Barrero L., Darío, 1979, Geology of the central Western Cordillera, west of Buga and Roldanillo, Colombia: Colombia Instituto Nacional de Investigaciones Geológico-Mineras, Publicaciones Geológicas Especiales del INGEOMINAS, no. 4, 75 p.

Bartok, P., Reijers, T.J.A., and Juhasz, I., 1981, Lower Cretaceous Cogollo Group, Maracaibo Basin, Venezuela: Sedimentology, diagenesis, and petrophysics: American Association of Petroleum Geologists Bulletin, v. 65–6, p. 1110–1134.

Bassinger, B. G., Harbison, R. N., and Weeks, L. A., 1971, Marine geophysical study northeast of Trinidad-Tobago: American Association of Petroleum Geologists Bulletin, v. 55, no. 10, p. 1730–1740.

Bateson, J. H., 1966, Some aspects of the geology of the Roraima Formation in British Guiana, *in* Transactions, Caribbean Geological Conference, III, Kingston, Jamaica, 1962, p. 144–150.

Bateson, J. H., and Hall, I.H.S., 1977, The Geology of the Maya Mountains, Belize: Great Britain Institute Geological Sciences, Overseas Memoir 3, 43 p.

Beck, C. M., 1977a, Geología de la Faja Piemontina y del frente de montaña en el noreste del Estado Guarico, Venezuela septentrional, *in* Memoria, Congreso Geologico Venezolano, V, Caracas, 1977, Venezuela Ministerio de Energía y Minas and Sociedad Venezolana de Geólogos, v. 2, p. 759–787.

——1978, Polyphase Tertiary tectonics of the Interior Range in the central part of the western Caribbean chain, Guarico State, northern Venezuela: Geologie en Mijnbouw, v. 57, no. 2, p. 99–104.

Beets, D. J., and MacGillavry, H. J., 1977, Outline of the Cretaceous and Early Tertiary history of Curaçao, Bonaire, and Aruba: *in* Guide to the Field Excursios on Curaçao, Bonaire, and Aruba, Netherlands Antilles: GUA Papers of Geology, Series 1, no. 10, Stichting GUA c/o Geologisch Instituut, 130 Nieuwe Prinsengracht, Amsterdam, The Netherlands, p. 1–6.

Bell, J. S., 1971, Tectonic evolution of the central part of the Venezuelan Coast Ranges, *in* Donnelly, T. W., ed., Caribbean geophysical, tectonic, and petrologic studies: Geological Society of America Memoir 130, p. 107–118.

Bell, J. S., 1972, Geotectonic evolution of the southern Caribbean area, *in* Shagam, R., and others, eds., Studies in earth and space sciences: Geological Society of America Memoir 132, p. 369–386.

Bell, J. S., and Peirson, A. L., 1971, Guia de la excursión geológica al frente de Montañas de Guárico (San Juan de los Morros-Camatagua-Altagracia de Orituco), *in* Memoria, Congreso Geológico Venezolano, IV, Venezuela Ministerio de Minas e Hidrocarburos, Boletín de Geología, Publicación Especial no. 5, v. 1, p. 329–356.

Bellizzia G., Alirio, 1972, Is the entire Caribbean Mountain belt of Venezuela allochthonous?, *in* Shagam, R., and others, eds., Studies in the earth and space sciences: Geological Society of America Memoir 132, p. 363–368.

Bellizzia G., A., Pimentel M., N., and Bajo O., R., compilers, 1976, Mapa geologico-estructural de Venezuela: Venezuela Ministerio de Minas e Hidrocarburos, scale 1:500,000.

Bellizzia G., Alírio, and Rodríguez, G., Domingo, 1968, Consideraciones sobre la estratigrafía de los Estados Lara, Yaracuy, Cojedes y Carabobo: Venezuela Ministerio de Minas e Hidrocarburos, Boletín de Geología, v. 9, no. 18, p. 515–563.

Bellizzia G., Alírio, Rodríguez, Domingo, and Graterol, Magali, 1972, Ofiolítas de Siquisique y Río Tocuyo y sus relaciones con la falla de Oca, *in* Transactions, Caribbean Geological Conference, VII, Isla de Margarita, Venezuela, 1971, p. 182–183.

Benaim, Ch., Nesin, 1974, Geología de la región El Dorado-Anacoco-Botanamo, Estador Bolívar, *in,* Memoria, Conferencia Geológica Inter-Guayanas, IX, Ciudad Bolívar, 1972, Venezuela Ministerio de Minas e Hidrocarburos, Boletín de Geología, Publicación Especial no. 6, p. 198–206.

Biju-Duval, B., Mascle, A., Montadert, L., and Wanneson, J., 1978, Seismic investigations in the Colombia, Venezuela, and Grenada Basins, and on the Barbados Ridge for future IPOD drilling: Geologie en Mijnbouw, v. 57, no. 2, p. 105–116.

Biju-Duval, B., Bizon, G., Mascle, A., and Muller, C., 1983, Active margin processes: Field observations in southern Hispaniola, *in* Watkins, J. S., and Drake, C. L., eds., Studies in continental margin geology: American Association of Petroleum Geologists Memoir 34, p. 325–344.

Birch, F. S., 1970, The Barracuda fault zone in the western North Atlantic: Deep-Sea Research, v. 17, p. 847–859.

Bishop, W. F., 1980, Petroleum geology of northern Central America: Journal of Petroleum Geology, v. 3, no. 1, p. 3–59.

Boiteau, Alice, and Michard, Andre, 1976, Donnees nouvelles sur le socle metamorphique de Cuba: Problemes d'application de la tectonique des plaques, *in* Transactions, Caribbean Geological Conference, VII, Saint-Francois, Guadeloupe, 1974, p. 221–226.

Boiteau, Alice, Michard, Andre, and Saliot, Pierre, 1972a, Metamorphisme de haute pression dans le complex ophiolitique du Purial (Oriente, Cuba): Comptes Rendus Academie des Sciences [Paris], ser. D, v. 274, p. 2137–2140.

Bonatti, Enrico, 1971, Rocks and sediments from the Barracuda fault zone, *in* Abstracts, Caribbean Geological Conference, V, St. Thomas, Virgin Islands, p. 45.

Bonini, W. E., Anomalous crust in the Eastern Venezuela Basin and the Bouguer gravity anomaly field of northern Venezuela and the Caribbean borderland: Geologie en Mijnbouw, v. 57, no. 2, p. 117–122.

Bonini, W. E., Acker, C., and Buzan, G., 1977a, Gravity studies across the western Caribbean Mountains, Venezuela, *in* Memoria, Congreso Latinoamericano de Geología, IV, Caracas, 1977, Venezuela Ministerio de Minas e Hidrocarburos, Boletín de Geología, Publicación Especial no. 7, v. 4, p. 2300–2323.

Bonini, W., Pimstein de Gaete, C., and Graterol, V., compilers, 1977b, Mapa de anomalías de Bouguer de la parte norte de Venezuela y areas vecinas: Venezuela Ministerio de Energía y Minas, scale 1:1,000,000.

Bosma, W., and de Roever, E.W.F., 1976, Results of recent geological studies in Suriname, *in* Memoria, Congreso Latinoamericano de Geología, II, Caracas, 1973, Venezuela Ministerio de Minas e Hidrocarburos, Boletín de Geología, Publicación Especial no. 7, v. 2, p. 687–708.

Bosma, W., and Groeneweg, W., 1970, Review of the stratigraphy of Suriname, *in* Guiana Geological Conference, VIII, Georgetown, Guyana, 1969, 29 p.

Botero A., Gerardo, 1963, Contribución al conocimiento de la geología de la zona central de Antioquia: Anales de la Facultad de Minas [Medellín, Colombia], no. 57, 101 p.

Bouysse, Philippe, and Martin, Pierre, 1979, Caractères morphostructuraux et évolution géodynamique de l'arc insulaire des Petites Antilles (Campagne ARCANTE 1): Bulletin du Bureau de Recherches Géologiques et Minières (deuxième série), Section IV, nos. 3–4, p. 185–210.

Bovenko, V. G., Shcherbakova, B. E., and Hernandes, H., 1980, 1982, Novyye geofizicheskiye dannyye o glubinnour stroyenii vostochnoy kuby (new geophysical data on the deep structure of eastern Cuba): Sovetskaya Geologiya, no. 9, p. 101–109: translation *in* International Geology Review, v. 24, no. 10, p. 1155–1162.

Bowen, J. M., 1972, Estratigrafía del precretáceo en la parte norte de la Sierra de Perijá, *in* Memoria, Congreso Geológico Venezolano, IV, Caracas, 1969, Venezuela Ministerio de Minas e Hidrocarburos, Boletín de Geología, Publicación Especial no. 5, v. 2, p. 729–761.

Bowen, Carl, 1966, Geology of central Dominican Republic (A case history of part of an island arc), *in* Hess, H. H., ed., Caribbean geological investigations: Geological Society of America Memoir 98, p. 11–84.

—— 1975, The geology of Hispaniola, *in* Nairn, A.E.M., and Stehli, F. G., eds., The Ocean basins and margins, v. 3, The Gulf of Mexico and the Caribbean: New York, Plenum Press, p. 501–552.

—— 1976, Caribbean gravity field and plate tectonics: Geological Society of America Special Paper 169, 79 p.

Bowin, C. O., and Nagle, F., 1982, Igneous and metamorphic rocks of northern Dominican Republic: an uplifted subduction zone complex, *in* Transactions, Caribbean Geological Conference, IX, Santo Domingo, Dominican Republic, 1980, v. 1, p. 39–50.

Boynton, C. H., Westbrook, G. K., Bott, M.H.P., and Long, R. E., 1979, A seismic refraction investigation of crustal structure beneath the Lesser Antilles island arc: Geophysics Journal Royal Astronomical Society, v. 58, p. 371–393.

Bracey, D. R., 1968, Structural implications of magnetic anomalies north of the Bahama-Antilles Islands: Geophysics, v. 33, no. 6, p. 950–961.

Brezsnyánszky, Karoly, and Iturralde-Vinent, M. A., 1978, Paleogeografía del Paleógeno de Cuba Oriental: Geologie en Mijnbouw, v. 57, no. 2, p. 123–133.

Briceño-Guarupe, L. A., 1978, The crustal structure and tectonic framework of the Gulf of Panama [M.S. thesis]: Corvallis, Oregon State University, 71 p.

Briden, J. C., Rex, D. C., Faller, A. M., and Tomblin, J. F., 1979, K-Ar geochronology and paleomagnetism of volcanic rocks in the Lesser Antilles island arc: Philosophical Transactions of the Royal Society of London, v. 291, no. 8 1393, p. 485–528.

Bryant, W. R., Meyerhoff, A. A., Brown, N. K., Jr., Furrer, M. A., Pyle, T. E., and Antoine, J. W., 1969, Escarpments, reef trends, and diapiric structures, eastern Gulf of Mexico: American Association of Petroleum Geologists Bulletin, v. 53, p. 2506–2542.

Buffler, R. T., Watkins, J. S., Shaub, F. J., Worzel, J. L., 1980, Structure and early geologic history of the deep central Gulf of Mexico basin, *in* Pilger, R. H., Jr., ed., The origin of the Gulf of Mexico and the early opening of the Central North Atlantic Ocean—Proceedings of a symposium, March 3-5, 1980, Baton Rouge, La.: Baton Rouge, Louisiana State University, School of Geoscience, 103 p.

Buffler, R. T., and others, 1981, A model for the early evolution of the Gulf of Mexico basin, *in* Colloquium on the Geology of Continental Margins, 26th International Geological Congress, Paris, July, 1980: Oceanologica Acta, Supplement to v. 4, p. 129–136.

Bunce, E. T., and Hersey, J. B., 1966, Continuous seismic profiles of the outer ridge and Nares Basin north of Puerto Rico: Geological Society of America Bulletin, v. 77, no. 8, p. 803–811.

Bunce, E. T., Phillips, J. D., Chase, R. L., and Bowin, C. O., 1970, The Lesser Antilles arc and the eastern margin of the Caribbean Sea, *in* Maxwell, A. E., ed., The Sea, v. 4, pts. 2-3,: New York, Wiley-Interscience, p. 359–385.

Bunce, E. T., Phillips, J. D., and Chase, R. L., 1974, Geophysical study of Antilles Outer Ridge, Puerto Rico Trench, and northeast margin of Caribbean Sea: American Association of Petroleum Geologists Bulletin, v. 58, no. 1, p. 106–123.

Bürgl, H., 1961, Historia geológica de Colombia: Revista de la Academia Colombiana de Ciencias Exactas, Fisicas, e Naturales, v. 11, no. 43, p. 137–193.

—— 1973, Precambrian to middle Cretaceous stratigraphy of Colombia: Translated by Allen, C. G., and Rowlinson, N. R., privately published by N. R. Rowlinson, Bogotá, Colombia, 214 p.

Burkart, Burke, 1978, Offset across the Polochic fault of Guatemala and Chiapas, Mexico: Geology, v. 6, p. 328–332.

Burke, Kevin, Grippi, Jack, and Sengor, A.M.C., 1980, Neogene structures in Jamaica and the tectonic style of the northern Caribbean plate boundary zone: Journal of Geology, v. 88, no. 4, p. 375–386.

Butterlin, J., 1977, Géologie structurale de la région des Caraïbes (Mexique—Amérique Centrale—Antilles—Cordillère Caraïbe): Paris, Masson, 259 p.

Campbell, C. J., 1968, The Santa Marta wrench fault of Colombia and its regional setting, *in* Transactions, Caribbean Geological Conference, IV, Port-of-Spain, Trinidad, W. I., 1965, p. 247–261.

—— 1974, Structural classification of northwestern South America, *in* Saunders, J. B., and others, eds., Contributions to the geology and paleobiology of the Caribbean and adjacent areas [Kugler volume]: Verhandlungen der Naturforschenden Gesellschaft in Basel, v. 84, no. 1, p. 68–79.

Campbell, C. J., and Bürgl, H., 1965, Section through the eastern cordillera of Colombia, South America: Geological Society of America Bulletin, v. 76, p. 567–590.

Case, J. E., 1974a, Oceanic crust forms basement of eastern Panamá: Geological Society of America Bulletin, v. 85, p. 645–652.

—— 1974b, Major basins along the continental margin of northern South America, *in* Burk, C. A., and Drake, C. L., eds., The geology of continental margins: New York, Springer-Verlag, p. 733–742.

—— 1975a, Geophysical studies in the Caribbean Sea: *in* Nairn, A.E.M., and Stehli, F. G., eds., The Ocean basins and margins, v. 3, The Gulf of Mexico and Caribbean: New York, Plenum Press, p. 107–180.

Case, J. E., Durán, L. G., López, R. A., Alfonso, and Moore, W. R., 1971, Tectonic investigations in western Colombia and eastern Panamá: Geological Society of America Bulletin, v. 82, p. 2685–2712.

Case, J. E., and Holcombe, T. L., 1980, Geologic-tectonic map of the Caribbean region: U.S. Geological Survey Miscellaneous Investigations Map I-1100, scale 1:2,500,000.

Case, J. E., and MacDonald, W. D., 1973, Regional gravity anomalies and crustal structure in northern Colombia: Geological Society of America Bulletin, v. 84, p. 2905–2916.

Chase, R. L., 1965, El Complejo de Imataca, la Anfibolita de Panamo, y la Tronjemita de Guri: rocas Precambrian del Cuadrilátero de Las Adjuntas-Panamo-Edo. Bolívar, Venezuela: Venezuela Ministerio de Minas e Hidrocarburos, Boletín de Geología, v. 7, no. 13, p. 105–216.

Chase, R. L., and Bunce, T. T., 1969, Underthrusting of the eastern margin of the Antilles by the floor of the western North Atlantic Ocean, and origin of the Barbados ridge: Journal of Geophysical Research, v. 74, no. 6, p. 1413–1420.

Christman, R. A., 1953, Geology of Saint-Martin, Saint-Bartholomew and Anguilla, Lesser Antilles: Geological Society of America Bulletin, v. 64, p. 65–96.

Coney, P. J., Jones, D. L., and Monger, J.W.H., 1980, Cordilleran suspect terranes: Nature, v. 288, p. 329–333.

Coronel, G. R., 1967, A geological outline of the Gulf of Venezuela, *in* Proceedings, World Petroleum Congress, VII, Mexico City, March, 1967, v. 2, p. 799–812.

—— 1970, Igneous rocks of the central Falcó: Associacion Venezolana Geología, Mineralogía, y Petrolera, Boletín Informativo v. 13, p. 155–162.

Couch, Richard, and Woodcock, S. F., 1981, Gravity and structure of the continental margins of southwestern Mexico and northwestern Guatemala: Journal of Geophysical Research, v. 86, no. B3, p. 1829–1840.

Dallmeyer, R. O., 1982, Age and character of pre-Mesozoic basement drilled in southeastern Gulf of Mexico: Tectonic implications: Geological Society of America Abstracts with Programs 1982, v. 14, nos. 1 and 2, p. 3.

Damuth, J. E., 1973, The western equatorial Atlantic-Morphology, Quaternary sediments, and climatic cycles [Ph.D. thesis]: New York, Columbia University, 602 p.

—— 1977, Late Quaternary sedimentation in the western equatorial Atlantic: Geological Society of America Bulletin, v. 88, p. 695–710.

Damuth, J. E., and Kumar, Naresh, 1975a, Amazon cone: Morphology, sediments, age, and growth pattern: Geological Society of America Bulletin, v. 86, p. 863–878.

—— 1975b, Late Quaternary depositional processes on the continental rise of the western equatorial Atlantic: Comparison with the western North Atlantic and implications for reservoir-rock distribution: American Association Petroleum Geologists Bulletin, v. 59, p. 2172–2181.

de Boer, J., 1979, The outer arc of the Costa Rican orogene (oceanic basement complexes of the Nicoya and Santa Elena Peninsulas [Costa Rica]): Tectonophysics, v. 56, p. 221–259.

de la Fuente, Luis, and Ellis, G. M., 1982, Informe sobre la investigación geológica de la Cordillera Oriental, sector El Rancho, *in* Transactions, Caribbean Geological Conference, IX, Santo Domingo, Dominican Republic, 1980, v. 2, p. 670–673.

del Giudice, D., and Recchi, G., 1969, Geología del area del proyecto minero de Azuero [Panamá]: Technical report prepared for the Government of the Republic of Panamá by the United Nations, 48 p.

Dengo, Gabriel, 1953, Geology of the Caracas region, Venezuela: Geological Society of America Bulletin, v. 64, no. 1, p. 7–40.

—— 1960, Notas sobre la geología de la parte central del litoral pacifico de Costa Rica: Instituto Geográfico de Costa Rica, Informe Semestral, Julio a Diciembre, San Jose Costa Rica, p. 43–58.

—— 1962a, Tectonic-igneous Sequence in Costa Rica: Petrologic Studies: A volume to honor A. F. Buddington: Geological Society of America, p. 133–161.

—— 1962b, Estudio geológico de la región de Guanacaste, Costa Rica: Instituto Geográfico de Costa Rica, Informe Semestral, 112 p.

—— 1973, Estructura geológica, historia tectónica y morfología de America Central: Instituto Centroamericano de Investigaciones Tecnologia Industrial (ICAITI), 2nd Edition, p. 1–52.

Diaz de Gamaro, M. L., 1977a, Estratigrafía y micropaleontología del Oligocene y Miocene inferior del centro de la Cuenca de Falcón, Venezuela: GEOS, v. 22, Escuela de Geología y Minas, Universidad Central de Venezuela, Caracas, p. 3–60.

—— 1977b, Revisión de la edades de las unidades litoestratigráficas en Falcón Central en base a su contenido de foraminíferos planctonicos, *in* Memoria, Congreso Geologico Venezolano, V, Caracas, 1977, Ministerio de Energía y Minas and Sociedad Venezolano de Geólogos, v. 1, p. 81–106.

Diebold, J. B., Stoffa, P. L., Buhl, P., and Truchan, M., 1981, Venezuela Basin crustal structure: Journal of Geophysical Research, v. 86, no. B9, p. 7901–7923.

Dillon, W. P., and Vedder, J. G., 1973, Structure and development of the continental margin of British Honduras: Geological Society of America Bulletin, v. 84, no. 8, p. 2713–2732.

Dillon, W. P., Vedder, J. G., and Graf, R. J., 1972, Structural profile of the northwestern Caribbean: Earth and Planetary Science Letters, v. 17, p. 175–180.

Donnelly, T. W., 1975, The geological evolution of the Caribbean and Gulf of Mexico—some critical problems and areas, *in* Nairn, A.E.M., and Stehli, F. G., eds., The Ocean basins and margins, v. 3, The Gulf of Mexico and the Caribbean: New York, Plenum Press, p. 663–685.

—— 1977, Metamorphic rocks and structural history of the Motagua suture zone, eastern Guatemala: Abstracts, Caribbean Geological Conference, VIII, Curaçao, 1977, Stichting GUA c/o Geologisch Instituut, 130, Nieuwe Prinsengracht, Amsterdam, The Netherlands, p. 40–41.

Donnelly, T. W., and Rogers, J.J.W., 1978, The distribution of igneous rocks throughout the Caribbean: Geologie en Mijnbouw, v. 57, no. 2, p. 151–162.

—— 1980, Igneous series in island arcs: The northeastern Caribbean compared with worldwide island-arc assemblages: *in* Smith, A. L., ed., Special issue on circum-Caribbean volcanism: Bulletin Volcanologique, v. 43-2, p. 347–382.

Dougan, Thomas, 1972, Origen y metamorfísmo de los gneises de Imataca y Los Indios, Rocas Precambricas de la Región de Los Indios-El Pilar, Estado Bolívar Venezuela, *in* Memoria, Congreso Geológico Venezolano, IV, Caracas, 1969: Venezuela Ministerio de Minas e Hidrocarburos, Boletín de Geología, Publicación Especial no. 5, v. 3, p. 1337–1548.

Draper, Grenville, and Lewis, J. F., 1982, Petrology, deformation, and tectonic significance of the Amina Schist, northern Dominican Republic, *in* Transactions, Caribbean Geological Conference, IX, Santo Domingo, Dominican

Republic, 1980, p. 53–64.

Duque-Caro, H., 1976, Características estratigráficas y sedimentarias del Terciario marino de Colombia, *in* Memoria, Congreso Latinoamericano de Geologia, II, Caracas, 1973, Venezuela Ministerio de Minas e Hidrocarburos, Boletín de Geología Publicación Especial no. 7, v. 2, p. 945–964.

—— 1979, Major structural elements and evolution of northwestern Colombia, *in* Watkins, J. S., and others, eds., Geological and geophysical investigations of continental margins: American Association of Petroleum Geologists Memoir 29, p. 329–351.

Eberle, W., Hirdes, W., Muff, R., and Pelaez, M., 1982, The geology of the Cordillera Septentrional (Dominican Republic), *in* Transactions, Caribbean Geological Conference, IX, Santo Domingo, Dominican Republic, 1980, v. 2, p. 619–632.

Edgar, N. T., Ewing, J. I., and Hennion, John, 1971, Seismic refraction and reflection in the Caribbean Sea: American Association of Petroleum Geologists Bulletin, v. 55, no. 6, p. 833–870.

Edgar, N. T., Holcombe, Troy, Ewing, John, and Johnson, William, 1973, Sedimentary hiatuses in the Venezuelan Basin, *in* Edgar, N. T., and others, eds., Initial Reports of the Deep Sea Drilling Project, v. 15: U.S. Government Printing Office, Washington, p. 1051–1062.

Elvir, A. R., 1974, Geología de Honduras: Republic of Honduras, Tegucigalpa, 44 p.

Espejo C., Aníbal, 1974, Geología de la región El Manteco-Guri, Estado Bolívar, Venezuela, *in* Memoria, Conferencia Geológica Inter-Guayanas, IX, Ciudad Guayana, 1972, Venezuela Ministerio de Minas e Hidrocarburos, Boletín de Geología Publicación Especial no. 6, p. 207–248.

Etayo-Serna, Fernando, 1976, Contornos sucesivos de mar Cretáceo en Colombia, *in* Etayo-Serna, Fernando, and Caceres-Giron, Carlos, eds., Memoria, Congreso Colombiano de Geología, I, Bogota, August 4-8, 1969: Universidad Nacional de Colombia, Bogotá, p. 217–252.

Etayo-Serna, Fernando, and Caceres-Giron, Carlos, eds., 1976, Memoria, Congreso Colombiano de Geología, I, Bogotá, August 4-8, 1969: Universidad Nacional de Colombia, Bogotá, 438 p.

Everett, J. R., 1970, Geology of the Comayagua quadrangle, Honduras, Central America [Ph.D. thesis]: Austin, University of Texas, 152 p.

Everett, J. R., and Fakundiny, R. H., 1976, Structural geology of El Rosario and Comayagua Quadrangles, Honduras, Central America: Publicaciones Geológicas del ICAITI, v. 5, p. 31–42.

Ewing, John, and Ewing, Maurice, 1970, Seismic reflection, *in* Maxwell, A. E., ed., The Sea, v. 4, pt. 1: New York, Wiley-Interscience, p. 1–51.

Ewing, J. I., Officer, C. B., Johnson, H. R., and Edwards, R. S., 1957, Geophysical investigations in the eastern Caribbean: Trinidad shelf, Tobago trough, Barbados Ridge, Atlantic Ocean: Geological Society of America Bulletin, v. 68, no. 7, p. 897–912.

Ewing, John, Antoine, John, and Ewing, Maurice, 1960, Geophysical measurements in the western Caribbean Sea and in the Gulf of Mexico: Journal of Geophysical Research, v. 65, no. 12, p. 4087–4126.

Ewing, J. I., Worzel, J. L., and Ewing, M., 1962, Sediments and oceanic structural history of the Gulf of Mexico: Journal of Geophysical Research, v. 67, p. 2509–2527.

Ewing, John, Talwani, Manik, and Ewing, Maurice, 1968, Sediment distribution in the Caribbean Sea, *in* Transactions, Caribbean Geological Conference, IV, Port-of-Spain, Trinidad and Tobago, 1965, p. 317–323.

Ewing, John, Talwani, Manik, and Ewing, Maurice, and others, 1967, Sediments of the Caribbean, *in* Studies in tropical oceanography: Miami, University of Miami, v. 5, p. 88–102.

Ewing, J. I., Edgar, N. T., and Antoine, J. W., 1971, Structure of the Gulf of Mexico and Caribbean Sea, *in* Maxwell, A. E., ed., The Sea, v. 4, pts. 2-3: New York, Wiley-Interscience, p. 321–358.

Ewing, M., and Heezen, B. C., 1955, Puerto Rico Trench topography and geophysical data: Geological Society of America Bulletin, v. 62, p. 255–268.

Ewing, M., and Worzel, J. L., 1954, Gravity anomalies and structure of the West Indies, Pt. I: Geological Society of America Bulletin, v. 65, no. 2, p. 165–173.

Ewing, M., Worzel, J. L., Ericson, D. B., and Heezen, B. C., 1955, Geophysical and geological investigations in the Gulf of Mexico, Part I: Geophysics, v. 20, p. 1–18.

Ewing, M., Worzel, J. L., and Ewing, J. I., 1962, Sediments and oceanic structural history of the Gulf of Mexico: Journal of Geophysical Research, v. 67, p. 2509–2527.

Ewing, M., Lonardi, A. G., and Ewing, J. I., 1968, The sediments and topography of the Puerto Rico Trench and Outer Ridge, *in* Transactions, Caribbean Geological Conference, IV, Port-of-Spain, Trinidad, 1965, p. 325–334.

Fahlquist, D. A., and Davies, D. K., 1971, Fault-block origin of the western Cayman Ridge, Caribbean Sea: Deep-Sea Research, v. 18, no. 2, p. 243–253.

Feininger, Tomas, 1970, The Palestina fault, Colombia: Geological Society of America Bulletin, v. 81, no. 4, p. 1201–1216.

Feo-Codecido, G., 1971a, Geología y recursos naturales de la Península de Paraguaná, Venezuela, *in* Proceedings, Symposium on investigations and resources of the Caribbean Sea and adjacent regions, Willemstad, Curaçao, Netherlands Antilles, 1968: UNESCO, Paris, p. 231–240.

—— 1972, Contribución a la estratigrafía de la cuenca Barinas-Apure, *in* Memorias, Congreso Geológico Venezolano, IV, Caracas, 1969: Venezuela Ministerio de Minas e Hidrocarburos, Boletín de Geología, Publicación Especial no. 5, v. 2, p. 773–795.

—— 1977b, Un esbozo geológico de la plataforma continental Margarita-Tobago, *in* Memoria, Congreso Latinoamericano de Geología; II, Caracas, 1973: Venezuela Ministerio de Minas e Hidrocarburos, Boletín de Geología, Publicación Especial no. 7, v. 3, p. 1923–1945.

Finch, R. C., 1981, Mesozoic stratigraphy of central Honduras: American Association of Petroleum Geologists Bulletin, v. 65, no. 7, p. 1320–1333.

Fink, L. K., Jr., 1971, [1970], Field guide to the Island of La Désirade with notes on the regional history and development of the Lesser Antilles island arc: American Geological Institute-National Science Foundation International Field Institute Guidebook to the Caribbean island-arc system, 1970, 17 p.

—— 1972, Bathymetric and geologic studies of the Guadeloupe region, Lesser Antilles Island Arc: Marine Geology, v. 12, no. 4, p. 267–268.

Fink, L. K., Jr., 1974, Geologie de la Désirade, Excursion B2, *in* Livretguide [Field guide] d'excursions dans les Antilles francaises, Caribbean Geological Conference, VII, Guadeloupe, 1974, p. 77–92.

Foote, R. Q., and Martin, R. G., 1981, Petroleum geology of the Gulf of Mexico Maritime Boundary Assessment area, *in* Powers, R. B., ed., Geologic framework, petroleum potential, petroleum-resource estimates, mineral and geothermal resources, geologic hazards, and deep-water drilling technology of the maritime boundary region in the Gulf of Mexico: U.S. Geological Survey Open-File Report 81-265, p. 68–79.

Fox, P. J., and Heezen, B. C., 1975, Geology of the Caribbean crust: *in* Nairn, A.E.M., and Stehli, F. G., eds., The Ocean basins and margins, v. 3, The Gulf of Mexico and the Caribbean: New York, Plenum Press, p. 421–466.

Fox, P. J., Heezen, B. C., and Johnson, G. L., 1970, Jurassic sandstone from the tropical Atlantic: Science, v. 170, p. 1402–1404.

Galavis, S., J. A., and Louder, L. W., 1970, Preliminary studies on geomorphology, geology and geophysics on the continental shelf and slope of northern South America (preprint): Eight World Petroleum Congress, Caracas, Venezuela, September 1970, 26 p.

Galli-Olivier, Carlos, 1979, Ophiolite and island-arc volcanism in Costa Rica: Geological Society of America Bulletin, v. 90, no. 5, p. 444–452.

Gansser, August, 1954, The Guiana Shield (S. America): Geological Observations: Eclogae Geologicae Helvetiae, v. 47, no. 1, p. 77–112.

—— 1955, Ein Beitrag zur Geologie und Petrographie der Sierra Nevada de Santa Marta (Kolumbien, Südamerika): Schweizer Mineralogische und Petrographische Mitteilungen, v. 35, no. 2, p. 209–279.

—— 1973, Facts and theories on the Andes: Journal of the Geological Society of London, v. 129, p. 93–131.

—— 1974, The Roraima problem (South America), *in* Jung, P., and others, eds., Contributions to the geology and paleobiology of the Caribbean and adjacent areas [Kugler volume]: Verhandlungen der Naturforschenden Gesellschaft in

Basel, v. 84, no. 1, p. 80–97.

Garrison, L. E., and others, 1972a, USGS-IDOE Leg. 3: Geotimes, v. 17, no. 3, p. 14–15.

—— 1972b, Acoustic reflection profiles—Eastern Greater Antilles: U.S. Department of Commerce National Technical Information Service PB2-07596.

Gaudette, H. E., Hurley, F. M., Espejo C., Anibal, and Dahlberg, E. H., 1977, Basamento Arqueano al sur del Complejo de Imataca en Venezuela y Surinam, in Memoria, Congreso Geológico Venezolano, V, Caracas, Venezuela Ministerio de Minas e Hidrocarburos and Sociedad Venezolana de Geologos, v. 2, p. 493–508.

Gaudette, H. E., Mendoza, V., Hurley, P. M., Fairbairn, H. W., 1978, Geology and age of the Paraguaza rapakivi granite, Venezuela: Geological Society of America Bulletin, v. 89, p. 1335–1340.

Gibbs, A. K., 1980, Geology of the Barama-Mazaruni Supergroup of Guyana [Ph.D. thesis]: Cambridge, Massachusetts, Harvard University, 446 p.

Gibbs, A. K., and Barron, C. N., 1983, The Guiana Shield reviewed: Episodes, v. 1983, no. 2, p. 7–14.

González de Juana, C., Iturralde, J., and Picard, X., 1980, Geología de Venezuela y de sus cuencas petroliferas: Caracas, Ediciones Fonives, 2 volumes, 1031 p.

González Silva, L. A., 1977, Geología de la Sierra del Interior (región central) y parte de los Llanos de Venezuela (incluyeno parte de los Estados Carabobo, Aragua, Guarico, y Cojedes), in Memoria, Congreso Latinoamericano de Geología, II, Caracas, 1973, Venezuela Ministerio de Minas e Hidocarburos, Boletín de Geología, Publicación Especial no. 7, v. 3, p. 1629–1650.

Gose, W. A., and Sanchez-Barreda, 1981 [1982], Paleomagnetic results from southern Mexico: Geofisica International v. 20, no. 3, p. 163–175.

Gose, W. A., Scott, G. R., and Schwartz, D. K., 1980, The aggregation of Mesoamerica: Paleomagnetic evidence, in Pilger, R. H., Jr., ed., The origin of the Gulf of Mexico and the early opening of the central North Atlantic Ocean, Proceedings of a symposium at Louisiana State University, Baton Rouge, March 3-5, 1980: School of Geoscience, Louisiana State University, p. 51–54.

Grauch, R. I., 1971, Geology of the Sierra Nevada south of Mucuchies, Venezuelan Andes: an aluminum-silicate-bearing metamorphic terrain [Ph.D. thesis]: Philadelphia, Pennsylvania University, 180 p.

—— 1972, Preliminary report of a late(?) Paleozoic metamorphic event in the Venezuelan Andes, in Shagam, R., and others, eds., Studies in Earth and Space Sciences, Geological Society of America Memoir 132, p. 465–473.

—— 1975, Geología de la Sierra Nevada al sur de Mucuchies, Andes Venezolanos: una region metamórfica de alumino-silicatos: Venezuela Ministerio de Minas e Hidrocarburos, Boletin de Geología, v. 12, no. 23, p. 339–441.

Griscom, Andrew, and Geddes, W. H., 1966, Island-arc structure interpreted from aeromagnetic data near Puerto Rico and the Virgin Islands: Geological Society of America Bulletin, v. 77, no. 2, p. 153–162.

Gunn, Bernard, and Roobol, John, 1976, The geochemistry of the limestone Caribbees, in Transactions, Caribbean Geological Conference, VII, Saint-Francois (Guadeloupe), 1974, p. 385–391.

Hamilton, W. H., 1969, Mesozoic California and the underflow of Pacific mantle: Geological Society of America Bulletin, v. 80, p. 2409–2429.

Hantke, G., and Parodi, I. A., 1966, Catalogue of the active volcanoes of the world including solfatara fields: part XIX, Colombia, Ecuador, and Peru: International Volcanological Association, Naples, 73 p.

Hargraves, R. B., 1968, Paleomagnetism of the Roraima dolerites: Geophysical Journal Royal Astronomical Society, v. 16, p. 147–160.

Hargraves, R. B., and Shagam, R., 1969, Paleomagnetic study of La Quinta Formation, Venezuela: American Association of Petroleum Geologists Bulletin, v. 53, no. 3, p. 537–552.

Hargraves, R. B., and Skerlec, G. M., 1982, Paleomagnetism of some Cretaceous-Tertiary igneous rocks on Venezuelan offshore Islands, Netherlands Antilles, Trinidad and Tobago, in Transactions, Caribbean Geological Conference, IX, Santo Domingo, Dominican Republic, 1980, p. 509–517.

Harvey, S.R.M., 1972, Origin of the southern Caribbean mountains, in Shagam, R., and others, eds., Studies in the Earth and Space Sciences, Geological

Society of America Memoir 132, p. 387–400.

Hay, John, and Aymard, Richard, 1977, El Cretáceo en el subsuelo de Anzoátegui y parte de Monagas, Cuenca de Venez uela Oriental, in Memoria, Congreso Geológico Venezolano, V, Caracas, 1977, Venezuela Ministerio de Energía y Minas and Sociedad Venezolana de Geológos, v. 4, p. 1557–1574.

Hayes, D. E., and others, 1970, Sites 143 and 144 [Demerara Rise and vicinity], in Hayes, D. E., and others, eds., Initial Reports of the Deep Sea Drilling Project, v. 14: U.S. Government Printing Office, Washington, p. 283–338.

Hedberg, H. D., 1950, Geology of the Eastern Venezuela Basin: Geological Society of America Bulletin, v. 61, p. 1173–1216.

Hedberg, H. D., Soss, L. C., and Funkhouser, H. J., 1947, Oil fields of Greater Oficina area, central Anzoátegui, Venezuela: American Association of Petroleum Geologists Bulletin, v. 31, p. 2089–2169.

Herd, D. G., 1974, Glacial and volcanic geology of the Ruiz-Tolima volcanic complex, Cordillera Central, Colombia [Ph.D. thesis]: Seattle, University of Washington, 78 p. (also 1982, Publicaciones Geologicas Especiales de Ingeominas No. 8, 48 p.

Hey, Richard, 1977, Tectonic evolution of the Cocos-Nazca spreading center: Geological Society of America Bulletin, v. 88, p. 1404–1420.

Holcombe, T. L., 1977, Caribbean bathymetry and sediments, in Weaver, J. D., ed., Geology, geophysics, and resources of the Caribbean, Report of the IDOE Workshop on the geology and marine geophysics of the Caribbean region and its resources, Kingston, Jamaica, 1975: University of Puerto Rico, Mayagüez, p. 27–62.

—— 1979, Geomorphology and subsurface geology west of St. Croix, U.S. Virgin Islands, in Watkins, J. S., and others, eds., Geological and geophysical investigations of continental margins: American Association of Petroleum Geologists Memoir 29, p. 353–362.

Holcombe, T. L., Vogt, P. R., Matthews, J. E., and Murchison, R. R., 1973, Evidence for sea-floor spreading in the Cayman Trough: Earth and Planetary Science Letters, v. 20, p. 357–371.

Hood, Peter, and Tyl, I., 1977, Residual magnetic anomaly map of Guyana and its regional geological interpretation, in Memoria, Congreso Latinoamericano de Geología, II, Caracas, 1973, Venezuela Ministerio de Minas e Hidrocarburos, Boletín de Geología, Publicación Especial no. 7, v. 3, p. 2219–2235.

Houtz, R. E., and Ludwig, W. J., 1977, Structure of Colombian Basin, Caribbean Sea, from profiler-sonobuoy measurements, in Abstracts, Caribbean Geological Conference, VIII, Curaçao, 1977, Stichting GUA c/o Geologisch Instituut, 130 Nieuwe Prinsengracht, Amsterdam, p. 74.

—— 1977, Structure of Colombia Basin, Caribbean Sea, from profiler-sonobuoy measurements: Journal of Geophysical Research, v. 82, no. 30, p. 4861–4867.

Huguett, Alcides, Galvis, Jaime, and Ruge, Primitivo, 1979, Geología, in La Amazona Colombiana y sus recursos: Proyecto Radargramétrico del Amazonas, Bogotá, Chapter 2, p. 29–92.

Hurley, P. M., Fairbairn, H. W., Gaudette, H. E., Mendoza, V., Martín B., C., and Espejo, A., 1973, Progress report on Rb-Sr age dating in the northern Guyana Shield: Massachusetts Institute of Technology Geochronology Laboratory Progress Report 20, 1972-1973, p. 1–7.

Ibrahim, A. K., Latham, G. V., and Ladd, John, 1979, Seismic refraction and reflection measurements in the Middle America Trench offshore Guatemala: Journal of Geophysical Research, v. 84, no. B10, p. 5643–5649.

Ibrahim, A. K., Carye, J., Latham, G., and Buffler, R. T., 1981, Crustal structure in Gulf of Mexico from OBS refraction and multichannel reflection data: American Association of Petroleum Geologists Bulletin, v. 65, no. 7, p. 1207–1229.

Instituto Cubano de Geodesía y Cartografía, 1978, Atlas de Cuba [especially geologic map, p. 24–25 and tectonic map, p. 26–27]: Instituto Cubano de Geodesía y Cartografía, Habana.

Interoceanic Canal Study Commission, 1968, Geology, final report, Route 17: Panama, Office Interoceanic Canal Studies, v. 1, 32 p.

Irving, E. M., 1975, Structural evolution of the northernmost Andes: U.S. Geological Survey Professional Paper 846, 47 p.

Iturralde-Vinent, M. A., 1977, Los movimientos tectónicos de la etapa de desarrollo platafórmico en Cuba: Academia de Ciencias de Cuba, Informe Científico-Tecnico no. 20, 24 p.

—— 1978, Los movimientos tectónicos de la etapa de desarrollo platafórmico en Cuba: Geologie en Mijnbouw, v. 57, no. 2, p. 205–212.

Jones, D. L., Silberling, N. J., Berg, H. C., and Plafker, George, 1981, Map showing tectonostratigraphic terranes of Alaska, columnar sections, and summary descriptions of terranes: U.S. Geological Survey Open-File Report 81-792, scale 1:2,500,000, 2 sheets, 19 p.

Joyce, James, 1982, The lithology and structure of the eclogite and glaucophanite-bearing rocks on the Samaná Peninsula, Dominican Republic, *in* Transactions, Caribbean Geological Conference, IX, Santo Domingo, Dominican Republic, v. 2, p. 417–421.

Joyce, James, and Nagle, F., 1980, Road Log August 26, Fourth day-morning Samaná Peninsula, *in* Field Guide, Caribbean Geological Conference, IX, Santo Domingo, Dominican Republic, 1980, p. 255–258.

Julivert, Manuel, 1970, Cover and basement tectonics in the Cordillera Oriental of Colombia, South America, and a comparison with some other folded chains: Geological Society of America Bulletin, v. 81, no. 12, p. 3623–3646.

Kalliokoski, J., 1965a, Geology of north-central Guayana shield, Venezuela: Geological Society of America Bulletin, v. 76, no. 9, p. 1027–1050.

—— 1965b, The metamorphosed iron-ore of El Pao, Venezuela: Economic Geology, v. 60, no. 1, p. 100–116.

Kearey, Philip, 1974, Gravity and seismic reflection investigations into the crustal structure of the Aves Ridge, eastern Caribbean: Geophysics Journal Royal Astronomical Society, v. 38, p. 435–448.

—— 1976, Gravity and seismic reflection investigation into the crustal structure of the Aves ridge, Eastern Caribbean, *in* Transactions, Caribbean Geological Conference, VII, Guadeloupe, 1974, p. 311–320.

Kearey, Philip, Peter, George, and Westbrook, G. K., 1975, Geophysical maps of the eastern Caribbean: Journal Geological Society of London, v. 131, p. 311–321.

Keats, W., 1976, The Roraima Formation in Guyana: A revised stratigraphy and a proposed environment of deposition, *in* Memoria, Congreso Latinoamericano de Geología, II, Caracas, 1973, Venezuela Ministerio de Minas e Hidrocarburos, Boletín de Geología, Publicación Especial no. 7, v. 2, p. 901–940.

Kellogg, J. N., 1980, Cenozoic basement tectonics of the Sierra de Perijá, Venezuela and Colombia, *in* Abstracts, Caribbean Geological Conference, IX, Santo Domingo, Dominican Republic, p. 35.

Kellogg, J. N., and Bonini, W. E., 1982, Subduction of the Caribbean plate and basement uplifts in the overriding South American plate: Tectonics, v. 1, no. 3, p. 251–276.

Kesler, S. E., 1971a, Petrology of the Terre-Neuve igneous province, northern Haiti: *in* Donnelly, T. W., ed., Caribbean geophysical, tectonic, and petrologic studies: Geological Society of America Memoir 130, p. 119–137.

—— 1980a, Geology and geochemistry of the Los Ranchos Formation, Dominican Republic, *in* Abstracts, Caribbean Geological Conference, IX, Santo Domingo, Dominican Republic, 1980, p. 35–36.

Kesler, S. E., Sutter, J. F., Issigonis, M. J., Jones, L. M., and Walker, R. L., 1977a, Evolution of porphyry copper mineralization in an oceanic island arc: Panama: Economic Geology, v. 72, p. 1142–1153.

Kesler, S. E., Sutter, J. F., Jones, L. M., and Walker, R. L., 1977, Early Cretaceous basement rocks in Hispaniola: Geology, v. 5, no. 4, p. 245–247.

Khudoley, K. M., and Meyerhoff, A. A., 1971, Paleogeography and geological history of Greater Antilles: Geological Society of America Memoir 129, 199 p.

Kim, J. J., Matumoto, Tosimatu, and Latham, G. V., 1982, A crustal section of northern Central America as inferred from wide-angle reflections from shallow earthquakes: Seismological Society of America Bulletin, v. 72, no. 3, p. 925–940.

Klar, G., 1979, Geochronology of the El Monteco-Guri and Guasipati areas, Venezuelan Guiana Shield [Ph.D. thesis]: Cleveland, Ohio, Case Western Reserve University, 177 p.

Kloosterman, J. B., 1976, Giant ring volcanoes of the Guiana Shield, *in* Memoria, Congreso Latinoamericano de Geología, II, Venezuela Ministerio de Minas e Hidrocarburos, Boletín de Geología, Publicación Especial no. 7, v. 2, p. 713–722.

Krause, D. C., 1971, Bathymetry, geomagnetism, and tectonics of the Caribbean Sea north of Colombia, *in* Donnelly, T. W., ed. Caribbean geophysical, tectonic, and petrologic studies: Geological Society of America Memoir 130, p. 35–54.

Krijnen, J. P., and Lee Chin, A. C., 1978, Geology of the northern, central, and south-eastern Blue Mountains, Jamaica, with a provisional compilation map of the entire inlier: Geologie en Mijnbouw, v. 57, no. 2, p. 243–250.

Krook, L., 1969, Investigations on the mineralogical composition of the Tertiary and Quaternary sands in northern Surinam, *in* Proceedings, Guiana Geological Conference, VII, Paramaribo, 1966: Verhandlingen van het Koninklijk Nederlands geologisch Mijnbouwkundig Genootschap, v. 27, p. 89–100.

Krook, L., and Mulders, M. A., 1971, Geological and pedological aspects of the Upper Coesewijne Formation: *in* Contributions to the geology of Suriname, 2: Geologisch Mijnbownkundige Dienst van Suriname, v. 21, p. 183–208.

Kroonenberg, S. B., 1982, A Grenvillian granulite belt in the Colombian Andes and its relation to the Guiana Shield: Geologie en Mijnbouw, v. 61, no. 4, p. 325–333.

Krushensky, R., and Elston, D., 1983, Caribbean plate tectonics: New evidence, new conclusions [abs.]: Abstracts, Caribbean Geological Conference, X, Cartagena, Colombia, 1983, p. 45–46.

Kugler, H. G., 1956, Trinidad, *in* Jenks, W. F., ed., Handbook of South American Geology: Geological Society of America Memoir 65, p. 355–365.

—— 1968, Sedimentary volcanism, *in* Transactions, Caribbean Geological Conference, IV, Port-of-Spain, Trinidad, 1965, p. 11–20.

Ladd, J. W., Ibrahim, A. K., McMillen, K. J., Matumoto, T., Latham, G. V., von Huene, R. E., Watkins, J. S., Moore, J. C., and Worzel, J. L., 1978, Tectonics of the Middle America Trench offshore Guatemala, *in* Proceedings, International Symposium on the February 4th, 1976, Guatemalan earthquake and the reconstruction process: Guatemala City, May 1978, 18 p.

Ladd, J. W., Shih, T.-C., and Tsai, C. J., 1981, Cenozoic tectonics of central Hispaniola and adjacent Caribbean Sea: American Association of Petroleum Geologists Bulletin, v. 65, no. 3, p. 466–489.

Ladd, J. W., Worzel, J. L., and Watkins, J. S., 1977, Multifold seismic reflection records from the northern Venezuela Basin and the north slope of the Muertos Trench, *in* Talwani, Manik, and Pitman, W. C., III., eds., Island arcs, deep sea trenches, and back-arc basins: American Geophysical Union, Maurice Ewing Series 1, p. 41–56.

Ladd, J. W., and Watkins, J. S., 1978, Active margin structures within the north slope of the Muertos Trench: Geologie en Mijnbouw, v. 57, no. 2, p. 255–260.

—— 1979, Tectonic development of trench-arc complexes on the northern and southern margins of the Venezuela Basin, *in* Watkins, J. S., and others, eds., Geological and geophysical investigations of continental margins: American Association of Petroleum Geologists Memoir 29, p. 363–371.

—— 1980, Seismic stratigraphy of the western Venezuela Basin: Marine Geology, v. 35, p. 21–41.

Lagaay, R. A., 1969, Geophysical investigations of the Netherlands Leeward Antilles: Verhandelingen der Koninklijke Nederlandse Akademie van Wetenschappen, Afd. Natuurkunde, Eerste Reeks-Deel 25, no. 2, North-Holland Publishing Company, 86 p.

Lamb, J. L., and Sulek, J. A., 1968, Miocene turbidites in the Carapita Formation of eastern Venezuela, *in* Transactions, Caribbean Geological Conference, IV, Port-of-Spain, Trinidad, 1965, p. 111–119.

Lewis, G. E., and Straczek, J. A., 1955, Geology of south-central Oriente, Cuba: U.S. Geological Survey Bulletin 975-D, p. 171–336.

Lewis, J. F., 1980, Resume of the geology of Hispaniola, *in* Field Guide, Caribbean Geological Conference, IX, Santo Domingo, Dominican Republic, 1980, p. 5–31.

—— 1980b, The south slope of the Cordillera Central, Four Day Field Trip F, First Day, *in* Field Guide, Caribbean Geological Conference, IX, Santo

Domingo, Dominican Republic, 1980, p. 169–171.

—— 1982a, Cenozoic tectonic evolution and sedimentation in Hispaniola, *in* Transactions, Caribbean Geological Conference, IX, Santo Domingo, Dominican Republic, 1980, v. 1, p. 65–73.

—— 1982b, Granitoid rocks in Hispaniola, *in* Transactions, Caribbean Geological Conference, IX, Santo Domingo, Dominican Republic, 1980, v. 2, p. 391–401.

—— 1982c, Ultrabasic and associatead rocks in Hispaniola, *in* Transactions, Caribbean Geological Conference, IX, Santo Domingo, Dominican Republic, v. 2, p. 403–408.

Lewis, John, and Robinson, Edward, 1976, A revised stratigraphy and geological history of the Lesser Antilles, *in* Transactions, Caribbean Geological Conference, VII, Saint-Francois (Guadeloupe), 1974, p. 339–344.

Llinas, Romeo, Gil, Nelson, Seward, Michael, Taveres, Ivan, and Snow, William, eds., 1982, Transactions, Caribbean Geological Conference, IX, Santo Domingo, Dominican Republic, 1980, 2 volumes, 716 p.

Lockwood, J. P., 1965, Geology of the Serrania de Jarara area, Guajira Peninsula, Colombia, [Ph.D. thesis]: Princeton, New Jersey, Princeton University, 237 p.

Lonsdale, Peter, and Klitgord, K. D., 1978, Structure and tectonic history of the eastern Panama Basin: Geological Society of America Bulletin, v. 89, p. 981–999.

López Ramos, E., 1974, Geología general y de México, Privately printed, Mexico City, D. F., 509 p.

—— 1975, Geological summary of the Yucatan Peninsula, *in* Nairn, A.E.M., and Stehli, F. G., eds.: The Ocean basins and margins, v. 3, The Gulf of Mexico and the Caribbean: New York, Plenum Press, p. 257–282.

—— 1981, Geología de Mexico, v. 3, second edition: Privately printed, Mexico City, 446 p.

Lowrie, A., 1978, Buried trench south of the Gulf of Panama: Geology v. 6, p. 434–436.

Lowrie, Allen, Aitken, Thomas, Grim, Paul, and McRaney, Linda, 1979, Fossil spreading center and faults within the Panama fracture zone: Marine Geophysical Researches, v. 4, p. 153–166.

Lu, R. S., McMillen, K. J., and Phillips, J. D., 1983, Multichannel seismic survey of the Colombia Basin and adjacent margin, *in* Watkins, J. S., and Drake, C. L., eds., Studies in continental margin geology, American Association of Petroleum Geologists Memoir 34, p. 395–410.

Ludwig, W. J., Houtz, R. E., and Ewing, J. I., 1975, Profiler-sonobuoy measurements in Colombia and Venezuela Basins, Caribbean Sea: American Association of Petroleum Geologists Bulletin, v. 59, no. 1, p. 115–123.

MacDonald, K. C., and Holcombe, T. L., 1978, Inversion of magnetic anomalies and sea-floor spreading in the Cayman trough: Earth and Planetary Science Letters, v. 40, p. 407–414.

MacDonald, W. D., 1964, Geology of the Serrania de Macuira area, Guajira Peninsula, Colombia [Ph.D. thesis]: Princeton, New Jersey, Princeton University, 167 p.

—— 1972b, Continental crust, crustal evolution, and the Caribbean: Geological Society of America Memoir 132, p. 351–363.

—— 1980, Anomalous paleomagnetic directions in Late Tertiary andesitic intrusions of the Cauca depression, Colombian Andes: Tectonophysics, v. 68, p. 339–348.

MacDonald, W. D., and Opdyke, N. D., 1972, Tectonic rotations suggested by paleomagnetic results from northern Colombia, South America: Journal of Geophysical Research, v. 77, no. 29, p. 2720–2730.

—— 1974, Triassic paleomagnetism of northern South America: American Association of Petroleum Geologists Bulletin, v. 58, no. 2, p. 208–215.

MacDonald, W. D., Doolan, B. L., and Cordani, U. G., 1971, Cretaceous-Early Tertiary metamorphic K-Ar values from the south Caribbean: Geological Society of America Bulletin, v. 82, p. 1381–1388.

McCann, W. R., and Sykes, L. R., 1981, Subduction of aseismic ridges in the northeastern Caribbean: Effects on seismicity and arc morphology [abs.]: EOS, American Geophysical Union Transactions, v. 62, no. 17, p. 323–324.

Maresch, W. V., 1972b, Guide of Excursion L-1, Field trip to the Rinconada Group, *in* Transactions, Caribbean Geological Conference, VI, Isla de Margarita, Venezuela, 1971, p. 20–21.

—— 1972c, Ecologitic-amphibolitic rocks on Isla Margarita, Venezuela: a preliminary report, *in* Shagam, R., and others, eds., Studies in Earth and space sciences: Geological Society of America Memoir 132, p. 429–437.

—— 1975, The geology of northeastern Margarita Island, Venezuela: a contribution to the study of Caribbean plate margins: Geologischen Rundschau, v. 64, no. 3, p. 846–883.

Marlow, M. S., Garrison, L. E., Martin, R. G., Trumbull, J.V.A., and Cooper, A. K., 1974, Tectonic transition in the northeastern Caribbean: U.S. Geological Survey Journal of Research, v. 2, no. 3, p. 289–302.

Martín Bellizzia, Cecilia, 1961, Geología del Macizo de el Baul, Estado Cojedes, *in* Memoria, Congreso Geológico Venezolano, III, Caracas, 1969, Venezuela Ministerio de Minas e Hidrocarburos, Boletín de Geología, Publicación Especial no. 3, v. 4, p. 1453–1530.

—— 1974, Paleotectónica del Escudo de Guayana, *in* Memoria, Conferencia Geológica Inter-Guayanas, IX, Ciudad Guayana, 1973, Venezuela Ministerio de Minas e Hidrocarburos Boletín de Geología, Publicación Especial no. 6, p. 251–305.

Martín F., Cecilia, 1978, Mapa tectónico, Norte de América del Sur: Venezuela Ministerio de Energía y Minas, scale 1:2,500,000.

Martín-Bellizzia, Cecila, and de Arozena, J.M.I., 1972, Complejo ultramáfico zonado de Tausabana-El Rodeo, gabro zonado de Siraba-Capuana, y complejo subvolcanico estratificado de Santa Ana, Paraguaná, Estado Falcón, *in* Transactions, Caribbean Geological Conference, VI, Isla de Margarita, Venezuela, 1971, p. 337–356.

Martin-Kaye, P. H., 1969, A summary of the geology of the Lesser Antilles: Overseas Geology Mineral Resources, Great Britain, v. 10, no. 2, p. 172–206.

Martin, R. G., 1980, Distribution of salt structures in the Gulf of Mexico: Map and descriptive text: U.S. Geological Survey Miscellaneous Field Studies Map MF-1213, 2 plates, 8 p.

Martin, R. G., and Foote, R. Q., 1981a, Geology and geophysics of the Maritime Boundary assessment areas, *in* Powers, R. B., ed., Geologic framework, petroleum potential, petroleum-resource estimates, mineral and geothermal resources, geologic hazards, and deep-water drilling technology of the Maritime Boundary region in the Gulf of Mexico: U.S. Geological Survey Open-File Report 81-265, p. 30–67.

Martínez, A. R., 1970, Giant fields of Venezuela: American Association of Petroleum Geologists Memoir 14, p. 326–336.

—— 1972a, Los recursos de hidrocarburos de Venezuela, *in* Memoria, Congreso Geológico Venezolano, IV, Venezuela Ministerio de Minas e Hidrocarburos, Boletín de Geología, Publicación Especial no. 5, v. 5, p. 2687–2727.

—— 1977a, El progreso de la exploración en las cuencas tradicionales, *in* Memorias, Congreso Geológico Venezolano, V, Ministerio de Energía y Minas and Sociedad Venezolano de Geólogos, v. 4, p. 1333–1348.

—— 1977b, Los recursos de hidrocarburos en las areas nuevas, *in* Memorias, Congreso Geológico Venezolano, V, Ministerio de Energía y Minas and Sociedad Venezolana de Geólogos, v. 4, p. 1349–1358.

Mascle, A., Montadert, L., Biju-Duval, B., Bizon, G., Muller, C., and Eva, A., 1980, Evolution paleobathymétrique et paleotectonique au sud d'Hispaniola durant le Tertiaire, *in* Abstracts, Caribbean Geological Conference, IX, Santo Domingo, Dominican Republic, 1980, p. 44–45.

Matthews, J. E., and Holcombe, T. L., 1976, Regional geological/geophysical study of the Carribean Sea (Navy Ocean Area NA-9): 1. Geophysical maps of the Eastern Caribbean: Washington, D.C., U.S. Naval Oceanographic Office, 43 p.

Mattson, P. H., 1973a, Middle Cretaceous nappe structures in Puerto Rican ophiolites and their relation to the tectonic history of the Greater Antilles: Geological Society of America Bulletin, v. 84, no. 1, p. 21–38.

Maurrasse, Florentin, 1980a, New data on the stratigraphy of the Southern Peninsula of Haiti: Transactions du 1st Colloque sur la Geologie D'Haïti, Port-au-Prince, Haïti, 1980, p. 184–199.

—— 1980b, Les marges continentales d'Haïti: Transactions du 1st Colloque sur la

Geologie D'Haïti, Port-au-Prince, Haïti, 1980, p. 200–206.

Maurrasse, Florentin, Husler, John, Georges, Gaston, Schmitt, Roman, and Damond, Paul, 1979, Upraised Caribbean sea floor below acoustic reflector B" at the Southern Peninsula of Haiti: Geologie en Mijnbouw, v. 58, no. 1, p. 71–83.

Maurrasse, Florentin, J-M.R., Pierre-Louis, Fritz, and Rigaud, J.-G., 1982, Cenozoic facies distribution in the Southern Peninsula of Haiti and the Barahona Peninsula, *in* Transactions, Caribbean Geological Conference, IX, Santo Domingo, Dominican Republic, 1980, v. 1, p. 161–174.

McBirney, A. R., and Williams, H., 1965, Volcanic history of Nicaragua: University of California Publications in Geological Sciences, v. 55, p. 1–65.

McCann, W. R., and Sykes, L. R., 1981, Subduction of aseismic ridges in the northeastern Caribbean: Effects on seismicity and arc morphology [abs.]: EOS (American Geophysical Union Transactions), v. 62, no. 17, p. 323–324.

Menéndez, A. V. de V., 1965, Geología del area del Trináco, centro norte de Estado Cojedes, Venezuela: Venezuela Ministerio de Minas e Hidrocarburos, Boletín de Geología, v. 6, no. 12, p. 417–543.

——1966, Tectónica de la parte central de las montañas occidentales del Caribe, Venezuela: Venezuela Ministerio de Minas e Hidrocarburos, Boletín de Geología, v. 8, no. 15, p. 116–139.

——1967, Tectonics of the central part of the western Caribbean Mountains, Venezuela: *in* Studies in tropical oceanography: Miami, University of Miami Press, v. 5, p. 103–130.

——1968, Revisión de la estratigrafía de la Provincia de Pastora segun el estudio de la region de Guasipati, Guayana Venezolana: Venezuela Ministerio de Minas e Hidrocarburos Boletín de Geología, v. 10, no. 9, p. 309–338.

——1974, Guía de la excursión geológica Guasipati-El Callao-Canaima, *in* Memoria, Conferencia Geológica Inter-Guayanas, IX, Ciudad Guayana, Venezuela, 1972, Venezuela Ministerio de Minas e Hidrocarburos, Boletín de Geología, Publicación Especial no. 6, p. 49–67.

Menéndez, Alfredo, Benaim, Nesin, and Espejo, Aníbal, 1974, Estratigrafía Precambrica de la provincia geológica de Pastora al este del Río Caroni y su correlacion tentativa interGuayana [abs.]: *in* Memoria, Conferencia Geológica Inter-Guayanas, IX, Ciudad Guayana, Venezuela, 1972, Venezuela Ministerio de Minas e Hidrocarburos, Boletín de Geología, Publicación Especial no. 6, p. 339–341.

Mendoza, Vincente, 1974, Geología del area del Río Suapure, parte Noroccidental del Escudo de Guayana, Estado Bolívar, Venezuela, *in* Memoria, Conferencia Geológica Inter-Guayanas, IX, Ciudad Guayana, Venezuela, 1972, Ministerio de Minas e Hidrocarburos, Boletín de Geología, Publicación Especial, no. 6, p. 306–338.

Mendoza, S., Vicente, 1977, E volucion tectónica del Escudo de Guayana, *in* Memoria Congreso Latinoamericano de Geología, II, Caracas, 1973, Venezuela Ministerio de Minas e Hidrocarburos, Boletín de Geologí, Publicación Especial No. 7, v. 3, p. 2237–2270.

Meyer, R. P., Mooney, W. D., Hales, A. L., Helsley, C. E., Woollard, G. P., Hussong, D. M., Kroenke, L. W., and Ramirez, J. E., 1976, Project Nariño III: Refraction observation across a leading edge, Malpelo Island to the Colombian Cordillera Occidental: *in* Sutton, G. H., and others, eds., The geophysics of the Pacific Ocean Basin and its margin, American Geophysical Union Monograph 19, p. 105–132.

Meyerhoff, A. A., 1973, Diapirlike features offshore Honduras: Implications regarding tectonic evolution of Cayman Trough and Central America: Discussion: Geological Survey of America Bulletin, v. 84, no. 6, p. 2147–2152.

Meyerhoff, A. A., and Hatten, C. W., 1974a, Bahamas Salient of North America: Tectonic framework, stratigraphy and petroleum potential: American Association of Petroleum Geologists Bulletin, v. 58, no. 6, pt. II of II, p. 1201–1239.

——1974b, Bahamas salient of North America, *in* Burk, C. A., and Drake, C. L., eds., The geology of continental margins: New York, Springer-Verlag, p. 429–446.

Meyerhoff, A. A., and Kreig, E. A., 1977a, Jamaican Petroleum potential-1: Future Jamaican exploration is justified, Part 1: Oil and Gas Journal v. 75,

——1977b, Jamaican Petroleum potential-2: Five major cycles make up Jamaican tectonic and structural history [part 2]: Oil and Gas Journal, v. 75, no. 37, p. 141–146.

Michelson, J. E., 1976, Miocene deltaic oil habitat: American Association of Petroleum Geologists Bulletin, v. 60, p. 1502–1519.

Miller, J. B., 1962, Tectonic trends in Sierra de Perijá and adjacent parts of Venezuela and Colombia: American Association of Petroleum Geologists Bulletin, v. 46, no. 9, p. 1565–1595.

Mills, R. A., and Hugh, K. E., 1974, Reconnaissance geologic map of Mosquitia region, Honduras and Nicaraguan Caribbean Coast: American Association of Petroleum Geologists Bulletin, v. 58, p. 189–207.

Montgomery, C. W., 1979, Uranium-Lead geochronology of the Archean Imataca Series, Venezuelan Guayana Shield: Contributions to Mineralogy and Petrology, v. 69, p. 167–176.

Montgomery, C. W., and Hurley, P. M., 1978, Total-rock U-Pb and Rb-Sr systematics in the Imataca Series, Guayana Shield, Venezuela: Earth and Planetary Science Letters, v. 39, p. 281–290.

Mooney, W. D., 1980, An East Pacific-Caribbean ridge during the Jurassic and Cretaceous and the evolution of western Colombia, *in* Pilger, R. H., Jr., ed., The origin of the Gulf of Mexico and the early opening of the central North Atlantic Ocean, Proceedings of a symposium at Louisiana State University, Baton Rouge March 3-5, 1980: School of Geoscience, Louisiana State University, Baton Rouge, p. 55–73.

Moore, G. T., and Fahlquist, D. A., 1976, Seismic profile tying Caribbean DSDP Sites 153, 151, and 152: Geological Society of America Bulletin, v. 87, p. 1609–1614.

Moreno, Luis, Lira M., Pedro, Mendoza, Vicente, and Ríos, J. H., 1977, Analysis de edades radiométricas en la parte oriental de la Guayana Venezolana y eventos tectónico-termales registrades, *in* Memorias, Congreso Geológico Venezolano, V, Caracas, 1977, Venezuela Ministerio de Energía y Minas and Sociedad Venezolana de Geólogos, v. 2, p. 509–518.

Morgan, B. A., 1969, Geología de la region de Valencia, Carabobo, Venezuela: Venezuela Ministerio de Minas e Hidrocarburos Boletín de Geología, v. 10, no. 20, p. 3–136.

Muessig, K. W., 1978, The central Falcón igneous suite, Venezuela: Alkaline basaltic intrusions of Oligocene-Miocene age: Geologie en Mijnbouw, v. 57, no. 2, p. 261–266.

——Structure and Cenozoic tectonics of the Falcón Basin, Venezuela and adjacent area: Geological Society of America Memoir 162 (this volume).

Murany, E. E., 1972a, Tectonic basis for Anaco fault, Eastern Venezuela: American Association of Petroleum Geologists Bulletin, v. 56, no. 5, p. 860–870.

——1972b, Structural analysis of the Caribbean coast eastern interior range of Venezuela, *in* Transactions, Caribbean Geological Conference, VI, Isla de Margarita, Venezuela, 1971, p. 295–298.

Murray, C. G., 1972, Zoned ultramafic complexes of the Alaskan type: Feeder pipes of andesitic volcanoes: *in* Shagam, R., and others, eds., Studies in the Earth and Space Sciences, Geological Society of America Memoir 132, p. 313–335.

Nagle, Frederick, 1974, Blueschist, eclogite, paired metamorphic belts, and the early tectonic history of Hispaniola: Geological Society of America Bulletin, v. 85, p. 1461–1466.

——1979, Geology of the Puerto Plata area, Dominican Republic, *in* Lidz, Barbara, and Nagle, Frederick, eds., Hispaniola: Tectonic focal point of the northern Caribbean—Three geologic studies in the Dominican Republic: Miami Geological Society, p. 1–28.

Nagle, F., Fink, L. K., Bostrom, K., and Stipp, J. J., 1973: Copper in pillow basalts from La Désirade, Lesser Antilles island arc: Earth and Planetary Science Letters, v. 19, p. 193–197.

Nagle, F., and Redmond, B., 1980, Road Log, August 25, Santiago to Samaná, Third day, *in* Field Guide, Caribbean Geological Conference, IX, Santa Domingo, Dominican Republic, 1980, p. 241–246.

Nairn, A.E.M., and Stehli, F. G., eds., 1975, The Ocean basins and margins, v. 3, The Gulf of Mexico and the Caribbean: New York, Plenum Press, 706 p.

Nelson, H. W., 1957, Contribution to the geology of the Central and Western Cordillera of Colombia in the sector between Ibagué and Cali: Leidse Geologische Mededelingen, Deel 22, 75 p.

Officer, C. B., Ewing, J. I., Hennion, J., F., Harkrider, D. G., and Miller, D. E., 1959, Geophysical investigations in the eastern Caribbean: Summary of 1955 and 1956 cruises, *in* Ahrens, L. H., and others, eds., Physics and Chemistry of the Earth: London, Pergamon Press, v. 3, p. 17–109.

Olszewski, W. J., Jr., Gaudette, H. E., and Mendoza, Vicente, 1977, Rb-Sr geochronology of the basement rocks, Amazonas Territory, Venezuela, A progress report: *in* Memoria, Congreso Geológico Venezolano, V, Ministerio de Energía y Minas and Sociedad Venezolana de Geólogos, v. 2, p. 519–526.

Onstott, T. C., and Hargraves, R. B., 1982, Paleomagnetic data and the Proterozoic apparent polar wander curve for the Venezuelan Guayana Shield, *in* Transactions, Caribbean Geological Conference, IX, Santo Domingo, 1980, v. 2: Universidad Católica Madre y Maestre, Santiago de los Caballeros, República Dominicana, p. 475–508.

Oxburgh, E. R., 1965, Geología de la región del Estado Carabobo, Venezuela: Venezuela Ministero de Minas e Hidrocarburos, Boletín de Geología v. 6, no. 6, no. 2, p. 113–208.

—— 1966, Geology and metamorphism of Cretaceous rocks in Eastern Carabobo State, Venezuelan coast ranges, *in* Hess, H. H., ed., Caribbean Geological Investigations: Geological Society of America Memoir 98, p. 241–310.

Palmer, H. C., 1979, Geology of the Moncion-Jarabacoa area, Dominican Republic, *in* Lidz, Barbara, and Nagle, Frederick, eds., Hispaniola: Tectonic focal point of the northern Caribbean—Three geologic studies in the Dominican Republic: Miami Geological Society, p. 29–68.

Pardo, Georges, 1975, Geology of Cuba: *in* Nairn, A.E.M., and Stehli, F. G., eds., The Ocean basins and margins, v. 3, The Gulf of Mexico and the Caribbean: New York, Plenum Press, p. 553–615.

Pennington, W. D., 1981, Subduction of the eastern Panamá Basin and seismotectonics of northwestern South America: Journal of Geophysical Research, v. 86, no. B11, p. 10753–10770.

Pérez, O. J., and Aggarwal, Y. P., 1981, Present-day tectonics of the southeastern Caribbean and northeastern Venezuela: Journal of Geophysical Research, v. 86, no. B11, p. 10791–10804.

Perfit, M. R., and Heezen, B. C., 1978, The geology and evolution of the Cayman Trench: Geological Society of America Bulletin, v. 89, no. 8, p. 1155–1174.

Perfit, M. R., Heezen, B. C., Rawson, M. R., and Donnelly, T. W., 1980, Chemistry, origin, and significance of metamorphic rocks from the Puerto Rico Trench: Marine Geology, v. 34, p. 125–156.

Persad, K. M., 1978, Hydrocarbon potential of the Trinidad area-1977: Geologie en Mijnbouw, v. 57, no. 2, p. 277–285.

Peter, George, 1972, Geologic structure offshore north-central Venezuela, *in* Transactions, Caribbean Geological Conference, VI, Isla de Margarita, Venezuela, 1971, p. 283–294.

Peter, George, and Westbrook, G. K., 1976, Tectonics of southwestern North Atlantic and Barbados Ridge complex: American Association of Petroleum Geologists Bulletin, v. 60, no. 7, p. 1078–1106.

Petersen, E. U., 1980, The Oxéc copper deposit, Guatemala: An ophiolite copper occurrence: Economic Geology, v. 75, p. 1053–1065.

Piburn, M. D., 1968, Metamorfísmo y estructúra del Grupo Villa de Cura, Venezuela Septentrional: Venezuela Ministerio de Minas e Hidrocarburos, Boletín de Geología, v. 9, no. 18, p. 184–290.

Pinet, P. R., 1971, Structural configuration of the northwestern Caribbean plate boundary: Geological Society of America Bulletin, v. 82, no. 7, p. 2027–2032.

—— 1972, Diapirlike features offshore Honduras: Implications regarding tectonic evolution of Cayman Trough and Central America: Geological Society of America Bulletin, v. 83, no. 7, p. 1911–1921.

—— 1973, Diapirlike features offshore Honduras: Implications regarding tectonic evolution of Cayman Trough and Central America: Reply: Geological Society of America Bulletin, v. 84, no. 6, p. 2153–2158.

—— 1975, Structural evolution of the Honduras continental margin and the sea floor south of the western Cayman Trough: Geological Society of America Bulletin, v. 86, p. 830–838.

Pinson, W. H., Jr., Hurley, P. M., Mencher, E., and Fairbairn, H. W., 1962, K-Ar and Rb-Sr ages of biotite from Colombia, South America: Geological Society of America Bulletin, v. 73, p. 907–910.

Polson, I. I., and Heñao, D., 1968, The Santa Marta wrench fault—a rebuttal, *in* Transactions, Caribbean Geological Conference, IV, Port-of-Spain, Trinidad, 1965, p. 263–266.

Poole, E. G., and Barker, L. T., 1982, The geology of the Scotland district, Barbados, *in* Transactions, Caribbean Geological Conference, IX, Santo Domingo, Dominican Republic, 1980, v. 2, p. 641–656.

Priem, H.N.A., 1978–1979, Isotopic dating in the Complejo migmatitica de Mitú, interim report: Proradam/Zwo Laboratorium Foor Istopengeologie, 5 p.

Pszczólkowski, A., 1977a, Stratigraphic-facies sequences of the Sierra del Rosário (Cuba): Bulletin de l'Académe Polonaise des Sciences, Série des Sciences de la Terre, v. 24, no. 3/4, p. 193–203.

—— 1977b, Nappe structure of Sierra del Rosário (Cuba): Bulletin de l'Académe Polonaise des Sciences, Série des Sciences de la Terre, v. 24, no. 3/4, p. 205–215.

Pyle, T. E., Meyerhoff, A. A., Fahlquist, D. A., Antoine, J. W., McCrevey, J. A., and Jones, P. C., 1973, Metamorphic rocks from northwestern Caribbean Sea: Earth and Planetary Science Letters, v. 18, p. 339–344.

Ramírez, J. E., 1968, Los volcanes de Colombia: Revista de la Academia de Colombia de Ciencias Exactas, Físicas y Naturales, v. 13, p. 227–235.

Reid, A. R., 1974, Stratigraphy of the type area of the Roraima Group, Venezuela, *in* Memoria, Conferencia Geológica Inter-Guayanas, IX, Ciudad Guayana, 1972, Venezuela Ministerio de Minas e Hidrocarburos, Boletín de Geología, Publicación Especial no. 6, p. 343–353.

Renz, O., 1960, Geología de la parte sureste de la Península de la Guajira (República de Colombia), *in* Memoria, Congreso Geológico Venezolano, III, Venezuela Ministerio de Minas e Hidrocarburos, Boletín de Geología, Publicación Especial no. 3, v. 1, p. 317–350.

—— 1977, The lithologic units of the Cretaceous in western Venezuela, *in* Memoria, Congreso Geológico Venezolano, V, Caracas, 1977, Ministerio de Energía y Minas and Sociedad Venezolana de Geólogos, p. 45–58.

Rezak, Richard, Antoine, J. W., Bryant, W. R., Fahlquist, D. A., and Bouma, A. H., 1972, Preliminary results of Cruise 71-A-4 of the R/V ALAMINOS in the Caribbean, *in* Transactions, Caribbean Geological Conference, VI, Margarita, Venezuela, 1971, p. 441–449.

Ríos, J. H., 1972, Geología de la región de Caicara, Estado Bolívar, *in* Memoria, Congreso Geológico Venezolano, IV, Venezuela, Ministerio de Minas e Hidrocarburos, Boletín de Geología, Publicación Especial no. 5, v. 3, p. 1759–1782.

Rodríguez, S. E., 1972, Analysis metalogénico del Yaracuy Occidental, *in* Transactions, Caribbean Geological Conference, VI, Isla de Margarita, Venezuela, 1971, p. 133–138.

Roemer, L., Bryant, W., and Fahlquist, D., 1976, A geophysical investigation of the Beata Ridge, *in* Transactions, Caribbean Geological Conference, VII, Guadaloupe, 1974, p. 115–125.

Rollins, J. F., 1965, Stratigraphy and structure of the Goajira Peninsula, northeastern Colombia [Ph.D. thesis]: University of Nebraska Studies, New Series, no. 30, 103 p.

Roper, P., 1978a, Stratigraphy of the Chuacús Group on the south side of the Sierra de la Minas Range, Guatemala: Geologie en Mijnbouw, v. 57, no. 2, p. 309–313.

—— 1978b, Structural fabric of serpentinite and amphibolite along the Motagua fault zone in El Progreso quadrangle, Guatemala: Gulf Coast Association of Geological Societies, Transactions, v. 28, pt. 2, p. 449–458.

Rosáles, Hugo, 1976, Excursion No. 5-Venezuela nororiental-Serranía del Interior, Primera parte: Maturin a Muelle de Cariáco, *in* Memoria, Congreso LatinoAmericano de Geología, II, Venezuela Ministerio de Minas e Hidrocarburras, Boletín de Geología, Publicación Especial no. 7, v. 2, p. 467–494.

Rosenfeld, J. H., 1981, Geology of the western Sierra de Santa Cruz, Guatemala, Central America: an ophiolite sequence [Ph.D. thesis]: Binghamton, New

York, State University of New York at Binghamton, 313 p.

Salvador, A., and Stainforth, R. M., 1968, Clues in Venezuela to the geology of Trinidad and vice versa, *in* Transactions, Caribbean Geological Conference, IV, Port-of-Spain, 1965, Transactions, p. 31–40.

Santamaria, Francisco, and Schubert, Carlos, 1974, Geochemistry and geochronology of the southern Caribbean-northern Venezuela plate boundary: Geological Society of America Bulletin, v. 85, no. 7, p. 1085–1098.

Santiago Acevedo, Jose, and Mejia Dautt, Octavio, 1980, Giant fields in the southeast of Mexico: *in* Transactions, Gulf Coast Association of Geological Societies, v. 30, p. 1–31.

Saunders, J. B., 1968, Field Trip Guide: Barbados, *in* Transactions, Caribbean Geological Conference, IV, Port-of-Spain, Trinidad, 1965, p. 443–449.

Saunders, J. B., and others, 1973, Cruise synthesis, *in* Edgar, N. T., Saunders, J. B., and others, Initial reports of the Deep Sea Drilling Project: Washington, D.C., U.S. Government, v. 15, p. 1077–1111.

Saunders, J. B., Jung, P., Geister, J., and Biju-Duval, B., 1982, The Neogene of the south flank of the Cibao Valley, Dominican Republic: a stratigraphic study, *in* Transactions, Caribbean Geological Conference, IX, Santo Domingo, Dominican Republic, 1980, v. 1, p. 151–160.

Schmidt-Effing, Reinhard, 1979a, Alter und Genese des Nicoya-Komplexes, einer ozeanischen Paläokruste (Oberjura bis Eozän) in südlichen Zentralamderika: Geologischen Rundschau, v. 68, no. 2, p. 457–494.

Schubert, Carlos, 1974, Seafloor tectonics east of the northern Lesser Antilles, *in* Abstracts, Caribbean Geological Conference, VII, Guadeloupe, 1974, p. 62–63.

—— 1982, Origin of Cariaco Basin southern Caribbean Sea: Marine Geology, v. 47, p. 345–360.

—— 1982, Neotectonics of Boconó fault, western Venezuela: Tectonophysics, v. 85, p. 205–220.

Schwartz, D. P., 1977, Active faulting along the Caribbean-North American plate boundary in Guatemala, *in* Abstracts, Caribbean Geological Conference, VIII, Curaçao, 1977, GUA Papers in Geology, Series 1, no. 9, p. 180–181.

Scientific Party DSDP Leg 77, 1981a, Off Mexico, *Challenger* probes past in the Gulf: Geotimes, v. 26, no. 11, p. 20–21.

Scientific Party DSDP Leg 78A, 1981b, Near Barbados Ridge, scraping off, subduction scrutinized: Geotimes, v. 26, no. 10, p. 24–25.

Scientific Party DSDP Leg 84, 1982, On Leg 84, Challenger drills again off Guatemala: Geotimes, v. 27, no. 7, p. 23–25.

Seely, D. R., 1979, The evolution of structural highs bordering major forearc basins, *in* Watkins, J. S., and others, eds., Geological and geophysical investigations of continental margins: American Association of Petroleum Geologists Memoir 29, p. 245–260.

Seely, D. R., Vail, P. R., and Walton, G. G., 1974, Trench slope model, *in* Burk, C. A., and Drake, C. L., eds., The geology of continental margins: New York, Springer-Verlag, p. 249–260.

Shagam, Reginald, 1960, Geology of central Aragua, Venezuela: Geological Society of America Bulletin, v. 71, p. 249–302.

—— 1972b, Geología de Los Andes centrales de Venezuela, *in* Memoria, Congreso Geológico Venezolano, IV, Venezuela Ministerio de Minas e Hidrocarburos, Boletín de Geología, Publicación Especial no. 5, p. 935–938.

—— 1972c, Evolución tectónica de los Andes Venezolanos, *in* Memoria, Congreso Geologico Venezolano, IV, Venezuela Ministerio de Minas e Hidrocarburos, Boletín de Geología, Publicación Especial no. 5, p. 1201–1261.

—— 1975, The northern termination of the Andes, *in* Nairn, A.E.M., and Stehli, F. G., eds., The ocean basins and margins, v. 3, The Gulf of Mexico and the Caribbean: New York, Plenum Press, p. 325–420.

—— 1977, Stratigraphic models for the northern Venezuelan Andes, *in* Memoria, Congreso Geológico Venezolano, V, Ministerio de Energía y Minas and Sociedad Venezolana de Geología, v. 2, p. 855–877.

—— 1977, Maracaibo microcraton and bordering orogenic belts . . . A region of enduring intra-plate tectonism?, *in* Abstracts, Caribbean Geological Conference, VIII, Curaçao, 1977, GUA Papers in Geology, Series 1, No. 9-1977, p. 185–186.

Shcherbakova, B. Ye., Bovenko, V. G., and Hernandez, H., 1977, Stroyeniye zemnoy kory Zapadnoy Kuby (Structure of the crust beneath western Cuba): Sovetskaya Geologya, no. 8, p. 138–143.

Shcherbakova, B. E., Bovenka, B. G., and Hernandez, H., 1978, 1980: Re'yeF poverkhnosti Mokhorovichicha v predelakh zapadnoy chasti o. Kuba: Doklady Akademii Nauk SSSR, v. 238, no. 3, p. 561–564; translation *in* Doklady Earth Science Section s, v. 238, no. 1, p. 7–9.

Shein, V. S., Ivanov, S. S., Kleshchev, K. A., Khain, V. Ye., Marrero, M., and Socorro, R., 1978, Tektonika Kuby i yeye shel'fa (Tectonics of Cuba and the surrounding shelf): Sovetskaya Geologya, no. 2, p. 104–119.

Shepard, F. P., 1973, Sea floor off Magdalena delta and Santa Marta area, Colombia: Geological Society of America Bulletin, v. 84, no. 6, p. 1955–1972.

Sheridan, R. E., Crosby, J. T., Bryan, G. M., and Stoffa, P. L., 1981, Stratigraphy and structure of southern Blake Plateau, northern Florida Straits, and northern Bahama Platform from multi-channel seismic reflection data: American Association of Petroleum Geologists Bulletin, v. 65, no. 12, p. 2571–2593.

Silver, E. A., and others 1972b, Acoustic-reflection profiles Venezuela continental borderland: U.S. Geological Survey Pub. No. USGS-GD-72-005.

Silver, E. A., Case, J. E., and MacGillavry, H. J., 1975, Geophysical study of the Venezuelan Borderland: Geological Society of America Bulletin, v. 86, no. 2, p. 213–226.

Skerlec, G. M., and Hargraves, R. B., 1980, Tectonic significance of paleomagnetic data from northern Venezuela: Journal of Geophysical Research, v. 85, No. B10, p. 5303–5315.

Smith, A. L., ed., 1980, Circum-Caribbean volcanism: Bulletin Volcanologique, v. 43, no. 2, p. 277–452.

Solomiac, H., 1974, La géologie et la métallogénie des îles de Saint-Martin (zone Francaise) et de Saint-Barthélémy, *in* Livret-guide [Field Guide] d'Excursions dans les Antilles Francaises, Caribbean Geological Conference, VII, Guadeloupe, 1974, p. 93–124.

Somin, M. L., and Millan, G., 1972, The metamorphic complexes of the Isle of Pines, Escambray, and Oriente of Cuba and their age [in Russian]: Akademia Nauk., U.R.S.S. geology series, no. 5, p. 48–57.

—— 1977, Sobre la edad de las rocas metamórficas Cubanas: Academía de Ciencias de Cuba, Informe Científico-Tecnico no. 2, p. 1–11.

Soviet Geophysical Committee, Academy of Sciences of the USSR, 1979, The tectonic and geodynamic[s] of the Caribbean region: Moscow, Nauka, 148 p.

Speed, R. C., 1981, Geology of Barbados: Implications for an accretionary origin: *in* Colloquium on the Geology of Continental Margins, 26th International Geological Congress, Paris, July, 1980: Oceanologica Acta, supplement to v. 4, p. 259–267.

Speed, R. C., and LaRue, D. K., 1982, Barbados: Architecture and implications for accretion: Journal of Geophysical Research, v. 87, no. B5, p. 3633–3643.

Stearns, Carola, Mauk, F. J., and Van der Voo, Rob, 1982, Late Cretaceous-Early Tertiary paleomagnetism of Aruba and Bonaire (Netherlands Leeward Antilles): Journal of Geophysical Research, v. 87, no. B2, p. 1127–1141.

Stephan, J. F., 1977, El contacto cadena Caribe-Andes Merideños entre Carora y El Tocuyo (Edo. Lara): Observacions en el Occidente Venezolano, *in* Memoria, Congreso Geológico Venezolano, Ministerio de Energía y Minas and Sociedad Venezolana de Geología, v. 2, p. 789–817.

Stoiber, R. E., and Carr, M. J., 1974, Quaternary volcanic and tectonic segmentation of Central America: Bulletin Volcanologique v. 37, p. 304–325.

Talwani, Manik, Windisch, C. C., Stoffa, P. W., Buhl, Peter, and Houtz, R. E., 1977, Multichannel seismic study in the Venezuelan Basin and the Curaçao Ridge, *in* Talwani, Manik, and Pitman, W. C., III, eds., Island arcs, deep sea trenches, and back-arc basins: American Geophysical Union, Maurice Ewing Series 1, p. 83–98.

Tarasiewicz, George, Eva, A., Wassal, Harry, and Nagle, Frederick, 1980, Metamorphic rocks and stratigraphy of central Tortue Island, Haiti *in* Abstracts, Caribbean Geological Conference, IX, Santo Domingo, Dominican Republic, 1980, p. 72–73.

Taylor, G. C., 1960, Geología de la Isla de Margarita, Venezuela, *in* Memoria, Congreso Geológico Venezolano, III, Caracas, 1960, Venezuela Ministerio de Minas e Hidrocarburos, Boletín de Geología, Publicación Especial no. 3, v. 2, p. 838–893.

Terry R. A., 1956, A geological reconnaissance of Panamá: California Academy of Sciences Occasional Papers, no. 23, 91 p.

Testarmante, M. M., and Gose, W. A., 1981, Jamaica-Paleomagnetic results from sedimentary rocks [abs.]: EOS, (American Geophysical Union Transactions), v. 62, no. 45, p. 854.

Thomas, D. J., 1972, Tertiary geology and paleontology (phyllum Mollusca) of the Guajira Peninsula, Colombia [Ph.D. thesis]: Binghamton, New York, State University of New York at Binghamton, 150 p.

Thomas, D. J., and MacDonald, W. E., 1976, Summary of the Tertiary stratigraphy and structure of the Guajira Peninsula, *in* Memoria, Congreso Colombiano de Geología, I, Etayo-Serna, Fernando, and Caceres-Giron, Carlos, eds., 1969, Universidad Nacional de Colombia, Bogotá, p. 207–216.

Thompson, A. V., 1966, Guide book of a geological section from Bogotá to the Central Cordillera: Colombian Society of Petroleum Geologists and Geophysicists, 20 p.

Tobisch, O. T., 1968, Gneissic amphibolite at Las Palmas, Puerto Rico, and its significance in the early history of the Greater Antilles island arc, Geological Society of America Bulletin, v. 79, p. 557–574.

Tomblin, J. F., 1975, The Lesser Antilles and Aves Ridge, *in* Nairn. A. E. M., and Stehli, F. G., eds., The ocean basins and margins, v. 3, The Gulf of Mexico and the Caribbean: Plenum Press, New York, p. 467–500.

Tschanz, C. M., Jimeno V., Andres, Cruz B., Jaime, and others, 1969, Mapa geológico de reconocimiento de la Sierra Nevada de Santa Marta—Colombia: Colombia Instituto Nacional de Investigaciones Geológico-Mineras, scale 1:200,000.

Tschanz, C. M., Marvin, R. F., Cruz B. Jaime, Mehnert, H. H., and Cebula, G. T., 1974, Geologic evolution of the Sierra Nevada de Santa Marta, northeastern Colombia: Geological Society of America Bulletin, v. 85, p. 273–284.

Tucholke, B. E., and Ewing, J. I., 1974, Bathymetry and sedimentary structure of the Greater Antilles Outer Ridge and vicinity: Geological Society of America Bulletin, v. 85, no. 11, p. 1789–1802.

Uchupi, Elazar, 1973, Eastern Yucatan continental margin and western Caribbean tectonics: American Association of Petroleum Geologists Bulletin, v. 57, no. 6, p. 1075–1085.

Uchupi, Elazar, Milliman, J. D., Luyendyk, B. P., Bowin, C. O., and Emery K. O., 1971, Structure and origin of southern Bahamas: American Association of Petroleum Geologists Bulletin, v. 55, no. 5, p. 687–704.

van Andel, Tj. H , Heath, G. R., Malfait, B. T , Heinrichs, D. F., and Ewing, J. I., 1971, Tectonics of the Panama Basin, eastern equatorial Pacific: Geological Society of America Bulletin, v. 82, p. 1489–1508.

van Andel, T. H., Heath, G. R., and others, 1973, Initial Reports of the Deep Sea Drilling Project, v. 16, Washington, D.C. (U.S. Government Printing Office), 949 p.

Van de Putte, H. W., 1972, Contribution to the stratigraphy and structure of the Roraima Formation, State of Bolívar, Venezuela, *in* Memoria, Conferencia Geológica Inter-Guayanas, Ciudad Guayana, 1972, Venezuela, Ministerio de Minas e Hidrocarburos, Boletín de Geología, Publicación Especial no. 6, p. 372–394.

Van Houten, F. B., 1976, Late Cenozoic volcaniclastic deposits, Andean foredeep, Colombia: Geological Society of America Bulletin, v. 87, p. 481–495.

Vásquez, E. E., and Dickey, P. A., 1972, Major faulting in north-western Venezuela and its relation to global tectonics, *in* Transactions, Caribbean Geological Conference, VI, Margarita, Venezuela, 1971, p. 191–202.

Vásquez, Enrique, and Masroua, L. F., 1973, Notas sobre la exploración en la Ensenda de la Vela, Estado Falcón: Abstracts, Congreso Latinoamericano de Geología, II, Caracas, Venezuela, 1973, p. 217–218.

—— 1976, Notas sobre la exploración en la Ensenada de la Vela, Estado Falcón, *in* Abstracts, Congreso Latinoamericano de Geología, II, Caracas, 1973, Venezuela, Ministerio de Minas e Hidrocarburas, Boletín de Geología, Publicación Especial no. 7, v. 2, p. 1137–1139.

Vedder, J. G., and others, 1972, Acoustic reflection profiles—East margin Yucatan Peninsula: U.S. Department of Commerce, National Technical Information Service PB2-07595.

Vedder, J. G., MacLeod, N. S., Lanphere, M. A., and Dillon, W. P., 1973, Age and tectonic implications of some low-grade metamorphic rocks from the Yucatan Channel: Journal of Research of the U.S. Geological Survey, v. 1, no. 2, p. 157–164.

Veldkamp, J., Mulder, F. G., and Zijderveld, J. D. A., 1971, Palaeomagnetism of Suriname dolerites: Physics of Earth and Planetary Interiors, v. 4, p. 370–380.

Vierbuchen, R. C., Jr., Tectonics of northeastern Venezuela and southeastern Caribbean: Geological Society of America Memoir 162 (this volume).

Vincenz, S. A., and Dasgupta, S. N., 1978, Paleomagnetic study of some Cretaceous and Tertiary rocks on Hispaniola: Pure and Applied Geophysics, v. 116, p. 1200–1210.

Vincenz, S. A., Steinhauser, P., and Dasgupta, S. N., 1973, Paleomagnetism of Upper Cretaceous ignimbrites on Jamaica: Zeitschrift für Geophysik, v. 39, p. 727–737.

Viniegra O., F., 1971, Age and evolution of salt basins of southeastern Mexico: American Association of Petroleum Geologists Bulletin, v. 48, p. 70–84.

—— 1981, Great carbonate bank of Yucatan, southern Mexico: Journal of Petroleum Geology, v. 3, no. 3, p. 247–278.

von Huene, Roland, Auboin, Jean, and others, 1980, Leg 67: The Deep Sea Drilling Project Mid-America Trench Transect off Guatemala: Geological Society of America Bulletin, Part 1, v. 91, p. 421–432.

Wadge, G., and Draper, G., 1978, Structural geology of the southeastern Blue Mountains, Jamaica: Geologie en Mijnbouw, v. 57, no. 2, p. 347–352.

Weaver, J. D., ed., 1977, Geology, geophysics, and resources of the Caribbean, Report of the IDOE Workshop on the geology and marine geophysics of the Caribbean Region and its resources, Kingston, Jamaica, Feb. 17-22, 1975: Mayagüez, University of Puerto Rico Press, 150 p.

Weeks, L. A., Lattimore, R. K., Harbison, R. N., Bassinger, B. G., and Merrill, G. F., 1971, Structural relations among Lesser Antilles, Venezuela, and Trinidad-Tobago: American Association of Petroleum Geologists Bulletin, v. 55, no. 10, p. 1741–1752.

Westbrook, G. K., 1975, The structure of the crust and upper mantle in the region of Barbados and the Lesser Antilles: Royal Astronomical Society Geophysics Journal, v. 43, p. 201–242.

—— 1982, The Barbados Ridge complex: tectonics of a mature forearc system: Geological Society of London Special Publication no. 8, p. 275–290.

Westbrook, G. K., Bott, M. H. P., and Peacock, J. H., 1973, The Lesser Antilles subduction zone in the region of Barbados: Nature Physical Sciences, v. 244, p. 18–20.

Weyl, Richard, 1980, Geology of Central America, 2nd edition: Berlin, Gebrüder Borntraeger, 371 p.

Wheeler, C. B., 1963, Oligocene and lower Miocene stratigraphy of western and northeastern Falcón Basin, Venezuela: American Association of Petroleum Geologists Bulletin, v. 47, p. 35–68.

Whetten, J. T., 1966, Geology of St. Croix, Virgin islands, *in* Hess, H. H., eds., Caribbean Geological Investigations: Geological Society of America Memoir 98, p. 177–239.

Wing, R. S., and MacDonald, H. C., 1973, Radar geology-petroleum exploration technique, eastern Panamá and northwestern Colombia: American Association of Petroleum Geologists Bulletin, v. 57, no. 5, p. 825–840.

Wong, Th. E., and Van Lissa, R. V., 1978, Preliminary report on the occurrence of Tertiary gold-bearing gravels in Surinam: Geologie en Mijnbouw, v. 57, no. 2, p. 365–368.

Wright, R. M., ed., 1974, Field guide to selected Jamaican geological localities: Jamaica Ministry of Mining and Natural Resources, Mines and Geology Division Special Publication no. 1, 59 p.

Zambrano, Elías, Vásquez, Enrique, Duval, Bernard, Latreille, Michel, and Coffinieres, Bernard, 1971, Sintesis paleogeográfica y petrolera del occidente de Venezuela, *in* Memoria, Congreso Geológico Venezolano, IV, Venezuela Ministerio de Minas e Hidrocarburos, Boletín de Geología, Publicacíon Especial no. 5, v. 1, p. 483–552.

Zeil, Werner, 1979, The Andes, a geological review: Berlin, Gebrüder Borntraeger, 260 p.

Manuscript Accepted by the Society September 1, 1983

Geological Society of America
Memoir 162
1984

Caribbean tectonics and relative plate motions

Kevin Burke*
Calvin Cooper**
John F. Dewey*
Paul Mann**
James L. Pindell**
Department of Geological Sciences
State University of New York
Albany, New York 12222

ABSTRACT

Alternative fits of the continents around the future site of the Caribbean about 200 Ma ago and alternative relative motions since then of North and South America and of Africa with respect to each other allow a wealth of information, including data tabulated here on the distribution of rift systems; early ocean floor; obducted ocean floor fragments and dated plutons to be assessed in relation to a history of Caribbean development.

After an early rift phase, the Gulf of Mexico formed by divergence mainly before the Caribbean itself. Convergence on what are now the northern and southern Caribbean margins during the Cretaceous produced arc-systems and carried the present Caribbean ocean floor, which represents an oceanic plateau, out of the Pacific. Cenozoic convergence in the Lesser Antilles and Central America has been contemporary with more than 1000 km of roughly eastward motion, distributed in wide plate boundary zones, of the Caribbean with respect to both North and South America. Moderate internal deformation of the Caribbean plate is perhaps attributable to its oceanic plateau character because it behaves mechanically in a way that is intermediate between that of normal ocean floor and continent. Although numerous problems remain in Caribbean geology, a framework into which many of them can be accommodated is beginning to emerge.

INTRODUCTION

Three different ways of interpreting the tectonic evolution of the Gulf of Mexico and the Caribbean have been advocated during the last century (Table 1). The seminal concept of continental drift was introduced for the Caribbean by Wegener (1922), and with the recognition of the plate structure of the earth, later interpretations involving continental displacement have come to dominate. The other two approaches interpret the Gulf of Mexico and the Caribbean either as largely subsided continental material (e.g., Suess, 1885; and Schuchert, 1935) or as the site of a permanent ocean basin (e.g., Willis, 1929, p. 328).

We here outline an interpretation, following that of Pindell and Dewey (1982), summarizing in maps and tables the data on which our interpretation is based. We attempt to point out ways in which the hypotheses advanced can be tested and draw attention to some of the many questions that are as yet unresolved.

THE FIT OF AFRICA, NORTH AMERICA, AND SOUTH AMERICA

The Caribbean formed by separation of North and South

---

*Also at: Burke, Lunar and Planetary Institute, NASA Road 1, Houston, Texas 77058; Dewey, Department of Geological Science Laboratories, University of Durham, South Road, Durham SH1 3LE, England.

**Present address: Cooper, Department of Geology, Rice University, Houston, Texas 77001; Pindell, Pennzoil Exploration and Production Company, Houston, Texas 77252; Mann, Institute for Geophysics, University of Texas at Austin, Austin, Texas 78712.

TABLE 1. SCHOOLS OF THOUGHT ON GULF OF
MEXICO/CARIBBEAN EVOLUTION

A. The Bailey Willis School of a permanent pre-Jurassic deep sea basin.

 1. Willis (1929), p. 328.
 2. Meyerhoff and Meyerhoff (1972).
 3. Cebull and Shurbet (1980).

B. The Edward Suess School of a subsided continental terrain.

 1. Suess (1885), p. 709.
 2. Schuchert (1935).
 3. Skvor (1969).
 4. Beloussov (1972).

C. The Alfred Wegener School of continental separation.

 1. Wegener (1929).
 2. Bullard et al. (1965).
 3. Dietz and Holden (1970).
 4. Le Pichon and Fox (1971).
 5. Freeland and Dietz (1972).
 6. Malfait and Dinkelman (1972).
 7. Walper and Rowett (1972).
 8. Moore and Castillo (1974).
 9. Van der Voo and French (1974).
 10. Ladd (1976a).
 11. Sclater et al. (1977).
 12. Pilger (1978).
 13. Dickinson and Coney (1980).
 14. Klitgord and Schouten (1980).
 15. Salvador and Green (1980).
 16. Walper (1980).
 17. White and Burke (1980).
 18. White (1980).
 19. Klitgord, Popenoe, and Schouten (in press).
 20. Pindell and Dewey (1982).

    and others.

America from Africa and from each other. Interpretations of how the Caribbean and Gulf of Mexico have evolved depend strongly on exactly how the three continents, as well as three small pieces of continental crust which we call Yucatan, Chortis (an area south of the Motagua suture in Guatemala, Honduras, and Nicaragua, Dengo, 1969), and the Isle of Pines, fitted together before the separation.

In Figure 1, we show a selection from a large number of published reconstructions that vary most significantly in their tightness of continental fit on the site of the Gulf of Mexico. Bullard and others (1965, Fig. 1a) matched Atlantic shorelines at 2 km depth leaving a wide gulf and some overlap in Central America. Le Pichon and Fox (1971, Fig. 1b) matched fracture zones on either side of the Atlantic and left a smaller Gulf of Mexico with more Central American overlap. Van der Voo and French (1974, Fig. 1c), using Permian paleomagnetic data, obtained a tight fit on the site of the gulf and further suggested that rotation about a pole within Saharan Africa might have changed the fit from that of Figure 1c to that of Figure 1a between the Early Permian and the Late Triassic. Pindell and Dewey (1982)

used paleomagnetism, Central Atlantic fracture zones, and a close fit between Brazil and the coast of the Gulf of Guinea to obtain the continental fit of Figure 1d which leaves room for Yucatan but not for other small continental objects in the Gulf of Mexico. These authors broke Mexico along three faults whose offsets they estimated by restoring the Huastecan fold belt (De Cserna, 1976) to linearity. They also placed Chortis south of and abutting Mexico, in this way solving the problem of Central American overlap. White (1980) and other authors who have used a looser fit have been able to place Chortis and the Isle of Pines as well as Yucatan within the Gulf of Mexico.

In this paper, we use the Pindell and Dewey fit of Figure 1d as a starting point for considering Caribbean evolution, emphasizing that it is sufficiently specific to be tested by additional geologic and possibly even paleomagnetic studies and that on larger scales than that of Figure 1 questions such as How much continental stretching is there in and around the Gulf of Mexico? and How much Cretaceous and Cenozoic deformation has there been along the northern coast of South America? become significant.

RELATIVE MOTION BETWEEN NORTH AND SOUTH AMERICA SINCE THE EARLY JURASSIC

Because magnetic anomalies and fracture zones nearly as old as the times of continental rupture have been mapped in the Central and South Atlantic, the motions of North and South America with respect to Africa can be fully described using the vectorial closure condition required by a three-plate system. Similar three-plate analyses have been attempted in the Alpine system, where, as in the Caribbean, smaller plates have been episodically deformed as a result of movement of their larger neighbors (Dewey and others, 1973). The finite difference method is fundamentally limited in accuracy by knowledge of ocean floor magnetic anomalies. Ladd (1976) was the first to use two sets of ocean floor data from the South and the Central Atlantic to establish relative motion between North and South America. Refinement of the motions he inferred will come mainly from better knowledge of the disposition of fracture zones and magnetic anomalies in the Atlantic which must remain a topic of continuing interest for Caribbean geologists. As yet, anomalies and fracture zones in the South Atlantic have been mapped much less thoroughly than those in the Central Atlantic.

We show, in Figure 2, two interpretations of the motion of South America with respect to a fixed North America that postdate Ladd's (1976) pioneer study. In Figure 3, we present a summary diagram indicating the style of deformation at specific times on the Caribbean site that we infer from the calculated motions. The *direct* dependence of Caribbean deformational style on the relative motion of North and South America (Fig. 3) assumes that motion is taken up along a single plate boundary zone, such as during Jurassic separation. Later, relative plate motions following separation acted across at least two plate boundaries (northern and southern Caribbean arcs or transforms)

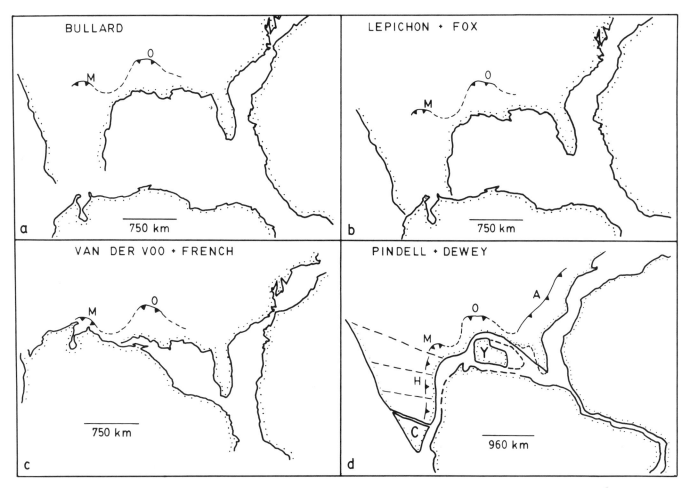

Figure 1. Four proposed Mesozoic continental reconstructions. a) Bullard, Everett and Smith, 1965, based on matching of opposing continental margins. b) LePichon and Fox, 1971, based on alignment of opposing marginal fracture zones. c) Van der Voo and French, 1974 (also Van der Voo, Mauk, and French, 1976), based on paleomagnetic data from circum-Atlantic continents. d) Pindell and Dewey, 1982, based on marginal fracture zone alignment, paleomagnetism and an improved fit between northern Brazil and the Guinea margin of Africa. Foldbelts: O - Ouachitas, M - Marathon, A - Appalachians, H - Huastecan; Continental blocks: C - Chortis, Y - Yucatan.

and therefore may be only *indirectly* manifested in Caribbean deformations. Figure 2a represents the path that we use (Pindell and Dewey, 1982); and Figure 2b, for comparison, is a path based on the results of Sclater, Hellinger, and Tapscott (1977) which was kindly prepared for us by Tapscott. Both paths show South America moving away from North America during the Jurassic, a process that led to the formation of ocean floor on the sites of the Gulf of Mexico and the Caribbean (EXTENSION and SINISTRAL SLIP in Fig. 3). The orientation of Mesozoic grabens is generally perpendicular to this direction (Fig. 4). This interval of extension is shown in Figure 2b as finishing with the end of the Jurassic, but in Figure 2a it is shown as continuing until 80 Ma ago. The difference between the two interpretations follows from the treatment of Africa as one or as two plates during the Early Cretaceous. Pindell and Dewey (1982) chose to accommodate the early opening of the South Atlantic by extension within the Benue trough (Burke and Dewey, 1974). This

keeps the northern coast of South America close against the coast of the Gulf of Guinea during the Early Cretaceous. On the other hand, Sclater and others (1977) regarded Africa as one plate throughout the Cretaceous and therefore required separation between South America and the Guinea coast as early as Neocomian times ending extension at the site of the Caribbean at the end of the Jurassic.

Between 80 and 65 Ma ago during the Late Cretaceous, both the plots of Figure 2 show South America moving to the ENE with respect to North America (SINISTRAL TRANSPRESSION in Fig. 3). This move is followed by an episode of EXTENSION and DEXTRAL SLIP (Fig. 3) between 65 and 36 Ma ago. Both Figure 2a and Figure 2b again show a similar pattern, in this case DEXTRAL TRANSPRESSION (Fig. 3) for the time since 38 Ma ago.

In summary, the motion of South-America with respect to North America consists of an initial, almost longitudinal, anti-

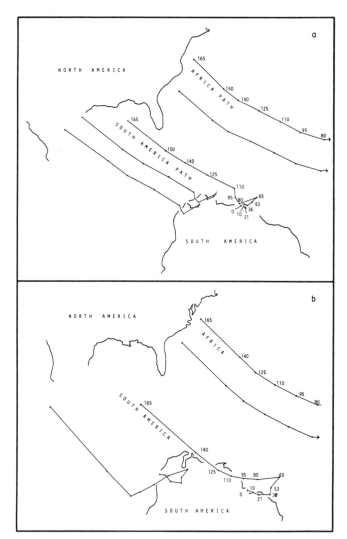

Figure 2. Relative motion vectors of several points of South America with respect to a fixed North America; a) based on data presented in Pindell and Dewey (1982); b) based on data presented in Sclater and others (1977). The differences between (a) and (b) result from different initial fits and different models for the opening of the South Atlantic Ocean. The position of South America with respect to North America at various times since the Jurassic provides a framework in which to base Caribbean evolution. Plate boundary configurations at each time must accord with the relative motions of the continents for the next interval.

clockwise rotation, an episode of southward latitudinal motion and finally longitudinal motion to the west with an element of clockwise rotation. Pindell and Dewey (1982) infer smaller overall displacements of South America after the initial eastward rotation than either Ladd (1976) or Sclater and others (1977) because of the accommodation of displacement in the Benue trough and because of a tighter fit of Africa and South America along the Guinea coast.

Refinement of understanding of the relative motion of South America with respect to North America is one area where continuing progress can be predicted. This progress is likely to indicate appropriate modifications of existing interpretations.

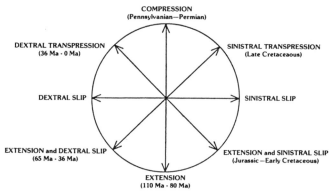

Figure 3. Diagram illustrating possible tectonic modes along a single plate boundary between the North and South American plates. Intervention of the Caribbean plate interferes with the application of this simple model. Predicted relative motions have resulted in extremely complex deformation to the Caribbean plate and the Northern and Southern Caribbean Plate Boundary Zones.

A FRAMEWORK FOR RECONSTRUCTING CARIBBEAN PLATE EVOLUTION

Once a choice of initial fit on the site of the Caribbean has been made and relative motions of the continents around the ocean have been calculated, a basis exists for interpreting sequential changes in plate configuration. The results of the many thousands of man years of geological research that has been conducted in and around the Caribbean during the last century and especially during the last 50 years provide a framework within which reconstructions for sequential time intervals can be prepared. Descriptions of rocks and structures in published work permit recognition, though not always unambiguous definition of environments such as: rifts, ocean floor, arc-systems, strike-slip systems, that are represented at particular times in particular areas. In Figures 4, 5, and 6 and in Tables 2, 3, 4, and 5, we present in summary form some of the information that has proved most useful to us in interpreting Caribbean evolution. Using the information summarized in these maps and tables, we depict six major stages of Caribbean evolution (Fig. 7). In Figures 8 and 9 and in Tables 6, 7, and 8, we present summaries of developments while the Caribbean plate has been moving eastward with respect to the Americas during the latest and the presently active phase of Caribbean evolution.

Rift and Early Ocean Floor and Related Facies

Features related to the early development of the Caribbean and Gulf of Mexico are shown in Figure 4 and are tabulated in Table 2. The early history of the Central Atlantic Ocean with: a Jurassic quiet zone (no. 1, Fig. 4, and Table 2), the Blake Spur anomaly (no. 2), marking a ridge-jump, the East Coast Anomaly (no. 3) close to the ocean margin offshore but turning onshore where it may mark the course of an end Paleozoic suture, and (no. 4, no. 5, and no. 6) complex basin development across the

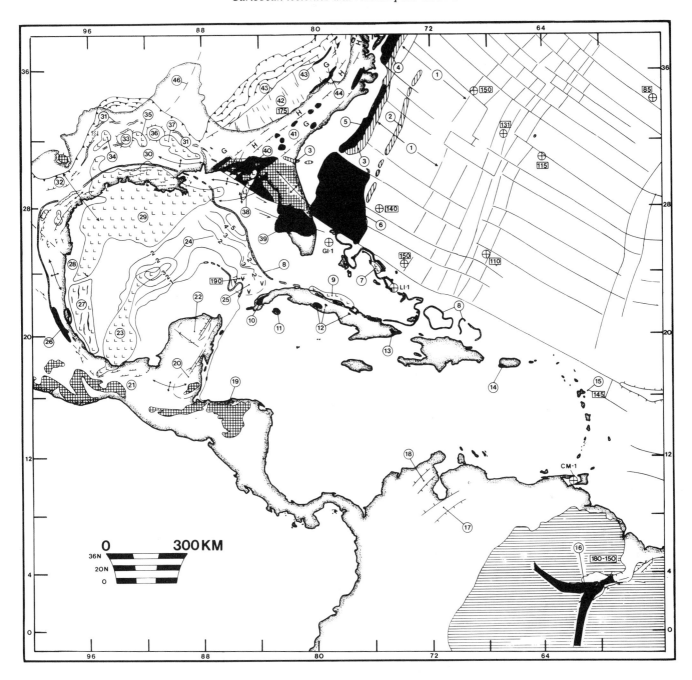

Figure 4. Compilation map of Jurassic-Early Cretaceous features developed during the early opening of the Caribbean and Central Atlantic area. Grey areas indicate rifts formed during the rupture of northern Africa from the eastern United States and South America from Mexico and the southern United States. Horizontally ruled area is the Guyana shield, disrupted by a failed triple arm rift system. Rifting began in Late Triassic-Early Jurassic approximately along the Pennsylvanian-Permian suture, marking the site of a Paleozoic ocean between North America and Africa-South America. Although the Newark group of rifts along the eastern seaboard are well exposed and studied (nos. 43 and 44) and those beneath the coastal plain (nos. 31, 40, 41) and the continental shelf (nos. 4, 5, 6) are known in the subsurface, those of northern South America (nos. 16, 17, 18) are relatively poorly known. Mesozoic rift deposits of northern South America have been disrupted by post-Eocene east-west South America-Caribbean relative plate motion; vestiges of graben have been identified in the upthrusted Perija and Merida Andes (nos. 17 and 18) and the Couva Marine One well in the Gulf of Paria (CM-1). See Table 2 for complete explanation.

TABLE 2.   MESOZOIC RIFTS, RIFT FACIES AND IGNEOUS ACTIVITY RELATED TO SEPARATION OF NOAM AND SOAM
(Keyed to Figure 4)

Nos. on Fig. 4	Name & Location	Type of Structure	Age & Method if Applicable	Thickness of Sediments	Associated Volcanism	Interpretation & Comments	Key References
1	Pre-M series ocean floor	Ocean crust, no magnetic anomalies	Plate accretion, 165-146 Ma	—	Oceanic crust	Produced during Jurassic "Quiet Period." Ridge jump occurred during this time	Klitgord and Grow, 1980 Vogt, 1973 Grow and Sheridan, 1981
2	Blake Spur Anomaly	Magnetic high	Middle Jurassic	—	Oceanic crust	Related to a major scarp in basement, possibly associated with a ridge jump	Klitgord and Grow, 1980 Klitgord and Schouten, 1980 Klitgord, Popenoe, and Schouten, in press
3	East Coast Magnetic Anomaly	Magnetic high	Jurassic	—	May be a partial cause for the anomaly	Uplifted block of oceanic crust(?) at ocean-continent boundary. On shore may be Hercynian suture	Klitgord and Grow, 1980 Grow and Sheridan, 1981
4	Baltimore Trough	Marginal graben complex	Triassic-Jurassic, active rifting	10-13 km	Low Cretaceous mafic intrusion	Extension during early opening of Atlantic	Grow and Sheridan, 1981 Klitgord and Behrendt, 1979 Grow, 1980
5	Carolina Trough— off SE U.S. coast	Marginal graben complex	Triassic-Jurassic, active rifting	Up to 11 km	?	Extension during early opening of Atlantic	Dillon and Paull, 1978 Klitgord and Behrendt, 1979 Grow and Sheridan, 1981
6	Blake Plateau Basin —off east Florida coast	Extended marginal graben complex	Triassic-Jurassic, active rifting	As thick as 12 km thickest in the west	Inferred (gravity modeling) mafic intrusions	Basin represents stretched, transitional crust	Dillon and Paull, 1978 Klitgord and Behrendt, 1979 Grow and Sheridan, 1981 Kent, 1979 Shipley et al., 1978
7	Exuma Sound Diapir	Salt diapir	May be correlative with Louann, Alegre and Casamance	?	—	Atlantic rifting salt	Lidz, 1973 Meyerhoff and Hatten, 1974 Burke, 1975
8	Probable Late Jurassic-Early Cretaceous reef trend	Reef	Late Jurassic-Early Cretaceous	Variable	—	Bahamian existence is questionable	Meyerhoff and Hatten, 1974 Antoine et al., 1974
9	Punta Alegre Salt Domes	Salt diapirs	Pre-Tithonian, Jurassic	?	—	Atlantic rifting salt	Meyerhoff et al., 1969 Meyerhoff and Hatten, 1968 Moore and del Castillo, 1974
10	San Cayetano Fm. of Pinar del Rió	Nonmarine(?) and marine fossiliferous beds	Early to medial Jurassic	?	—	Accreted to or thrust onto Cuba arc during its pre-Eocene evolution	Meyerhoff et al., 1969 Khudoley and Meyerhoff, 1971
11	Isle of Pines, crystalline metamorphics	Low- to high-grade metamorphosed rocks	Paleozoic(?)	?	—	Continental fragments(?)	Meyerhoff et al., 1969 Kuman and Gavilán, 1965 Mattson, 1979
12	Sierra de Trinidad and other metamorphics	Metamorphosed rocks	Jurassic(?)	?	—	Mesozoic Arc rocks(?)	Mattson, 1979 Meyerhoff et al., 1969 Meyerhoff and Hatten, 1968
13	Eastern Oriente Province Metamorphics	Metamorphosed rocks	Jurassic(?)	—	—	Mesozoic Arcs	Meyerhoff et al., 1969 Mattson, 1979
14	Bermeja Complex	Metamorphosed dismembered Ophiolite	~126 Ma	?	(Ophiolite)	Ophiolitic sliver in accretionary wedge	Mattson, 1960, 1973, 1979 Mattson and Pesagno, 1979
15	Desirade Ophiolite	Uplifted oceanic crust	Late Jurassic K-Ar dates	?	Spilite-Keratophyre	In accretionary wedge	Fink, 1972
16	Uraricoera-Rio Blanco-North Savannas rift system	Rift at high angle to margin	Jurassic	Up to 6000 meters, contains evaporites	Basaltic-Tholeiitic	May represent failed arm of a ridge-ridge-ridge triple junction	Berrangé and Dearnley, 1975 McConnell et al., 1969
17	Merida Andes	Upthrust Jurassic Rift	Jurassic Pollen	>1.5 km	Vitric tuff	Late Cenozoic upthrust Jurassic rifts (ticks downthrown side)	Schubert et al., 1979 Green, 1976
18	Perija Andes	Upthrust(?) Jurassic Rift	—	>1.0 km	?	Similar to 17(?)	Kellogg and Bonini, 1982

TABLE 2. MESOZOIC RIFTS, RIFT FACIES AND IGNEOUS ACTIVITY RELATED TO SEPARATION OF NOAM AND SOAM
(Keyed to Figure 4) *Continued*

Nos. on Fig. 4	Name & Location	Type of Structure	Age & Method if Applicable	Thickness of Sediments	Associated Volcanism	Interpretation & Comments	Key References
19	Metamorphic rocks	—	Variable, Precambrian to Late Paleozoic	?	Cretaceous-Tertiary intrusions	Basement complex of Chortis block	King, 1969
20	Yucatan Redbeds	—	Jurassic	Variable	—	Related to continental rifting	Lopez-Ramos (1975)
21	Isthmus of Tehuantepec	Suture(?)	Late Jurassic	—	—	May have been differential movement between Yucatan block and Mexican blocks	Pindell and Dewey, 1982 Gose and Sanchez-Barreda (in press)
22	Subsurface of North Yucatan	?	Upper Cretaceous	~400 m	Submarine Andesitic flows	?	Weidie et al., 1980
23	Campeche Knolls	Salt domes	Inferred Jurassic -Cretaceous	?	—	"Challenger" unit containing the evaporites pinches out against older basement of Campeche shelf	Ladd et al., 1976b Moore and Castillo, 1974 Watkins et al., 1978 Antoine et al., 1974
24	Pre-Middle Cretaceous Isopachs (km)	Surface of Pre-Middle Cretaceous	Pre-Middle Cretaceous	As indicated	—	—	Buffler et al., 1980
25	Jordan, Pinar del Rio, Catoche, and an unnamed Knoll	Igneous(?) intrusions	Pre-Lower Cretaceous(?)	—	?	Unclear	Antoine et al., 1974 Bryant et al., 1969
26	Huayacocotla Aulacogen	Graben at a high angle to inferred rifting. (Aulacogen)	~185 Ma Sinemurian-Hettangian	3 km	Sinemurian basic volcanism	May contain oldest (dated) marine incursion into a circum-gulf graben	Schmidt-Effing, 1980
27	Mexican Ridges	Folds in sedimentary section	Post-Jurassic	3 km sediment involved in folding	—	Anticlinal folding of sedimentary section related to salt tectonics	Garrison and Martin, 1973 Antoine et al., 1974 Bryant et al., 1968
28	Texas-Sonora Megashear	Continental transform	No later than early Jurassic	—	?	Aided with post-rifting emplacement of Mexico	Silver and Anderson, 1974 Pindell and Dewey, 1982
29	Gulf Coast Salt Dome Province	Bedded and diapiric salts	Middle to Late Jurassic	Variable	—	Occur in diapirs that pierce Tertiary strata	Kirkland and Gerhard, 1971 Antoine et al., 1974
30	Wiggins Arch	Horst	Triassic-Jurassic	—	—	Horst in Gulf Coast thinned basement	White, 1980 Antoine et al., 1974
31	Luling-Mexia-Talco South Arkansas-Pickens-Gilbertown graben system	Grabens and half grabens, collectively forming a fairly continuous belt of extension	Triassic-Jurassic	Variable, up to 3000 m	Basic dykes and sills	Separates thick continental crust of NOAM from rifts and horsts beneath Gulf Coastal Plain	King, 1969 Halbouty, 1967 Bishop, 1973 Hager and Burnet, 1960 Walthall and Walper, 1967
32	San Marcos Uplift	Horst	Triassic-Jurassic	—	—	Horst(?)	Antoine et al., 1974
33	Sabine Uplift	Horst (Stranded block)	Triassic-Jurassic	—	No Jurassic	Drilling has recovered Mississippian volcaniclastics Horst in thinned basement	Pindell and Dewey, 1982 Antoine et al., 1974 Bishop, 1973
34	East Texas Salt Basin	Intra-Horst Basin	Jurassic	?	?	Diapirs	Antoine et al., 1974 Fig. 1
35	North Louisiana Salt Basin	Intra-Horst Basin	Jurassic	?	?	Diapirs	Antoine et al., 1974 Fig. 1
36	Monroe Uplift	Horst	Triassic-Jurassic	—	—	Horst in thinned basement	Antoine et al., 1974 Bishop, 1973
37	Mississippi Salt Basin	Intra-Horst Basin	Jurassic	?	?	Diapirs	Antoine et al., 1974
38	Eastern extension of Louann Salts	Diapirs	Jurassic	?	—	Diapirs	Meyerhoff and Hatten, 1974

TABLE 2.   MESOZOIC RIFTS, RIFT FACIES AND IGNEOUS ACTIVITY RELATED TO SEPARATION OF NOAM AND SOAM
(Keyed to Figure 4) *Continued*

Nos. on Fig. 4	Name & Location	Type of Structure	Age & Method if Applicable	Thickness of Sediments	Associated Volcanism	Interpretation & Comments	Key References
39	South Florida Volcanic Province	Volcanics and Sediments	Triassic-Jurassic	?	Diabases, basalts, and rhyolites	Rift-related volcanism	Klitgord, Popenoe and Schouten, in press
40	West Florida Basin	Block-faulted Paleozoic basement, overlain by Arkosic sediments	Triassic(?)	Triassic sediments, 3700 ft. Total basin depth, about 10,000 ft.	Diabase sills	Extension related to breakup of Pangea	Klitgord, Popenoe and Schouten, in press Arden, 1974; Barnett, 1975
41	Atlantic Coastal Plain	Onlapped coastal plain deposits	Cretaceous to present	Variable	—	—	Cook et al., 1981
42	Piedmont Province	Metamorphic thrust sheet(s)	Mid to Late Paleozoic	—	Granitic to basic intrusions	Westward migrating thrust sheet(s) during Paleozoic (Alleghenian) Orogeny	Cook et al., 1981 Cook et al., 1979
43	Danville Basin	Half-grabens, down to east	Late Triassic-Early Jurassic	~3200 meters	Tholeiitic sills and diabase	Graben axis is parallel to basement fabric	Cornet, 1977; Bain, 1957; DeBoer and Snider, 1979 Olsen et al., 1982
44	Wadesboro, Sanford, Durham Basins	Half-grabens, down to east	Late Triassic-Early Jurassic	2000-3000 meters	Tholeiitic sills and diabase	Graben axis parallel to basement fabric	Cornet, 1977 DeBoer and Snider, 1979 Olsen et al., 1982
45	Brevard Zone	Complex fault zone involving much shear	Several episodes of motion, late Paleozoic and early Mesozoic	—	—	Fault motion may have allowed continental escape during Paleozoic Orogeny	Bobyarchick, 1981 Cook et al., 1981 Cook et al., 1979
46	Mississippi Embayment	Reactivated Precambrian Rift	Jurassic-Cretaceous	1000-1500 meters (Jurassic-present)	Cretaceous syenite and Lampophyre	Precambrian structural feature reactivated during Mesozoic rifting	Ervin and McGinnis, 1975 King, 1969

continental margin perhaps coincides with the opening of the Gulf of Mexico which was apparently complete by 140 Ma ago. The isolated salt diapir of Exuma sound in the Bahamas (no. 7) and those of the north coast of Cuba (no. 9) can be interpreted as links between contemporary evaporites in and around the Gulf of Mexico and those at the mouth of the Casamance River in Senegal (Burke, 1975). The postulated extension of a Late Jurassic and Cretaceous reef trend over the Bahamas (no. 8) emphasizes the occurrence of unusually elevated Jurassic ocean floor in this area. Explanations of the elevation as due to transpression on a fracture zone associated with a long transform (Pindell and Dewey, 1982) and as overlying hotspot volcanoes (Dietz, 1973) have both been advanced. The alternative idea that the Bahamas overlie continent has seemed unlikely because Africa and North America fit too closely together (Fig. 1).

The Cayetano formation of Cuba (no. 10), although now thrust-bound in the western part of the island's suture zone, contains sediments of facies reminiscent of early Atlantic-type margin deposition and as such represents an environment not elsewhere exposed in either the Gulf of Mexico or the coast of the Atlantic although possibly encountered in Great Isaac no. 1 (east of the tip of Florida on Fig. 4).

The Isle of Pines (no. 11) with its kyanite-bearing crystalline siliceous rocks is of uncertain origin although apparently continental. A likely possibility is that it represents a fragment of

Paleozoic basement from Yucatan clipped off as Cuba moved past during the Late Cretaceous (Fig. 7c and d). The other metamorphic rocks of Cuba (no. 12 and 13) can be interpreted as related either to emplacement of subvolcanic granodiorite and tonalite plutons during arc development or to subduction of Atlantic ocean floor and ultimate suturing of the Cuban arc against the Bahamas.

Jurassic ophiolitic material within the Caribbean has been recognized only locally. The Bermeja complex of southwestern Puerto Rico (no. 14) contains ocean-floor rocks formed during the Late Jurassic although its incorporation into an accretionary wedge took place later and the ophiolite of Desirade (no. 15), incorporated in the Lesser Antillean accretionary wedge during the Cenozoic, consists of ocean floor made at the mid-Atlantic ridge, during the Late Jurassic.

Rifts of Late Triassic to Early Jurassic age extend from Guyana and Brazil (no. 16) to Svalbard in an area which, before the Atlantic opened, was comparable in extent to that of the active East African Rift system (Burke, 1976). Although the Newark group of rifts along the eastern seaboard of North America are well-exposed and studied and those of the Gulf of Mexico are well-known in the subsurface, those of northern South America (no. 16, 17, 18) are relatively poorly known. Two of these rifts, the Sierra Merida (no. 17) and Sierra Perija (no. 18), have been involved in Neogene thrusting related to strike-slip movements and their Andean setting has perhaps been responsible for delay in the recognition of their rift character (but see Green, 1976, Fig. 14, column 22). Ocean floor began to form later in the Caribbean than in the Gulf of Mexico and the Triassic/Early Jurassic rift sediments of Venezuela do not pass up rapidly into Atlantic-type margin sediments. Perhaps the strongest evidence of the onset of oceanic conditions in the Caribbean itself lies in more than 1 km of Late Jurassic or very Early Cretaceous evaporites encountered in the Couva Marine 1 Cm-1 on Fig. 4) well in the Gulf of Paria (cf. Potter 1965).

Rocks forming the isthmus of Panama and the southern part of Central America consist mainly of arc and accretionary terrains of Cretaceous and Cenozoic age. These rocks do not at present appear to add much to understanding of the early development of the Caribbean; but outcrops of Paleozoic rocks in Honduras farther north (no. 19) form part of the Chortis block whose position before ocean-opening and whose relationship to the Nicaraguan Rise, which stretches northeastward from Chortis toward Jamaica, are critical in considering early Caribbean evolution. We use a reconstruction placing Chortis adjacent to southern Mexico (Fig. 1d) before rifting began in the Gulf of Mexico and the Caribbean. Jurassic red beds in Yucatan (Fig. 4, no. 20) show that the rifting event was felt in that area and the displacement of Yucatan from the northern side of the Gulf of Mexico, as indicated by the evaporite distribution on both sides of the gulf (White, 1980), requires a suture or strike-slip boundary to pass through the isthmus of Tehuantepec (no. 21). This structure would not have been active after 140 Ma ago by which time the present structure of the Gulf of Mexico was established. The only evidence suggesting that tectonic activity in Yucatan persisted later is the occurrence of submarine andesites, apparently of Cretaceous age, in the subsurface of northern Yucatan (no. 22), an as yet unexplained phenomenon.

Structures of the gulf itself (no. 23 to no. 29) include evidence of marginal rifting in Huayacocotla (no. 26); thick evaporites (no. 23, no. 27, no. 29) that we interpret as overlying rift crust (i.e., stretched continental crust) in some areas and oceanic crust in other areas as well as thick pre-mid Cretaceous sediments (no. 24). The projection of the Texas-Sonora megashear (28) is shown on Figure 4 to indicate that Gulf of Mexico structures which are Late Triassic to medial-Jurassic in age postdate all megashear movement.

The Gulf Coast of the United States of America (no. 28-no. 38) records a broad zone of extension between a fault system bordering continental crust of normal thickness (no. 31) and ocean floor (beneath much of no. 29) with horsts generally marked by an absence of evaporites and rifts marked by thick, diapiric evaporite bodies.

On the eastern side of the Gulf of Mexico, continental crust forming the basement of the Florida peninsula consists of a northern area, underlain by crystalline rocks and Paleozoic cover which is interpreted as sutured against North America at the end of the Pennsylvanian, the suture lying beneath the west Florida basin (no. 40), and a southern area (no. 39), consisting of Triassic to Jurassic volcanic rocks which are interpreted as of rift facies.

North of Florida, the Atlantic coastal plain (no. 41) represents Cretaceous to Recent onlap sediments across the rifted Atlantic ocean margin. Landward of the coastal plain, rocks of the Paleozoic southern Appalachian Mountains are exposed in the Piedmont (no. 42) metamorphic and igneous belt with major structural breaks such as the Brevard Zone (no. 45). Rifts associated with the initiation of the Atlantic Ocean outcrop within the Piedmont (no. 43, 44). Farther west the Mississippi embayment (no. 46) is the most spectacular of these as it reaches farther into the North American continent than other structures associated with the formation of the Atlantic and the Gulf of Mexico. This extension and also continuing seismicity are attributed to a position overlying a Late Precambrian rift structure.

Ocean Floor and its Obducted Fragments

A major limitation in interpreting the tectonic evolution of the Caribbean is that no area of the ocean floor, formed during the initial separation of Africa and South America from North America, is preserved except as slivers in accretionary prisms. Most of the Caribbean is presently underlain by ocean floor that was carried in from the Pacific (Wilson, 1966) while material that was earlier on the site was being subducted or obducted at Late Mesozoic convergent boundaries. Some writers have suggested that the Caribbean ocean floor grew in place (e.g., Mooney, 1980), and magnetic anomalies have been discerned by other workers in both the Colombian basin (Christofferson 1976) and the Venezuelan basin (Ghosh and others, this volume). The greater part of the present Caribbean ocean floor, in both the Colombian and the Venezuelan basins (that containing the B" reflector) is about a kilometer shallower than is to be expected from the generally Late Cretaceous age that it has yielded in Deep Sea drill holes (Table 3). This anomaly has led to comparisons with the oceanic plateaus of the western Pacific (Burke and others, 1978) which are also unusually shallow for their age. These plateaus are known from Deep Sea drilling and from outcrops on land (Hughes and Turner, 1977) to be composed, at least in their shallower parts, of alternations of basaltic flows and sills with pelagic limestones. Seismic refraction measurements in the Caribbean suggest a general depth to the M discontinuity of 15-20 km (e.g., Edgar and others, 1971) with much of the thick oceanic crust yielding velocities appropriate for basalt. The association of pelagic limestones and basaltic rocks recorded in DSDP holes from the Caribbean (Table 3) further suggests a close resemblance to the plateaus of the western Pacific. Failure as yet to recognize a coherent pattern of magnetic anomalies on the Caribbean ocean floor has led to the suggestion (Burke and others, 1978) that the Caribbean represents an off-the-ridge accumulation like much of the mid-Pacific mountains rather than a plateau formed at a ridge like the Ontong Java plateau.

In Figure 5 we show the circum-Caribbean distribution of rock associations which have been interpreted as fragments of oceanic floor. They occur mostly in accretionary wedges asso-

TABLE 3. SUMMARY OF DSDP HOLES AND DREDGE HAULS OF THE CARIBBEAN CRUST (Keyed to Figure 5)

Location	Description	References
DSDP 146	Dolerite sills beneath mixed Santonian sediments, dominantly limestones, turbidites, and basaltic ash flow. No pillows suggests magmas partly intrusive	Donnelly and others, 1973
DSDP 150	Basaltic ash in Santonian limestones, dolerite beneath Coniacian sediment. No pillows, magmas may be partly intrusive	Donnelly and others, 1973
DSDP 151	Santonian beds probably contain basaltic ash	Donnelly and others, 1973
DSDP 152	Basalt beneath Campanian limestone contains marble inclusions, one of which has Campanian foraminifera	Donnelly and others, 1973
DSDP 153	Mixed lithologies containing Santonian and Coniacian basaltic ash beds, above basalt	Donnelly and others, 1973
DSDP 154	Hole penetrated only Neogene pelagic and terrestrial sediments	Donnelly and others, 1973
Aves Ridge	Biomicrites, granodiorite, and diabase. Intrusions date between Middle Cretaceous and Paleocene	Fox and Heezen, 1975
Beata Ridge	Basalts, dolerites, oldest paleontological sample Middle Eocene carbonate	Fox and Heezen, 1975
Puerto Rico Trench	Serpentinites, carbonates, greenschist facies basalts. Mona Passage black marble	Perfit et al., 1980
La Désirade (escarpment)	Metamorphosed basalts to gabbros (relict textures); plutonic gabbros	Fox and Heezen, 1975

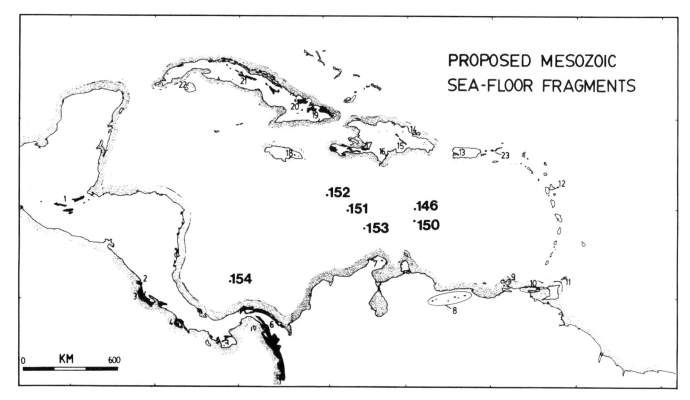

Figure 5. Proposed fragments of Mesozoic seafloor in the Caribbean region. Fragments include both normal seafloor and B" exposures. Numbers refer to explanations in Tables 3 and 4. Heavily printed numbers are DSDP holes.

ciated with subduction zones. It has been suggested (Burke and others, 1978) that attempts to subduct the thick oceanic crust of the Caribbean have proved relatively unsuccessful and that subduction has stopped with the arrival at the subduction zones of buoyant Caribbean ocean floor. An attempt has been made (Table 4) to distinguish between dismembered ophiolites that consist of fragments of ocean floor of normal crustal thickness and those associations which result from attempts to subduct the thick crust that now occupies much of the Caribbean.

Accretionary prisms containing fragments of ocean floor of normal thickness can be expected to include substantial proportions of ultramafic material, gabbros, and locally, perhaps, recognizable sheeted dikes. On the other hand, the thicker oceanic crust is likely to be represented dominantly by basaltic material, perhaps commonly in sills rather than extrusives with siliceous and calcareous pelagic sediments and little or no ultramafic material. The absence of ultramafic material in the thrust masses of the accretionary prisms results from the great thickness of the basalt in the thicker oceanic crust.

Ocean-floor fragments occur in a great variety of environments within the area shown in Figure 5. If the Wilson cycle presently operating in the Caribbean develops to a conclusion, if for example South America collides in the future with North America, then the complete tectonic history of the fragments of ocean floor already incorporated in islands and continental margins may prove very difficult to decipher.

For example: in the Sierra de Santa Cruz of Guatemala (no. 1 on Fig. 5 and Table 4), the El Tambor ophiolite occupies a suture zone between the Yucatan and Chortis continental blocks; but on the interpretation we use here, there has never been a large ocean between Chortis and North America (see Fig. 1d). The ophiolitic material represents slivers of ocean floor picked up as Chortis moved eastward along the Cayman trough-Motagua transform system during the Later Cenozoic to occupy an area where the western end of the Greater Antillean arc had collided to form a suture zone against North America 80 Ma ago. The B" ocean floor does not appear to be represented in the El Tambor ophiolite because it contains a substantial proportion of ultramafic and gabbroic rocks. The width of ocean in this area remains poorly constrained and further work may prove our present interpretation invalid.

Ophiolite complexes on the Pacific coast of Central America and northern South America are of particular interest as they may preserve evidence of the passage of the thick ocean floor from the Pacific into the Caribbean. Although the Santa Elena complex (no. 2) contains too much ultramafic material to be likely to represent B", the Nicoya, Osa, and Azuero complexes (no. 3, 4, and 5) as well as those of eastern Panama and western Colombia (no. 6) lie within accretionary prisms mostly of Late Cretaceous and Early Cenozoic age which contain relatively little ultramafic material and are much more likely to preserve pieces of the thickened ocean floor. The komatiite bearing accretionary prism of Isla Gorgona (Echeverría, 1980) may possibly also represent material from the thickened Caribbean ocean floor.

Along the northern coast of South America, ophiolite slices have been recognized in an accretionary wedge in the Guajira peninsula (no. 7). Farther east in the Villa de Cura Nappe system (no. 8), ophiolites appear to have been obducted perhaps in one piece (Skerlec and Hargraves, 1980) from a marginal basin between an arc (largely represented by granodioritic intrusives in the offshore Venezuelan islands) and the South American continent (Maresch, 1974). Fragments of ophiolite preserved along the El Pilar fault system (no. 10) perhaps represent parts of the marginal basin that is better preserved in the Villa de Cura Nappe. Whether the ophiolite of Tobago (no. 11) also represents part of this system is presently unresolved.

The ophiolite material in La Desirade (no. 12) differs from those considered so far in having come from the Atlantic spreading center. It is likely that there are large quantities of comparable material in the accretionary wedge of the Lesser Antilles although no other outcrops have been recognized.

Ophiolitic bodies of the Greater Antilles (no. 13-no. 23) fall into three groups. One group which includes those of northern Cuba (no. 19, 20, and 21) and northern Hispaniola (no. 14) consists of slices of Atlantic ocean floor involved in the suturing of the Greater Antilles against the Bahama Bank which have been thrust over the Bahamian carbonate succession in the way first depicted by Wassall (1957). A second group outcropping on the southern side of the Greater Antilles consists of Caribbean ocean floor slices in accretionary wedges. The Bermeja complex of Puerto Rico (no. 13) which is the best developed of these, with its abundant ultramafic bodies, is not thought to contain B" material but to represent remnants of the original Caribbean ocean floor. In some interpretations (White and Burke, 1980), the slivers of dismembered ophiolite in Jamaica (no. 18) represent old ocean floor in a similar accretionary wedge while the dismembered ophiolites in accretionary wedges of the southern peninsula of Haiti (no. 17) and the La Cienaga complex of the Dominican Republic (no. 16) contain only B" material. The nature of the third group of ophiolitic bodies, such as those of central Hispaniola (no. 19), is very uncertain. It is conceivable that they are representatives of either the northern or the southern belt, but they could be of a completely different origin. Some (e.g., Bowin, 1975) have considered that they do not occur in an accretionary wedge but represent ocean floor on top of which the Greater Antillean arc has been built.

Dated Plutons

The dated plutons of the Caribbean and its surroundings constitute an important part of the framework for interpretation of Caribbean plate evolution and some information about them is summarized in Figure 6 and in Table 5. A clear distinction can be made between a set of ages that are older than the formation of the Caribbean, most of which are of Permian age and can be interpreted as related to convergence and final collision between North and South America and a set of Cretaceous and Paleogene ages which record the evolution of the convergent boundaries of

K. Burke and Others

TABLE 4.   PROPOSED FRAGMENTS OF MESOZOIC SEAFLOOR IN

	Name and Location	Age of Formation	Rock Types and Thicknesses		
			Ultramafics	Gabbro	Sheeted Diabase Dikes
1	Sierra de Santa Cruz, Guatemala	Aptian basalts early Cenomanian limestone	Serpentinized peridotite, harzburgite	Massive and cumulate late gabbro section >2 km thick	section is >300 m thick
2	Santa Elena Peridotite, Costa Rica	Latest Jurassic to Santonian	Harzburgite, lherzolite	Rare pegmatite gabbros	Very abundant and intrudes peridotites
3	Nicoya Complex, Costa Rica	Latest Jurassic to Late Cretaceous	Small outcrops have been reported by Galli-Olivier, 1979	Gabbroic to dioritic intrusions of Culebra are emplaced into Nicoya arc presumably of arc-tholeiitic affinity	2 km including interlayered pillow basalts
4	Osa Peninsula, Costa Rica	Early Cretaceous	Small outcrops	Abundant	Abundant
5	Azuera Peninsula, Panama	Early Cretaceous	Small areas of ultramafics	None reported	None reported
6	Eastern Panama and Western Colombia	Late Cretaceous or older in west; pre-Eocene in east	Isolated outcrops in major fault zones	Fault blocks	Fault blocks
7	Guajira Peninsula Colombia; Paraguana, Venezuela	Early Cretaceous	Serpentinites	Meta-gabbros	None reported
8	Villa de Cura; Tinaquillo, Venezuela	Early Cretaceous	Pyroxenite	Gabbro bodies	None reported
9	Isla de Margarita, Venezuela	Early Cretaceous	Serpentinite	Metagabbros	Diabase dikes 66 ± 5 Ma (K-Ar)
10	Araya-Paria Peninsula, Venezuela	Early Cretaceous	Serpentinite	Scattered gabbros	None reported
11	Tobago	Early Cretaceous	Serpentinized peridotite	High-Al gabbros	Some diorite intrusions
12	La Desirade, Lesser Antilles	145 ± 5 m.y. Upper Jurassic K-Ar age on Trondjhemite	None reported	Gabbroic rocks from dredge hauls off submarine scarp	None reported
13	Bermeja Complex. Puerto Rico	Early Tithonian to late Aptian radiolarians in chert	3000 m of serpentinized peridotite interpreted from gravity studies	Possibly amphibolized gabbro	diabase in pebbles in conglomerate overlying Mayaguez group
14	Samana Peninsula and Northern Dominican Republic, Hispaniola	Early Cretaceous	Eclogite serpentinite blue amphibole	None reported	None reported
15	Duarte Formation, Dominican Republic, Hispaniola	Early Cretaceous	Serpentinized peridotite, harzburgite, dunite	Extensive gabbro	Diorite intrusive

AND AROUND THE CARIBBEAN PLATE (Keyed to Figure. 5)

Basalt	Overlying Sediments	Comments	References
Aptian flows; 2 pillow lava sequences are >600 m thick	Pelagic chert limestone; volcanoclastic conglomerates	Occurs in a suture zone; appears to be normal ocean crust and not B″	Peterson and Zantop, 1980 Rosenfeld, 1980, 1981 Bertrand et al., 1978
pillow basalts cover peridotite with apparent angular unconformity	Cenomanian Radiolarian chert	Separates continental crust of the Chortis block from oceanic crust overlain by arc material to the south. Probably not B″	de Boer, 1979 Galli-Olivier, 1979 Dengo, 1962 Schmitt-Effing, 1979 Azema and Tournon, 1980
massive units of pillow lavas	Pelagic chert and lava breccia	Nicoya complex is typical seafloor (ridge) emplaced between late Santonian to early Campanian (Galli-Olivier, 1979); possibly represents B″ crust	deBoer, 1979 Galli-Olivier, 1979 Kuijpers, 1980 Schmitt-Effing, 1979
undetermined thickness	Pelagic chert	Continuation of Nicoya complex. Possibly represents B″ crust	References same as Nicoya Complex
Pillow basalts	Campanian limestone	Similar to Nicoya, Osa complex. Early Cretaceous ophiolitic basement overlain by Campanian volcanic and plutonic complex	Ferencic et al., 1971 Galli-Olivier, 1979
Relationship unclear	Relationship unclear	Complete ophiolites not recognized	Case, 1974 Case and Holcombe, 1980 Bourgeois et al., 1982
Relationship unclear	Relationship unclear	Occurs in fault zones	Bellizzia, 1972 Alvarez, 1971 Skerlec and Hargraves, 1980 MacDonald, 1964
Pillow basalts	3000 m metamorphosed volcano-sedimentary rocks	Scattered ophiolitic members not seen as a continuous body. Interpreted as obducted marginal basin floor and not B″	Maresch, 1974 MacKenzie, 1960 Gealey, 1980 Skerlec and Hargraves, 1980 Vierbuchen, 1978
Basalts	Relationships unclear	Similar to Villa de Cura	Vignali, 1979 Maresch, 1975
Pillow lavas	Radiolarian cherts	Along El Pilar Fault Zone in accretionary prism probably part of Villa de Cura	Virbuchen, 1978 Vignali, 1979
Relationship unclear	Relationship unclear	In accretionary prism not B″	Maxwell, 1948 Rowley and Roobol, 1978
Pillow basalts	Deep marine radiolarian cherts interbedded and overlying pillow basalts	Probably occurs in accretionary prism and does not represent B″	Mattinson et al., 1973 Mattinson et al., 1980 Fink, 1972 Fink et al., 1972
~300 m pillow basalts	200-250 m chert	Slivers of ocean floor not B″ in accretionary prism	Mattson, 1960, 1973 Mattson and Pessagno, 1979
None reported	Highly faulted schistose limestone and marble	Slivers in accretionary prism; not B″	Nagle, 1979, 1974 Joyce, 1980
Relationship unclear	Relationship unclear	Bounded by high-angle faults in strike-slip zone; not B″	Bowin, 1966 Kesler et al., 1977 Haldemann, 1980

TABLE 4. PROPOSED FRAGMENTS OF MESOZOIC SEAFLOOR IN

Name and Location	Age of Formation	Rock Types and Thicknesses		
		Ultramafics	Gabbro	Sheeted Diabase Dikes
16 La Cienaga Complex, Dominican Republic, Hispaniola	Cretaceous (early)	None reported	Gabbro occurs as fault-bounded blocks	None reported
17 Dumisseaux Fm. Southern Haiti, Hispaniola	Early Cretaceous	None reported	Rare occurrences	None reported
18 Bath-Dunrobin Fm.; Arntully, Green Bay Areas, Jamaica	Associated Campanian limestone	Serpentine occurs in fault zones	150-200 m (maximum)	None reported
19 Eastern Oriente Province Cuba —Purial (easternmost) —Sierra de Nipe (south central)	Pre-Aptian	Serpentinized peridotite, harzburgite and gabbro section is 4 km thick	Gabbroic intrusions in ultramafics	None reported
20 Gibara Area, NW Oriente Province, Cuba	Early Cretaceous	Slightly serpentinized peridotite, harzburgite	Brecciated gabbros associated with ultramafics	None reported
21 Central and Western Cuba	Early Cretaceous	>2300m serpentinite encountered in well, amphibolites	None reported	None reported
22 Isla de Pinos, Cuba	Jurrasic (?)	Isolated serpentinite outcrops	None reported	None reported
23 Water Island Fm., Virgin Islands	Late early Cretaceous	None reported	None reported	None reported

the Caribbean itself. Early Paleozoic and Precambrian ages have not been plotted because the scope of this review is limited.

HOW DID THE CARIBBEAN EVOLVE?

We here pose the question of how the Caribbean has evolved since the beginning of the Jurassic and attempt to suggest some possible ways of testing the plate evolution sequence of Pindell and Dewey (1982). An abbreviated version of this sequence is shown in Figure 7 and contains many elements of earlier models belonging to the Wegener school of continental separation (Table 1c).

By 150 Ma ago (Fig. 7a), the Gulf of Mexico was largely formed. There had been major stretching under the Gulf Coast with horsts formed in the Wiggins Arch and the Sabine Uplift, Yucatan had rotated away from North America, and evaporites had been deposited both on rift floor and on oceanic crust in the Gulf. The youngest ocean floor in the gulf (which may be about 140 Ma old) was formed after the evaporites ceased to be depos-

ited. Mexico had attained something closer to its present shape than it had in Late Permian times by processes which are approximated by left-lateral strike-slip displacement on three major faults (see Fig. 7a). This movement was probably completed by the end of Triassic time (Pindell and Dewey 1982), but it seems likely that it will prove very difficult to establish exactly what happened in Mexico around the end of the Paleozoic and the beginning of the Mesozoic because Permian and Triassic outcrops in Mexico are small and isolated. Preliminary paleomagnetic evidence indicates that a significant part of Mexico rotated 130° counterclockwise during Early Mesozoic time, coeval with the opening of the Gulf of Mexico, and that a major structural discontinuity must exist between Mexico and cratonic North America (Gose, and others, 1982). In Figure 7a, a spreading center is shown as beginning to develop between South America and Yucatan while motion in the Gulf of Mexico continued. An alternative possibility is that the gulf had completely formed before motion began between Yucatan and South America. Completion of the establishment of the Gulf of Mexico involves

AND AROUND THE CARIBBEAN PLATE (Keyed to Figure. 5)

Basalt	Overlying Sediments	Comments	References
Basalt; no pillow structures	Intensely deformed micritic limestones and cherts	Slivers in accretionary prism; probably B″	Pindell et al., 1981
Both pillowed and massive basalt	Pelagic limestones cherts, siltstones overlain unconformably by upper Eocene limestone	probably sub B″ in accretionary prism	Maurasse et al., 1979 Maurasse, 1980 Waggoner, 1978
2.5 km max. of basalt and gabbro	Thin siliceous sediment	Bath-Dunrobin and Arntully are slivers in an accretionary prism; probably not B″	Wadge et al., 1982 Chubb, 1954 Draper et al., 1976
None reported	Local metasediments and chert	High pressure metamorphic aureole around ultramafic sedimentary cover limits age to pre-Camp. Maastrictian. In accretionary prism; not B″	Boiteau et al., 1972a, b Boiteau et al., 1976 Kozary, 1968 Lewis and Straczek, 1955
Relationship unclear	<100 m volcanic siltstone	Occur as thrust sheets not massif-like uplifts as in eastern Oriente province in accretionary prism; not B″	Kozary, 1968 Thayer and Guild, 1947 Wassal, 1956
Occurs locally	Necomanian abyssal limestone enclosed in serpentine	Extensive trend of ultramafics. Many thrust bounded blocks or sheets in accretionary prism; not B″	Meherhoff and Hitten, 1968 Hill, 1959 Wassal, 1956 Boyanov et al., 1975
None reported	Relationship unclear	Isla de Pinos is likely to be true continental crust, perhaps with a fragment of oceanic crust thrust · onto it; not B″	Trujillo, 1975 Meyrehoff and Hatten, 1968
Spillites keratophyres pillow structures	Radiolarian tuff	Structures suggest occurrences in accretionary prism; probably not B″	Donnelly, 1966

rotation of Yucatan against Mexico at the isthmus of Tehuantepec. This area is covered with thick, later sequences and the structure beneath is as yet unresolved. Good evidence of when a rifted margin formed on the north coast of South America has not been recognized. If the evaporites of the Gulf of Paria are about 135 Ma old and represent this event, then the beginning of ocean-floor formation in the Caribbean would approximate the time of ending of ocean-floor formation in the Gulf of Mexico.

The relationship between the timing of formation of the Gulf of Mexico and of the Caribbean ocean floors may also be related to the timing of a major ridge jump in the Atlantic (no. 2, Table 2).

The position of the southern Florida volcanic rocks (no. 39, Table 2) is similar to that of the nodal volcanic concentrations which occur in the East African Rift system at places where rifts join or make angular bends. On some interpretations (e.g., Dewey and Burke, 1974), the places where these nodal volcanic rocks occur are hotspots, and the Bahamas might be considered to overlie volcanics erupted on the track of such a hotspot. Dis-

crimination between this and the alternative possibility, that the Bahamas are carbonate material deposited on a fracture zone elevated through Late Jurassic transpression, does not at present seem possible but represents an interesting problem.

Separation of Africa from South America is shown (in Fig. 7b) as starting at 110 Ma but on other interpretations could have happened about 20 Ma earlier. The Greater Antilles perhaps having nucleated on a Central Atlantic fracture zone occupied at this time a boundary between the North American plate and a plate in the Pacific (perhaps the Farallon plate). Over about the next 50 Ma, the Greater Antilles crossed the space between Mexico and South America on a generally northward track to collide with the Bahamas during the Paleogene. This interpretation of conditions at about 110 Ma ago is considered most attractive because of its simplicity, but it leaves some problems less than fully resolved. For example, evidence considered to represent mid-Cretaceous subduction toward the north on the south side of the Greater Antilles (e.g., no. 13, Table 4) and evidence thought to indicate northward subduction in the offshore islands of Vene-

TABLE 5.   SELECT DATED PLUTONS IN THE CARIBBEAN REGION (Keyed to Figure 6)

Number	Location/Name	Composition	Age (my)	References
	MAYA MOUNTAINS, BELIZE		290 ± 10 K-Ar	
1	Mt. Pine Ridge	True granite	(discordant) 390	Kesler et al., 1974
2	Hummingbird and Lockscomb	Granodiorite	210, Rb-Sr	Bateson, 1972
3	Western Altos Cuchumatanes, NW Guatemala	Granitic plutons	Post E. Permian, pre-L. Jurassic by stratigraphic relationships. Post-Early Cretaceous plutons as well	Anderson et al., 1973
4	**SAN MARCOS STOCK**	Granodiorite	150 ± 13 Rb-Sr whole rock	Horne et al., 1976
5	Chiquimula pluton	Granodiorite with some gabbro and diorite	84 K-Ar	Clemons and Long, 1971
	Southern Guatemala		95 Rb-Sr	
	Northern Honduras		215 Rb-Sr whole rock	
6	**SIERRA de OMOA**	Metamorphosed intermediate plutons	305 ± 12 Rb-Sr whole rock	Horne et al., 1976
	NW Honduras	Unmetamorphosed intermediates	150 ± 13 Rb-Sr whole rock	Horne et al., 1974
	Upua Valley, Honduras			
7	San Pedro Sula	Granodiorite	36 K-Ar biotite (minimum reset)	Horne et al., 1974, 1976
	Mezapa	Tonalite	72 K-Ar biotite	
8	Tela	Tonalite	73 K-Ar biotite, 93 K-Ar hornblende	
	CORDILLERA NOMBRE de DIOS			
	N. Honduras			
9	Las Mangas	Tonalite	57 K-Ar biotite	Horne et al., 1974
10	Piedras Nagras	Tonalite	57 K-Ar biotite, 72 K-Ar biotite	
	CENTRAL HONDURAS			
11	Minas de Oro	Granodiorite	61 K-Ar biotite	Horne et al., 1974
12	San Francisco	Granodiorite	59 K-Ar replicate analysis	
13	San Ignacio	Granodiorite	114 K-Ar biotite	
14	Dipilto Batholith	Granitic adamellite to biotite rich tonalite	140 ± 15 Rb-Sr whole rock, possible early tertiary thermal event as well	Horne et al., 1976
	Honduras-Nicaragua border			
	SIERRA NEVADA de SANTA MARTA, Colombia			
15	Santa Marta batholith	Granodiorite	44/49 K-Ar biotite/hornblende	Tschanz et al., 1974
16	Buritaca pluton	Granodiorite	49 K-Ar biotite/hornblende	
17	Latal and Tucurinquita plutons	Granodiorite	54/58 K-Ar biotite/hornblende	
	NORTHERN ANDES, Colombia			
18	Anacué River	Granitic pluton	160 ± 30	Irving, 1975
19	near Virginias	Diorite	160 ± 7 K-Ar hornblende	
20	Paraguaná Peninsula, Venezuela	Granitic pluton intrudes quartz-feldspar micaschist and gneiss	262-265 U-Pb (Permian)	Santamaria and Schubert, 1974
21	Aruba	Quartz diorite batholith with gabbroic intrusions	67-73 ± 4 K-Ar 70 ± 2 Rb-Sr	Santamaria and Schubert, 1974 Priem et al., 1978
22	Curacao	Sills and dykes intruding Zeolite facies metamorphics	74 ± 5 K-Ar whole rock 84 ± 6	Santamaria and Schubert, 1974 Beets, 1972
23	Gran Roque	Quartz diorite	65 ± 3.6 K-Ar	Santamaria and Schubert, 1974
24	Margarita Island	Matasiete granite	72 ± 6 K-Ar amphibole	Santamaria and Schubert, 1974
25	Los Testigos	Metagranitic rocks	47 ± 6 K-Ar amphibole, extension of Lesser Antilles?	Santamaria and Schubert, 1974
	CARIBBEAN MTS., Venezuela			
26	Guaremal	Granite	33-35 ± 2 K-Ar	Santamaria and Schubert, 1974
		diorite	79 K-Ar	
27	Choroni	Granite	30 ± 2 K-Ar	Santamaria and Schubert, 1974
28	Cantagallo	Gabbroic	65-67 K-Ar	Santamaria and Schubert, 1974
29	Oritapo, Northern Venezuela	Diorite	K-Ar 76 ± 4 biotite/hornblende	Santamaria and Schubert, 1974
30	El Baul	Granodiorite	270 K-Ar	Santamaria and Schubert, 1974
31	Ares Ridge	Granodiorite dredge	65 K-Ar biotite	Sutter, in Santamaria and Schubert, 1974
32	Paria Peninsula	Igneous-metamorphic association	53 ± 3 K-Ar whole rock 128 ± 11 hornblende concentrate analysis, same rock. Emplacement pre-Late Jurassic	Santamaria and Schubert, 1974
33	Tobago	Gabbro diorite granodiorite	127 ± 7 K-Ar whole rock 113 ± 6 K-Ar whole rock 102 ± 4 K-Ar whole rock	Rowley and Roobol, 1978
34	St. Croix, Virgin Islands	Hornblende gabbro Southgate pluton dykes	66 ± 3 K-Ar hornblende	Speed et al., 1979
35	Virgin Gorda batholith	Granodiorite narrows quartz diorite	35 K-Ar biotite/hornblende 24 K-Ar hornblende	Cox et al., 1977 Kesler and Sutter, 1979
	PUERTO RICO			
36	San Lorenzo batholith	Granodiorite; quartz diorite	71-78 K-Ar biotite, altered(?)	Cox et al., 1977 Kesler and Sutter, 1979
37	Morovis Stock	Granodiorite	88 K-Ar hornblende	Cox et al., 1977 Kesler and Sutter, 1979

TABLE 5.   SELECT DATED PLUTONS IN THE CARIBBEAN REGION (Keyed to Figure 6) *Continued*

Number	Location/Name	Composition	Age (my)	References
38	Utuado batholith	Dioritic; granodiorite	73-68 K-Ar biotite	Cox et al., 1977
	Stocks south of Utuado	quartz diorite	38-48 K-Ar biotite	Kesler and Sutter, 1979
39	Terre-Neuve, Haiti	Granodiorite	62-66 K-Ar biotite	Kesler, 1971
	(continues into Dom. Rep.)			
	CORDILLERA CENTRAL, Dominican Republic			
40	Medina Stock	Tonalite	79 ± 2 K-Ar biotite/hornblende	Bowin, 1975, Table 3
41	El Rio batholith	Tonalite	86 K-Ar hornblende	Kesler et al., 1977
	SIERRA MAESTRA, Cuba			
42	Nima-Nima	Plagiogranite, diorite	49 ± 6, 58 ± 8 K-Ar whole rock	Meyerhoff et al., 1969
43	Daiquiri	quartz diorite	49 ± 6 K-Ar whole rock	Meyerhoff et al., 1969
44	Northern Oriente Province, Cuba	Pegmatite dyke in metamorphics	119 ± 10% whole rock	Meyerhoff et al., 1969
45	Las Villas Province, Cuba	Tres quanos quartz monzonite	61 K-Ar biotite	Meyerhoff et al., 1969
46	Sierra de Trinidad, Cuba	Granodiorite	180 K-Ar whole rock	Meyerhoff et al., 1969
	JAMAICA			
47	Guava River	Granodiorite	80 ± 5 K-Ar	Wadge et al., 1982
48	Above rocks	Granodiorite	65 ± 3 K-Ar, Rb-Sr, U-Pb	Chubb and Burke, 1963
49	Central Inlier (Ginger Ridge)	Granodiorite	85 ± 9 K-Ar isochron intercept.	Lewis et al., 1973
50	Nicaraguan Rise	Granodiorite	Late Cretaceous	Arden, 1975

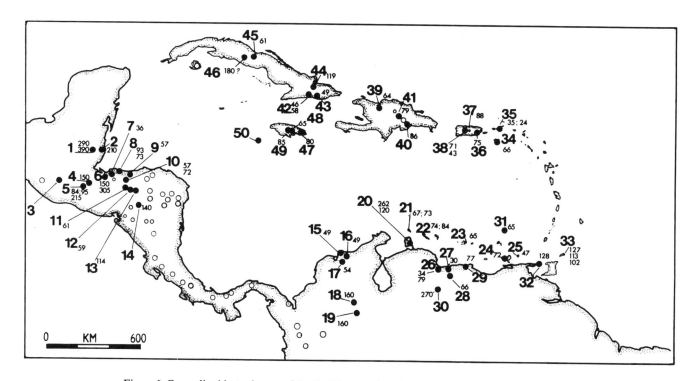

Figure 6. Generalized isotopic ages of the Caribbean region. Open circles represent a selection of as yet undated plutons and closed circles, dated bodies. Small numbers refer to generalized ages of plutons. Large numbers correlate with the explanation in Table 5. The plutons can generally be regarded as the roots of volcanic edifices. In Central America, the wide variety of ages reflects the circum-Pacific accretionary style active in the Pacific region as far back as late Paleozoic. Ages in the Greater Antilles indicate subduction from Cretaceous to the end of Eocene. On the mainland of South America, plutons west of #20 include circum-Pacific material of Jurassic age. The majority of the others are Cretaceous and Cenozoic. Ten Permian ages record a Hercynian collision between North and South America.

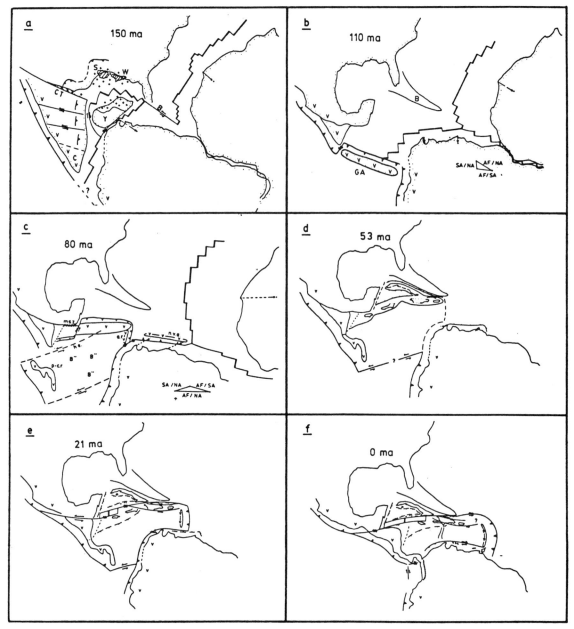

Figure 7. General outline of Caribbean/Gulf of Mexico evolution in the framework of relative motion rectors (see Fig. 2a and Pindell and Dewey, 1982). Three general phases of geologic development are predicted here from consideration of North America-South America relative plate motion; a fourth is not predicted and is shown in Figure 9. The three phases are: 1) *Jurassic-Lower Cretaceous extension sinistral slip* and development of predominately stable shelf margins (crosses=evaporites; W=Wiggins Arch; S=Sabine Arch; Y=Yucatan block; C=Chortis block; B=Bahama Platform; CT=Chihuahua Trough; V=volcanic arc; barbed lines=thrusts or subduction zones; barb on hanging wall; arrows=a displacement of points relative to a fixed North America—Fig. 7a); 2) *Cretaceous-Paleogene convergence* of South America with respect to North America with development of several arc terrains (GA= Greater Antilles; p-cr=Panama-Costa Rica arc) with at least two episodes of arc-continent collisions western Greater Antilles with Yucatan block at Motagua suture zone (MZ) and the Netherlands Antilles (n.a.) with northern South America; B" oceanic plateau may have undergone sinistral shear along faults like the Hess escarpment (h.e.) as a response to oblique motion of the Americas (see vector triangle) (Fig. 7b, c, d); 3) post-Eocene eastward relative motion of the Caribbean plate with respect to the Americas and development of the complex North and South Caribbean strike-slip plate boundary zones (Fig. 7e, f). Shapes of landmasses, particularly northern South America, have been severely altered by this strike-slip phase, as well as a subsequent collisional phase (Fig. 9) and are shown in earlier time frames only as approximate geographical references.

zuela and in Tobago older than 80 Ma ago (e.g., no. 33, Table 5) have to be reinterpreted. An alternative interpretation sees the Greater Antilles at this time as an arc-system looking like a capital letter *Y* on its side with the single limb on the east, representing subduction to the north in the Venezuelan Antilles, and two limbs on the west representing subduction to the north in the southern Greater Antilles and subduction to the south on the northern Greater Antilles (Burke and others, 1980, Fig. 2; White and Burke, 1980). However, it is not surprising that there is as yet no consensus on this issue because intense strike-slip faulting in the Neogene has offset Greater Antillean Cretaceous outcrops huge distances and restoration of their original disposition is as yet a long way from being possible.

At 80 Ma ago (Fig. 7c), we show radical changes in Caribbean development. An oceanic plateau that now occupies much of the Caribbean ocean floor was moving in between Mexico and South America perhaps along transforms, one of which may be represented by the Hess escarpment (Fig. 9), and the Greater Antilles were sliding past Yucatan with their western end (Jamaica and the Nicaraguan Rise) colliding with that peninsula along the Motagua suture zone. The eastward component of motion of the buoyant ocean floor is accommodated by the beginning of convergence along an arc now marked by the Aves swell, for which subduction is shown to the west although it could have been to the east. On the southwestern side of the buoyant ocean floor, convergence is shown as beginning to build the Panama-Costa Rica arc. The arc of offshore Venezuela had formed by subduction to the north and was very young at this time on the interpretation used here. It is shown as about to collide with Venezuela, an event which involves obduction of ocean floor from between the arc and the continent to form the Villa de Cura Nappe (no. 8, Table 4).

At 53 Ma ago (Fig. 7d), the Greater Antilles are shown colliding with the Bahamas, an event which was completed earlier in Cuba, than farther east in Hispaniola and possibly also in Puerto Rico. We depict the splitting of the Greater Antilles arc at this time to form the Yucatan basin as a back-arc basin (Gealey, 1980), the timing being compatible with the depth of the basin and with rifting events on its margins. The Panama and Costa Rican arc system is shown as having joined Chortis along a suture marked by the Santa Elena ophiolite (no. 2, Fig. 5). The Hess escarpment, extending ENE from the suture, is interpreted as ceasing strike-slip motion at this time. There was little motion between the Caribbean and North America at about 53 Ma ago and the Aves arc may have ceased activity. Motion of plates in the Pacific with respect to the Americas was convergent with a major strike-slip system (or its equivalent) linking the arcs of the North and South American cordilleras.

From about the end of the Eocene (38 Ma), what motion there is of South America with respect to North America is westward with a small component of clockwise rotation, but movement eastward of the Caribbean plate with respect to both North and South America became much more important in

structural development. This motion was accommodated on the great strike-slip systems of both the northern and southern Caribbean margins with their broad plate boundary zones and has resulted in convergence along the Lesser Antillean arc. Quantitative assessment of the minimum amount of Caribbean eastward movement is made possible by the existence of the Cayman spreading center (a short ridge segment within the Montagua-Swan-Oriente transform system, Holcombe and others, 1973) which, we suggest, began to operate about 38 Ma ago and which has accommodated about 1400 km of motion.

Although Cuba appears to have behaved as an integral part of North America throughout the time of strike-slip motion on the northern Caribbean boundary, the other islands of the Greater Antilles have been sliced to shreds by strike-slip faults. For example, in Figure 7e (at 21 Ma ago), Jamaica and the southern peninsula of Hispaniola are shown as moving east with respect to the greater part of that island; and another major strike-slip fault, roughly in the Cibao valley, is shown as active farther north. Strike-slip motion along the various members of the Motagua fault system has carried Chortis to its present position from its former location on the southern coast of Mexico (see Fig. 7d, e, and f). This motion accounts, at least in part, for the much narrower forearc-system of the middle America trench north of the Caribbean, North American, and Cocos plates triple junction. No part of the forearc north of the triple junction to the mouth of the Gulf of California is older than Miocene because Chortis had occupied this position prior to that time. However, the forearc south of the triple junction includes much older material and is therefore wider because this margin had been convergent for a longer time. Paleomagnetic studies (Gose and Swartz, 1977) indicate complex rotations of the Chortis block from the Aptian to the present.

Oligocene and later strike-slip motion on the southern border of the Caribbean is even greater than on the northern border because a few hundred kilometers of westward motion of South America with respect to North America have to be added to the approximately 1400 km eastward motion of the Caribbean with respect to North America. Most of this motion is offshore from Venezuela, and only small offsets have been mapped on faults on shore. The most spectacular consequence of this strike-slip motion has been to carry the island of Barbados with its quartz-rich Eocene accretionary wedge, acquired when close to South America in the Guajira area, about 1500 km eastward to its present, rather isolated position (Dickey, 1980).

PLATE BOUNDARY ZONES OF THE NORTHERN AND SOUTHERN CARIBBEAN

Since the Caribbean began to move eastward with respect to the Americas about 38 Ma ago, it has not behaved entirely rigidly. Along both northern and southern boundaries, complex plate boundary zones have developed. These are shown in Figure 8 with 33 of the individual basins associated with the plate boundary zones identified on the map and tabulated in

TABLE 6.   POST-EOCENE STRIKE-SLIP BASINS AROUND THE CARIBBEAN PLATE

Name	Age	Type	Sediment Thickness (meters)	Known Volcanism	References
1 Central American	?-present	Between NE-trending secondary extensional faults	—	Quaternary alkaline basalts along NE faults	Plafker, 1976 Muehlberger, 1976
2 Caymen	Post-Eocene-present	Short spreading ridge between long transforms	Variable—dependent on distance from spreadng and landmasses	Mid-ocean ridge basalt	Holcombe et al. 1973 Perfit and Heezen, 1978
3 79°W 18°20′N	Post-Pliocene	Pull-apart basin	—	—	Holcombe, 1981 Holcombe and Sharman, in press
4 78°W 20°N	?-present	—	—	—	Case and Holcombe, 1980
5 Cauto	?-post-Eocene-present	Shape suggests triangular opening behind rotating block	1100 m of post-Lower Miocene	—	Iturralde-Vinent, 1969, 1978; Breznyansky and Iturralde-Vinent, 1978
6 Yallahs	?-present	Between NE-trending, secondary extensional faults	Minimum 500 m	—	Burke, 1967
7 West Navassa	?-present	Pull-apart	—	Alkali basalt	Burke et al., 19080
8 Central Navassa	?-present	Pull-apart	—	—	Mann et al., 1981
9 East Navassa	?-present	Pull-apart	—	—	Mann et al., 1981
10 Southern Haitian (2)	Quat.	Pull-apart	—	—	Mann and Burke, 1982
11 Port au Prince	?-present	Between NE-trending, secondary extensional faults	—	Pleistocene basalts aligned along NE-bounding faults	Woodring, et al., 1924; Wadge and Wooden, 1982
12 Cul de Sac-Enriquillo	? Medial Eocene-present	Between strike-slip faults from Eocene to Late Oligocene Transpressional faults to present	3 km of Late-Miocene sediments	Pleistocene basalts closeby on NE-extensional faults	Bowin, 1975; Wadge and Wooden, 1982 Cooper, unpubl. field data
13 Low Layton	Miocene	Pull-apart	—	Late Miocene alkali basalt 9.5 m.y.	Manna and Burke, 1980; Wadge, 1981
14 76°W 19°30′N	?-present	Pull-apart	—	—	Case and Holcombe, 1980
15 Northern and Southern Gonave	?-present	Between strike-slip and transpressional faults	—	—	Goreau, 1980
16 Windward Passage	Post-Pliocene?	Pull-apart	—	—	Goreau, 1980
17 Hispaniola (Old Bahama Channel)	?-present	Along transpressional faults	—	—	Goreau, 1980
18 Cibao	Late Miocene to to present	Along transpressional fault	4.6 km of post-late Miocene sediment	—	Bowin, 1975 Saunders et al., 1980
19 Tavera	Early-Late Oligocene	Parallel to strike-slip faults	Minumum 1.5 km	—	Palmer, 1979; Groetsh, 1980; Mann et al., 1981

TABLE 6.   POST-EOCENE STRIKE-SLIP BASINS AROUND THE CARIBBEAN PLATE Continued

Name	Age	Type	Sediment Thickness (meters)	Known Volcanism	References
20 San Juan	Medial Eocene to present	Parallel to s.s. faults from Eocene to Late Oligocene; transpressional to present	Minimum of 3 km of post-Lower Miocene	Plio-Pleistocene alkaline basalts; some along NE fissures	Michael and Millar, 1977 Nemec, 1980
21 Azua	?-present	Between NE-trending secondary extensional faults	Minimum 3 km of post-Lower Miocene	—	Bowin, 1975; Mann, unpubl. field data
22 San Pedro	Oligocene-present	Parallel to transpressional faults	3 km	—	Ladd et al., 1981
23 Muertos	?-present	Parallel to transpressional overthrust		—	Ladd et al., 1977
24 Puerto Rico	Post-Late Eocene	Parallel to transpressional overthrust	3.5 km of post-Miocene subsidence	—	Perfit et al., 1980
25 Anegada	Post-Eocene	Pull-apart	—	—	Donnelly, 1964
26 Cariaco	?-present	Pull-apart	—	—	Schubert, 1982; Silver et al., 1975; Case, 1974
27 Orchila	?-present	Pull-apart	—	—	Case, 1974; Peter, 1972
28 Los Rogues	?-present	Parallel to strike-slip faults (?)	—	—	Silver et al., 1975
29 Rancheria	?-present	Parallel to transpressional faults (?)	—	—	Case and Holcombe, 1980
30 Falcon and Bonaire	Oligocene-Lower Miocene	Incipient short ridge between long transforms (?)	3.5 km at 50-70 cm/1000 yrs	Alkali basalts (28-23 m.y.)	Muessig, 1978; Silver et al., 1975; Case, 1974
31 Yaracuy	Late Pliocene-present	Pull-apart	—	—	Shubert, 1980
32 Mucuchies	Late Pliocene-present	Pull-apart	—	—	Schubert, 1980
33 Gonzalez	Late Pliocene-present	Pull-apart	—	—	Schubert, 1980

Table 6. Bucher (1948) was the first to point out that secondary structures on the two sides of the Caribbean had orientations comparable to those that might be expected in a glacier flowing from west to east; that is, extensional structures trend NE-SW on the northern side and NW-SE on the southern side. Compressional structures trend in the complementary directions on the two sides. This topic has been elaborated on by later authors (for example, Burke and others, 1980; Kafka and Weidner, 1981; Sykes and others, 1983); but as the summary in Table 6 indicates, development of secondary structure associated with the major strike-slip systems on the northern and southern sides of the Caribbean is very complex and has as yet been worked out in very few areas. Because some of the basins in the plate boundary zones are oil producing (for example, the Maracaibo basin and southern Trinidad) and others have oil potential, the plate boundary zones of the Caribbean are likely to be areas in which there will be future opportunities through oil exploration for learning about the processes of plate interaction at transform boundaries over time scales varying up to 30 million years.

The geology of the two plate boundary zones is so complex that it has not been possible yet to establish offsets on many of the individual strike-slip faults within either the northern or the southern transform systems. This problem arises partly because on both sides of the Caribbean the transform boundaries roughly parallel Cretaceous and Paleogene east-west trending convergent boundaries, and thus makes offset markers difficult to find. A second consideration has been that until lately many workers thought there were no large offsets on the northern and southern

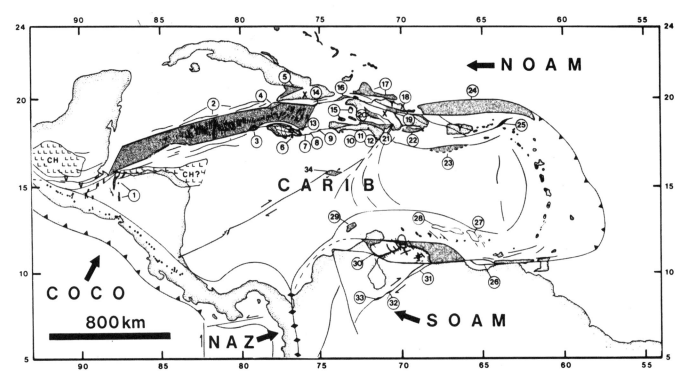

Figure 8. Post-Eocene eastward movement of the Caribbean relative to North and South America has produced about 29 strike-slip basins along both the northern sinistral plate boundary zone and southern dextral plate boundary zone. Most prominent along the northern boundary is the 1400 km long Cayman trough (no. 2), floored by oceanic and thinned island arc crust produced at the Mid-Cayman spreading center, shown in black. The Cayman trough provides firm evidence for large and steady post-Eocene eastward displacement of the Caribbean plate relative to the America plates. The Falcon-Bonaire Basin (no. 30) in northern South America is a pull-apart basin, although it contains only scattered alkali basalts. Smaller pull-aparts are developed at left (northern boundary) and right (southern boundary) steps along throughgoing strike-slip faults. The large number of basins in the Hispaniola area is related to the disruption of Cretaceous-Eocene island arc belts that lay across the strike of the proto-Cayman Trough Fault System. Some of these are deeps related to ongoing underthrusting of Jurassic Atlantic crust (24) and buoyant Cretaceous Caribbean seafloor (23) beneath the arc material of the Greater Antilles. A second generation of basins (nos. 1, 31, 32, 33, and 34) began developing on northeast-trending faults three to five million years ago following the collision of the Panama arc with South America in Colombia. The arc-continent collision drove a complex wedge, intruding the Santa Marta block, northward into the Caribbean plate causing dextral movement along the Bocono Fault (basins no. 31, 32, 33) and dextral reactivation of the Hess escarpment as inferred from the direction of stepping of offsets (basins no. 34 and 82W). Northeast movement of the Nicaraguan Rise relative to the Colombian basin along the Hess escarpment causes Quaternary rifting and basaltic volcanism in Central America (no. 1) and compresses the earlier generation of east-west basins in Hispaniola (nos. 10, 11, 12, 20, 21).

sides of the Caribbean. Recognition of the extent of ocean floor in the Cayman trough has made this position no longer tenable, and we present a summary of strike-slip offsets that have been claimed in Table 7. It seems likely that the establishment of strike-slip offsets within the plate boundary zones is an area in which much progress can be expected.

Paleomagnetic studies of the Caribbean have not been extensively considered in this review because they have been discussed in other papers in the memoir. We would interpret most of the presently available paleomagnetic azimuths as products of rotation in the northern and southern strike-slip plate boundary zones (Fig. 8). In the sinistral northern plate boundary zone, the

paleomagnetic results (see, for example, a summary in Skerlec and Hargraves, 1980, their Fig. 11) are compatible with anticlockwise rotations. On the dextral southern border, rotations are clockwise (Skerlec and Hargraves, 1980; Stearns, and others, 1982).

ACTIVE DEFORMATION OF THE CARIBBEAN PLATE

Recorded seismic events enabled Molnar and Sykes (1969) to establish the identity of the Caribbean as a rigid lithospheric plate. There are, however, significant departures from rigidity in

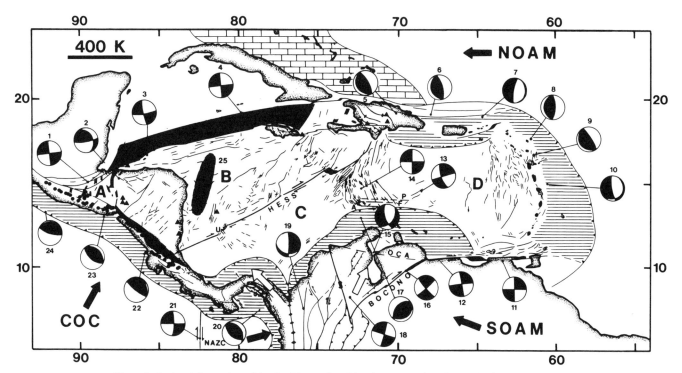

Figure 9. Active deformation of the Caribbean plate. Numbers on earthquake mechanisms refer to Table 8. Horizontally ruled areas are accretionary prisms including those within the Caribbean largely related to strike-slip motion. Shaded areas are extensional, the seismically active elliptical area (25) close to letter "B" is interpreted as reactivating the Paleogene San Andres trough. Round dots are active volcanoes in arcs and are mainly andesitic. Triangles are either intraplate or within plate boundary zone pull-apart structures (see Wadge and Wooden, 1982). Thin lines within the plate represent active faults, mainly structures that cut sediments on reflection profiles. The suture zone on the southern edge of the map at 77° West marks the collision of the Panama arcs with South America, a critical event in establishing the Neotectonic zones A, B, C, and D of the plate and in moving the Bonaire Block (open triangle) toward the Caribbean.

Caribbean plate behavior not only in the existence of the wide plate boundary zones of the northern and southern margins but also in indications of active deformation within the plate interior. These indications include: 1) numerous intraplate earthquakes; 2) intraplate faulting affecting young sediments, as seen on reflection profiles, and shown on Figure 9; and 3) strike-slip mechanisms for intraplate earthquakes (no. 13 and no. 14 in Fig. 9 and Table 8) and intraplate volcanism (black triangles on Fig. 9).

The internal deformation of the Caribbean plate and its apparent subduction between the North and South American plates (Fig. 9) have been interpreted as an ongoing response to Late Tertiary convergence of: 1) the South American and Caribbean plates (Jordan, 1975; Pindell and Dewey, 1982); 2) the South and North American plates (Ladd, 1976; Pindell and Dewey, 1982; Sykes, and others, 1983); or 3) a small continental fragment "escaping" from either the collision of the Carnegie Ridge (Pennington, 1981) or the Panama arc (Pindell and Dewey, 1982; Burke and Mann, 1982). Late Neogene uplift and basin formation and initiation of strike-slip faults (Case and Holcombe, 1980) approximately coincide with closure of the Pacific-

Caribbean Seaway (Keigwin, 1978) and suggests to us that Panama arc collision with Colombia drove a complex wedge (the Bonaire Block of Silver, and others, 1975) northward into the Caribbean (Fig. 9). Older strike-slip faults like the Oca (Fig. 9), that had previously accommodated east-west Caribbean-South America relative plate motion, were rendered inactive over the last few million years by the northward movement of the wedge along the dextral Bocono and the sinistral Santa Marta Fault Zones (an open broad arrow in Fig. 9 shows the direction of motion of the Bonaire Block).

The response of the Caribbean plate to the northward motion of the block is more typical of continental areas (for example, China) than oceanic areas and may be attributed to the unusually thick B" oceanic crust of the Caribbean (Burke, and others, 1978). The internal deformation of the plate permits subdivision into four zones (A to D on Fig. 9; Burke and Mann, 1982), which we suggest have developed distinctively since the collision of the Panama arc with Colombia: 1) a Central American zone bounded by the middle America arc, the Honduras Depression, and the Motagua Fault System (Zone A in Fig. 9); 2) the Nicaraguan Rise north of the Hess escarpment (Zone B); 3) the Vene-

TABLE 7. SUMMARY OF STRIKE-SLIP OFFSETS RELATED TO THE MOTION OF THE CARIBBEAN PLATE RELATIVE TO THE AMERICA PLATES

No.	Name of Fault and Location	Sense of Offset*	Max. Amount of Offset	Type and Age of Rocks Offset	Slip Rate	Comments and Interpretations	Key References
1	Motagua, Guatemala	LL	58.3	Quarternary stream terraces (40,000 to 10,000 yrs.)	Avg. .6 cm/yr 0.45 to 1.8 cm/yr	Presently Motagua FZ is main strand of the PBZ in Central America	Schwartz et al., 1979
2	Polochic Fault	LL	132 ± 5 km	Late Cretaceous structure and stratigraphy	Movement-mid Miocene to mid Pliocene	Post Pliocene displacement shifted to other faults in W. Guatemala, Chiapas and the Motagua FZ	Burkart, 1978
3	Polochic Fault	LL	—	Offset of recent river valleys	Quat.	—	Kupfer and Godoy, 1967
4	Motagua-Polochic System	LL	-1000 km?	Pre-Neogene accretionary prism and pre-Mesozoic rocks of Chortis and Southwestern Mexico	1000 km in < 40 m.y.	Narrow Neogene accretionary prism north of plate boundary formed on strike-slip fault; wide accretionary prism to the south not affected	Watkins et al., 1981; King, 1969
5	Swan-Motagua System, Central America	LL	1000 km	salt(?) diapirs (Honduras); salt deposits (Guatemala)	1-2 cm/yr	Offset and rate based on salt deposit offsets	Pinet, 1972
6	Swan-Oriente (Cayman Trough)	LL	~1400 km.	Paleogene arc rocks are offset to extent of water of oceanic depths in trough	3.7 cm/yr for 37 m.y. closes the Cayman Trough	Cayman Trough pull-apart probably initiated in Late Eocene-Early Oligocene	Holcombe et al., 1973; MacDonald, 1976; Mann and Burke, 1981; Sykes et al., 1983
7	Nine Faults in Jamaica	LL	40 km (cumulative)	Most faults appear Neogene in age	4 mm/yr for 10 m.y. (est)	Uncertainty in the age of some of the mapped offsets; no offsets mapped for some faults in this zone	Burke et al., 1980
8	Enriquillo-Plantian Garden System	LL	~10 km	Cretaceous rocks in Jamaica	—	Total offset in area is probably much greater	Burke et al., 1981
9	Oriente-Northern Hispaniola system	LL	~200 km	Granodiorite plutons and ophiolites in S. Cuba and Northern Hispaniola; dissimilar basement juxtaposed in Hispaniola	~200 km from late Eocene	Lateral offset accompanied by large amounts of post-Oligocene uplift	Hass and Maxwell, 1953; Malfait and Dinkelman 1972; Mann and Burke, 1981; Burke et al., 1981
10	Septentrional-Puerto Rican Trench System	LL	~400 km	Northern Hispaniola forearc offset from Puerto Rican arc	<400 km from mid Eocene	Strike-slip, probably accompanied by rotation	Mann and Burke, 1981; Burke et al., 1981
11	Puerto Rico-St. Croix area	LL	80 km	Small closed basin 50 km NE of St. Croix is the faulted extension of trough between Saba and Anguilla range	—	These depressions may be strike-slip related but much younger	Donnelly, 1964

TABLE 7. SUMMARY OF STRIKE-SLIP OFFSETS RELATED TO THE MOTION OF THE CARIBBEAN PLATE RELATIVE TO THE AMERICA PLATES

No.	Name of Fault and Location	Sense of Offset*	Max. Amount of Offset	Type and Age of Rocks Offset	Slip Rate	Comments and Interpretations	Key References
12	El Pilar, Venezuela	RL	100 km	Faults and sediments	1-2 cm/yr	Major fault of South Caribbean PBZ; accomodates perhaps 10% of all motion	Schubert, 1979 Vierbuchen, 1978
13	La Victoria, Venezuela	RL	Several km	Displaced drainage	—	Part of N. Venezuela Fault system	Smith, 1953 Menendez, 1966 Schubert, 1981
14	Bocono, Venezuela	RL	60-100m (Late Pleistocene) 25-40 km inferred since Pliocene	Late Pleistocene glacial moraines ($\sim 12{,}000\ ^{14}C$ yrs.)	0.5-0.8 cm/yr.	Fault bounds Bonaire Block driven NE at 5 m.y. arc. cont. collision (cf. Fig. 9)	Schubert and Sifontes, 1970
15	Border Faults of Falcon-Bonaire Basin	RL	~ 500 km	Sediments in basin are Oligocene and younger. Intrusion ages 28-23 m. y.	500 km over ~ 30 m.y.	Shape of F-B basin suggests a similar history of E-W movement to Cayman Trough. N. movement of Bonaire Block has placed it in compression	Muessig, 1978
16	Oca Fault, Venezuela, Colombia	RL	20 km	Subsurface data (Tertiary sediments in Falcon Basin?)	< 20 km in 5 m.y.	Practically seismically inactive at present; inactivity may have begun 5 m.y. ago and be related to NE movement of Bonaire Block	Schubert, 1981

*LL=Left-lateral
RL=Right-lateral

TABLE 8. FIRST MOTIONS USED TO CONSTRAIN THE MOTIONS OF THE CARIBBEAN PLATE
(Keyed to Figure 9; figure numbers listed next to author are the source of mechanisms shown in Figure 9)

Keyed Number	Source	Comments and Interpretation
1	Langer and Bollinger, 1979, Fig. 4	Composite focal mechanism for after shock of Feb. 4 1976 earthquake (M = 7.5). Sinistral slip on Motagua FZ
2	Dean and Drake, 1978, Fig. 4, Solution 51	Intraplate event (H = 24 km) of a thrust nature with a large component of strike-slip (P axis plunges 20°-roughly in the same direction as plate motion)
3	Molnar and Sykes, 1969, Fig. 12, Solution 103	Sinistral on Swan FZ (H = 12 km)
4	Molnar and Sykes, 1969, Solution 101	Sinistral slip on Oriente FZ (H = 0 km)
5	Molnar and Sykes, 1969, Solution 105	Underthrusting of NOAM beneath Hispaniola in dominantly strike-slip environment (H = 39 km)
6	Kafka, 1980, Solution 2	Underthrusting of NOAM beneath Puerto Rico Trench (H = 28 km)
7	Kafka, 1980, Solution 108	Same as number 6 (H = 51 km)
8	Kafka, 1980, Solution 109	Underthrusting of NOAM beneath Lesser Antilles Arc (H = 27 km)
9	Kafka, 1980, Solution 23	Same as number 8 (H = 57 km)
10	Kafka, 1980, Solution 110	Same as number 8 (H = 79 km)
11	Molnar and Sykes, 1969, Fig. 12, Solution 115	Dextral slip on El Pilar FZ (H = 20 km)
12	Rial, 1978, Fig. 1, Solution 6	Interpreted by Rial as NIOW sinistral slip; may be related to northward movement of the Bonaire Block
13	Kafka and Weidner, 1979, Fig. 8	Left-lateral intraplate event studied by Rayleigh wave method; sense of shear opposite to that predicted by Prandtl cell model; may reflect internal deformation caused by northward movement of Bonaire Block
14	Molnar and Sykes, 1969	Intraplate event in the vicinity of inactive strike-slip faults near DSDP site known from reflection profiles. Interpretation same as number 13
15	Pennington, 1981, Fig. 4, Solution 39	Normal faulting may reflect active subvsidence of Maracaibo Bay and internal deformation of Bonaire Block
16	Pennington, 1981, Fig. 4, Solution 36	
17	Vierbuchen, 1978, Fig. 23, Solution V5	Underthrusting of Caribbean plate beneath Bonaire Block
18	Pennington, 1981, Fig. 4, Solution 33	Orientations of planes, poles, and axes determined by Pennington from figure in Dewey, (1972)—(nodal plane ambiguity). Complex strike-slip movements in collisional zone
19	Pennington, 1981, Fig. 4, Solution 16	Combination thrust and strike-slip in complex Panama-SOAM collisional area
20	Pennington, 1981, Fig. 4, Solution 11	Poles to planes presented in Molnar and Sykes, 1969; the axes were computed by Pennington. Underthrusting of Nazca plate beneath SOAM
21	Pennington, 1981, Fig. 4, Solution 6	Dextral slip on Nazca-Cocos transform boundary
22	Dean and Drake, 1978, Fig. 3, Solution 28	Underthrusting of Cocos plate beneath Caribbean
23	Dean and Drake, 1978, Fig. 3, Solution 23	Same as 22
24	Dean and Drake, 1978, Fig. 3, Solution 22	Same as 22
25	Sykes et al., 1983, Fig. 2	Western Nicaraguan Rise seismic zone—may reflect extension and incipient rifting of area B as a result of convergence of the Bonaire Block

zuelan basin between the Beata Ridge and the Lesser Antilles (Zone C); and 4) the Venezuelan basin between the Beata Ridge at the Lesser Antilles (Zone C).

Uplift of the Beata Ridge at the place where the Caribbean is narrowest between Guajira and Hispaniola is suggested to be, at least in part, a consequence of northward movement of the Bonaire Block; and the deformation pattern of the eastern Caribbean (area D on Fig. 9), which has been compared to that of a modified Prandtl cell implying extrusion toward the east (Burke and others, 1978), can also be ascribed to the influence of the Bonaire Block. West of the Beata Ridge, movement of the Bonaire Block has been responsible for reactivation of the Hess escarpment interpreted as taking place in a right lateral sense from the direction of stepping of offsets (between areas B and C on Fig. 9). This reactivation has accommodated motion of the Colombian basin (area C) toward Panama with the initiation of the foreland thrust belt-like deformation on the northern side of the Panama isthmus, the shutting off of arc volcanism on the isthmus and uplift of ophiolites on the southern side. Internal deformation in area B north of the Hess escarpment is in the opposite sense to that to the south. Compressional structures dominate close to Hispaniola and extension is most obvious in the Honduras Depression, an area where Wadge and Wooden (1982) have identified a province of alkaline volcanic activity. Active structures in area A on the Central American mainland have been related to segmentation of the underlying slab of Cocos plate but are dominantly compatible with E-W extension like that of area B.

Interpretation of earthquake mechanisms in the Caribbean is somewhat hazardous because many events are on faults in plate boundary zones and are unrelated to overall plate motion. We have plotted 24 mechanisms from published sources on Figure 9 which we regard as compatible with nonseismological information on the tectonics over the last 5 Ma. For example, in the islands of Hispaniola and Jamaica, there are many young faults which strike roughly E-W; we have observed that faults in this area trending north of west are associated with thrusting and folding while faults striking south of west are extensional with pull-apart segments (Burke, and others, 1981). This difference leads us to interpret the motion of North America with respect to the Caribbean as being that shown in Figure 9, from east to west.

CONCLUSION

The broad outlines of Caribbean history are beginning to emerge with the recognition that the pre-Caribbean continental fits and the motions of the boundary plates can be generally established and are compatible with much of what is known about Caribbean geology. It is our hope that our discussion will prove useful in emphasizing aspects of Caribbean studies in which the record is less than completely understood. A limitation, as in all geological research, is that the processes that have operated in producing the Caribbean of today have destroyed a great part of the rocks and structures that recorded what happened in the past. With this reservation, it appears that the time is now ripe to

attempt understanding of many aspects of Caribbean geology within a plate tectonic framework.

ACKNOWLEDGMENTS

Caribbean studies at the State University of New York at Albany have been supported by NSF (Geology Program Grants 7614754 and 7803319), by NASA (Geodynamics Program Grant NAG5-155), as well as by support from the Phillips, Cities Services, and Weeks Petroleum Companies, and by Robertson Research, Inc. Former colleagues who have worked with us include Jack Grippi and Gary White, and we have profited from discussions with fellow students of the Caribbean in the United States and especially in Jamaica, Haiti, and the Dominican Republic. We greatly appreciate the assistance and cooperation we have received from government and university authorities in those three countries in furthering our studies. Very helpful reviews were provided by Rob Van der Voo, Jim Case, and Bob Hargraves. Computational facilities at the Lamont Doherty Geological Observatory used in the preparation of Figure 7 were kindly made available by W. C. Pitman, III. Participation in the Caribbean Research Project of the Lunar and Planetary Institute is also gratefully acknowledged. A portion of the research reported in this paper was done while Burke was a Visiting Scientist at the Lunar and Planetary Institute, which is operated by the Universities Space Research Association under Contract No. NASW-3389 with the National Aeronautics and Space Administration. This paper is Lunar and Planetary Institute Contribution No. 503.

REFERENCES CITED

Alvarez, W., 1971, Fragmented Andean Belt of Northern Colombia, *in* T. W. Donnelly, ed., Caribbean Geophysical, Tectonic and Petrologic Studies: Geological Society of America Memoir 130, p. 77–96.

Anderson, T. H., Burkart, B., Clemons, R. E., Bohnenberger, O. H., and Blount, D. N., 1973, Geology of the western Altos Cuchumatanes, northwestern Guatemala: Geological Society of America Bulletin, v. 84, p. 805–826.

Antoine, J. W., Martin, R., M Pyle, T., and Bryant, W. R., 1974, Continental Margins of the Gulf of Mexico *in* Burk, C. A., and Drake, D. L., eds., The Geology of Continental Margins: New York, Springer-Verlag, p. 683–693.

Arden, D. D., Jr., 1974, Geology of the Suwannee Basin interpreted from geophysical profiles: Gulf Coast Association of Geological Society Transactions, v. 24, p. 223–230.

——1975, Geology of Jamaica and the Nicaragua Rise *in* Nairn, A.E.M., and Stehli, F. H., eds., The Ocean Basins and Margins, v. 3, New York, Plenum Press, p. 617–661.

Azema, J., and Tournon, J., 1980, La Peninsule de Santa Elena, Costa Rica: un massif ultrabasic charrie en marge Pacific de l'Amerique centrale: Comptes Rendus Académie des Sciences [Paris], Ser. D, v. 290, p. 9–12.

Bain, G. W., 1957, Triassic Rift structure in eastern North America: Transactions of the New York Academy of Sciences, Ser. 2, v. 19, p. 409–502.

Barnett, R. S., 1975, Basement structure of Florida and its tectonic implications: Transactions, Gulf Coast Association of Geological Societies, v. 25, p. 122–142.

Bateson, J. H., 1972, New interpretation of geology of Maya Mountains, British Honduras: American Association of Petroleum Geologists Bulletin, v. 55, p. 936–963.

Beets, D. J., 1972, Lithology and stratigraphy of the Cretaceous and Danian succession of Curaçao: Uitgaven "Natuurwetenschappelijke Studiekring voore Suriname en de Nederlandse, Antillen," Utrecht, no. 70, p. 165.

Bellizzia, A. G., 1972, Is the entire Caribbean mountain belt of Northern Venezuela allochthonous?, in Shagam, R., and others, eds., Studies in Earth and space sciences: Geological Society of America Memoir 132, p. 363–368.

Beloussov, V. V., 1972, Basic trends in the evolution of continents: Tectonophysics, 13(1-4), p. 95–117.

Berrangé, J. P., and Dearnley, R., 1975, The Apoteri volcanic formation—tholeiitic flows in the North Savannas Graben of Guyana and Brazil: Geologische Rundschau, v. 64, p. 883–899.

Bertrand, J., Delaloye, M., Fontignie, D., and Vuagnat, M., 1978, Ages (K-Ar) sur diverses ophiolitiques et roches associées de la Cordillère Centrale du Guatémala: Schweizerische Mineralogische und Petrographische Mitteilungen, v. 58, p. 405–413.

Biju-Duval, B., and others, 1980, Tertiary Sequences South of the Cordillera Central, in Field Guide to the Ninth Caribbean Geological Conference: Santo Domingo, p 107–138.

Bishop, W. V., 1973, Late Jurassic Contemporaneous Faults in North Louisiana and South Arkansas: American Association of Petroleum Geologists Bulletin, v. 57, p. 858–877.

Bobyarchick, A., 1981, The Eastern Piedmont Fault System and its Relationship to Alleghanian Tectonics in the Southern Appalachians: Journal of Geology, v. 89, p. 335–347.

Boiteau, A., Michard, A., and Saliot, P., 1972, Metamorphisme de haute pression dans le complexe ophiolitique du Purial (Oriente, Cuba): Comptes Rendus Académie des Sciences [Paris], Ser. D, v. 274, p. 2137–2140.

Boiteau, A., Butterlin, J., Michard, A., and Saliot, P., 1972, Le complexe ophiolitique du Purial (Oriente, Cuba) et son metamorphisme de haute pression: problems et datation et de correlation: Comptes Rendus Académie des Sciences [Paris], Ser. D, v. 275, p. 895–898.

Boiteau, A., and Michard, A., 1976, Données nouvelles sur le socle métamorphique de Cuba, problèmes d'application de la tectoniques des plaques: Transactions of the Caribbean Geological Conference, 7th, Guadeloupe, French Antilles, p. 221–226.

Bourgois, J., Calle, B., Tournon, J., and Toussaint, J. F., 1982, The Andean ophiolitic megasutures on the Buga-Buenaventura traverse (Western Cordillera-Valle Colombia): Tectonophysics, v. 82, p. 207–230.

Bowin, C. O., 1966, Geology of Central Dominican Republic: a case history of part of an island arc: Geological Society of America Bulletin, v. 98, p. 11–84.

—— 1975, The Geology of Hispaniola, in Nairn, A.E.M., and Stehli, F. G., eds., The Gulf of Mexico and the Caribbean, Ocean Basins and Margins, v. 3: New York, Plenum Press, p. 501–552.

Boyanov, I., Goranov, G., and Cabrera, R., 1975, Algunos nuevos datos sobre la geologia de los complejos de anfibolitas y granitoides en la parte sur de Las Villas: Serie Geologica, Academia de Ciencias de Cuba, Instituto de Geologia y Paleontologia, v. 19, 14 p.

Brezsnyánszky, K., and Iturralde-Vinent, M. A., 1978, Paleogeografia del Paleogeno de Cuba: Transactions of the Caribbean Geological Conference, 8th, Willemstad, 1977, Geologie en Mijnbouw, v. 57:2, p. 123–132.

Bryant, W. R., Antoine, J., Ewing, M., and Jones, B., 1968, Structure of Mexican continental shelf and slope, Gulf of Mexico: American Association of Petroleum Geologists Bulletin, v. 52, no. 7, p. 1204–1228.

Bryant, W. R., Meyerhoff, A. A., Brown, N. K., Jr., Furrer, M. A., Pyle, T. E., and Antoine, J. W., 1969, Escarpments, reef trends, and diapiric structures, eastern Gulf of Mexico: American Association of Petroleum Geologists Bulletin, v. 53, no. 12, p. 2506–2542.

Bucher, W. H., 1947, Problems of Earth deformation illustrated by the Caribbean Sea basin: Transactions of the New York Academy of Sciences, Ser. 2, v. 9, p. 98–116.

Buffler, R. T., Watkins, J. S., Shaub, F. J., and Worzel, J. L., 1980, Structure and early geologic history of the deep central Gulf of Mexico basin, Pilger, R. H., Jr., ed., in Symposium on the origin of the Gulf of Mexico and the early opening of the Central North Atlantic Ocean, Louisiana State University,

Baton Rouge, Louisiana, p. 3–16.

Bullard, E. C., Everett, J. E., and Smith, A. G., 1965, The fit of the continents around the Atlantic: Royal Society of London Philosophical Transactions, Ser. A, v. 258, p. 41–51.

Burkart, B., 1978, Offset across the Polochic fault of Guatemala and Chiapas Mexico: Geology, v. 6, p. 38–332.

Burke, K., 1967, The Yallahs basin: a sedimentary basin southeast of Kingston, Jamaica: Marine Geology, v. 5, p. 45–60.

Burke, K., 1975, Atlantic evaporites formed by evaporation of water spilled from Pacific, Tethyan and Southern Oceans: Geology, v. 3, p. 613–616.

Burke, K., 1976, Development of graben associated with the initial ruptures of the Atlantic Ocean: Tectonophysics, v. 36, p. 93–111.

Burke, K., and Dewey, J. F., 1974, Two plates in Africa during the Cretaceous?: Nature, v. 249, p. 313–316.

Burke, K., and Mann, P., 1982, Did the Pliocene Panama-South America collision cause internal deformation of a previously rigid Caribbean plate?: Geological Society of America Abstract with Programs, v. 14, no. 7, p. 456.

Burke, K., Fox, P. J., and Sengor, A.M.C., 1978, Buoyant ocean floor and the evolution of the Caribbean: Journal of Geophysical Research, v. 83, no. B8, p. 3949–3954.

Burke, K., Grippi, J., and Sengor, A.M.C., 1980, Neogene structures in Jamaica and the tectonic style of the Northern Caribbean Plate Boundary Zone: Journal of Geology, v. 88, p. 375–386.

Burke, K., Mann, P., Dixon, T., and Nemec, M., 1981, Neotectonics of Hispaniola [abs.]: EOS (American Geophysical Union Transactions), v. 62, no. 45, p. 1051.

Case, J. E., 1974, Oceanic crust forms basement of Eastern Panama: Geological Society of America Bulletin, v. 85, p. 645–652.

—— 1974, Major basins along the continental margin of northern South America, in Burk, C. A., and Drake, C. L., eds., Geology of the continental margins: New York, Springer-Verlag, p. 733–741.

Case, J. E., and Holcombe, T. L., 1980, Geologic-Tectonic map of the Caribbean Region: U.S. Geological Survey Miscellaneous Investigation ser. I-1100, scale 1:2,500,000.

Cebull, S. E., and Shurbet, D. H., 1980, The Ouachita Belt in the Evolution of the Gulf of Mexico, Pilger, R. H., Jr., ed., in Symposium on the origin of the Gulf of Mexico and the early opening of the Central North Atlantic Ocean, Louisiana State University, Baton Rouge, Louisiana, p. 17–26.

Christofferson, E., 1976, Colombian Basin magnetism and Caribbean plate tectonics: Geological Society of America, v. 87, p. 1255–1258.

Chubb, L. J., 1954, The Lazaretto Section Jamaica: Colonial Geology and Mineral Resources, v. 4, p. 233–247.

Chubb, L. J., and Burke, K., 1963, Age of the Jamaican granodiorite: Geological Magazine, v. 100, p. 524–532.

Clemons, R. E., and Long, L. E., 1971, Petrologic and isotopic study of the Chiquimula pluton, Southeastern Guatemala: Geological Society of America Bulletin, v. 82, 2729–2740.

Cook, F. A., Albough, D. S., Brown, L. D., Kaufman, S., Oliver, J. E., and Hatcher, R. D., Jr., 1979, Thin-skinned tectonics in the crystalline Southern Appalachians, COCORP seismic-reflection profiling of the Blue Ridge and Piedmont: Geology, v. 7, p. 563–567.

Cook, F. A., Brown, L. D., Kaufman, S., Oliver, J. E., and Petersen, T. A., 1981, COCORP seismic profiling of the Appalachian Orogen beneath the coastal plain of Georgia: Geological Society of America Bulletin, Part 1, v. 92, p. 738–748.

Cornet, B., 1977, The palynostratigraphy and age of the Newark Supergroup [Ph.D. thesis]: University Park, Pennsylvania State University, 505 p.

Cox, D. P., Marvin, R. F., M'Gonigle, J. W., McIntyre, D. H., and Rogers, C. L., 1977, Potassium-Argon geochronology of some metamorphic, igneous, and hydrothermal events in Puerto Rico and the Virgin Islands: Journal of Research of the U.S. Geological Survey, v. 5, p. 689–703.

Dean, B. W., and Drake, C. L., 1978, Focal mechanism solutions and tectonics of the Middle America Arc: Journal of Geology, v. 86, p. 111–128.

De Boer, J., 1979, The outer arc of the Costa Rican Orogen (oceanic basement complexes of the Nicoya and Santa Elena Peninsulas): Tectonophysics, v. 56, p. 221–254.

De Boer, J., and Snider, F. G., 1979, Magnetic and chemical variations of Mesozoic diabase dikes from eastern North America: Evidence for a hotspot in the Carolinas?: Bulletin of Geological Society of America, Part I, v. 90, p. 185–198.

DeCserna, Z., 1976, Mexico geotectonics and mineral deposits: New Mexico Geological Society Special Publication, no. 6, p. 18–25.

Dengo, G., 1969, Problems of tectonic relations between Central America and the Caribbean, Transactions, Gulf Coast Association of Geological Societies, v. 19, p. 311–320.

—— 1972, Review of Caribbean serpentinites and their tectonic implications, *in* Shagam, R., and others, eds., Studies in earth and space sciences: Geological Society of America Memoir 132, p. 303–312.

Dewey, J. F., Pitman, W. C., III, Ryan, W.B.F., and Bonnin, J., 1973, Plate tectonics and the evolution of the Alpine system: Geological Society of America Bulletin, v. 84, p. 3137–3180.

Dewey, J. F., and Burke, K., 1974, Hotspots and continental breakup, some implications for collisional orogeny: Geology, v. 2, p. 57–60.

Dickey, P., 1980, Barbados as a fragment of South America ripped off by continental drift: Transaction of the Caribbean Geological Conference, 9th, Santo Domingo, p. 51–52.

Dickinson, W. R., and Coney, P. F., 1980, Plate tectonic constraints on the origin of the Gulf of Mexico, Pilger, R. H., Jr., ed., *in* Symposium on the origin of the Gulf of Mexico and the early opening of the Central Atlantic Ocean, Louisiana State University, Baton Rouge, Louisiana, p. 27–36.

Dietz, R. S., 1973, Morphologic fits of North America/Africa and Gondwana, Review, *in* Tarling, D. A., and Runcorn, K., eds., Implications of continental drift to the Earth science: London Academic Press, v. 2, p. 865–872.

Dietz, R. S., and Holden, J. C., 1970, Reconstruction of Pangea: Journal of Geophysical Research, v. 75, p. 4939–4956.

Dillon, W. P., and Paull, C. K., 1978, Interpretation of multichannel seismic reflection profiles of the Atlantic continental margin off the coasts of South Carolina and Georgia: U.S. Geological Survey Miscellaneous Field Studies Map MF-936.

Donnelly, T. W., 1963, Genesis of albite in early orogenic volcanic rocks: American Journal of Science, v. 261, p. 957–972.

—— 1964, Evolution of Eastern Greater Antillean arc: American Association of Petroleum Geologists Bulletin, v. 48, p. 680–696.

—— 1966, Geology of St. Thomas and St. John, U.S. Virgin Islands, *in* Hess, H. H., ed., Caribbean geological investigations: Geological Society of America Memoir 98, p. 85–176.

Donnelly, T. W., Melson, W., Kay, R., and Rogers, J.J.W., 1973, Basalts and dolerites of Late Cretaceous age from the Central Caribbean, *in* Edger, N. T., and others, Initial Reports of the Deep Sea Drilling Project, v. 15, Washington, D.C., U.S. Government Printing Office, p. 989–1011.

Donnelly, T. W., and Rogers, J.J.W., 1978, The distribution of igneous rocks throughout the Caribbean: Geologie en Mijnbouw, v. 57, p. 151–162.

Draper, G., Harding, R. R., Horsfield, W. T., Kemp, A. W., and Tresham, A. E., 1976, Low grade metamorphic belt in Jamaica and its tectonic implications: Geological Society of America Bulletin, v. 87, p. 1283–1290.

Echeverría, L. M., 1980, Tertiary or Mesozoic komatiites from Gorgona Island, Columbia: field relations and geochemistry: Contributions to Mineralogy and Petrology, v. 73, p. 253–266.

Edgar, N. T., Ewing, J. I., and Hennion, J., 1971, Seismic refraction and reflection in Caribbean Sea: American Association of Petroleum Geologists Bulletin, v. 55, p. 833–870.

Ervin, P. C., and McGinnis, L. D., 1975, Reelfoot rift: reactivated precursor to the Mississippi embayment: Geological Society of America Bulletin, v. 86, p. 1287–1295.

Ferencic, A., del Giudice, D., and Recchi, G., 1971, Tectomagmatic and metallogenic relationships of the region central Panama-Costa Rica: Transactions of the Caribbean Geological Conference, 5th, St. Thomas, Virgin Islands, p. 189–195.

Fink, L. K., Jr., 1972, Bathymetric and geologic studies of the Guadeloupe region, Lesser Antilles Island arc: Marine Geology, v. 12, p. 267–288.

Fink, L. K., Jr., Harper, C. T., Stipp, J. J., and Nayle, F., 1972, The tectonic significance of La Desirade—Possible relict seafloor crust, *in* Petzell, C., ed., Transactions of the Caribbean Geological Conference, 6th, Margarita, Venezuela, p. 302.

Fox, P. J., and Heezen, B. C., 1975, Geology of the Caribbean Crust, *in* The Ocean Basins and Margins, v. 3: Nairn, A.E.M., and Stehli, F. G., eds., Plenum Press, New York, p. 421–466.

Freeland, G. L., and Dietz, R. S., 1972, Plate tectonic evolution of the Caribbean-Gulf of Mexico Region, Transactions of the Caribbean Geological Conference, 6th, Margarita, Venezuela, p. 259–264.

Galli-Olivier, C., 1979, Ophiolite and island arc volcanism in Costa Rica: Geological Society of America Bulletin, v. 90, p. 445–452.

Garrison, L. E., and Martin, R. G., 1973, Geologic structures in the Gulf of Mexico basin: U.S. Geological Survey Professional Paper, v. 773, 85 p.

Gealey, W. K., 1980, Ophiolite obduction mechanism, *in* Ophiolites: Proceedings International Ophiolite Symposium, Cyprus, 1979, Cyprus Geological Survey Department, p. 288–243.

Goreau, P., 1981, Tectonic evolution of the northern Caribbean plate margin, [Ph.D. thesis]: Cambridge, Massachusetts Institute of Technology and Woods Hole, Woods Hole Oceanographic Institute, 245 p.

Gose, W. A., and Sanchez-Barreda, L. A., 1983, Paleomagnetic results from southern Mexico: Geofisica Internacional, v. 23 (in press).

Gose, W. A., and Swartz, D. K., 1977, Paleomagnetic results from Cretaceous sediments in Honduras: Tectonic implications: Geology, v. 5, p. 505–508.

Gose, W. A., Belcher, R. C., and Scott, G. R., 1982, Paleomagnetic results from northeastern Mexico: evidence for large Mesozoic rotations: Geology, v. 10, p. 50–54.

Green, A. R., 1976, The evolution of the earth's crust and sedimentary basin development, *in* Heacock, J. G., ed., The earth's crust: American Geophysical Union Monograph 20, p. 1–17.

Groetsch, G. J., 1980, Resedimented conglomerates and turbidites of the Upper Tavera Group, Dominican Republic: Transactions of the Caribbean Geological Conference, 9th, Santo Domingo, p. 191–198.

Grow, J. A., 1980, Deep structure and evolution of the Baltimore Canyon Trough in the vicinity of the COST B-3 well, *in* Scholle, P. A., ed., Geologic Studies of the COST B-3 well, U.S. Geological Survey, Circular, v. 833, p. 117–125.

Grow, J. A., and Sheridan, R. E., 1981, Deep structure and evolution of the continental margin off the eastern United States: Oceanologia Acta, No. SP, Proceedings of the International Geological Congress, 26th, Geology of Continental Margins Symposium, Paris, July 1980, p. 11–19.

Hager, D. S., and Burnett, C. M., 1960, Mexia-Talco fault line in Hopkins and Delta counties, Texas: American Association of Petroleum Geologists Bulletin, v. 44:3, p. 316–356.

Halbouty, M. T., 1967, Salt domes: Gulf region, United States and Mexico (second edition): Houston, Gulf Publishing Company, 425 p.

Haldemann, E. G., Brouwer, S. B., Blowes, J. H., and Snow, W. E., 1980, Lateritic nickel deposits at Bonao: *in* Field Guide to the Ninth Caribbean Geological Conference, Santo Domingo, p. 69–80.

Hess, H. H., and Maxwell, J. C., 1953, Caribbean research project: Geological Society of America Bulletin, v. 64, p. 1–6.

Hill, P. A., 1959, Geology and structure of the northwest Trinidad mountains, Las Villas Province, Cuba: Geological Society of America Bulletin, v. 70, p. 1459–1478.

Holcombe, T. L., 1981, Effects of changes in North American-Caribbean plate motion on development of the Cayman Trough: Geological Society of America Abstracts with Programs, v. 13, no. 7, p. 474.

Holcombe, T. L., and Sharman, G. E., 1983, "Post-Miocene North American Caribbean Motion: Evidence from the Cayman Trough", Geology, v. 11 (in press).

Holcombe, T. L., Vogt, P. R., Matthews, J. E., and Murchison, R. R., 1973, Evidence for sea-floor spreading in the Cayman Trough: Earth and Planetary Science Letters, v. 20, p. 357–371.

Hopkins, H. R., 1973, Geology of the Aruba Gap Abyssal Plain near DSDP Site 153, in Edgar, N. T., and others: Initial Reports of the Deep Sea Drilling Project, v. 15, Washington, D.C., U.S. Government Printing Office, p. 1039–1050.

Horne, G. S., Pushkar, P., and Shafiquallah, M., 1974, Laramide plutons on the landward continuation of the Bonacca ridge, Northern Honduras: Transactions of the Caribbean Geological Conference, 7th, Guadeloupe, p. 583–588.

Horne, G. S., Clark, G. S., Pushkar, P., 1976, Pre-Cretaceous rocks of northwestern Honduras: basement terrane in Sierra de Omoa, American Association of Petroleum Geologists, v. 60, no. 4, p. 566–583.

Hughes, G. W., and Turner, C. C., 1977, Upraised Pacific floor, southern Malaita, Solomon Islands: Geological Society of America Bulletin, v. 88, p. 412–414.

Irving, E. M., 1975, Structural evolution of the northernmost Andes, Colombia: U.S. Geological Survey Professional Paper, v. 846, p. 1–47.

Iturralde-Vinent, M., 1969, Principal characteristics of Cuban Neogene stratigraphy: American Association of Petroleum Geologists Bulletin, v. 53, p. 1938–1955.

—— 1978, Los Movimientos tectonicos de la etapa de desarrollo plataformico en Cuba, in Transactions of the Caribbean Geological Conference, 8th, Willemstad, 1977: Geologie en Mijnbouw, v. 57:2, p. 205–212.

Jordan, T. H., 1975, The present-day motions of the Caribbean Plate: Journal of Geophysical Research, v. 80, p. 4433–4439.

Joyce, J., 1980, Geology of Samana Peninsula: Field guide to the Ninth Caribbean Geological Conference, Santo Domingo, p. 247–254.

Kafka, A. L., 1980, Caribbean tectonic processes: seismic surface wave source and path property analysis [Ph.D. thesis]: State University of New York, Stony Brook, New York, 276 p.

Kafka, A. L., and Weidner, D. J., 1979, The focal mechanisms and depths of small earthquakes as determined from Rayleigh-wave radiation patterns: Bulletin of the Seismological Society of America, v. 69, no. 5, p. 1379–1390.

—— 1981, Earthquake focal mechanisms and tectonic processes along the southern boundary of the Caribbean plate, Journal of Geophysical Research, v. 86, 2877–2888.

Keigwin, L. D., 1978, Pliocene closing of the Isthmus of Panama: Geology, v. 6, p. 630–634.

Kent, K. M., 1979, Two-dimensional gravity model of the Southeast George embayment—Blake Plateau [Ms. thesis]: University of Delaware, Newark, Delaware, 89 p.

Kesler, S. E., 1971, Petrology of the Terre-Neuve igneous province, northern Haiti: Geological Society of America Memoir 130, p. 119–137.

Kesler, S. E., and Sutter, J. F., 1979, Compositional evolution of intrusive rocks in the eastern Greater Antilles island arc: Geology, v. 7, p. 197–200.

Kesler, S. E., Kienle, O. F., and Bateson, J. H., 1974, Tectonic significance of intrusive rocks in the Maya Mountains, British Honduras: Geological Society of America Bulletin, v. 85, p. 549–552.

Kesler, S. E., Sutter, J. F., Jones, L. M., and Walker, R. L., 1977, Early Cretaceous basement rocks in Hispaniola: Geology, v. 5, p. 245–247.

Khudoley, K. M., and Meyerhoff, A. A., 1971, Paleogeography and geologic history of Greater Antilles: Geological Society of America Memoir 129, 199 p.

King, P. B., 1969, Tectonic map of North America: U.S. Geological Survey Monograph 69-1, scale 1:5,000,000.

Kirkland, D. W., and Gerhard, J. E., 1971, Jurassic salt, central Gulf of Mexico, and its temporal relation to circum-Gulf evaporites: American Association of Petroleum Geologists Bulletin, v. 55, p. 680–686.

Klitgord, K. D., and Behrendt, J. D., 1979, Basin Structure of the U.S. Atlantic Margin, in Investigations of continental margins: American Association of Petroleum Geologists Memoir 29, p. 85–112.

Klitgord, K. D., and Schouten, H., 1980, Mesozoic evolution of the Atlantic, Caribbean and Gulf of Mexico, Pilger, R. J., Jr., ed., in Symposium on the origin of the Gulf of Mexico and the early opening of the Central North Atlantic, Louisiana State University, Baton Rouge, Louisiana, p. 100–101.

Klitgord, K. D., and Grow, J. A., 1980, Jurassic seismic stratigraphy and basement structure of Western Atlantic Magnetic Quiet Zone: American Associ-

ation of Petroleum Geologists Bulletin, v. 64, no. 10, p. 1658–1680.

Klitgord, K. D., Popenoe, P., and Schouten, H., 1983, Florida: A Jurassic transform plate boundary: Journal of Geophysical Research, v. 88 (in press).

Kozary, M. T., 1968, Ultramafic rocks in the thrust zones of northwestern Oriente Province, Cuba: American Association of Petroleum Geologists Bulletin, v. 52, p. 2298–2317.

Kuijpers, E. P., 1980, The geologic history of the Nicoya ophiolite complex, Costa Rica, and its geotectonic significance: Tectonophysics, v. 68, p. 233–255.

Kuman, V. E., and Gavilan, R. R., 1965, Geologia de Isla de Pinos: Revista de Tecnologica, Ministerio de Industrias, La Habana, v. 3, no. 4, p. 20–38.

Kupfer, D. H., and Godoy, J., 1967, Strike-slip faulting in Guatemala [abs.]: EOS (American Geophysical Union Transactions), v. 48, p. 215.

Ladd, J. W., 1976a, Relative motion of South America with respect to North America and Caribbean tectonics: Geological Society of America Bulletin, v. 87:60702, p. 969–976.

Ladd, J. W., et al., 1976b, Deep seismic reflection results from the Gulf of Mexico: Geology, v. 4, p. 365–368.

Ladd, J. W., Worzel, J. L., and Watkins, J. S., 1977, Multifold seismic reflection records from the north slope of the Muertos Trench, in Talwani, M., and Pitman, W. C., eds., Island Arcs, deep sea trenches and back arc basins: Washington, D.C., American Geophysical Union, p. 41–58.

Ladd, J. W., Shih, T. C., and Tsai, C. J., 1981, Cenozoic tectonics of Central Hispaniola and adjacent Caribbean Sea: American Association of Petroleum Geologists Bulletin, v. 65, no. 3, p. 466–489.

Langer, C. J., and Bollinger, G. A., 1979, Secondary faulting near the terminus of a seismogenic strike-slip fault: Aftershocks of the 1976 Guatemala earthquake: Bulletin Seismological Society of America, v. 69, no. 2, p. 427–444.

LePichon, X., and Fox, P. J., 1971, Marginal offsets, fracture zones and the early opening of the North Atlantic: Journal of Geophysical Research, v. 76, p. 6294–6308.

Lewis, G. E., and Straczek, J. A., 1955, Geology of south-central Oriente, Cuba: U.S. Geological Survey Bulletin, v. 975-D, p. 171–336.

Lewis, J. F., Harper, C. T., Kemp, A. W., and Stipp, J. J., 1973, Potassium argon retention ages for some Cretaceous rocks from Jamaica: Geological Society of America Bulletin, v. 84, p. 335–340.

Lidz, B., 1973, Biostratigraphy of Neogene cores from Exuma Sound diapirs, Bahama Islands: American Association of Petroleum Geologists Bulletin, v. 57, no. 5, p. 841–857.

MacDonald, W. D., 1976, Cretaceous-Tertiary evolution of the Caribbean: Transactions of the Caribbean Geological Conference, 7th, Guadeloupe, p. 69–78.

—— 1964, Geology of the Serrania de Macuira area, Guajira Peninsula, Columbia [Ph.D. thesis]: Princeton University, Princeton, New Jersey, 167 p.

MacKenzie, D. B., 1960, High-temperature alpine-type peridotite from Venezuela: Geological Society of America Bulletin, v. 17, p. 303–318.

Malfait, B. T., and Dinkelman, M. G., 1972, Circum-Caribbean tectonic and igneous activity and the evolution of the Caribbean plate: Geological Society of America Bulletin, v. 83, p. 251–272.

Mann, P., and Burke, K., 1980, Neogene wrench faulting in the Wagwater Belt, Jamaica: Transactions of the Caribbean Geological Conference, 9th, Santo Domingo, p. 95–97.

—— 1981, Post-Eocene NOAM-CARIB relative plate motion and strike-slip offset of the Greater Antilles Island Arc [abs.]: Geological Society of America Abstracts with Programs, v. 13, no. 7, p. 503.

—— 1982, Basin formation at intersections of conjugate strike-slip faults: examples from southern Haiti [abs.]: Geological Society of America Abstracts with Programs, v. 14, no. 7, p. 555.

Mann, P., Burke, K., Draper, G., and Dixon, T., 1981, Tectonics and sedimentation at a Neogene strike-slip restraining bend, Jamaica [abs.]: EOS (American Geophysical Union Transactions), v. 62, no. 45, p. 1051.

Maresch, W. V., 1974, Plate tectonics origin of the Caribbean mountain system of Northern South America: Discussion and proposal: Geological Society of America Bulletin, v. 85, p. 669–682.

—— 1975, The geology of northeastern Margarita island, Venezuela: A contribution to the study of Caribbean plate margins: Geologische Rundschau, v. 64, p. 847–883.

Mattinson, J. M., Fink, L. K., Jr., and Hopson, C. A., 1973, Age and origin of ophiolitic rocks on La Desirade Island, Lesser Antilles Island Arc: Carnegie Institution of Washington, Yearbook, v. 72, p. 616–623.

—— 1980, Geochronologic and isotopic study of the La Desirade Island basement complex: Jurassic Oceanic Crust in the Lesser Antilles?: Contributions to Mineralogy and Petrology, v. 71, p. 237–245.

Mattson, P. H., 1960, Geology of the Mayaguez area, Puerto Rico: Geological Society of America Bulletin, v. 71, p. 319–362.

—— 1973, Middle Cretaceous nappe structures in Puerto Rican ophiolites and their relation to the tectonic history of the Greater Antilles: Geological Society of America Bulletin, v. 84, no. 1, p. 21–37.

—— 1979, Subduction, buoyant breaking, flipping and strike-slip faulting in the Northern Caribbean: Journal of Geology, v. 87, p. 293–304.

Mattson, P. H., and Pessagno, E. A., Jr., 1979, Jurassic and Early Cretaceous radiolarians in Puerto Rican ophiolite—tectonic implications: Geology, v. 7, p. 440–444.

Maurasse, F.J.-M.R., 1980, Relations between the geologic setting of Hispaniola and the evolution of the Caribbean: Presentations/Transactions du 1er collogue sur la geologie d'Haiti, Port-au-Prince, p. 246–264.

Maurasse, F.J.-M.R., Husler, J., Georges, G., Schmitt, R., and Damond, P., 1979, Upraised Caribbean sea-floor below acoustic reflector B" at the southern peninsula of Haiti: Geologie en Mijnbouw, v. 58, p. 71–83.

Maxwell, J. C., 1948, Geology of Tobago, British West Indies: Geological Society of America Bulletin, v. 59, p. 801–854.

McConnell, R. B., Masson Smith, D., and Berrangé, J. P., 1969, Geological and geophysical evidence for a rift valley in the Guyana Shield: Geologie en Mijnbouw, v. 48, p. 189–200.

Menendez, A., 1966, Tectónica de la parte central de las Montanas del Caribe occidentales, Venezuela: Boletin de Geologico (Caracas, Venezuela), v. 8, no. 15, p. 3–72.

Meyerhoff, A. A., and Hatten, C. W., 1968, Diapiric structures in central Cuba, *in* Diapirism and diapirs: American Association of Petroleum Geologists Memoir 8, p. 315–357.

—— 1974, Bahamas Salient of North America, *in* Burk, C. A., and Drake, C. L., eds., The geology of continental margins: New York, Springer-Verlag, p. 429–446.

Meyerhoff, A. A., Khudoley, K. M., and Hatten, C. W., 1969, Geologic significance of radiometric dates from Cuba: American Association of Petroleum Geologists Bulletin, v. 53, p. 2494–2500.

Meyerhoff, A. A., and Meyerhoff, H. A., 1972, "The new global tectonics": major inconsistencies: American Association of Petroleum Geologists Bulletin, v. 56, no. 2, p. 269–336.

Michael, R. C., and Millar, G., 1977, Tertiary geology of the southwestern flank of the Cordillera Central, Dominican Republic [abs.]: Caribbean Abstract for the Geological Conference, 8th, Willemstad, 1977, p. 123–124.

Molnar, P., and Sykes, L. R., 1969, Tectonics of the Caribbean and Middle America regions from focal mechanisms and seismicity: Geological Society of America Bulletin, v. 80, p. 1639–1684.

Mooney, W. D., 1980, An East Pacific-Caribbean Ridge during the Jurassic and Cretaceous and the evolution of western Colombia, *in* Pilger, R. H., Jr., Symposium on the origin of the Gulf of Mexico and the early opening of the central North Atlantic Ocean: Louisiana State University, Baton Rouge, Louisiana, p. 55–73.

Moore, G. W., and del Castillo, L., 1974, Tectonic evolution of the southern Gulf of Mexico: Geological Society of America Bulletin, v. 85, p. 607–618.

Muehlberger, W. R., 1976, The Honduras depression: Publicaciones Geologicas del ICAITI, no. V: 4th Reunion Geologicas de America Central, Guatemala, p. 43–51.

Muessig, K. W., 1978, The central Falcon igneous suite, Venezuela: alkaline basaltic intrusions of Oligocene-Miocene age, *in* Transactions of the Caribbean Geological Conference, 8th, Williamstad, 1977: Geologie en Mijnbouw, v. 57:2, p. 261–266.

Nagle, F., 1974, Blueschist, eclogite, paired metamorphic belts, and the early tectonic history of Hispaniola: Geological Society of America Bulletin, v. 85, p. 1461–1466.

—— 1979, Geology of the Puerto Plata area, Dominican Republic, *in* Lidz, B., and Nagle F., eds., Hispaniola: tectonic focal point of the northern Caribbean three geologic studies in the Dominican Republic: Miami, Miami Geological Society, p. 1–28.

Nemec, M. C., 1980, A two phase model for the tectonic evolution of the Caribbean [abs.]: Transactions of the Caribbean Geological Conference, 9th, Santo Domingo, p. 23–34.

Olsen, P. E., McCune, A. R., and Thomson, K. S., 1982, Correlation of the Early Mesozoic Newark Supergroup by vertebrates, principally fishes: American Journal of Science, v. 282, p. 1–44.

Palmer, H. C., 1979, Geology of the Moncion-Jarabacoa area, Dominican Republic, *in* Lidz, B., and Nagle, F., eds., Hispaniola: Tectonic Focal Point of the Northern Caribbean: Miami, Miami Geological Society, p. 29–68.

Pennington, W. D., 1981, The subduction of the eastern Panama Basin and the seismotectonics of northwestern South America: Journal Geophysical Research, v. 86, no. B11, p. 10753–10770.

Perfit, M. R., and Heezen, B. C., 1978, The geology and evolution of the Cayman Trench: Geological Society of America Bulletin, v. 89, p. 1155–1174.

Perfit, M. R., Heezen, B. C., Rawson, M. R., and Donnelly, T. W., 1980, Chemistry, origin and significance of metamorphic rocks from the Puerto Rico Trench: Marine Geology, v. 34, p. 125–156.

Peter, G., 1972, Geologic structure offshore, north-central Venezuela: Transactions of the Caribbean Geological Conference, 6th, Margarita, Venezuela, 1971, p. 283–294.

Petersen, E. U., and Zantop, H., 1980, The Oxec deposit, Guatemala: an ophiolite copper occurrence: Economic Geology, v. 75, p. 1053–1065.

Pilger, R. H., Jr., 1978, A closed Gulf of Mexico, Pre-Atlantic ocean plate reconstruction and the early rift history of the Gulf and North Atlantic: Transactions of the Gulf Coast Association of Geological Societies, v. 28, p. 385–393.

Pindell, J., Cooper, C., and Burke, K., 1981, Late Cretaceous Accretionary Prism in the Sierra Bahoruco, southwest Dominican Republic: EOS (American Geophysical Union Transactions), v. 62, no. 17, p. 405.

Pindell, J., and Dewey, J. F., 1982, Permo-Triassic reconstruction of western Pangea and the evolution of the Gulf of Mexico/Caribbean Region: Tectonics, v. 1, Part 2, p. 179–211.

Pinet, R. R. , 1972, Diapir-like features offshore Honduras: implications regarding tectonic evolution of Cayman Trough and Central America: Geological Society of America Bulletin, v. 83, p. 1911–2158.

Plafker, G., 1976, Tectonic aspects of the Guatemala earthquake of 4 February, 1976: Science, v. 193, p. 1201–1208.

Potter, H. C., 1965, Discussion of the papers in the first session, *in* Salt Basins around Africa, Institute of Petroleum, London, Elsevier, p. 77.

Priem, H.N.A., Beets, D. J., Boelrijk, N.A.I.M., Hebeda, E. H., Verdurmen, E.A.Th., and Verschure, R. H., 1978, Rb-Sr evidence for episodic intrusion of the Late Cretaceous tonalitic batholith of Aruba, Netherlands Antilles, *in* MacGillary, H. J., and Beets, D. J., eds., Transactions of the Caribbean Geological Conference, 8th, Willemstad, 1977: Geologie en Mijnbouw, v. 57, p. 293–296.

Ramos Lopez, E., 1975, Geological Summary of the Yucatan Peninsula, *in* Nairn, A.E.M., and Stehli, F.G.Y., The ocean basins and margins, v. 3, The Gulf of Mexico and the Caribbean: New York, Plenum Press, p. 257–280.

Rial, J. A., 1978, The Caracas, Venezuela earthquake of July 1967: A multiple-source event: Journal of Geophysical Research, v. 83, p. 5405–5414.

Rosenfeld, J. H., 1980, The Santa Cruz ophiolite, Guatemala, Central America: Transactions of the Caribbean Geological Conference, 9th, Santa Domingo, p. 451–452.

—— 1981, Geology of the Western Sierra de Santa Cruz, Guatemala, Central America: An ophiolite sequence [Ph.D. thesis]: Binghampton, New York, State University of New York at Binghamton, Binghampton, New York, 313 p.

Rowley, K. C., Roobol, J. M., 1978, Geochemistry and age of the Tobago igneous rocks, Caribbean Geological Conference, 8th, Willemstad, 1977, *in* Transactions of the Geologie en Mijnbouw, v. 57, p. 315–318.

Salvador, A., and Green, A. R., 1980, Opening of the Caribbean Tethys, *in* Geology of the Alpine Chain born of the Tethys: International Geological Congress, 26th, Colloquium C5, Bureau de Recherces, Geologiques et Minieres Memoir 115, p. 224–229.

Santamaria, F., and Schubert, C., 1974, Chemistry and geochronology of the southern Caribbean-Northern Venezuelan Plate Boundary: Geological Society of America Bulletin, v. 85, p. 1085–1098.

Saunders, J. B., Jung, P., Geister, J., and Biji-Duval, B., 1980, The Neogene of the south flank of the Cibao Valley, Dominican Republic: a stratigraphic study, *in* Transactions of the 9th Caribbean Geological Conference, Santo Domingo, p. 151–160.

Schmidt-Effing, R., 1979, Alter und Genese des Nicoya-Komplexes, einer ozeanischen Päläokruste (Oberjura bis Eozän) im südliche Zentralamerika: Geologische Rundschau, v. 68, p. 457–494.

—— 1980, The Huayacoctla Avlacogen in Mexico (lower Jurassic) and the origin of the Gulf of Mexico, *in* Symposium on the origin of the Gulf of Mexico and the early opening of the central North Atlantic, Louisiana State University, Baton Rouge, Louisiana, p. 79–86.

Schubert, C., 1980, Late-Cenozoic pull-apart basins, Bocono fault zone, Venezuelan Andes: Journal Structural Geology, v. 2:4, p. 463–468.

—— 1981, Are the Venezuelan fault systems part of the southern Caribbean plate boundary?: Geologische Rundschau, v. 70, p. 542–551.

—— 1982, Origin of Cariaco Basin, southern Caribbean Sea: Marine Geology, v. 47, p. 345–360.

Schubert, C., and Sifontes, R. S., 1970, Bocono Fault, Venezuelan Andes: Evidence of postglacial movement: Science, v. 170, p. 66–69.

Schubert, C., Sifontes, R. S., Padron, V. E., Velez, J. R., and Loaiza, P. A., 1979, Formacion La Quinta (Jurasico) Andes Meridenos: Geologia de la seccion tipo: Acta Cientifica Venezolana, v. 30, p. 42–55.

Schubert, C., 1935, Historical geology of the Antillean-Caribbean region or the lands bordering the Gulf of Mexico and the Caribbean Sea: New York, Wiley, p. 811.

Schwartz, D. P., Cluff, L. S., and Donnelly, T. W., 1979, Quaternary faulting along the Caribbean-North Plate boundary in Central America, *in* Whitten, C. A., Green, R., and Meade, B. K., eds., Recent crustal movements, 1977: Tectonophysics, v. 52, p. 431–445.

Sclater, J. B., Hellinger, S., and Tapscott, C., 1977, The paleobathymetry of the Atlantic Ocean from the Jurassic to the present: Journal of Geology, v. 85, p. 509–552.

Shipley, T. H., Buffler, R. T., and Watkins, J. S., 1978, Seismic stratigraphy and geologic history of Blake Plateau and adjacent western Atlantic continental margin: American Association of Petroleum Geologists Bulletin, v. 62, p. 792–812.

Silver, L. T., and Anderson, T. L., 1974, Possible left-lateral early to middle Mesozoic disruption of the southwestern North American craton margin [abs.]: Geological Society of America Abstracts with Programs, v. 6, p. 955.

Silver, E. A., Case, J. E., and MacGillavry, H. J., 1975, Geophysical study of the Venezuelan borderland: Geological Society of America Bulletin, v. 86, p. 213–226.

Skerlec, G. M., and Hargraves, R. B., 1980, Tectonic significance of paleomagnetic data from northern Venezuela: Journal of Geophysical Research, v. 85, no. B10, p. 5303–5315.

Skvor, V., 1969, The Caribbean area: a case of destruction and regeneration of a continent: Geological Society of America Bulletin, v. 80, p. 961–968.

Smith, R. J., 1953, Geology of the Los Teques-Cua Region, Venezuela, Geological Society of America, Bulletin, v. 64, p. 41–64.

Speed, R. C., Gerhard, L. C., and McKee, E. H., 1979, Ages of deposition, deformation, and intrusion of Cretaceous rocks, eastern St. Croix, Virgin Islands: Geological Society of America Bulletin, v. 90, p. 629–632.

Stearns, C., Mauk, F. J., and Van der Voo, R., 1982, Late Cretaceous-Early Tertiary paleomagnetism of Aruba and Bonaire: Netherlands Leeward Antilles, Journal of Geophysical Research, v. 87, no. B2, p. 1127–1141.

Suess, E., 1885, Das Antlitz der Erde: Prag, F. Tempsky, v. 1, 779 p.

Sykes, L. R., McCann, W. R., and Kafka, A. L., 1982, Motion of Caribbean plate during last 7 million years and implications for earlier Cenozoic movements: Journal of Geophysical Research, v. 87, no. B13, p. 10,656–10,676.

Thayer, T. P., and Guild, P. W., 1947, Thrust faults and related structures in Eastern Cuba: EOS (Transactions American Geophysical Union), v. 28, p. 919–930.

Trujillo, G. M., 1975, El complejo cristalino mesozoico de Isla de Pinos. Su Metamorfismo: Academia de Ciencias de Cuba, Instituto de geologia y Paleontologia, Serie Geologica, no. 23, 16 p.

Tschanz, C. M., Marvin, R. F., Cruz, B. J., Mehnert, H., and Cebula, G. T., 1974, Geologic evolution of the Sierra Nevada de Santa Marta, northeastern Colombia: Geological Society of America Bulletin, v. 85, p. 273–284.

Van der Voo, R., and French, R. B., 1974, Apparent polar wandering for the Atlantic bordering continents: Late Carboniferous to Eocene: Earth Science Reviews, v. 10, p. 99–119.

Van der Voo, R., French, R. B., and Mauk, F. J., 1976, Permian-Triassic continental configurations and the origin of the Gulf of Mexico: Geology, v. 4, p. 177–180.

Vierbuchen, R. C., Jr., 1978, The tectonics of northeastern Venezuela and the southeastern Caribbean Sea [Ph.D. thesis]: Princeton, New Jersey, Princeton University, 175 p.

Vignali, M., 1979, Estratigrafia y estructura de las cordilleras metamorficas de Venezuela oriental (peninsula de Araya-Paria y isla de Margarita): Geos (Venezuela), no. 25, Escuela de Geologica y Minos, Caracas, p. 19–66.

Vogt, P. R., 1973, Early events in the opening of the North Atlantic, *in* Tarling, D. H., and Runcorn, S. K., eds., Implications of continental drift to the earth sciences, v. 2: London, Academic Press, p. 693–712.

Wadge, G., 1982, A Miocene submarine volcano at Low Layton, Jamaica: Geology Magazine, v. 119 (2), p. 183–199.

Wadge, G., Jackson, T. A., Isaacs, M. C., and Smith, T. E., 1982, The ophiolitic Bath-Dunrobin Formation, Jamaica: significance for Cretaceous plate margin evolution in the northwestern Caribbean, Journal of the Geological Society, London, 139, p. 321–333.

Wadge, G., and Wooden, J. L., 1982, Late Cenozoic alkaline volcanism in the northwestern Caribbean: tectonic setting and Sr isotopic characteristics: Earth and Planetary Science Letters, v. 57, p. 35–46.

Waggoner, D. G., 1978, Sr and Nd-isotope study of Caribbean basalts belonging to the B" event [abs.]: EOS (Transactions American Geophysical Union), v. 59, p. 404.

Walper, J. L., 1980, Tectonic evolution of the Gulf of Mexico, *in* Pilger, R. H., Jr., ed., Symposium on the origin of the Gulf of Mexico and the early opening of the Central North Atlantic Ocean: Louisiana State University, Baton Rouge, Louisiana, p. 87–98.

Walper, J. L., and Rowett, C. L., 1972, Plate tectonics and the origin of the Caribbean and the Gulf of Mexico: Transactions of the Gulf Coast Association of Geological Societies, v. 22, p. 105–116.

Walthall, B. H., and Walper, J. L., 1967, Peripheral Gulf rifting in northeast Texas: American Association of Petroleum Geologists Bulletin, v. 51, no. 1, p. 102.

Wassal, H., 1956, The relationship of oil and serpentine in Cuba: International Geological Congress of Mexico, 20th, sect. 3, p. 67–77.

Watkins, J. S., Ladd, J. W., Buffler, R. T., Shaub, F. J., Houston, M. H., Worzel, J. L., 1978, Occurrence and evolution of salt in deep Gulf of Mexico, *in* Bouma, A. H., Moore, G. T., and Coleman, J. M., eds., Framework, facies and oil trapping characteristics of the Upper Continental Margin: American Association of Petroleum Geologists Studies in Geology, no. 7, p. 43–65.

Watkins, J. S., McMillen, K. J., Bachman, S. B., Shipley, T. H., Moore, J. C., and Angevine, C., 1981, Tectonic synthesis, leg 66: transect and vicinity, *in* Watkins, J. S., et al.: Initial Reports of the Deep Sea Drilling Project, v. 66, Washington, D.C., U.S. Government Printing Office, p. 837–849.

Wegener, A., 1922, Die Enstehung der Kontinente und Ozeane: Braunschweig, Friedrich Vieweg und Sohn, 144 p.

Weidie, A. E., Ward, W. C., and Marshall, R. H., 1980, Geology of the Yucatan Platform, *in* Thorsen, W. G., chairperson, Geology of Cancun, Quintana Roo, Mexico: W. Texas Geological Society, p. 50–68.

White, G. W., 1980, Permian-Triassic continental reconstruction of the Gulf of Mexico-Caribbean Area: Nature, v. 283, p. 823–826.

White, G. W., and Burke, K. C., 1980, Outline of the tectonic evolution of the Gulf of Mexico and Caribbean Region: Houston Geological Society Bulletin, June, 1980, p. 8–13.

Wilhelm, O., and Ewing, M., 1972, Geology and history of the Gulf of Mexico: Geological Society of America Bulletin, v. 83, p. 575–599.

Williams, M. D., 1975, Emplacement of Sierra de Santa Cruz, Eastern Guatemala: American Association of Petroleum Geologists Bulletin, v. 59, p. 1211–1216.

Willis, B., 1929, Continental Genesis: Geological Society of America Bulletin, v. 40, p. 281–336.

Woodring, W. P., Brown, J. S., and Burbank, W. S., 1924, Geologie de la Republique d'Haiti: Baltimore, Lord Baltimore Press, 631 p.

MANUSCRIPT ACCEPTED BY THE SOCIETY SEPTEMBER 1, 1983

Printed in U.S.A.

Geological Society of America
Memoir 162
1984

Seafloor spreading magnetic anomalies in the Venezuelan Basin

N. Ghosh
S. A. Hall
J. F. Casey
Department of Geosciences
Central Campus
University of Houston
Houston, Texas 77004

ABSTRACT

A compilation of existing magnetic data clearly demonstrates the presence of extensive, NE-SW trending, linear anomalies over the central Venezuelan Basin. These long wavelength, small amplitude anomalies are truncated in the east by a series of N-S linear anomalies over the Aves Ridge, and in the south by E-W trending anomalies over the Aruba Basin, Curacao Ridge, and Los Roques Basin. In the southeastern corner of the basin, there is a magnetic quiet zone similar to that observed in the North Atlantic and Pacific Oceans.

Analysis of the NE-SW anomalies reveals an axis of symmetry which crosses the basin from north of the Guajira Peninsula to near the Muertos Trough at 68° W. Modelling indicates that the linear anomalies are the result of a phase of seafloor spreading between 153 and 127 m.y. at a half rate of 0.4 to 0.5 cm y^{-1}. The quiet zone is therefore believed to correspond to a period in the Middle Jurassic which may be characterized by frequent short reversals.

The magnetic study together with other geologic and geophysical evidence suggests that the Venezuelan Basin formed in the Pacific region as a western extension of the N. Atlantic in Middle-Late Jurassic. Spreading appears to have ceased when, in the early Cretaceous, the South Atlantic began to open. As a result of these changes in plate motion, the Venezuelan Basin became trapped behind the juvenile Antilles arc-trench system. The Venezuelan Basin was then gradually inserted into the Caribbean region as this system migrated eastward with respect to North and South America.

INTRODUCTION

It is now well established that studies of the regular patterns of magnetic anomalies over the world's major ocean basins can be used to successfully reconstruct the former positions of continents and the associated plate motions (e.g., Sclater and others, 1977). While the seafloor spreading history of the South and Central Atlantic Oceans is relatively well known, the early spreading history between the North and South American plates has remained obscure. This is largely because much of the oceanic crust that was created between North and South America has since been subducted beneath the Caribbean Arc system during a period from the Mesozoic to the present. As a result, the age of the oceanic crust and its spreading history cannot, in general, be determined. The only deep basin floored by oceanic crust near the Caribbean region that can be directly related to the early separation of North and South America is the present Gulf of Mexico (Shepherd and others, 1982); however, small amplitude, smooth magnetic anomalies have rendered its seafloor spreading history difficult to interpret.

Most models for the evolution of the region suggest that the bulk of the Caribbean plate has been transported to its present position from a more westerly location (e.g., Malfait and Dinkleman, 1972; Sykes and others, 1982; Pindell and Dewey, 1982). Much of the plate tectonic interpretation for the development of the Caribbean is based upon the geological history of mainland South and Central America, the various islands, and their margins. A study of magnetic anomalies within the deep ocean basins of the Caribbean is therefore of particular importance in that it provides an independent means for evaluating the tectonic evolution of the region. The major deep ocean basins that are currently within the Caribbean plate are the Colombian and Venezuelan Basins: in this paper we concentrate on the Venezuelan Basin.

PREVIOUS WORK

Interpretations of magnetic anomaly information available for the Venezuelan Basin are relatively sparse, and consequently, its age and spreading history are poorly known. There have been few detailed magnetic studies; possibly because of early reports of the presence of smooth magnetic anomalies (Ewing and others, 1960). In a study of several aeromagnetic profiles from Project MAGNET, Donnelly (1973) described a series of NE-SW trending linear anomalies. Unfortunately, the widely spaced nature of the profiles limited the reliability of the correlations and therefore the observed trends. A more detailed marine magnetic survey of the north central part of the basin was carried out by R/V Eastward (Watkins and Cavanaugh, 1974) in which a total of 1550 nautical miles of data were collected. The survey comprised seven NW-SE profiles spaced approximately 32 km apart. On the basis of these data, Watkins and Cavanaugh were able to identify three separate zones of anomalies: a striped zone over the central area, a smooth or quiet zone towards the southeast, and a short wavelength anomaly zone in the eastern part of the survey area. The striped zone corresponds to the linear pattern of Donnelly (1973) with the dominant trend being NE-SW. A number of detailed surveys which deal with structures along the coastlines and nearshore areas have also been carried out (e.g., Silver and others, 1975).

Although there have been several studies of localized surveys, none has so far made a detailed synthesis of all available data. In this paper we attempt this synthesis by studying the characteristic shapes and prevalent trends of the magnetic anomalies throughout the basin. This information is used to construct seafloor spreading models for the evolutionary history of the Venezuelan Basin. We present evidence that recognizable seafloor spreading anomalies exist within the basin and attempt to show that these anomalies may be related to the early opening between the North and South American plates. In addition, evidence for a major reorganization of the plates in this region is presented which is best explained in terms of a change in the relative motion of North and South America in response to the rifting of South America and Africa in the Early Cretaceous.

DATA SOURCES AND REDUCTION PROCEDURES

A compilation of available marine magnetic measurements collected in the Venezuelan Basin (Fig. 1) by various vessels between 1961 and 1977 has been made using data from the National Geophysical and Solar Terrestrial Data Center, the Marine Science Institute of the University of Texas, and the Office of Naval Research (including the detailed compilation of Caribbean data by J. E. Matthews (1976)). Airborne data, especially those from Project MAGNET (Donnelly, 1973), were not used directly in the compilation because of their variable elevations. Aeromagnetic profiles were used, however, to supplement information on anomaly trends in those areas where the seaborne data gave insufficient control. The available track coverage is shown in Figure 2a. As can be seen, the central and southern portions of the basin are fairly well covered but data density in the eastern and western areas, including the Beata Ridge, is relatively sparse.

In order to directly compare the data from diverse sources, a uniform data reduction procedure has been adopted. Two main corrections are necessary: (1) the removal of the Earth's main field component (i.e., the "normal" correction), and (2) subtraction of the influence of diurnal variations of the magnetic field. In dealing with data which cover a 17-year period, adequate correction for secular changes becomes of critical importance. The first correction was therefore accomplished by using the International Geomagnetic Reference Field (I.G.R.F.) as modified by Barraclough and others (1975) which appears to better predict the secular variations than previous models of the geomagnetic field (Hall, 1979). The data were corrected for diurnal changes using records from the geomagnetic observatory at San Juan, Puerto Rico. Hourly mean values of the geomagnetic field were available for the interval 1961-1977. The magnetic data were corrected by subtracting the appropriate hourly value for the day of measurement. A summary of the intervals of heightened magnetic activity associated with magnetic storms was used to identify data gathered during such periods. No attempt was made to correct such data which were simply removed from the data base. An analysis of the mis–ties associated with the magnetic values at the crossing points of ships' tracks demonstrates that the correction for diurnal effects significantly reduces the crossing errors. Crossing errors after correction for diurnal effects are all <70 nT.

DESCRIPTION OF ANOMALIES

The anomalies over the central area are rather subdued with peak to trough amplitudes of 75 to 200 nT and wavelengths of 50 to 100 km while those along the margins of the basin are somewhat larger (200 to 300 nT) with shorter wavelengths. Linear trends are observed in various parts of the basin with a smooth magnetic zone (quiet zone) in the southeastern corner. This smooth zone is the same as that previously reported by Watkins and Cavanaugh (1974). The linear anomalies may be separated into three distinct zones on the basis of the location and orientation of the correlated anomalies: (1) a dominant NE-SW pattern over the central, western, and northwestern portions of the basin; (2) a predominantly E-W trend along the southern rim of the basin and over the Aruba Basin and Curacao Ridge; and (3) a much less well-documented N-S pattern over the eastern edge of

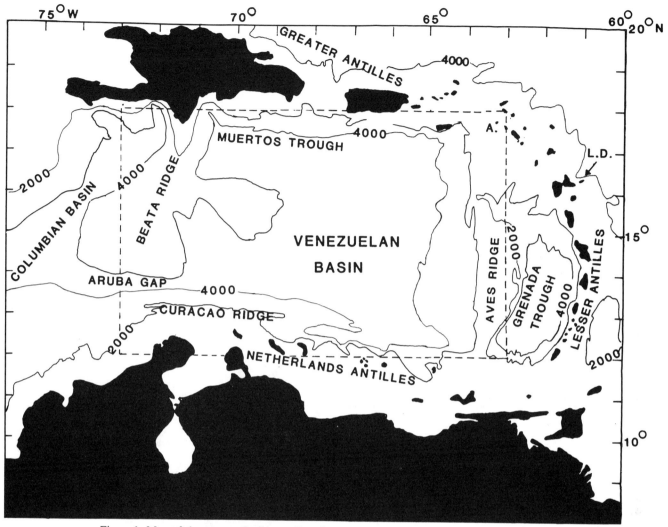

Figure 1. Map of the eastern Caribbean. Boxed area shows the location of the study area and the geographic limits of Figures 3, 6, and 9.

the basin and adjacent parts of the Aves Ridge where the track coverage is relatively sparse.

NE-SW ANOMALIES

Clearly within the basin proper the dominant features are the NE-SW linear anomalies and the magnetic quiet zone in the southeastern corner of the basin. Figure 3 shows several profiles which cross the central area from NW to SE, i.e., approximately normal to the NE-SW lineations. The profiles display several peaks and troughs which can be correlated over more than 500 km (Fig. 3). The anomalies have amplitudes of typically 150 to 200 nT although profile 1 anomalies are somewhat smaller (~75 nT). Peak to trough distances are typically 25 to 50 km. In the northwestern corner of the basin, over and adjacent to the Beata Ridge, the anomalies (profiles 10, 13, and 14, Fig. 3) also form a northeasterly trend. Amplitudes here are slightly larger (200 to 300 nT) suggesting shallower magnetic sources which

may be related to the ridge. This may also account for the somewhat better correlation of features at the northwestern end of these profiles (Fig. 3).

QUIET ZONE

As seen in Figure 3, the NE-SW linear features are difficult to recognize towards the southeastern end of profiles 1 to 7 which cross a zone characterized by low amplitude (<100 nT), longer wavelength (~100 km) anomalies. This zone broadens southwards from 80 km (profile 1) to more than 250 km (profile 7) forming a triangular area. The area is bounded on the east by N-S anomalies over the Aves Ridge and to the south by E–W anomalies over the Curacao Ridge. In general, within this zone it is difficult to correlate anomalies from one profile to another. The correlations shown (Fig. 3) are tenuous at best and probably reflect a bias towards the neighboring dominantly NE-SW anomalies. The anomalies in the quiet zone appear to be trun-

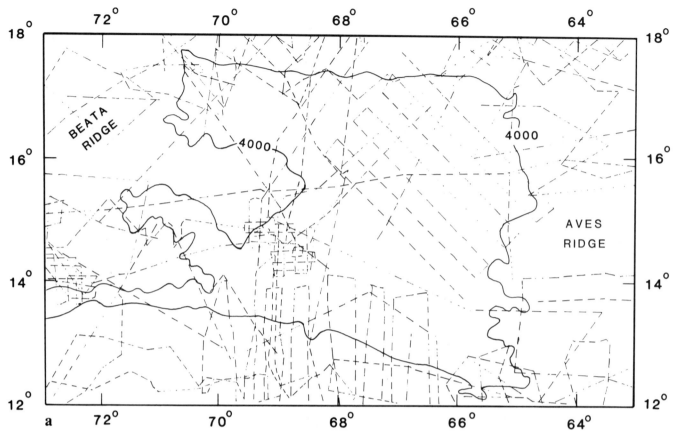

Figure 2 (this and facing page). (a) Location of ships' tracks in the Venezuelan Basin for which magnetic data are available. Bathymetric contours in meters. (b) Location of numbered magnetic anomaly profiles shown in Figures 3, 4, and 5. Dotted lines show the locations of Project MAGNET profiles 0040 and 230 previously correlated by Donnelly (1973).

cated towards the northeastern corner of the basin by the N-S linear anomalies. Track control in this region is poor, however, and consequently continuation of either trend (i.e., N-S or NE-SW) is questionable.

E-W ANOMALIES

Along the southern periphery of the basin, there are a series of E-W trending anomalies over the Aruba Basin, Curacao Ridge, Los Roques Basin, and southern edge of the Venezuelan Basin. These anomalies, previously described by Silver and others (1975), have variable amplitudes (0 to 300 nT) and, in general, shorter wavelengths than those over the central basin area (Fig. 4). There is a significant reduction in their amplitude over the Curacao Ridge and Los Roques Basin which reflects a substantial increase in basement depth in this area. Refraction studies show up to 14 km of low-velocity, possibly sedimentary, material in the Los Roques Basin (Ewing and others, 1970). The boundary between the dominant NE-SW anomalies and these E-W anomalies is not well constrained by the data but coincides approximately with the northern margin of a separate block, the Bonaire

block, identified by Silver and others (1975). The exact relationship between the magnetic anomalies and this small block is unknown. More detailed surveys of the area, especially across the boundary between the major anomaly trends, are needed to clarify the significance of the E-W anomalies.

N-S ANOMALIES

Over the extreme eastern edge of the basin and the adjacent Aves Ridge, there is a series of poorly defined N-S linear anomalies. Some profile to profile correlations (Fig. 5) are quite reasonable, but others because of the ship track distribution (Fig. 2a) and anomaly amplitudes (25 to 150 nT) must be considered tenuous. In particular, the anomalies over the Aves Ridge have shorter wavelengths and more variable amplitudes (Fig. 5), making it difficult to correlate individual features. The short wavelength anomalies over the Aves Ridge may be related to basement topography, but those anomalies to the west are more difficult to explain in terms of basement relief. It appears more likely that these anomalies are due to lateral changes in the magnetization of

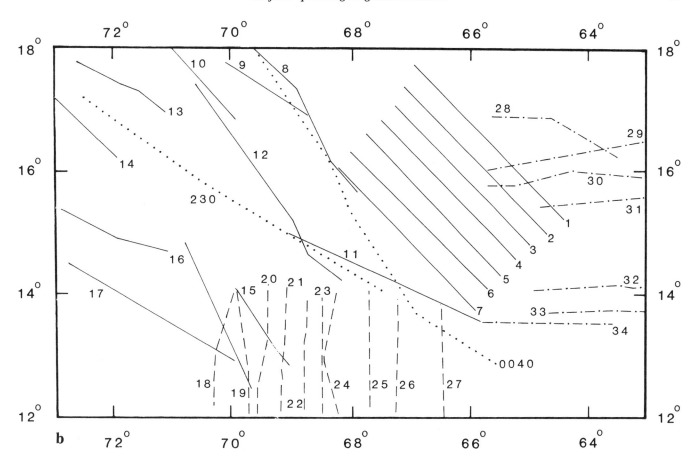

the basement. The origin of the N-S anomalies is currently unknown although they are probably related to the evolution of the Antilles Arc system.

BASIS FOR CORRELATIONS

The correlations of individual peaks and troughs for all parts of the basin are shown in Figure 6. Where possible these correlations were obtained by identifying anomalies using the intersections of the ships' tracks. Where tracks do not cross, the characteristic shape of individual anomalies has been used to make the correlations. For example, there is a twin peaked anomaly (feature A, Fig. 3) which can be recognized on profiles 8, 0040, 9, 12, 16, and 17. In addition, there is a broad positive (feature B, Fig. 3) which can be traced across several profiles. Some of these features have been previously correlated by Donnelly (1973) and Watkins and Cavanaugh (1974). Where there are no intersections of tracks and characteristically shaped anomalies are not easily identified—for example, profiles 31, 32, 33 (Fig. 5)—the overall trend of the adjacent anomalies has been

used to project the correlations across ships' tracks; where such correlations are tentative they are labelled with a "?" (Figs. 3, 4, and 5).

DATA ANALYSES

Because of their central location in the Venezuelan Basin, the dominantly NE-SW anomalies were selected for detailed analysis and modelling studies. These NE-SW linear anomalies have similar wavelengths but slightly smaller amplitudes than those observed over the abyssal areas of the major ocean basins. Anomaly amplitudes in the Pacific Ocean are generally 200 to 300 nT (Larson and Hilde, 1975) while those in the Indian Ocean are commonly less than 250 nT (Fisher and others, 1971). In the South Atlantic, amplitudes range from 150 to typically 400 nT but average less than 300 nT (Rabinowitz and LaBrecque, 1979). In the case of the Venezuelan Basin, the depth of the magnetized layer is believed to be slightly greater than in these ocean basins and, consequently, the somewhat lower anomaly amplitudes are considered to be consistent with a sea-

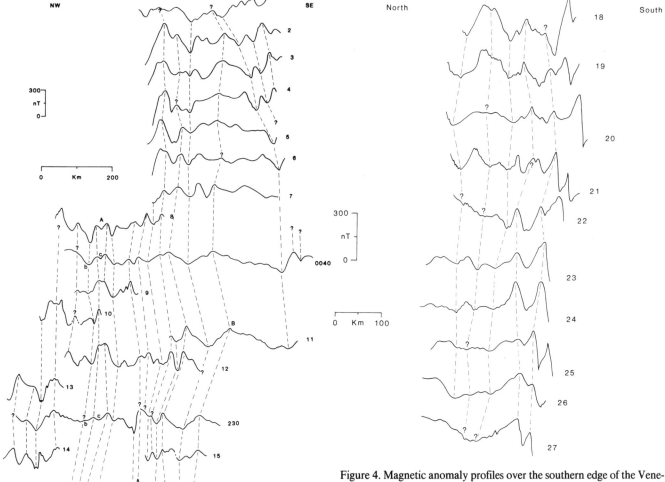

Figure 3. Magnetic anomaly profiles over the central and northern Venezuelan Basin. Location of profiles shown in Figure 2b. Correlations are shown as dashed lines. Tentative correlations indicated by "?". A denotes distinctive double peak feature, B denotes a broad positive anomaly. b and c are anomalies correlated by Donnelly (1973) on Project MAGNET profiles 0040 and 230.

Figure 4. Magnetic anomaly profiles over the southern edge of the Venezuelan Basin. Location of profiles shown in Figure 2b. Dashed lines indicate correlations. Tentative correlations denoted by "?".

floor spreading origin. Seismic reflection data (e.g., Ludwig and others, 1975; Stoffa and others, 1981), magnetic data (Donnelly, 1973), and depth to magnetic basement studies (Ghosh, 1982) indicate that the B″ reflector is neither true acoustic nor magnetic basement throughout most of the basin but the exact nature of the crust in the region is currently unknown. In order to determine whether the linear magnetic anomalies are compatible with a seafloor spreading origin, two techniques were employed: (a) de-skewing of the anomalies in a search for any axes of symmetry, and (b) modelling of the anomalies using synthetic seafloor spreading profiles.

DE-SKEW

The shape of marine magnetic anomalies depends, in part, upon the latitude and orientation of the source body at the time it acquires its magnetization. For bodies which do not become magnetized at the magnetic poles, the shape of the anomalies will not be, in general, symmetric but rather skewed and contain both positive and negative parts. Thus by measuring the amount of skewness, information regarding the latitude and/or orientation at which the body became magnetized may be obtained. This information is obtained by de-skewing the anomalies by progressive amounts until a symmetrical anomaly is produced. The amount of de-skewing required is a measure of the angular distance to the paleomagnetic pole. In the case of seafloor spreading magnetic anomalies, it is also possible to determine the location of the spreading axis. Magnetic anomalies symmetrically placed on either side of this axis will have similar or "identical" shapes.

Using the correlations shown in Figure 3, a composite anomaly profile was constructed from profiles oriented more or less orthogonal to the NE-SW linear anomalies. This profile was

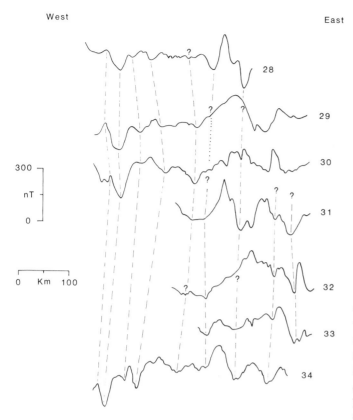

Figure 5. Magnetic anomaly profiles over the eastern Venezuelan Basin and Aves Ridge. Location of profiles shown in Figure 2b. Dashed lines indicate correlations. Tentative correlations denoted by "?".

then de-skewed using the method of Schouten and McCamy (1972). The amount of de-skewing required suggests the anomalies formed close to their present latitude. On the basis of the de-skew analysis, an axis of symmetry has been identified which lies close to the center of the basin (Fig. 7). Folding the composite profile about this axis reveals an excellent correlation for many of the features. Because of their low amplitudes, it is difficult to obtain meaningful de-skew results for those anomalies which lie in the quiet zone. There is no corresponding quiet zone at the northwest end of the profile as the basin is limited here by Hispaniola. In addition, the magnetic signature to the west may be influenced by the Beata Ridge in which case a match for these anomalies would not be expected on the de-skewed profile.

SEAFLOOR SPREADING MODEL

Synthetic seafloor spreading anomaly profiles were generated and compared with the various profiles shown in Figure 3. As noted above, the age of the crust beneath the Venezuelan Basin is unknown and, consequently, a variety of models were generated. Results of DSDP studies in the Caribbean (Saunders and others, 1973) indicate the age of the B″ reflector is approximately 80 m.y. The underlying basement must therefore be older than this. Using this as a lower bound for the age of the crustal

material, models covering the period 84 to 162 m.y. were generated for a range of spreading rates. The upper bound of 162 m.y. was chosen as this represents the earliest period for which a reasonably detailed magnetic polarity scale exists (Larson and Hilde, 1975). If the crust is older than this, then meaningful information regarding its age and rate of formation cannot currently be obtained from a study of the magnetic anomalies. Between about 80 and about 110 m.y., there is a long period of normal polarity (Mercanton Normal), possibly interrupted by a number of short intervals of reverse polarity (Lowrie and others, 1980). This period gives rise to a magnetic quiet zone which is observed in both the South Atlantic (Rabinowitz and LaBrecque, 1979) and Pacific Oceans (Larson and Chase, 1972) and provides a possible explanation for zone of similar character observed in the southeastern corner of the Venezuelan Basin. The results of the de-skew analysis indicate an axis of symmetry crossing the center of basin, northwest of the quiet zone. This axis would form the spreading center in any seafloor spreading origin of the crust in the Venezuelan Basin. Its location with respect to the quiet zone requires that the crust beneath the axis of symmetry be younger than that beneath the quiet zone. As the upper age of the Cretaceous long normal polarity interval is approximately 80 m.y., it follows that this crust would be younger than the overlying B″ material. Consequently, the quiet zone is not believed to be due to spreading during the Cretaceous long normal interval.

The anomalies must therefore be due to polarity reversals older than 110 m.y. Synthetic seafloor spreading anomalies were generated for the time interval 110 to 162 m.y. using a wide range of spreading rates. The lowest half spreading rate was 0.3 cm y^{-1} based upon producing the observed anomalies over the entire time interval. Setting a maximum spreading rate is difficult as the entire crust present could have been formed in a relatively brief period. However, using the broad kinematic constraints provided by the relative motions of the major plates over this time interval, a maximum half spreading rate of 2 cm y^{-1} was used.

A uniform depth of 6.5 km below sea level was chosen for the upper surface of the magnetized layer on the basis of detailed depth to basement studies (Ghosh, 1982) using an autocorrelation technique developed by Phillips (1979). This depth is compatible with that estimated by Donnelly (1973) from the comparison of seaborne and airborne magnetic profiles in the Venezuelan Basin. The depth is also consistent with the results of seismic reflection surveys (e.g., Ludwig and others, 1975). The best fit to the observed composite profile was obtained using a model in which spreading took place between 153 and 127 m.y. at a rate of 0.4 to 0.5 cm y^{-1} (Fig. 8). The axis of symmetry of the de-skewed profile corresponds exactly with the proposed spreading axis.

The spreading rate obtained, viz. ~0.5 cm y^{-1}, is low when compared with rates in the major ocean basins. However, comparable rates have been reported from the Labrador Sea (Kristoffersen and Talwani, 1977) and more recently for the Arctic Ocean (Jackson and Reid, 1983). In both the Labrador Sea and the Arctic Ocean, identifiable magnetic anomalies are present.

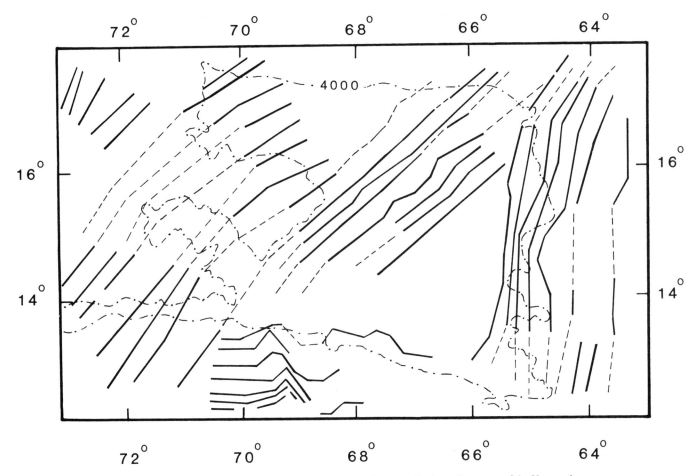

Figure 6. Map showing correlation of individual magnetic anomalies in various parts of the Venezuelan Basin. Dashed lines are used to indicate less reliable correlations.

Although the simple model of Vine and Matthews (1963) provides an excellent explanation for the magnetic anomalies over most spreading ridges, there is some concern that as the spreading rates become small the existence of a finite, possibly 1-3 km, intrusion zone will result in the absence of identifiable magnetic anomalies over such ridges. As noted above, this does not appear to be the case. A finite intrusion zone results in a smoothing of the magnetic signature such that many of the fine-scale features are effectively filtered out. To demonstrate that the broader scale features are preserved, the seafloor modelling was also carried out using an intrusion width of 8 km (Fig. 8). As can be seen, even with a wide intrusion zone the major features, though smoothed, may still be identified.

At either end of the profiles, the correlation is much poorer. Towards the northwest, feature C (Fig. 8) does not correlate well with the synthetic profile. This anomaly is believed to be due to an edge effect at the southern margin of Hispaniola and therefore does not correspond to a seafloor spreading anomaly. At the southeastern end of the profiles, the anomalies are those of the magnetic quiet zone which may correspond to a long normal polarity interval in the Jurassic (McElhinny and Burek, 1971; Irving and Pullaiah, 1976). Recent studies of the Jurassic

(Steiner, 1980) suggest the lack of anomalies may in fact be due to many short reversals rather than a single long period of normal polarity. Isochrons based upon this modelling are shown in Figure 9.

THE MAGNETIC QUIET ZONE

The magnetic quiet zone in the southeastern corner of the Venezuelan Basin corresponds to a zone where the B″ reflector has a rough or irregular appearance and the crust has nearly normal oceanic thickness. The area to the northwest is characterized by well-developed magnetic anomalies, a thicker crust, and corresponds to a zone where B″ appears smooth (Diebold and others, 1981). Where the B″ reflector appears smooth, it is thought to be due to extensive basaltic sills and flows of Coniacian age (Saunders and others, 1973). The implication of the seismic study of Diebold and others (1981) is that the area of rough acoustic basement to the southeast may not contain the sill layer, but simply a normal oceanic crustal sequence. If correct, this rough basement may reflect crustal accretion at a slow, as opposed to a fast, spreading center as supported by the seafloor spreading rates proposed on the basis of our magnetic interpreta-

NW

SE

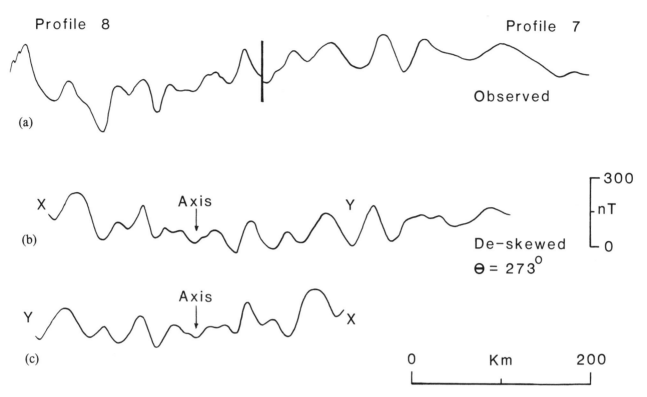

Figure 7. (a) Composite observed magnetic anomaly profile across the Venezuelan Basin, (b) de–skewed composite profile (Θ = 273°), and (c) northwestern half of de–skewed composite profile rotated about the axis of symmetry.

tion. The boundary between the rough and smooth zone is also the approximate position of the Central Venezuelan Basin Fault. It appears that the motion along this fault locally occurred both prior to and after Coniacian age (i.e., the sill event) (Diebold and others, 1981). This structure is obviously of concern in presenting a seafloor spreading model in that the fault zone could represent a zone of large lateral displacement along which formerly distinct parts of the Venezuelan Basin have been juxtaposed. Alternatively, as speculated by Diebold and others (1981), the motion on the fault zone could be normal dip-slip and could have occurred because of decoupling along the rough-smooth boundary in response to increased subsidence caused by more sedimentary loading in the bathymetrically deeper rough zone. Sedimentary rocks above the B″ reflector appear to be significantly thicker southeast of the fault zone. It is unclear at present whether the Central Venezuelan Basin Fault represents a large lateral dislocation zone or a much smaller dip-slip fault zone.

In our modelling we have assumed that the rough-smooth boundary in the Venezuelan Basin does not represent a large lateral dislocation zone, but an isochron of Jurassic age (Fig. 9). Because the boundary between rough and smooth areas has a NE-SW trend parallel to magnetic lineations, any lateral motion

parallel to the trend of these lineations may not severely affect the seafloor spreading model. It should be emphasized that our model assumes structural continuity across the rough-smooth boundary. The coincident location of the acoustically rough-smooth boundary, the magnetically quiet versus higher amplitude boundary, and the Central Venezuelan Basin Fault is a peculiar feature. Clearly, further study of this area is required before any definite conclusions can be reached as to the continuity of the basin. The good correlation of anomalies with the seafloor spreading model to the northwest of the boundary would suggest, however, that the smooth zone to the southeast was contiguous with the central part of the basin.

TECTONIC EVOLUTION OF THE VENEZUELAN BASIN

Since a major part of the oceanic crust originally created between North and South America has since been subducted beneath the Caribbean Arc system, it is difficult to determine the early tectonic history of the area. One of the most useful techniques for examining the evolution of the Caribbean is that of determining the relative motions of the North American, South

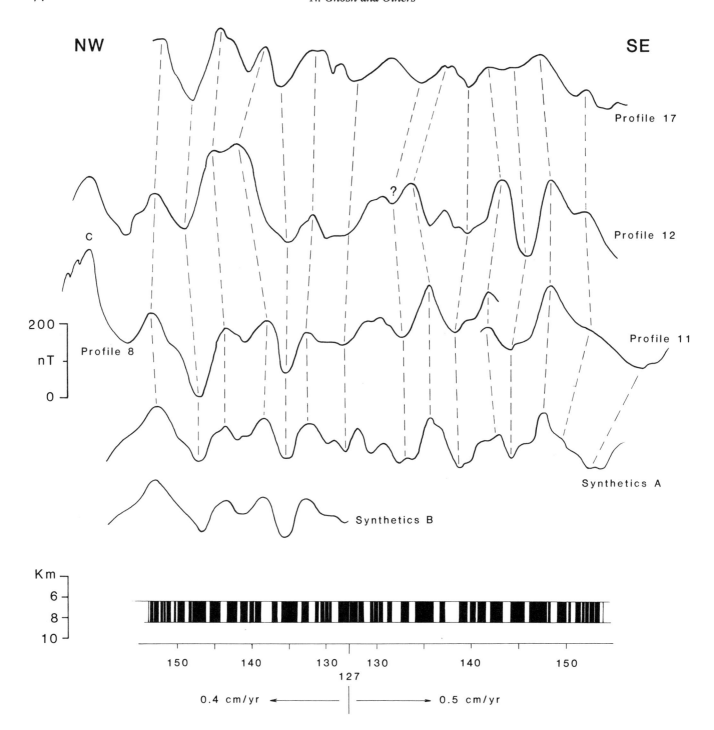

Figure 8. Synthetic seafloor spreading model for the NE-SW linear anomalies across the Venezuelan Basin. Correlations shown by dashed lines. Depth of magnetized layer = 6.5 km, thickness of layer = 2 km. Synthetics A has been generated using the model of Vine and Matthews (1963). Synthetics B has been generated using a broad intrusion zone (8 km) which produces a smooth transition from one polarity to the other.

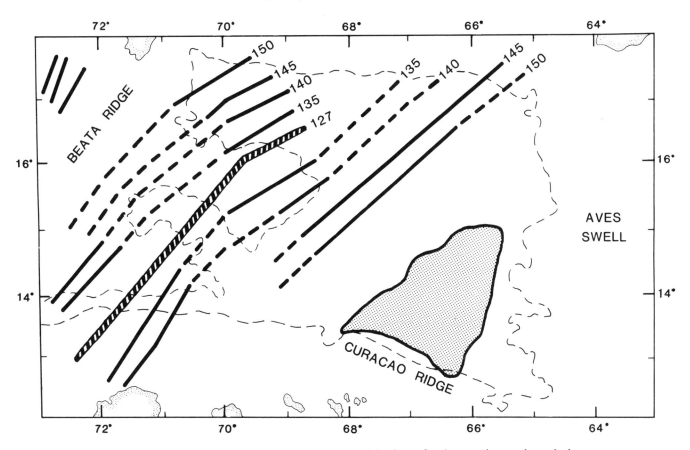

Figure 9. Map showing the location of the proposed isochrons for the oceanic crust beneath the Venezuelan Basin. Stippled area is the magnetic quiet zone. Hashed line shows location of fossil spreading ridge.

American, and African plates. This three-plate system provides a tectonic framework into which the relative positions of North and South America at various times in the past can be fitted. These positions may then be used to estimate the size of the Caribbean Basin at various times since the initial breakup of Pangea. In the following analysis, we use the reconstruction of Sclater and others (1977). This reconstruction differs from that of Pindell and Dewey (1982) in that they propose no significant motion between South America and Africa in the equatorial Atlantic region prior to 110 m.y. We note, however, that the opening of the South Atlantic between 130 and 125 m.y. (Schlater and others, 1977; Rabinowitz and LaBrecque, 1979) requires that there is relative motion between Africa and South America during this same time period if Africa and South America behave as rigid plates. The reconstruction of Pindell and Dewey (1982) assumes that Africa does not behave as a single rigid plate, but that Early Cretaceous extension across the Benue Trough and associated Early Cretaceous rifts was significant enough to allow northwest and southwest Africa to behave as two independent plates. In this model, North Africa and South America behave as a single rigid plate without extensional motion between the northern Brazilian margin and North Africa. Sial (1976), however, indicates that extensive dike swarms in northern Brazil are associated with the

passive margin history in that region and that ages of these characteristically tholeiitic diabases range from about 200 to 120 m.y. They are interpreted to be related to a long protracted history of crustal extension and magmatism accompanying the breakup of northern South America and Africa in the region. This interpretation appears to be corroborated by similar isotopic ages for dike swarm fields in the Liberia-Nigeria region which complement those of northern Brazil and by continuation of the Brazilian dike fields of similar age into neighboring French Guiana to the north (Behrendt and Wotorson, 1970; May, 1971). Dates reported on basalts derived from drill sites that are 125 km offshore of northern Brazil within the Amapa Basin similarly show ages as old as 210 m.y. (Asmus, 1982). Because there is abundant evidence that the northern Brazilian margin was tectonically and magmatically active between 130-125 m.y. and because the evidence for the amount of Cretaceous deformation accommodated by the motion across the Benue Trough cannot be confirmed quantitatively, we prefer at this stage to adopt the previous reconstruction of Sclater and others (1977) which predicts that rifting of northern South America and Africa occurred about 125 m.y. B.P. This is supported by evidence that a major plate reorganization and changes in motion occurred in the Caribbean region at this time (see following discussion).

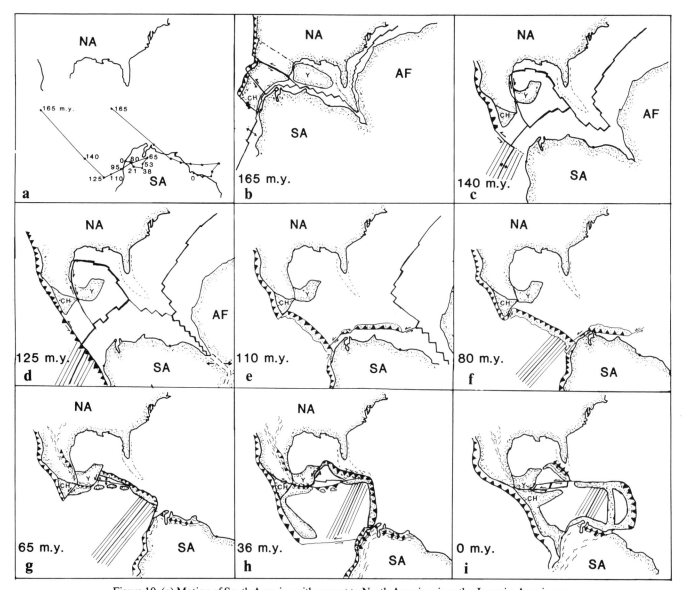

Figure 10. (a) Motion of South America with respect to North America since the Jurassic. Ages in m.y. compiled by Sclater and others, 1977. Points on the South American plate at the location today (0) have moved along a path indicated over the last 165 m.y. (figure after Burke and others, 1984). (b-i) Generalized schematic diagram showing the evolution of the Venezuelan Basin from Mid-Jurassic to Present (modified after Pindell and Dewey, 1982; Dickenson and Coney, 1980). Note Mid-Jurassic rifting and divergence between North and South America (b, c); initiation of subduction, cessation of spreading in the Venezuelan Basin and its entrapment at 127 m.y. (d); insertion of the Venezuelan Basin into the Caribbean region and eventual collision of Cuba and the Bahama Bank in Eocene time (e, f, g) and post-Eocene east-west strike-slip motion of the Venezuelan Basin with respect to North and South America (h, i). CH = Chortis Block; Y = Yucatan. Lineated area corresponds to the Venezuelan Basin through time.

Using points now situated on the northern coast of Venezuela, the motion from 165 m.y. to the present of South America relative to North America is shown in Figure 10a (after Burke and others, 1984, compiled from Sclater and others, 1977). Between 165 and 125 m.y. B.P., this part of South America and North America diverged along an approximately NW-SE direction. After rifting, new ocean floor formed simultaneously at two spreading centers: one in the Gulf of Mexico (i.e., North of Yucatan) (Shepherd and others, 1982), the other located between Yucatan and South America (i.e., the area now occupied by the present Caribbean plate) (Fig. 10b, c). This period of seafloor spreading corresponds to the time during which the crust of the Venezuelan Basin was created (i.e., 153-127 m.y.).

About 125 m.y. B.P., there was a major change in the relative motion of North and South America (Fig. 10d) associated with the rifting of Africa and South America and the opening of the South Atlantic (Sclater and others, 1977; Rabinowitz and LaBrecque, 1979). This change in relative plate motion corresponds approximately to the time of cessation of spreading in the Venezuelan Basin where the youngest magnetic anomaly located over the proposed fossil spreading axis is interpreted to be 127 m.y. in age. Modelling of magnetic anomalies within the Gulf of Mexico suggests that seafloor spreading between Yucatan and North America also ceased at about 127 m.y. (Shepherd and others, 1982). We believe the initiation of subduction zones of the ancestral Lesser and Greater Antilles was coincident with the cessation of both of these spreading centers at ~127 m.y. and occurred at a time of major plate reorganization in the Caribbean region.

It appears that the Venezuelan Basin was created during a Mid-Jurassic to Early Cretaceous phase of seafloor spreading along a ridge system which extended from the North Atlantic to the Pacific between Yucatan and South America (Fig. 10a, b, c). The similarly slow spreading rates in the Venezuelan Basin and the North Atlantic support this interpretation (see previous discussion). We believe that the Venezuelan Basin was originally part of the Pacific Basin and that upon Early Cretaceous rifting of Africa and South America in response to changes in plate motion, a westward-dipping subduction zone was initiated within the Pacific Basin. The Venezuelan Basin, at this stage, became a "trapped" back arc basin west of an ensimatic arc. The oldest isotopic ages of arc-related autochthonous igneous and metamorphic rocks recorded in the Greater and Lesser Antilles, 123 ± 2 m.y. and 127 ± 6 m.y. (Kesler and others, 1977; Bowin, 1975; Nagle, 1974), probably approximate the time of initiation of this arc activity and are essentially identical to the proposed time of ridge "shut off." Ages of the oldest arc-related tholeiitic rocks in the Venezuelan Antilles, the Netherlands Leeward Antilles, and Caribbean mountain system show similar ages (~130-114 m.y. B.P.) (Santamaria and Schubert, 1974). Except for ages from a trondhjemite from La Desirade (see discussion of Dinkleman and Brown, 1977; Mattinson, 1980), we are not aware of any pre-Early Cretaceous isotopic ages or pre-Albian fossil ages from autochthonous arc-related igneous, metamorphic, or sedimentary rocks.

The older ages recorded from La Desirade (Fig. 1) have been interpreted in a number of ways (e.g., see Mattinson and others, 1973; Mattinson, 1980; Fox and Heezen, 1975; Donnelly and Rogers, 1978, Donnelly, 1980; and Bouysse and others, 1983). Mattinson and others (1973) and Mattinson (1980) have suggested that the older oceanic crustal basement upon which the Lesser Antilles island arc has been built may be exposed on La Desirade. If this interpretation is correct, isotopically dated trondhjemitic rocks from La Desirade can be interpreted as part of an ophiolitic suite trapped within the forearc of the Lesser Antilles upon initiation of a westward-dipping intra-oceanic subduction zone in the Mesozoic. This ophiolitic basement should be similar in age to the oceanic crust (i.e., the Venezuelan Basin

crust) trapped within the back arc region upon initiation of subduction. This model assumes that the Aves Ridge represents a detached Mesozoic to Early Cenozoic remnant arc which ceased to be active in Mid-Eocene (Fox and Heezen, 1975) as the result of back arc spreading within the Grenada Trough.

$^{206}Pb/^{238}U$ and $^{207}Pb/^{206}Pb$ age determinations on zircon separates from trondhjemites thought to be part of the ophiolitic assemblage of La Desirade yield crystallization age of 145-150 m.y. (Mattinson and others, 1973; Mattinson, 1980). These ages are consistent with the interpretation that this ophiolitic assemblage represents an oceanic basement complex to the Lesser Antilles Arc that originated at the same spreading center as the crust in the central Venezuelan Basin. Not only does this age lie within the Late Jurassic to Early Cretaceous period of seafloor spreading proposed for the "trapped" Venezuelan Basin, but the projection of the 150 m.y. isochron proposed on the basis of magnetic correlations (Fig. 9) intersects the northern part of the Aves Ridge-Lesser Antilles Arc system near the position of La Desirade. The ophiolite assemblage in turn has apparently been intruded and succeeded by younger Mid-Cretaceous (112-119 m.y.) to Cenozoic arc-related basaltic, rhyolitic and andesitic volcanics locally intercalated with pelagic cherts (Bouysse and others, 1983).

The ophiolites of Guatemala and Cuba which are interpreted to represent the forearc basement and foundation of the ancestral Greater Antilles Arc are at least as old as Early Cretaceous (Wadge and others, in press). These ophiolites were obducted during the Late Cretaceous and Eocene arc-continent collisions between the northeastward moving ancestral Greater Antilles Arcs and Yucatan and the Bahama Platform, respectively. No younger oceanic basement foundation to the ancestral Antilles Arc (other than possible back arc ophiolites opened after the initiation of subduction) are known from the Caribbean. For these reasons, we believe that the Venezuelan Basin was generated prior to the initiation of subduction and forms, in part, the older oceanic basement foundation of the arc. The timing of initial arc volcanism indicates that the cessation of spreading along the fossil ridge axis within the Venezuelan Basin is essentially synchronous with the initiation of subduction.

Our schematic models assume only minor tectonic dislocation and disruption of the main oceanic part of the Caribbean plate which includes both the Venezuelan Basin and Colombian Basin. These dislocations within the Caribbean plate (e.g., the Hess Escarpment and the Beata Ridge) are probably the result of the non-rigid behavior of the Caribbean plate during Late Cretaceous and Cenozoic convergence of the North American and South American plates (Burke and others, 1978). The origin of the adjacent Colombian Basin directly to the west of the Venezuelan Basin is unclear at present and beyond the scope of this paper. The proposed magnetic correlations and lineation trends throughout the Colombian Basin, however, have been based on sparse and often widely spaced magnetic profiles. The basin is probably older than the Late Cretaceous age proposed on the basis of these magnetic data (Christofferson, 1973) since DSDP drilling information and seismic work indicate the presence of seismic layer B″ (or sill layer) within the Colombian Basin (Donnelly and others, 1973; Burke and others, 1978). These data would suggest that the oceanic crust within the Colombian Basin is older than Santonian-Coniacian in age (Donnelly and others, 1973). We are in the preliminary stages of compiling and reprocessing all existing magnetic data from the Colombian Basin to evaluate more precisely its age and origin. Until such analysis of these data are completed, the tectonic relationships between the Colombian and Venezuelan Basins will be poorly understood. In the opinion of the authors, there seem to be two possible scenarios. Either the Colombian Basin evolved simply as a once contiguous part of the Venezuelan Basin subjected to a later, mild tectonic disruption (e.g., formation of the Beata Ridge Fault Zone), or it evolved independently and was tectonically juxtaposed with the Venezuelan Basin. If the former is the case, we would predict that our analysis of magnetic data and further drilling would show the age of the Colombian Basin to be Late Jurassic or older.

From ~125 to 65 m.y. B.P., the motion between northern South America and North America was slightly convergent along an ENE-WSW direction (Fig. 10e, f). The juvenile Antillean Arc system migrated north and eastward into the narrowing oceanic tract between North and South America. A separate arc system which initiated to the east of the Antilles Arc collided with the north-facing stable continental margin of Venezuela resulting in the emplacement of the Villa de Cura nappes (Maresch, 1974). Late Cretaceous collisions of segments of a once more continuous Antilles may be recorded in Colombia (Goossens and others, 1977) and in Central America (Pindell and Dewey, 1982) as the arc was inserted between North and South America.

During the Late Cretaceous to Eocene (Fig. 10g, h), continued migration of the Antilles Arc northward with respect to North America and subduction of Atlantic crust resulted in the opening of the Yucatan Basin (Hall and Yeung, 1980) possibly by back arc spreading and the eventual collision of Cuba with the Bahama Bank in Eocene time (Burke and others, 1984).

Relative motion between North America and South America during the Cenozoic has been relatively minor with a slight divergence between 65-36 m.y. B.P. and a small westward movement of South America during the last 36 m.y. B.P. Since the Eocene (Fig. 10i), the Caribbean plate (including the Venezuelan Basin) has undergone a dominantly eastward strike-slip motion with respect to North America with a minimum displacement along the Motagua-Swan-Oriente Fault System of 1200-1400 km (Pindell and Dewey, 1982; Sykes and others, 1982; Burke and others, 1984). Because post-Eocene motion between North and South America is small, a similar amount of Caribbean-South American right-lateral strike-slip motion must have occurred parallel with the Venezuelan Antilles. Therefore, the Lesser Antilles arc and the trapped Venezuelan Basin must have migrated past the coast of northern Venezuela largely by post-Eocene strike-slip motion. The Late Cretaceous emplacement of the Villa de Cura nappes (Maresch, 1974) would thus have preceded this passage and oblique collision of the Lesser Antilles with the northern coast of Venezuela throughout the later part of the Cenozoic.

The model we invoke for the insertion of the Venezuelan Basin into the Caribbean region is not unlike those presented in other discussions (e.g., Malfait and Dinkleman, 1973; Pindell and Dewey, 1982; Burke and others, 1984). The strong evidence for Late Jurassic to Early Cretaceous seafloor spreading in the Venezuelan Basin, we feel, provides new constraints on these models and allows, as presented here, further refinements for the evolution of the Caribbean plate.

ACKNOWLEDGMENTS

The technical assistance of A. Shepherd and discussions with K. Burke, J. Dewey, and G. Wadge are greatly appreciated. We also appreciate the many constructive comments on earlier drafts of the manuscripts by the G.S.A. reviewers. S. Hall acknowledges support from a Research Initiation Grant awarded by the University of Houston.

REFERENCES CITED

Asmus, H. E., 1982, Geotectonic significance of Mesozoic-Cenozoic magmatic rocks in the Brazilian continental margin and adjoining emerged area: Fifth Congress of Latin America Geology, Argentina, Actos III, p. 761–779.

Barraclough, D. R., Harwood, J. M., Leaton, B. R., and Malin, S.R.C., 1975, A model of the geomagnetic field at epoch 1975: Geophysical Journal Royal Astronomical Society, v. 43, p. 645–659.

Behrendt, J. C., and Wotorson, C. S., 1970, Aeromagnetic and gravity investigations of the coastal area and continental shelf of Liberia, West Africa, and their relation to continental drift: Geological Society of America Bulletin, v. 81, p. 3563–3574.

Bouysse, P., Schmidt-Effiny, R., and Westercamp, D., 1983, La Desirade Island (Lesser Antilles) revisited: Lower Cretaceous radiolarian cherts and arguments against ophiolitic origin for the basal complex: Geology, v. 11, p. 244–247.

Bowin, C., 1975, The geology of Hispaniola, *in* Nairn, A.E.M., and Stehli, F. G.,

eds., The Gulf of Mexico and the Caribbean, ocean basins and margins, v. 3: New York, Plenum Press, p. 501–552.

Burke, K., Fox, P. J., and Sengor, A.M.C., 1978, Bouyant ocean floor in the evolution of the Caribbean: Journal of Geophysical Research, v. 83, p. 3949–3954.

Burke, K., Cooper, C., Dewey, J. F., Mann, W. P., and Pindell, J. L., 1984, Caribbean tectonics and relative plate motions, *in* Bonini, W. E., and others, eds., The Caribbean-South American Plate Boundary and Regional Tectonics: Geological Society of American Memoir 162 (this volume).

Christofferson, E., 1973, Linear magnetic anomalies in the Colombian Basin, central Caribbean Sea: Geological Society of America Bulletin, v. 84, p. 3217–3230.

Dickenson, W. R., and Coney, P. F., 1980, Plate tectonic constraints on the origin of the Gulf of Mexico, *in* Pilger, R. H., Jr., ed., The origin of the Gulf of Mexico and the early opening of the central Atlantic: Baton Rouge, Louisiana State University, p. 27–36.

Diebold, J. B., Stoffa, P. L., Buhl, P., and Truchan, M., 1981, Venezuelan Basin crustal structure: Journal of Geophysical Research, v. 86, no. B9, p. 7901–7923.

Dinkleman, M. G., and Brown, J. F., 1977, K-Ar geochronology and its significance to the geological setting of La Desirade (Lesser Antilles): Caribbean Geological Conference, 8th, Curacao, Abstracts, p. 38–39.

Donnelly, T. W., 1973, Magnetic anomaly observations in the eastern Caribbean Sea, *in* Edgar, N. T., and others, eds., Initial Reports of the Deep Sea Drilling Project, v. 15, U.S. Government Printing Office, Washington, D.C., p. 1023–1030.

——1980, Igneous series in island arcs: the northeastern Caribbean compared with worldwide island arc assemblages: Bulletin Volcanologique, v. 43-2, p. 347–382.

Donnelly, T. W., Melson, W., Kay, R., and Rogers, J.J.W., 1973, Basalts and dolerites of Late Cretaceous age from the central Caribbean, *in* Edgar, N. T., and others, eds., Initial Reports of the Deep Sea Drilling Project, v. 15, U.S. Government Printing Office, Washington, D.C., p. 989–1011.

Donnelly, T. W., and Rogers, J.J.W., 1978, The distribution of igneous rock suites throughout the Caribbean: Geologie en Mijnbouw, v. 57, p. 151–162.

Ewing, J. I., Antoine, J., and Ewing, M., 1960, Geophysical measurements in the western Caribbean Sea and in the Gulf of Mexico: Journal of Geophysical Research, v. 65, p. 4087–4126.

Ewing, J. I., Edgar, N. T., and Antoine, J. W., 1970, Structure of the Gulf of Mexico and Caribbean Sea, *in* Maxwell, A. E., ed., The Sea, v. 4, part 2: New York, Wiley-Interscience, p. 321–358.

Fisher, R. L., Sclater, J. G., and McKenzie, D. P., 1971, Evolution of the Central Indian Ridge, Western Indian Ocean: Geological Society of America Bulletin, v. 82, p. 553–562.

Fox, P. J., and Heezen, B. C., 1975, Geology of the Caribbean crust, *in* Nairn, A. E., and Stehli, F. G., eds., The ocean basins and margins, v. 3: New York, Plenum Press, p. 421–466.

Ghosh, N., 1982, Magnetic anomaly studies and determination of the depth to magnetic basement in the Venezuelan Basin [M.S. thesis]: Houston, University of Houston.

Goossens, P. J., Rose, W. I., Jr., and Flores, D., 1977, Geochemistry of tholeiites of the Basic Igneous Complex of northwestern South America: Geological Society of America Bulletin, v. 88, p. 1711–1720.

Hall, S. A., 1979, Geomagnetic secular variation and secular acceleration in the Red Sea area: Geophysical Journal Royal Astronomical Society, v. 58, p. 583–592.

Hall, S. A., and Yeung, T., 1980, A study of magnetic anomalies in the Yucatan Basin, *in* Transactions of the Caribbean Geological Conference, IX, Santo Domingo, 1980: Universidad Catolica Madre y Maestra, Santiago de los Caballeros, Republica Dominicana.

Irving, E., and Pullaiah, G., 1976, Reversals of the geomagnetic field, magneto-stratigraphy, and relative magnitude of the paleosecular variation in the Phanerozoic: Earth Science Review, v. 12, p. 35–64.

Jackson, H. R., and Reid, I., 1983, Oceanic magnetic anomaly amplitudes: variation with sea-floor spreading rate and possible implications: Earth and Planetary Science Letters, v. 63, p. 368–378.

Kesler, S. E., Sutter, J. F., Jones, L. M., and Walker, R. L., 1977, Early Cretaceous basement rocks in Hispaniola: Geology, v. 5, p. 245–247.

Kristoffersen, Y., and Talwani, M., 1977, Extinct triple junction south of Greenland and the Tertiary motion of Greenland relative to North America: Geological Society of America Bulletin, v. 88, p. 1037–1049.

Larson, R. L., and Chase, C. G., 1972, Late Mesozoic evolution of the Western Pacific Ocean: Geological Society of America Bulletin, v. 83, p. 3627–3644.

Larson, R. L., and Hilde, T.W.C., 1975, A revised time scale of magnetic reversals for the Early Cretaceous and Late Jurassic: Journal of Geophysical Research, v. 80, p. 2586–2594.

Lowrie, W., Channell, J.E.T., and Alvarez, W., 1980, A review of magnetic stratigraphy investigations in Cretaceous pelagic carbonate rocks: Journal of Geophysical Research, v. 85, p. 3597–3605.

Ludwig, W. J., Robert, E. H., and Ewing, J. I., 1975, Profiler sonobuoy measurements in Colombia and Venezuelan Basins, Caribbean Sea: American Association of Petroleum Geology Bulletin, v. 59, p. 115–124.

Malfait, B. T., and Dinkleman, M. G., 1972, Circum Caribbean tectonics and igneous activity and the evolution of the Caribbean Plate: Geological Society of America Bulletin, v. 83, p. 251–271.

Maresch, W. V., 1974, Plate tectonics origin of the Caribbean Mountain system of northern South America: Geological Society of America Bulletin, v. 85, p. 669–682.

Matthews, J. E., 1976, Total intensity geomagnetic field of the Caribbean region: U.S. Geological Survey, Miscellaneous Studies Map MF-742.

Mattinson, J. M., 1980, Geochronologic and isotopic study of the La Desirade Island basement complex: Jurassic oceanic crust of the Lesser Antilles: Contributions to Mineralogy and Petrology, v. 71, p. 237–245.

Mattinson, J. M., Fink, L. K., Jr., and Hopson, C. A., 1973, Age and origin of ophiolitic rocks on La Désirade Island, Lesser Antilles Island Arc: Annual Report of the Director of the Geophysical Laboratory, Carnegie Institution, p. 616–623.

May, P. R., 1971, Pattern of Triassic-Jurassic diabase dikes around the North Atlantic in the context of pre-drift position of the continents: Geological Society of America Bulletin, v. 82, p. 1285–1292.

McElhinny, M. W., and Burek, P. B., 1971, Mesozoic paleomagnetic stratigraphy: Nature, v. 232, p. 98–102.

Nagle, F., 1974, Blueschist, eclogite, paired metamorphic belts, and the early tectonic history of Hispaniola: Geological Society of America Bulletin, v. 85, p. 1461–1466.

Phillips, J. D., 1979, ADEPT: A program to estimate depth to magnetic basement from sampled magnetic profiles: U.S. Geological Survey Open File Report 79-367, 35 p.

Pindell, J., and Dewey, J. F., 1982, Permo-Triassic reconstruction of western Pangea and the evolution of the Gulf of Mexico/Caribbean region: Tectonics, v. 1, p. 179–212.

Rabinowitz, P. D., and LaBrecque, J., 1979, The Mesozoic South Atlantic Ocean and evolution of the continental margins: Journal of Geophysical Research, v. 84, p. 5973–6002.

Santamaria, F., and Schubert, C., 1974, Geochemistry and geochronology of the southern Caribbean-northern Venezuelan plate boundary: Geological Society of America Bulletin, v. 85, p. 1085–1098.

Saunders, J. B., Edgar, N. T., Donnelly, T. W., and Hay, W. W., 1973, Cruise synthesis, *in* Edgar, N. T., and others, eds., Initial Reports of the Deep Sea Drilling Project, v. 15, U.S. Government Printing Office, Washington, D.C., p. 1077–1111.

Schouten, H., and McCamy, K., 1972, Filtering marine magnetic anomalies: Journal of Geophysical Research, v. 77, p. 7089–7099.

Sclater, J. G., Hellinger, S., and Tapscott, C., 1977, The paleobathymetry of the Atlantic Ocean from the Jurassic to the Present: Journal of Geology, v. 85, p. 509–533.

Shepherd, A. V., Hall, S., and Snow, R., 1982, Magnetic and gravity fields of the eastern Gulf of Mexico (82° - 90° W): Geological Society of America Abstracts with Programs, v. 14, no. 7, p. 615.

Sial, A. N., 1976, The post-Paleozoic volcanism of northeast Brazil and its tectonic significance: Annaes de la Academia, Brasilerra, de Ciencias 48, p. 299–311.

Silver, E. A., Case, J. E., and MacGillavry, H. J., 1975, Geophysical study of the Venezuelan borderland: Geological Society of America Bulletin, v. 86, p. 213–226.

Steiner, M. B., 1980, Investigation of the geomagnetic field polarity during the Jurassic: Journal of Geophysical Research, v. 85, p. 3572–3586.

Stoffa, P. L., Mauffret, A., Truchan, M., and Buhl, P., 1981, Sub B″ layering in the southern Caribbean: the Aruba gap and Venezuelan Basin: Earth and Planetary Science Letters, v. 53, p. 131–146.

Sykes, L. R., McCann, W. R., and Kafka, A. L., 1982, Motion of Caribbean Plate during last 7 million years and implications for earlier Cenozoic movements: Journal of Geophysical Research, v. 87, p. 10656–10676.

Vine, F. J., and Matthews, D. H., 1963, Magnetic anomalies over ocean ridges: Nature, v. 199, p. 947–949.

Wadge, G., Draper, G., and Lewis, J. F., 1983, Ophiolites of the northern Caribbean: reappraisal of their roles in the evolution of the Caribbean Plate Boundary: Special Publication of the Geological Society of London on Ophiolites and Oceanic Lithosphere (in press).

Watkins, J., and Cavanaugh, T., 1974, Implications of magnetic anomalies in the Venezuelan Basin: Transactions of the VII Caribbean Geological Conference, Guadaloupe, p. 129–138.

MANUSCRIPT ACCEPTED BY THE SOCIETY SEPTEMBER 1, 1983

Geological Society of America
Memoir 162
1984

Plate tectonic evolution of the Caribbean region in the mantle reference frame

R. A. Duncan
School of Oceanography
Oregon State University
Corvallis, Oregon 97331

R. B. Hargraves
Department of Geological and Geophysical Sciences
Princeton University
Princeton, New Jersey 08540

ABSTRACT

The mantle reference frame defined by stationary hotspots has been used to determine the positions and motions of continental and oceanic plates surrounding the Caribbean region from late Jurassic time (140 m.y.) to the present. First, the position of the Pacific plate and the Pacific-Farallon spreading ridge has been reconstructed using the ages and geometry of island and seamount chains emanating from Pacific hotspots. Then, by assuming symmetric spreading across the Pacific-Farallon ridge, the motion of the Farallon plate relative to the mantle has been calculated. This shows that the postulated oceanic plateau which may form the core of the present Caribbean plate could have been erupted onto late Jurassic to early Cretaceous oceanic lithosphere as the Farallon plate passed over the Galapagos hotspot, hypothesized to have been initiated in mid- to late Cretaceous time (100 to 75 m.y. B.P.).

The thickened volcanic plateau collided with the Greater Antilles Arc, then filling the gap between South America and nuclear Central America, in late Cretaceous time (80 to 70 m.y. B.P.) and was not subducted; instead, subduction of the Farallon plate commenced behind the plateau. This buoyant, indigestible piece of oceanic lithosphere drove the Greater Antilles Arc northeastwards, accompanied by subduction of proto-Caribbean crust, until it collided with the Bahama platform in late Eocene time. Concomitantly, the trench and island arc which developed behind (southwest of) the plateau generated what is now a part of Central America. Subsequent westward subduction of Atlantic lithosphere beneath the Lesser Antilles Arc and continuing eastward subduction of oceanic lithosphere beneath Central America, together with transform faulting (left-lateral Cayman transform fault in the north, right-lateral strike-slip motion in and off Venezuela in the south) defined the present boundaries of the Caribbean plate.

INTRODUCTION

The Caribbean region has evolved at the boundary between two continental plates, North and South America, since early Jurassic time (Dickinson and Coney, 1980; Pindell and Dewey, 1982). Its tectonic history is a Gordian Knot left by seafloor spreading, plate convergence, and large transform fault displacements. It is also possible that a large portion of the Caribbean plate is unusually thick ocean floor, formed in the Pacific and transported into the region from the west (Malfait and Dinkelman, 1972, Burke and others, 1978). Much of the geologic record has been destroyed by subduction and so must be inferred from indirect evidence. The geological complexity and missing evidence allow many possible models of Caribbean evolution. In this paper the circum-Caribbean plate motions are described with respect to the hotspot (= mantle) reference frame. From these the convergence rate and direction between oceanic lithosphere from the west (Pacific basin) and North and South America are determined. In particular, the case for collision of an oceanic plateau with the Greater Antilles Arc to form the Caribbean plate is examined.

UNUSUAL SEAFLOOR IN THE CARIBBEAN

Understanding of the tectonic evolution of the Caribbean plate has been particularly frustrated by the lack of easily identifiable seafloor magnetic anomalies. This obscurity has been presumed to be related in some way to the excess thickness of the crust. Seismic refraction studies have revealed that the crust is from 10 to 15 km thick (Officer and others, 1959; Edgar and others, 1971; Houtz and Ludwig, 1977). The oceanic floor is also 1-2 km shallower than would be predicted from thermal subsidence (Burke and others, 1978). Houtz and Ludwig (1977) emphasized that the oceanic crust is not uniformly thick but consists of basins and ridges, perhaps corresponding to normal ocean floor and that thickened by later volcanic eruption, respectively.

As a result of the drilling performed on Leg 15 of the Deep Sea Drilling Program, the additional thickness of the Caribbean seafloor is identified with widespread occurrence of basaltic sills and flows overlain by sediments of late Cretaceous age (Coniacian to Campanian). These volcanic rocks in turn overlie sediments (Hopkins, 1973) and oceanic crust of presumed Cretaceous or earlier age (Burke and others, 1978). The oceanic basement has not been penetrated by drilling. Recently, however, magnetic anomalies recorded from the Venezuela basin have been interpreted by Ghosh and others (1984) to have an ENE trend and to be of late Jurassic to early Cretaceous age. The presence of these anomalies (Donnelly, 1973), no matter their precise age, indicates that the oceanic crust beneath the volcanic and sedimentary carapace must be older than the Cretaceous magnetic quiet zone (118 to 83 m.y. [Cox, 1983]).

Removal of 1 to 2.5 km of basaltic sills and flows would restore the crust to very nearly a normal oceanic thickness (Houtz and Ludwig, 1977). This volcanic overlay is considered to be responsible for the B″ reflector (Donnelly and others, 1973) seen on reflection records throughout the central Caribbean. The restricted age range of overlying sediments (~80-75 m.y. [Donnelly and others, 1973]) indicates that this great eruptive episode terminated rather abruptly. Petrologically, the layer B″ basalts comprise plagioclase tholeiites of two geochemically distinct types (Bence and others, 1975). One is a low-K_2O, low-TiO_2, LREE-depleted tholeiite indistinguishable from the majority of basalts formed at spreading ridges. The other (at DSDP site 151) is a higher K_2O, higher TiO_2, LREE-enriched tholeiite typical of Icelandic and Hawaiian tholeiites and a few ocean floor basalts which may have erupted away from the ridges (Bence and others, 1975).

Various basalt exposures on islands surrounding the Caribbean have been tentatively correlated with layer B″ and the underlying oceanic crust (Donnelly, 1975; Burke and others, 1978; Beets and others, 1984). In the western Cordillera of Colombia and Ecuador (Mooney, 1980), the stratigraphic column comprises circa Jurassic to early Cretaceous pillow basalts of ocean ridge composition, overlain by deep water sediments and, in turn, by late Cretaceous basaltic flows and sills. A similar sequence constitutes the Nicoya Complex of Central America, although the volcanics and sediments overlying the deep water sediments are more clearly of trench/island arc affinity (Schmid-Effing, 1979). In the Dutch Antilles, pillowed volcanic rocks and intercalated sediments also occur. The extensive igneous rocks are basaltic (MORB type) on Aruba and Curacao, while the basaltic andesite-rhyolites of the Washikemba series of Bonaire are more distinctly of island arc affinity (Beets and others, 1984; Primitive Island Arc series of Donnelly and Rogers [1978]). Possible analogues of these stratigraphic sections are overthrust onto the mainland of Venezuela as the blueschist metavolcanics of the Villa de Cura series. The basement of the islands constituting the Greater Antilles represents a complex composite of oceanic crust, sediment, and arc-related volcanics and sediments; all are of late Jurassic to late Cretaceous age.

It is apparent, then, that throughout the Caribbean there is a wide areal extent of penecontemporaneous volcanism ending in the late Cretaceous. This has prompted others to draw analogies between the Caribbean and anomalously shallow oceanic plateaus in the Pacific, such as the Ontong Java and the Manihiki plateaus (Mattson, 1969; Burke and others, 1978). Burke and others (1978) claim that the Caribbean plate exists because the buoyancy of such a plateau renders it virtually non-subductible. They suggest that the borders of the plateau have been obducted and intermixed with arc volcanic rocks at the surrounding plate boundaries—hence the B″ basaltic rocks exposed on circum-Caribbean islands and on the western margins of northwestern South and Central America. The bulk of the plateau which is purportedly indigestible, in the plate-tectonic sense, remains as the Caribbean Sea.

The pivotal question with regard to the present Caribbean seafloor, however, concerns its paleogeographic position. At one extreme it may be claimed that the eruption of layer B″ basalts took place essentially in the site of the present Caribbean, onto crust formed in Jurassic to early Cretaceous time by spreading between North and South America. Alternatively, it came in from somewhere in the Pacific. This latter model, in one form or another, is attractive to many authors (Mattson, 1969; Burke and others, 1978; Schmid-Effing, 1979; Pindell and Dewey, 1982; Sykes and others, 1982; Beets and others, 1984). The present Caribbean is bounded on the west and east by active subduction zones—the Central America Trench and the Lesser Antilles Trench, respectively. The northern and southern boundaries show evidence of previous subduction, but there the most recent motion is transcurrent, moving Caribbean crust eastwards with respect to both North and South America (e.g., Sykes and others, 1982).

The problem now exists of estimating how far from the west the Caribbean plate has travelled. If it has come from the Pacific, this oceanic lithosphere could have been generated at the Pacific-Farallon spreading ridge (roughly north-south orientation) or at the Farallon-Phoenix spreading ridge (roughly northeast-southwest orientation). As discussed earlier, there is some evidence that the oceanic basement of the Caribbean is mid-Jurassic to early Cretaceous in age. Based on the age difference between crystallization and emplacement, Mooney (1980) has concluded that the lowermost basaltic rocks of the western Cordillera of Ecuador-Colombia were formed at a spreading ridge probably within 450 to 900 km of the western margin of South America. Although the position of the Farallon-Phoenix spreading ridge is completely unknown because subduction has removed it, the Pacific-Farallon ridge was much too far away, so Mooney (1980) favors a more east-westerly Farallon-Phoenix ridge for the generation of these rocks. The identification of northeast-southwest trending magnetic anomalies (Ghosh and others, 1984; see also Christofferson, 1976) also supports generation at a ridge with such an orientation.

PLATE MOTIONS IN THE HOTSPOT REFERENCE FRAME

Pindell and Dewey (1982) have calculated the relative motions between North and South America from Jurassic time to the present by referencing each plate to Africa and obtaining finite difference rotations. They demonstrated that Jurassic through Cretaceous separation between North and South America would generate oceanic lithosphere but would also provide a gap between the continents large enough to accommodate "exotic" oceanic blocks colliding from the west. The motion between North and South America and oceanic plates in the Pacific basin has not yet been estimated. This may be done by adding the relative plate motions from Africa through Antarctica to the Pacific. Such a solution, however, incurs additive errors at each plate boundary and is subject to the large uncertainty of possible plate motions within Antarctica (Jurdy, 1979; Gordon and Cox, 1980; Duncan, 1981; Dalziel, 1982).

An alternative estimate of the motion between the western

TABLE 1. MOTIONS OF PLATES SURROUNDING THE CARIBBEAN, 140 m.y. TO PRESENT

Time (m.y.)	Latitude (°N)	Longitude (°E)	Angle (°ccw)
Pacific Plate over Hotspot Frame			
42-0	68.0	-75.0	-34.0
65-42	17.0	-107.0	-14.0
74-65	22.0	-95.0	-7.5
100-74	36.0	-76.0	-15.0
140-110	85.0	-165.0	14.0
Farallon (Nazca) Plate-Pacific Plate			
12-0	70.0	-80.0	18.0
25-12	70.0	-90.0	20.0
38-25	82.0	65.0	12.2
74-38	73.0	60.0	31.5
119-74	60.0	50.0	44.0
140-119	44.0	36.0	18.5
Farallon (Nazca) Plate over Hotspot Frame			
12-0	71.5	-86.8	9.0
25-12	67.2	-102.2	10.6
38-25	32.9	141.6	6.1
42-38	31.9	-45.8	-3.9
48-42	31.8	129.6	5.7
65-48	36.4	137.1	18.6
74-65	20.9	154.7	9.7
100-74	31.0	162.2	25.7
119-100	43.1	140.4	13.6
149-119	20.6	146.8	15.0
South America over Hotspot Frame			
21-0	65.4	-17.8	-2.9
38-21	63.4	100.1	-2.5
65-38	79.4	91.0	-8.8
86-65	16.0	106.1	-8.6
100-86	62.3	-21.1	-6.3
119-100	26.6	-10.4	-7.3
140-119	26.5	36.4	-5.0
North America over Hotspot Frame			
38-0	37.8	107.9	-5.5
65-38	54.0	98.9	-7.7
86-65	28.7	114.9	-12.6
100-86	63.3	63.0	-5.9
119-100	28.7	24.4	-6.2
149-119	34.8	14.5	17.1

Atlantic plates and the Pacific basin plates may be made using the hotspot reference frame (Morgan, 1981; Duncan, 1981). The motion of the Pacific plate is given by the geometry and documented age progressions within island and seamount chains (Table 1). Motion has been west-northwest, along the trend of the Hawaiian Islands, from about 42 m.y. until the present. Pacific plate velocity over the hotspots during this period is well constrained by island and seamount ages in many volcanic lineaments. Between 42 m.y. and middle Cretaceous time (100 to 110 m.y.), motion was almost due north, following the trend of the Emperor Seamounts and Line Islands. From late Jurassic to middle Cretaceous time, the direction of Pacific plate motion is defined by east-west volcanic lineaments such as the Mid-Pacific Mountains and Magellan Seamounts. Pacific plate velocity during this earliest interval was probably slow (Henderson and Gordon, 1981) although age information is sparse.

From this basis of Pacific plate motion in the hotspot reference frame, we reconstruct the position of the Pacific-Farallon spreading ridge through time using identified magnetic anomalies on the Pacific plate (Hilde and others, 1976; Handschumacher, 1976; Klitgord and Mammerickx, 1982). The ridge positions are illustrated in Figures 1 to 8. We calculate the motion of the Farallon plate in the hotspot reference frame by adding the relative motion between the Pacific and Farallon plates to the Pacific

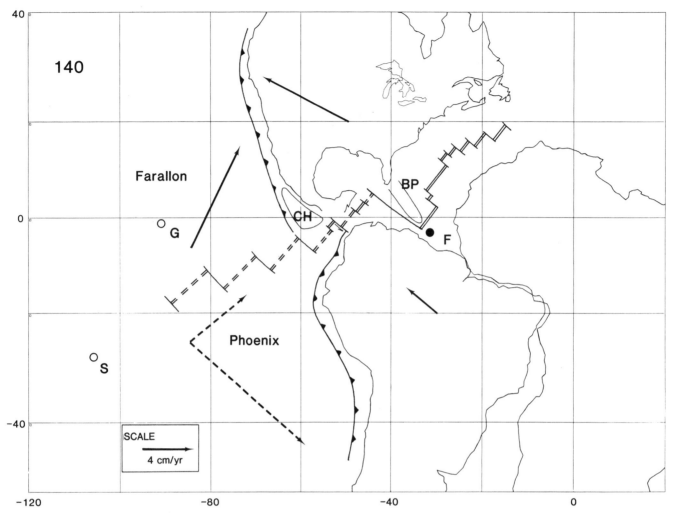

Figure 1. 140 m.y. B.P. plate positions in the hotspot reference frame. Bold arrows indicate plate motions with respect to stationary hotspots. The hypothetical position of the Farallon-Phoenix spreading ridge is shown by dashed lines. The Phoenix plate moved either to the northeast or to the southeast (bold, dashed lines), depending on whether Farallon-Phoenix spreading was slow or fast, respectively. Fernando de Noronha (F), Galapagos (G), and Sala y Gomez (S) are presently active hotspots. Only the Fernando hotspot was active (solid circle) in late Jurassic time, forming the volcanic base of the Bahama Plateau (BP).

motion already described. Then the Pacific-Farallon relative motions given by Handschumacher (1976) and Klitgord and Mammerickx (1982) are used for 38 m.y. (magnetic anomaly 13, by the time scale of Cox [1983]) to the present. Pieces of the Farallon plate older than Eocene (magnetic anomaly 21: 48 m.y.) have been subducted beneath North and South America, so we must assume that spreading was symmetric about the Pacific-Farallon ridge in order to calculate poles of rotation from magnetic anomalies and transform faults recorded on the Pacific plate. This is probably a good assumption since half-spreading rates are moderate to fast (4-8 cm/yr) from Jurassic to Eocene time.

Pacific-Farallon relative motions and our calculated Farallon plate motions for times of prominent magnetic anomalies appear in Table 1. The motions of North and South America in

the hotspot reference frame were determined in a similar fashion by adding Atlantic relative motions (those selected by Pindell and Dewey [1982]) to African plate motion in the hotspot reference frame (Morgan, 1981; Duncan, 1981). We then used the rotation data in Table 1 to reconstruct plate positions and trajectories for various times from late Jurassic to the present (Figs. 1 to 8).

In the following discussion, we briefly summarize the major features of a model of Caribbean tectonic evolution as we propose they relate to the plate motions derived from the hotspot reference frame.

140 m.y. (anomaly M14). The Gulf of Mexico has been formed by seafloor spreading in Jurassic time, as North America moved northwestward away from Gondwanaland (Dickinson and Coney, 1980; Pindell and Dewey, 1982). We concur with Mooney (1980) in connecting the Central Atlantic spreading

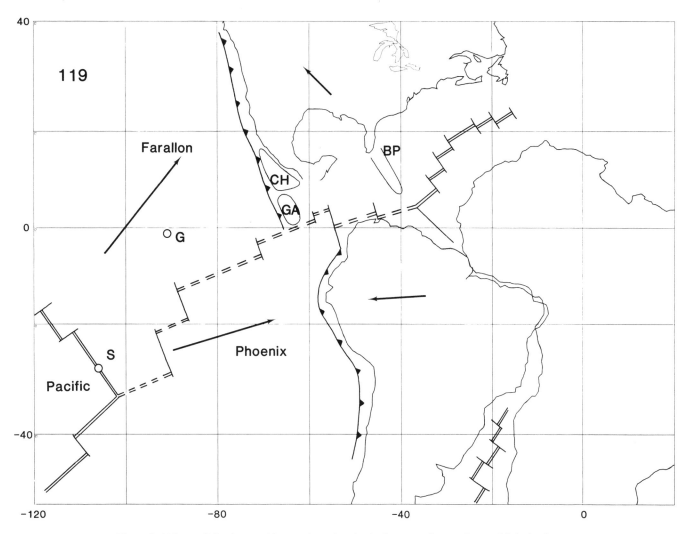

Figure 2. 119 m.y. B.P. plate positions and motions in the hotspot reference frame. GA is the Greater Antilles Arc.

ridge with spreading between the Farallon and Phoenix plates in the Pacific basin. This spreading ridge then passes through the Caribbean region, resulting from northwest-southeast separation of North and South America. The Farallon plate is subducting beneath North America, but the nature of the interaction between the Phoenix plate and South America is not clear. Because the Farallon plate is moving rapidly northeastwards, this interaction will depend upon the rate of spreading on the Farallon-Phoenix ridge, which is unknown. If spreading is fast, the Phoenix plate will be subducting beneath South America; but if it is slow, primarily right-lateral transcurrent motion will occur along the northwestern margin of South America. The Bahama Plateau is thought to have a volcanic foundation beneath the carbonate platform (Dietz and Holden, 1973; Burke and others, 1978; Dickinson and Coney, 1980). We agree with Dietz and Holden (1973) and Dickinson and Coney (1980) who postulate an origin by hotspot volcanism. Morgan (1981) proposed specifically the Fernando de Noronha hotspot whose position falls at the south-

east end of the Bahama Plateau at 140 m.y. in our reconstruction. The present positions of the Galapagos and Sala y Gomez hotspots are also shown (Fig. 1).

119 m.y. (anomaly MO). Spreading is active in the Central and South Atlantic and begins at this time in the Equatorial Atlantic (Pindell and Dewey, 1982), forming a triple junction north of South America. We propose that spreading on the ridge running through the Caribbean (Mooney, 1980) essentially terminates when the Equatorial Atlantic spreading begins. Until this happens, three spreading ridges meet in the southeast Pacific basin, separating the Pacific, Farallon, and Phoenix plates. The Sala y Gomez hotspot, if active, would lie very close to the Pacific-Farallon spreading ridge.

100 m.y. (Quiet zone, interpolated). Spreading through the Atlantic is underway. The Farallon-Phoenix spreading is dead. Subduction of the Farallon plate is now continuous along the western side of the Americas, forming what will become the Greater Antilles Arc adjacent to the proto-Caribbean. The Villa

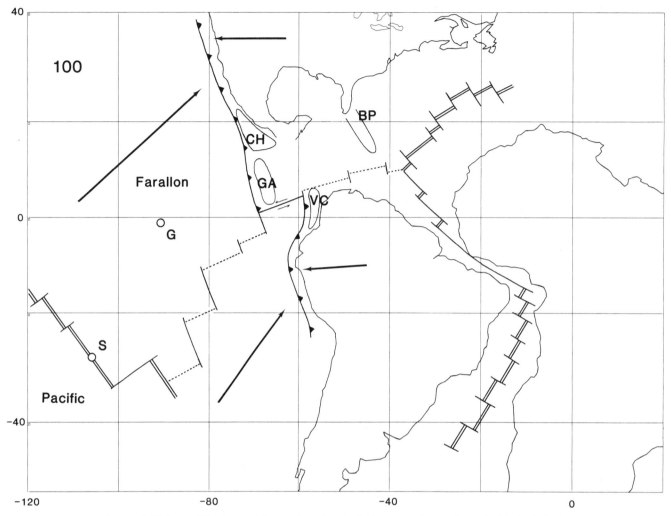

Figure 3. 100 m.y. B.P. plate positions and motions in the hotspot reference frame. VC is the Villa de Cura complex. Spreading on the Farallon-Phoenix ridge stopped as spreading began in the Equatorial Atlantic.

de Cura complex is produced by east-facing subduction beneath the northwestern margin of South America. Late Jurassic and early Cretaceous Caribbean ocean floor lies passive behind the Greater Antilles Arc. We hypothesize that at about 100 m.y. the Galapagos hotspot burst into activity, erupting voluminous basalts onto oceanic crust formed earlier at the Farallon-Phoenix spreading ridge in late Jurassic to early Cretaceous time; a substantial oceanic plateau results.

80 m.y. (anomaly 33). By 80 m.y. the Galapagos plateau encounters and clogs the Greater Antillean trench/arc, pushing it to the northeast with counterclockwise rotation. A southern segment of the arc, destined to become the Dutch Antilles and Villa de Cura belt of northern Venezuela, is translated eastward and rotated clockwise. Extensions of this segment are accreted to Ecuador and Colombia. The Galapagos hotspot continues to erupt basalts onto northeastwardly moving ocean floor of early Cretaceous age formed at the Pacific-Farallon ridge.

60 m.y. (anomaly 26). The geological history of the southern Caribbean borderland (Curacao, Aruba, Bonaire, and the Caribbean Mountains of Venezuela [Beets and others, 1984]) places the timing of the insertion from the west of this oceanic plateau between late Cretaceous and Paleocene. Paleomagnetic evidence (Skerlec and Hargraves, 1980; Hargraves and Skerlec, 1980; Stearns and others, 1982) demonstrates that clockwise rotation of these borderlands must have continued after ~90 m.y. (the age of the batholiths on Aruba [Beets and others, 1984]) and pre-Paleocene (the age of the unrotated limestone Morros overlying the Villa de Cura complex [Hargraves and Skerlec, 1980]). Translation of the Villa de Cura onto the Venezuelan mainland may have continued until Eocene time. Similarly, beginning in the late Cretaceous but continuing until late Eocene time (Mattson, 1979; Dickinson and Coney, 1980), the Greater Antilles Arc (including Cuba and Hispaniola) moves northeastward, consuming Caribbean ocean floor of Mesozoic age beneath it. This

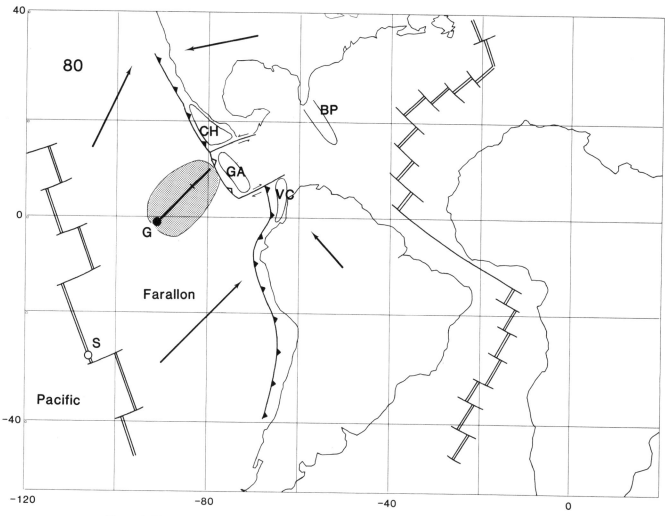

Figure 4. 80 m.y. B.P. plate positions and motions in the hotspot reference frame. Predicted hotspot tracks are shown as bold lines emanating from active hotspots (solid circles). Large volumes of basalts erupted at the Galapagos hotspot formed an oceanic plateau (stippled pattern) between about 100 and 75 m.y., which collided with the Greater Antilles Arc in the late Cretaceous.

motion ceases when the subduction zone collides with the Bahama platform (Burke and others, 1978; Dickinson and Coney, 1980). Some underthrusting of Bahaman lime-stones beneath the arc rocks in Cuba has occurred, but the bulk of the Bahama platform was apparently too buoyant to be subducted. Following the proposal of Burke and others (1978), we believe the northeastward motion of the Greater Antilles Arc was caused by collision of an oceanic plateau from the west. Specifically, the Jurassic to Cretaceous ocean floor thickened by eruption of basalts over the Galapagos hotspot between 100 and 75 m.y. arrives from the west at the correct time and in the correct position to clog the trench and push the Greater Antilles Arc northeastward through the Caribbean with simultaneous counterclockwise rotation. Perhaps somewhat before, but certainly by 60 m.y., the Sala y Gomez hotspot became active at or very close to a segment of the Pacific-Farallon ridge.

38 m.y. (anomaly 13). With the docking of the oceanic plateau against the Bahama platform, further northeastward motion is impeded. Subduction of the Farallon plate behind the plateau continues, however, forming the Central America arc. The present northern and southern transform boundaries of the Caribbean plate begin to develop at this time (Burke and others, 1978; Sykes and others, 1982). Farallon plate motion away from the Sala y Gomez hotspot forms the Nazca Ridge between 60 and 38 m.y. An abrupt change in Farallon plate motion relative to the hotspots at about 40 m.y. is reflected by the change in orientation from the Nazca Ridge to the Sala y Gomez Ridge, mirroring the change in Pacific plate motion that produced the Hawaiian-Emperor bend.

21 m.y. (anomaly 6). Left-lateral strike-slip motion of about 1200 km along the northern boundary (Cayman transform, Montagua-Polochic fault system) continues to allow eastward motion of the Caribbean plate relative to North America and eastward motion of the Chortis block from the Pacific side of

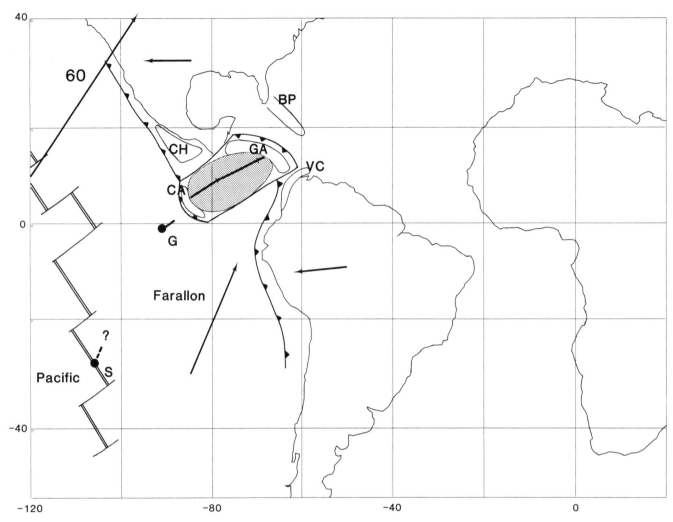

Figure 5. 60 m.y. B.P. plate positions and motions in the hotspot reference frame. CA is the Central America arc, formed behind the northeastwardly moving oceanic plateau. The Sala y Gomez hotspot (S) is active by this time.

nuclear Central America into the gap between Central and South America (Dickinson and Coney, 1980; Pindell and Dewey, 1982). In the east, renewed subduction of Atlantic crust beneath the Caribbean plate, which began earlier in response to this eastward motion, leads to the development of the volcanic arc of the Lesser Antilles. On the southern boundary, transcurrent motion is much more distributed.

At about 25 m.y. (between the time of magnetic anomalies 6 and 7), a reorganization of plate boundaries in the eastern Pacific leads to the ridge jump from the Galapagos Rise (interior to the present Nazca plate) to the present East Pacific Rise and splitting of the Farallon plate at the east-west trending Galapagos spreading ridge to form the Cocos plate (north) and the Nazca plate (south) (Handschumacher, 1976). From its position beneath a spreading ridge, the Galapagos hotspot now produces two volcanic lineaments, the Cocos Ridge and the Carnegie Ridge (Hey and others, 1977). The Sala y Gomez hotspot continues to pro-

duce the Sala y Gomez Ridge with gaps due to spreading ridge jumps.

Present. The present Caribbean plate is bounded to the north and south by transform faults (highly distributed on the south) and to the east and west by subduction zones. Its motion in the hotspot reference frame is now very slow (~1 cm/yr) (Minster and Jordan, 1978), although there is significant relative motion between it and North and South America and the Cocos plate (Sykes and others, 1982). Volcanic lineaments on the Cocos and Nazca plates emanating from the Galapagos and Sala y Gomez hotspots match well with those predicted from the plate motions we calculate.

In Figures 9a, 9b, and 9c, we illustrate calculations of plate interactions between the western margins of North and South America and ocean plates to the west. For the time interval 140 to 120 m.y., the Farallon plate converged with North America, but the motion of the Phoenix plate with respect to South Amer-

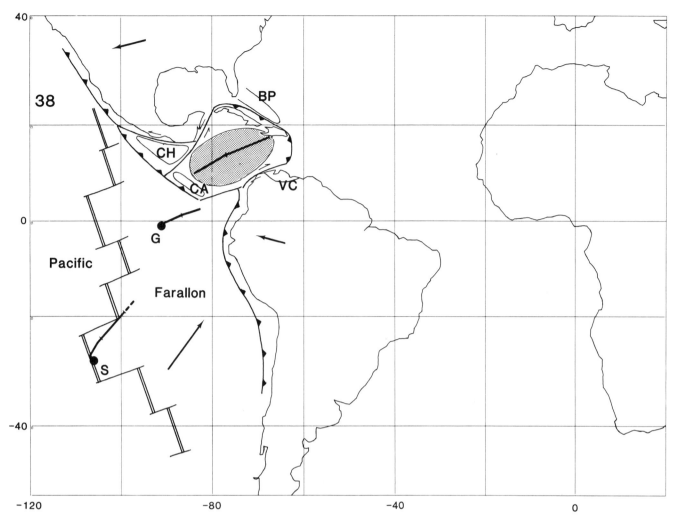

Figure 6. 38 m.y. B.P. plate positions and motions in the hotspot reference frame. Northeastward motion of the oceanic plateau is blocked by the Bahama Plateau (BP).

ica is uncertain. Relative motion vectors for Farallon-North America interaction are shown in Figure 9a. The vector lengths give the velocity of convergence, and their orientations give the direction of approach of the Farallon plate relative to the south-western margin of North America. Northeasterly convergence at about 10 cm/yr characterized the relative motion between these two plates during this period.

Calculation of the interaction of the Phoenix plate with South America depends critically on the Farallon-Phoenix spreading rate—which is unknown. Since the Farallon plate moved rapidly to the northeast during this period, a slow spreading rate (1 to 2 cm/yr) would imply that Phoenix plate motion was also moving to the northeast, but at a somewhat slower velocity. Such a situation requires northward motion of the Farallon-Phoenix ridge and primarily right-lateral transcurrent motion between the Phoenix and South America plates. A fast Farallon-Phoenix spreading rate (10 to 15 cm/yr), on the other

hand, would require southward motion of the Phoenix plate. Both alternatives are illustrated in Figure 9b.

Since the cessation of spreading on the Farallon-Phoenix ridge at about 119 m.y., the South America plate has converged with the Farallon plate (becoming the Nazca plate after 25 m.y.). In Figure 9c, vectors show the Farallon plate's motion with respect to northwestern South America from 110 m.y. to the present in 10-m.y. increments. A change in Pacific plate motion at about 100 m.y. led to rapid northeasterly convergence in the Caribbean region which linked the active subduction zones on the western margins of North and South America.

Convergence was rapid, 16 to 10 cm/yr, between 110 and 50 m.y. with only minor changes in direction. The Villa de Cura complex (northern Venezuela) and the Greater Antilles Arc formed in the early stages of this subduction. Between 50 and 40 m.y., another major change in Pacific plate motion (at the Hawaiian-Emperor bend) caused a slowing in convergence rate

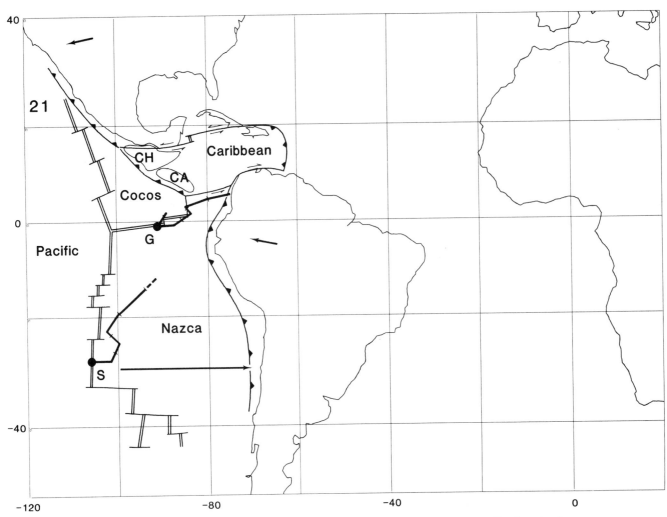

Figure 7. 21 m.y. B.P. plate positions and motions in the hotspot reference frame. The Farallon plate has broken into the Cocos and Nazca plates at about 25 m.y.

to 6 to 8 cm/yr. Subsequent breakup of the Farallon plate and spreading ridge reorientations have caused a swing to a more easterly direction of convergence.

CONCLUSIONS

We have calculated the motions of plates bordering the Caribbean region in the hotspot (= mantle) reference frame for Jurassic time to the present. As shown earlier by Pindell and Dewey (1982), the size of the Caribbean through time is given directly by the amount of North America-South America separation or closure, each referred to African plate motion by seafloor spreading. Using the hotspot reference frame, we quantify convergence rates and directions for subduction of oceanic plate formed in the Pacific basin beneath North and South America. This obviates many of the errors attending the relative motion circuit through Antarctica.

As previously proposed (Burke and others, 1978), the unus-

ually thick core of the Caribbean plate probably formed as an oceanic plateau in the Pacific basin and was thrust into the Caribbean region from the west. We find that such a plateau could have formed over the Galapagos hotspot in late Cretaceous time (100-75 m.y.) on oceanic crust formed earlier at the Farallon-Phoenix ridge. Subsequent northeastward motion of the Farallon plate delivered this plateau to the Caribbean where it pushed the Greater Antilles Arc northeastward into collision with the Bahama platform in Eocene time. Deep-sea drilling in the Caribbean through layer B" to sediments and oceanic basement would provide a test of the proposed age relationships.

While some basalts from the Caribbean ocean crust exhibit attributes of hotspot volcanism (Bence and others, 1975), Batiza and others (1980) have pointed out chemical similarities between many Caribbean B" basalts and sills and lava flows in the Nauru Basin, and at the Manihiki and Ontong Java plateaus, in the western Pacific. On the basis of trace element concentrations, these basalts are chemically distinct from both normal MORB's

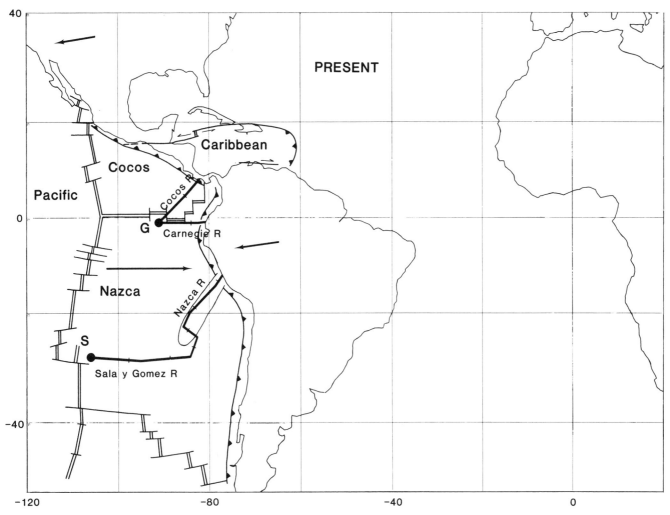

Figure 8. Present plate motions around the Caribbean region. Predicted hotspot tracks emanating from the Galapagos (G) and Sala y Gomez (S) hotspots match well with the known aseismic ridges on the Cocos and Nazca plates.

and tholeiites related to hotspot activity (Hawaii, Iceland) and have been termed ocean-plateau tholeiites (Tokuyama and Batiza, 1981). The crustal age and geophysically inferred structure of the Pacific oceanic plateaus are also similar to the central Caribbean (Burke and others, 1978; Larson and Schlanger, 1981). Batiza and others (1980) speculate that large portions of the western Pacific Ocean, formed in the southeast Pacific in late Jurassic/early Cretaceous time, may have been constructed by additions of basaltic sills and flows of similar compositions, erupted in an off-ridge tensional setting. These plateaus, including the Caribbean, may owe their origin to a widespread mid-Cretaceous volcanic "event" such as that proposed by Haggerty and others (1982) or to plate motion over hotspots as modeled by Henderson and Gordon (1981).

Further work on the geochemical composition of layer B" basalts and comparison with rocks erupted at the Galapagos Islands is needed to test that predicted correlation with hotspot activity. Additional research on ophiolitic sections in Caribbean borderlands is required to define the possible correlation of these rocks and the timing of their emplacement on continental margins.

Our plate motions, based on hotspots underlying the African and Pacific plates, predict hotspot tracks which match well with observed volcanic lineaments. As proposed by Morgan (1981), the Bahama platform may be reasonably related to latest Jurassic volcanic activity at the Fernando de Noronha hotspot. Drilling through the thick carbonate cap at several sites would reveal the nature of the pedestal rocks. The Sala y Gomez hotspot could have produced the Nazca Ridge and the Sala y Gomez Ridge between roughly 60 m.y. and the present. Radiometric dating of dredged material from these lineaments would test the predicted ages. And the Galapagos hotspot has been producing paired volcanic ridges since the breakup of the Farallon plate into the Cocos and Nazca plates.

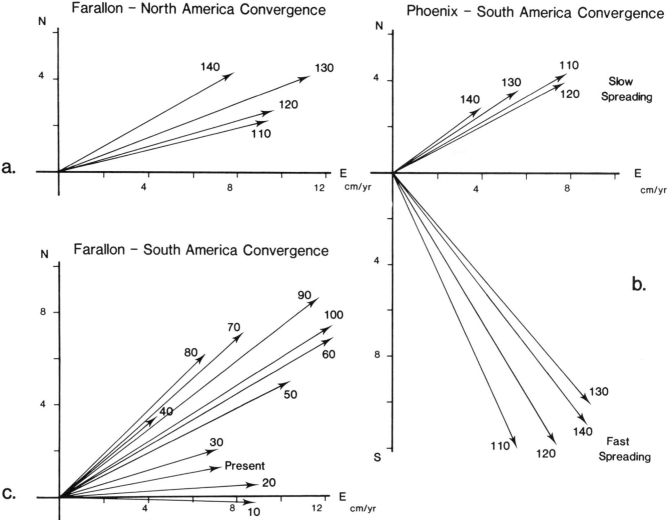

Figure 9. The relative motion between the eastern Pacific oceanic lithosphere and the Central America region from 140 m.y. to the present is represented by vectors whose orientations give the direction of motion and whose lengths give the rate of that motion. From 140 to 110 m.y., the relative motion between the Farallon plate and southwestern North America was primarily convergent (a). During the same period, relative motion between the Phoenix plate and northwestern South America could have been convergent or right-lateral transcurrent, depending on whether spreading on the Farallon-Phoenix ridge was fast or flow, respectively (b). From 110 to 40 m.y. the Farallon plate converged with northwestern South America in northeasterly directions, and from 40 m.y. to the present in easterly directions (c).

ACKNOWLEDGMENTS

This research was partially supported by NSF grant OCE80-24284 and the Office of Naval Research (R.A.D.) and by NSF grant EAR80-08207 (R.B.H.).

REFERENCES CITED

Batiza, R., Larson, R. L., Schlanger, S. O., Shcheka, S. A., and Tokuyama, H., 1980, Trace element abundances in basalts of Nauru Basin: Nature, v. 286, p. 476–478.

Beets, D. J., Maresch, W. V., Klaver, G. Th., Mottana, A., Bocchio, R., and

Beunk, F. F., 1984, Magmatic rock series and high-pressure metamorphism as constraints on the tectonic history of the southern Caribbean: *in* Bonini, W. E. and others, eds., The Caribbean-South American Plate Boundary and Regional Tectonics: Geological Society of America Memoir 162 (this volume).

Bence, A. E., Papike, J. J., and Ayuso, R. A., 1975, Petrology of submarine basalts from the Central Caribbean: DSDP Leg 15: Journal of Geophysical Research, v. 80, p. 4775–4804.

Burke, K., Fox, P. J., and Sengor, A.M.C., 1978, Buoyant ocean floor and the evolution of the Caribbean: Journal of Geophysical Research, v. 83, p. 3949–3954.

Christofferson, E., 1976, Colombian basin magnetism and Caribbean plate tectonics: Geological Society of America Bulletin, v. 87, p. 1255–1258.

Cox, A., 1983, Magnetostratigraphic time scale, in Harland, B., and others eds.,

The Geological Time Scale: Cambridge University Press (in press).

Dalziel, I.W.D., 1982, The early (pre-middle Jurassic) history of the Scotia Arc region: A review and progress report, in Craddock, C., ed., Antarctic Geoscience: Madison, Wis., University of Wisconsin Press, p. 111–126.

Dickinson, W. R., and Coney, P. J., 1980, Plate tectonic constraints on the origin of the Gulf of Mexico, in Pilger, R. H., Jr., ed., The origin of the Gulf of Mexico and the early opening of the central North Atlantic Ocean: Houston Geological Society Continuing Education Series, School of Geoscience, Louisiana State University, Baton Rouge, La., p. 27–36.

Dietz, R. S., and Holden, J. C., 1973; Geotectonic Evolution and subsidence of Bahama Platform: Reply; Geological Society of America Bulletin, v. 84, p. 3477–3482.

Donnelly, T. W., 1973, Magnetic anomaly observations in the eastern Caribbean Sea, in Edgar, N. T., and others, eds., Initial Reports of the Deep Sea Drilling Project, v. 15: Washington, D.C., U.S. Government Printing Office, p. 1023–1030.

—— 1975, The geological evolution of the Caribbean and the Gulf of Mexico, in Nairn, A.E.M., and Stehli, F. G., eds., The Ocean Basins and Margins, v. 3, The Gulf of Mexico and the Caribbean: New York, Plenum Press, p. 663–689.

Donnelly, T. W., and Rogers, J., 1978, The distribution of igneous suites throughout the Caribbean: Geologie en Mijnbouw, v. 57, p. 151–162.

Donnelly, T. W., and others, 1973, Basalts and dolerites of Late Cretaceous age from the Central Caribbean, in Edgar, N. T., and others, eds., Initial Reports of the Deep Sea Drilling Project, v. 15: Washington, D.C., U.S. Government Printing Office, p. 989–1011.

Duncan, R. A., 1981, Hotspots in the southern oceans—an absolute frame of reference for motion of the Gondwana continents: Tectonophysics, v. 74, p. 29–42.

Edgar, N. T., Ewing, J. I., and Hennion, J., 1971, Seismic refraction and reflection in Caribbean Sea: American Association of Petroleum Geologists Bulletin, v. 55, p. 833–870.

Ghosh, N., Hall, S. A., and Casey, J. F., 1984, Seafloor spreading magnetic anomalies in the Venezuelan Basin: in Bonini, W. E. and others, eds., The Caribbean-South American Plate Boundary and Regional Tectonics: Geological Society of America Memoir 162 (this volume).

Gordon, R. G., and Cox, A., 1980, Paleomagnetic test of the early Tertiary plate circuit between the Pacific basin plates and the Indian plate: Journal of Geophysical Research, v. 85, p. 6534–6546.

Haggerty, J. A., Schlanger, S. O., and Premoli-Silva, I., 1982, Late Cretaceous and Eocene volcanism in the southern Line Islands and implications for hotspot theory: Geology, v. 10, p. 433–437.

Handschumacher, D. W., 1976, Post-Eocene plate tectonics of the eastern Pacific, in Sutton, G. H., and others, eds., The Geophysics of the Pacific Ocean Basin and its margin: Washington, D.C., American Geophysical Union Monograph 19, p. 177–202.

Hargraves, R. B., and Skerlec, G. M., 1980, Paleomagnetism of some Cretaceous-Tertiary igneous rocks on Venezuelan offshore islands, Netherlands Antilles, Trinidad and Tobago: Ninth Caribbean Geological Congress, v. 1, p. 509–518.

Henderson, L. J., and Gordon, R. G., 1981, Oceanic plateaus and the motion of the Pacific plate with respect to the hotspots: Transactions American Geophysical Union, v. 62, p. 1028.

Hey, R., Johnson, G. L., and Lowrie, A., 1977, Recent plate motions in the Galapagos: Geological Society of America Bulletin, v. 88, p. 1385–1403.

Hilde, T.W.C., Uyeda, S., and Kroenke, L., 1976, Tectonic history of the western Pacific, in Drake, C. L., ed., Geodynamics: Progress and Prospects: Washington, D.C., American Geophysical Union, p. 1–15.

Hopkins, H., 1973, Geology of the Aruba Gap abyssal plain near DSDP site 153, in Edgar, N. T., and others, eds., Initial Reports of the Deep Sea Drilling Project, v. 15: Washington, D.C., U.S. Government Printing Office, p. 1039–1050.

Houtz, R., and Ludwig, W. J., 1977, Structure of the Colombia basin, Caribbean Sea, from profiler sonobuoy measurements: Journal of Geophysical Research, v. 82, p. 4861–4867.

Jurdy, D. M., 1979, Relative plate motions and the formation of marginal basins: Journal of Geophysical Research, v. 84, p. 6796–6802.

Klitgord, K. D., and Mammerickx, J., 1982, Northern East Pacific Rise: Magnetic anomaly and bathymetric framework: Journal of Geophysical Research, v. 87, p. 6725–6750.

Larson, R. L., and Schlanger, S. O., 1981, Cretaceous volcanism and Jurassic magnetic anomalies in the Nauru basin, western Pacific Ocean: Geology, v. 9, p. 480–484.

Malfait, B. T., and Dinkelman, M. G., 1972, Circum-Caribbean tectonic and igneous activity and the evolution of the Caribbean plate. Geological Society of America Bulletin, v. 83, p. 251–272.

Mattson, R. M., 1969, The Caribbean: a detached relic of Darwin Rise: Transactions American Geophysical Union, v. 50, p. 317.

—— 1979, Subduction, buoyant breaking, flipping and strike-slip faulting in the northern Caribbean: Journal of Geology, v. 87, p. 293–304.

Minster, J. B., and Jordan, T. H., 1978, Present day plate motions: Journal of Geophysical Research, v. 83, p. 5331–5354.

Mooney, W. D., 1980, An East Pacific-Caribbean ridge during the Jurassic and Cretaceous and the evolution of western Colombia, in Pilger, R. H., Jr., ed., The origin of the Gulf of Mexico and the early opening of the central North Atlantic Ocean: Houston Geological Society Continuing Education Series, School of Geoscience, Louisiana State University, Baton rouge, La., p. 55–74.

Morgan, W. J., 1981, Hotspot tracks and the opening of the Atlantic and Indian Oceans, in Emiliani, C., ed., The Sea, v. 7: New York, Interscience-Wiley, p. 443–475.

Officer, C., Ewing, J., Hennion, J., Harkinder, D., and Miller, D., 1959, Geophysical investigations in the eastern Caribbean—Summary of the 1955 and 1956 cruises, in Ahrens, L. M., Press, F., Rankama, K., and Runcorn, S. K., eds., Physics and Chemistry of the Earth, v. 3: New York, Pergamon, p. 17–109.

Pindell, J., and Dewey, J. F., 1982, Permo-Triassic reconstruction of western Pangea and the evolution of the Gulf of Mexico/Caribbean region: Tectonics, v. 1, p. 179–211.

Schmid-Effing, R., 1979, Geodynamic history of the oceanic crust in Southern Central America, Fourth Latin American Geological Congress: Geologische Rundschau, v. 68, p. 457–492.

Skerlec, G. M., and Hargraves, R. B., 1980, Tectonic significance of paleomagnetic data from northern Venezuela: Journal of Geophysical Research, v. 85, p. 5303–5315.

Stearns, C., Mauk, F. J., and Van der Voo, R., 1982, Late Cretaceous-Early Tertiary paleomagnetism of Aruba and Bonaire (Netherlands Leeward Antilles): Journal of Geophysical Research, v. 87, p. 1127–1141.

Sykes, L. R., McCann, W. R., and Kafka, A. L., 1982, Motion of the Caribbean plate during the last 7 million years and implications for earlier Cenozoic movements: Journal of Geophysical Research, v. 87, p. 10656–10676.

Tokuyama, H., and Batiza, R., 1981, Chemical composition of igneous rocks and origin of the sill and pillow basalt complex of Nauru Basin, southwest Pacific, in Larson, R. L., Schlanger, S. O. et al., eds., Initial Reports, Deep Sea Drilling Project, v. 61; Washington, D.C., U.S. Government Printing Office, p. 673–687.

MANUSCRIPT ACCEPTED BY THE SOCIETY SEPTEMBER 1, 1983

Printed in U.S.A.

Geological Society of America
Memoir 162
1984

Magmatic rock series and high-pressure metamorphism as constraints on the tectonic history of the southern Caribbean

Dirk J. Beets*
Geologisch Instituut
Universiteit van Amsterdam
Nieuwe Prinsengracht 130
Amsterdam, The Netherlands

Walter V. Maresch
Institut für Mineralogie
Ruhr-Universität Bochum
463 Bochum, F.R. Germany

Gerard Th. Klaver
Geologisch Instituut
Universiteit van Amsterdam
Nieuwe Prinsengracht 130
Amsterdam, The Netherlands

Annibale Mottana
Istituto di Mineralogia e Petrografia
Università
Piazzale delle Scienze 5
00185 Roma, Italy

Rosangela Bocchio
Istituto di Mineralogia
Petrografia e Geochimica
Università
Via Botticelli 23
20133 Milano, Italy

Frank F. Beunk
Geologisch Instituut
Universiteit van Amsterdam
Nieuwe Prinsengracht 130
Amsterdam, The Netherlands

Hendrik P Monen
Geologisch Instituut
Universiteit van Amsterdam
Nieuwe Prinsengracht 130
Amsterdam, The Netherlands

ABSTRACT

The contact relationships between metabasalts (eclogites, glaucophanites, prasinites, etc.) and the enclosing mica schists in the Venezuelan Coast Ranges favour a common petrological history for both. Mineralogical disequilibria, such as replacement textures and mineral zoning in the metabasic assemblages, can all be related to a single metamorphic cycle. Relict deuteric/late magmatic hornblende epitaxially overgrown by the barroisitic amphibole of this metamorphic event shows that the latter has been the only regional metamorphic episode to have affected these rocks. The high-pressure character of the metamorphism is a logical consequence of overthrusting related to collision of the Aruba-Blanquilla island arc with the South American continental margin. The well-defined stratigraphy and detailed radiometric dating of intrusive rocks in the Netherlands Antilles indicate that this collision took place in the Coniacian/Campanian interval.

The igneous rocks of the Netherlands Antilles, which are considered to form part of the colliding arc, consist largely of submarine volcanics, as well as a tonalite/gabbro batholith. These rocks range in age from middle Albian to Coniacian. The volcanics of Curaçao and Aruba are composed of basalts with a MORB chemistry and are oceanic in origin. Nevertheless, these sequences differ from "normal" oceanic crust by their thickness, chemical homogeneity and the non-depleted nature of the source, thus suggesting that they were fed by a prolific chondritic mantle plume. The volcanics of Bonaire range from basalt to rhyolite in composition and are chemically related to the primitive island arc series. An important characteristic is the high initial water content of these magmas,

---

*Present address: Geological Survey of the Netherlands, Spaarne 17, Haarlem, The Netherlands.

as shown by the geometry and mineralogy of the flows. The tonalite/gabbro batholith on Aruba is of calc-alkaline composition. The similarity in chemistry between the volcanics of Bonaire and the Villa de Cura Group of the Venezuelan mainland supports the view that the latter is an overthrust remnant of the colliding arc. Comparison with the metabasalts of the La Rinconada Group of Margarita Island is equivocal.

Existing paleomagnetic evidence indicates that the arc underwent a 90° north-south to east-west rotation shortly before collision, and that the colliding arc extended via the Aves Ridge into the Greater Antilles. The age of the oldest volcanics of this arc, and thus the age of its origin, is uncertain. Most data favour formation in the Early Cretaceous, but a Late Jurassic age is also possible. Consequently, two alternative models for the evolution of the arc are proposed: one in which the arc forms as a lengthening Central American "isthmus" in response to opening of the Caribbean in the Late Jurassic, and a second in which the arc originates in the Pacific in the Early Cretaceous. In order to collide with the northern as well as southern margin of the Caribbean, the arc must have lengthened to more than twice the width of the gap. Lengthening probably occurred at a trench-trench-ridge triple junction, thus explaining the large volume of MORB-volcanics in the arc. The high water content of the primitive island arc series is attributed to a high rate of subduction due to the worldwide surge in spreading rates in the mid-Cretaceous. In the second model, a strong causal connection between the mid-Cretaceous thermal event in the Pacific as well as the origin and evolution of the arc/trench system is postulated.

INTRODUCTION

A consideration of the present-day, right lateral strike-slip nature of the boundary between the Caribbean and South American plates, and the Mesozoic high-pressure metamorphic assemblages found there, makes it clear that there must be a fundamental difference between the present tectonic regime and the one active in the Mesozoic. While geophysical methods can be used to ascertain present-day tectonics, the Mesozoic situation can be reconstructed only by concise study of the sedimentary, metamorphic, and volcanic rocks of that age. Such studies in the southeast Caribbean are hampered by the fact that the relative positions of various critical rock units have been drastically changed both during Mesozoic tectonic activity and in the Tertiary. The unfossiliferous and petrologically often monotonous nature of the rocks involved makes the correlation and differentiation necessary for reconstructing the Mesozoic situation difficult and has given rise in recent years to a number of interpretations (e.g., Bell, 1971, 1972; Bellizzia, 1972; Maresch, 1974; Beets, 1975; Stephan and others, 1980; Skerlec and Hargraves, 1980; Talukdar and Loureiro, 1982).

The basic premise of a compressive plate margin along the Caribbean/South American juncture is the common theme linking most of the above-mentioned paleotectonic proposals. A recurring motif is the proposition that one or more island arcs collided with the South American craton during the Late Mesozoic, and that large-scale southward thrusting and/or gravity sliding have occurred along the cratonic margin during and after these collisions. Nevertheless, despite broad agreement on the basic paleotectonic framework, there is considerable debate on a number of important aspects. How many island-arc collisions occurred and when? Where did these arcs originate? Is the chain

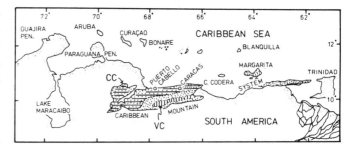

Figure 1. Location map of the Caribbean-South American borderland. Aruba, Curaçao, and Bonaire comprise the Netherlands Antilles. Heavy stippling and ruling: Caribbean Mountain System. Subdivisions are Coast Ranges Tectonic Belt (CC, horizontal ruling) and Villa de Cura Tectonic Belt (VC, vertical ruling).

of predominantly volcanic islands from Aruba to Blanquilla (Fig. 1) a remnant of such an arc, or is it a welt of upthrust Caribbean ocean floor? How can the huge volumes of basic to acid volcanics represented by the Villa de Cura Group of the Caribbean Mountain System and the La Rinconada Group of Margarita Island (Fig. 1) be related to such paleotectonic schemes?

In the present paper, we propose to work backwards by reexamining first the problem of high-pressure metamorphism in the area and then discussing the geochemistry of the magmatic rocks of Aruba, Bonaire, and Curaçao, as well as those of the Villa de Cura and La Rinconada Groups. Regional high-pressure metamorphism requires anomalous geothermal conditions such as are to be expected in subduction zones as well as collision zones characterized by overthrusting and crustal thickening. The geochemistry of magmatic rocks is strongly influenced by the tectonic environment of magma generation, differentiation, and crystallization. Thus the nature and timing of a high-pressure metamorphic imprint and the type, distribution, and age of original magmatic activity must have been dictated by the type of plate interaction that has occurred.

Because this study deals in large part with metavolcanics, the problems inherent in chemical analysis and radiometric age dating of such rocks are of concern. It is well known that seafloor weathering and hydrothermal activity, as well as metamorphic processes, can alter the relative proportions of major and trace elements. Thus the metavolcanic rocks analyzed may no longer reflect accurately the original magma compositions, and changes in daughter/parent ratios in radiometric systems can lead to spurious ages. Detailed discussions of the extent of possible alteration in the analyses discussed in this paper are given in Klaver and others (in preparation) and Mottana and others (1979). Enough analyses are available from the Dutch Leeward Islands and from Margarita Island so that those presented here may be accepted with confidence. Radiometric age determinations, on the other hand, remain a grave problem in the Southern Caribbean. Detailed work on the Aruba batholith and paleontological control on Curaçao, to be discussed more fully below, confirm that many of the Rb/Sr and K/Ar determinations available are probably distorted to varying degrees. Such dates are used, albeit with caution, in the following discussions because it seems unrealistic to ignore them completely.

Much of the evidence discussed is, by its very nature, of permissive and supporting, rather than of unequivocally determinative character. Our purpose is to outline the boundary constraints to which the solution of the problem must conform. Beyond this, only sound geological reasoning will lead to new insights. This paper is an attempt to find a common paleotectonic framework for the detailed geological, petrological, mineralogical, geochemical, and geophysical knowledge of three independent working groups in the Caribbean/South American borderland. A positive aspect of this endeavour has been that, despite the need to compromise on a number of interpretations, we could reach basic agreement on the main theme presented here. We are convinced that periodic assessments of this type are necessary in order to outline those directions of future research that will most profitably lead to further progress in our understanding of the area.

LATE MESOZOIC METAMORPHISM

Two discrete belts of high-pressure metamorphic rocks may be recognized in the Caribbean Mountain System of northern Venezuela. One, the Villa de Cura Belt (Fig. 1), is an allochthonous unit of metavolcanics and intercalated metasediments, approximately 250 km by 30 km in areal extent, enclosed by sediments and very low-grade metasediments. The alkali amphiboles glaucophane and crossite, as well as lawsonite, attest to the low-grade blueschist nature of regional metamorphism in the Villa de Cura Group (Piburn, 1967). The second belt is more difficult to define in terms of areal extent. It is typified mainly by metamorphosed mafic dykes, sills, and irregular lensoid blocks that are found within the volumetrically dominant metasediments of the Coast Ranges scattered in a narrow, 5-20 km-wide coastal strip from Cabo Codera (and Margarita Island) in the east to approximately Puerto Cabello in the west. Typical high-pressure assemblages include eclogite and barroisite- (less commonly glaucophane-, crossite-, or winchite-) bearing amphibolites (e.g., Morgan, 1970; Maresch and Abraham, 1981; Talukdar and Loureiro, 1982; Maresch and others, 1982). This coastal strip has been termed the "frange côtière-Margarita" by Stephan and others (1980), and interpreted to be a melange zone by Talukdar and Loureiro (1982).

In recent years, the concept of "polymetamorphism" of the high-pressure assemblages found in the Coast Ranges and Margarita Island has gained wide acceptance (e.g., Blackburn and Navarro, 1977; Martín, 1978; Stephan and others, 1980; Talukdar and Loureiro, 1982). For Stephan and others, as well as Talukdar and Loureiro, this "polymetamorphism" is important evidence for their postulate of a regional high-pressure metamorphic event in the Late Jurassic to Early Cretaceous, followed by greenschist metamorphism in the Late Cretaceous. Repeatedly, the presence of "high-grade" eclogite or garnet amphibolite in "low-grade" greenschists is taken as evidence for the present-day allochthonous position of the former (e.g., Vignali, 1979; Stephan and others, 1980). Because these concepts have a decided influence on Late Mesozoic tectonic models in the region, their justification must be considered in more detail.

Most workers at present favour an Early Cretaceous or older age of metamorphism for the Villa de Cura Belt (e.g., Maresch, 1974; Stephan and others, 1980; Talukdar and Loureiro, 1982). This interpretation is based on a single K-Ar whole-rock date of 100 ± 10 Ma on actinolite-rich metatuff reported by Piburn (1967), but recent determinations of 107 ± 3 Ma, 99.2 ± 3 Ma, and 97.5 ± 3 Ma on hornblende of an unmetamorphosed intrusive into the Villa de Cura Group (Hebeda and others, this volume) have provided corroboration.

The inferred age of metamorphism thus would coincide with the Late Jurassic to Early Cretaceous event postulated by Stephan and others, as well as Talukdar and Loureiro, for the Coast Ranges. We agree with a pre-mid-Cretaceous metamorphic age in the Villa de Cura Belt. As will be developed more fully in this paper, however, we postulate that the Villa de Cura suite represents an overthrust remnant of a colliding island arc, and that the blueschist metamorphic imprint observed stems from pre-collision subduction processes along this arc. Moreover, it is our intent to show in the following section that the Coast Ranges have been subjected to *only one* regional metamorphic episode, which took place in the Late Cretaceous.

"Allochthonous" Origin of High-Pressure Assemblages in Metabasic Rocks.

An important dispute centers on the applicability of pressure and temperature estimates obtained on high-pressure metabasic assemblages to the enclosing, mineralogically non-determinative mica schists. Both geological and petrological lines of reasoning can be followed. The former involve the nature of the contact between eclogite/glaucophanite/garnet amphibolite/prasinite and the enclosing mica schist, the latter the compatibility in metamorphic grade between metabasalt and enclosing metasediment.

There is no doubt that many of the high-pressure assemblages are now found as irregular "blocks" dispersed in the ubiquitous mica schist of the Coast Ranges, but it is highly debatable as to whether this can be taken as evidence either for the synsedimentary deposition or the tectonic injection of such blocks already in metamorphosed form (cf. Vignali, 1979; Talukdar and Loureiro, 1982; Stephan and others, 1980). The differential response of interlayered basalt and sediment to intense folding and deformation can easily lead to an "exotic block" aspect. In fact, where folding has been somewhat less intense, contact relations typical of sills and crosscutting dykes from 15 to 200 cm or more thick are not uncommon (e.g., Punta Corey and Playa Caribe on Margarita Island). The contact between the metabasic La Rinconada Group of Margarita Island and overlying metasedimentary units is clearly gradational and involves interlayering of metavolcanics and metasediments (Maresch, 1975).

Petrological arguments sustain an *in situ* interpretation. Morgan's (1970) elegant petrological analysis showed that the eclogites, garnet amphibolites, and epidote amphibolites of Puerto Cabello are in fact isofacial (i.e., metamorphosed under the same load-pressure and temperature conditions). A similar conclusion has been reached by Maresch (1977) on Margarita Island. The observed differences in mineralogy are due to variations in the composition of the metamorphic fluid phase and/or the bulk rock (e.g., Newton and Fyfe, 1976; Schliestedt, 1979).

With only minor exceptions, all eclogites from the Coast Ranges that have been studied in detail are typical of Smulikowski's (1969, 1972) "ophiolite" group, with some tendency toward his "common" group (e.g., Morgan, 1970; Navarro, 1974; Maresch, 1971; Maresch and Abraham, 1981). Such eclogites are found on a worldwide scale to be associated typically with rocks of no higher grade than blueschist, greenschist, or epidote-amphibolite. The greenschist to epidote-amphibolite grade of the Venezuelan Coast Ranges is thus completely in accord with the type of eclogite found there. Tectonic juxtaposition as suggested by Talukdar and Loureiro (1982) and Stephan and others (1980) need not be invoked.

Thus, there is good evidence for interpreting most of the small bodies of eclogites, glaucophanites, garnet amphibolites, epidote amphibolites, and prasinites (greenschists) of the Venezuelan Coast Ranges as basaltic sills and dykes intruded into sediment, which were later deformed and metamorphosed together with the enclosing sediment to the assemblages found today.

This conclusion does not rule out the presence of allochthonous units in the Caribbean Mountain System, but it requires that such units be much larger than the small blocks of metabasaltic rocks in question here.

"Polymetamorphism"

This term, though often employed, is actually not very useful for the problem under discussion. In any metamorphic episode, a rock must follow a cycle in which it is heated and compressed to some maximum value and subsequently cooled and decompressed to surface conditions where it is examined. It is commonly assumed (see Miyashiro, 1973, p. 56) that the resultant mineral assemblage reflects an equilibrium frozen in at the highest temperature reached. The reason is that dehydration reactions during prograde metamorphism will be rapid as compared to retrograde rehydration, which suffers from the lack of the rate-enhancing H_2O that has been lost during the prograde path. However, as Norris and Henley (1976) and Wood (1979) have pointed out, prograde high-pressure, low-temperature metamorphism need not involve dehydration. Furthermore, eclogites, representing an "H_2O-poor" equilibrium compared with most non-eclogitic country rocks, will serve as active sinks for available H_2O during both prograde and retrograde paths. They will be particularly susceptible to continuing recrystallization during the whole metamorphic cycle. Disequilibrium and overprint textures will then simply reflect the P/T path of the eclogite during a single cycle of metamorphism. It is important to differentiate this *single-cycle overprinting* from the effects caused by *two or more, discrete metamorphic* cycles.

For this purpose, petrological and mineralogical disequilibrium features in high-pressure metamorphic rocks on Margarita Island—a geological continuation of the Coast Ranges—have been analyzed (Maresch and Abraham, 1981; Maresch and others, 1979). Both *in situ* occurrences in the >2000 m metavolcanic sequence of the La Rinconada Group (Maresch, 1975), as well as isolated occurrences in the mica schists of the overlying Juan Griego Group (Rudolph, 1981; Maresch and others, 1982) have been studied. The latter are deemed to be petrologically analogous to the Coast Range examples.

Well-studied examples of overstepped, low-variance reactions are the following (Maresch and Abraham, 1981):

$$\text{omphacite + paragonite + garnet + rutile} \rightarrow$$
$$\text{albite + epidote + barroisite + sphene} \quad (1)$$
$$\text{barroisite + paragonite} \rightarrow \text{albite + epidote + chlorite} \quad (2)$$

which lead to spectacular disequilibrium textures such as those shown in Figure 2. Such textures are ubiquitous in the eclogites of Margarita (Maresch, 1972; Navarro, 1974, 1977; Maresch and Abraham, 1981; Rudolph, 1981), and probably in those of the Coast Ranges (Martín, 1978; Talukdar and Loureiro, 1982) and Guajira Peninsula (Green and others, 1968) as well. Reaction (2) may also be observed in banded amphibolites (Maresch, 1971).

As discussed in detail by Maresch and Abraham (1981), all evidence points to the fact that these reaction relations, which ultimately would lead to an epidote-amphibolite facies assemblage without omphacite and paragonite, are due to late-stage depressurization effects during exhumation of the deeply buried

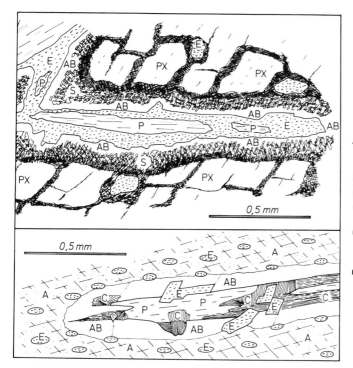

Figure 2. Thin section sketches showing prominent replacement textures caused by reactions (1) (top) and (2) (bottom). *Top*: paragonite (P) crystals replaced by epidote (E) and albite (AB). Adjacent omphacite (PX) altered to epidote and symplectite (S) of amphibole and albite. *Bottom*: paragonite (P) replaced by epidote (E), chlorite (C), and separated from amphibole (A) by secondary albite (AB). Diablastic amphibole + epidote matrix surrounding paragonite porphyroblast is schematic.

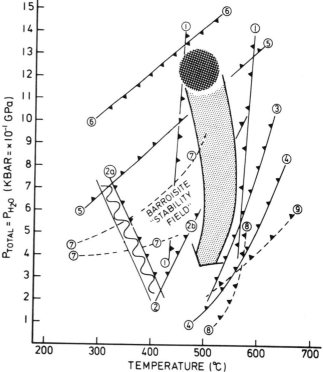

Figure 3. Estimate of P_{total}-T conditions of crystallization and late-stage depressurization path of an eclogite exhibiting reaction textures of Figure 2 caused by reactions (1) and (2) (see text). Saw teeth on various constraints show direction compatible with observed assemblage. Curves: (1) *P-T* interval indicated by Fe^{2+}/Mg distribution between omphacite and garnet; (2) estimates based on muscovite-paragonite solvus; (3) upper stability of antigorite (Evans and others, 1976) found in associated ultramafics; (4) upper stability of paragonite + quartz (Chatterjee, 1972); (5) low-*P* stability of omphacite according to Currie and Curtis (1976); (6) low albite = pure jadeite + quartz; (7) stability field of barroisite amphibole (Ernst, 1979); (8) upper stability limit of chlorite in basaltic system (Liou and others, 1974; QFM buffer); (9) upper stability limit of epidote + quartz (Liou, 1973; QFM buffer). Figure taken from Maresch and Abraham (1981).

material during a single metamorphic cycle. Figure 3 summarizes the late-stage load-pressure-temperature-time (P-T-t) path deduced by Maresch and Abraham (1981) for an eclogite from the north coast of Margarita. The temperature increase during depressurization is probably due to heating during initial uplift caused by concurrent reestablishment of a normal geothermal gradient.

Amphibole zoning in the La Rinconada Group of Margarita Island supports the P-T-t path shown in Figure 3. It has been found on a regional scale (Maresch and others, 1979; Maresch and Abraham, 1981), that the evolution of amphibole composition is from cores of relatively high-pressure barroisite to rims of lower-pressure hornblende, both in the eclogites and in the enclosing metabasalt. The barroisite-hornblende transition is relatively sharp and usually less than 5 μm wide. The increase in the temperature-sensitive tschermakite and edenite components (the pargasite trend), coupled with the decrease in the pressure-sensitive glaucophane/crossite components from core to rim (Fig. 4), again suggests a significant pressure decrease concurrent with still rising temperatures during the time interval between core and rim growth (e.g., Holland and Richardson, 1979; Laird and Albee, 1981). There is no evidence to suggest that the P—T conditions favouring barroisitic core growth were *not* directly

followed by the P-T conditions favouring hornblende rim growth, as would otherwise be expected in multi-cycle metamorphism.

In a few samples, where the late-stage kinetics were particularly favourable for amphibole growth, further evolution of the lower-pressure hornblende towards actinolite mirrors the ultimate return to near-surface conditions and completion of the metamorphic cycle.

Talukdar and Loureiro (1982) also describe zoned amphiboles from Coast Range eclogites and amphibolites. No analytical data are given, but "glaucophane" is reported to be rimmed by "barroisite" or "actinolite." In broad terms, this sequence also reflects evolution from high-pressure/low-temperature through high-pressure/higher-temperature to low-pressure/low-temperature conditions (cf. Holland and Richardson, 1979), but perhaps at lower absolute temperatures than on Margarita (cf. Ernst, 1979). Nevertheless, further studies on these latter occurrences

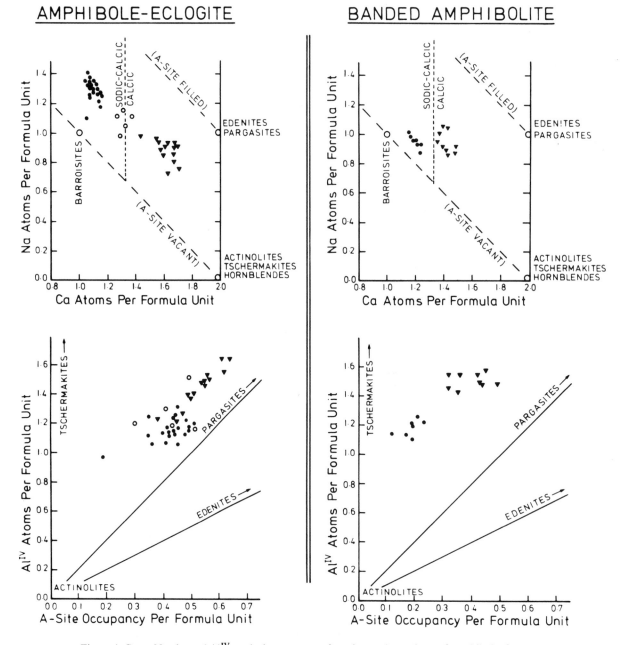

Figure 4. Ca *vs.* Na plot and Al^{IV} *vs.* A-site occupancy for microprobe analyses of amphiboles from an amphibole-eclogite (see Maresch and Abraham, 1981) and a banded amphibolite (sample location IB of Mottana and others, 1979) from the La Rinconada Group (Margarita Island) to show variations in tschermakite, edenite, and glaucophane/crossite content. Latter plots at Na = 2.0, Ca = 0.0. End-member terminology after Leake (1978). Filled circles: cores; filled triangles: rims; open circles: core-rim transition zone where amenable to analyses. Fe^{3+}/Fe^{2+} calculated as described by Maresch and Abraham (1981).

are necessary, for amphibole compositions are notoriously susceptible to the influence of local bulk chemistry, especially when replacing pyroxene in eclogites (e.g., Ungaretti and others, 1981).

Many garnets in metabasites from Margarita Island and the Coast Ranges exhibit distinct growth zones. Commonly an inclusion-rich core abruptly gives way to an inclusion-free rim (Blackburn and Navarro, 1977; Rudolph, 1981; Talukdar and

Loureiro, 1982). Blackburn and Navarro (1977) have correlated this zone boundary with a small compositional discontinuity in garnets from several eclogites and amphibole eclogites from Margarita Island. However, the corresponding Fe/Mg fractionation data between garnet zones and coexisting pyroxene also given by these authors yield a difference in equilibration temperature across the zone boundary of at most 40°C (see discussion by

Maresch and Abraham, 1981). The discontinuity may mark a small growth hiatus during the prograde P-T path, perhaps due to a discontinuous silicate reaction, although admittedly chemical and textural zoning patterns in eclogite garnets are still problematical (e.g., Ghent, 1982).

An important step towards clarifying the prograde path has recently been made (Maresch, work in progress) with the identification of relict, often Cr-bearing, deuteric hornblende from the original gabbroic rock in amphibolite of the La Rinconada Group (Fig. 5, Table 1). In contrast to the blue-green barroisite of regional metamorphism, which is oriented in the plane of the foliation, these greenish-brown hornblendes also occur in small "augen," around which the planar fabric has been bent. Besides the Cr-content inherited from the magmatic pyroxene that has been replaced (Table 1), the typical, fine-grained ilmenite exsolution texture of the pyroxene (e.g., Maresch, 1971, Fig. 2-13) is often preserved (Fig. 5), although ilmenite may be replaced by sphene or rutile in some samples. These deuteric hornblendes are altered to and directly overgrown by the barroisite of regional metamorphism. It is therefore now possible to conclude that the first regional metamorphic episode to which the gabbroic/basaltic material of the La Rinconada Group of Margarita was subjected was of the intermediate- to high-pressure type, and, as discussed above, all subsequent overprinting can be related to the same metamorphic episode. The impact of this mineralogical evidence could be strengthened if detailed microfabric analyses were available on the metamorphic rocks of the Venezuelan Coast Ranges, in order to link the sequence of changing P-T conditions to the regional deformational phases observed (e.g., Vignali, 1972, 1979; Talukdar and Loureiro, 1982; Maresch, 1971). However, very little conclusive work has been done in this regard; microfabric studies may be recommended as a fruitful direction of future research.

We conclude that only one metamorphic cycle has affected the La Rinconada Group, and, by induction, all the metabasic rocks of the high-pressure coastal strip in the Coast Ranges (or "frange côtière-Margarita" of Stephan and others, 1980) from Puerto Cabello to Margarita Island as well.

The Orogenic Mechanism Responsible for Metamorphism

The existence of high-pressure/low-temperature metamorphism logically raises the question of how such conditions of deep burial, coupled with anomalously low temperatures, can be achieved. Maresch and Abraham (1981) estimated the conditions of equilibration for an eclogite from Margarita Island to be >11.5 kbar (1.15 Gpa) and 450-525°C. This implies an overburden of at least 40 km. Morgan (1967), in his study of the eclogites of the Puerto Cabello region, first recognized the geotectonic problems involved. Simple stratigraphic burial, as envisioned by Morgan as well as Piburn (1967), does not solve the enigma, because this would lead to temperatures in excess of 600°C for a burial depth of 40 km (cf. Thompson, 1981).

Bell (1971) noted that high-pressure metamorphism could

TABLE 1. REPRESENTATIVE MICROPROBE ANALYSES OF RELICT DEUTERIC/LATE MAGMATIC HORNBLENDE (ANALYSIS 5663-52) AND BARRIOSITE OVERGROWTH OF REGIONAL HIGH-PRESSURE METAMORPHISM (ANALYSIS 5663-45)

	5663-52	5663-45
SiO_2	49.5	48.5
TiO_2	0.42	0.25
Al_2O_3	8.3	11.9
Cr_2O_3	0.43	0.00
FeO*	13.8	14.4
MgO	13.3	11.1
CaO	11.4	8.5
Na_2O	1.4	3.1
K_2O	0.11	0.28
	98.7	98.0
Cation contents for 23 oxygens		
Si	7.12	7.00
Al	0.88	1.00
	8.00	8.00
Al	0.53	1.02
Ti^{4+}	0.046	0.028
Cr^{3+}	0.049	0.000
Fe^{2+}*	1.66	1.74
Mg	2.84	2.38
	5.13	5.17
Ca	1.75	1.31
Na	0.40	0.88
K	0.020	0.052
	2.17	2.24

* Total Fe as FeO.

have been contemporaneous with and related to overthrusting of the Villa de Cura klippe from north to south. In the following years, a number of authors (e.g., Bell, 1971; Maresch, 1972; Bellizzia, 1972; Murray, 1972; Santamaría and Schubert, 1974) attempted to explain high-pressure metamorphism and Late Mesozoic orogenic activity along the Caribbean/South American borderland in terms of a steady-state, southward dipping subduction zone. Nevertheless, as Maresch (1974) pointed out, both the distribution of volcanic rocks and the type and age of metamorphic rocks in the Caribbean Mountain System of northern Venezuela fail to support this concept. Maresch (1974, 1976) summarized evidence which led him to postulate that high-pressure metamorphism was due to the collision of an island arc (part of which is now exposed as the Aruba to Blanquilla island chain) with the South American continental margin in the Late Cretaceous. Partial overthrusting, or obduction, of this volcanic welt could then account for the high-pressure/low-temperature conditions of metamorphism experienced by the continental edge. Because the collision impeded further north-south convergence between island arc and continent, Maresch (1974) suggested that a new, south-dipping subduction zone formed north of the collision zone. Consequently, in the partially obducted arc isostatic readjustment led to decoupling between the arc and the continental margin (now considerably foreshortened) along the primary thrust zone. While the continental edge rebounded upward, most of the island arc subsided north of the continent to form a region of extensive block-faulting (e.g., Ball and others, 1971). However, some of the thrust slices remained linked to the

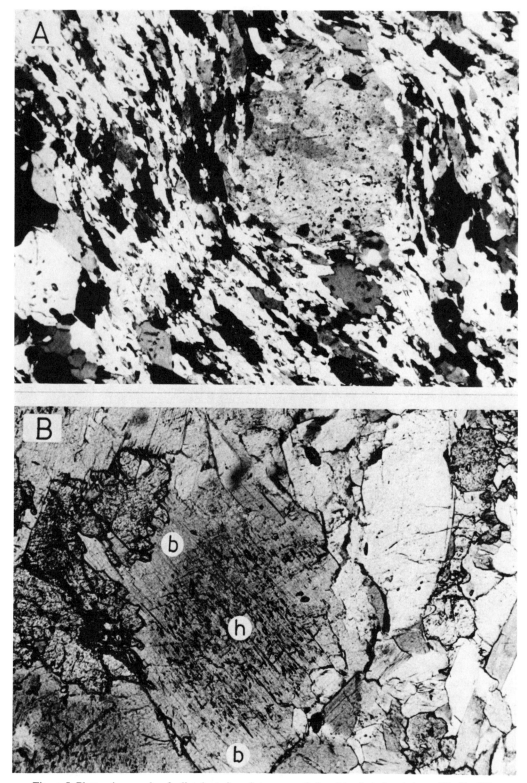

Figure 5. Photomicrographs of relict deuteric or late-magmatic hornblende in banded amphibolite (A) and coarse-grained metaintrusive (B). A: "augen" of hornblende with thin rim of barroisite distorting planar fabric of rock (crossed polarizers); B: amphibole crystal in amphibole-sphene-epidote intergrowth with oriented ilmenite platelets clouding hornblende core (h); the clear rims are barroisite (b) (plane light). Long dimensions of photomicrographs represent 2.8 and 1.3 mm in (A) and (B), respectively.

continent and shared the same subsequent tectonic history. The Villa de Cura Belt was suggested to have been such an overthrust remnant that slid, or was thrust, southward during post-collision deformational events, such as the isostatic rebounding noted above. Such a model harmonizes well with the general concepts of large-scale, southward-directed thrusting and gravity sliding independently developed by a number of workers (e.g., Bell, 1971; Bellizzia, 1972; Stephan and others, 1980), although the details may vary considerably from author to author.

In recent years, the thermal perturbations of large-scale overthrusting and associated high-pressure metamorphism have been investigated in considerable detail (see summary by Thompson, 1981; also Richardson and England, 1979; Draper and Bone, 1981). There are a number of parameters in these models that may be varied, viz. thermal conductivity, existence of heat sources or sinks in the crust involved, magnitude of heat flow from below the crust, thickness of overthrust slab, erosion rate or rate of tectonic unroofing, and the elapsed time after thrusting and before erosion or unroofing begins. The results, schematically summarized in Figure 6, all show that P-T conditions conducive to eclogite formation are reached in the lower slab, i.e., the continental margin, for almost all logical values of these variables. Rather, the problem is to exhume the high-pressure assemblages before they are obliterated by the return to a normal temperature distribution. Depending on the parameters of the model chosen, limits on the maximum burial time range from at least 20 m.y. to 100 m.y. or more, time intervals that appear geologically attainable.

The model presented in Figure 6 assumes gradual unroofing through erosion. Tectonic unroofing through isostatic rebound or tilting, as assumed in the model suggested by Maresch (1974), would of course considerably hasten this process. It is interesting to note that in the upper slab, i.e., the volcanic arc, there is initial *cooling* (path of rock parcel "C") after overthrusting (shear heating is considered to be restricted to the immediate vicinity of the thrust zone). Such "quenching" could be instrumental in preserving the metamorphic character of the lower part of the overthrust slab, i.e., island arc. Although for simplicity, identical, homogeneous geothermal gradients are assumed in Figure 6 for both island arc and continental margin in order to demonstrate the basic principles involved, the leading edge of the arc complex should involve blueschist metamorphism in the trench environment. Thus, a ready explanation presents itself for the origin and preservation of the blueschist metamorphism in the Villa de Cura Belt.

We conclude that the island-arc/continental margin collision model is a reasonable working hypothesis in that it complies not only with the tectonic and petrological evidence available, but also with the geophysical parameters predicted for such a collision.

Figure 6 also points out an extremely important feature of the overthrust process. It shows the P-T-t paths of three parcels of rock for a model with constant thickness of overthrust slab and constant values of the parameters outlined above. A study of the

distribution of mineral assemblages at a contemporary erosion surface will not yield information on the shape of a paleogeotherm unless the relative *time* of equilibration of each of the regionally distributed assemblages is known. If we assume that assemblages are "frozen in" at the highest temperature reached (e.g., Miyashiro, 1973, p. 56), then in Fig. 6 parcel of rock "A" and "B" will equilibrate last at 40 and 60 m.y. after overthrusting, respectively. The apparent erosion surface P-T gradient deduced from the study of these two rock samples would bear no relationship to any of the paleogeotherms that actually existed. If we further consider that the overthrust plate will not have been uniformly thick, that the various parameters affecting heat flow may be non-uniform, and that the rate of erosional or tectonic unroofing can have varied from place to place, it becomes obvious that the present erosional surface of the Caribbean Mountain System must be an extremely complex record of partial sets of variable geotherms that have affected the rock units exposed.

LATE MESOZOIC TO EARLY TERTIARY MAGMATISM

An island-arc/continent collision event will be reflected in the regional metamorphic imprint of the collision zone, as discussed in the previous section. Implicit in such a model is a specific sequence of tectonic environments that should constrain the geochemistry of the magmatic rocks generated, differentiated, and crystallized during the plate-tectonic processes leading to and post-dating the collision.

Maresch (1974) attempted to fit the magmatic rocks of the region into four petrotectonic complexes with distinct age ranges (Fig. 7):

1) Because the La Rinconada Group of Margarita Island appears to underlie shelf-type sediments (Maresch, 1972), this thick (>2000 m) succession of metamorphosed basic flows and intrusives was interpreted to have accumulated as a consequence of the Jurassic or older rifting episode that formed the proto-Caribbean.

2) The basic to intermediate, mainly extrusive magmatic rocks of Aruba, Curaçao, Bonaire, the Venezuelan islands, and the Villa de Cura Group of the mainland (as an overthrust remnant) were considered to be the relics of the oceanic island arc that collided with the South American continent.

3) To explain the volumetrically minor, predominantly basic volcanic activity in a broad belt parallel to the coast in the central part of the western Caribbean Mountains of Venezuela (represented by units such as the Tacagua Formation, Pilancones Formation, Conoropa rocks, Los Naranjos Member, and Aragüita Formation), Maresch assumed that a short-lived, south-dipping subduction zone formed along the craton in the mid-Cretaceous, and that subduction beneath the oceanic island arc ceased as the intervening ocean was consumed.

4) Late Cretaceous and Early Tertiary calc-alkaline batholiths and minor volcanics of the Aruba-Blanquilla island chain and the Venezuelan Coast Ranges were ascribed to post-collision

PRE-THRUSTING STAGE

SCHEMATIC CROSS-SECTION   CORRESPONDING DEPTH-T SECTION

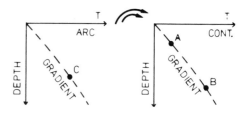

POST-THRUSTING STAGE

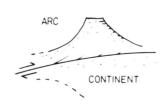

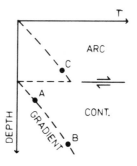

EROSION AND UPLIFT

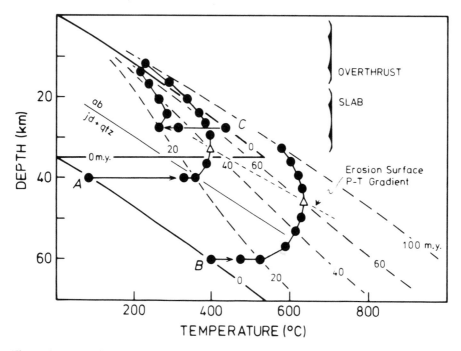

Figure 6. Schematic example of the thermal effects accompanying large-scale overthrusting (based essentially on Figure 2 of Thompson, 1981, where detailed parameter values may be found). The model assumes an elapsed time of 20 m.y. after overthrusting and before erosion sets in. Depth-temperature coordinates for three parcels of rock are shown at 10 m.y. intervals during uplift, paleogeotherms at 20 m.y. intervals. Location of albite = jadeite + quartz equilibrium shown for orientation. For further discussion, see text.

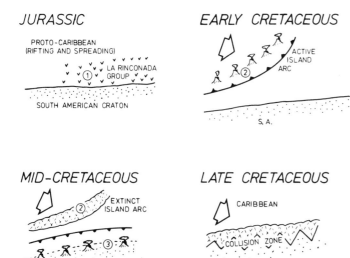

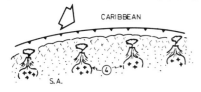

Figure 7. Plan-view sketches illustrating island-arc/continent collision sequence as suggested by Maresch (1974). Numbers refer to the four petrotectonic complexes discussed in the text.

subduction along a newly formed southward-dipping subduction zone north of the collision zone.

This outline of magmatic events accompanying collision was based on scanty geochemical data and poor age control. New data have accumulated to such an extent that a review and revision of these proposals is necessary.

Relative Timing of Magmatism and Collision

Maresch (1974) assumed a Late Cretaceous age for the island arc/continent collision. By contrast, Beets (1975) arrived at a Late Jurassic to Early Cretaceous age based on the available stratigraphy of the Netherlands Antilles. However, the discovery of new fossil localities on Bonaire shortly thereafter showed that Beets' conclusion was based on a misinterpretation of the stratigraphic relationship between the Washikemba and Rincon Formations. These new data also favoured a Late Cretaceous age for the collision event (Beets and others, 1977; MacGillavry and Beets, 1977). Nevertheless, a Late Jurassic to Early Cretaceous age for the collision has been revived by Stephan and others (1980) *in addition* to a Late Cretaceous event. We are not in a position to judge the stratigraphic arguments for their conclusion. However, such an early age for the collision and concomitant

high-pressure metamorphism of the Coast Ranges necessitates that:

1) The high-pressure metamorphic rocks of the Coast Ranges and Margarita Island be "polymetamorphic" (in the sense of having been subjected to more than one metamorphic cycle). As has been argued in a previous section, petrological evidence speaks against this conclusion.

2) The basic to intermediate volcanism of the Netherlands Antilles would post-date the first collision. It could then be related causally to Donnelly and other's (1973) Late Cretaceous "flood basalt event," rather than to an oceanic island arc; but, as will be shown in subsequent discussions, such an alternative is untenable on geochemical grounds.

We base our conclusions concerning the age of the collision on the detailed stratigraphy of the Netherlands Antilles, where pre-collision, igneous sequences of Albian to Coniacian age are unconformably overlain by post-collision Campanian and younger sediments (Fig. 8).

The oldest sequence is exposed on the island of Curaçao and comprises a roughly 5 km thick succession of predominantly pillow basalts, the Curaçao Lava Formation. A pelagic intercalation in mid-formation contains ammonites yielding a middle Albian age (Wiedmann, 1978). Whole-rock K/Ar dating of two basalt samples of this unit by Santamaria and Schubert (1974) gave ages of 118 ± 10 and 126 ± 12 Ma, and of three diabase samples of one outcrop by Priem (personal communication, 1982) yielded 93 ± 3 Ma for one of the samples and much younger, 73 and 69 Ma, ages for the other two.

Based on the scarcity of pelagic intercalations in the Curaçao Lava Formation, Klaver (in preparation) estimates that piling up of the basalts of this unit took place in less than 10 million years. Taking Odin and Kennedy's (1982) dates of 107 ± 1 and 95 ± 1 Ma for the base and the top of the Albian, respectively, the 118 ± 10 and 126 ± 12 dates of Santamaría and Schubert seem a little too old. Perhaps the reason is that these lavas extruded under high hydrostatic pressure, so that excess ^{40}Ar was trapped in the glassy rims of the pillows (Dalrymple and Moore, 1968). The 73 and 69 Ma dates of Priem are reset ages. As shown by Beets (1972), the Cretaceous and Danian succession of the Netherlands Antilles was metamorphosed under low-grade conditions (zeolite and prehnite-pumpellyite facies) in the Late Cretaceous and Early Tertiary. This event has reset most of the radiometric systems (Priem and others, 1977).

The Curaçao Lava Formation is separated from the overlying Campanian/Maastrichtian Knip Group and Danian Midden-Curaçao Formation by an angular unconformity (Klaver, in preparation; Beets, 1972, 1977). The Knip Group and Midden-Curaçao Formation are sedimentary units of up to 2 km in thickness. The former consist of pelagic, silica-rich rocks and clastic sediments, the latter of clastic sediments (turbidites) only. In both units local volcanic debris is associated with detritus derived from a sialic source. This is the first indication in the Curaçaoan section of the proximity of the South American continent. Volcanic rocks form a minor constituent in these units—a few basaltic flows and

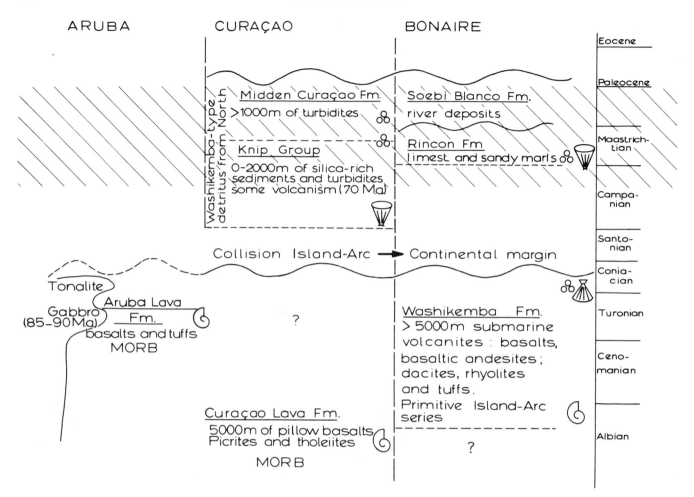

Figure 8. Correlation chart of the Cretaceous and Early Tertiary sequences of Curaçao, Aruba, and Bonaire. Except for the tonalite-gabbro batholith of Aruba, ages of all units are based on fossils: ammonites for the Curaçao and Aruba Lava Formations, ammonites, inoceramids, and planktonic foraminifera for the Washikemba Formation, and rudists and planktonic foraminifera for the late Senonian units. An extensive radiometric dating program on rocks of the Aruban batholith has shown that emplacement of the batholith occurred in the 85-90 Ma interval and that the published age of 70-75 Ma (Priem and others, 1968, 1977; Santamaría and Schubert, 1975) is a reset age due to a low-grade thermal event in the late Senonian and Early Tertiary (hatched field).

rhyodacitic tuffs in the Knip Group, and a few sills and plugs of andesitic composition in both (Beets, 1972). The unconformity between the Curaçao Lava Formation and the Knip Group in our opinion is the time interval into which the collision of the island arc with the continental margin of Venezuela must be placed. The minor volcanics in the Knip Group belong to the post-collision, calc-alkaline arc of Maresch (1974).

On Bonaire the Washikemba Formation is the oldest sequence exposed. It ranges in age from late Albian to Coniacian (Beets and others, 1977; Smit, 1977). The unit is more than 5 km thick, consisting mainly of submarine volcanics (flows of basalts, basaltic andesites, and andesites; flows, sills, and subaqueous flows of rhyodacite and rhyolite) with intercalations of pelagic sediments, mainly black shales, cherts, radiolarites, and cherty limestones. Volcaniclastic sandstones and boulder-beds, indicative of the emergence of volcanic islands, are restricted to the

upper part of the formation. Except for this volcanic debris, the Washikemba Formation contains no detritus.

The unit is unconformably overlain by two small erosional remnants of the Rincon Formation (MacGillavry and Beets, 1977), consisting of rudist- and pseudorbitoid-bearing limestones, marls, and calcareous sandstones. The marls contain a middle to late Maastrichtian fauna of planktonic foraminifera. The detritus in the sandstones is all volcanic debris, probably derived from the Washikemba Formation. Sialic detritus from the mainland first appears in the ?Paleocene/Eocene Soebi Blanco Formation, an alluvial deposit of river conglomerates (Beets and others, 1977). Here again, the unconformity between the Washikemba Formation and the Rincon Formation must mark the collision. The data from both unconformities indicate that the island-arc/continent collision at the site of the Netherlands Antilles must have taken place in the Coniacian to Campanian interval.

The oldest unit of Aruba is the Aruba Lava Formation, a succession about 3 km thick of sills, flows, and pyroclastics of basaltic composition with intercalations of volcaniclastic sediments. MacDonald (1968) found ammonites in one of these intercalations, leading to a provisional Turonian age. Newly collected material confirms this age (Wiedmann, personal communication, 1980). The Aruba Lava Formation is intruded by a composite tonalite batholith (Westermann, 1932). The history of dating of this batholith is interesting and instructive. In 1966, Priem and others proposed an age of 73.5 ± 3 Ma for the tonalite, on the basis of a whole-rock/biotite Rb-Sr age (72.5 ± 4 Ma) and a biotite K/Ar age (74.5 ± 4 Ma) obtained from a single sample (ages recalculated with modern decay constants). This age was roughly confirmed by Santamaría and Schubert (1974), who reported 67 ± 4 Ma (K/Ar on biotite of a quartz diorite). In a progress report on newly collected samples presented at the 8th Caribbean Geological Conference, Priem and others (1977) gave a Rb/Sr isochron based on 9 whole-rock and 3 biotite determinations, indicating an age of 72 ± 2 Ma for the batholith. However, an isochron age solely based on the whole-rock data yielded a large error (72 ± 29 Ma), because of the low Rb/Sr ratios and some scatter of the data-points. Moreover, a whole-rock/biotite pair from the northwestern part of the batholith gave a much too high Rb/Sr date of 85 Ma. It was subsequently found (Priem and others, 1978) that another whole-rock/biotite pair from northwest Aruba also yielded the higher age. The whole-rock data are inconclusive and could fit both an 85 Ma and a 70 Ma isochron. Priem and others (1978) suggested that these data might be interpreted as reflecting episodic intrusion: an older 85 Ma event and younger intrusive activity around 70 Ma. Considering the inferred age of the collision (the interval Coniacian/Campanian) and the close association and similarity of the dated igneous rocks, this solution did not appear very satisfactory. Therefore, considering the retentivity of hornblende for argon, K/Ar dating of hornblende from all samples containing sufficient amphibole was carried out. Invariably, ages of 85-90 Ma were obtained, whether from the northwestern or central part of the batholith (Priem, personal communication, 1982). This fact and the Turonian (= 91-88 Ma according to Odin and Kennedy, 1982) age of the country rock, lead to the final conclusion that emplacement of the batholith took place in the 85-90 Ma interval and that the low-grade thermal event around 70-60 Ma has reset most of the radiometric systems, except for the K/Ar in amphibole. This conclusion has some important implications. In the first place, it shows that the emplacement of the batholith occurred prior to collision, so that it, contrary to all earlier interpretations, forms part of the colliding oceanic island arc. As will be discussed below, this conclusion is consistent with the initial $^{87}Sr/^{86}Sr$ ratio of the batholith and the paleomagnetic data as given by Stearns and others (1982) and Hargraves and Skerlec (1980). It furthermore shows that calc-alkaline volcanism is not restricted to the late Senonian to Tertiary cordilleran-type arc, but that it also occurs in the older, colliding arc. In the third place, intrusion of this calc-alkaline complex occurred shortly after deposition of the

Aruba Lava Formation, which consists of volcanics with a chemistry similar to that of mid-ocean ridge basalt (Donnelly and Rogers, 1978). And finally, the history of dating studies on the batholith shows with what care radiometric dates in such a complicated collision-type boundary must be interpreted.

The Aruba Lava Formation and the tonalite batholith are unconformably overlain by Eocene limestones. Late Senonian sediments are absent on Aruba, so that the section yields no additional constraints on the timing of the collision.

The data from the Netherlands Antilles modify the collisional model of Maresch (1974), while confirming its main features. Contrary to what Maresch assumed, the oceanic island arc activity did not cease in the mid-Cretaceous but was active up to the very moment of the collision. Although this fact does not rule out the existence of an arc-trench system along the South American craton in the mid- to Late Cretaceous (Maresch's second arc), it questions its necessity. Considering the small volume of volcanics of this alleged arc and the structurally and stratigraphically still ill-defined position of these units, we propose to omit this arc-trench system and to simplify the model to a collision of oceanic island arc and continental margin in the Coniacian/Campanian interval, followed by renewed subduction north of the collision zone in the late Senonian. The latter is responsible for some minor calc-alkaline magmatism and thermal activity leading to low-grade metamorphism of the magmatic sequences and a resetting of the radiometric ages of most of the igneous rocks to between 75 and 60 Ma.

The Curaçao Lava Formation, the Washikemba Formation, the Aruba Lava Formation, and the Aruban batholith are considered by us to be remnants of the colliding oceanic arc. In the next section, their geochemistry will be discussed and compared with magmatic sequences from mainland Venezuela and the Island of Margarita. Of particular interest is the geochemical relationship between the volcanics of the Netherlands Antilles and those of the Villa de Cura Tectonic Belt on the mainland, inasmuch as the latter has been regarded as an overthrust remnant of the colliding island arc (Skerlec, 1976; Skerlec and Hargraves, 1980; Maresch, 1974, 1976). The Coniacian/Campanian age of the collision, however, shows that final emplacement of the Villa de Cura Belt in the Paleocene (Bell, 1971, 1972) must be due to post-collision deformational events.

Geochemistry and Geology of the Oceanic Island Arc

Curaçao: Curaçao Lava Formation. The more than 5 km thick Curaçao Lava Formation consists almost exclusively of basic volcanic rocks extruded and deposited in a submarine environment. The only exception is an intercalation of pelagic sediment a few metres in thickness. The main rock type is pillow basalt, which is virtually the only rock-type found in the lower half of the formation. In the upper half, pillow lavas still predominate, but alternate with reworked hyaloclastites and diabase sills and dykes (Fig. 9).

The stratigraphy of this basalt sequence is reflected largely in

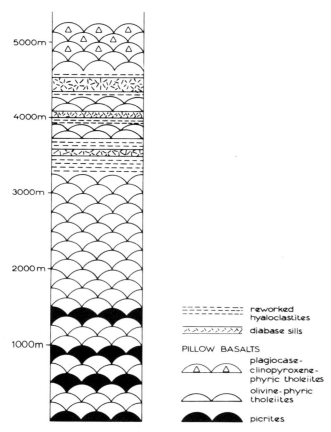

Figure 9. Simplified lithologic column of the Curaçao Lava Formation. The lower 3000 m of this unit consists almost entirely of pillow basalts. In the upper 2000 m, pillow basalts alternate with reworked hyaloclastites and diabase sills. Pelagic intercalations are extremely scarce because of continuous piling-up of basalts. Density of the magma controls the distribution of basalt types in the formation.

the density of the lavas (Klaver, in preparation). The lower half of the formation consists of picrites and olivine-phyric tholeiites. The MgO percentage of these rocks varies from 31 (olivine cumulate) to 8 (olivine-phyric tholeiites). The upper half consists of olivine-phyric tholeiites and plagioclase-pyroxene-phyric tholeiites with 10 to 5 percent MgO (Fig. 10). Picrites, olivine-phyric tholeiites and plagioclase-pyroxene-phyric tholeiites form a consanguineous suite. They are derived from a picritic parent with at least 17.5 percent MgO (Beets and others, 1982)—the picrites and olivine-phyric tholeiites by addition or fractionation of olivine (Fo 87), and the plagioclase-pyroxene-phyric tholeiites by further fractionation of olivine, plagioclase and pyroxene, the latter in about equal amounts. Model calculations suggest that the parental magma resulted from about 30 percent batch partial melting of a mantle of pyrolitic composition under dry conditions.

As shown by Donnelly and others (1973), Beunk and Klaver (1977) and Donnelly and Rogers (1978), major and trace element chemistry of the majority of the basalts are identical to those of mid-ocean ridge basalt. The chondrite-normalized REE patterns (Fig. 11) of the rocks are basically flat (La/Sm$_N$ = 0.8 -

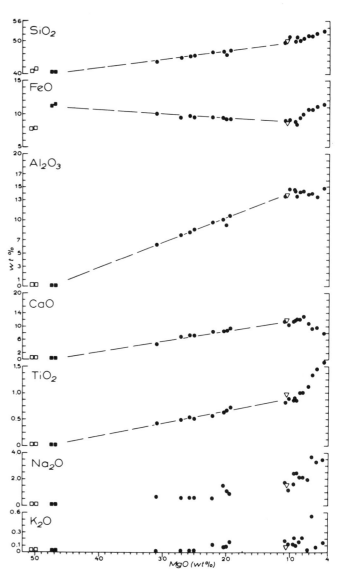

Figure 10. MgO-variation diagram of picrites and tholeiites of the Curaçao Lava Formation (solid dots). Closed squares: olivine phenocrysts (Fo 87 and Fo 88) of picrites; open squares: olivine megacrysts (Fo 91.9 and Fo 92) of picrites; triangle: composition of glass of picrite. As the composition of the glass of the picrites plots within the field of the olivine-phyric tholeiites (8–11% MgO) on the Fo 87/88 control line, picrites and olivine-phyric tholeiites are related by equilibrium crystallization of olivine (Beets and others, 1982). The change in slope of the control line between 8 and 9% MgO indicates that fractionation of plagioclase and pyroxene defines the composition of the more evolved basalts.

1.0), indicating that the suite belongs to the T-type MORB. A similar conclusion may be drawn from Th-Ta-Hf plots (Klaver, in preparation) and the relatively undepleted K-content of the basalts (Beets and others, 1982). Nevertheless, the formation differs in a number of important respects from normal oceanic crust. Most obvious is its thickness, which is several times that of layer 2, but more or less similar to that of the mid-Cretaceous oceanic plateaus in the Pacific, such as the Ontong Java and Manihiki plateaus (Hussong and others, 1979; Carlson and oth-

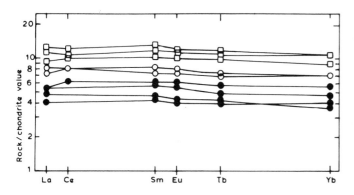

Figure 11. Chondrite-normalized REE-patterns of representative picrites and tholeiites of the Curaçao Lava Formation. Closed dots: picrites; open dots: olivine-phyric tholeiites; open squares: plagioclase-pyroxene-phyric tholeiites. MgO-content of the samples are, from bottom to top: 27.5; 25.1; 21.0; 20.7; 11.0; 10.3; 8.3; 6.0; 4.9%. Similarity of patterns of the various basalts, despite strongly divergent MgO-content, indicates that olivine is the major fractionating phase. The patterns are consistent with consanguinity of the suite.

ers, 1980), and the smooth, thick Caribbean crust of the western part of the Venezuela Basin (Edgar and others, 1973; Diebold and others, 1981; Stoffa and others, 1981). A second major difference is the wide range in MgO content of the Curaçaoan basalts, from picrites with 31-19 wt% to evolved tholeiites with about 5 wt%. Whereas picrites and evolved basalts are quite common in the Curaçao Lava Formation, they are absent or rare in oceanic crust (Melson and others, 1976; Bryan, 1979; Bryan and others, 1979). As shown by Huppert and Sparks (1980), Stolper and Walker (1980), and Sparks and others (1980), the restricted compositional range of mid-ocean ridge basalt is due to the occurrence of a density minimum in the liquid line of descent, which defines the composition of the melt leaving the steady-state magma chamber underlying the ridge. The common occurrence of picritic flows in the lower half of the Curaçao Lava Formation indicates that at this stage of development of the volcanic unit no large magma chamber was present. In this respect the Curaçao Lava Formation is comparable to some flood basalt sequences, in particular those associated with the initiation of spreading, as for instance the Tertiary volcanic province of Baffin Island-West Greenland (Clarke, 1970; Clarke and Pederson, 1976) and the Nuanetsi Province of Rhodesia (Cox, 1972; Cox and Jamieson, 1974) of Karoo age. The absence of picrites in the upper half of the Curaçao Lava Formation suggests that, due to piling up of the lavas on the seafloor, the height of the magma column became such that the dense picritic liquid could no longer extrude and only olivine-phyric tholeiitic liquid reached the surface. In most of these tholeiites, olivine is the only liquidus mineral; only occasionally is it joined by plagioclase. Most of the plagioclase and pyroxene fractionation resulting in the evolved basalts of the upper part of the unit took place in the sills intruding the formation, and not in the magma chambers underlying it. This suggests a relatively short residence time of the magma in the deeper reservoirs due to a high rate of supply of the parental liquid.

The third difference from normal oceanic crust is the homogeneous nature of the Curaçao Lava Formation. All basalts of the unit are related by low-pressure fractionation of olivine ± plagioclase ± pyroxene to a single parent.

In conclusion, it can be stated that the Curaçao Lava Formation is basically a MORB sequence formed by shallow (< 70 km) melting of a non-depleted pyrolitic source under essentially dry conditions in a tensional environment. However, it is not just another piece of ocean floor. The thickness of the unit, its chemical homogeneity, and the non-depleted nature of the source suggest that the Curaçao Lava Formation was fed by a prolific, chondritic mantle source supplying magma at a rate surpassing that of the lateral removal of newly formed crust, rather than being the result of partial melting of mantle diapirs passively rising because of tension and pressure release in a ridge environment. The inferred absence of a large magma reservoir in the early stage of development of the Curaçao Lava Formation suggests that the unit represents the initial tapping of this mantle source.

Considering the close association in space and time of the Curaçao Lava Formation with magmatic sequences of island arc origin on Bonaire and Aruba, we conclude that the formation is part of the island arc succession, despite its MORB nature. As will be discussed below, it may owe its origin to lengthening of the arc at a trench-trench-ridge triple junction.

Bonaire, Washikemba Formation. The Washikemba Formation is a submarine island arc succession consisting of lavas, shallow intrusions, and subaqueous pyroclastic flows, alternating with pelagic sediments and, in the upper part of the unit, volcaniclastic sandstones and boulder-beds. Neither top nor base of the formation is exposed. The most complete succession is situated in northwest Bonaire (Fig. 12), where a gently northeastward-dipping section of more than 5 km thickness is exposed. Imprints of ammonites in cherty shales in the lower half of the section point to a ?late Albian age; cherty limestones of the upper half of the section contain faunas of inoceramids and planktonic foraminifera of Turonian and Coniacian age (Beets and others, 1977; Smit, 1977).

An SiO_2-histogram for the volcanics of the Washikemba Formation, including the subaqueous pyroclastic flows, shows a bimodal distribution with peaks in the group of basaltic andesite (50-60 percent SiO_2) and rhyolite (70-75 percent SiO_2) (Fig. 13). As a rough estimate, lavas, sills and subaqueous pyroclastic flows of dacitic to rhyolitic composition form 55 percent of the total volume of volcanic rocks exposed in Bonaire. The remaining 45 percent consists of pillow lavas, sheet flows, and laccoliths of basaltic to andesitic composition. In the northwestern section (Fig. 12), the proportion is about 35 percent basaltic andesite and andesite, 45 percent dacite and rhyolite, including subaqueous pyroclastic flows (25 percent), and 20 percent sediments (pelagic and volcaniclastic).

All sediments in the formation are indicative of accumulation in deeper water, including the volcaniclastic sandstones and boulder-beds, which are turbidites and channel facies debris

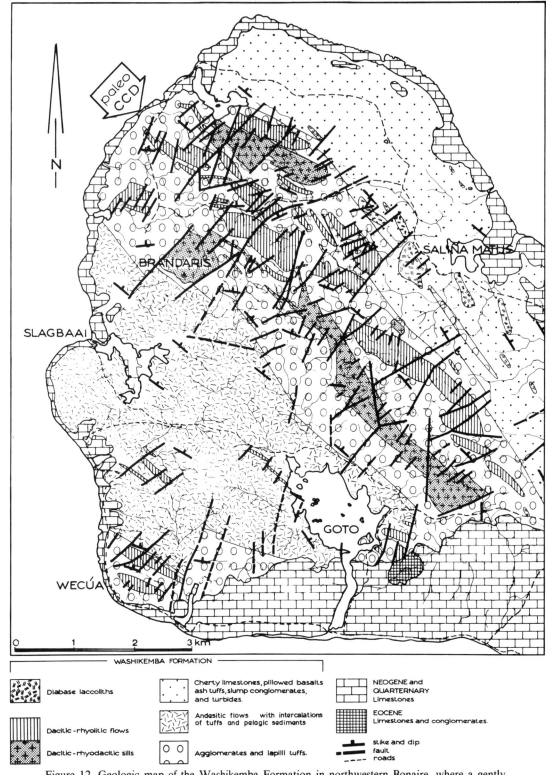

Figure 12. Geologic map of the Washikemba Formation in northwestern Bonaire, where a gently northeastward-dipping, continuous section of over 5000 m thickness is exposed. The Washikemba Formation is unconformably overlain by Eocene and younger limestones, which fringe the coastline. Arrow in upper left corner gives the stratigraphic level of the inferred Turonian paleo-CCD in the Washikemba Formation.

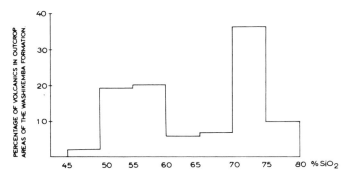

Figure 13. SiO$_2$-histogram of the volcanics of the Washikemba Formation. At a rough estimate, lavas, sills, and subaqueous pyroclastic flows of dacitic to rhyolitic composition form 55% of the total volume of igneous rocks; the remaining 45% is of basaltic to andesitic composition.

flows. Rounding, and the nature of the pebbles in the latter rocks, point to derivation from volcanic islands. The debris in the subaqueous pyroclastic flows consists largely of angular pumice fragments; and considering the absence of rounded fragments, it is believed that these represent redeposited ejecta from submarine vents. All pelagic sediments of the upper half of the formation are cherty limestones and cherty marls; those in the lower half are cherts, radiolarites, and black, cherty shales without any carbonate. The absence of carbonate in sediments of the lower half of the section suggests that they have been deposited below the paleo-Carbonate Compensation Depth (CCD). Data from Thierstein and Okada (1979) on the depth of the CCD in the Cretaceous Pacific Ocean—as will be discussed below, that is the realm where we think the island arc originated—yield about 3500 m for the Albian CCD and a shallowing to 2700 m in the Turonian. Considering the age of the faunas found in the cherty limestones of the Washikemba Formation, the boundary between carbonate-free and carbonate-rich sediments in the section (Fig. 12) corresponds to the Turonian CCD at 2700 m depth. The ammonites of late Albian age (imprints, no carbonate remains) in cherty shales about 1000 to 1500 m lower in the section may have been deposited close to the Albian CCD at 3500 m. If our assumption on the paleo-CCD in the Washikemba Formation is correct, it implies that the northwestern sequence has been formed between a depth of about 3500 m near the base and 1000 m near the top of the section. No carbonate has been found in the sediments of the formation exposed in southeastern Bonaire, which suggests that this succession has been entirely deposited below the Turonian CCD.

The conclusion on the depth of deposition of the Washikemba Formation has important implications for the nature of its volcanism, as all lavas, including those from the lower half, are vesicular, indicating a high initial volatile content of the magma. The amount of vesicles is variable, but may become as much as 20 vol%. This amount of vesicles suggests that water is the major gas phase and that the melts are saturated with respect to H$_2$O (Klaver and others, in preparation). Extrapolating Hamilton and others' (1964) curves for the solubility of water to the inferred

pressure range of eruption of the Washikemba Formation lavas gives about 1.5 wt% H$_2$O at 350 bars and 0.5 wt% at 100 bars in basalt, and 2.5 to 1 wt% in andesite. An initial high water content of the Washikemba magma is corroborated by the following evidence:

1. Magnetite and hornblende replace clinopyroxene as liquidus minerals in the restmelts of sheet-flows of basaltic andesite occurring at about 1000-2000 m below the presumed paleo-CCD in the northwestern facies succession (Fig. 12). The minerals appear after crystallization of 65-70 wt% plagioclase and clinopyroxene (Klaver and others, in preparation). For hornblende to crystallize, the melt must contain about 3 wt% water. As magnetite crystallization precedes that of the hornblende, the composition of the restmelt shifts to that of dacite. Even at this composition, a pressure of 300 to 400 bars is necessary to keep 3 wt% water in solution in the melt (Hamilton and others, 1964). Such a pressure corresponds to 3-4 km water depth. The minerals of the restmelt have been quenched and often enclose amygdaloidal vesicles indicative of second-stage boiling, showing that when hornblende reaches the liquidus, the melt is saturated with water. Sheet-flows of basaltic andesite with hornblende occur only in that part of the Washikemba Formation below the presumed paleo-CCD.

2. Shape and size of the dacitic to rhyolitic sills and flows mentioned above (Fig. 12) suggest a much lower viscosity than would be expected for dry melts of this composition. Using flow directions based on the shape of vesicles in sections parallel to the bedding, a minimum "aspect ratio" in the sense of Walker (1973) was calculated for 15 flows of the northwestern outcrop area ranging in SiO$_2$ content from 68 to 75 percent. These varied between 10 and 35, with a mean at 25, which implies a viscosity of about 10^4-10^5 Poise. Using the methods of Bottinga and Weill (1972) and Shaw (1972) to calculate viscosity from the oxide-percentages of the major elements, it was shown that a dry melt of this composition (71 percent SiO$_2$) would have a viscosity of 6.5×10^7 Poise at an eruption temperature of 900°C, but that addition of 2-3 wt% H$_2$O would lower this to a range compatible with an aspect ratio of 25. A similar conclusion is reached when the aspect ratio principle is applied to the rhyodacitic sill between Brandaris and Goto (Fig. 12). In itself, a sill of this composition (Appendix 1) seems a contradiction. Yet, this intrusive body with subparallel roof and floor, a length of over 5 km, and a thickness of about 250 m has an SiO$_2$ content varying between 68 and 73 wt%. The conclusion is that the viscosity of this magma must have been lowered considerably by the presence of water, which, at the confining pressure at this depth, escaped during crystallization but not during flow. However, in addition to these obviously low-viscosity dacites and rhyolites, rhyolites also occur which form steep-sided domes and small lobate flows suggestive of high viscosity at the time of eruption.

3. Low phenocryst content of the volcanics, as water reduces not only the viscosity but also the density of a melt, and hence greatly decreases its ability to carry crystals in suspension (Mysen, 1977).

4. Quench textures in the restmelts of basaltic andesite sheet-flows and in the groundmass of the low-viscosity rhyodacitic sills in the northwestern outcrop area, which are due to a shift of the solidus because of water loss at second-stage boiling.

Considering the large amount of dacite and rhyolite (Fig. 13) in the Washikemba Formation, an island-arc origin for this sequence seems obvious. A more precise classification of the volcanics in terms of partial melting and fractionation paths is complicated by the alteration of the rocks to low-temperature mineral assemblages. Alteration is probably due to the combined effects of degassing during and after crystallization of the flows, hydrothermal cells of cooling seawater (ocean-floor metamorphism), and low-grade metamorphism during the late Senonian to Early Tertiary thermal event (Fig. 8). Alteration strongly affected the relatively mobile elements Ca, Na, K, Rb, Sr, and Ba, and to a much lesser extent Al, Fe, Mg, Ti, Zr, Y, Nb, Hf, Th, and the REE.

The most primitive basalt of the Washikemba Formation—KV554 (Appendix 1), with 7.7 percent MgO and an Mg-number of 57—is very close in composition to Ewart's (1976) average island-arc basalt. KV554 differs from Curaçaoan MORB by its lower content in the high field-strength ions Ti, Zr, Nb, and Hf, less so in REE-content, and the higher content of Al_2O_3 and the alkalies at equivalent Mg-number. The depletion in high field-strength ions relative to MORB is a distinctive, though problematical feature of island-arc basalts and might be due to fractionation of minor titanium phases at an early stage (Wood and others, 1979; Saunders and others, 1980; Perfit and others, 1980). Liquidus phase relations of basalt under hydrous conditions (Holloway and Burnham, 1972) show that plagioclase is not a liquidus phase, while clinopyroxene, olivine, and magnetite are important. Although KV554 and all other basalts and basaltic andesites of the Washikemba Formation contain both plagioclase and clinopyroxene phenocrysts, it may be assumed, on the basis of the relatively low Ni- and Cr-contents, that they derive from more primitive parents. The higher content in Al_2O_3 relative to MORB suggests that in these more primitive parents plagioclase is not a liquidus phase. Fractionation of clinopyroxene and plagioclase in various proportions defines the main trend from basalt to andesite in the formation. Within this range REE-patterns vary from horizontal, about 10 times chondritic (KV554), to LREE enriched with La about 25 times chondritic (79KV330/340; Fig. 14); the high field-strength ions Zr, Ti, Y, and Nb show a general increase (Fig. 15) as the distribution coefficients of plagioclase and clinopyroxene are low for these elements (Pearce and Norry, 1979). Whether the dacites and rhyolites of the Washikemba Formation are derived from the more basic rocks by crystal fractionation is questionable. Two main groups of acid rocks can be distinguished: low-viscosity dacites and rhyolites, and high-viscosity rhyolites. Chemically, the two groups are distinguished by a different behaviour of the high field-strength ions Zr, Ti, Nb, and Y and by their REE-patterns. The high-viscosity rhyolites show a continuous increase of Zr and Y, stabilization of Nb at 5-6 ppm, and a sharp decrease

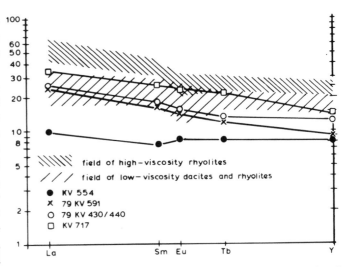

Figure 14. Chondrite-normalized REE-patterns of selected volcanics of the Washikemba Formation, Bonaire. KV554 is the most primitive basalt of the Washikemba Formation; 79KV591 is a basaltic andesite; 79KV430/440, is the mean of 11 analyses of an andesitic sheet flow associated with the low-viscosity dacites and rhyolites; KV717 is an andesite associated with the high-viscosity rhyolites (see appendix 1).

in TiO_2 (Fig. 15), suggesting that clinopyroxene and plagioclase are joined by titano-magnetite as the main fractionating phases (Pearce and Norry, 1979). The low-viscosity dacites and rhyolites, on the other hand, have concentrations of incompatible elements, including Hf and Th, which are scarcely higher than those of the basaltic andesites with which they are associated. Figure 15 shows that the Zr-content of these rocks is similar to that of the basaltic andesites and andesites, Nb decreases to 1 ppm or less, and only Y shows a slight increase. The high-viscosity rhyolites have a higher REE-content with La 40 to 70 times chondritic and La/Yb$_N$ ratios of 2-3 (Fig. 14); the low-viscosity dacites and rhyolites have flatter patterns (La/Yb$_N$ = 1-2) and are less enriched (La is 15-35 times chondritic). Both groups have distinct negative Eu anomalies. Although the behaviour of the high field-strength ions suggests derivation of the high-viscosity rhyolites from the basalts by fractionation of clinopyroxene, plagioclase, and magnetite, the Hf/Th ratios of the rhyolites differ from those of the associated basalts and basaltic andesites, suggesting a different mantle source. Hf/Th ratios of the low-viscosity dacites and rhyolites are similar to those of the basalts, but no phases are known which explain the behaviour of the trace elements. On the basis of the sharp decrease in Ti and Nb (Fig. 15), it is suggested that hornblende could be an important fractionating phase, but the REE-patterns as well as the low Hf and Th content do not confirm this. A derivation by shallow melting of basalts underlying the section does not give a satisfactory solution for the trace element behaviour of these rocks. Although such melting could easily produce dacites (Helz, 1976), with a similar Hf/Th ratio as the original basalt, the REE patterns of such melts would differ considerably from those of the low-viscosity dacites and rhyolites (Hanson, 1980).

The predominance of clinopyroxene as a liquidus phase in

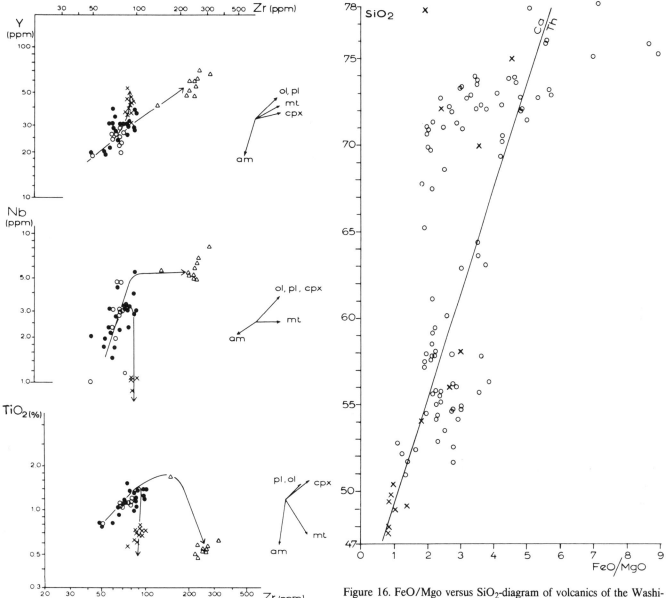

Figure 15. Zr versus TiO₂-, Nb-, and Y-diagram of volcanics of the Washikemba Formation. Closed dots: basaltic andesites and andesites of the northwestern outcrop area; open dots: basalts to andesites of southeastern Bonaire; triangles: high-viscosity rhyolites and one associated andesite; crosses: low-viscosity dacites and rhyolites. Vector diagrams at right give the behaviour of the high field-strength ions by fractionation of am (= amphibole), pl (= plagioclase), ol (= olivine), cpx (= clinopyroxene), and mt (= magnetite). Partition coefficients are from Pearce and Norry (1979).

Figure 16. FeO/Mgo versus SiO₂-diagram of volcanics of the Washikemba Formation (open dots) and Villa de Cura Group (crosses). Basalts and basaltic andesite of the Washikemba Formation show some iron enrichment and tend to a tholeiitic trend. However, most of the dacites and rhyolites of the formation plot in the calc-alkaline field as defined by Miyashiro (1974).

the basalt to andesite range, the absence of hypersthene, and the relatively low La/Yb$_N$ ratios suggest that the Washikemba volcanics are related to Kuno's (1968) pigeonitic series and Jakeš and Gill's (1970) island-arc tholeiite series. However, they differ from these series by the large amount of siliceous rocks. Moreover, when plotted on a FeO/MgO versus SiO₂ diagram (Fig. 16), the rocks show little Fe-enrichment and most of the

andesites and the more siliceous rocks plot in Miyashiro's (1974) calc-alkaline field. The AFM-diagram (Fig. 17) shows the same trend: some Fe-enrichment in the more basic rocks and a strong "calc-alkaline" Fe-depletion in the andesite-rhyolite range. The siliceous volcanics differ, however, from "normal" calc-alkaline rocks by the much lower content in incompatible elements. Donnelly and Rogers (1978) considered the Washikemba volcanics to belong to their "primitive island-arc" suite (PIA) and stressed the resemblance with the volcanics of the Louisenhoj Formation of the Virgin Islands. The PIA series is distinguished from the calc-alkaline series by its lower content of LIL elements,

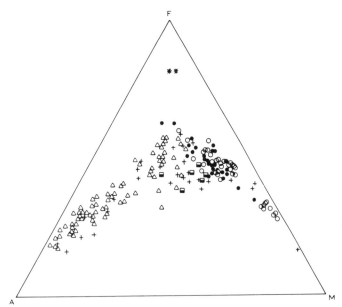

Figure 17. AFM-diagram of the volcanics of the Curaçao Lava Formation (open dots), Aruba Lava Formation (closed dots), Aruban batholith
(half-filled squares), Washikemba Formation (triangles), and Villa de
Cura Group (crosses). Curaçao and Aruba Lava Formations show a
distinct tholeiitic trend, the Aruban batholith, and most of the Washikemba Formation and Villa de Cura Group volcanics a calc-alkaline
trend. Asterisks represent microprobe analyses of amphibole of a basaltic
andesite of the Washikemba Formation ($A = Na_2O + K_2O$; $F = FeO$;
$M = MgO$).

including REE, Th, U, Zr, K, and Ba. According to Donnelly
and Rogers (1980), "The difference is most striking in the
abundant siliceous differentiates, in which these elements are
scarcely higher in abundance than in mafic examples." A basic
feature of the series is, moreover, the intrinsically hydrated nature
of its magma. The original definition of the PIA series was based
on the earliest volcanic rocks in Puerto Rico and the Virgin
Islands. Later it was extended to volcanic rocks in the Dominican
Republic, Bonaire, Tobago, Venezuela, and La Désirade (Donnelly and Rogers, 1978). Because of its early position in the
stratigraphy of the volcanic rocks in the Caribbean, Donnelly
and Rogers (1980) suggested that PIA magma was generated by
massive melting of hydrated mantle, which rises in front of the
descending edge of a newly subducted slab at the initial stage of
island-arc genesis. Although we do not understand how and why
this magma reaches the surface instead of freezing deeper in the
crust because of water loss, we agree with Donnelly and Rogers
(1978, 1980) that the initial high water content is one of the main
characteristics of this series.

The exposed section of the Washikemba Formation spans at
least 7 million years, using Odin and Kennedy's (1982) ages for
the top of the Albian and the base of the Coniacian. Throughout
the entire section evidence is found for the high initial water
content of the magma, implying that this cannot be related only
to such a relatively short-lived event as initiation of subduction,
but must be an intrinsic feature of the system. The source for this

water probably is the subducting slab. As there seems to be no
reason to assume that Mesozoic oceanic crust stored more water
than its recent equivalent, the solution must be found in the rate
of introduction of water in the subduction zone. Based on the age
of the Washikemba Formation (Albian-Coniacian = Cretaceous
magnetic quiet zone), we suggest that the exceptionally water-rich
nature of the volcanics of the sequence is related to high rates of
subduction in response to the worldwide surge in spreading rates
postulated for this period (Larson and Pitman, 1972; Hays and
Pitman, 1973; Lowrie, 1979). Whether this interpretation also
holds for the other PIA series in the Caribbean is difficult to
prove, as the ages of these units are still ill-established.

***Aruba: Aruba Lava Formation (= diabase-schist-tuff
formation of Westermann, 1932).*** The Aruba Lava Formation
is over 3 km thick and consists of basalts, diabases, pyroclastic
and volcaniclastic sediments and phyllites, all exposed in the
central part of Aruba. The outcrop area of this unit is truncated at
its northeastern side by the Caribbean Sea, and elsewhere by the
composite batholith which intrudes it. The structure of the unit is
complicated, mainly due to the overprinting of an early phase of
folding by a system of E-W-striking faults in the southern part of
the outcrop area. This fault system post-dates intrusion of the
batholith, and is responsible for the common occurrence of
strongly developed foliation in the rocks bordering the faults.

The formation has been metamorphosed by the intrusion of
the batholith. Metamorphism ranges from hornblende-hornfels
facies at the contact with the batholith to prehnite-quartz facies at
a distance of about 4 km from this contact. The majority of the
rocks have metamorphic parageneses characteristic of the albite-
epidote hornfels facies. Rocks bordering the E-W-oriented late
faults are altered to chlorite-talc schists.

Basalt flows and diabase sills form the main components of
the formation (Fig. 18). The former occur both as pillow lavas
and sheet-flows. Diabase sills are found at various levels in the
formation and may attain a thickness of up to 300 m. The basalts
alternate with pyroclastic and volcaniclastic sediments of varying
nature. Most conspicuous is a level of up to 100 m thickness of
well-sorted conglomerates about 750 m above the base of the
formation (Fig. 18). This conglomerate, which consists of well-
rounded pebbles, cobbles and boulders of basalt and diabase, is
associated with accretionary lapilli tuffs, and locally overlies a
paleosol of weathered basalt with well-preserved exfoliation
structures. Rounding and sorting of the conglomerates strongly
suggest current or wave action. As accretionary lapilli are formed
in eruptive clouds of subaerial volcanoes, we think that this part
of the section was entirely formed on land. However, pillow
basalts with associated marine sediments occur both below and
above this part of the section, so that the conglomerates must be
seen as a short emergent interval in an otherwise marine section.
Emergence is probably due to the local building-up of a volcano
above sealevel. Accretionary lapilli tuffs were occasionally found
embedded between marine sediments (Fig. 18), indicating that
emergent volcanoes were still nearby, that the depth of deposition
was moderate, and that a large part of the section probably

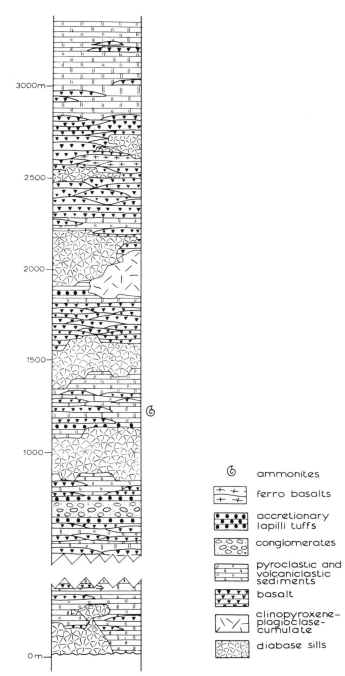

Figure 18. Lithologic columns of the Aruba Lava Formation, a more than 3 km thick alternation of basalts, diabases, and pyroclastic and volcaniclastic sediments. As discussed in the text, the main part of the section is considered to represent the submarine flank of an emerged volcano.

represents the submarine flank of an emergent volcano. An environment similar to that of the Cayman Trench, as suggested by Donnelly and Rogers (1978) for this unit, is therefore considered to be very unlikely, despite the fact that the entire volcanic sequence has a MORB-chemistry. The majority of the volcanic sediments are ash and lapilli tuffs of basaltic composition, hyaloclastites, and fine-grained turbidites consisting of volcanic debris.

A pebbly mudstone intercalated with these turbidites contains a shallow marine fauna, among which are ammonites of Turonian age.

The chemistry of the Aruban basalts closely resembles that of the basalts of the Curaçao Lava Formation, except for the absence of picrites and the presence of more evolved ferrobasalts in the upper part of the formation. The MgO-content of the volcanics varies between 10 and 4 percent, with the exception of a clinopyroxene-plagioclase cumulate with almost 12 percent MgO (Fig. 19). Plagioclase and clinopyroxene are the liquidus minerals within this range; because of metamorphism, they are commonly partly or completely replaced by albite + epidote and a light green amphibole. Olivine was not recognized. Fractionation of plagioclase dominates over clinopyroxene, eventually leading to ferrobasalts in the upper part of the formation (Appendix 1). As in Curaçao, fractionation occurs, at least in part, in the diabase sills: the clinopyroxene-plagioclase cumulate (Fig. 18) at about 2 km above the base of the section is probably the cumulate of the ferrobasalts several hundreds of metres higher in the section. Incompatible element ratios, as for instance Hf/Th (Fig. 20), of the Aruban basalts are identical to those of the Curaçaoan ones, indicating a similar mantle source. The chondrite-normalized REE-patterns of the basalts and diabases of the Aruba Lava Formation (Fig. 21) differ mainly from those of the Curaçao Lava Formation by the presence of positive Eu anomalies in the diabases (cumulates) and negative Eu anomalies in the evolved basalts, stressing the greater importance of plagioclase fractionation in the Aruba Lava Formation. However, La/Yb and La/Sm ratios are similar.

On the basis of the thickness of the basalt succession, the wide range in MgO-content, the composition and the homogeneous nature of the source, it was suggested that the Curaçao Lava Formation was fed by a prolific, chondritic mantle source. The same arguments hold for the Aruba Lava Formation. If the age assignments for the two units are correct (middle Albian for the Curaçao Lava Formation and Turonian for the Aruba Lava Formation), then this source lasted for a period of more than 10 million years.

Aruba: Tonalite/Gabbro batholith. The main member of this batholith is a hornblende tonalite with schlieren of trondhjemite and granitic pegmatite (Westermann, 1932; Helmers and Beets, 1977). The (quartz)-norite to quartz-hornblende gabbro is a slightly older, separate intrusion and forms roof pendants in the tonalite. As has been discussed at length, the age of the batholith ranges between 85 and 90 Ma. Other than a few major element analyses published by Van Tongeren (1934), little chemical work has been done as yet. In order to determine whether there is a direct relationship between the tonalite/gabbro batholith and the basalts of the Aruba Lava Formation, we analysed several samples. The data are plotted on an MgO-variation diagram (Fig. 19). In Figure 21 chrondrite-normalized REE-patterns of two diorites are also shown. The MgO-variation diagram shows that contrary to the tholeiitic Fe-enrichment trend of the Aruba Lava Formation, the gabbro and tonalitic dykes show a decrease

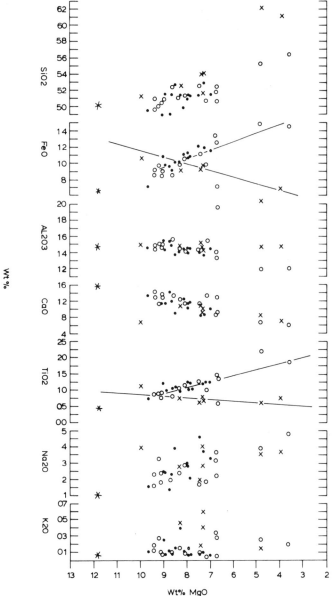

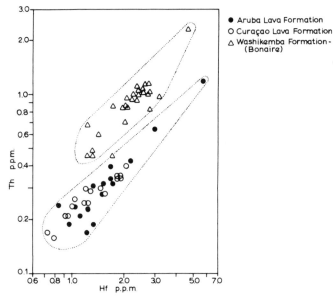

Figure 20. Hf versus Th diagram of basalts and diabases of the Curaçao Lava Formation (open dots) and Aruba Lava Formation (closed dots), and of basalts, basaltic andesites, and andesites of the Washikemba Formation (triangles). Aruban and Curaçaoan basalts and diabases plot in the same field, which suggests a similar mantle source.

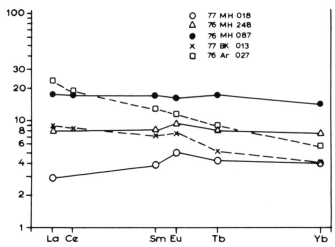

Figure 19. MgO-variation diagram of Aruba Lava Formation and the tonalite/gabbro batholith of Aruba. Open dots: diabases; closed dots: basalts; asterisk: clinopyroxene-plagioclase cumulate; crosses: batholith samples. The different fractionation trends of Aruba Lava Formation and batholith are most obvious in the MgO versus FeO and MgO versus TiO_2 plots, the volcanics having a distinct tholeiitic trend, whereas the batholith samples show a calc-alkaline trend.

Figure 21. Chondrite-normalized REE-patterns of representative igneous rocks of the Aruba Lava Formation and the tonalite/gabbro batholith, 77MH018: clinopyroxene-plagioclase cumulate; 76MH248; pillow basalt; 76MH087: ferrobasalt (all three from the Aruba Lava Formation); 77BK013 and 76Ar027: diorite samples of batholith. For chemical analyses, see appendix 1.

in Fe with decreasing MgO and increasing SiO_2, in other words, a calc-alkaline trend. The different fractionation path of batholith and basalts is accentuated by the TiO_2 versus MgO plot: in the Aruba Lava Formation Ti is incompatible and remains in the melt up to low MgO-contents, whereas the trend of gabbro and dykes shows that throughout the suite a Ti-bearing phase (titano-magnetite or hornblende) effectively fractionates TiO_2. The light-REE-enriched and heavy-REE-depleted patterns of gabbro and dykes are consistent with a calc-alkaline trend for the batholith.

The initial $^{87}Sr/^{86}Sr$ value of 0.703-0.704 reported by Priem and others (1978) suggests a mantle origin for the batholith.

Venezuela: Villa de Cura Group. The Villa de Cura Group is a 4-5 km thick sequence of metamorphic volcanic and volcaniclastic rocks. The unit occurs in a roughly E-W-trending belt of approximately 250 km length, bounded by the Agua Fría Fault in the north and by thrust faults in the south. As discussed above, the rocks have been metamorphosed under high-pressure/low-temperature conditions (Piburn, 1967). The group yielded no fossils; one K-Ar whole-rock radiometric age of 100 ± 10 Ma for an actinolitic metatuff was given by Piburn. Alaskan-type ultramafic intrusions from the Villa de Cura Group were studied by Murray (1972), who considered them to be the deep, subvolcanic cumulates below an andesitic volcanic arc. Hornblendes of two samples of hornblendite and one gabbro from different localities within the El Chacao ultramafic intrusion were dated by Hebeda and others (see this volume). The results of K-Ar dating of these hornblendes are 107 ± 3 Ma, 99.2 ± 3 Ma, and 97.5 ± 3 Ma, thus roughly confirming the Piburn date. Four K-Ar whole-rock age determinations of gabbro in the Villa de Cura Group by Santamaria and Schubert (1974) gave ages of 67-65 Ma. In view of Priem's results and considering the Late Cretaceous-Paleocene thermal event in rocks of the Netherlands Antilles, these latter ages seem to be reset.

The Villa de Cura Group is an allochthonous mass derived from the north and emplaced in Late Cretaceous to Paleocene time (Menéndez, 1967; Bell, 1972). The suggestion of Maresch (1974) and later workers that the Villa de Cura Group formed part of the colliding arc, and hence came from north of the continental margin of South America, was confirmed by paleomagnetic work on the Alaskan-type ultramafic intrusions and other igneous rocks in this belt (Skerlec and Hargraves, 1980). These authors show that the entire belt underwent a 90° clockwise megatectonic rotation from an original N-S trend to the present E-W trend. The K-Ar age of the hornblendes of the ultramafic rocks implies that rotation occurs between about 95 Ma and the final emplacement of the group in Maastrichtian-Paleocene time.

The lithology of the Villa de Cura Group strongly suggests an island-arc origin for most of the volcanics in this sequence, despite the fact that it has been metamorphosed under HP/LT conditions. In most of the units, metabasalts are associated with volcanics of intermediate and acid composition; and in one of the units, the Santa Isabel Formation, metamorphic, siliceous volcanics, the quartz-albite granulites of Shagam (1960) and Piburn (1967), are the prevailing rock-type. Moreover, volcaniclastic and pyroclastic rocks are important constituents of the group, and suggest emergent volcanoes and/or high vapor pressures. Based on major element analyses given by Shagam (1960), Donnelly and Rogers (1978) consider the Villa de Cura volcanics to belong to the primitive island-arc series.

Samples from the Villa de Cura Group were obtained by us through the courtesy of Princeton University and were derived from the original Ph.D. collections of Shagam, Oxburgh, Seiders, Bell, and Konigsmark. None was originally collected for geochemical purposes, and many are strongly foliated. Although relics of an originally porphyritic texture are common, it is often difficult to decide whether the rocks are lavas or tuffs. The latter could involve a possibly biased chemistry because of reworking. We have attempted to select with care the best samples available.

The Villa de Cura Group is subdivided into four formations with transitional boundaries (Shagam, 1960). A short description of each formation and the samples analysed in the present study follows.

1. *El Caño Formation*—In its type-section, the unit consists largely of basic metatuffs. According to Piburn (1967), most of these represent reworked deposits. Conglomerates occur in the basal part. Lavas and sills are found throughout the formation, although in small amounts.

Three samples were analyzed. Two are reworked tuffs, and one, Ar 220 (Appendix 1) is a plagioclase-clinopyroxene phyric flow of andesitic composition. Although metamorphosed, the texture and chemistry of this rock is similar to the sheet-flows of basaltic andesite and andesite in the Washikemba Formation. The chondrite-normalized REE-pattern (Fig. 22) is identical to that of 79KV430/440 (Fig. 14).

2. *El Chino Formation*—According to Piburn (1967), the unit consists of metatuffs alternating with black phyllites and glaucophane and lawsonite schists. Near the top, clinopyroxene-phyric lavas predominate. Sample S762 (Appendix 1) is a strongly foliated metaporphyrite with up to 40 percent relic phenocrysts of clinopyroxene. Except for these, no original igneous textures have been preserved. Glaucophane, a green amphibole, chlorite, and albite are the main metamorphic minerals. Chemically, the rock is a high Mg-basalt. Low Ti, Zr, and Nb and large amounts of clinopyroxene phenocrysts suggest affinity with island-arc tholeiites. The REE-pattern (Fig 22) is about 7 times chondritic, except for Yb, which is depleted with respect to the light and middle REE. This might be due to the large numbers of clinopyroxene phenocrysts in this sample. Sample S609 (Appendix 1) is a strongly foliated and metamorphic rhyolite flow or tuff, with deformed relics of plagioclase and quartz phenocrysts. Low Ti, Y, Zr, and Nb, and the light REE-enriched and middle REE-depleted pattern (Fig. 22), suggest that hornblende was an important fractionating phase. Low concentration of REE and of the high field-strength ions suggests an affinity with the primitive island arc series, although the REE-pattern is different from that of the rhyolites of the Washikemba Formation.

3. *El Carmen Formation*—This 1200 m thick unit consists largely of metamorphic clinopyroxene-phyric basalts, volcanic conglomerates and tuffs (Piburn, 1967). One picrite and two basalts of this unit were analyzed (Appendix 1). All have been metamorphosed; our samples show no distinct foliation.

Picrite and basalt are clinopyroxene-phyric; irregular chlorite/serpentine patches are probably sheared and replaced olivine phenocrysts. Plagioclase joins clinopyroxene and olivine as phenocrysts in the least Mg-rich basalts Ar 1211 and Ar 724. Most basalts are vesicular; in some, the vesicles have rims of quench crystals reminiscent of similar textures in Bonairan basalts. The chemistry of picrite and basalt is very similar to that of MORB, except for the lower Ti-content. REE-patterns show slight depletion of light and heavy REE. Despite chemical similarity to MORB, the early appearance of clinopyroxene on the liquidus suggests that the magma was water-bearing (Holloway and Burnham, 1972), which is not in accordance with the

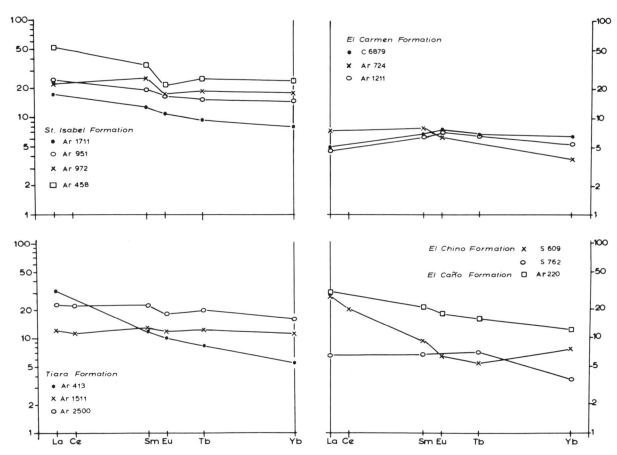

Figure 22. Chondrite-normalized REE-patterns of rocks of the various formations of the Villa de Cura Group. Ar220: metamorphic plagioclase-clinopyroxene-phyric andesite; S762: metamorphic, foliated clinopyroxene-phyric basalt; S609: metamorphic, foliated rhyolite flow or tuff; C6879: metamorphic clinopyroxene-olivine-phyric picrite; Ar 724 and Ar 1211: metamorphic clinopyroxene plagioclase phyric basalt; Ar 1711: Crossite-schist (?andesite); Ar 972, Ar 951, and Ar 458: metamorphic foliated rhyolite flows or tuffs.

MORB-chemistry.

4. *Santa Isabel Formation*—The Santa Isabel Formation consists mainly of quartz-albite granulites, the strongly foliated, crenulated and metamorphic equivalents of tuffs and flows of rhyolitic and dacitic composition (Shagam, 1960; Piburn, 1967). They are associated with metabasalts and metasediments.

Sample Ar 1711 (Appendix 1) is a strongly crenulated crossite and Mg-riebeckite-bearing schist with the chemistry of an andesite. The concentration of the high field-strength ions Zr, Nb, Ti, and Y and that of the REE (Fig. 22) is similar to that of the andesites of the Washikemba Formation and sample Ar 220 of the El Caño Formation. On the basis of chemistry, two groups can be distinguished among the quartz-albite granulites. Predominant are siliceous rocks with very low concentrations of Zr, Nb, and REE (samples Ar 972 and Ar 951; Appendix 1), and moderate concentration of Ti and Y. In a Zr versus TiO_2, Nb, and Y diagram (Pearce and Norry, 1979), this group plots near the field of low-viscosity rhyolites and dacites of the Washikemba Formation. REE-patterns of the low-viscosity rhyolites and these rocks are similar (Fig. 22 and 14). The second group, represented by

sample Ar 458 (Appendix 1), has higher Nb, Zr, and REE concentrations, and plots near the field of the high-viscosity rhyolites of the Washikemba Formation on the Pearce and Norry diagram; REE-patterns are identical.

The chemistry of the igneous rocks of the Villa de Cura Group supports an island-arc environment for this 4-5 km thick sequence. The chemistry and petrology of the Villa de Cura volcanics show similarities as well as differences with respect to the Washikemba volcanics. A difference is the more primitive nature of the basalts of the Villa de Cura Group. However, as discussed, the low Ni- and Cr-content of the Bonairan basalts indicates derivation by low-pressure fractionation of olivine and pyroxene from more primitive parents, which could well be similar to the basalts of the El Carmen Formation. The large amount of siliceous differentiates in the Santa Isabel Formation and the resemblance in chemistry of these rocks to the Washikemba Formation rhyolites and dacites indicate that similar, though still poorly understood, processes led to their origin. These facts, and the corresponding age, suggest that the two units derive from the same island-arc, although their post-depositional history was very different. As shown by Skerlec and Hargraves (1980), the HP/LT metamorphism of the group occurs before the 90° clockwise rotation, and is therefore due to subduction-related processes in which the Villa de Cura part of the island-arc was involved.

Venezuela: La Rinconada Group, Margarita Island.

The metamorphic terrane of Margarita Island, Venezuela (Fig. 1), can be divided into a metavolcanic unit defined as the La Rinconada Group (Maresch, 1975), and the metasedimentary units Juan Griego Group and Los Robles Group (Venezuela, 1970). Ultramafic rocks metamorphosed together with these units and usually strongly serpentinized are especially common in the La Rinconada Group. Non-metamorphic sediments overlying this basal complex are Eocene or younger in age. As discussed in an earlier section, a Coniacian-Santonian age of metamorphism appears likely. Remnants of microfossils in the metamorphic Los Robles Group yield ?Cenomanian ages (Furrer, personal communication to Vignali, 1972).

Several types of intrusive magmatic rocks are found (e.g., González de Juana and Vignali, 1972; Maresch, 1971). The most important of these are sheared post-metamorphic granites (such as the El Salado Granite of Maresch, 1971), which are relatively minor in extent, as well as intermediate intrusives, such as tonalites and trondhjemites. Although this fact has not been stressed in recent discussions of Margarita Island (e.g., Vignali, 1979; Stephan and others, 1980), the high-K_2O granites and the low-K_2O tonalites and trondhjemites form two distinct groups of intrusives. The tonalites are intimately associated, often in lit-par-lit fashion, with the metavolcanic rocks of the La Rinconada Group, and are predominantly metamorphosed to quartz-albite-white mica orthogneiss.

The stratigraphy of the metamorphic units of Margarita Island has undergone a number of drastic revisions, the details of which were described by Maresch (1971, 1975) and Vignali (1979). We interpret the La Rinconada Group to be a basal unit that is overlain via a transitional contact by the metasedimentary rocks of the Juan Griego Group, which in turn is reported to be conformably overlain by the Los Robles Group (Vignali, 1972). The stratigraphy adopted is important because it constrains the age of the La Rinconada Group. Correlation of the Juan Griego Group with the metasedimentary rocks of the Venezuelan mainland (e.g., Venezuela, 1970, p. 314; Maresch, 1972) and the Cenomanian fossil age from the Los Robles Group indicate that the underlying La Rinconada Group must be of Late Jurassic age or older. Thus a complete Upper Jurassic-Cenomanian section is inferred to exist on Margarita Island, a conclusion that argues against the Early Cretaceous metamorphic event postulated by Stephan and others (1980).

The metavolcanic rocks of the La Rinconada Group crop out as a sequence of banded amphibolites with paraconformable layers and lenses, from decimetres to tens of metres thick, of massive unfoliated amphibolite. The structure-corrected thickness of the exposed section has been estimated as exceeding 2000 m (Maresch, 1971). The trace element spectrum of the banded amphibolites (Mottana and others, 1979) does not support Ma-

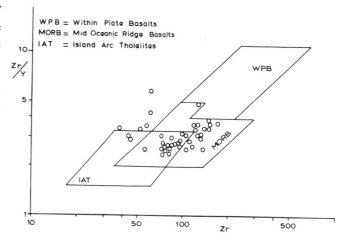

Figure 23. Zr/Y versus Zr-diagram of rocks of the La Rinconada Group, Margarita Island.

resch's (1971, 1975) initial interpretation, based on macroscopic aspect, that these rocks originated mainly as pyroclastics. Rather, they must represent former flows and possibly stretched, sheared pillows. The massive amphibolites are probably gabbroic or diabasic intrusives into this volcanic pile (Maresch, 1971, 1977; Mottana and others, 1979). No volcanic structures are now preserved. The grade of metamorphism reached may be roughly described by the classical epidote-amphibolite facies. In rocks of suitable bulk and fluid composition in the northern part of the outcrop area, eclogites and related omphacite-bearing assemblages are common (Maresch, 1977; Maresch and Abraham, 1981).

A geochemical study (major and minor elements, REE) has been undertaken on a suite of metabasaltic samples from the La Rinconada Group, in order to clarify the tectonic setting in which these magmatic rocks originated. A progress report of this study has been presented by Mottana and others (1979). An obvious uncertainty is the problem of possibly extensive mobility of many rock components during the metamorphic process, although the discovery that REE-elements display low mobility during progressive high-pressure metamorphism of Alpine mafic rocks up to the blueschist/eclogite facies (Morten and others, 1979) gives cause for optimism.

The geochemical evidence obtained points to a tholeiitic character for the La Rinconada Group, although SiO_2 may reach 54 wt %. A definitive distinction between MORB and island-arc tholeiite (IAT), or the PIA ("primitive island-arc") classification of Donnelly and Rogers (1978), is not possible with the data at hand. Elements such as Zr and Y, which are considered relatively immobile, favour a MORB character (Fig. 23). On the other hand, REE-patterns and enrichment factors (Fig. 24) are quite variable and display characteristics between those typical of MORB and IAT or PIA.

A geochemical distinction between the La Rinconada Group and the basic volcanics of the Netherlands Antilles and the Villa de Cura Group/Tiara Formation is not possible with the

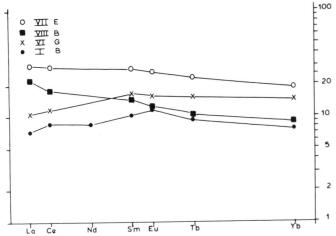

Figure 24. Chondrite-normalized REE-patterns of metabasalts of the La Rinconada Group of Margarita Island. For chemical analyses, see Appendix 1.

data at hand. The inferred age of the La Rinconada Group suggests that it formed during rifting of the proto-Caribbean, in which case a MORB character for the basalts would appear more logical.

No geochemical data have been obtained on the acid intrusives. However, the contact relationships and the low-K_2O character of the tonalitic orthogneisses (Maresch, 1971) are typical of many so-called plagiogranites (e.g., Brown and others, 1979). Their origin must be related to the problems of the origin of plagiogranites in ophiolites as a whole. The unmetamorphosed intermediate intrusives and the granites could be related to the post-collisional calc-alkaline volcanic arc of Maresch (1974). The high-K_2O nature of these granites deserves attention. Walker and others (1972) have also described K-feldspar-rich granitic rocks from the Aves Ridge 200 km to the north in the Caribbean Sea. Initial $^{87}Sr/^{86}Sr$ ratios as low as 0.7038 obtained for these rocks are suggestive of a mantle origin.

CONCLUSIONS

General Statement

In the preceding part, we have shown that only one regional metamorphic event, of high-pressure character, has affected the Caribbean/South American plate margin in Late Mesozoic time. This event can be related to the collision of an island-arc with the northern continental margin of South America in the Coniacian to Campanian interval. The blueschist metamorphism of the Villa de Cura Group pre-dates this event and is due to subduction-related processes in an island-arc environment.

We have studied remnants of this island-arc exposed on the islands of Aruba, Curaçao, and Bonaire, as well as in the Villa de Cura Tectonic Belt of northern Venezuela. The origin of the La Rinconada Group of Margarita Island is still equivocal, but the evidence at hand leads us to believe that this unit belongs to the continental margin association of northern South America, overridden during collision with the island-arc.

Island-arc magmatism is represented by the Washikemba Formation of Bonaire, the tonalite/gabbro batholith of Aruba and the Villa de Cura Group of Venezuela. The tonalite/gabbro batholith is a calc-alkaline complex. Volcanics of the Washikemba Formation and the Villa de Cura Group have the characteristics of the primitive island-arc series. High initial water-content of the magma, which according to Donnelly and Rogers (1978) is an important feature of this series, can be demonstrated for the Washikemba volcanics. This island-arc volcanism is associated with basalt-sequences of MORB-affinity on the islands of Curaçao and Aruba. These sequences differ from normal oceanic crust by a much thicker crustal section (more than 5 and 3 km, respectively) and the homogeneous and chondritic nature of their mantle source.

Origin and Evolution of the Island-Arc

In order to define the site of origin of the Mesozoic island-arc and the cause of its formation, the extent of the volcanic sequences thought to belong to it and the age of the oldest rocks in the arc must be known. Although much geological and geophysical work has been done in the Caribbean, these two basic questions are still difficult to answer; consequently, the models presented here are still speculative.

When Maresch (1974) proposed a collision-type model for the Cretaceous history of the southern Caribbean margin, the Caribbean as a whole was ignored. In a later paper, Maresch (1976) proposed a continuation of this Mesozoic arc into the Aves Ridge and Lesser Antilles arc. However, different models were proposed for the Greater Antilles (Mattson, 1977, 1979), and it was far from clear how the various models could be fitted together. A strong plea for an island-arc collision with both the northern and southern margin of the Caribbean was given by Dickinson and Coney (1980), based on plate tectonic constraints and geological data. The most convincing, though not conclusive, evidence for a continuation of the colliding arc from northern South America, via the Aves Ridge and Lesser Antilles arc, into Puerto Rico, Hispaniola, and Cuba is based on paleomagnetic data (Skerlec and Hargraves, 1980). As mentioned previously, the anomalous easterly declinations of the island-arc volcanics in the Caribbean Mountain System of Venezuela suggest a 90° clockwise rotation of the arc relative to stable South America in the Late Cretaceous. Stearns and others (1982) report the same results from rocks of the Washikemba Formation of Bonaire and from the Aruba Lava Formation and the batholith of Aruba. As discussed by Skerlec and Hargraves, the majority of the paleomagnetic data from Mesozoic rocks along the northern margin of the Caribbean is characterized by anomalous westerly declinations, and points to an anticlockwise megatectonic rotation with regard to stable North America. Such a pattern would be expected if an island-arc originally oriented N-S were to move into the Caribbean and collide with its northern and southern margins, provided that the natural remanent magnetization (NRM) of the rocks was acquired in periods of normal polarity. Skerlec and Hargraves (1980) suggest that both sets of NRM were acquired during one long-lasting polarity epoch. The only polarity interval which qualifies is the Cretaceous magnetic quiet period, which lasted from the Barremian/Aptian boundary to the Santonian (Lowrie and others, 1980), and which has an uninterrupted normal polarity for 30 (Odin and Kennedy, 1982) or 40 Ma (Larson and Hilde, 1975; Lowrie and others, 1980). Skerlec and Hargraves' (1980) suggestion would imply that most of the island-arc volcanics incorporated in the collision were formed (or at least magnetized) within this period. Although the ages of the volcanic sequences in the Caribbean are still not precisely known, our present knowledge seems to confirm their suggestion in main lines. The ages of all volcanic units of the island-arc discussed in the preceding section fall within this interval. Moreover, in all volcanosedimentary, fossil-bearing units in the Caribbean, faunas older than Early Cretaceous have never been encountered, except for the possible Late Jurassic Radiolaria of the Bermeja complex of Puerto Rico (Mattson and Pessagno, 1979). However, radiometric dating of igneous and metamorphic volcanics in the Caribbean also yields ages older than the Barremian-Aptian boundary. Most of these fall in the Early Cretaceous, as for instance, 123 ± 2 Ma for amphibolite (K-Ar on hornblende and plagioclase; Kesler and others, 1977); 127 ± 5 Ma for hornblendite (K-Ar on hornblende; Bowin, 1975) in the Dominican Republic; 114-130 ± 14 Ma for amphibolites, dolerites, and gabbros from Venezuela and the Venezuelan islands Los Monjes and Gran Roque (whole-rock K-Ar; Santamaría and Schubert, 1974); and 127 ± 7 Ma for gabbro of Tobago (whole-rock K-Ar; Rowley and Roobol, 1978). For the amphibolites, these are metamorphic ages that post-date the igneous event by an unknown amount. Older ages are scarce. Khudoley and Meyerhoff (1971) report a date of 180 Ma for a granodiorite from Las Villas Province, Cuba, and Mattinson and others (1973) give ages of 142 ± 10 Ma (U-Pb on zircon) for quartz-keratophyre of La Désirade. The 180 Ma is probably incorrect, as the Caribbean was still closed at that time (Ladd, 1976). The Jurassic age for the La Désirade volcanics has been criticized by Dinkelman and Brown (1977), who find an age of 87 ± 12 Ma (argon isochrons) and consider this to be an upper limit for the ophiolite suite of the island. Consequently, the data on the maximum age of the volcanics in the Caribbean still leave room for speculation concerning the timing of the origin of the island-arc. Most data suggest a late Early Cretaceous time for its formation, but an earlier, Jurassic age cannot be excluded.

The presence of metamorphosed basic and ultrabasic rocks, including eclogites, in a broadly arcuate band across south-central Guatemala, from near the Mexican border to the Caribbean margin (Dengo, 1972; Donnelly, 1977; Roper, 1978), suggests that the island-arc volcanics extend from Cuba into Central America. A continuation of the arc along the western coast of Mexico is still open to question. The Sierra Madre del Sur was thought to be underlain by a Precambrian and Paleozoic metamorphic basement (De Cserna, 1965; Kesler, 1973) intruded by Mesozoic batholiths. Recent work (Campa-Uranga, 1979; Rangin, 1978; Gastil and others, 1978), however, suggests the existence of a Mesozoic, oceanic island-arc succession in Peninsular California and western Mexico, which was welded to the continent in late Albian (Gastil and others, 1978) or Cenomanian time. Although this latter age coincides rather well with the time of collision in the Caribbean, the volume of Jurassic (and even Triassic) volcanics in the Peninsular Arc seems to be much larger than that found in the Caribbean. A much better case for a continuation of the colliding arc beyond the Caribbean can be made for the western Cordillera of Colombia and its extension in Ecuador. Based on similarities in gross structure, a continuation between the western end of the Caribbean Mountain System of Venezuela and the northeastern end of the Colombian Andes was suggested by Alvarez (1971). Now that more data on the geology of the western Cordillera of Colombia have become available (Case and others, 1971; Barrero, 1977; Goossens and Rose, 1973; Goossens and others, 1977), the fact emerges that the huge volumes of submarine volcanics include both MORB-sequences and island-arc tholeiites and andesites (Barrero, 1977; Goossens and others, 1977). The ages reported by the authors mentioned are all late Early Cretaceous and Late Cretaceous, and consequently, are in the same range as those of the Netherlands Antilles.

In order to be objective, we propose two alternative models: one in which the island-arc originates as a Central American "isthmus" when the Caribbean begins to open in the Late Jurassic, and one in which the trench-arc system forms in the Pacific in Early Cretaceous time.

In model I (Fig. 25), it is assumed that Pacific oceanic crust was consumed in a Jurassic and older arc-trench system along the margin of the then-still-united Americas (Dickinson and Coney, 1980). Opening of the Caribbean in the Late Jurassic (Ladd, 1976) resulted in extension of this arc across the gap which formed between Guatemala and South America. At that time the Nicaragua Rise was still outside the Caribbean (Gose and Swartz, 1977). The Jurassic Mid-Atlantic/Caribbean spreading ridge would have connected with this trench at a trench-trench-ridge (TTR) triple junction. Subduction started as west-facing, down-to-the-east, but it is envisaged that this sense reversed in the late Early Cretaceous to east-facing, down-to-the-west, so that from then on models I and II of Figure 25 are roughly similar. Whether the arc originated in the "straits" between Guatemala and Colombia (model I), or swept through this gap (model II), it is clear that, in order to form the large loop that collided with the Bahamas Bank as well as the South American margin, the arc had to lengthen to more than twice the width of the then-existing gap. This fact and the presence of MORB-sequences of similar age as the PIA series in the island-arc (e.g., Aruba Lava Formation) suggest that lengthening of the arc occurred at a TTR triple junction. The ballooning of the arc to allow it to collide both with the northern and southern Caribbean margins necessitates the formation of new crust in the back-arc basin. A connection of the TTR triple junction with an RRR junction in the back-arc basin,

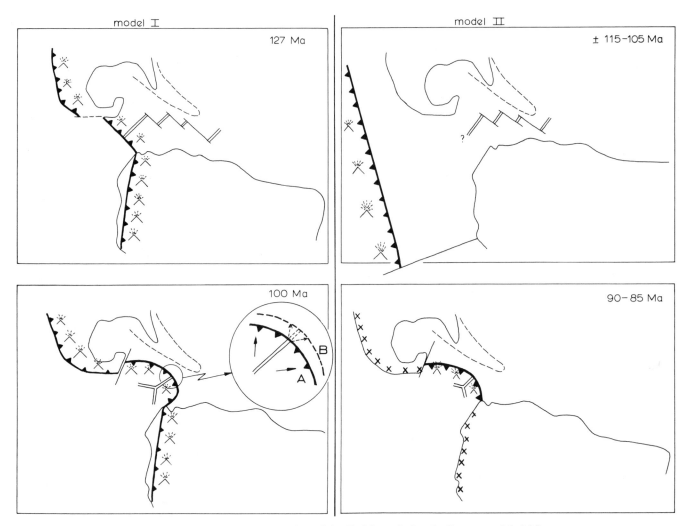

Figure 25. Two alternative reconstructions of the Caribbean during the Cretaceous. Model I assumes origin of the island arc in the Central American isthmus in the Late Jurassic with a reversal of the sense of subduction in the mid-Cretaceous. Model II assumes initiation of subduction in the Early Cretaceous in the Pacific and subsequent movement of the arc towards the east. The arc collides with the western margin of South and North America in Cenomanian or Turonian time and with the Bahama bank and the northern South American margin in the Santonian. Relative position of the Americas after Ladd (1976).

much the same as has been suggested for the evolution of the New Hebrides-Vityas, Fiji, and Lau-Tonga island-arcs in the Pacific in the Late Tertiary (Falvey, 1978), could provide a simplified model for the evolution of the island-arc and the Caribbean basin in the Cretaceous (Fig. 25).

Although older (Jurassic?) oceanic crust could have been moved into the Caribbean behind the island-arc, the balloon-like expansion of the arc indicates that at least part of the Caribbean crust must have formed concomitant with the arc. The Coniacian age of the top of the basement in the Venezuelan Basin (Edgar, Saunders, and others, 1973) does not contradict this. Seismic work and deep sea drilling results have shown that the crust of the central and northwestern part of the Venezuelan Basin differs from "normal" oceanic crust by a greater thickness, smooth surface, the presence of sub-basement reflectors and a bathymetry too shallow for crust of Cretaceous age (Edgar and others, 1971; Edgar, Saunders, and others, 1973; Diebold and others, 1981). To explain these features, Edgar, Saunders, and others (1973) and Donnelly (1975) proposed that the drilled Coniacian flows and sills belonged to a Late Cretaceous "flood basalt event" superimposed upon an older (?Jurassic) oceanic crust. Diebold and others (1981) point to the resemblance between the Venezuelan Basin crust and that of the Nauru Basin in the western Pacific, where a mid-Cretaceous volcanic complex of more than 500 m thickness overlies a Late Jurassic oceanic basement (Larson and Schlanger, 1981). This off-ridge volcanic complex forms part of the Early to mid-Cretaceous thermal event in the western Pacific (Schlanger and others, 1981).

An alternative, which would better fit our models, is that the thick Venezuelan Basin crust is formed by back-arc spreading in the Cretaceous. As shown by Sleep and Windley (1982), the thickness of the oceanic crust mainly depends on the depth at which partial melting to produce basalt starts, and this in turn is a function of the heat of the mantle source: the hotter the source, the thicker the crust. As discussed earlier, the characteristics of the Curaçao and Aruba Lava Formations—chemical homogeneity; thickness of basalt pile building up to above sealevel in the case of the Aruba Lava Formation; large amount of picrites in the Curaçao Lava Formation; virtual absence of pelagic intercalations—point to a prolific and hot mantle plume. These units are considered to have formed at the TTR triple junction. A similar hot source behind the arc does not seem unlikely.

Whether formed by off-ridge volcanism or back-arc spreading, the resemblance between the Venezuelan Basin crust and that of the western Pacific, where thick and buoyant Cretaceous crust is commonly found, has been noted by many workers (e.g., Donnelly, 1975; Burke and others, 1978; Batiza and others, 1980; and Schlanger and others, 1981). Some of these western Pacific plateaus are considered to have formed at spreading centers, mostly triple junctions, as for instance, Ontong Java (Andrews, Packham, and others, 1975), Manihiki (Schlanger, Jackson, and others, 1976), and Hess Rise (Vallier and others, 1981). Others, such as the Nauru Basin (Larson and Schlanger, 1981) and the Mid-Pacific Mountains (Thiede and others, 1981), represent off-ridge volcanism. Because of the buoyant nature of these plateaus, their oceanic origin has been questioned. However, recent seismic work (Hussong and others, 1979; Carlson and others, 1980) shows that they consist of a 6-15 km thick, high-velocity basal layer overlain by units having velocities typical of oceanic layers 2 and 3, which at Ontong Java are up to five times thicker than in the oceanic crust of the surrounding basin. The nature of the basal layer is difficult to assess, but it is certainly not of continental origin. Western Pacific plateaus as well as off-ridge volcanism date from the 120 to 90 Ma interval and bear witness to an event with a much greater than normal heat output from the mantle (Turcotte and Burke, 1978; Watts and others, 1980; Sprague and Pollack, 1980).

These considerations lead to model II, in which the arc-trench system originated in the Early Cretaceous in the Pacific Ocean as east-facing, down-to-the-west, moved eastwards, collided with the western margin of North and South America in mid-Cretaceous time, and invaded the Caribbean in the Late Cretaceous. This model renders it possible to include the volcanics of the western Cordillera of Colombia and those of eastern Mexico in this arc system. Moreover, the model suggests a link between the mid-Cretaceous thermal event in the Pacific and the initiation of a trench-arc system by plate rupture along a pre-existing weakness zone (a transform fault) and down-flexing under the load of off-ridge volcanism (Cloetingh and others, 1982).

The cause of the mid-Cretaceous thermal event is not known. As it occurs just before and during the initial stage of the Cretaceous magnetic quiet period, a connection between the two events is suggested. A coupling of processes in the upper mantle

and outer core was inferred by Vogt (1975) on the basis of the synchronism in variations in geomagnetic reversal frequency and major plate-tectonic changes at 42-45, 70-80, and 110-120 Ma. McElhinny (1979) suggested that convection caused by heat crossing the core-mantle boundary might be intermittent, consisting of rising plumes which are occasionally emitted from a thermal boundary layer at the base of the mantle. According to him, this could give rise to variations in the Rayleigh number of the core; such fluctuations are likely to disturb the flow field in the core and thus affect the frequency of reversals. Perhaps the extensive outpouring of basalt in the mid-Cretaceous of the Pacific realm represents such an episode of heat transfer from the core-mantle boundary. Plumes and plume systems rising from deeper parts of the mantle could well explain the prolific, homogeneous, and chondritic nature of the mantle source of the volcanics of the Curaçao Lava Formation and Aruba Lava Formation.

The mid- and Late Cretaceous worldwide rise in sealevel is intimately associated with the thermal event in the mid-Cretaceous. Schlanger and others (1981) ascribe this rise to regional uplift in the Pacific because of heating of the Pacific lithospheric plates in the mid-Cretaceous, thus discarding the earlier idea of Hays and Pitman (1973) and Pitman (1978) that this rise is due to an expansion of the spreading ridges because of a worldwide surge in spreading rates (Larson and Pitman, 1972). As discussed by Berggren and others (1975), Schlanger and others (1981), and Lowrie and others (1980), the ages of the boundaries of the Cretaceous magnetic quiet zone used by Larson and Pitman (1972) to calculate spreading rates proved to be wrong. Using paleomagnetic results of well-dated pelagic sections of Hauterivian to Paleocene age from Umbria and the southern Alps, Lowrie and others (1980) conclude that the Cretaceous magnetic quiet interval lasted from 115 to 78 Ma, thus reducing spreading rates to about 70% of the values given by Larson and Pitman (1972). However, this implies that they are still appreciably higher than during formation of the preceding M sequence anomalies. As mentioned earlier, the water-rich nature of the Washikemba Formation volcanics and other primitive island-arc sequences in the Caribbean (Donnelly and Rogers, 1978), in our opinion, can be explained only by assuming high rates of subduction below the arc, as the water must come from the subducted slab. High subduction rates imply a high rate of introduction of water in the subduction zone and a depression of the isotherms. As most of the water is released before the slab reaches a depth of 100 km and the temperature probably is too low to allow partial melting at this depth, the water will cause mantle diapirs to rise to shallower depths where partial melting takes place and magma of the tholeiitic island-arc series is generated (Ringwood, 1975).

In summary, it can be stated that the present data on the geology of the Cretaceous island-arc sequences in the Caribbean still leave much room for speculation. We have acknowledged this fact by proposing *two* models for the origin of the island-arc that collides with the Bahamas Bank and the northern continental margin of South America in the Late Cretaceous. In both models, the magmatic evolution of the arc is strongly influenced by the mid-Cretaceous thermal event responsible for the huge oceanic plateaus in the Pacific, a general increase in spreading rates, and a worldwide rise in sealevel. The origin of the arc in the first model is a conventional one and is due to extension of an existing subduction zone across the Central American isthmus when North and South America separated. In the second model, the mid-Cretaceous event is considered to be responsible for the origin as well as the evolution of the island-arc and its back-arc basin. Although the concept is speculative and unconventional, it best explains the present data, and we consider it an important basis for future research.

ACKNOWLEDGMENTS

Many colleagues contributed in one way or another to the collected results presented here. We are especially grateful to Dr. Harry N. A. Priem and his coworkers of the ZWO laboratory for isotope geology for permission to use all their unpublished radiometric age data on the Cretaceous rocks of the Netherlands Antilles and for specially undertaking work on samples from the Villa de Cura Group of Venezuela (Hebeda and others, this volume). We are also grateful to the faculty of the Department of Geological and Geophysical Sciences, Princeton University, for providing us with material on the Villa de Cura Group from thesis collections; to T. C. Onstott and J. Wolinski for cutting them and shipping this material to us; and to Professor R. B. Hargraves for sending us samples of the El Chacao intrusion.

We sincerely appreciate the time-consuming reviews of our lengthy manuscript provided by Th. W. Donnelly, R. B. Hargraves, B. A. Morgan, and R. Shagam, which led to an improved presentation and clearer logic.

We would also like to thank the following persons for their generous help: S. Judson, G. Liborio, L. Morten, D. Falvey, M. de Bruin and coworkers, H. J. MacGillavry, C. Kieft, P. Maaskant, Th. B. Roep, F. Kievit, A. van Eunen, and J. Boer.

Part of the work reported here was financially supported by grants MA 689/1 (Deutsche Forschungsgemeinschaft, Bonn) to W.V.M.; CT75.00881,05 (Consiglio Nazionale delle Ricerche, Roma) to A. M.; WR75-211, 212, and 213 (Netherlands Foundation for the Advancement of Tropical Research—WOTRO) to D.J.B., G.Th.K., and F.F.B.; and the Molengraaff Fund to G. Th. K. and H. M.

APPENDIX 1. CHEMICAL ANALYSES OF IGNEOUS ROCKS OF THE CURACAO LAVA FORMATION, WASHIKEMBA FORMATION, ARUBA LAVA FORMATION, ARUBAN BATHOLITH, VILLA DE CURA GROUP, AND LA RINCONADA GROUP.
(main elements in wt%, trace elements in ppm)

	Curacao Lava Formation			Washikemba Formation						
	1	2	3	4	5	6	7	8	9	10
SiO_2	47.38	50.68	52.81	50.95	52.18	58.10	56.19	71.07	70.92	73.37
TiO_2	0.70	1.04	1.48	0.81	1.13	1.37	1.69	0.57	0.73	0.69
Al_2O_3	10.32	14.34	13.62	16.51	18.11	15.53	16.23	15.47	13.96	13.12
FeO	10.27	10.36	12.20	10.27	7.78	9.38	7.56	3.65	4.63	3.58
MnO	0.18	0.18	0.17	0.18	0.13	0.11	0.14	0.03	0.15	0.07
MgO	20.68	8.56	6.00	7.69	6.30	4.15	2.72	0.40	2.63	1.19
CaO	8.49	12.46	9.99	7.53	8.30	5.45	6.41	0.14	0.49	0.66
Na_2O	1.16	2.17	3.42	5.03	4.46	4.44	7.13	7.11	6.04	5.57
K_2O	0.10	0.16	0.08	0.94	1.39	1.17	1.53	0.85	0.51	1.54
P_2O_5	0.05	0.07	0.11	0.09	0.19	0.22	0.39	0.06	0.20	0.20
Sc	31.8	46.-	46.-	34.99	32.39	29.20	23.00	12.00	18.00	16.25
Cr	1847.-	387.-	104.-	145.20	158.60	34.30		3.10	4.44	3.60
Ni	870.-	109.-	57.1	36.0	51.3	17.7	2.48	2.57		3.34
Y	13.5	20.5	22.1	18.5	25.8	32.8	41.6	72.6	43.0	45.0
Zr	45.6	66.8	81.1	49.1	77.5	92.2	152.0	271.0	94.4	89.2
Nb	3.4	6.53	7.60	1.00	3.07	3.40	5.68	6.92	1.07	
La	2.03	3.05	4.56	3.62	9.01	9.27	12.60	24.40	8.66	10.79
Ce	6.13	7.18	11.30		16.40		27.90		17.35	18.18
Sm	1.43	2.18	2.86	2.03	3.78	4.32	5.89	9.79	4.89	5.27
Eu	0.55	0.74	0.91	0.74	1.26	1.37	2.00	2.70	1.45	1.44
Tb	0.31	0.47	0.67	0.48	0.69	0.80	1.26	1.62	1.02	1.02
Yb	1.31	1.91	2.65	2.03	2.19	3.08	3.50	6.70	3.96	4.13
Hf	0.92	1.22	1.75	1.31	1.96	2.63	3.70	6.90	3.13	3.22
Th	0.23	0.28		0.49	0.85	1.17	1.00	2.19	1.24	1.25

APPENDIX 1 (continued)

	Aruba Lava Formation			Aruban Bath.		Villa de Cura Group				
	11	12	13	14	15	16	17	18	19	20
SiO_2	49.99	51.49	55.36	62.22	61.29	56.07	48.03	77.88	49.49	49.90
TiO_2	0.45	1.04	2.18	0.58	0.76	1.63	0.61	0.23	0.61	0.64
Al_2O_3	13.56	14.97	11.87	14.66	14.83	16.75	13.33	12.14	10.95	16.29
FeO	6.84	9.17	14.86	5.23	6.87	8.76	10.21	1.49	10.36	9.11
MnO	0.13	0.14	0.19	0.07	0.12	0.17	0.18	0.01	0.19	0.18
MgO	11.87	8.59	4.78	4.79	3.96	3.21	11.98	0.77	12.16	9.23
CaO	15.84	12.03	6.39	8.55	7.03	5.58	13.12	0.34	12.79	10.77
Na_2O	1.03	2.35	3.95	3.61	3.74	5.65	1.58	5.31	2.11	3.00
K_2O	0.05	0.08	0.26	0.16	1.16	1.70	0.73	1.70	1.05	0.75
P_2O_5	0.03	0.10	0.17	0.12	0.25	0.52	0.16	0.05	0.16	0.07
Sc	52.79	50.31	49.24	26.72	24.7	18.50	46.-	4.6	44.70	48.0
Cr	1476.-	356.5	12.51	127.2	70.58	5.90	564.-	4.9	844.0	340.0
Ni	217.-	111.-	43.5	34.9	16.7	6.10	229.-	2.78	313.0	86.0
Y	12.4	21.4	38.1	14.1	18.5	33.5	13.40	20.1	17.0	18.8
Zr	31.3	64.4	124.-	48.5	62.3	105.-	37.80	105.-	44.4	50.2
Nb	2.47	5.78	11.5	1.60	3.15	4.34	2.75	<0.56	3.15	2.21
La	1.06	3.04	6.44	3.12	8.55	11.35	2.48	10.45	1.83	1.78
Ce		12.16	16.41	8.00	18.48	22.30		19.39	8.65	5.59
Sm	0.88	1.84	3.93	1.72	2.98	5.00	1.59	2.28	1.61	1.84
Eu	0.45	0.83	1.43	0.65	0.99	1.59	0.62	0.58	0.66	0.66
Tb	0.25	0.49	0.99	0.29	0.49	0.93		0.33	0.38	0.36
Yb	1.00	1.91	3.49	0.97	1.43	3.08	0.94	1.94	1.65	1.64
Hf	0.62		<2.98	1.37	2.02	2.80		3.02	0.81	1.02
Th			<0.65	0.23	1.38	1.00		1.78		

APPENDIX 1 (continued)

	Villa de Cura Group					La Rinconda Group			
	21	22	23	24	25	26	27	28	29
SiO_2	48.96	58.22	70.0	72.14	75.04	50.59	49.97	53.78	51.56
TiO_2	0.70	0.92	0.81	0.74	0.34	2.05	1.40	0.90	1.94
Al_2O_3	14.62	14.65	14.26	12.89	13.84	12.54	15.26	13.25	13.07
Fe_2O_3						6.90	4.34	5.58	5.13
FeO	10.55	10.05	3.66	4.49	1.91	6.10	5.11	6.86	6.21
MnO	0.18	0.08	0.05	0.09	0.08	0.23	0.16	0.20	0.22
MgO	10.09	3.37	1.03	1.84	0.42	5.45	7.42	5.24	6.83
CaO	11.45	5.25	1.76	1.12	0.51	11.14	11.08	8.50	9.65
Na_2O	2.49	6.21	8.10	6.45	6.89	1.44	2.78	2.85	3.01
K_2O	0.20	1.08	0.05	0.04	0.91	0.37	0.05	0.11	0.12
P_2O_5	0.19	0.17	0.25	0.22	0.06	0.30	0.23	0.20	0.42
Sc	42.6	33.6	16.0	14.60	9.11				
Cr	366.0	21.9	5.40	4.70			157.-	53.-	72.-
Ni	117.0	23.0	4.13	6.00	3.75	10.0	53.-	34.-	45.-
Y	14.4	22.9	47.30	39.0	57.2	44.-	39.-	30.-	47.-
Zr	38.5	64.6	127.-	100.-	223.-	117.-	111.-	80.-	174.-
Nb	2.84	2.38	1.79	<0.65	3.30				
La	1.75	6.35	8.35	8.92	19.50	2.6	3.9	8.30	11.40
Ce		13.99	21.39	18.70	41.70	8.26	11.3	17.14	28.00
Nd						5.95	8.22	9.27	
Sm	1.54	2.99	5.85	4.49	8.14	2.36	3.84	3.34	6.57
Eu	0.63	0.96	1.55	1.45	1.91	1.02	1.38	1.11	2.29
Tb	-.-	0.56	1.09	0.89	1.45	0.54	0.89	0.59	1.37
Yb	1.37	2.01	4.64	3.76	6.16	1.90	3.48	2.20	4.68
Hf			3.84	3.09	6.71				
Th			0.84	0.80	1.74				
Lu						0.32	0.70	0.41	0.75

1. 79KV615/620/621 - Mean of three samples of one pillow of picrite; 2. 79Bel83 - olivine-phyric pillow basalt; 3. Be 2665 - plagioclase-pyroxene-phyric basalt; 4. KV554 - Basalt sheet flow; 5. 79KV591 - basaltic andesite, pillow lava; 6. 79KV430/440 - andesitic sheet flow, mean of 11 analyses; 7. KV717 - andesite associated with high-viscosity rhyolites; 8. KV663 - high-viscosity rhyolite; 9. 79KV163 - Brandaris-Goto rhyodacitic sill; 10. KV526 - low-viscosity rhyolite; 11. 77MH018 - clinopyroxene-plagioclase cumulate; 12. 76MH248 - pillow basalt; 13. 76MH087 - ferrobasalt; 14. 77BK013 - diorite; 15. 76AR027 - diorite; 16. Ar220 - metamorphic plagioclase-clinopyroxene-phyric andesite, El Cano Formation; 17. S762 - metamorphic, foliated clinopyroxene-phyric basalt, El Chino Formation; 18. S609 - metamorphic, foliated rhyolite flow or tuff, El Chino Formation; 19. C6879 - metamorphic clinopyroxene-?olivine-phyric picrite, El Carmen Formation; 20. AR 724 - metamorphic clinopyroxene-plagioclase-phyric basalt, El Carmen Formation; 21. Ar 1211 - metamorphic clinopyroxene-plagioclase-phyric basalt, El Carmen Formation; 22. Ar 1711 - crossite schist (?andesite), Santa Isabel Formation; 23. Ar972, 24. Ar951, 25. Ar458 - metamorphic, foliated rhyolite flows or tuffs, Santa Isabel Formation; 26. IB, 27. VIG, 28. VIIB, 29. XIIE - metabasic rocks (data from Mottana and others, in press). All iron as FeO, except for samples 26-28.

APPENDIX 2. ANALYTICAL PROCEDURES

The samples for chemical analyses of the Curaçao Lava Formation, Aruba Lava Formation, Washikemba Formation, and the Aruban Batholith were selected after an extensive field and petrographic study of each formation. We have selected the samples in such a way that they represent both the vertical as well as the lateral variation of rock type found in each formation. In total, 80 samples of the Curaçao Lava Formation, 150 samples of the Washikemba Formation, 40 samples of the Aruba Lava Formation, and 10 samples of the Aruban Batholith were analysed. Details of sampling procedures and analytical techniques for the La Rinconada Group specimens have been described in detail by Mottana and others (in press).

The samples were prepared in a tungsten carbide crusher and a quartz agate mill. Immediately after the final grinding, three splits of the ground sample were separated for the XRF major element, SRF trace element, and the instrumental neutron activation analysis, respectively. Major elements were analysed using standard XRF techniques, and total precision is generally on the order of 1%, except for Na2O and Mg, for which the estimated precision is 10% at 1 wt% concentration. The accuracy is of the same order of magnitude as the precision. The trace elements Ni, Cu, Zn, Rb, Sr, Zr, Y, and Nb in whole-rock samples were measured using the XRF techniques developed by Vie le Sage and others (1979). The analytical precision of the XRF trace element data varies from about 20% at the 1 ppm level to a few percent at 100 ppm. The estimated accuracy is better than 5% for Ni, Cu, Zn, Rb, and Sr, 5 to 10% for Y, and 10% for Zr and Nb.

The determination of the REE, Sc, Co, Cr, Th, Hf, and Ta was carried out by instrumental neutron activation analysis (INAA) at the Inter-university Reactor Institute (IRI) in Delft (The Netherlands). The method used is a slightly modified version of the procedure described by De Bruin and Korthoven (1972) and Korthoven and de Bruin (1977). For purposes of analytical quality, U.S.G.S. standards were regularly analysed together with the samples. The mean of 15 measurements of U.S.G.S. standard BHVO-1, together with the standard deviation (1σ), is given in Table 1, as well as the values of BHVO-1 measured by Potts and others (1981) for comparison. Because of the low concentrations of Ta, Th, Tb, and Yb in some samples these were also analysed using an epithermal irradiation facility (ENAA). Again BHVO-1 was analysed together with these samples for quality control. The values obtained by the ENAA method of 18 measurements of BHVO-1 are given in Table 1. The concentration of Ta in the samples could not be used because of contamination of Ta from the WC-crusher (about 0.1 ppm).

APPENDIX 2: TABLE 1. INSTRUMENTAL NEUTRON ACTIVATION ANALYSIS OF BHVO-1

Method of analysis No. of samples, 1σ	INAA (Delft) 15	1σ	ENAA (Delft) 18	1σ	INAA*	Chondrites[†]
Sc	31.5	2.7	30.2	4.6		
Cr	286	2.4				
Co	44.9	3.1	45.1	2.7	44.4	
La	16.9	5.3	15.6	3.1	15.5	0.367
Ce	37.5	6.4	-		38.9	0.957
Sm	6.2	4.5	7.3	6.4	6.11	0.231
Eu	2.07	3.8			2.11	0.087
Tb	0.82	24	0.97	3.4	0.97	0.058
Yb	1.82	9.8	2.15	10	2.09	0.248
Lu	0.35	20			0.30	0.0381
Hf	4.4	3.8			4.4	
Ta			1.14	2.3	1.2	
Th	1.13	9.8	1.36	3.3	1.3	

*Potts and others, 1981.
[†]Chondrite abundances are from Evensen and others (1978) multiplied by 1.5 to allow for removal of volatiles.

REFERENCES CITED

Alvarez, W., 1971, Fragmented Andean Belt of Northern Colombia, *in* Donnelly, T. W., ed., Caribbean geophysical, tectonic and petrologic studies: Geol. Soc. America Mem. 130, p. 77–96.

Andrews, G., Packham et al., 1975, Initial Reports of the Deep Sea Drilling Project, v. 30 (U.S. Government Printing Office, Washington, D.C.), 753 p.

Ball, M. M., Harrison, C.G.A., Supko, P. R., Bock, W., and Maloney, N. J., 1971, Marine geophysical measurements on the southern boundary of the Caribbean Sea, *in* Donnelly, T. W., ed., Caribbean geophysical, tectonic, and petrologic studies: Geol. Soc. America Mem. 130, p. 1–33.

Barrero, L. Darío, 1977, Geology of the central Western Cordillera, west of Buga and Roldanillo, Colombia [Ph.D. thesis]: Col. School of Mines, Golden, Colo., 148 p.

Batiza, R., Larson, R. L., Schlanger, S. O., Shcheka, S. A., and Tokuyama, H., 1980, Trace element abundances in basalts of Nauru Basin: Nature, v. 286, p. 476–479.

Beets, D. J., 1972, Lithology and stratigraphy of the Cretaceous and Danian succession of Curaçao [Ph.D. thesis]: Publication of Natuurwet. Studiekring Sur. and Neth. Antilles, no. 70, 153 p.

—— 1975, Superimposed island arcs along the southern margin of the Caribbean, *in* Borradaile, G. J., ed., Progress in Geodynamics, Geodynamics Project, Sc. Rep. 13, p. 218–234.

—— 1977, Cretaceous and Early Tertiary of Curaçao, *in* 8th Caribbean Geological Conference Guide to Field Excursions, GUA Papers of Geology, Amsterdam, no. 10, p. 7–17.

Beets, D. J., Klaver, G. Th., and Mac Gillavry, H. J., 1977, Geology of the Cretaceous and Early Tertiary of Bonaire, *in* 8th Caribbean Geological Conference Guide to Field Excursions, GUA Papers of Geology, Amsterdam, no. 10, p. 18–28.

Beets, D. J., Klaver, G. Th., Beunk, F. F., Kieft, C., and Maaskant, P., 1982, Picrites as parental magma of MORB-type tholeiites: Nature, v. 296, p. 341–343.

Bell, J. S., 1971, Tectonic evolution of the central part of the Venezuelan Coast Ranges, *in* Donnelly, T. W., ed., Caribbean geophysical, tectonic, and petrologic studies: Geol. Soc. America Mem. 130, p. 107–118.

—— 1972, Geotectonic evolution of the southern Caribbean area, *in* Studies in earth and space sciences (Hess Vol.): Geol. Soc. America Mem. 132, p. 367–386.

Bellizzia, G. A., 1972, Is the entire Caribbean Mountain Belt of northern Venezuela allochthonous?, *in* Studies in earth and space sciences (Hess Vol.): Geol. Soc. America Mem. 132, p. 363–368.

Berggren, W. A., McKenzie, D. P., Sclater, J. G., and Van Hinte, J. E., 1975, World-wide correlation of Mesozoic magnetic anomalies and its implication, discussion and reply: Geol. Soc. America Bull., v. 86, p. 267–269.

Beunk, F. F., and Klaver, G. Th., 1977, Geochemistry of the igneous rocks of Bonaire, Aruba and Curaçao, *in* 8th Caribbean Geological Conference Abstracts, GUA Papers of Geology, Amsterdam, no. 9, p. 17–18.

Blackburn, W. H., and Navarro, E., 1977, Garnet zoning and polymetamorphism in the eclogitic rocks of Isla de Margarita, Venezuela: Canadian Mineralogist, v. 15, p. 257–266.

Bottinga, Y., and Weill, D. F., 1972, The viscosity of magmatic silicate liquids: a model for calculation: Am. Jour. Sci., v. 272, p. 438–475.

Bowin, C., 1975, The Geology of Hispaniola, *in* Nairn, A.E.M., and Stehli, F. G., eds., The Ocean Basins and Margins, Plenum Press, New York, v. 3, p. 501–550.

Brown, E. H., Bradshaw, J. Y., and Mustoe, G. E., 1979, Plagiogranite and keratophyre in ophiolite on Fidalgo Island, Washington: Geol. Soc. America Bull., v. 90, p. 493–507.

Bryan, W. B., 1979, Regional variation and petrogenesis of basalt glasses from the FAMOUS Area, Mid-Atlantic Ridge: Jour. Petrol., v. 20, p. 293–327.

Bryan, W. B., Thompson, G., and Frey, F., 1979, Petrologic character of the Atlantic crust from DSDP and IPOD drill sites, *in* Talwani, M., Harrison, C., and Hayes, D. E., eds., Deep Drilling Results in the Atlantic Ocean:

Ocean Crust, Maurice Ewing Series 2, Am. Geoph. Union, Washington, D.C., p. 273–285.

Burke, K., Fox, P. J., and Sengör, A.M.C., 1978, Buoyant ocean floor and the evolution of the Caribbean: Jour. Geophys. Res., v. 83, p. 3949–3954.

Campa-Uranga, M. F., 1979, La Sierra Madre del Sur: Modelo de borde oceanico, *in* Memoria Congreso Latinoamericano de Geología, IV, Trinidad and Tobago (in press).

Carlson, R. L., Christensen, N. I., and Moore, R. P., 1980, Anomalous crustal structures in ocean basins: continental fragments and oceanic plateaus: Earth Planet. Sci. Lett., v. 51, p. 171–180.

Case, J. E., Durán S., L. G., López, R. A., and Moore, W. R., 1971, Tectonic investigations in western Colombia and eastern Panama: Geol. Soc. America Bull., v. 82, p. 2685–2711.

Chatterjee, N. D., 1972, The upper stability limit of the assemblage paragonite + quartz and its natural occurrences: Contrib. Mineral. Petrol., v. 34, p. 288–303.

Clarke, D. B., 1970, Tertiary basalts of Baffin Bay: Possible primary magma from the mantle: Contrib. Mineral. Petrol., v. 25, p. 203–224.

Clarke, D. B., and Pedersen, A. K., 1976, Tertiary volcanic province of West Greenland, *in* Escher, A., and Watts, W. S., eds., Geology of Greenland, Geol. Surv. Greenland, p. 365–385.

Cloetingh, S.A.P.L., Wortel, M.J.R., and Vlaar, N. J., 1982, Evolution of passive continental margins and initiation of subduction zones: Nature, v. 297, p. 139–142.

Cox, K. G., 1972, The Karroo volcanic cycle: J. Geol. Soc. London, v. 128, p. 311–336.

Cox, K. G., and Jamieson, B. G., 1974, The olivine-rich lavas of Nuanetsi: a study of polybaric magmatic evolution: Jour. Petrol., v. 14, p. 269–301.

Currie, K. L., and Curtis, L. W., 1976, An application of multicomponent solution theory to jadeitic pyroxenes: Jour. Geology, v. 84, p. 179–194.

Dalrymple, G. B., and Moore, J. G., 1968, Argon 40: Excess in submarine pillow basalts from Kilauea volcano, Hawaii: Science, v. 161, p. 1132–1135.

De Bruin, M., and Korthoven, P.J.M., 1972, Computer oriented system for non-destructive Neutron Activation Analysis: Analytical Chemistry 44 (14), p. 2382–2385.

De Cserna, Z., 1965, Reconocimiento geológico en la Sierra Madre del Sur de Mexico, entre Chilpancingo y Acapulco, estado de Guerrero: Mexico Univ. Nac. Autonoma Inst. Geologia Bol., v. 62, p. 76.

Dengo, G., 1972, Review of Caribbean serpentinites and their tectonic implications, *in* Studies in earth and space sciences (Hess Vol.): Geol. Soc. America Mem. 132, p. 303–312.

Dickinson, W. R., and Coney, P. J., 1980, Plate tectonic constraints on the origin of the Gulf of Mexico, *in* Pilger, R. H., ed., The origin of the Gulf of Mexico and the early opening of the central North Atlantic Ocean, Louisiana State University, p. 27–36.

Diebold, J. B., Stoffa, P. L., Buhl, P., and Truchan, M., 1981, Venezuela Basin crustal structure: Jour. Geophys. Res., v. 86, p. 7901–7924.

Dinkelman, M. G., and Brown, J. F., 1977, K-Ar Geochronology and its significance to the geologic setting of La Désirade, Lesser Antilles, *in* 8th Caribbean Geological Conference Abstracts, GUA Papers of Geology, Amsterdam, no. 9, p. 38–39.

Donnelly, T. W., 1975, The geological evolution of the Caribbean and Gulf of Mexico—some critical problems and areas, *in* Nairn, A.E.M., and Stehli, F. G., eds., The Ocean Basins and Margins, v. 3, Plenum Press, New York, p. 663–689.

—— 1977, Metamorphic rocks and structural history of the Motagua Suture Zone, eastern Guatemala, *in* 8th Caribbean Geological Conference Abstracts, GUA Papers of Geology, Amsterdam, no. 9, p. 40–41.

Donnelly, T. W., and Rogers, J.J.W., 1978, The distribution of igneous rock suites throughout the Caribbean: Geologie en Mijnbouw, v. 57, p. 151–162.

—— 1980, Igneous series in Island Arcs: the northeastern Caribbean compared with worldwide island-arc assemblages: Bull. Volcanologique, v. 43,

p. 347–383.

Donnelly, T. W., Kay, R., and Rogers, J.J.W., 1973, Chemical petrology of Caribbean basalts and dolerites, Leg. 15, Deep Sea Drilling Project: EOS (Trans. Amer. Geophys. Union), v. 54, p. 1002–1004.

Donnelly, T. W., Melson, W., Kay, R., and Rogers, J.J.W., 1973, Basalts and dolerites of Late Cretaceous age from the central Caribbean, in Edgar, N. T., Saunders, J. B. et al., Initial Reports of the Deep Sea Drilling Project, v. 15, (U.S. Government Printing Office, Washington, D.C.), p. 989–1005.

Draper, G., and Bone, R., 1981, Denudation rates, thermal evolution, and preservation of blueschist terrains: Jour. Geology, v. 89, p. 601–613.

Edgar, N. T., Ewing, J. I., and Hennion, J., 1971, Seismic refraction and reflection in Caribbean Sea: Bull. Am. Assoc. Petr. Geol., v. 55, p. 833–870.

Edgar, N. T., Saunders, J. B. et al., 1973, Initial Reports of the Deep Sea Drilling Project, v. 15, (U.S. Government Printing Office, Washington, D.C.), 1137 p.

Ernst, W. G., 1970, Tectonic contact between the Franciscan melange and the Great Valley sequence—crustal expression of a Late Mesozoic Benioff zone: Jour. Geophys. Res., v. 75, p. 886–901.

—— 1979, Coexisting sodic and calcic amphiboles from high-pressure metamorphic belts and the stability of barroisitic amphibole: Mineral. Mag., v. 43, p. 269–278.

Evans, B. W., Johannes, W., Oterdoom, H., and Trommsdorf, V., 1976, Stability of chrysotile and antigorite in the serpentine multisystem: Schweiz. Mineral. Petrograph. Mitt., v. 56, p. 79–93.

Evensen, N. M., Hamilton, P. J., and O'Nions, R. K., 1978, Rare-earth abundances in chondritic meteorites: Geochimica et Cosmochimica Acta, v. 42, p. 1199–1212.

Ewart, A., 1976, Mineralogy and chemistry of modern orogenic lavas—Some statistics and implications: Earth Planet. Sci. Lett., v. 31, p. 417–432.

Falvey, D. A., 1978, Analysis of Palaeomagnetic Data from the New Hebrides: Bull. Aust. Soc. Explor. Geophys., v. 9, p. 117–123.

Gastil, G., Morgan, G. J., and Krummenacher, D., 1978, Mesozoic history of peninsular California and related areas east of the Gulf of California, in Howell, D. G., and McDougall, K. A., eds., Mesozoic Paleogeography of the Western United States, Pacific Section Soc. Ec. Pal. Min., Los Angeles, Cal., p. 107–117.

Ghent, E. D., 1982, Chemical zoning in eclogite garnets—a review [abs.]: Terra cognita, v. 2, p. 301.

González de Juana, C., 1968, Guía de la excursión geológica à la parte oriental de la Isla Margarita: Asoc. Venezolana Geol. Min. y Petro, 42 p.

González de Juana, C., and Vignali, M., 1972, Rocas metamórficas e ígneas en la Península de Macanao, Margarita, Venezuela: Transactions, Caribbean Geological Conference, VI, Margarita, Venezuela, 1971, p. 63–68.

Goossens, P. J., and Rose, W. I., 1973, Chemical composition and age determination of tholeiitic rocks in the Basic Igneous Complex, Ecuador: Geol. Soc. America Bull., v. 84, p. 1043–1052.

Goossens, P. J., Rose, W. I., and Flores, D., 1977, Geochemistry of tholeiites of the Basic Igneous Complex of northwestern South America: Geol. Soc. America Bull., v. 88, p. 1711–1720.

Gose, W. A., and Swartz, D. K., 1977, Paleomagnetic results from Cretaceous sediments in Honduras: Tectonic implications: Geology, v. 5, p. 505–508.

Green, D. H., Lockwood, J. P., and Kiss, E., 1968, Eclogite and almandine-jadeite-quartz rock from the Guajíra Peninsula, Colombia, South America: Am. Mineralogist, v. 53, p. 1320–1335.

Hamilton, D. C., Burnham, C. W., and Osborn, E. F., 1964, The solubility of water and effects of oxygen fugacity and water content on crystallization in mafic magmas: Jour. Petrol., v. 5, p. 21–39.

Hanson, G. N., 1980, Rare Earth Elements in petrogenetic studies of igneous systems: Ann. Rev. Earth Planet. Sci., v. 8, p. 371–406.

Hargraves, R. B., and Skerlec, G. M., 1980, Paleomagnetism of some Cretaceous-Tertiary igneous rocks on Venezuelan offshore islands, Netherlands Antilles, Trinidad and Tobago, in Transactions, Caribbean Geological Conference, IX, Santo Domingo, 1980, p. 509–513.

Hays, J. D., and Pitman, W. C., 1973, Lithospheric plate motion, sea level changes and climatic and ecological consequences: Nature, v. 246, p. 18–22.

Hebeda, E. H., Verdurmen, E. A. Th., and Priem, H.N.A., K-Ar hornblende ages from the El Chacao complex north-central Venezuela (this volume).

Helmers, H., and Beets, D. J., 1977, Cretaceous of Aruba, in 8th Caribbean Geological Conference Guide to Field Excursions, GUA Papers of Geology, Amsterdam, no. 10, p. 29–35.

Helz, R. T., 1976, Phase relations of basalt in their melting ranges at $P_{H20} = 5$ Kb. Part II. Melt compositions: Jour. Petrol., v. 17, p. 139–193.

Holland, T.J.B., and Richardson, S. W., 1979, Amphibole zonation in metabasites as a guide to the evolution of metamorphic conditions: Contrib. Mineral. Petrol., v. 70, p. 143–148.

Holloway, J. R., and Burnham, C. W., 1972, Melting relations of basalt with equilibrium water pressure less than total pressure: Jour. Petrol., v. 13, p. 1–19.

Huppert, H. E., and Sparks, R. S., 1980, Restrictions on the compositions of mid-ocean ridge basalts: a fluid dynamic investigation: Nature, v. 286, p. 46–48.

Hussong, D. M., Wipperman, L. K., and Kroenke, L. W., 1979, The crustal structure of the Ontong Java and Manihiki oceanic plateaus: Jour. Geophys. Res., v. 84, p. 6003–6010.

Jakeś, P., and Gill, J., 1970, Rare Earth elements and the island arc tholeiitic series: Earth Planet. Sci. Lett., v. 9, p. 17–28.

Kesler, S. E., 1973, Basement rock structural trends in southern Mexico: Geol. Soc. America Bull., v. 84, p. 1059–1064.

Kesler, S. E., Sutter, J. F., Jones, L. M., and Walker, R. L., 1977, Early Cretaceous basement rocks in Hispaniola: Geology, v. 5, p. 145–147.

Khudoley, K. M., and Meyerhoff, A. A., 1971, Paleogeography and geologic history of the Greater Antilles: Geol. Soc. America Mem. 129, p. 1–199.

Korthoven, P.J.M., and De Bruin, M., 1977, Automatic interpretation of γ-ray data obtained in non-destructive Activation Analysis: Journal of Radioanalytical Chemistry, 35, p. 127–137.

Kuno, H., 1968, Origin of andesite and its bearing on the island arc structure: Bull. Volcanologique, v. 29, p. 195–222.

Ladd, J. W., 1976, Relative motion of South America with respect to North America and Caribbean tectonics: Geol. Soc. America Bull., v. 87, p. 969–976.

Laird, J., and Albee, A. L., 1981, Pressure, temperature, and time indicators in mafic schist: their application to reconstructing the polymetamorphic history of Vermont: American Journal of Science, v. 281, p. 127–175.

Larson, R. L., and Pitman, III, W. C. , 1972, World-wide correlation of Mesozoic magnetic anomalies, and its implications: Geol. Soc. America Bull., v. 83, p. 3645–3661.

Larson, R. L., and Hilde, T.W.C., 1975, A revised time scale of magnetic reversals for the Early Cretaceous and Late Jurassic: Jour. Geophys. Res., v. 80, p. 2586–2594.

Larson, R. L., and Schlanger, S. O., 1981, Cretaceous volcanism and Jurassic magnetic anomalies in the Nauru Basin, western Pacific Ocean: Geology, v. 9, p. 480–484.

Leake, B. E., 1978, Nomenclature of amphiboles: Can. Mineralogist, v. 16, p. 501–520.

Leake, B. E., Hendry, G. L., Kemp, A., Plant, A. G., Harrey, P. K., Wilson, J. R., Coats, J. S., Aucott, J. W., Lunel, T., and Howarth, R. J., 1969, The chemical analysis of rock powders by automatic X-ray fluorescence. Chem. Geol., v. 5, 7–86.

Liou, J. G., 1973, Synthesis and stability relations of epidote, $Ca_2Al_2FeSi_3O_{12}$ (OH): Jour. Petrol., v. 14, p. 381–413.

Liou, J. G., Kuniyoshi, S., and Ito, K., 1974, Experimental studies of phase relations between greenschist and amphibolite in a basaltic system: Am. Jour. Science, v. 274, p. 613–632.

Lowrie, W., 1979, Geomagnetic reversals and ocean crust magnetization, in Talwani, M., Harrison, C. G., and Hayes, D. E., eds., Deep drilling results in the Atlantic Ocean: Ocean crust, Maurice Ewing Series 2, p. 135–151.

Lowrie, W., Channell, J.E.T., and Alvarez, W., 1980, A review of magnetic stratigraphy investigations in Cretaceous pelagic carbonate rocks: Jour.

Geophys. Res., v. 85, p. 3597–3605.

Ludden, J. N., and Thompson, G., 1979, An evaluation of the behaviour of rare earth elements during the weathering of sea-floor basalt: Earth Plan. Sci. Lett., v. 43, p. 85–92.

MacDonald, W. D., 1968, *in* Status of geological research in the Caribbean 14, Un. Puerto Rico, p. 40.

MacGillavry, H. J., and Beets, D. J., 1977, The Rincon problem or how to confuse the geologist, *in* Abstracts, Caribbean Geological Conference, VIII, Curaçao, Netherlands Antilles, GUA Papers of Geology, Series 1, v. 9, p. 103–105.

Maresch, W. V., 1971, The metamorphism and structure of northeastern Margarita Island, Venezuela [Ph.D. thesis]: Princeton, N. J., Princeton Univ., 278 p.

—— 1972, Eclogitic-amphibolitic rocks on Isla Margarita, Venezuela: A preliminary account, *in* Studies in earth and space sciences (Hess volume): Geol. Soc. America Mem. 132, p. 429–437.

—— 1974, Plate tectonic origin of the Caribbean Mountain System of northern South America: Discussion and proposal: Geol. Soc. America Bull., v. 85, p. 669–682.

—— 1975, The geology of northeastern Margarita Island, Venezuela: A contribution to the study of Caribbean plate margins: Geol. Rundschau, v. 64, p. 846–883.

—— 1976, Implications of a Mesozoic to Early Tertiary collision-type plate-tectonic model in northern Venezuela for the southern Caribbean Region, *in* Transactions, Caribbean Geological Conference, VII, Guadeloupe, 1974, p. 485–491.

—— 1977, Similarity of metamorphic gradients in time and space during metamorphism of the La Rinconada Group, Margarita Island, Venezuela, *in* Abstracts, Caribbean Geological Conference, VIII, Curaçao, Netherlands Antilles, GUA Papers of Geology, Series 1, v. 9, p. 110–111.

Maresch, W. V., and Abraham, K., 1981, Petrography, mineralogy, and metamorphic evolution of an eclogite from the Island of Margarita, Venezuela: Jour. Petrol., v. 22, p. 337–362.

Maresch, W. V., Medenbach, O., and Rudolph, A., 1982, Winchite and the actinolite-glaucophane miscibility gap: Nature, v. 296, p. 731–732.

Maresch, W. V., Abraham, K., Bocchio, R., and Mottana, A., 1979, Systematic compositional variations in amphiboles from the La Rinconada Group metabasalts, Margarita Island, and their paleotectonic implications, *in* Memoria, Congreso Latinoamericano de Geología, IV, Trinidad and Tobago (in press).

Martín, F. C., 1978, Ideas sobre diversas asociaciones metamórficas y complejos ofiolíticos en la evolución del borde sur de la placa tectónica del Caribe: Geologie en Mijnbouw, v. 57, p. 379.

Mattinson, J. M., Fink, L. K., and Hopson, C. A., 1973, Age and origin of ophiolitic rocks on La Désirade Island, Lesser Antilles Island Arc: Carnegie Inst. Wash. Yearbook 72, p. 616–623.

Mattson, P. H., 1977, Caribbean plate tectonic model, *in* Mattson, P. H., ed., West Indies Island Arcs, Benchmark Papers in Geology, v. 33, p. 353–359.

—— 1979, Subduction, buoyant braking, flipping, and strike-slip faulting in the northern Caribbean: Jour. Geol., v. 87, p. 293–304.

Mattson, P. H., and Pessagno, E. A., 1979, Jurassic and Early Cretaceous radiolarians in Puerto Rican ophiolite-Tectonic implications: Geology, v. 7, p. 440–445.

McElhinny, M. W., 1979, Palaeogeomagnetism and the Core-Mantle Interface, *in* McElhinny, M. W., ed., The Earth: Its Origin, Structure and Evolution, Academic Press, London, p. 113–136.

Melson, W. G., Vallier, T., Wright, T. L., Byerly, G., and Nelen, J., 1976, Chemical diversity of abyssal volcanic glass erupted along Pacific, Atlantic, and Indian Ocean sea-floor spreading centers: AGU Geophys. Mon. 19, p. 351–368.

Menéndez, A., 1967, Tectonics of the central part of the Western Caribbean Mountains, Venezuela: Studies in tropical oceanography no. 5, Univ. Miami, Inst. Marine Sci., p. 103–130.

Miyashiro, A., 1973, Metamorphism and metamorphic belts: Allen and Unwin, London, 492 p.

—— 1974, Volcanic rock series in island arcs and active continental margins: Am.

Jour. Sci., v. 274, p. 321–355.

Morgan, B. A., 1967, Geology of the Valencia area, Carabobo, Venezuela [Ph.D. thesis]: Princeton, N.J., Princeton Univ., 220 p.

—— 1970, Petrology and mineralogy of eclogite and garnet amphibolite from Puerto Cabello, Venezuela: Jour. Petrol., v. 11, p. 101–145.

Morten, L., Brunfelt, A. O., and Mottana, A., 1979, Rare earth abundances in superferrian eclogites from the Voltri Group (Pennidic Belt, Italy): Lithos, v. 12, p. 25–32.

Mottana, A., Bocchio, R., Liborio, G., Morten, L., and Maresch, W. V., 1979, Geochemistry of the La Rinconada Group metabasalts of Isla Margarita, Venezuela: an interim account, *in* Memoria, Congreso Latinoamericano de Geología, IV, Trinidad and Tobago (in press).

Murray, C. G., 1972, Zoned ultramafic complexes of the Alaskan-Ural type: Feeder pipes of andesitic volcanoes, *in* Studies in earth and space sciences (Hess volume): Geol. Soc. America Mem. 132, p. 313–335.

Mysen, B. O., 1977, The solubility of H_2O and CO_2 under predicted magma genesis conditions and some petrological and geophysical implications: Rev. Geophys. Space Phys., v. 15, p. 351–361.

Navarro, F. E., 1974, Petrogenesis of the eclogitic rocks of Isla de Margarita, Venezuela [Ph.D. thesis]: Lexington, Kentucky, Univ. Kentucky, 222 p.

—— 1977, Eclogitas de Margarita: evidencias de polimetamorfísmo: *in* Memoria, Congreso Geológico Venezolano, V, Caracas, p. 653–661.

Newton, R. C., and Fyfe, W. S., 1976, High pressure metamorphism, *in* Bailey, D. K., and Macdonald, R., eds., The evolution of the crystalline rocks: London, Academic Press, p. 101–186.

Norris, R. J., and Henley, R. W., 1976, Dewatering of a metamorphic pile: Geology, v. 4, p. 333–336.

Odin, G. S., and Kennedy, W. J., 1982, Mise à jour de l'échelle des temps mésozoíques: C. R. Acad. Sc. Paris, t. 294, p. 383–387.

Oxburgh, E. R., 1966, Geology and metamorphism of Cretaceous rocks in eastern Carabobo State, Venezuelan Coast Ranges, *in* Hess, H. H., ed., Caribbean geological investigations: Geol. Soc. America Mem. 98, p. 241–310.

Pearce, J. A., and Norry, M. J., 1979, Petrogenetic implications of Ti, Zr, Y, and Nb variations in volcanic rocks: Contrib. Mineral. Petrol., v. 69, p. 33–47.

Perfit, M. R., Gust, D. A., Bence, A. E., Arculus, R. J., and Taylor, S. R., 1980, Chemical characteristics of island-arc basalts: implications for mantle sources: Chemical Geology, v. 30, p. 227–257.

Piburn, M. D., 1967, Metamorphism and structure of the Villa de Cura Group, Northern Venezuela [Ph.D. thesis]: Princeton, N.J., Princeton Univ., 148 p.

Pitman III, W. C., 1978, Relationship between eustacy and stratigraphic sequences of passive margins: Geol. Soc. America Bull., v. 89, p. 1389–1403.

Potts, P. J., Thorpe, O. W., and Watson, J. S., 1981, Determination of the rare-earth element abundances in 29 international rock standards by instrumental activation analysis: a critical appraisal of calibration errors: Chemical Geology, v. 34, p. 331–352.

Priem, H.N.A., Boelrijk, N.A.I.M., Verschure, R. H., Hebeda, E. H., and Lagaay, R. A., 1966, Isotopic age of the quartz-diorite batholith on the island of Aruba, Netherlands Antilles: Geologie en Mijnbouw, v. 45, p. 188–190.

Priem, H.N.A., Andriessen, P.A.M., Beets, D. J., Boelrijk, N.A.I.M., Hebeda, E. H., Verdurmen, E.A.Th., and Verschure, R. H., 1977, Isotopic dating of the quartz-diorite batholith of Aruba and the crystalline cores of Curaçao and Bonaire, *in* Abstracts, Caribbean Geological Conference, VIII, Curaçao, Netherlands Antilles, GUA Papers of Geology, Series 1, no. 9, p. 149–155.

Priem, H.N.A., Beets, D. J., Boelrijk, N.A.I.M., Hebeda, E. H., Verdurmen, E.A.Th., and Verschure, R. H., 1978, Rb-Sr evidence for episodic intrusion of the Late Cretaceous tonalitic batholith of Aruba, Netherlands Antilles: Geologie en Mijnbouw, v. 57, p. 293–296.

Rangin, C., 1978, Speculative model of Mesozoic geodynamics, central Baja California to northeastern Sonora (Mexico), *in* Howell, D. G., and McDougall, K. A., eds., Mesozoic Paleogeography of the Western United States, Pacific Section Soc. Ec. Pal. Min., Los Angeles, Cal., p. 85–107.

Richardson, S. W., and England, P. C., 1979, Metamorphic consequences of crustal eclogite production in overthrust orogenic zones: Earth Planet. Sci. Lett., v. 42, p. 183–190.

Ringwood, A. E., 1975, Composition and Petrology of the Earth's Mantle, McGraw-Hill, New York.

Roper, P. J., 1978, Stratigraphy of the Chuacus Group on the south side of the Sierra de las Minas range, Guatemala: Geologie en Mijnbouw, v. 57, p. 309–313.

Rowley, K. C., and Roobol, J. M., 1978, Geochemistry and age of the Tobago igneous rocks: Geologie en Mijnbouw, v. 57, p. 315–318.

Rudolph, A., 1981, Petrographie und Petrologie von eklogitischen Linsen in Glimmerschiefern der Insel Margarita, Venezuela [Dipl. thesis]: Bochum, F. R. Germany, Ruhr-Univ. Bochum, 137 p.

Santamaría, F., and Schubert, C., 1974, Geochemistry and geochronology of the Southern Caribbean-Northern Venezuela plate boundary: Geol. Soc. America Bull., v. 85, p. 1085–1098.

Saunders, A. D., Tarney, J., and Weaver, S. D., 1980, Transverse geochemical variations across the Antarctic Peninsula: implications for the genesis of calc-alkaline magmas: Earth Planet. Sci. Lett., v. 46, p. 344–360.

Schlanger, S. O., Jackson, E. D., et al., 1976, Initial Reports of the Deep Sea Drilling Project 33: (U.S. Government Printing Office, Washington, D.C.).

Schlanger, S. O., Jenkyns, H. C., and Premola-Silva, I., 1981, Volcanism and vertical tectonics in the Pacific Basin related to global Cretaceous transgressions: Earth Planet. Sci. Lett., v. 52, p. 435–449.

Schliestedt, M., 1979, Phasengleichgewichte in Eklogit-Glaukophanit-Wechsellagerungen der Insel Sifnos (Kykladen, Griechenland)—ein Beitrag zum Eklogitproblem: Fortschr. Mineralogie, v. 57, Beiheft, S. 140–141.

Shagam, R., 1960, Geology of central Aragua, Venezuela: Geol. Soc. America Bull., v. 71, p. 249–302.

Shaw, H. R., 1972, Viscosities of magmatic silicate liquids; an empirical method of prediction: Am. Jour. Sci., v. 272, p. 870–893.

Silver, E. A., Case, J. E., and MacGillavry, H. J., 1975, Geophysical study of the Venezuelan borderland: Geol. Soc. America Bull., v. 86, p. 213–226.

Skerlec, G. M., 1976, The western termination of the Caribbean Mountains of Venezuela: A progress report, *in* Transactions, Caribbean Geological Conference, VII, Guadeloupe, 1974, p. 493.

Skerlec, G. M., and Hargraves, R. B., 1980, Tectonic significance of paleomagnetic data from northern Venezuela: Jour. Geophys. Res., v. 85, p. 5303–5315.

Sleep, N. H., and Windley, B. F., 1982, Archean plate tectonics: constraints and inferences: Jour. of Geology, v. 90, p. 363–379.

Smit, J., 1977, Planktonic foraminiferal faunas from the upper part of the Washikemba Formation, Bonaire, *in* Abstracts, Caribbean Geological Conference, VIII, Curaçao, Netherlands Antilles, GUA Papers of Geology, Series 1, no. 9, p. 192–193.

Smulikowski, K., 1968, Differentiation of eclogites and its possible causes: Lithos, v. 1, p. 89–101.

—— 1972, Classification of eclogites and allied rocks: Krystalinikum, v. 9, p. 107–130.

Sparks, R.S.J., Meyer, P., and Sigurdsson, H., 1980, Density variation amongst mid-ocean ridge basalt: implications for magma mixing and the scarcity of primitive lavas: Earth Planet. Sci. Lett., v. 46, p. 419–430.

Sprague, D., and Pollack, H. N., 1980, Heat flow in the Mesozoic and Cenozoic: Nature, v. 285, p. 393–395.

Stearns, C., Mauk, F. J., and Van der Voo, R., 1982, Late Cretaceous-Early Tertiary paleomagnetism of Aruba and Bonaire (Netherlands Leeward Antilles): Jour. Geophys. Res., v. 87, p. 1127–1141.

Stephan, J.-F., Beck, C., Bellizzia, A., and Blanchet, R., 1980, La chaîne caraïbe du Pacifique à l'Atlantique: Bur. Recherches Geol. Min. Mem. 115 (Paris), p. 38–59.

Stoffa, P. L., Mauffret, A., Truchan, M., and Buhl, P., 1981, Sub B" layering in the southern Caribbean: the Aruba gap and Venezuela Basin: Earth Planet. Sci. Lett., v. 53, p. 131–146.

Stolper, E., and Walker, D., 1980, Melt density and the average composition of basalt: Contrib. Mineral. Petrol., v. 74, p. 7–12.

Talukdar, S., and Loureiro, D., 1982, Geología de una zona ubicada en el segmento norcentral de la Cordillera de la Costa, Venezuela: Metamorfismo y deformatión. Evolución del margen septentrional de Suramérica en el marco de la tectónica de placas: GEOS (Caracas), v. 27, p. 15–76.

Talwani, M., Windisch, C. C., Stoffa, P. L., Buhl, P., and Houtz, R. E., 1977, Multi-channel seismic study in the Venezuelan Basin and the Curaçao Ridge, *in* Talwani, M., and Pitman III, W. C., eds., Island Arcs, Deep Sea Trenches and Back-Arc Basins, Maurice Ewing Series I, Am. Geoph. Union, Washington, D.C., p. 83–99.

Taylor, G. C., 1960, Geology of the Island of Margarita, Venezuela [Ph.D. thesis]: Princeton, N.J., Princeton Univ., 121 p.

Thiede, J., Dean, W. E., Rea, D. K., Vallier, T. L., and Adelseck, C. G., 1981, The geologic history of the Mid-Pacific Mountains in the central North Pacific Ocean—a synthesis of Deep-Sea Drilling Studies, *in* Thiede, J., Vallier, T. L. et al., Initial Reports of the Deep Sea Drilling Project, v. 62, (U.S. Government Printing Office, Washington, D.C.), p. 1073–1120.

Thierstein, R., and Okada, H., 1979, The Cretaceous/Tertiary boundary event in the North Atlantic, *in* Tucholke, B. E., Vogt, P. R., et al., Initial Reports of the Deep Sea Drilling Project, v. 43, (U.S. Government Printing Office, Washington, D.C.), p. 601–617.

Thompson, A. B., 1981, The pressure-temperature (P, T) plane viewed by geophysicists and petrologists: Terra Cognita, v. 1, p. 11–20.

Turcotte, D. L., and Burke, K., 1978, Global sea-level changes and the thermal structure of the earth: Earth Planet. Sci. Lett., v. 41, p. 341–346.

Ungaretti, L., Smith, D. C., and Rossi, G., 1981, Crystal-chemistry by X-ray structure refinement and electron microprobe analysis of a series of sodic-calcic to alkalic-amphiboles from the Nybö eclogite pod, Norway: Bulletin Mineralogique, v. 104, p. 400–412.

Vallier, T. R., Rea, D. K., Dean, W. E., Thiede, J., and Adelseck, C. G., 1981, The geology of Hess Rise, central North Pacific Ocean, *in* Thiede, J., Vallier, T. L. et al., Initial Reports Deep Sea Drilling Project, v. 62 (U.S. Government Printing Office, Washington, D.C.), p. 1031–1071.

Van Tongeren, W., 1934, Chemical analyses of some rocks from Aruba: K. Akad. Wetens., Amsterdam, Sec. Sci., v. 37, p. 162–167.

Venezuela, Dirección de Geología, 1970, Lexico Estratigráfico de Venezuela, 2nd Ed.: Boletín de Geología, Publicación Especial no. 4, 756 p.

Vié le Sage, R., Quisefit, J. P., Dejean de la Batie, R., and Faucherre, J., 1979, Utilisation du rayonnement primaire diffusé par l'échantillon pour une détermination rapide et précise des élements traces dans les roches: X-Ray Spect., v. 8, p. 121–128.

Vignali, M., 1972, Analisis estructural y eventos tectónicos de la Península de Macanao, Margarita, Venezuela, *in* Transactions, Caribbean Geological Conference, VI, Margarita, Venezuela, 1971, p. 241–246.

—— 1979, Estratigrafía y estructura de las cordilleras metamórficas de Venezuela oriental (península de Araya-Paria e isla de Margarita): GEOS (Caracas), v. 25, p. 19–66.

Vogt, P. R., 1975, Changes in geomagnetic reversal frequency at times of tectonic change: evidence for coupling between core and upper mantle processes: Earth Planet. Sci. Lett., v. 25, p. 313–321.

Walker, B. M., Vogel, T. A., and Ehrlich, R., 1972, Petrogenesis of oceanic granites from the Aves Ridge in the Caribbean Basin: Earth Plan. Sci. Lett., v. 15, p. 133–139.

Walker, G.P.L., 1973, Lengths of lava flows: Royal Society of London Philosophical Transactions, ser. A, v. 274, p. 107–118.

Watts, A. B., Bodine, J. H., and Ribe, N. H., 1980, Observation of flexure and the geological evolution of the Pacific Ocean Basin: Nature, v. 283, p. 532–537.

Westermann, J. H., 1932, The geology of Aruba: Geograf. Geol. Meded. Physiogr. Geol. Reeks, Utrecht, no. 7.

Wiedmann, J., 1978, Ammonites from the Curaçao Lava Formation, Curaçao, Caribbean: Geologie en Mijnbouw, v. 57, p. 361–364.

Wood, D. A., Joron, J.-L., Treuil, M., Norry, M., and Tarney, J., 1979, Elemental and Sr isotope variations in basic lavas from Iceland and the surrounding ocean floor: Contrib. Mineral. Petrol., v. 70, p. 319–339.

Wood, R. M., 1979, A re-evaluation of the blueschist facies: Geol. Mag., v. 116, p. 21–33.

MANUSCRIPT ACCEPTED BY THE SOCIETY SEPTEMBER 1, 1983

Geological Society of America
Memoir 162
1984

Caribbean structural breaks and plate movements

Peter H. Mattson
Department of Earth and Environmental Sciences
Queens College
City University of New York
Flushing, New York 11367

ABSTRACT

The Caribbean Plate was created as the North and South American Plates began to separate about 140 m.y. ago, allowing the Phoenix/Farallon spreading ridge to extend eastward. The Caribbean Plate was separated from the Phoenix/Farallon spreading ridge, about 110-100 m.y. ago, by a subduction zone near present-day Central America, connected by transform faults to the older Greater Antilles subduction zone. Geologic data from the margins of the Caribbean Plate indicate six important discontinuities in the history of the plate. Near the beginning of the Albian (110 m.y. B.P.), northeast-dipping subduction in the Greater Antilles zone may have been blocked by underthrusting of part of the Chortis Block of Central America, causing subduction to flip to southwest-dipping, perhaps followed by the beginning of subduction beneath Central America. A Santonian (85 m.y. B.P.) discontinuity may be the result of thickened oceanic crust formed at the Galapagos hot spot reaching the Central American subduction zone and blocking or modifying subduction. At the beginning of the Tertiary (66 m.y. B.P.), the Caribbean Plate changed its relative motion from northeastward to eastward and began to underthrust northern South America. In the Late Oligocene (27 m.y. B.P.), the Farallon/Phoenix Plate separated into the Nazca and Cocos Plates. The present-day emergence (5-0 m.y. B.P.) has not yet been correlated with plate motion changes.

Major changes in plate arrangement and motion are thus reflected in the Caribbean Plate by major geologic discontinuities such as unconformities. Geologic structures along the plate margins are the resultants of the direction of relative plate motion, and of the type of lithosphere. Oblique plate motion produced horizontal slip accompanied by slow subduction without volcanism, by uplift and erosion, or by a combination of processes. Oceanic lithosphere forms relatively simple plate boundaries, as in the Greater Antilles, but continental lithosphere forms complex border zones composed of old structural blocks moving along ancient zones of weakness, as in northern South America.

INTRODUCTION

To examine the margins of the Caribbean Plate (Fig. 1), one must first place the Caribbean Plate into its position in the dynamic set of lithospheric plates forming the surface of the earth. Two hundred million years ago, with much contemporary detail lost, a small number of large plates formed the western hemisphere: the Pacific, Farallon, Phoenix, Americas, Eurasia/Africa, and Antarctic Plates. The Atlantic Ocean opened, splitting Pangea and disrupting Paleozoic orogenic belts. Duncan and Har-

graves (this volume) suggested that oceanic crust destined to become the Caribbean Plate was generated on an east-northeast-trending spreading ridge between the Farallon and Phoenix Plates, perhaps during Jurassic times. The Caribbean region has been shaped by the interaction of the Farallon, North American, and South American Plates; the small-plate or microplate behavior of the Caribbean between the larger plates in this region is apparent in the geological record of the last 100 million years.

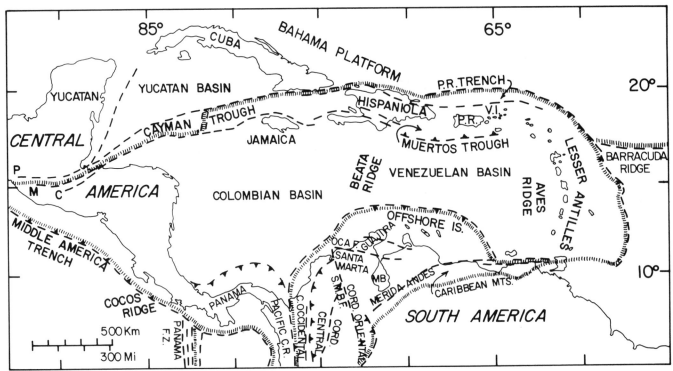

Figure 1. Geography and present-day plate boundaries in the Caribbean. Short hatching marks approximate plate boundaries or outlines of boundary zones; dashed lines are faults. Abbreviations are: F, fault; F.Z., fault zone; C.R., Coast Range; Cord., Cordillera; S.M.B.F., Santa Marta-Bucaramanga Fault; PR, Puerto Rico; VI, Virgin Islands; MB, Maracaibo Basin; NP, Nazca Plate. Central American faults are M, Motagua; P, Polachic; C, Chamelecon. Arrows show approximate direction of relative plate motion.

The Caribbean Plate is the northeasternmost of the three small plates (Caribbean, Cocos, and Nazca in Figs. 2 and 3) that lie between three major plates: the North American, South American, and Pacific Plates. North America and South America in the first plate reconstructions were considered part of a single plate (Morgan, 1968), but Ladd (1976) and other authors have shown that the plates moved independently since at least 180 m.y. B.P. The Americas Plates are now bordered on the west by the Pacific, Cocos, and Nazca Plates; but at an earlier time there was a simple system of Pacific, Farallon, Phoenix, and Americas Plates. Spreading at the East Pacific Rise and Mid-Atlantic Ridge gradually changed this simple pattern. Spreading between the Farallon and Phoenix Plates ceased between 119 and 100 m.y. B.P., when spreading began on the Mid-Atlantic Ridge in the equatorial Atlantic (Duncan and Hargraves, 1984), and the combined Farallon/Phoenix Plate moved as a single unit until 27 m.y. B.P. Atwater (1970) and Menard (1978) described the process of fragmentation of the Farallon/Phoenix Plate as parts of the East Pacific Rise spreading ridge approached the subduction zone bordering the Americas Plate. The East Pacific Rise spreading ridge moved eastward relative to the Americas Plates and began to be subducted about 26-27 m.y. B.P. (Handschumacher, 1976). Also about 27 m.y. B.P., the Cocos and Nazca Plates began their separate existence, perhaps resolving stresses created as the Farallon Plate was simultaneously subducted to the northeast beneath

the Caribbean Plate and subducted to the east beneath the South American Plate (Lonsdale and Klitgord, 1978; Hey, 1977). Van Andel and others (1971) suggest that the Coiba aseismic ridge on the Cocos Plate, as it entered the Middle American Trench, plugged that part of the subduction zone, causing the transform border between the Cocos and Nazca Plates to jump to the west.

This paper surveys the structure and tectonic history of selected areas at the northern and southern borders of the Caribbean Plate and attempts to relate the younger tectonic patterns to the character of the Caribbean Plate and its neighboring plates, and to present-day movements. The controlling factors, given the plate movement vectors, seem to be the oceanic nature of the southern part of the North American Plate, the continental character of the northwestern South American Plate, and the changing character of the Caribbean Plate near its margins.

The time intervals of 140-110, 110-85, 85-66, 66-45, 45-27, and 27-0 m.y. B.P. (Fig. 3) were modified from Pindell and Dewey (1982), by considering ages of unconformities and changes in tectonic style, tectonic intensity, volcanism, plutonism, or sedimentation recorded in the geology of the Caribbean and its borderlands.

CARIBBEAN PLATE BOUNDARIES

The northern border of the present Caribbean Plate extends

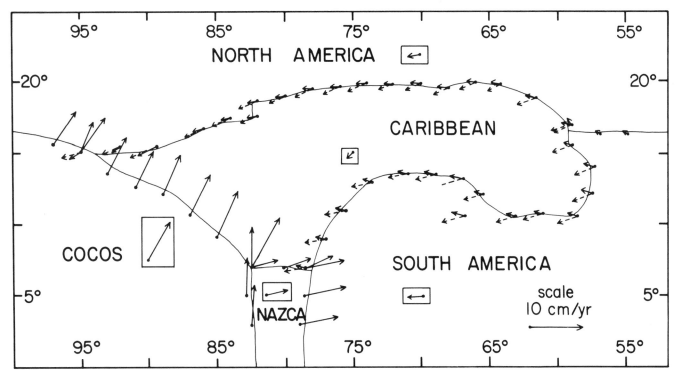

Figure 2. Relative and absolute motions of the Caribbean and adjoining plates for last 5 million years. Calculated using eulerian poles and relative angular velocities of Minster and Jordan (1978). Arrows begin at point of calculation, chosen near Caribbean Plate margins but on adjoining plates; they show motion of external plate relative to Caribbean Plate and hence are essentially maximum stress vectors. Dotted arrows are calculated using eulerian poles and angular velocities of Sykes and others (1982). Absolute motion vectors, in boxes, calculated from data of Minster and Jordan (1978) for points at tails of arrows. There is no marine seismic evidence for the position of the North American-Caribbean-Cocos triple junction.

from Guatemala in the west to the Lesser Antilles in the east (Figs. 1, 2). The border is complex in relative plate motion, crustal type, and modern topography. Its west end is the Caribbean-Cocos-North America triple junction in the Middle American Trench (Fig. 1). Eastward, the plate boundary passes beneath the volcanic pile in Guatemala and emerges as a large left-lateral fault zone: the Motagua, Polochic, and perhaps the Chamelecon (Fig. 1) are the largest faults in the zone (Schwartz and others, 1979). From Guatemala eastward, the boundary becomes the Cayman Trough, where it is also a zone composed of several faults (Sykes and others, 1982). From the eastern end of the Cayman Trough between Cuba and Hispaniola, it passes either north of Hispaniola (Molnar and Sykes, 1969; Case and Holcombe, 1980), through northern Hispaniola (Goreau, 1980), or as multiple strands through Hispaniola. The boundary continues in the Puerto Rico Trench eastward until both turn south; a separate strand may pass along the Muertos Trough south of Puerto Rico (Fig. 1). At some point in the Puerto Rico Trench opposite the Lesser Antilles, perhaps near the Barracuda Ridge, the boundary passes the Caribbean-North America-South America triple junction and continues southward, separating the Caribbean and south American Plates.

The Caribbean Plate is bounded on the south by the South American Plate (Fig. 1). It can be considered to end westward in a complex triple junction of the Caribbean, South American, and Nazca Plates (Fig. 2). The South American Plate is mostly exposed above sea level where it lies against the Caribbean Plate; the Caribbean Plate is mostly submarine, although other interpretations of the boundary are possible (compare Pennington, 1981). Shagam (1975), Bell (1974), and Campbell (1974) have published comprehensive geological summaries of these regions, and these papers provide the sources for most of the structural data presented here.

CARIBBEAN PLATE MOTIONS

The major characteristics shaping the margins of the Caribbean Plate are the magnitudes and directions of relative motion of the four plates that border the Caribbean Plate, and the crustal characteristics of the plates themselves. Figure 2 shows the vectors of relative motion at approximately two-degree intervals around the margins of the Caribbean Plate. The vectors were calculated using the rotation poles and angular velocities measured by Minster and Jordan (1978). Dashed vectors are calculated from poles and angular velocities measured by Sykes and others (1982), for the North American-Caribbean and South

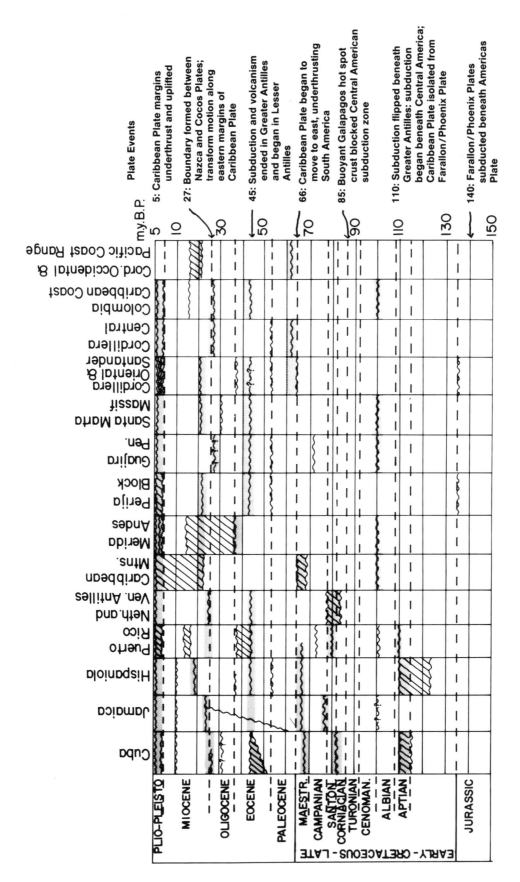

Figure 3. Tectonic breaks and plate events in the Caribbean Plate. Undulating lines show time position of unconformities; diagonal lines show hiatus in deposition or lack of preservation. Shading shows correlation of major discontinuities.

American-Caribbean plate boundaries. The two sets of vectors differ because Minster and Jordan used data only from Central America and the Cayman Trough, whereas the Sykes group used those data plus fault mechanisms and other seismic data from the northeastern Caribbean. Thus they differ most strongly in the eastern Caribbean. The Sykes group estimated a relative plate motion magnitude of 3.7 cm/yr, greater than the 2.0 cm/yr rate used by Minster and Jordan.

Where plate motions are predominantly parallel to the plate margins, as along the northern margin and along the eastern part of the southern margin, slip is smooth and topographic trends parallel plate movement. Minor convergent components of motion, best shown using the Sykes vectors, create minor, slow subduction and minor folding.

Where motion is oblique to plate margins, as along the western part of the southern margin (from 69 to 77 degrees west latitude), compressional stresses cause subduction of the Caribbean Plate beneath the South American Plate, forcing the South American Plate upward and fragmenting it into a mosaic of fault blocks. The complicated structural pattern in northern South America is due to this convergence superimposed on earlier Mesozoic and Paleogene convergent systems formed as South America moved against the Caribbean and Nazca Plates, and superimposed on the even earlier Paleozoic orogenic systems. Boundaries of mountain blocks tend to parallel older structural trends within the blocks themselves rather than paralleling modern plate margins.

Figure 2 also displays some of the complexity of the plate configuration near Panama where four plates and two triple junctions lie close together. Case (personal communication, 1983), because of data from reflection profiles and first motion studies, considered Panama as a fifth plate, moving independently of its neighbors; and this makes the configuration even more complex. A first suggestion is that the Cocos-Caribbean-Nazca triple junction should be unstable, with the Nazca Plate pulling away to the east as it subducts beneath South America, at the same time as the Cocos Plate subducts beneath Central America and moves northward along the Panama Fracture Zone. More careful study shows that the Panama Fracture Zone (Fig. 1) is unusual because it is not parallel to nearby plate motion directions. The Cocos Pate is actually subducting to the northeast and keeping the north-trending fracture zone closed. This effectively moves the triple junction slowly eastward.

The Cocos-Caribbean-Nazca triple junction is also unstable in another way. The nearby subduction of the aseismic Cocos Ridge (Lonsdale and Klitgord, 1978) causes pieces of the Cocos Plate to split off and join the Nazca Plate, thus causing the triple junction to appear to jump westward along the Middle American Trench. The relative velocities of this eastward drift and westward jumping have not yet been calculated.

The triple junction between the Nazca, Caribbean, and South American Plates is extremely complicated. South America is moving westward relative to the Nazca and Caribbean Plates, forming the mosaic of mountain and basin blocks of the northern

Andes. The northern convexity of the Panamanian isthmus might be due to this westward shoving, essentially compressing the southwestern corner of the Caribbean Plate between the South American and Cocos Plates. Continuation of this motion should eventually collapse the convex-to-the-north arc of the isthmus onto itself, against northeastern Colombia.

Minster and Jordan (1978) calculated absolute motions of the plates during the last 5 million years, using a hot-spot reference frame. Absolute motions for representative points on the Caribbean and adjoining plates have been calculated and are shown on Figure 2 as the boxed vectors positioned away from the plate boundaries. The resultant of absolute velocity vectors from adjoining plates is the vector of relative motion at the boundary, at any particular point. Note that the Caribbean Plate moves very slowly in the hot-spot reference frame, as pointed out by Minster and Jordan. Perhaps this is either because it lies above subducting plates, or because it moves laterally beside subducting plates but is only itself subducted near the west end of its southern margin, near the Santa Marta Massif. The slow convergence taking place within the Caribbean Plate, in the Muertos Trough, and beneath the Curacao ridge is the result of the small component of compressive stress across the Caribbean Plate caused by the minute convergence of the North and South American Plates (Fig. 2).

Kellogg (1980) and Kellogg and Bonini (1982) combined geophysical and geological methods in a study of the Perija block, Santa Marta Massif, Merida Andes, and the Cordillera Oriental. They suggested that Perija, Santa Marta, and the Merida Andes are thrust to the northwest along gently-dipping (15-20 degrees) faults as the Caribbean Plate is slowly subducted beneath northwestern South America (Fig. 2). They relate movements on other major faults in the Merida Andes and the Santander Massif to east-directed subduction of the Nazca Plate beneath South America.

Malfait and Dinkelman (1972), Ladd (1976), Pindell and Dewey (1982), Duncan and Hargraves (1984), and other writers have published map sequences showing the changing Caribbean Plate from 125-200 m.y. B.P. to the present. It is beyond the scope of this study to analyze these works in detail; as more becomes known, more accurate models have been put forward, and certainly improvements will continue to be made. The latest set of maps (Fig. 4, adapted from data in Pindell and Dewey, 1982) shows a Caribbean Plate created as a separate plate 125 m.y. ago, moving northeastward relative to a fixed North American Plate until about 65 m.y. B.P. when the subduction zone between the North American and Caribbean Plates consumed all the oceanic crust between island arc and continental lithosphere. The Caribbean Plate then moved to the east, from 65 m.y. B.P. to the present (Pindell and Dewey, 1982), consuming North American Plate ocean crust by subduction beneath the Lesser Antilles.

The shape of the Caribbean Plate has not changed greatly since its creation, except about 65 m.y. ago, when Cuba was severed from the Caribbean Plate and sutured onto the North American Plate, the Yucatan Basin began to open as a small

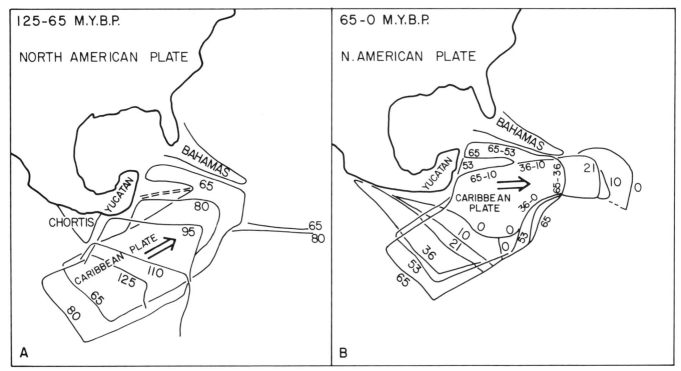

Figure 4. Position and shape of the Caribbean Plate since 125 m.y. B.P. Compiled from Pindell and Dewey (1982). A, 125 to 65 m.y. B.P.; B, 65 m.y. B.P. to the present. Small numbers show ages of plate boundary segments in m.y. B.P.; arrows, approximate direction of Caribbean Plate movement; double dashed lines, back-arc spreading axis active 80-65 m.y. B.P. to form the Yucatan Basin. East-west lines labeled 65 and 80 in A are connections to Mid-Atlantic Ridge spreading axis at those times. North America, the Bahamas, and Yucatan have been held fixed, in order to show the change in shape and position of the Caribbean Plate in its eastward movement along the southern margin of the North American Plate. Note the Caribbean Plate's change in direction of motion at 65 m.y. B.P. and its seeming decrease in surface area beginning at that time, during its apparent partial subduction beneath South America.

spreading center, and the Caribbean Plate began to move eastward. Pindell and Dewey's (1982) maps (Fig. 4) suggest that the eastward movement may have caused overlapping of the Caribbean and South American Plates in northern Colombia and Venezuela, via a process of slow, nonvolcanic subduction as described above (Kellogg and Bonini, 1982), and perhaps including the distortion of the Isthmus of Panama.

STRUCTURAL EVENTS IN THE CARIBBEAN PLATE

Late Jurassic-Early Cretaceous (Figure 5a: 140-110 m.y. B.P.)

Unmetamorphosed shelf sediments of middle Jurassic to Early Cretaceous age crop out in northern and western Cuba, showing only evidence of continental-margin, nonvolcanic erosion, sedimentation, and subsidence during this time in that part of Cuba. Lithological similarities to rocks of the same age in Honduras have been pointed out by Khudoley and Meyerhoff (1971).

Metamorphic rocks in Cuba are of two types. Metamorphosed siliceous and pelitic sedimentary rocks in central and western Cuba may be the metamorphosed equivalents of the Middle Jurassic to Early Cretaceous rocks just described, or they may be older. Metamorphosed basic volcanic rocks in eastern Cuba, cut by a 120-m.y. B.P. unmetamorphosed pegmatite (Meyerhoff and others, 1969) and associated with ultramafic rocks and high-pressure/low-temperature blueschist metamorphic rocks may represent an early subduction episode. In that case, rock distributions suggest that a southern plate subducted beneath what is now the North American Plate (Mattson, 1979). This subduction zone must have been uplifted and eroded in the 120-110-m.y. B.P. interval, because deposits in eastern Cuba of latest Aptian to Albian age (110-100 m.y. B.P.) contain volcanic rocks belonging to a younger subduction episode.

The metamorphic core of Hispaniola is cut by two intrusions dated at 123 and 127 m.y. B.P. (Kesler and others, 1977), and overlain unconformably by less metamorphosed rocks containing mid-Aptian to mid-Albian (about 112-105 m.y. B.P.) fossils (Bowin, 1966). Draper and Lewis (1980) and Lewis (1980) describe the northern portion of the metamorphic com-

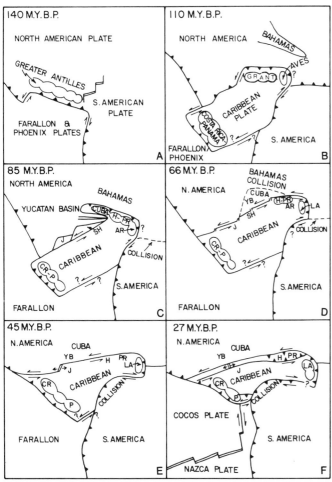

Figure 5. Structure of the Caribbean Plate since 140 m.y. B.P. Shapes of Caribbean Plate taken from Pindell and Dewey (1982). Symbols: toothed lines, subduction zones; double lines, spreading zones; single lines or lines with arrows, transform plate boundaries; dashed line, collision plate boundary. Active subduction-generated island arcs are outlined. Abbreviations: AR, Aves Ridge; CR, Costa Rica; GR ANT, Greater Antilles; H, Hispaniola; J, Jamaica; LA, Lesser Antilles; P, Panama; PR, Puerto Rico; SH, southern Haiti; YB, Yucatan Basin.

plex as the Amina Formation, which they interpret as a pyroclastic and epiclastic part of an accretionary prism of a Cretaceous island arc, metamorphosed to greenschist facies. The rocks are chemically basaltic and keratophyric (Lewis, 1980) and are perhaps coeval with a suite of greenschist-grade mafic volcanic rocks, the Duarte Formation, exposed to the south (Bowin, 1975). Between the Amina and Duarte Formations is a fault zone Fig. 6) containing large bodies of ultramafic rock.

This distribution of pre-Albian volcanic and ultramafic lithologies in Hispaniola suggests that a subduction zone dipping northward might have existed at that time beneath Hispaniola. The suture is now exposed as the fault zone containing ultramafic rocks, separating ocean crust on the south from island arc on the north (Bowin, 1975). Subduction must have ended between 123 and 105 m.y. ago.

In Puerto Rico (Fig. 7), serpentinized harzburgite and dunite, pillow lava, metatholeiite, and radiolarite form the oldest part of the Mesozoic core of the island (Mattson, 1973). The ultramafic rocks are probably ophiolite; the metatholeiite by its chemistry could correspond to either ocean crust or early island arc type (Lee and Mattson, 1976). The radiolarite contains Late Jurassic and Early Cretaceous radiolarians (Mattson and Pessagno, 1979). Three radiometric dates in the metatholeiite are 85, 110, and 112 m.y. B.P. (Tobisch, 1968; Mattson, 1966; Cox and others, 1977). The sequence is overlain unconformably by a volcanic series less deformed and metamorphosed than the older rocks. The volcanic series is as old as Albian by fossils, and contains 109 and 100 m.y. B.P. plutons (Cox and others, 1977). Thus the break between tectonic sequences may have happened about 110 m.y. ago, perhaps somewhat older if the metamorphic rocks have been reheated, as the single 85-m.y. B.P. number suggests.

The only evidence of vergence of the subduction zone in Puerto Rico is that the slightly younger plutonic rocks are located east-northeast of the ophiolite, suggesting an island arc north of subducting ocean crust. Common strike-slip faulting and a complicated structural history subsequent to this eposide may have confused these old geographic relations.

In summary, Cuba and Hispaniola have good evidence of a north-dipping subduction zone active until about 120-110 m.y. B.P., and evidence in Puerto Rico suggests a break in activity about 110 m.y. B.P., perhaps slightly older. The dip of the subduction zone was northward, based upon good evidence in Cuba and Hispaniola, and on weaker evidence in Puerto Rico. Ocean crust being subducted was of Early Cretaceous and Late Jurassic age in Puerto Rico.

There is no sedimentological or other direct evidence that the island arc was emergent, nor was there evidence of continental provenance for sediments in the arc, except perhaps in western Cuba during the Jurassic portion of the episode. In both Pindell and Dewey's (1982) and Duncan and Hargraves' (1984) reconstructions, the western Greater Antilles were situated at about this time near the Chortis Block of Central America, so Chortis may have been a source of continental detritus for the Cuban Jurassic deposits.

Uplift, erosion, or the beginning of new deposition marked the boundary between Early and Late Cretaceous (about 100 m.y. B.P.) throughout northern South America. During the 140-110-m.y. B.P. interval, northern South America was largely emergent, shedding clastic detritus from uplifted Paleozoic terranes. Marginal areas such as the Santa Marta Massif and Caribbean Mountains were warped downward, receiving pyroclastic and epiclastic material from offshore volcanoes to the north. A metamorphic episode may have taken place in the Santa Marta Massif at about this time, with emplacement of ultramafic rocks.

Convergence of the Phoenix and Americas Plates might have caused South America to rise during this interval. Subduction would have created a volcanic arc, and the beginning of eastward movement of the Greater Antilles segment of the sub-

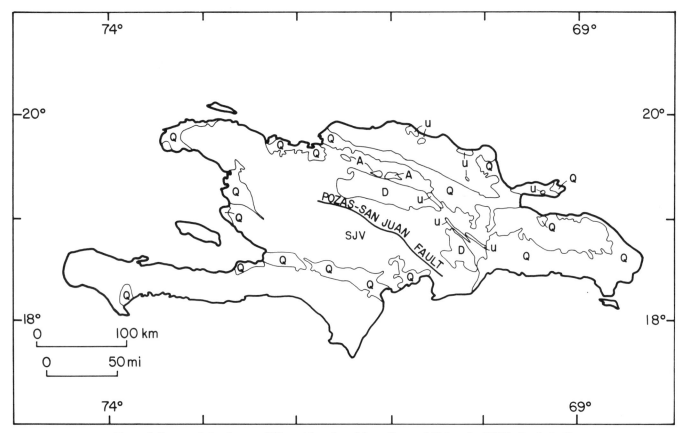

Figure 6. Structural sketch map of Hispaniola, adapted from Case and Holcombe (1980). Symbols: Q, alluvium; blank areas, Tertiary and Cretaeous strata younger than 110 m.y. B.P.; D, Duarte Formation; A, Amina Formation; U, ultramafic rocks (D, A, and U are 140-110 m.y. B.P.); SJV, San Juan Valley.

duction zone would have caused transform motion at the ends of the segment.

Middle Cretaceous (Figure 5b: 110-85 m.y. B.P.; Albian-Santonian)

The Middle Cretaceous time was one of the most important rock-forming intervals in the Greater Antilles, creating thick sequences of largely submarine volcanic rocks. In Cuba two volcanic sequences crop out, the Domingo and Cabaiguan belts of Pardo (1975). The older Domingo belt contains basic volcanic rocks, ultramafic rocks, spilitized basalts, and crystal tuffs in a thick, unfossiliferous series. Conformably above and south of the Domingo belt is the Cabaiguan belt of clastic and volcanic rocks of basalt, andesite, and more silicic composition, including lava flows, pillow lava, and tuff. Both series are dominantly submarine, although the southern unit contains Cenomanian limestones with oolite clasts derived from Upper Jurassic limestones apparently exposed to the south.

The older Cuban volcanic sequence contains no fossils, but volcanic-derived shales, clays, and cherts north of the volcanic belt are of latest Aptian to Albian age (110-100 m.y. B.P.; Pardo, 1975). The middle of the younger Cuban sequence contains Cenomanian and Turonian microfossils (100-87 m.y. B.P.). The top

1000 feet of the Cabaiguan section contains Maestrichtian fossils and is considered in the next interval. No fossils from the Coniacian to Campanian interval have yet been found, although an unfossiliferous 2000-foot volcanic unit stratigraphically between Turonian and Maestrichtian rocks may date from this interval. The character of volcanism seems to change in the Santonian, from deep-sea to shallow marine and from greater to lesser intensity. Uplift culminating in emergence, unaccompanied by folding or other obvious tectonism formed a disconformity in the Santonian in some parts of Cuba, and a slight unconformity in other parts (Pardo, 1975).

Some unmetamorphosed volcanic rocks crop out in Jamaica (Fig. 8) dating from this time, but most unmetamorphosed deposits were clastic or epiclastic (Arden, 1975). The Benbow inlier in north-central Jamaica contains volcanic rock and limestone formations of Barremian to Albian age (120-100 m.y. B.P.), including andesite and basalt lavas, pillow lavas, breccias, shales, sandstones, conglomerates, cherts, and limestones. The central inlier contains a thick Cenomanian to Turonian volcanic conglomerate and breccia with minor epiclastic rocks, basalt lavas and dikes, and conglomerates (Arthurs Seat formation), cut and metamorphosed by a granodiorite dated at 85 m.y. B.P. A Turonian to early Campanian mudstone (Peters Hill formation) overlies the Arthurs Seat formation. Campanian volcaniclastic rocks,

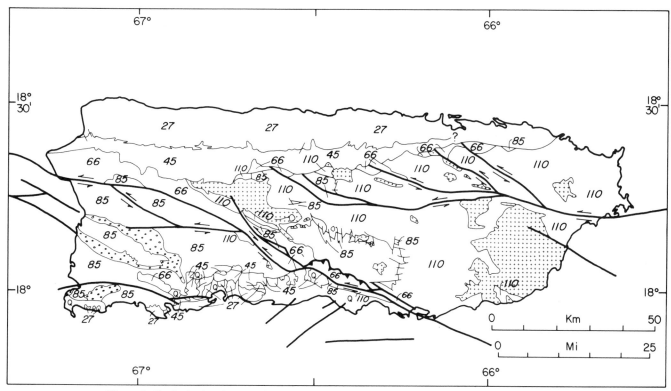

Figure 7. Geologic sketch map of Puerto Rico. Adapted from Garrison and others (1972), Glover (1971), Monroe (1980), Nelson and Monroe (1966), Berryhill (1965), and Nelson (1967a, 1967b). Only land-exposed structures are shown. Symbols: sawtooth lines, thrust faults or slides; lines with arrows, strike-slip faults; other heavy lines, normal faults. Patterns are: circles, ultramafic and other rocks of 140-110 m.y. B.P. interval; crosses, granitic rocks of several ages. Areas shown as 110, 85, 66, 45, and 27 are strata of 110-85, 85-66, 66-45, 45-27, and 27-0 m.y. B.P. ages, respectively.

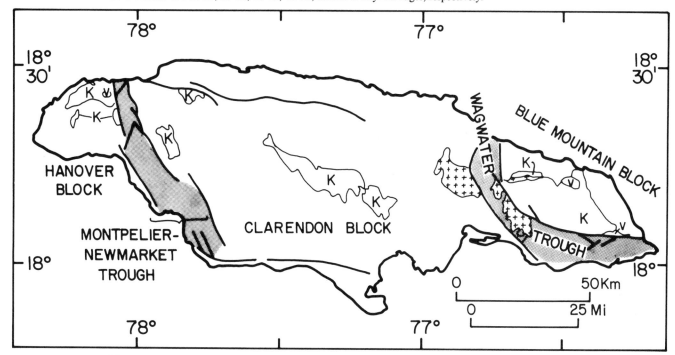

Figure 8. Structural sketch map of Jamaica. Adapted from Arden (1975). Symbols: blank areas, Tertiary strata; K, Cretaceous strata; v, volcanic rocks; +, granitic rocks; black areas, ultramafic rocks; heavy lines, faults; shading marks, structural troughs.

andesite flows, and volcanic breccias overlie these rocks unconformably. In other parts of the island, however, the unconformity separates Coniacian and basal Campanian rocks.

Two units of metamorphic rocks may also date from this time. The Westphalia and Mount Hibernia schist complexes are in fault contact with each other in the southwest Blue Mountain inlier in east-central Jamaica (Draper and others, 1976). The Westphalia schist consists of highly deformed quarzo-feldspsathic and amphibolite-grade rocks, including sedimentary rocks, basalt, andesite, granodiorite, and ultramafic rocks. The Mount Hibernia schist complex contains marble interbedded with schistose mafic volcanic rocks; greenschist mineralogy is prevalent with blueschist facies in the eastern exposures.

The Westphalia schist is overlain unconformably by Eocene rocks and contains a K-Ar retention age of 76 m.y. B.P., probably indicating Late Cretaceous metamorphism. The Mount Hibernia schist is overlain unconformably by (Arden, 1975) or in fault contact with (Krijnen and Chin, 1978) Campanian to Maestrichtian sedimentary and volcanic rocks. Draper and others (1976) consider the metamorphic rocks as pre-Early Cretaceous apparently based on one locality of tentatively identified benthonic rudists, unsupported by published mapping as is the estimate of Krijnen and Chin, which is accepted in this paper.

The stratigraphic relation between the two metamorphic units is unknown; the Westphalia schist is metamorphosed to a higher grade than is the Mount Hibernia schist, but both could represent different levels in the same episode. The higher metamorphic grade than that of other Jamaican rocks of this age suggests that the metamorphic rocks are older, but fossil evidence does not support more than minimum ages.

The metamorphic rocks represent a converging plate boundary complex. Mount Hibernia schist forms a high-pressure/low-temperature subduction zone, now uplifted; and the Westphalia schist is the volcanic island arc, of greenschist grade. If their original relative positions are still correctly shown, the subduction zone dipped to the south or southwest.

The tectonic position of Jamaica during 110-85 m.y. B.P. is not clearly shown in the geologic record. Reconstructions place the island at the west end of the Greater Antilles between Cuba and Yucatan (Pindell and Dewey, 1982; Fig. 5b, 5c) in a complex area of transform, subduction, and rift boundaries. The Nicaragua Rise during the Early Cretaceous seems to have been a site of volcanism (Arden, 1975), but the tectonic cause of the volcanism is not evident. Shallow-water deposition and erosional products are more common here than in other parts of the Greater Antilles, perhaps because of the proximity of the Chortis and Yucatan continental blocks.

Hispaniola contains three sequences of rocks that can be placed with varying certainty into the 110-85-m.y. B.P. interval. All are relatively unmetamorphosed, thus apparently younger than the strong metamorphic episode culminating about 110 m.y. B.P.

Rocks containing mid-Aptian to mid-Albian fossils of shallow-water types are mapped as the Los Ranchos, Hatillo, and Las Lagunas formations (Bowin, 1966, 1975), exposed north of the metamorphic complex in central Dominican Republic. Rock types include quartz keratophyre, keratophyre, dacite, andesite tuffs, cherts, and minor sedimentary rocks. The Tireo formation, tuffs and quartz keratophyres with minor andesite, basalt, and limestone, contains Cenomanian fossils in its lower part; it is exposed south of the metamorphic complex. On the southern peninsula of Haiti pillow basalts are interbedded with Aptian-Albian sedimentary rocks, and a volcanic unit similar in age to the Tireo occurs in the northern peninsula of Haiti.

Unmetamorphosed submarine basalts, tuffs, wackes, and some radiolarites (Siete Cabezas and Peravillas formations) rest unconformably upon the metamorphic complex in the central cordillera. These rocks are unconformably below Lower Eocene rocks; Bowin believes them to be pre-middle Aptian based on the age of the Los Ranchos formation; but they are nowhere in contact with the Los Ranchos, and it is possible that they are the southeastern lateral equivalents of the Los Ranchos.

The upper part of the Tireo formation, and volcanic sediments and limestones in northwestern Haiti, both contain Campanian-Maestrichtian fossils and belong in the next younger interval. More work, especially more fossil dating, may permit greater precision in marking the 85-m.y. B.P. discontinuity, for which the strongest evidence is in Puerto Rico.

Certainly there was volcanism during this interval in Hispaniola. There is some suggestion that submarine volcanism predominated in the early part of the interval, and that shoaling occurred near the end of the interval and perhaps continued into the next younger episode. No evidence for vergence of the 110-85-m.y. B.P. subduction zone has been found in Hispaniola.

Puerto Rico also contains two volcanic sequences within this episode, the Albian pre-Robles sequence and the Albian to Santonian Rio Orocovis Group (Berryhill and others, 1960; Mattson, 1974). The pre-Robles sequence is almost entirely marine volcanic rocks: lavas, breccias, tuffs, and volcanic sediments predominantly andesitic in composition. Two thin conglomerates contain cobbles of Albian limestone. The Rio Orocovis Group rests upon the pre-Robles sequence with slight unconformity. It includes pillow lavas, pyroclastic and epiclastic volcanic rocks of andesite and basaltic andesite composition. No detrital quartz or other continental erosional debris has been found. Volcanic deposition was probably in deep water, as the only evidences of shallow-water deposition are the two thin conglomerate members and a nonmarine tuff breccia unit near the top of the Rio Orocovis Group. At the top of the Rio Orocovis, faulting, uplift, and erosion formed a moderately angular unconformity.

In Puerto Rico, fossils in rocks from this episode range in age from Albian (108-100 m.y. B.P.) to Santonian (85-82 m.y. B.P.). The next younger unit, unconformable on the older unit, also contains Santonian fossils, so the end of this tectonic episode is tightly constrained to 85-82 m.y. B.P. in Puerto Rico.

During the 110-85 m.y. B.P. interval, the Greater Antilles developed thick volcanic sequences of subduction zone type, largely from submarine volcanoes, with only minor shallow-water deposition. Shallow-water Upper Jurassic limestones were exposed to erosion south of Cuba in the Cenomanian, perhaps on the then nearby Chortis or Yucatan blocks.

There is no evidence for dip of the subduction zone in Puerto Rico or Hispaniola. In Cuba deep-sea basalts and dolerites may form ocean crust in the Domingo belt; the belt was thrust northward onto sedimentary rocks of other belts during Eocene deformation. Southward-dipping major thrust and reverse faults suggest southward-dipping subduction, but most faults are Eocene in age. The lack of volcanic deposition on the shelf sediments north of the volcanic belt during this interval, where a volcanic arc would form if subduction dipped to the north, indicates a southward dip is more probable. Tectonic inclusions of older metamorphic rocks in the Domingo belt ultramafic rocks suggest that the Domingo belt is part of an active subduction zone rather than part of a passive upper plate, which it would be if subduction dipped northward. In Jamaica there may have been a subduction zone dipping to the south at this time, but the relevant rocks are poorly dated and may have had their relative positions changed by faulting.

Thus the 110-85 m.y. B.P. subduction zone in the Greater Antilles probably dipped to the south, based on evidence from Cuba and Jamaica.

During the 110-85 m.y. B.P. interval in northern South America, uplift of the Cordillera Central, Guajira Block, Santa Marta Massif, and other mountain blocks continued, with rapid clastic and carbonate deposition in adjacent basins such as the Cauca and Magdalena valleys at the north end of the Cordillera Central. Volcanism in Aruba and Curacao in the Netherlands Antilles is tholeiitic, perhaps originating at a mantle hot spot (Beets and others, 1984); Bonaire exposes a slightly younger series of primitive island arc and calc-alkaline volcanic rocks. Ammonites from the Curacao lavas are middle Albian in age (Beets and others, 1984). However, the same rock unit on Curacao and igneous rocks from two Venezuelan islands and the Paraguana Peninsula have nine K/Ar ages of 114-130 m.y. B.P. (Santamaria and Schubert, 1974); thus it is possible that the volcanic episode in the Netherlands and Venezuelan Antilles began in the Early Cretaceous, before the 110-85 m.y. B.P. interval. At the end of the interval, uplift and tilting of the Netherlands and Venezuelan Antilles arc may have marked its collision with the continent (Beets and MacGillavry, 1977; Beets and others, 1984).

Tholeiitic volcanism including the formation of basalt, spilite, and ultramafic rocks occurred in the Cordillera Occidental and the Pacific Coast Ranges at this time, and some volcanism was located to the east, on the west flank of the Cordillera Central.

Deformation and metamorphism are not as precisely located as in the Greater Antilles. Orogenic activity was recorded in the Cordillera Central: granite intrusion, ultramafic emplacement, and low-grade metamorphism. Possible blueschist-facies rocks are reported by R. Shagam (personal communication, 1983) in the Cauca valley. Formation of greenschist-facies phyllites in the

Santa Marta Massif may date from this time or earlier. Collision of the Netherlands and Venezuelan Antilles arc with South America created high-pressure blueschist- and eclogite-facies rocks at the end of the interval (Beets and others, 1984).

Late Cretaceous (Figure 5c: 85-66 m.y. B.P., Santonian-Maestrichtian)

The Campanian and Maestrichtian in Cuba were periods of island volcanism, deep- and shallow-water clastic, epiclastic, carbonate, and reef deposition, and moderate to intense deformation. Moderately active volcanism, less intense but more siliceous than in the Middle Cretaceous, produced basalt lava and beds of andesitic tuff and breccia. Deformation beginning in western Cuba caused uplift, folding, and largely southward thrusting in the early Maestrichtian. In general, volcanism in Cuba was distributed south of a carbonate bank during this time, suggesting that the subduction zone dipped beneath the volcanic area from along its north or south margin. Ultramafic rock distribution along the north margin favors a southward-dipping zone.

Jamaica contains strata that are predominantly epiclastic and clastic, largely Maestrichtian but perhaps also Campanian in age. Volcanic tuff beds have been mapped, and the Late Cretaceous section in the central inlier includes andesite flows and andesite and basalt dikes. Roobol (1972) has mapped several small volcanic centers in eastern Jamaica.

The island was emergent before the end of the Maestrichtian; granodiorite intrusions dated at 63 +/- 3 m.y. B.P. (Harland and others, 1964) and faulting probably accompanied the uplift; large graben such as the Wagwater Trough may have begun to form at this time (Arden, 1975).

Volcanism and nonvolcanic sedimentation both occurred in Hispaniola during this time. Limestones interbedded with basalt lavas and other volcanic rocks form the southern peninsula of Haiti and with andesitic tuffs are exposed in central Haiti. Shallow-water clastic sediments, tuff, and limestone are mapped as the Maestrichtian part of the Tireo formation in central Dominican Republic. Near the north coast of Hispaniola, in the Puerto Plata area, andesite lava, tuff, spilite, and keratophyre(?) crop out, associated with a melange deposit containing blocks of ultramafic and blueschist metamorphic rocks. These Puerto Plata rocks are undated but lie unconformably below Paleocene strata.

Deformation, probably uplift, folding, and erosion, occurred in Hispaniola within or at the end of the Maestrichtian. A few areas in central Dominican Republic show conformable relations between Cretaceous and Tertiary strata, so the unconformity may have been formed before the end of the Cretaceous, or may not have taken place everywhere.

Bowin (1975) suggests that a northern plate was subducting beneath the Caribbean Plate from Early Cretaceous to early Maestrichtian time. He (as well as Nagle, 1974) thought the melange in the Puerto Plata area to be a fragment of that subduction zone.

Puerto Rico exposes a moderate angular unconformity

within the Santonian, at the base of the Santonian-Maestrichtian sequence. The upper boundary is well defined in western Puerto Rico as an angular unconformity within the Maestrichtian, separating two series both containing Maestrichtian microfossil localities (Mattson, 1960).

Andesitic, basaltic andesitic, and some basaltic volcanism were widespread in Puerto Rico during this time. Activity was probably a series of island volcanoes with fringing reefs and epiclastic deposits. Thick, volcanic conglomerates were deposited with lavas in central Puerto Rico, and relatively thin sequences of limestones and epiclastic volcanic rocks in southwestern Puerto Rico.

Folding, faulting, uplift, erosion, and gravity sliding northward into the volcanic pile all took place in the early Maestrichtian. Hypabyssal granitic intrusions date from this time. Subduction probably dipped to the south from a subduction zone now exposed on the south wall of the Puerto Rico Trench north of the island (Perfit and others, 1980); in that case the southwestern carbonate sequence is part of a back-arc high.

The Greater Antilles during 85-66 m.y. B.P. formed a volcanic island arc, containing island volcanoes and reefs, with a subduction zone north of the Cuba-Puerto Rico trend. Jamaica was also near or on a volcanic island arc, but may have been separated from the other areas by transform faults west of Cuba. Carbonate deposition is preserved from northern Cuba and southwest Puerto Rico, off the trend of the island arc; and the actual subduction zone may be preserved in Cuba, Hispaniola, and north of Puerto Rico. The period ended with faulting, folding, intrusion of granitic rocks, uplift, and erosion. Subduction may have ended when the northern Cuba-Bahama carbonate-covered crust began to enter the subduction zone north of Cuba, a collision probably beginning in western Cuba in the early Maestrichtian and moving eastward.

In northern South America, the 85-66 m.y. B.P. interval is not well defined. Major tholeiitic volcanism ended in the Cordillera Occidental and Pacific Coast Ranges, but minor calc-alkaline volcanism continued in the Netherlands Antilles (Beets and others, 1984). In the Caribbean Mountains uplift, deformation, and southward gravity sliding probably began during the Maestrichtian, to culminate during the next interval. Elsewhere, block uplift and subsidence continued.

Early Tertiary (Figure 5d: 66-45 m.y. B.P.; Paleocene-Middle Eocene)

In central and western Cuba, uplift and erosion followed by deposition of clastic and carbonate sediments persisted throughout this interval. The only volcanic activity was in easternmost Cuba, forming thick volcanic and epiclastic deposits, including calc-alkaline basalt, andesite, dacite, and rhyolite volcanic breccias and some lavas. Deformation was intense during this time. Folding and faulting swept across Cuba, culminating in Early Eocene in the west and Middle Eocene in central Cuba. Activity

included folding, normal and reverse faulting, northward thrust faulting and gravity sliding with displacements of as much as 60 miles (Pardo, 1975). Left-lateral strike-slip faulting was common, and ultramafic rocks were emplaced tectonically in the fault zones. Intensity of deformation increased from south to north, and decreased to the east; in Sierra Maestra in eastern Cuba only a disconformity separates Middle from Upper Eocene strata. Strong erosion and uplift caused thick clastic and carbonate deposition on widespread peneplaned or eroded surfaces throughout Cuba beginning in the Late Eocene.

Tertiary nonvolcanic clastic and carbonate deposition began in Jamaica perhaps as early as Paleocene time; volcanism probably resumed in the Early Eocene. In the late Early Eocene (Butterlin, 1977) or Middle Eocene (Arden, 1975), normal and strike-slip faulting in two major zones created graben separating three uplifted structural blocks. The Wagwater belt between the Clarendon and Blue Mountain blocks is arcuate, convex to the southwest. Sedimentation in the Wagwater belt began with evaporites, continued as pyroclastic and epiclastic material, and included thick conglomerates. Jackson and Smith (1980) described volcanism in the belt as bimodal: tholeiitic basalts and spilites interlayered with calc-alkaline dacites, perhaps formed during creation of an inter-arc or back-arc basin. Island-arc-type granitic intrusions are exposed in central and eastern Jamaica.

Unconformities are mapped in the Paleocene to Early Eocene, and in Late Eocene in many places in Jamaica, but deposition in the Wagwater Trough is continuous from Paleocene through Oligocene (Arden, 1975).

In Hispaniola, the Paleocene and Early Eocene were times of erosion and volcanism in some places, and of limestone deposition in other places. Volcanic, epiclastic, and clastic rocks are exposed in central Haiti and the Dominican Republic. Elsewhere limestone and clastic deposits were common, except in the Puerto Plata area of northern Dominican Republic, where andesite and dacite tuffs have been mapped. In Middle Eocene times, wackes and tuffs were deposited in central Dominican Republic, and basalt pillow lavas formed with limestones in northwestern Haiti. Central Haiti contains Middle Eocene clastic rocks, limestones, and andesite tuff and lava intruded by Upper Eocene granodiorite and dolerite. A submarine slide or olistostrome containing Middle Eocene and older rock blocks moved downslope to the north, in the Puerto Plata area; it was uplifted and emergent before early Late Eocene time. Nagle (1974) considers it a fragment of a subduction zone.

The Late Eocene was a period of important change in Hispaniola. Volcanism ceased to be of major importance, volcanic rocks of this age found only in northwestern Haiti. Intrusive stocks of granodiorite, diorite, gabbro, and dolerite cut younger rocks in central Dominican Republic and Haiti. In eastern Hispaniola deformation produced east-trending folds, as in Puerto Rico to the east; in central and western Hispaniola folding retained earlier west-northwest trends. Overthrusts at this time were to the north (Bowin, 1975).

In Puerto Rico a tectonic break within the Maestrichtian is

more widespread and important than the break at the end of the Maestrichtian. Shallow-water calc-alkaline volcanic activity, probably forming island volcano and reef complexes, combined with erosional products of Cretaceous volcanic terranes to form the stratigraphic section. Fossil ages locate the interval boundaries closely: Maestrichtian at the base (perhaps 70 m.y. B.P.), Middle Eocene at the top of the interval (about 45 m.y. B.P.). Late Eocene rocks have not yet been discovered; Early Oligocene fossils are found in rocks near the base of the next interval. Green siliceous tuff layers in Middle Eocene strata are probably equivalent to Layer A and A" of the Caribbean and north Atlantic ocean floor (Mattson and Pessagno, 1971). Plutonic and hypabyssal intrusion of granitic rocks accompanied by metamorphism and mineralization both began and ended the episode; the younger intrusions have been dated as young as 38 m.y. B.P., Latest Eocene or Early Oligocene (Barabas, 1971).

Folding was more intense in the Maestrichtian than in the Eocene deformation, and faulting was more frequent in the Eocene. Northward-moving overthrusts and slides, normal faults, and strike-slip faults were common. Strata contain east-trending fold axes except in the regions of thickest deposition, where westnorthwest fold axes predominate. This may be the result of the control of folding in areas of thin deposition by older easttrending fractures in the pre-Eocene basement. Away from the basement shear zones, in deeper parts of the Paleogene basin, fold orientations may be controlled primarily by regional stresses.

In summary, the Greater Antilles during the 66-45 m.y. B.P. interval (perhaps beginning a few million years earlier according to the Puerto Rico data) was a time of island-arc volcanism, clastic and epiclastic sedimentation, and deformation. Overthrusting and sliding were northward; a subduction zone was located north of the islands. The volcanic arc was positioned south of thrust movements in Cuba, in the same general location as thrusting in Hispaniola, and north of thrusting in Puerto Rico. This oblique crossing of two tectonic features suggests that they are of different ages or at least produced by different stress fields. Jamaica does not seem clearly related to the other Greater Antilles during this time. Perhaps it represents an inter-arc or back-arc basin as Jackson and Smith (1980) suggested.

Northern South America contains scattered evidence of the 66-45 m.y. B.P. episode. Volcanism, epiclastic deposition, and intrusion continued in the Netherlands Antilles through the Paleocene. Uplift and gentle folding in the eastern Caribbean Mountains were accompanied by southward sliding of large tectonic blocks and low-grade metamorphism, while intense alpine-type folding and eastward thrusting occurred in the western Caribbean Mountains (Kellogg, 1984). In the Merida Andes, marine upper Cretaceous and clastic Paleocene to Eocene valley-fill deposits were deformed by block-uplift with little folding or thermal metamorphism, before deposition of Oligocene to Pliocene clastic, partly continental deposits (Shagam, 1975).

To the southwest in the Cordillera Oriental and Santander Massif, uplift and minor heating produced an outward overthrusting and overturning of folds, symmetrical in the Cordillera Orien-

tal but toward the west in Santander. Minor Paleocene volcanism is noted by Shagam (1975).

Shallow-water and nonmarine deposits of Oligocene and younger age in the Perija block, following marine Cretaceous and early Tertiary deposits, suggest rapid uplift and erosion in the Eocene, forming unconformities dated by Kellogg (1984) as Early Eocene (53 m.y. B.P.) and Middle Eocene (45 m.y. B.P.). In the Santa Marta Massif, Paleocene and Eocene granite intrusions and a lack of sedimentary deposits from that time also suggest uplift and erosion. In the Cauca Valley of the Cordillera Central, marine Cretaceous is overlain with probable unconformity by nonmarine Paleocene to Oligocene strata; and on the east flank of the mountain range, marine Cretaceous and Paleocene valley-fill deposits are overlain unconformably by molasse clastics of Eocene and Early Oligocene age. Granites of about this age are exposed in the Cauca Valley (Shagam, 1975).

In the Cordillera Occidental and the Pacific Coast Range of Colombia, volcanism with strong deformation occurred in the Maestrichtian to Eocene interval. Submarine volcanic rocks included calc-alkaline gabbro, spilite, diabase, and tuff. Activity probably began before the end of the Cretaceous and extended across the interval into the Oligocene. Regional metamorphism, rocks moderately to tightly folded and overturned to the west, abundant faults, and uplift characterized this period. Miocene clastic rocks in faulted graben cover these units unconformably.

Duque-Caro (1979; 1984) described intense deformation and diapirism in the San Jacinto belt, north of the Cordillera Occidental near the Caribbean coast of Colombia, in an accretionary prism of deep-water and shallow-water sediments of Cretaceous to Early Oligocene age. Narrow anticlines and broad synclines, mud volcanoes and intrusive diapirs, and common faults were formed from Paleocene through Early Oligocene time. Unconformities marked the late Middle Eocene and the end of the Early Oligocene. Duque-Caro suggested that the deformation, while directly caused by gravitational instability in the turbidite-pelagic accretionary prism, was ultimately the result of plate convergence along northern South America.

Thus northern South America shows evidence of two major tectonic processes active 66 to 45 m.y. ago. The Netherlands Antilles, Cordillera Occidental, San Jacinto belt, and Pacific Coast Range were sites of island-arc or submarine volcanic activity continuing throughout the interval but not well-defined by the interval boundaries, probably indicating subduction at the northern and northwestern edges of the South American Plate. The Caribbean Mountains were folded and thrust eastward and southward, but the other parts of northern South America behaved as rigid blocks, uplifted and intruded with little internal deformation.

Middle Tertiary (Figure 5e: 45-27 m.y. B.P.; Late Eocene-Late Oligocene)

In Cuba, widespread deposition of carbonate, evaporite, and clastic material beginning in Late or upper Middle Eocene time

(Khudoley and Meyerhoff, 1971) was followed by gentle folding and uplift in the Oligocene, and by another transgression beginning near the start of the Miocene. Moderate folding took place locally at the edge of the Bahama Platform in northeastern Cuba (Brezsnyanszky and Iturralde-Vinent, 1978). Iturralde-Vinent (1981) described block uplifts and the formation of graben and half-graben during this time. However, folding and erosion of older strata at the beginning of the interval were evident only near the margins of the subsiding basins; strata are conformable in the deep basins (Iturralde-Vinent, 1972). Almost no igneous activity was reported; Brezsnyanszky (personal communication, 1977) mentioned the existence of north-trending igneous dikes in the Sierra Maestra, cutting Upper Eocene rocks and covered by Miocene strata.

Jamaica preserves carbonate deposits with less than 2 percent terrigenous material from the 45-27-m.y. B.P. interval, and rare clay, shale, and lignite (Arden, 1975). Arden suggested that there was local emergence during the Late Eocene, followed by almost complete submergence except for a few low islands. A more widespread uplift and emergence ended the interval in Middle Oligocene to early Miocene time, forming an important unconformity (Arden, 1975).

In Hispaniola carbonate deposition predominated during this interval. Minor volcanism persisted: Upper Eocene basalt, dolerite, and microdiorite in northern Haiti; volcanic fragments in limestone in the Cordillera Central; and clastic rocks, some tuffaceous, in northern Dominican Republic.

The interval limits in Hispaniola differ somewhat from 45 and 27 m.y. B.P. Near the base of the interval in Haiti, a widespread unconformity separates Upper Eocene carbonates and minor volcanic rocks from Lower Oligocene carbonates and minor volcanic rocks; in Dominican Republic only the unconformity separating Middle and Upper Eocene strata exists. At the top of the interval, an unconformity separating Oligo-Miocene from Miocene rocks at about 20 m.y. B.P. is mapped in Haiti. In northern Hispaniola a significant unconformity separating older clastic rocks of the Tabera Formation from younger carbonate and clastic rocks of the Cercado Formation is considered by Bowin (1975) to be within the early Miocene, about 18 m.y. B.P., but by Butterlin (1977) to lie at the Oligocene-Miocene boundary, about 22 m.y. B.P.

Rocks of this interval in central and western Hispaniola are folded and faulted (Michael and Lewis, 1980) in a major deformation, but the eastern part of the island did not experience any such orogeny, nor did other islands in the Greater Antilles.

Puerto Rico had much the same history as eastern Hispaniola during 45-27 m.y. B.P. It experienced only uplift, erosion, and clastic and carbonate deposition. Hypabyssal porphyry intrusions accompanied by copper mineralization cut pre-Oligocene rocks during this time, but there is no evidence of other igneous activity. Basins north and south of the island accumulated detritus from the eroding uplands. No break is evident near 27 m.y. B.P., except a change in deposition in the Late Oligocene or early Miocene from dominantly clastic sediments with subordinate marls and limestones to dominantly limestone with calcarenite and marl (Mattson, 1973; Monroe, 1980), perhaps indicating subsidence or lateral movement of the eroding landmass in central Puerto Rico.

In the Greater Antilles, then, the 45-27-m.y. B.P. interval was characterized by uplift, erosion, and the deposition of clastic and carbonate rocks. Minor igneous activity was perhaps a waning phase of the extensive early Tertiary activity; perhaps the Oligocene dikes in eastern Cuba marked rifting along the Cayman Trough. The interval ended with uplift and emergence in Cuba and Jamaica, strong folding and faulting in central and western Hispaniola, and possible submergence in eastern Hispaniola and Puerto Rico. This suggests a stress field oblique to a transform boundary, with compression and extension in different places along the boundary.

Northern South America during 45-27 m.y. B.P. was composed of rising mountain blocks and subsiding basins. Erosion produced continental debris which filled basins flanking the mountains and grabens within them. Within the blocks, compressional deformation such as folding and faulting seem limited to the Caribbean Mountains during this time; in that area thrust faulting and sliding to the south were common. Deformation in the Caribbean Mountains increased in intensity from south to north. Volcanism and plutonism were reported only in the Cordillera Occidental and Pacific Coast Range (Shagam, 1975; Irving, 1971).

Uplift and tilting, forming unconformities, ended the interval in Late Oligocene in the Magdalena and Cauca valleys, between Early Oligocene and mid-Miocene in the Merida Andes, in about mid-Oligocene in the Santa Marta Massif and in Late Oligocene to mid-Miocene in the Cordillera Occidental and Pacific Coast Range. Deformation that began in the Paleocene continued in the San Jacinto belt north of the Cordillera Occidental, as described above; it decreased in intensity markedly after an unconformity ending the Early Oligocene (Duque-Caro, 1979; 1984), as deposition shoaled. In the Perija block, Oligocene to Pliocene deposition was broken by rapid uplift and erosion, forming a major unconformity, in the Late Oligocene (25 m.y. B.P.; Kellogg, 1984). In the Netherlands Antilles, uplift and erosion occurred without deformation or preserved deposition.

Shagam (1975) considered that true Andean uplift began at the end of this interval, in the early Miocene. Dip-slip movement began at this time on the Oca and Santa Marta faults (Tschanz and others, 1974), although strike-slip movement had already began on those faults in the early Tertiary (Shagam, 1975).

Major uplift and erosion filled the 45-27-m.y. B.P. interval in northern South America. Only the Caribbean Mountains were deformed internally by folding, thrust faulting, and sliding; the San Jacinto belt was deformed diapirically. The Cordillera Occidental and Pacific Coast Range were the sites of volcanism and intrusion, perhaps the result of the preceding episode's subduction. Increased uplift may be the result of more rapid movement of the Caribbean Plate past or beneath northern South America during the 36-21-m.y. B.P. interval (Fig. 4). At the same time, plate movement at the northern edge of the Caribbean Plate was largely transform, with local extension or convergence dependent on the relations between stresses and plate boundary orientations.

Late Tertiary and Quaternary (Figure 5f: 27-0 m.y. B.P.; Miocene to Recent)

Cuba exposes large areas of shallow-water and continental carbonate, evaporite, clay, and clastic sediments formed since 27 m.y. B.P., largely Miocene, deposited in basins or along the present coast. Episodic uplift formed unconformities at the top of the middle Miocene (11 m.y. B.P.), at the top of the upper Miocene (5 m.y. B.P.), and at the top of the Pliocene (about 2 m.y. B.P.) (Iturralde-Vinent, 1969). Gentle folds formed parallel to the island axis at the end of middle Miocene. During late Miocene, the island became essentially completely emergent. Faulting and shearing occurred in Miocene rocks near some major vertical fault zones, such as one trending north from the Bay of Pigs. Terraces as high as 300 m at the east end of Cuba are probably of Pleistocene age (Butterlin, 1977).

In Jamaica, uplift in mid-Miocene time began the "Alpine Orogeny" (Arden, 1975). The different structural blocks were uplifted and tilted, and faulting produced steep, straight north and south coastlines. A few Miocene alkalic lavas are exposed on the north coast of the island. Widespread unconformities date from the end of middle Miocene and the end of the Pliocene. Rocks are carbonates, marls, reefs, and clastics. Arden (1975) suggested that the last episode of lateral slip on the Cayman Trough began in the middle of Miocene time, tilting Jamaica southward and shifting the crest of the Nicaragua Rise northward.

Hispaniola contains a variety of clastic and carbonate middle Miocene and younger rocks exposed in several coastal and inland areas. A large basin of middle Miocene clastic continental and marine rocks including evaporites is folded in the Central Plateau of Haiti and southeastward in the Dominican Republic. Detrital marine middle Miocene rocks are tilted seaward on the flanks of the northwest peninsula of Haiti. Grabens containing deformed Miocene and Pliocene clastic rocks form the Cauca Valley in Dominican Republic and the Enriquillo/Cul-de-Sac graben in southern Hispaniola. Middle Miocene rocks are gently folded and uplifted above 1000 m on older normal faults near the north coast of Dominican Republic, and some gravity slides took place at that time (Nagle, 1979). Unconformable on other middle and upper Miocene rocks are Pleistocene reef and coastal deposits, uplifted as high as 640 m in the northwest peninsula (Butterlin, 1977). Pleistocene volcanic activity is alkalic near the Cul-de-Sac graben in Haiti, and calc-alkalic in the southern Dominican Republic (Vespucci, 1983); both types are probably related to oblique transform motion along the plate boundary.

Puerto Rico during this interval was the site of limestone and marl deposition, with a smaller clastic component than in older rocks. Most deposits are Miocene, with rare Pliocene and uplifted Pleistocene coastal deposits. Monroe (1980) described uplift of the central core of the island in middle Miocene time, about 15 m.y. B.P. Uplift, tilting, and rare faulting and folding

occurred in the early Miocene to Pliocene; terraces as high as 620 m (Weaver, 1960) are dated as early Miocene or younger. The island is tilted downward to the north, presumably since early Miocene. Hot springs, topographic alignments, and common horizontal slickensides in rocks as young as Miocene suggest that lateral faulting still continues.

During the last 27 m.y., the Greater Antilles has undergone mainly uplift, erosion, and tilting; folding has been important only in Hispaniola. Several large graben formed in Hispaniola in the Miocene and were uplifted in the Pliocene. Lateral strike-slip movements were evident in Hispaniola and Puerto Rico, and vertical movements occurred throughout the Greater Antilles but were greatest near the Cayman Trough. Jamaica and Puerto Rico tilted in opposite directions during this interval: Jamaica down to the south and Puerto Rico down to the north. Reflection seismology by Ladd and others (1977) indicates slow convergent underthrusting in the Muertos Trench south of Dominican Republic and Puerto Rico, and Sykes and others (1982) describe a component of convergence beneath the Puerto Rico Trench north of Puerto Rico. These convergence rates probably total no more than about 10 mm per year. Thus the geological evidence of uplift, lateral slip, and minor convergence is consistent with oblique stress along the northern Caribbean Plate boundary.

Mountain blocks in northern South America during the 27-0 m.y. B.P. interval were uplifted, shedding clastic sediments into graben, as in the Caribbean Mountains, and into flanking basins, as the Magdalena, Cauca, Atrato, and Cariaco basins. Schubert (1982, p. 357) described the Cariaco Basin, east of the Caribbean Mountains (Fig. 9), as "a pull-apart basin developed in the overlap region of two right-lateral *en echelon* faults." He measured horizontal slip of 25-100 km, and vertical slip between basin and margins of 2.5-7 km. The basin began to form at least 2.5 m.y. ago, perhaps earlier in the late Tertiary. From northwest of the Cariaco Basin, Ladd and others (1977) presented seismic profiles interpreted as showing present-day underthrusting in the Venezuelan Basin, beneath the Curacao Ridge of the Netherlands Antilles.

Most of the modern faults were active: vertical movement on the Oca and Santa Marta faults, for example, created 12 km of structural relief since early Miocene. Sediments were arkosic and clastic in areas such as the Santa Marta Block, carbonate in the Netherlands Antilles, and dominantly continental elsewhere. Pliocene to Recent calc-alkaline siliceous volcanism erupted on the crest of the Cordillera Central, mostly south of 5 degrees north latitude. Miocene and Pliocene andesite and rhyolite tuffs were interbedded with clastic sediments in the Magdalena Valley, and tonalite plutons dated Early Oligocene and early Miocene occur in the Cordillera Occidental (Shagam, 1975).

Uplift and tilting were intermittent throughout the interval. Unconformities are at the base, within, and at the top of the Miocene, and in and above the Pliocene. Campbell (1974) charts distinct phases of mobility during this time: early, middle, late, and latest Miocene, and Pleistocene. Most phases include uplift and tilting; folding, southward overthrusting, and gravity sliding

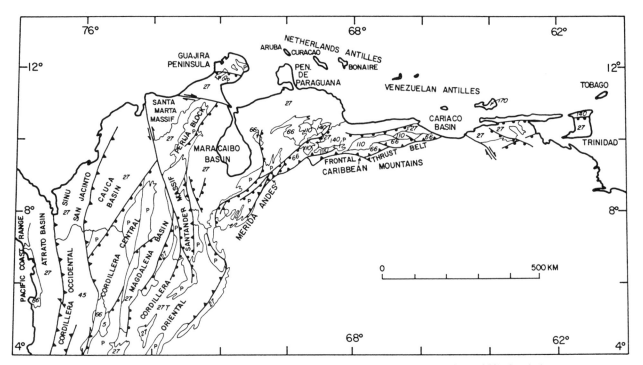

Figure 9. Southwestern corner of the Caribbean Plate. Adapted from de Alemeida (1978). Symbols: 27T, structures 0-27 m.y. B.P., some strata of same age; 27, strata 0-27 m.y. B.P.; 45, 66, 85, 110, 140 represent structures and some strata of 27-45, 45-66, 66-85, 85-110, and 110-140 m.y. B.P. respectively; P, pre-Mesozoic structures and strata; lines with solid dots, normal faults with dots on down side; lines with arrows, lateral faults showing displacement; lines with triangles, thrust faults with triangles on upper block. Some normal faults have probable components of lateral slip, but these components are not shown.

took place in the Caribbean Mountains until early Miocene. Also within this interval, moderate folding and compressional deformation reached a maximum in the Cordillera Occidental between middle and late Miocene, but not until the end of the late Miocene in the Cordillera Oriental (Campbell, 1974).

At the present time, uplift, lateral slip, and vertical slip are occurring along major faults in northern South America. Upper Oligocene to Holocene deep-sea and shallow-water sediments form the Sinu belt west of the San Jacinto belt near the Caribbean coast of Colombia (Duque-Caro, 1979; 1984). From middle Miocene through Early Pliocene, this accretionary prism was intensely deformed and diapirically intruded, forming the same type of folded, faulted, and intruded structures that formed earlier in the San Jacinto belt described above, but displaced later in age and westward in location. Intensity of deformation increased from south to north, as evidenced by steeper dips, more compaction, tighter folds, and more thrust faults. Northeast-trending structures continue along strike offshore (Case and Holcombe, 1980). Unconformities are recorded at the end of middle Miocene, the end of Early Pliocene, and in mid-Pleistocene. Uplift and emergence with little deformation followed the mid-Pliocene unconformity and continue to the present.

Kellogg and Bonini (1982) presented geologic, gravity, and seismic data indicating that the Caribbean Plate is being subducted beneath the South American Plate between 73 and 77

degrees west latitude (Fig. 9). They calculate a convergence rate of 1.9 cm/yr, with thrust displacements of 16-26 km laterally beneath the Perija block, about 10 km beneath the Santa Marta Massif, and 25 km beneath the Merida Andes.

Northern South America thus shows widespread evidence of block uplift but no volcanism or plutonism, except in the southwest in the Cordillera Occidental where South America lies opposite the subducting Nazca Plate. Along the Caribbean coast of Colombia, plate convergence caused deformation of sediments as young as Holocene, in the Sinu belt west of the San Jacinto belt. Slow convergence by underthrusting is taking place at the present time beneath the Perija block, the Santa Marta Massif, and the Merida Andes. Lateral slip near the Caribbean coast of Venezuela was probably caused by transform movement between the Caribbean Plate and the South American Plate. The movement was oblique, as it was at the northern margin of the Caribbean Plate, creating slow convergent thrusting in the Venezuelan Basin north of the Netherlands Antilles.

EVOLUTION OF THE CARIBBEAN PLATE

Geology

Figure 3 tabulates the unconformities and hiatuses described in the preceeding section, for structural blocks on the northern and southern margins of the Caribbean Plate. The figure illustrates the proposition that uplift, deformation, and change in sedimentary, igneous, or tectonic character occurred episodically along the boundaries of the Caribbean Plate. The times of maximum activity (110, 85, 66, 45, 27, and 0-5 m.y. B.P.) are probably times of change in plate motions.

The Aptian-Albian boundary (110 m.y. B.P.), near the end of the Early Cretaceous, is well recorded in Cuba, Hispaniola, and Puerto Rico as a time of metamorphism, deformation, and intrusion. In western Cuba there is a suggestion that an episode of subduction ended because of continental lithosphere entering the subduction zone (Mattson, 1979), perhaps part of the Chortis Block of Central America. Along the southern Caribbean plate boundary, unconformities are described at the end of the Early Cretaceous, but without enough detail to be sure that the break is clearly distinct from that in the northern Caribbean. Evidence seems best in the Caribbean Mountains, the Santa Marta Massif, and on the Colombia-Caribbean coast. In the Perija block and the Cordillera Oriental, the only break is at the base of the Cretaceous. Evidence from the southern Antilles is conflicting: fossils suggest a hot-spot lava sequence beginning after the break, but radiometric dates suggest an older age, and there is no evidence of a 110-m.y. B.P. break.

The 85-m.y. B.P. (Santonian) break is basal Santonian in Cuba, mid-Santonian in Puerto Rico, basal Campanian in Jamaica, and Coniacian to basal Campanian in the southern Antilles. It is not recorded in Hispaniola nor in the rest of northern South America. Volcanic activity decreased or changed character in the Greater Antilles and southern Antilles. Elsewhere clastic and carbonate deposition accompanied uplift and erosion. In Puerto Rico faulting formed grabens, suggesting a shift from compressional to tensional or lateral slip along the plate boundary.

The 66-m.y. B.P. (base of Paleocene) break is recorded everywhere except in the southern Antilles and from the Merida Andes to the Santa Marta Massif in northern South America. In Puerto Rico, there are two tectonic breaks at about this time (about 73 and 66 m.y. B.P.); and the earlier break is accompanied by stronger deformation, uplift, and erosion than the later break. In the Caribbean Mountains, the break may begin about 73 m.y. B.P. (early Maestrichtian) also, but no record of the 73-66 m.y. B.P. interval is preserved.

The 45-m.y. B.P. (between Middle and Late Eocene) break is seen almost everywhere in the Greater Antilles and northern South America. It is absent only in Jamaica, the Caribbean Mountains, Cordillera Central, Cordillera Occidental, and Pacific Coast Range, where uplift and erosion was more or less continuous throughout the Tertiary. This break marked the end of volcanism everywhere except Hispaniola and the Cordillera Occidental/Pacific Coast Range, suggesting an end to almost all subduction on the north and south edges of the Caribbean Plate; volcanism began about this time in the Lesser Antilles at the east edge of the plate, and continued at the west edge in Central America.

The 27-m.y. B.P. (Late Oligocene) break is found everywhere along the northern and southern borders of the Caribbean Plate, except in Hispaniola where it may correspond to an unconformity in the early Miocene (about 20 m.y. ago). In Puerto Rico there is no unconformity or hiatus until the early Miocene, but the source of clastic detritus disappeared about this time, as recorded in the preserved sediments; presumably subsidence or lateral displacement was the cause. In northern South America, the break is recorded at different levels in the Late Oligocene or early Miocene. Some areas such as the Merida Andes displayed continuous emergence across this time, rather than a distinct deformational event.

The present time (0-5 m.y. B.P.; Pliocene-Holocene) is certainly another major break, but it is too close to permit many conclusions to be drawn. Unconformities caused by uplift and erosion are widespread. Three unconformities formed in the Merida Andes since the beginning of the Pliocene, for example, four since late Miocene in the Cordillera Oriental, two since Miocene in the Perija block, and two since Miocene in Cuba. Uplift of 300-600 m has been described in Cuba and Hispaniola in the Pleistocene. Sykes and others (1982) suggest a change in plate motion at about 2.4 m.y. B.P.; but there may be other events in the last few million years which in the longer-term geologic record would blend together, but which we may eventually be able to decipher.

Plate Events

The two Americas plates have been moving independently for at least 180 m.y. The South American Plate moved to the southeast relative to the North American Plate from 180 to 127 m.y. B.P. and from 84 to 38 m.y. B.P. (Ladd, 1976). The space created between them must have been filled in the first instance by a spreading ridge, either a short independent ridge bounded by transforms, or a ridge connected with the existing Pacific or Atlantic spreading ridges. The earliest opening, beginning at 180 m.y. B.P., was around a rotation pole in the Atlantic (Ladd, 1976), so it is more likely that the East Pacific Rise extended an arm into the Caribbean (Fig. 5). Duncan and Hargraves (1984) suggested that this spreading ridge was the border between the Farallon and Phoenix Plates. Pindell and Dewey (1982), however, believed that this extension from the Pacific did not begin until about 95 m.y. B.P., and that a spreading ridge extended from the Atlantic into the Caribbean from 180 m.y. B.P. until then. The lack of exposures of Caribbean ocean crust of this age renders further speculation on these alternatives pointless.

The formation of Caribbean lithosphere, whether by extension from east or west (Figure 5 shows eastward extension), was not a simple passive filling of a gap. Modern island chains in the Caribbean expose relics of several compressional island arcs, including some fragments as old as Late Jurassic or Early Cretaceous (110-140 m.y. B.P.; Mattson, 1979). Clearly, these are remnants of a stage earlier than the 27-m.y. B.P. time of break-up of the Farallon Plate into Nazca and Cocos Plates. Malfait and

Dinkelman (1972) suggested that, in southern Central America, island-arc volcanism began in Eocene times and covered the entire present-day arc by Early Oligocene time. However, recent work by Schmidt-Effing (1980) indicated that island-arc volcanism and subduction began in Central America in the Cenomanian (92-100 m.y. B.P.). Aubouin and Azema (1980) suggested that Central America was the border between the Caribbean Plate and plates to the west in late Mesozoic times, because of the presence of Campanian and younger Cretaceous rocks of island-arc origin on the Central American flank of the Middle America Trench. Earlier oceanic rocks in Central America are as old as Middle Jurassic or Early Cretaceous (Dengo, 1969; Schmidt-Effing, 1980), but these might have formed at Pacific spreading centers farther to the west. Paleomagnetic studies of rocks from Caribbean islands (MacDonald and Opdyke, 1972; Vincenz and others, 1973; Stearns and others, 1982; Skerlec and Hargraves, 1980) indicate that the islands may have begun their formation southwest of their present positions, apparently west of the present position of South America (Fig. 4).

Movement of the Caribbean Plate relative to the North American Plate from 125 m.y. B.P. to the present was northeastward (Fig. 4a) until about 65 m.y. B.P., and eastward since then (Fig. 4b). The earlier motion probably began when the North and South American Plates began to act independently in the Early Jurassic, rotating to form a wedge-shaped gap widening to the west (Fig. 5a). The Farallon and Phoenix Plates filled that gap—in fact overfilled it, and the east-dipping subduction zone between the Farallon/Phoenix and Americas Plates was extended eastward by transform offsets. The subduction zone extended at least from Cuba through Hispaniola, Puerto Rico, speculatively into the Aves Ridge, and through the southern Netherlands and Venezuelan Antilles (Beets and others, 1984).

The Caribbean Plate may have been cut off from the Farallon/Phoenix Plates about 100 m.y. ago, by the extension of a subduction zone along Panama (Fig. 3, 4a, 5b), as Schmidt-Effing (1980) suggested (see also Pindell and Dewey, 1982; Aubouin and Azema, 1980; Case and others, 1971; Lloyd, 1963). The Caribbean Plate was then isolated from both the major Atlantic and Pacific spreading systems for the first time, by subduction zones extending beneath it on its northeast and southwest sides, and by transform faults on its northwest and southeast sides (Fig. 5b, 5c).

Several events took place at about this time, 110-66 m.y. B.P. in the Late Cretaceous (Fig. 5b, 5c, 5d). Perhaps about 110 m.y. ago, subduction flipped polarity or changed to transform motion in the Greater Antilles (Mattson, 1979). About 85 m.y. B.P., the Aves Ridge and southern Antilles part of the subduction zone ceased subduction, and thick basalt flows covered the floor of the Colombian and Venezuelan Basins (Donnelly and others, 1973). Subduction and volcanism began about 60-45 m.y. B.P. in the Lesser Antilles.

A conspicuous change in the Caribbean Plate boundary dynamics during 110-66 m.y. B.P. was the conversion of subduction boundaries to transform boundaries. Consumption zones were originally oriented normal to major plate slip vectors but were swung subparallel to them as the Caribbean Plate protruded eastward between the North and South American Plates. The east-trending subduction zones could be converted more easily to transform zones than reoriented again. Thus, most of the Greater Antilles ended vulcanism and became sliced by large strike-slip faults in Late Cretaceous to Eocene time. Vulcanism in the mid-Cayman Trough spreading center represents leakage along the generally transform boundary, and in Hispaniola perhaps local subduction, although the alkalic composition of some of the Hispaniola volcanic rocks suggests a rift source.

The Aves Ridge subduction zone, however, was oriented normal to the slip vectors. Perhaps as a first stage, it opened a small back-arc spreading center at about 85 m.y. B.P., flooding the Caribbean basins with basalt. Then in a second stage, it turned off consumption of Caribbean crust, flipped, and jumped to the present Lesser Antilles position. Alternatively, following the model of Duncan and Hargraves (1984), the thickened basaltic crust in the Colombian and Venezuelan Basins could have formed above the Galapagos hot spot in the Pacific, about 100-80 m.y. ago. By Late Eocene time, 38-45 m.y. B.P., a new plate pattern was stable and continued to the present: slip motion along north and south Caribbean Plate margins and subduction at the Lesser Antilles. A second-order stress across the northern and southern boundaries, caused by slight convergence of the North and South American Plates since 38 m.y. B.P. (Oligocene), caused slow subduction, thus far discovered north and south of the eastern Greater Antilles, near the southern plate margin north of the Netherlands West Indies (Ladd and others, 1977; Kellogg and Bonini, 1982), and off the north coast of Panama and Colombia (Fig. 1).

In the Pacific west of the Caribbean Plate, the Farallon Plate continued to fragment (Fig. 5), forming the Nazca and Cocos Plates about 27 m.y. B.P. (early Miocene or Late Oligocene). The Cocos Plate subducted beneath the Caribbean Plate but began to split off fragments to the Nazca Plate, as aseismic ridges on the Cocos Plate clogged the Middle American Trench subduction zone (Van Andel and others, 1971; Lonsdale and Klitgord, 1978). The Nazca Plate may have begun with a transform margin bordering the Caribbean Plate (Lonsdale and Klitgord, 1978); it now has that general relation (Minster and Jordan, 1978; Pindell and Dewey, 1982). Pennington (1981) described earthquake focal mechanisms from this border region showing maximum compression in an east-west direction, indicating movement on northeast- to northwest-trending faults. He suggested that subduction of the Nazca Plate beneath the South American Plate caused compression in the Caribbean Plate near Panama (Fig. 2, 9). The movement of the Caribbean Plate relative to the Nazca Plate would be essentially of the transform type, with internal deformation within the Caribbean Plate in Panama.

SUMMARY

The Caribbean Plate probably began its existence as an arm

of the Farallon and Phoenix Plates about 140 million years ago, during the spreading apart of the North and South American Plates. It became an identifiable small plate about 85 to 100 million years ago when it was separated from the Farallon/Phoenix Plate by subduction west of Central America.

The northern margin of the Caribbean Plate was an ocean-to-ocean margin for most of its existence. Subduction zones were continuously active along the northern margin until the Late Cretaceous or early Tertiary, when lateral slip motion became dominant. Compressional folds, emplacement of ophiolite, calc-alkalic vulcanism, and granitic plutonism occurred, and an extended history is partly known.

The southern margin of the Caribbean Plate adjoins the mostly continental South American Plate. In the eastern part of the margin, in the Caribbean Mountains of Venezuela and the offshore islands, a complicated history of subduction, uplift, and nappe formation extended from the Jurassic through the Eocene.

The western part of the southern margin has been rising along gently-dipping thrust faults beneath the mountain massifs. This uplift and fracturing is the result of convergence of the Caribbean and South American Plates, first partly resolved by subduction in the Mesozoic and Paleogene in the Cordillera Occidental, Pacific Coast Range, Santa Marta Massif, Guajira Peninsula, and southern Antilles, and later resolved by vertical movements throughout northern South America and by slow subduction of the Caribbean Plate beneath the South American Plate along the Caribbean coast of Colombia. The Caribbean Mountains, already oriented parallel to the plate movements, were preserved relatively intact although cut into strips by faults paralleling the plate margin.

Modern compressive movements have created mud vulcanism, diapirism, folding, and thrust faulting near the Caribbean coast of Colombia. Slow convergence is proceeding north of Panama, offshore of Caribbean Colombia, north of Venezuela, and south of Hispaniola and Puerto Rico. There is also a component of subduction beneath the Puerto Rico Trench north of the Greater Antilles. The eastern Panamanian isthmus is being bent into a great loop convex to the north and probably eventually will collapse upon itself, as South America pushes onto the Caribbean Plate.

The geologic structures along the margins of the Caribbean Plate and the width of the marginal zones themselves are functions of the vectors of relative plate motion at the margins, and of the type of lithosphere that forms the adjoining plates. Plate motion parallel to plate boundaries produces transform boundaries, as in the Greater Antilles; but oblique plate motion generates a component of compression which causes slow nonvolcanic sub-

duction accompanied by uplift, as in northern coastal Colombia. Where simple oceanic lithosphere forms both sides of a plate boundary, simple subduction or lateral slip along a single plane occurs. Where modified oceanic or continental lithosphere forms part of a plate boundary, lateral slip occurs along multiple planes in a boundary zone as near the Greater Antilles; or subduction takes place slowly, vulcanism is limited or absent, and the dominant physiographic effect may be vertical faulting and uplift, as in northern Colombia.

Geologic data indicate that important changes in tectonic and depositional styles at the margins of the Caribbean Plate took place at 110 m.y. B.P. (basal Albian), 85 m.y. B.P. (Santonian), 66 m.y. B.P. (basal Tertiary), 45 m.y. B.P. (Late Eocene), 27 m.y. B.P. (Late Oligocene), and 5-0 m.y. B.P. (Present). Near the beginning of the Albian (110 m.y. B.P.), part of the Chortis Block may have entered a northeast-dipping subduction zone south of the western Greater Antilles, causing subduction to flip to a northwest-dipping zone located north of the Greater Antilles. Subduction probably began at the west edge of the Caribbean Plate at this time, beginning to form Central America. The Santonian break (85 m.y. B.P.) may be the result of thickened oceanic crust formed at the Galapagos hot spot (Duncan and Hargraves, 1984) reaching the Central American subduction zone and blocking or modifying further subduction. At the beginning of the Tertiary (66 m.y. B.P.), the Caribbean Plate changed its relative motion from northeastward to eastward, and began to underthrust northern South America. In the Late Oligocene (27 m.y. B.P.), the Farallon/Phoenix Plate separated into the Nazca and Cocos Plates. The cause of the 5-0 m.y. B.P. emergence cannot yet be discerned.

Thus major changes in plate arrangement and motion are marked in the Caribbean Plate by major geologic discontinuities.

ACKNOWLEDGMENTS

This research was partiallly supported by C.U.N.Y.-F.R.A.P. Research Grant 13380, and by the Republic of Venezuela. I have benefited from discussions with Lynn Sykes, William McCann, Alan Kafka, Margaret Winslow, and Hannes Brueckner; from critical reviews of the manuscript by Jim Case, Robert Hargraves, Jim Kellogg, Fred Nagle, and Reg Shagam; and from programming advice from Scott Weaver. Robert Hargraves and Reg Shagam kindly provided preprints of several forthcoming papers. The vectors shown in Figure 2 were calculated by a HP-41C calculator program available from the writer. Mary Ann Luckman drafted the figures.

REFERENCES CITED

de Almieda, F.F.M., 1978, Tectonic map of South America: Geological Society of America, Map and Chart Series, MC-32, scale 1:5,000,000.

Arden, D. D., 1975, Geology of Jamaica and the Nicaragua Rise, *in* Nairn, A.E.M., and Stehli, F. G., editors, The Ocean Basins and Margins: the Gulf of Mexico and the Caribbean: New York, Plenum Press, p. 617–662.

Atwater, T., 1970, Implications of plate tectonics for the Cenozoic tectonic evolution of western North America: Geological Society of America Bulletin, v. 81, p. 3513–3536.

Aubouin, J., and Azema, J., 1980, A propos de l'origine de la plaque caraibe: la face pacifique de l'Amerique centrale: C. R. Acad. Sc. Paris, series D, v. 291, p. 33–37.

Barabas, A. H., 1971, K-Ar dating of igneous events and porphyry copper mineralization in west central Puerto Rico (abstract): Geological Society of America Abstracts with Programs, v. 3, no. 7, p. 498.

Beets, D. J., and MacGillavry, H. J., 1977, Outline of the Cretaceous and early Tertiary history of Curacao, Bonaire and Aruba: Caribbean Geological Conference, 8th, Curacao, Guide to the Field Excursions of Curacao, Bonaire, and Aruba, Netherlands Antilles, p. 1–6.

Beets, D. J., Maresch, W. V., Klaver, G. T., Mottana, A., Bocchio, R., Beunk, F. F., and Monen, H. P., 1984, Magmatic rock series and high-pressure metamorphism as constraints on the tectonic history of the southern Caribbean, *in* Bonini, W. E. and others, eds., The Caribbean-South American Plate Boundary and Regional Tectonics: Geological Society of America Memoir 162 (this volume).

Bell, J. S., 1974, Venezuelan Coast Ranges, *in* Spencer, A. M., editor, Mesozoic-Cenozoic Orogenic Belts: Geological Society, London, Special Publication 4, p. 683–703.

Berryhill, H. L., Jr., 1965, Geology of the Ciales quadrangle, Puerto Rico: United States Geological Survey, Bulletin 1184, 116 p.

Berryhill, H. L., Jr., Briggs, R. P., and Glover, L., III, 1960, Stratigraphy, sedimentation, and structure of Late Cretaceous rocks in eastern Puerto Rico— preliminary report: American Association of Petroleum Geologists, Bulletin, v. 44, no. 2, p. 137–155.

Bonini, W. E., Garing, J. D., and Kellogg, J. N., 1980, Late Cenozoic uplifts of the Maracaibo-Santa Marta block, slow subduction of the Caribbean Plate and results from a gravity study: Caribbean Geological Conference, 9th, Santo Domingo 1980, preprint, 2 p.

Bowin, C., 1966, Geology of central Dominican Republic (a case history of part of an island arc): Geological Society of America, Memoir 98, p. 11–84.

—— 1975, The geology of Hispaniola, *in* Nairn, A.E.M., and Stehli, F. G., editors, The Ocean Basins and Margins: The Gulf of Mexico and the Caribbean: New York, Plenum Press, p. 501–552.

Brezsnyanszky, K., and Iturralde-Vinent, M. A., 1978, Paleogeografia del Paleogeno de Cuba oriental: Geologie en Mijnbouw, v. 57, p. 123–133.

Butterlin, J., 1977, Geologie Structurale de la Region des Caraibes: Paris, Masson, 255 p.

Campbell, C. J., 1974, Colombian Andes, *in* Spencer, A. M., editor, Mesozoic-Cenozoic Orogenic Belts: Geological Society, London, Special Publication 4, p. 705–724.

Case, J. E., Duran S., L. G., Lopez R., Alfonso, and Moore, W. R., 1971, Tectonic investigations in western Colombia and eastern Panama: Geological Society of America Bulletin, v. 82, p. 2685–2712.

Case, J. E., and Holcombe, T. L., 1980, Geologic-tectonic map of the Caribbean region: U.S. Geological Survey Miscellaneous Investigations Map I-1100, scale 1:2,500,000.

Cox, D. P., Marvin, R. F., M'Gonigle, J. W., McIntyre, D. H., and Rogers, C. L., 1977, Potassium geochronology of some metamorphic, igneous, and hydrothermal events in Puerto Rico and the Virgin Islands: Journal of Research, United States Geological Survey, v. 5, p. 689–703.

Dengo, G., 1969, Problems of tectonic relations between Central America and the Caribbean: Gulf Coast Association of Geological Societies, Transactions, v. 19, p. 311–320.

Donnelly, T. W., Melson, W., Kay, R., and Rogers, J.J.W., 1973, Basalts and dolerites of Late Cretaceous age from the central Caribbean, *in* Edgar, N. T., Kaneps, A. G., and Herring, J. R., editors, Initial reports of the Deep Sea Drilling Project: Washington, National Science Foundation, v. 15, p. 989–1012.

Draper, G., Harding, R. R., Horsfield, W. T., Kemp, A. W., and Tresham, A. E., 1976, Low-grade metamorphic belt in Jamaica and its tectonic implications: Geological Society of America Bulletin, v. 87, p. 1283–1290.

Draper, G., and Lewis, J. F., 1980, Preliminary report on deformation and structure of the Amina schists, northern Dominican Republic: Caribbean Geological Conference, 9th, Santo Domingo, preprint, 4 p.

Duncan, R. A., and Hargraves, R. B., 1984, Tectonic evolution of the Caribbean region in the mantle reference frame, *in* Bonini, W. E. and others, eds., The Caribbean-South American Plate Boundary and Regional Tectonics: Geological Society of America Memoir 162 (this volume).

Duque-Caro, H., 1979, Major structural elements and evolution of northwestern Colombia, *in* Watkins, J. S., and others, editors, Geological and geophysical investigations of continental margins: American Association of Petroleum Geologists Memoir 29, p. 329–351.

—— 1984, Structural style, diapirism, and accretionary episodes of the Sinu-San Jacinto terrain, northwestern Colombia, *in* Bonini, W. E. and others, eds., The Caribbean-South American Plate Boundary and Regional Tectonics: Geological Society of America Memoir 162 (this volume).

Garrison, L. E., Martin, R. G., Jr., Berryhill, H. L., Jr., Buell, M. W., Jr., Ensminger, H. R., and Perry, R. K., 1972, Preliminary tectonic map of the eastern Greater Antilles region: U.S. Geological Survey, Miscellaneous Geologic Investigations Map I-732, scale 1:500,000.

Glover, L., III, 1971, Geology of the Coamo area, Puerto Rico, and its relation to the volcanic arc-trench association: United States Geological Survey, Professional Paper 636, 102 p.

Goreau, P. D., 1980, The geophysics and tectonic evolution of the eastern Cayman Trough and western Hispaniola (abstract): Caribbean Geological Conference, 9th, Santo Domingo, Resumenes, p. 27–28.

Handschumacher, D. W., 1976, Post-Eocene plate tectonics of the eastern Pacific, *in* Sutton, G. H., and others, editors, Geophysics of the Pacific Ocean and margin: American Geophysical Union Monograph 19, p. 177–202.

Harland, W. B., Smith, A. G., and Wilcock, B., editors, 1964, Summary of the Phanerozoic time scale, p. 260–262, *in* The Phanerozoic Time Scale: Geological Society London Quarterly Journal, v. 120-S, 458 p.

Hey, R., 1977, Tectonic evolution of the Cocos-Nazca spreading center: Geological Society of America Bulletin, v. 88, p. 1404–1420.

Irving, E. M., 1971, La evolucion estructural de los Andes mas septentrionales de Colombia: Instituto Nacional de Investigaciones Geologico-Mineras, Ministerio de Minas y Petroleos, Bogota, Bolotin Geologico, v. 19, no. 2, 90 p.

Iturralde-Vinent, M. A., 1969, Principal characteristics of Cuban Neogene stratigraphy: American Association of Petroleum Geologists, Bulletin, v. 53, no. 9, p. 1938–1955.

—— 1972, Principal characteristics of Oligocene and Lower Miocene stratigraphy of Cuba: American Association of Petroleum Geologists, Bulletin, v. 56, p. 2369–2379.

—— 1981, Nuevo modelo interpretativo de la evolucion geologica de Cuba: Ciencias de la Tierra y del Espacio, no. 3, p. 51–89.

Jackson, T. A., and Smith, T. E., 1980, Mesozoic and Cenozoic mafic magma types of Jamaica and their tectonic significance: Caribbean Geological Conference, 9th, Santo Domingo, preprint, 2 p.

Kellogg, J. N., 1980, Cenozoic basement tectonics of the Sierra de Perija, Venezuela and Colombia: Caribbean Geological Conference, 9th, Santo Domingo, preprint, 2 p.

—— 1984, Cenozoic tectonic history of the Sierra de Perija, Venezuela-Colombia, and adjacent basins, *in* Bonini, W. E. and others, eds., The Caribbean-South American Plate Boundary and Regional Tectonics: Geological Society of America Memoir 162 (this volume).

Kellogg, J. N., and Bonini, W. E., 1982, Subduction of the Caribbean plate and basement uplifts in the overriding South American plate: Tectonics, v. 1, p. 251–276.

Kesler, S. E., Lewis, J. F., Jones, L. M., and Walker, R. L., 1977, Early island-arc intrusive activity, Cordillera Central, Dominican Republic: Contributions to Mineralogy and Petrology, v. 65, p. 91–99.

Khudoley, K. M., and Meyerhoff, A. A., 1971, Paleogeography and geological history of Greater Antilles: Geological Society of America, Memoir 129, 199 p.

Krijnen, J. P., and Chin, A.C.L., 1978, Geology of the northern, central, and southeastern Blue Mountains, Jamaica, with a provisional compilation map of the entire inlier: Geologie en Mijnbouw, v. 57, p. 243–250.

Ladd, J. W., 1976, Relative motion of South America with respect to North America and Caribbean tectonics: Geological Society of America Bulletin, v. 87, p. 969–976.

Ladd, J. W., Worzel, J. L., and Watkins, J. S., 1977, Multifold seismic reflection records from the northern Venezuela basin and the north slope of the Muertos Trench: American Geophysical Union, Maurice Ewing Series, v. 1, p. 41–56.

Lee, V., and Mattson, P. H., 1976, Metamorphosed oceanic crust or early volcanic products in Puerto Rico basement rock association: Caribbean Geological Conference, 7th, Guadeloupe 1974, Transactions, p. 263–270.

Lewis, J. F., 1980, Overall summary of the tectonic evolution of Hispaniola: Caribbean Geological Conference, 9th, Santo Domingo, preprint, 6 p.

Lloyd, J. J., 1963, Tectonic history of the south Central-American orogen, *in* Childs, O. E., and Beebe, B. W., editors, Backbone of the Americas: American Association of Petroleum Geologists Memoir 2, p. 88–100.

Lonsdale, P., and Klitgord, K. D., 1978, Structure and tectonic history of the eastern Panama Basin: Geological Society of America Bulletin, v. 89, p. 981–999.

MacDonald, W. D., and Opdyke, N. D., 1972, Tectonic rotations suggested by paleomagnetic results from northern Colombia, South America: Journal of Geophysical Research, v. 77, p. 5720–5730.

Malfait, B. T., and Dinkelman, M. G., 1972, Circum-Caribbean tectonic and igneous activity and the evolution of the Caribbean plate: Geological Society of America Bulletin, v. 83, p. 251–272.

Mattson, P. H., 1960, Geology of the Mayaguez area, Puerto Rico: Geological Society of America Bulletin, v. 71, p. 319–362.

—— 1966, Geological characteristics of Puerto Rico, p. 124–138, *in* Poole, W. H., editor, Continental Margins and Island Arcs: Geological Survey of Canada, Paper 66-15.

—— 1973, Middle Cretaceous nappe structures in Puerto Rican ophiolites and their relation to the tectonic history of the Greater Antilles: Geological Society of America Bulletin, v. 84, p. 21–37.

—— 1974, Puerto Rico-Virgin Islands, *in* Spencer, A. M., editor, Mesozoic-Cenozoic orogenic belts: Geological Society, London, Special Publication 4, p. 639–661.

—— 1979, Subduction, buoyant braking, flipping, and strike-slip faulting in the northern Caribbean: Journal of Geology, v. 87, p. 293–304.

Mattson, P. H., and Pessagno, E. A., Jr., 1971, Caribbean Eocene volcanism and Atlantic Ocean Layer A: Science, v. 174, p. 138–139.

—— 1979, Jurassic and early Cretaceous radiolarians in Puerto Rican ophiolite—tectonic implications: Geology, v. 7, p 440–444.

Menard, H. W., 1978, Fragmentation of the Farallon plate by pivoting subduction: Journal of Geology, v. 86, p. 99–110.

Meyerhoff, A. A., Khudoley, K. M., and Hatten, C. W., 1969, Geologic significance of radiometric dates from Cuba: American Association of Petroleum Geologists Bulletin, v. 53, no. 12, p. 2494–2500.

Michael, R., and Lewis, J. F., 1980, Structure and tectonics of south-central flank of the Massif du Nord-Cordillera Central and adjacent portions of the Plateau Central-San Juan Valley, Hispaniola: Caribbean Geological Conference, 9th, Santo Domingo, preprint, 4 p.

Minster, J. B., and Jordan, T. H., 1978, Present-day plate motions: Journal of Geophysical Research, v. 83, p. 5331–5354.

Molnar, P., and Sykes, L. R., 1969, Tectonics of the Caribbean and Middle America region from focal mechanisms and seismicity: Geological Society of America Bulletin, v. 80, p. 1639–1684.

Monroe, W. H., 1980, Geology of the Middle Tertiary formations of Puerto Rico: United States Geological Survey, Professional Paper 953, 93 p.

Morgan, J. W., 1968, Rises, trenches, great faults, and crustal blocks: Journal of Geophysical Research, v. 73, p. 1959–1982.

Nagle, F., 1974, Blueschist, eclogite, paired metamorphic belts and the early tectonics of Hispaniola: Geological Society of America Bulletin, v. 85, p. 1461–1466.

—— Geology of the Puerto Plata area, Dominican Republic, p. 1–28, *in* Lidz, B., and Nagle, F., editors, Hispaniola: tectonic focal point of the northern Caribbean: Miami, Florida, Miami Geological Society, 96 p.

Nelson, A. E., 1967a, Geologic map of the Corozal quadrangle, Puerto Rico: United States Geological Survey, Miscellaneous Geologic Investigations, Map I-473, 1:20,000 scale.

—— 1967b, Geologic map of the Utuado quadrangle, Puerto Rico: United States Geological Survey, Miscellaneous Geologic Investigations, Map I-480, 1:20,000 scale.

Nelson, A. E., and Monroe, W. H., 1966, Geology of the Florida quadrangle, Puerto Rico: United States Geological Survey, Bulletin 1221-c, 22 p.

Pardo, G., 1975, Geology of Cuba, *in* Nairn, A.E.M., and Stehli, F. G., editors, The ocean basins and margins: volume 3, the Gulf of Mexico and the Caribbean: New York, Plenum Press, p. 553–615.

Pennington, W. D., 1981, Subduction of the eastern Panama Basin and seismotectonics of northwestern South America: Journal of Geophysical Research, v. 86, p. 10753–10770.

Perfit, M. R., Heezen, B. C., Rawson, M., and Donnelly, T. W., 1980, Chemistry, origin and tectonic significance of metamorphic rocks from the Puerto Rico Trench: Marine Geology, v. 34, p. 125–156.

Pindell, J., and Dewey, J. F., 1982, Permo-Triassic reconstruction of western Pangea and the evolution of the Gulf of Mexico/Caribbean region: Tectonics, v. 1, p. 179–211.

Roobol, M. J., 1972, The volcanic geology of Jamaica: Sixth Caribbean Geological Conference, Margarita, Venezuela 1971, Transactions, p. 100–107.

Santamaria, F., and Schubert, C., 1974, Geochemistry and geochronology of the southern Caribbean-northern Venezuela plate boundary: Geological Society of America Bulletin, v. 85, p. 1085–1098.

Schmidt-Effing, R., Gursky, H. J., Strebin, M., and Wildberg, H., 1983, The ophiolites of southern Central America with special reference to the Nicoya Peninsula (Costa Rica): Ninth Caribbean Geological Conference, Santo Domingo 1980, Transactions, p. 423–429.

Schubert, C., 1982, Origin of the Cariaco basin, southern Caribbean Sea: Marine Geology, v. 47, p. 345–360.

Schwartz, D. P., Cluff, L. S., and Donnelly, T. W., 1979, Quaternary faulting along the Caribbean-North American plate boundary in Central America: Tectonophysics, v. 52, p. 431–445.

Shagam, R., 1975, The northern termination of the Andes, *in* Nairn, A.E.M., and Stehli, F. G., editors, The ocean basins and margins: volume 3, the Gulf of Mexico and the Caribbean: New York, Plenum Press, p. 325–420.

Skerlec, G. M., and Hargraves, R. B., 1980, Tectonic significance of paleomagnetic data from northern Venezuela: Journal of Geophysical Research, v. 85, p. 5303–5315.

Stearns C., Mauk, F. J., and Van der Voo, R., 1982, Late Cretaceous-early Tertiary paleomagnetism of Aruba and Bonaire (Netherlands Leeward Islands): Journal of Geophysical Research, v. 87, p. 1127–1141.

Sykes, L. R., McCann, W. R., and Kafka, A. L., 1982, Motion of Caribbean plate during last 7 million years and implications for earlier Cenozoic movements: Journal of Geophysical Research, v. 87, p. 10,656–10,676.

Tobisch, O. T., 1968, Gneissic amphibolite at Las Palmas, Puerto Rico, and its significance in the early history of the Greater Antilles island arc: Geological Society of America Bulletin, v. 79, p. 557–574.

Tschanz, C. M., Marvin, R. F., Cruz B., J., Mehnert, H. B., and Cebula, G. T., 1974, Geologic evolution of the Sierra Nevada de Santa Marta, northeastern

Colombia: Geological Society of America Bulletin, v. 85, p. 273–284.

Van Andel, T. H., Heath, G. R., Malfait, B. T., Heinrich, D. F., and Ewing, J. I., 1971, Tectonics of the Panama Basin, eastern equatorial Pacific: Geological Society of America Bulletin, v. 82, p. 1489–1508.

Vespucci, P., 1983, Preliminary account of the petrology of the late Cenozoic volcanic province of Hispaniola: Ninth Caribbean Geological Conference, Dominican Republic 1980, Transactions, p. 379–389.

Vincenz, S. A., Steinhauser, P., and Dasgupta, S. N., 1973, Paleomagnetism of Upper Cretaceous ignimbrites on Jamaica: Zeitschrift fur Geophysik, v. 39, p. 727–737.

Weaver, J. D., 1960, Notes on higher level erosion surfaces of Puerto Rico: Second Caribbean Geological Conference, Mayaguez, Puerto Rico 1959, p. 90–98.

Manuscript Accepted by the Society September 1, 1983

Geological Society of America
Memoir 162
1984

Seismic reflection profiles across the southern margin of the Caribbean

John W. Ladd
Lamont-Doherty Geological Observatory
of Columbia University
Palisades, New York 10964

Marek Truchan
Gulf Research and Development Company
11111 South Wilcrest
Houston, Texas 77099

Manik Talwani
Gulf Science and Technology
11111 South Wilcrest
Houston, Texas 77099

Paul L. Stoffa
Department of Geology
The University of Texas
Austin, Texas 78712

Peter Buhl
Lamont-Doherty Geological Observatory
of Columbia University
Palisades, New York 10964

Robert Houtz
Gulf Research and Development Company
11111 South Wilcrest
Houston, Texas 77099

Alain Mauffret
Département de Géologie Dynamique
Université Pierre et Marie Curie
4 Place Jussieu
75230 Paris
Cedex 05
France

Graham Westbrook
Department of Geological Sciences
University of Durham
Durham DH1 3LE, England

ABSTRACT

Seismic sections across the southern margin of the Caribbean reveal structures related to the convergence of the Caribbean and South American plates. The South Caribbean Deformed Belt and its eastward extension, the Curacao Ridge, is a zone of intensely deformed Cretaceous and Tertiary sediments that lies along the southern edge of the Colombia and Venezuela Basins. Undeformed sediments of the Caribbean basins abut the deformed belt abruptly to the north. To the south, the South Caribbean Deformed Belt gives way to older deformed belts of the Netherlands and Venezuelan Antilles Ridge and to the continental margin of Colombia and Venezuela containing pre-Tertiary structures. Along most of the South Caribbean Deformed Belt an apron of sediments prograges northward across the deformed belt suggesting active deformation at the northern edge of the belt and progressively older Tertiary deformation to the south. Caribbean oceanic crust extends southward beneath the deformed belt and southward-dipping reflections occur within the deformed belt possibly indicating slices of oceanic crust incorporated within it. Bottom simulating reflectors along parts of the deformed belt indicate the presence of gas hydrates. The chemical phase relationships of gas hydrates and the depth of the bottom simulating reflections indicate a thermal gradient of approximately 0.04 degrees/meter.

INTRODUCTION

The southern margin of the Caribbean from Panama to Trinidad is a complex zone of convergence and transcurrent faulting associated with relative motion between the Caribbean and South American plates. The deformation zone varies in width and structural expression—off Panama the zone is about 160 km wide and is dominated by compressive features. In Colombia and western Venezuela, the zone extends about 600 km from the deformation front in the Colombia Basin to the southeastern boundary of the Merida Andes. Here strike-slip faulting accompanies thrusting. In western Venezuela, the zone extends some 450 km from the southern flank of the Cordillera de la Costa to the northern flank of the Curacao Ridge. Here again thrusting and strike-slip faulting are dominant. In eastern Venezuela, the zone narrows dramatically to about 200 km from the southern flank of the Serrania del Interior to the southern flank of the Grenada Basin (Case and Holcombe, 1980).

Numerous papers (Krause, 1971; Case, 1974; Silver and others, 1975; Talwani and others, 1977; Ladd and Watkins, 1978; Kellogg and Bonini, 1982) have described the offshore structures that occur along this margin. Other papers (Malfait and Dinkelman, 1972; Maresch, 1974; Bowin, 1976; Ladd, 1976; and Pindell and Dewey, 1982) have discussed the tectonics of the region in plate tectonic terms. Discussions based on offshore data have been rather general compared with those of onshore data because our knowledge of offshore geology has been limited by the rather modest resolution of the nonproprietary seismic data that have been available and by the almost total lack of nonproprietary well data from offshore. We have a general idea that since the late Eocene, this margin has experienced a combination of strike-slip motion and convergence—a pattern that continues to this day (Ladd, 1976; Jordan, 1975). However, we have very little knowledge of how this oblique convergence has varied during this time interval in both space and time. Yet the tenets of plate tectonics and complexities of the deformed belt suggest that there must have been continual temporal and spatial changes (Dewey, 1975).

In an attempt to obtain a better understanding of the structure and stratigraphy of this margin, a set of multichannel seismic reflection lines was shot along this margin (Figure 1 and Plate 1 in pocket inside back cover) (Talwani and others, 1977; Diebold and others, 1981).

TECTONICS

This data set shows compressive deformation localized in a narrow zone some tens of kilometers wide at the northern edge of the southern Caribbean margin. This is not to say that all compressive strain occurs in this narrow zone, for there is a great deal of Tertiary faulting and folding to the south. But the accompanying profiles emphasize the concentration of intense deformation in the Curacao Ridge and its westward extension, the South Caribbean Deformed Belt (Case, 1974).

There is an ambiguity in the width of the South Caribbean Deformed Belt because we do not know the nature of acoustic basement beneath much of the margin or the age of the undeformed sediments that prograde across the margin from the south. If we knew more about acoustic basement and the age of prograding sediments, we would be able to address more directly the question of synchroneity of deformation among the South Caribbean Deformed Belt and the subparallel belts to the south. On line 130 sediments are ponded in the Rancheria Basin behind the deformed belt. The same is true of lines 8 and 97 where terrigenous sediments are ponded in the Los Roques Trough south of the Curacao Ridge. Refraction studies (Edgar and others, 1971; Diebold and others, 1981) indicate that the Curacao Ridge is indeed a sediment ridge. The generally incoherent nature of the reflection data from this ridge, similar to many convergent margin accretionary complexes, suggests the intensely deformed nature of the sediments that compose the ridge. The other seismic lines, however, have sediment aprons that prograde out over the deformation complex. It is generally not clear where igneous basement ends laterally and sedimentary deformation complex begins beneath the undeformed apron.

The localization of the deformation suggests that it occurs in response to a localized set of stresses such as those that would occur along a zone of crustal thrusting (Cowan and Silling, 1978; Chapple, 1978). In essence, basement thrusting will cause a set of shear stresses along the base of the overlying sediment body close to the subcrop of the basement fault. Finite strength and yielding of the overlying sediments will cause the stress to decrease away from the thrust zone. The sharp deformation front seen at the northern and southern margins of the South Caribbean Deformed Belt may be due to the existence of a deformation threshold: If the basal shear stress is less than some finite amount, deformation will not occur. The appearance of the deformed belt in the seismic reflection records is similar in general to other accretionary complexes associated with subduction zones. We have hesitated in calling this region a subduction zone, however, because it does not have the associated volcanics and well-developed earthquake Benioff zone generally associated with a subduction zone. A component of thrusting is consistent with the relative plate motions deduced by Jordan (1975), Ladd (1976), and Kellogg and Bonini (1982), but not with motions deduced by Sykes and others (1982).

On several of the seismic sections, one can see reflections continuing southward under the deformation complex from the oceanic basins to the north (see sections 8, 97, 132, 134, 136). On line 136, there occurs within the deformation complex above the southward continuation of Venezuela Basin reflections a southward-dipping reflection labeled R1, suggesting a tectonically emplaced piece of Venezuela Basin crust. Farther up the slope beneath shotpoint 600 is another landward-dipping reflection labeled R2. Ladd and Watkins (1978) found disconnected pieces of reflections apparently indicating pieces of Venezuela Basin crust within the deformation complex about 100 km to the east of line 136. The landward-dipping reflections are reminiscent of

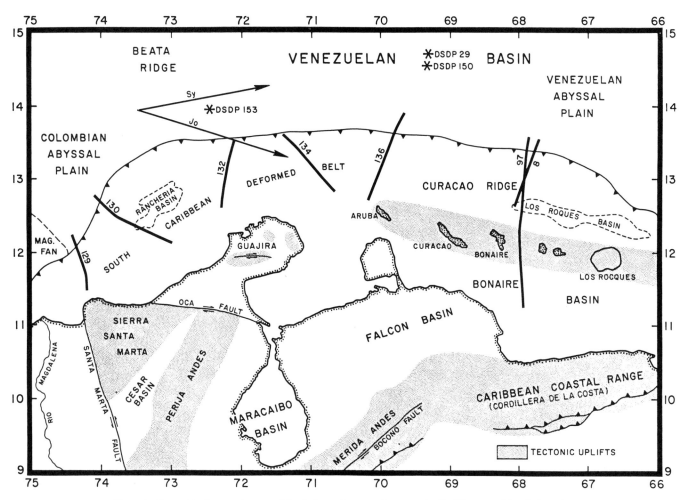

Figure 1. Generalized tectonic map of the deformation zone between the Caribbean and South America plates. Seismic lines illustrated in plate 1 are indicated by heavy black lines with line numbers. The stippled regions are tectonic uplifts containing pre-Tertiary rock after Case (1974). The two arrows in the region of the Beata Ridge labeled Sy and Jo indicate the relative motion of the Caribbean with respect to South America according to Sykes and others (1982) and according to Jordan (1975). The Jordan relative motion vector seems more consistent with the component of compression indicated by the compressive features along the southern Caribbean margin.

landward-dipping reflections seen in the Middle America Trench offshore Guatemala (Ibrahim and others, 1979; Ladd and others, 1982) which were interpreted as landward-dipping slices of oceanic crust.

Because of the inability to clearly distinguish deformation complex from other kinds of acoustic basement, it is difficult to make general statements about spatial trends in the size of the deformation complex. On line 130, the deformation complex is restricted to the region between the Rancheria Basin and the Colombia Basin—a zone about 30 km wide. On lines 8 and 97, we know that the deformation complex is limited by the Venezuela Plain to the north and Los Roques Trough to the south, giving a width of 60 to 70 km for the deformation complex. It is difficult to judge the lateral extent of the complex on the other lines. On line 136, the deformation complex may extend as far

south as shotpoint 600 where reflection R2 indicated landward-dipping lithic slices. Along parts of the southern Caribbean margin, there may be two or more episodes of formation of deformation complex. In the region of the Curacao Ridge and the Netherlands and Venezuelan Antilles, the Antilles ridge may represent an older deformation complex in which Venezuelan Basin crust was uplifted exposing the basalts of B" on the island of Curacao. Orogenic deformation along the Netherlands Antilles Ridge is for the most part pre-Eocene (MacGillavry, 1977). The Curacao Ridge may be a younger late Tertiary deformation complex. This lateral gradient in time of deformation may account for landward-dipping reflectors such as R2 on line 136 beneath a relatively thick, mildly deformed slope apron. (The base of the slope apron is labeled B on the section). A similar situation was encountered offshore Guatemala in the Middle America Trench where a late Tertiary sediment apron progrades seaward over an early Tertiary accretionary complex (Aubouin and others, 1982; Ladd and others, 1982; von Huene and others, 1982).

SEISMIC STRATIGRAPHY

We are handicapped in our attempts to analyze the stratigraphy of the South Caribbean Deformed Belt by a lack of nonproprietary wells within the belt and only a few nonproprietary holes to the north in the Colombian and Venezuelan Basins. The lone exceptions to well information from the deformed belt are the generalized descriptions by Duque-Caro (1979, and this volume) and Buquero (1983) for the continental shelf section off the Guajira Peninsula. Our best information for the sedimentary section to the north of the deformed belt is from DSDP drill sites in the Venezuela Basin. In spite of this paucity of data, a few general comments can be made about the age of the seismic units within and adjacent to the South Caribbean Deformed Belt.

The top of oceanic crust in the Colombian Basin is seen dipping apparently southeastward between shotpoints 1500 and 1800, and between 9 and 10 seconds of two-way reflection time on line 130. Farther landward toward the South Caribbean Deformed Belt the top of oceanic crust is lost in the multiple that begins just below 10 seconds; but if the dip of the top of the oceanic crust is continued to the seaward toe of the accretionary prism from shotpoint 1800, the top of oceanic crust projects to about 12 seconds of two-way traveltime at shotpoint 3200. With an average sediment velocity of 2 to 3 km/sec for the sediment overlaying oceanic crust, the depth from sea surface to the top of the oceanic crust will be between 11 and 15 km.

In the Rancheria Basin, an irregular surface is seen at seven seconds of two-way time that looks like the top of oceanic crust. We have labeled this surface "top of basement." The depth to this irregular surface is approximately 7 km below the sea surface. If this indeed is a piece of oceanic crust similar in composition and tectonic origin to the crust seen in the Colombia Basin, then there must be a structural discontinuity in oceanic crust beneath the South Caribbean Deformed Belt. This basement may be an oceanic remnant trapped at the initiation of thrusting as described by Seely (1979). Thrusting probably began in late Eocene or early Oligocene times when the present regime of Caribbean plate motions began (Ladd, 1976).

DSDP Site 154 on a low ridge in the Colombia Basin reached Miocene terrigenous sands between 250 and 278 m subbottom in a section that is older than the turbidites that are ponded around the ridge; but it is difficult to tie the stratigraphy of Site 154 to seismic line 130. The age of the oceanic crust in the Colombian Basin is unknown though drilling on the lower flank of the Nicaragua Rise at DSDP Site 152 suggests a possible Late Cretaceous age for at least some of the crustal material. A pre-Tertiary age for Colombian Basin ocean crust is also suggested by the lack of any obvious spreading center for the formation of oceanic crust in the Colombia Basin coupled with the isolation of the Colombian Basin from other oceanic areas by Late Cretace-

ous and early Tertiary volcanic terrains of Central America and the Greater Antilles. All we can say about the age of the sediments northward of the South Caribbean Deformed Belt is that they are probably Cretaceous and younger.

Beneath the continental slope between shotpoints 5000 and 5500 on line 130, we see acoustic basement labeled "B" rising from 7 seconds to 2.5 seconds. A sonobuoy shot to the northeast from the end of line 130 indicates the top of a crustal layer with velocity of 6.2 km/sec at 2.5 seconds. For the top of a crustal layer to be so shallow, it probably is underlain by crust of continental thickness. Consequently, acoustic basement beneath the upper continental slope on line 130 may be the top of landward-thickening continental crust. There is probably a transition from oceanic to continental crust at the landward end of the Rancheria Basin.

Several unconformities exist within the section landward of the accretionary prism. The oldest unconformity is the onlapping of sediments onto acoustic basement "B" beneath the continental slope. Two unconformities are indicated beneath the upper continental slope which are stratigraphically beneath the unconformity marked "U" but stratigraphically higher than the deep unconformity that dips to the northwest between shotpoints 4400 and 4000. The youngest unconformity indicated on the section lies 0.8 sec subbottom in the Rancheria Basin.

There is very little published information about the stratigraphy of the offshore region along the Colombian coast. Duque-Caro (1979) refers to this region as the Sinu belt composed of late Oligocene-early Miocene pelagic mudstones, late Miocene-Pliocene abyssal turbidites, and Pleistocene-Holocene shallow water carbonate facies composed of shales, reef limestones, sandstones, and conglomerates. He describes a high-pressure shale of late Oligocene-early Miocene age underlying Miocene turbidites. The shale appears on seismic profiles as a "transparent or blank zone of no reflections which permits relatively easy identification." There may be a problem, however, differentiating on seismic reflection profiles between acoustic basement caused by this shale interval and acoustic basement which is continental crust.

If we have correctly concluded that acoustic basement in the Rancheria Basin is a piece of trapped Cretaceous oceanic crust, we may be looking at a Cretaceous to Recent section in the Rancheria Basin. This conclusion is based in part on correlation of the top of acoustic basement with unpublished sonobuoy results indicating crustal velocities at 2.5 seconds at the end of line 130 and in part on the similarity in appearance of the top of acoustic basement with the top of oceanic crust in the Colombian Basin. However, acoustic basement on line 130 in the Rancheria Basin could conceivably be the top of the shale layer; we may then be looking only at a Miocene and younger section. This uncertainty could be resolved with a good seismic refraction or wide-angle reflection line shot in the Rancheria Basin to determine a velocity for the material beneath the top of acoustic basement at 7 seconds.

Farther east our clues to the seismic stratigraphy come

mainly from the geology of the islands of Aruba, Curacao, and Bonaire. Line 97 crosses the Netherlands Antilles Ridge just to the east of Bonaire. The Cretaceous to Danian volcanic-sedimentary sequence on Bonaire is strongly folded and overlain unconformably by weakly folded Eocene limestones. The Eocene is overlain by Neogene limestones that have not been affected by folding (Beets, 1977). Possibly then acoustic basement labelled B on line 97 over the Netherlands Antilles Ridge is the top of the Cretaceous sequence. The section above B could be the Eocene to Quaternary sequence. The southern flank of the Los Rocques Basin is probably the northward limit of this Cretaceous "basement." The highly deformed sediments beneath the Curacao Ridge to the north are undoubtedly Late Cretaceous and Tertiary sediments deformed to form the Curacao Ridge in the Neogene.

Line 136 terminates close to the island of Aruba where unfolded Neogene limestones and clastics overlie an igneous Cretaceous "basement." The top of acoustic "basement" marked B on line 136 may be the top of the Cretaceous igneous sequence. Although acoustic "basement" continues northward to the Venezuela Basin, the sedimentary nature of the Curacao Ridge further to the east argues that the northern portion of acoustic basement between approximately shotpoints 800 to 1600 is sediments, not volcanics. The boundary between sedimentary basement and igneous basement is not at all apparent on these records.

The landward-dipping reflectors R1 and R2 may be caused by landward-dipping slices of oceanic crust emplaced by an accretionary process associated with underthrusting or subduction along this margin. The Cretaceous volcanics observed in outcrop on Aruba may be just another slice of such rocks; i.e., Aruba may be a subaerial exposure of the material that causes reflections R1 and R2 in the seismic data. Possibly the Netherlands Leeward Antilles were uplifted into their present position by processes of convergence and southward underthrusting not unlike the processes that formed the South Caribbean Deformed Belt to the north. Possibly there has been a northward migration through time of the locus of deformation. Another possibility is that acoustic basement all the way from shotpoint 1600 to the south end of the line is a mixture of slices of oceanic crust and oceanic sediments emplaced in pre-Neogene time.

Drilling on Leg 15 of the Deep Sea Drilling Project clarified the stratigraphy in the Venezuela Basin to the north of the deformation front on lines 136, 8, and 97. On line 136 the reflection labeled B" can be correlated with the top of Coniacian basalts penetrated at DSDP drill sites 146/149 and 150. In the western Venezuela Basin in the region of the drill sites and line 136, B" is the top of material with seismic velocities greater than 5 km/sec. Diebold and others (1981) have shown that Venezuela Basin crust in the eastern Venezuela Basin near lines 97 and 8 is different in character from the western basin crust. B" cannot be traced to line 97 as a continuous reflection from line 136. On line 97 we have labeled B" the reflection which occurs at the top of material with seismic velocity greater than 5 km/sec (Diebold and others, 1981). On line 97 B" is an irregular reflection unlike planar B" to the west. The strong planar reflection above B" on line 97 is more

like B" to the west on line 136 and on other seismic lines from the western Venezuela Basin (Ladd and Watkins, 1980).

On line 136 we have labeled a reflection A" in the Venezuela Basin. A" can be correlated with drilling results to indicate that A" is the boundary between soft, semiconsolidated Eocene and younger pelagic oozes above and Eocene and older lithified cherts and chalks below.

The horizontally stratified ponded sediments between shotpoints 1600 and 1950 on line 136, and between shotpoints 4250 and the end of the line on line 97 are turbidites of the Venezuela Abyssal Plain. These turbidites are Pliocene and Pleistocene in age.

GAS HYDRATES

Several of the seismic sections show a well-developed bottom simulating reflection (BSR) about 0.5 seconds subbottom within the deformation complex. Similar reflections have been noted in the accretionary complexes of Circum-Pacific subduction zones (Shipley and others, 1979; Yamano and others, 1982), and have been attributed to reflections from the base of a gas hydrate. BSRs are seen along the South Caribbean Deformed Belt on lines 130, 132, 8, and 97. They are seen in water depths ranging from 2.3 to 4.2 kilometers and at subbottom depths of 400 to 500 meters.

The BSR occurs at a phase boundary in the chemical system of methane and water. Above the BSR, temperatures are low enough and pressures are high enough that methane combines with water to form a solid hydrate. Below the BSR, methane occurs either as free gas or dissolved in water in the sediment pore space. Since the position of the phase boundary is governed by temperature and pressure, we can use the presence of the BSR to estimate the temperature gradient in the section since we can estimate pressure gradients fairly well independent of the BSR. Three thousand meters of water exerts a pressure of about 300 bars. Another 500 meters of sediment exerts at least another 50 bars. Therefore, the pressure at the BSR in 3 km of water would be approximately 350 bars. Using the phase diagram from Shipley and others (1979), we find that at a pressure of 350 bars the temperature at the BSR should be approximately 24°C. Since the sea bottom temperature is about 4°C, we have a temperature gradient in the deformation complex of .04 deg/meter. Similar gradients have been determined for the accretionary prism of the Middle America Trench and for the Blake Outer Ridge (Uyeda and others, 1982). Gonzalez de Juana and others (1980) report similar gradients in the Lake Maracaibo region. The above thermal gradient assumes that the BSR always lies at the phase boundary between hydrate and free gas. However, the formation of a hydrate requires a large quantity of gas relative to the gas that could be dissolved in water just below the hydrate. Therefore, if insufficient gas were present, the BSR might actually lie above the location of the phase boundary and the thermal gradient calculated on the basis of the BSR position would be too high. The BSR may not always be located on the phase boundary as suggested by the lack of consistent increase in subbottom depth with water bottom depth. A linear relationship between water depth and subbottom depth of the BSR has been seen along the Pacific coast of Central America but was not clearly seen on the Blake Outer Ridge off the east coast of the United States. A program of thermal gradient measurements using a bottom penetrating probe is necessary to calibrate the South Caribbean BSR.

CONCLUSION

The South Caribbean Deformed Belt and its eastward extension, the Curacao Ridge, are Neogene accretionary complexes formed in a compressional regime as South America is thrust northward over the Caribbean basins. This deformation zone is part of the broader boundary zone between the South American and Caribbean plates that has experienced deformation almost continually in one form or another since early Cretaceous time. The South Caribbean Deformed Belt is probably just the last element to be added to the northern margin of South America after the emplacement of the Netherlands Antilles and other structural elements to the west. During the Neogene, the compressional strain between South America and the Caribbean has been a second order component of the overall relative motion between the two plates. Strike-slip shear is the dominant component of the relative motion, but strike-slip shear strain is difficult or impossible to detect in seismic reflection records. The compression along the South Caribbean Deformed Belt is consistent with the relative motions of South America with respect to the Caribbean determined by Jordan (1975) but not with the relative motion determined by Sykes and others (1982).

Structurally, the South Caribbean Deformed Belt is a thick wedge of semiconsolidated, highly deformed sediments overlying a southward extension of Caribbean oceanic crust. The deformation of much of the belt has progressed to the extent that reflection profiling does not generally detect continuous stratigraphic horizons within the accretionary complex. The highly deformed accretionary complex is overlain by a northward prograding undeformed sediment apron. Drilling offshore Guajira and the geology of the Netherlands Leeward Antilles suggest that the sediment apron is of Neogene age. The northward progradation of the apron suggests that the base of the apron becomes progressively younger to the north indicating progressively younger deformation of the complex to the north.

We are envisioning here a steady-state process in which the Deformed Belt has grown progressively northward throughout the Neogene and the sediment apron has followed the northward migration of deformation. A similar scenario had been envisioned by Seely and others (1974) for the Middle America Trench offshore Guatemala. Subsequent drilling on Leg 84 of the Deep Sea Drilling Project, however, indicated that there was an early Tertiary phase of tectonic accretion followed by a Neogene phase of nonaccretion or tectonic erosion. Off Guatemala the Neogene apron overlies an accretionary complex of Cretaceous rocks that was probably emplaced in the late Paleocene. Drilling has shown the Guatemala margin to be a non-steady-state convergent feature. Possibly the same is true of the South Caribbean Deformed Belt. At present we are just too ignorant of the age and lithology of material that composes the belt.

A prominent feature on many of the seismic lines is a bottom simulating reflector that is the base of a gas hydrate zone. This feature, which is common in accretionary complexes, permits an estimate of the thermal gradient within the complex of 0.04 degrees C/meter. Similar gradients have been detected farther south in the Lake Maracaibo region and in other accretionary complexes.

ACKNOWLEDGMENTS

This paper was reviewed internally by Marcus Langseth and Steve Cande. The paper has also benefited from reviews by Thomas Donnelly, John Grow, and the editors of this memoir. Our thanks to all of them. The data presented in this paper were collected and processed with funds from National Science Foundation Grant OCE76-82328. Lamont-Doherty Geological Observatory Contribution Number 3474.

REFERENCES CITED

Aubouin, J., and others, 1982, Leg 84 of the Deep Sea Drilling Project: Subduction without accretion: Middle America Trench off Guatemala: Nature, v. 297, p. 458–460.

Baquero, E., 1983, Estructura y litologia de un area al NW de la Peninsula Guajira un estudio de estratigrafia sismica [Thesis]: Bogata, Universidad Nacional de Colombia, Facultad de Ciencias, Departamento de Geociencias, 117 p.

Beets, D. J., 1977, Cretaceous and early Tertiary of Curacao: Guide to the Field Excursions on Curacao, Bonaire, and Aruba, Netherlands Antilles: Caribbean Geological Conference, VIII, Curacao, 1977, p. 7–17.

Bowin, C., 1976, Caribbean gravity field and plate tectonics: Geological Society of America Special Paper 169, 79 p.

Case, J. E., 1974, Major basins along the continental margin of northern South America: in Burk, C. A., and Drake, C. L., eds., The Geology of Continental Margins: Springer-Verlag, New York, p. 733–742.

Case, J. E., and Holcombe, T. L., 1980, Geologic-tectonic map of the Caribbean region: in U.S. Geological Survey Miscellaneous Investigation Series, Map I-1100, scale 1:2,500,000.

Chapple, W. M., 1978, Mechanics of thin-skinned fold-and-thrust belts: Geological Society of America Bulletin, v. 89, p. 1189–1198.

Cowan, W. M., and Silling, R. M., 1978, A dynamic, scaled model of accretion at trenches and its implications for the tectonic evolution of subduction complexes: Journal of Geophysical Research, v. 83, p. 5389–5396.

Dewey, J. F., 1975, Finite plate rotations: Some implications for the evolution of rock masses at plate margins: American Journal of Science, v. 275A, p. 371–412.

Diebold, J. B., Stoffa, P. L., Buhl, P., and Truchan, M., 1981, Venezuela Basin Crustal Structure: Journal of Geophysical Research, v. 86, p. 7901–7923.

Duque-Caro, H., 1979, Major structural elements and evolution of northwestern Colombia, in Watkins, J. S., Montadert, L., and Dickerson, P., eds., Geological and Geophysical Investigations of Continental Margins: American Association of Petroleum Geologists Memoir 29, p. 329–351.

—— 1984, Structural style and orogenic phases of the folded belt, northwestern Colombia, *in* Bonini, W. E. and others, eds., The Caribbean-South American Plate Boundary and Regional Tectonics: Geological Society of America Memoir 162 (this volume).

Edgar, N. T., Ewing, J. I., Hennion, J., 1971, Seismic refraction and reflection in Caribbean Sea: American Association of Petroleum Geologists Bulletin, v. 55, p. 883–870.

Gonzalez de Juana, C., Iturralde de Arozena, J. M., and Picard C. X., 1980, Geologia de Venezuela y de sus cuencas petroliferas: Caracas, Foninves, p. 895.

Ibrahim, A. K., Latham, G. V., and Ladd, J., 1979, Seismic refraction and reflection measurment in the Middle America Trench offshore Guatemala: Journal of Geophysical Research, v. 84, p. 5643–5650.

Jordan, T. H., 1975, The present-day motions of the Caribbean plate: Journal of Geophysical Research, v. 80, p. 4433–4439.

Kellogg, J. N., and Bonini, W. E., 1982, Subduction of the Caribbean plate and basement uplifts in the overriding South American plate: Tectonics, v. 1, no. 3, p. 251–276.

Krause, D. C., 1971, Bathmetry, geomagnetism and tectonics of the Caribbean Sea north of Colombia: *in* Donelly, T. W., ed., Caribbean Geophysical Tectonic, and Petrologic Studies: Geological Society of America Memoir 130, p. 35–54.

Ladd, J. W., 1976, Relative motion of South America with respect to North America and Caribbean tectonics: Geological Society of America Bulletin, v. 87, p. 969–976.

Ladd, J. W., Ibrahim, A. K., McMillen, K. M., Latham, G. V., von Huene, R. E., 1982, Interpretation of seismic reflection data of the Middle America Trench off Guatemala: *in* Initial Reports of the Deep Sea Drilling Project, Leg 67, p. 675–689.

Ladd, J. W., and Watkins, J. S., 1978, Tectonic development of trench-arc complexes on the northern and southern margins of the Venezuela Basin, *in* Watkins, J. S., Montadert, L., Dickerson, P., eds., Geological and Geophysical Investigations of Continental Margins: Amererican Association of Petroleum Geologists Memoir 29, p. 363–371.

—— 1980, Seismic stratigraphy of the western Venezuela Basin: Marine Geology, v. 35, p. 21–41.

MacGillavry, H. J., 1977, Tertiary formations: Guide to the Field Excursions on Curacao, Bonaire, and Aruba, Netherlands, Antilles: Caribbean Geological Conference, VIII, Curacao, 1977, p. 36–38.

Malfait, B. T., and Dinkelman, M. G., 1972, Circum-Caribbean tectonic and igneous activity and the evolution of the Caribbean plate: Geological Society of America Bulletin, v. 83, p. 251–272.

Maresch, W. V., 1972, Plate-tectonics origin of the Caribbean mountain system of northern South America: Discussion and Proposal: Geological Society of America Bulletin, v. 85, p. 669–682.

Pindell, J., and Dewey, J. F., 1982, Permo-Triassic reconstruction of western Pangea and the evolution of the Gulf of Mexico/Caribbean region: Tectonics, v. 1, p. 179–212.

Seely, D. R., 1979, The evolution of structural highs bordering major forearc basins; *in* Watkins, J. S., Montadert, L., Dickerson, P. W., eds., Geological and Geophysical Investigations of Continental Margins; American Association of Petroleum Geologists Memoir 29, p. 245–260.

Seely, D. R., Vail, P. R., and Walton, G. G., 1974, Trench slope model: *in* Burk, C. A. and Drake, C. L., eds., The Geology of Continental Margins: New York, Springer-Verlag, p. 249–260.

Shipley, T. H., Houston, M. H., Buffler, R. T., McMillen, F. J., Ladd, J. W., and Worzel, J. L., 1979, Seismic evidence of widespread possible gas hydrate horizons on continental slopes and rises: American Association of Petroleum Geologists Bulletin, v. 63, p. 2204–2213.

Silver, E. A., Case, J. E., MacGillavry, H. J., 1975, Geophysical study of the Venezuelan borderland: Geological Society of American Bulletin, v. 86, p. 213–226.

Sykes, L. R., McCann, W. R., and Kafka, A. L., 1982, Motion of the Caribbean plate during last 7 million years and implications for earlier Cenozoic movements: Journal of Geophysical Research, v. 87, p. 10656–10676.

Talwani, M., Windisch, C. C., Stoffa, P. L., Buhl, P., and Houtz, R. E., 1977, Multichannel seismic study in the Venezuelan Basin and the Curacao Ridge, *in* Talwani, M., and Pitman, W. C. III, eds., Island Arcs, Deep Sea Trenches and Back-Arc Basins: American Geophysical Union, Maurice Ewing Series I, p. 83–98.

von Huene, R., Ladd, J., and Norton, I., 1982, Geophysical observations of slope deposits, Middle America Trench off Guatemala: *in* Initial Reports of the Deep Sea Drilling Project, Leg 67, p. 71–732.

Yamano, M., Uyeda, A., Aoki, Y., and Shipley, T. H., 1982, Estimates of heat flow derived from gas hydrates: Geology, v. 10, p. 339–343.

MANUSCRIPT ACCEPTED BY THE SOCIETY SEPTEMBER 1, 1983
LAMONT-DOHERTY CONTRIBUTION NUMBER 3474

Geological Society of America
Memoir 162
1984

Magnetic provinces in western Venezuela

William E. Bonini
Department of Geological and Geophysical Sciences
Princeton University
Princeton, New Jersey 08544

ABSTRACT

Magnetic provinces, recognized by trend geometry and anomaly character, are delineated in western Venezuela utilizing aeromagnetic data from the 1950s oil company surveys. In northwestern Venezuela, the Guajira-Paraguaná province (east-west anomalies) lies north of a proposed east-west fault zone extending from the southern Paraguaná Peninsula across the Gulf of Venezuela and south of the pre-Tertiary outcrops on the Guajira Peninsula. South of this fault zone and on the west is the Perijá province (northeasterly trends) and on the east the Coro province (northwesterly trends). The Oca zone province (east-west trends) separates the northern and southern parts of the Perijá province. Geologic features which can result in these magnetic anomalies are fault blocks, east-west faults, some in sets, faults oblique to east-west shear, and probably intrusions into the crust parallel to these features. These have been produced by differential motion between the major zones of dislocation resulting from right-lateral offsets on east-west transcurrent faults during extension in the Tertiary. Thus, the magnetic anomalies locate a complex zone of Cenozoic interaction on this margin of the Caribbean-South American plate boundary.

South of the Oca fault, the Perijá province trends are related to the geologic trends of the Sierra de Perijá, and the Jurassic La Quinta graben complex and the Central Lake Graben. There are only modest, trendless anomalies in the eastern Maracaibo Basin, the Falcón Basin, and the area to the south. Magnetic basement rocks here are probably igneous and metamorphic Paleozoic units, similar to those exposed in the Venezuelan Andes. Southeast of the Andes, a major lineation between the Barinas and Río Meta provinces is correlated with the projection of the Altamira fault. Unmetamorphosed lower Paleozoic sedimentary rocks are preserved in a block dropped down to the north. This is possibly an extension of the Espino Graben complex, which is 300 km to the northeast and in which Jurassic basalt is preserved. Strong magnetic anomalies west and south of the El Baúl Uplift are probably associated with Jurassic volcanics, similar to those exposed in the uplift, perhaps preserved in the same graben complex. These appear as major features of the interior plains (llanos) north of the Guayana Shield.

The primary subdivision of the magnetic provinces is two-fold. North of and including the Oca fault-Coro area, magnetic anomalies document the complex zone of interaction related to east-west shear in Cenozoic time. South of this, magnetic anomalies are related to the distribution of pre-Cretaceous rocks, some of which were involved in the plate boundary interactions, but others to the south are related to the development of the pre-Cretaceous terrane north of the Guayana Shield.

161

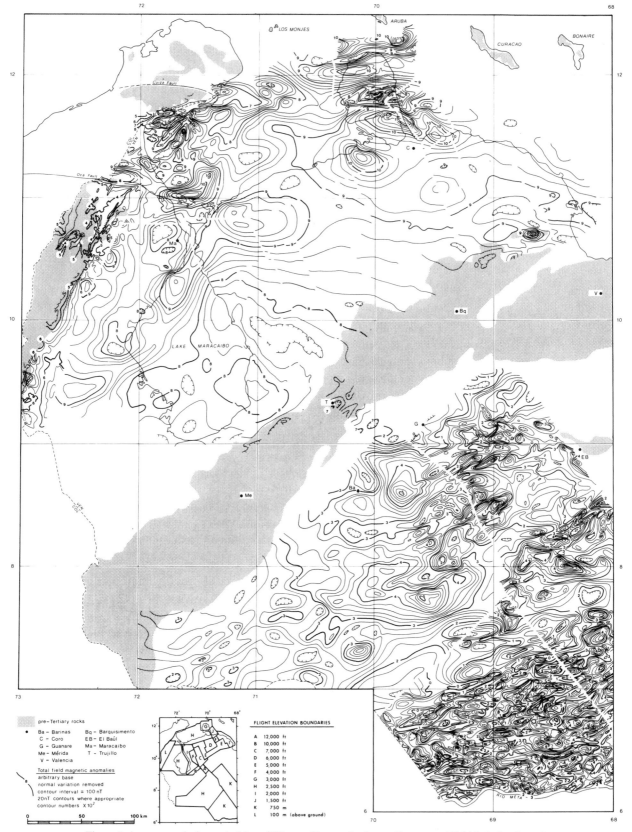

Figure 1. Aeromagnetic Anomaly Map of Western Venezuela. Anomalies are total field (relative datum) compiled from 12 surveys at flight elevations shown. See text for sources of data. Major contours are 100 nT with 20 nT contours in areas of low-gradient anomalies.

INTRODUCTION

The upper Phanerozoic tectonic development, nature, and location of the plate "boundary" between the Caribbean and South American plates has been the subject of many studies (e.g., Rod, 1956; Vasquez and Dickey, 1972; Case, 1974; Silver and others, 1975; Jordan, 1975; Case and Holcombe, 1980; and Bonini and others, 1980). It is treated also in some detail by many papers in this volume; yet still there is no agreement on the subject. At one extreme the present boundary has been placed north of the Curacao Ridge [South Caribbean Deformed Belt (Kellogg and Bonini, 1982)] along the La Orchila Basin and into the Urica fault trend (Vierbuchen, 1979) or the Los Bajos-El Soldado trend (Perez and Aggarwal, 1981) in eastern Venezuela and Trinidad, and at the other extreme along the Boconó-San Sabastian-El Pilar fault trends from the Venezuelan Andes into eastern Venezuela (Vasquez and Dickey, 1972; Malfait and Dinkelman, 1972). The purpose here is not to review published ideas concerning the nature of that plate boundary except to note that it is apparent that it is the product of a complicated late Mesozoic and Cenozoic history, involving rotation of blocks and rifting [Netherland and Venezuelan Antilles, Falcón Basin (Silver and others, 1975; Muessig, 1984)]; transform, transverse or strike-slip faulting [El Pilar and Urica (Vierbuchen, 1979, 1984), Oca and Boconó (Vasquez and Dickey, 1972), and Santa Marta faults (Ward and others, 1973; Tschanz and others, 1974)], and subduction or underthrusting at a slow rate with low seismicity [Curacao Ridge area, South Caribbean Deformed Belt (Silver and others, 1975; Kellogg and Bonini, 1982)].

It is clear that no simple description of the plate boundary is possible at this stage. What we see, can best be described as the result of the interplay of relative motions between the Caribbean and South American plates, influenced on the west by Nazca plate motions. This is a region where the boundary is diffuse and these relative motions are distributed in a variety of ways on the continental crustal margin of South America and its borderland.

Many geological and geophysical studies have dealt with aspects of the problem using results of geologic mapping, fault displacements, gravity and paleomagnetic data, seismic activity, first-motion studies, and seismic refraction and offshore reflection surveys (see Case and others, 1984, for an extensive bibliography). Although some widely spaced magnetic track lines are available on the Venezuelan offshore borderland (Peter, 1972; Silver and others, 1975), the aeromagnetic data onshore in Venezuela, which might be useful in studies of the boundary, have resided in confidential oil company files. The purpose of this paper is to present and analyze on a qualitative basis recently released magnetic data of the 1950s for the study and delineation of elements of the complex Caribbean-South American plate boundary south to the Guayana Shield in western Venezuela.

AEROMAGNETIC SURVEY DATA

Aeromagnetic surveys were made of potential oil-producing areas of Venezuela using the fluxgate magnetometer in the 1950s. These surveys were made for several oil companies and apparently data were traded among groups, such that composite maps were commonly available within the petroleum industry in Venezuela. A. Bellizzia, then director, Dirección de Geología, Ministerio de Energía y Minas (MEM), obtained from the petroleum company, Maraven, release of these maps for use on the cooperative program between MEM and Princeton University. These maps were obtained from Gulf Research and Development Corporation, Houston. Data from a 1976 survey in the Sierra de Perijá, flown for the Corpozulia company, were made available through MEM.

In 1981-82 absolute high-resolution aeromagnetic surveys were flown over north-central and eastern Venezuela and offshore areas for Petróleos de Venezuela. Inasmuch as this most recent new survey does not cover much area west of Longitude 70°W in Venezuela, it is appropriate to present the 1950s data for this volume (Fig. 1).

Flight spacing on most of the 1950s surveys was from 1.6 to 5 km (1 to 3 mi) and the flight elevation ranged from 750 m to 3.7 km (Fig. 1). The Perijá survey was flown at 100 m above terrain and at 1 km spacing over much of the area. The composite map (Fig. 1) uses a 100 nanotesla (nT) contour interval (1 nT = 1 gamma), but 20 nT contours are provided to fill selected low-gradient areas. Survey boundaries are delineated by narrow blank regions on the map, and each survey is specified in the legend by flight elevation (Fig. 1). With minor adjustments, continuous contours can be interpolated for an eye-smooth map. All have had the core-derived field removed, although its precise magnitude is unknown. Fluxgate magnetometer measurements are relative, so the maps show only the anomalies relative to an arbitrary datum. The fact that these are relative anomalies does not affect the geologic interpretation of the data. A cross-section (Fig. 2) is given with the line of section and the magnetic provinces discussed indicated on Figure 3.

MAGNETIC ANOMALY CONSIDERATIONS

Rock Magnetization

Induced magnetization (J_i) is that induced in the Earth's magnetic field (F). The magnetization vector is parallel to and is the product of that field and the susceptibility (k) [$J_i = k \cdot F$]. On the other hand, the natural remanent magnetization (NRM) is the magnetization remaining when the ambient field is removed (i.e., magnetization in field-free space). It is also a vectorial quantity and itself may be the resolution of several NRM components.

The total magnetization of a rock is a vector, which is the resultant of J_i and NRM. Depending upon the direction and intensity of each vector, J_i and NRM can reinforce each other (if in the same general direction), or cancel or reduce (if opposed), or give a resultant magnetization significantly different from each (e.g., at 90° to each other). The importance of each vector de-

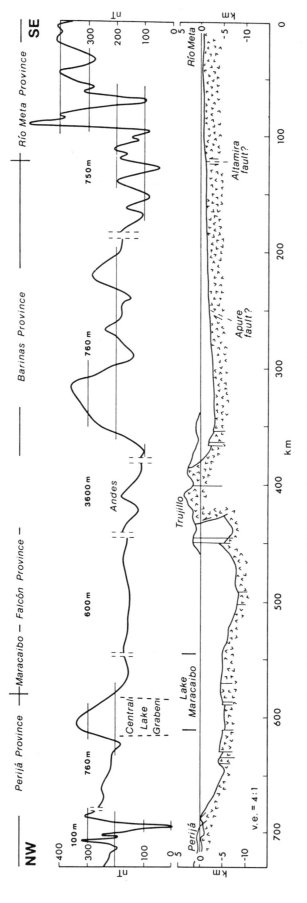

Figure 2. Cross-section from the Río Meta in the southeast to the Colombian-Venezuela border in the Sierra de Perijá. Magnetic anomalies and flight elevations are given and the pre-Cretaceous basement elevation is shown (Bellizzia and others, 1976). Sedimentary cover is undifferentiated. See Figure 3 for line of section.

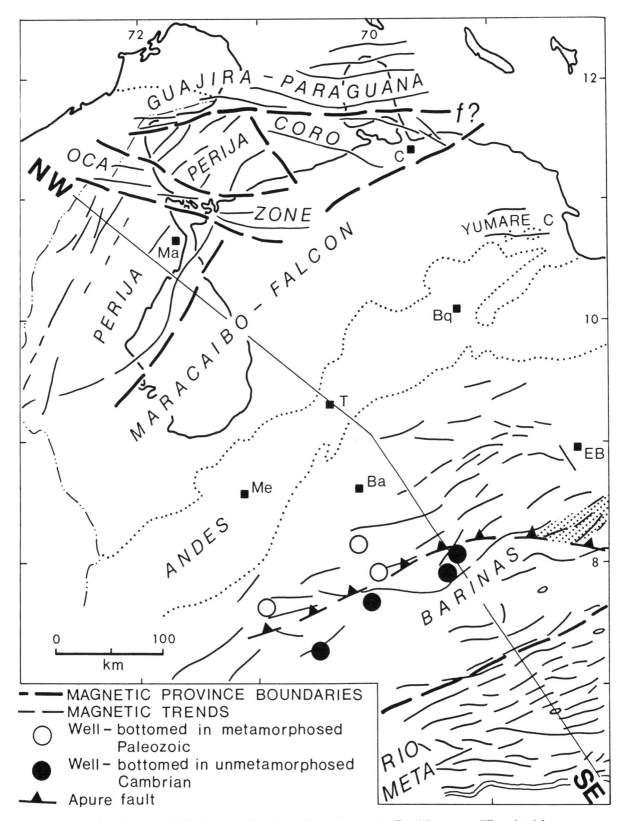

Figure 3. Magnetic Provinces and Trends in Western Venezuela. Trend lines are undifferentiated for positive and negative trends. Locations of the proposed Apure fault and significant wells in the Barinas-Apure Basin (Barinas province) are from Feo-Codecido and others (1984). Proposed location for the Altamira fault is along the Barinas-Río Meta province boundary.

pends upon their magnitude; and the Koenigsberger ratio (Q) is generally calculated to relate those intensities (but not direction), [$Q = NRM / J_i$].

It is unfortunate that in making and reporting paleomagnetic results, as well as in sampling basement rocks in wells, few researchers give complete or indeed any, bulk magnetic parameters for the rocks. To interpret magnetic field anomalies one must predict, calculate, or assume reasonable total magnetization (intensity as well as direction) for the suspected source rocks. This requires estimates of the susceptibility to determine induced magnetization, as well as the NRM intensity and direction to properly assess the importance of remanence.

Magnetization of Venezuelan Rocks

Magnetic anomalies of geologic interest result from the magnetic field produced by the total magnetization of rocks, more specifically the magnetic minerals contained in those rocks. As a generalization, most large amplitude magnetic anomalies are produced by magnetite-rich mafic igneous rocks making up the basement or that occur within the sedimentary column, because potentially these rocks have the highest susceptibilities (Table 1) and NRM. Acidic igneous and metamorphic basement rocks may on occasion produce noticeable, and in cases strong, anomalies of interest.

Whereas Table 1 summarizes these general considerations, Table 2 lists magnetic parameters available for Venezuelan rocks of interest here. Most are taken from paleomagnetic studies, but the data are incomplete for use in analyzing magnetic field anomalies. However, several points can be made:

1. High susceptibilities have been measured, and specific units can be placed in the high to intermediate ranges:

Susceptibilities greater than 1.0×10^{-3} emu/cm³
($J_i = 0.4 \times 10^{-3}$ emu/cm³)

Jurassic La Quinta Formation, Sierra de Perijá, andesites and lavas, alkalai basalt dikes
Isla de Toas basalt
Eastern Venezuela Basin quartz diorite
Precambrian Cuchivero Group granite, Guayana Shield
Precambrian Imataca Complex, Guayana Shield, mafic granulite, charnockite, quartz monzonite
Precambrian Roraima quartz dolerite, Guayana Shield

Susceptibilities between 1.0 and 0.4×10^{-3} emu/cm³
($J_i = 0.4$ and 0.15×10^{-3} emu/cm³)

Palmar granite, Sierra de Perijá
Precambrian Imataca Complex, Guayana Shield, amphibolite, biotite gneiss
Precambrian Roraima noritic dolerite, Guayana Shield

2. NRM intensities similar to the range of induced intensities (J_i) listed above have been measured and could be particularly important in total magnetization. However, the NRM of different

TABLE 1. PERCENTAGE DISTRIBUTION OF MAGNETIC SUSCEPTIBILITIES IN MAJOR ROCK TYPES

Susceptibility range (emu/cm³)	Rock type				
	Mafic effusive rocks (%)	Mafic plutonic rocks (%)	Granites & allied rocks (%)	Gneisses schists slates (%)	Sedimentary rocks (%)
Less than 10^{-4}	5	24	60	71	73
Between 10^{-4} and 10^{-3}	29	27	23	22	19
Between 10^{-3} and 4×10^{-3}	47	28	16	7	4
Greater than 4×10^{-3}	19	21	1	0	4

Note: The number of samples ranges from 45 to 97 per rock type. Data compiled by P. M. Hurley and reproduced from Table 25-2 in Lindsley and others (1966). Values given are in the cgs system (1 emu/cm³ = 0.0796 S. I. units).

samples may be scattered by secondary overprint, lightening strikes, or multicomponent magnetization, and the resultant NRM may be very small.

NRM intensities greater than 0.4×10^{-3} emu/cm³

Isla de Toas basalt
Jurassic La Teta volcanics, Guajira
Eocene Parashi dikes and diorite, Guajira
El Baúl Jurassic Guacamayos volcanics
Jurassic Bolívar diabase dikes, Guayana Shield
Eastern Venezuela Basin quartz diorite
Precambrian Parguaza rapakivi granite, Guayana Shield
Precambrian Cuchivero granite, Guayana Shield
Precambrian Imataca Complex, Guayana Shield, mafic granulite, charnockite, amphibolite, quartz monzonite
Precambrian Roraima noritic and quartz dolerites

NRM intensities between 0.4 and 0.15×10^{-3} emu/cm³

Jurassic La Quinta andesites, lavas, alkalai basalt dikes
Precambrian Cuchivero metavolcanics

3. Q ratios for the Sierra de Perijá La Quinta Formation rocks are less than 1, and thus the induced magnetization should dominate.

4. Except for one, all Q ratios for Guayana shield rocks are above 1, significantly so in many cases. Caution, therefore is needed in quantitative interpretation of anomalies resulting from these rocks, because remanence may dominate.

5. Core samples from the Eastern Venezuela Basin show Q ratios significantly below 1 for all but one unit. Most, if not all, of these wells are believed to bottom within the metamorphosed Paleozoic province (Feo-Codecido and others, 1984). Although generally low magnetization is expected for metamorphic and

TABLE 2. AVERAGE MAGNETIC SUSCEPTIBILITIES, NATURAL REMANENT MAGNETIZATION
INTENSITIES, AND KOENIGSBERGER RATIOS FOR VENEZUELAN ROCKS

Area, units, rock types	n	Susceptibilities $\times 10^{-3}$ emu/cm^3 (range)	NRM intensities $\times 10^{-3}$ emu/cm^3 (range)	Q ratio (range)	Source
Sierra de Perijá					
La Quinta Fm. (Jurassic)					
Andesites and lavas (51)*	30	1.36 (0.03-2.42)	0.25 (0.05-0.5)	0.49 (0.11-6.67)	1
Alkali basalt dikes (35)*	20	2.28 (0.66-3.10)	0.33 (0.1-0.6)	0.39 (0.09-1.21)	1
Redbeds (17)*	23	0.16 (0.03-0.50)	0.03 (0.01-0.09)	0.54 (0.44-1.13)	1
Andesitic flows and tuff	7	7.9 (0.2-16.0)	..	..	2
Palmar granite	4	0.82 (0.48-1.16)	..	..	2
Cretaceous and Paleocene					
Sandstone, limestone	4	0.04 (0.02-0.26)	..	..	2
Isla de Toas					
Basalt (7)*	4	1.5	0.9	1.6	1
Guajira Peninsula					
La Teta volcanics (Jurassic)	44	..	1-0.1	..	3
Siapana (Permian-Triassic?)					
Dikes	17	..	0.01	..	3
Diorite	26	..	0.001	..	3
Parashi (Eocene)					
Dikes and diorite	6	..	1-0.1	..	3
Paraguaná Peninsula					
Santa Ana Complex (Cretaceous)					
Gabbro and volcanics	33	0.10 (0.03-0.3)	0.08	2.1	4
El Baúl Uplift					
Guacamayos Group (Jurassic)					
Volcanic flow units	46	..	4.0	..	5
Eastern Venezuela Basin					
Oil well core samples from Anzoátegui and Monagas+					
El Tigre 1, granite	..	0.07	0.004	0.1	6
Pilón 1, granite	..	0.03	0.001	0.07	6
Yopales 1, granitoid	..	0.12	0.013	0.2	6
Merey 2, quartz diorite	..	3.4	1.7	1.0	6
Uracoa 1, biotite schist	..	0.24	0.07	0.6	6
Temblador 4, biotite schist	..	0.026	0.005	0.4	6
Guayana Shield					
Bolívar diabase dikes (Jur.)	41	0.2	1	13.8	5
Parguaza batholith (Precamb.)					
Rapakivi granite	26	0.31 (0.02-0.70)	3.0 (0.06-12.7)	27.6 (2.4-74)	7
Cuchivero Group (Precambrian)					
Metavolcanics	29	0.18 (0.04-0.52)	0.33 (0.52-1.04)	5.2 (1.2-31)	7
Granite	5	1.44	6.33	12.6	7
Imataca Complex (Precambrian)					
Mafic granulite	35	4.7 (0.2-8.9)	4.26 (1.09-9.5)	2.6 (1.0-15.7)	7
Charnockite	35	1.4 (0.72-2.2)	3.00 (0.46-12.4)	6.1 (0.7-28)	7
Amphibolite	15	0.52 (0.26-0.78)	0.62 (0.40-0.84)	3.4 (1.5-9.0)	7
Quartz monzonite	7	6.9	4.6	1.9	7
Biotite gneiss	6	0.4	0.08	0.54	7
Roraima (Precambrian)					
Noritic dolerite	33	0.57 (0.35-0.81)	0.53 (0.32-5.2)	2.51 (1.1-39.6)	7
Quartz dolerite	12	4.6 (4.0-5.2)	20.3 (15.6-25.1)	11.9 (11.0-13.8)	7

Note: n = number of samples reported by author; NRM = natural remanent magnetization; Q = Koenigsberger ratio. In many cases authors have reported site averages based on a number of samples with no range of sample values for a given site. The average values of susceptibilities and NRM intensities given here are numerical averages of site values for each rock type in a given area irrespective of vector direction. Average Q values were calculated from the average susceptibilities and NRM intensities. The ranges given in parentheses are the low and high site values reported by authors.
Sources: 1 = Maze and Hargraves, 1984; 2 = Kellogg, 1981; 3 = MacDonald and Opdyke, 1972; 4 = Skerlec and Hargraves, 1980; 5 = MacDonald and Opdyke, 1974; 6 = Slichter, 1942; 7 = Onstott and Hargraves, 1980.
*Susceptibilities not reported by Maze and Hargraves (1984). Values reported here were measured by C. Rine, Princeton University, on same cores or representative samples used by them. The number measured is indicated in parentheses following the rock type name.
+Vacquier (in Slichter, 1942) noted that the NRM data are too scattered to be of practical value in terms of geologic structure, and suggested that the magnetizations of these cores have been disturbed by drilling. If scatter alone was the criterion for this suggestion, it is quite probable that in light of present knowledge these data are a reasonable representation of susceptibilities and NRM intensities. This, of course, does not rule out drilling disturbance.

granitic rocks (Table 1), the low magnetization and Q ratio for rocks of this province suggest there might be significant differences in magnetic anomalies produced over this metamorphic province and the Guayana Shield terrane.

Other Factors Affecting Magnetic Field Anomalies

The characteristics of magnetic anomalies vary with a number of factors in addition to the total magnetization (see Reford, 1964). These are:

Orientation of body in the Earth's field
Inclination of the Earth's field
Body shape, width, and length
Extent in depth of the body
Distance between the top of the body and the observation level (elevation of the aircraft).

There are usually too many variables to determine a unique solution for a causative body. However, depth estimates to a magnetic source can be made, and in many cases, this is the most important use of magnetic anomaly analyses.

The surveys flown over western Venezuela involved considerable variation in "depth" to magnetic basement. With increase in depth, r, the width (or wavelength) of an anomaly will increase. At the same time for a simple situation and depending upon orientation and other factors, the magnitude of the anomaly tends to fall off as: (a) $1/r$ for a long, wide, deep body; (b) as $1/r^2$ for a long, concentrated body; and (c) as $1/r^3$ for a concentrated body. Thus, the shape and depth of the body provide controls on the size and shape of the anomaly.

The trend of the anomalies, or trend geometry, can be a useful tool in qualitative interpretation. In identifying magnetic provinces, the trends or shape of the magnetic anomalies, their orientation and extent, can be useful because it is assumed that they are related to structural trends or geologic body shapes, and thus provide geological strike patterns. Trend geometry is frequently observed even though anomalies weaken and broaden with increased depth.

MAGNETIC PROVINCES

Magnetic provinces (Fig. 3) are outlined on the basis of the character of magnetic anomalies, their wavelength, amplitude, spatial density, and trend orientation. The magnetic anomaly character presumably reflects trends of geologic origin, geometry and orientation of bodies, magnetization of the igneous and metamorphic basement, and of units, such as volcanics, within the sedimentary column, and relief on or depth to magnetic basement. In the last named case this could be related to gradual deepening of the basement, or an abrupt change, such as where the basement is displaced by faulting. A magnetic body could produce a high only, a low only, or a high and a low of varying proportion, depending upon factors such as body strike and dip, NRM, etc. Trends on Figure 3 are the trends of highs and lows, not an interpretation of specific details of the causative bodies.

Obviously no unique interpretation is possible, but important features of the map permit designation of the following provinces:

I. Perijá Province.

This region is characterized by northeasterly trending anomalies. The Sierra de Perijá has many northeasterly geologic elements, e.g., the Sierra itself, the Perijá and Tigre faults, and the Lajas granitic pluton (Miller, 1962; Bellizzia and others, 1976; Kellogg, 1984). There are a series of circular high closures east of the range front which are aligned on a trend more or less northeasterly. Along the western margin of Lake Maracaibo there is another prominent northeasterly trend with an amplitude of about 200 nT, which at its southern end (at about Latitude 9° 30′N) is more east-northeasterly, and which at its northern end (at about Latitude 10°45′N) merges with an east-west trend in the Oca fault magnetic province in the western Falcón Basin. Pumpin (in Bartok and others, 1981) identifies the Central Lake Graben as a Jurassic graben, 30 km wide and 5 km deep, based on seismic sections and well data, and suggests that the magnetic anomaly (Fig. 2) is derived from a deep source related to the graben formation.

Candidate sources for the magnetic anomalies in this province are the basaltic andesitic flows of the Jurassic La Quinta formation exposed in the Sierra de Perijá (Table 2; Maze, 1984). The Q ratio calculated for these rocks (Table 2) is less than 0.6, suggesting that induced magnetization is dominant. From these data (Table 2) the expected contributions from the Palmar granite would be an order of magnitude less, and the sedimentary rocks almost two orders of magnitude less than that of the La Quinta volcanics.

Other exposed rocks in the Sierra de Perijá are pre-Devonian metasedimentary rocks, varying from sub-greenschist facies in exposures in Colombia and in places amphibolite facies on the Venezuelan side, and the Lajas granite complex (Kellogg, 1984). Unmetamorphosed sedimentary rocks of Devonian, Carboniferous, Permian, Mesozoic, and Cenozoic age, consists of shales, sandstones, quartzites, and limestones. They would all be expected to have low magnetization and to contribute little to the magnetic anomalies. Exceptions to this, in addition to the Jurassic La Quinta volcanics, are the pseudo "basaltic tholeiites" of the Paleocene Marcelina formation reported by Moticska (1977). Moticska reports that these are produced by combustion of coal in that formation. The total rock analysis, including abundant glass, is that of a dacite, but it contains up to 8 percent titaniferous magnetite. Because the Perijá aeromagnetic survey was flown only 100 m above ground level, these "volcanics" produce a series of sharp, narrow, linear magnetic features (Fig. 1) confined to exposures of the Marcelina formation.

The northeasterly trends continue to the north of the east-west Oca fault magnetic zone (Figs. 1 and 3) to a point in the western Gulf of Venezuela just south of the coast of the Guajira Peninsula about 30 km south of the Cuiza fault (Fig. 1). The identification of these northeast trends as a northern subprovince of the Perijá province is based solely on the trend geometry. However, northeast-trending faults are shown on the map by Case and Holcombe (1980).

II. Oca Zone Province.

The Oca zone, about 30 km wide, straddles the Oca fault and appears to continue into a broad east-west high in western Falcón. This parallels a change in structure documented by Muessig (1984). Intense anomalies over the Isla de Toas, which is along the Oca fault, possibly reflect the high magnetization of the basalts exposed there (Table 2). Maze and Hargraves (1984) report that the NRM vector is close to the present field direction. Displacement relief across the Oca fault is about 1 km, down to the north (Feo-Codecido, 1972). This is insufficient in itself to produce such a wide trend of anomalies. However, there are other east-west structural trends in the area, e.g., the Montes de Oca, south of the fault, and possibly parallel faults and fault blocks now covered, which could produce basement relief north of the Oca fault trace. In addition there are possible igneous bodies, or rotated blocks associated with right-lateral east-west Cenozoic movements on the Oca and parallel faults.

III. Guajira-Paraguaná Province.

The strong northeasterly trends of the northern part of the Perijá province terminate against east-west oriented anomalies, at or just south of the Guajira Peninsula coastline (Figs. 1 and 3) in the western Gulf of Venezuela. These east-west trends continue across the Gulf of Venezuela into the prominent east-west trends on the Paraguaná Peninsula over the Santa Ana Complex and on the northern part of the peninsula. Although there is little magnetic data over the Colombian Guajira, the east-west magnetic trends appear to be associated with units shown as Jurassic in age by Bellizzia and others (1976). On the Venezuela-Colombia border, the anomalies are on strike with outcrop trends of the Guajira trough, Cocinas Range, and Cuiza fault (Case and Mac-Donald, 1973). The southern boundary east of 71° longitude is less pronounced.

Candidate source rocks for the magnetic anomalies in this province are the mafic and ultra-mafic units of the Lower Cretaceous Santa Ana Complex, south-central Paraguaná. The Jurassic metamorphic units and upper Paleozoic acid intrusives exposed in north-central Paraguaná show no associated anomalies. Skerlec and Hargraves (1980) report paleomagnetic data from the Santa Ana Complex gabbros and volcanics. Average susceptibilities are low as are the NRM average intensities (Table 2). However, the Q ratio is about 2, indicating that the remanence dominates. Since the NRM direction is southerly and down (about 25°), the nature of the aeromagnetic field anomalies (a broad low with local highs) probably reflects the NRM dominance in the Santa Ana Complex area.

Magnetic field anomalies could be produced by the La Teta volcanics on southeastern Guajira. Part of these volcanics proba-

bly underlies the strong east-west aeromagnetic anomalies. NRM intensities for the La Teta lavas in a Jurassic redbed sequence are comparable to the NRM intensities of the La Quinta flows in the Sierra de Perijá (Table 2). No susceptibilities are reported for the La Teta volcanics, but if they are similar to the La Quinta volcanics magnetically, susceptibilities could be high enough to produce the field anomalies. A broad east-west low north of the La Teta outcrop area trends eastward into the Gulf of Venezuela, and is on strike with the Santa Ana Paraguaná low. Although of unknown origin, it is on the Cuiza fault trend (Fig. 1).

The continuity of magnetic trends from eastern Guajira across the northern Gulf of Venezuela, across the Paraguaná Peninsula and into the western Bonaire Basin, strongly suggests continuity of geologic trends and that these areas today constitute a single tectonic unit. If indeed there has been relative motion between Guajira and Paraguaná implied by the contrasting paleomagnetic vectors (Skerlec and Hargraves, 1980), these east-west trends require additional explanation. It is suggested that the strong northwesterly anomalies in the Coro area and the east-west anomalies in northern Paraguaná are related to features resulting from extensional tectonics during the Oligocene and Miocene (Muessig, 1984) associated with the right-lateral motion between the Caribbean and South American crustal plates.

IV. Coro Province.

There is a large, high closure, elongate in the northwesterly direction west of Coro (C on Fig. 1 and 3), which appears to continue northwesterly into the Gulf of Venezuela. Case and Holcombe (1980) also show northwesterly trending faults in that area. Similarly, the trends north and northeast of Coro and on the southern coast of the Paraguaná Peninsula appear more as northwesterly trends, and could represent a magnetic-trend boundary between the Guajira-Paraguaná province and the Coro province. That line of demarcation is mapped as an east-west fault in southern Paraguaná (Bellizzia and others, 1976) and this or a parallel fault could extend across the Gulf of Venezuela as another east-west fault system between the Oca and Cuiza faults (Fig. 3).

V. Maracaibo Basin-Falcón Basin Province

This is a region of generally low magnetic relief covering the area south of the Coro province extending into the states of Lara and Yaracuy, south and southwest of Coro. Exceptions to this are the several in-line east-west anomalies (Fig. 3) associated with the Precambrian(?) Yumare complex (Bellizzia and others, 1976). The Maracaibo Basin-Falcón Basin province is, however, an area containing a sedimentary sequence as thick as 10 km (Bellizzia and others, 1976), but in general there are no significant trends south and southwest of the Coro region. This province could be an area consisting of basement units characterized by low susceptibility (felsic metamorphic rocks?); or the anomalies are low because of simple decrease in amplitude with depth; or a combination of both. The anomalies observed on the crossing of the Andes near Trujillo (Figs. 1, 2) show only modest magnetic relief (80 nT) even though the flight distance above "basement" of metamorphosed Paleozoic units was only 1 to 2.25 km. This suggests that magnetic basement in this province could be an extension of the Paleozoic metamorphic basement province recognized by Feo-Codecido and others (1984) east and southeast of the Andes and exposed in the Andes.

VI. Barinas Province.

This area, southeast of the Venezuelan Andes, covers a large portion of the Barinas-Apure Basin. Magnetic trends are generally easterly and northeasterly, with some nearly circular anomalies. The basement is at a depth of less than 0.3 km along the Río Meta (at the southern edge of the map, Figs. 1, 3) and deepens to about 3.5 km along the Andean front (Bellizzia and others, 1976). The anomalies change character along a northeasterly line, which is proposed as the boundary between this and the Río Meta province to the south (Fig. 3). North of this line anomalies are broader and, with exceptions, lower in relative value than those to the south. Exceptions are the high values near El Baúl and some high amplitude, tighter anomalies in a zone 50 to 100 km due south of El Baúl (see stippled area, Fig. 3). The northern edge of this area is a lineation, which could be a fault.

Feo-Codecido and others (1984) report on basement wells and significant, deep holes which did not reach basement. The wells of importance and the Apure fault location from their paper are shown on Figure 3. They describe the Apure fault zone as the boundary between the Paleozoic orogenic belt to the north and the older, yet unmetamorphosed Paleozoic sedimentary rocks to the south. They propose that the fault is a thrust to explain the proximity of the upper Paleozoic schists and granites about 30 km north of the subcrop of the unmetamorphosed Cambrian Carrizal Formation.

West of Longitude 69°W (Fig. 1) there does not appear to be a significant difference in magnetic character of anomalies between the area north and south of the projection of the Apure fault, although the basement south of the fault should be Precambrian rather than metamorphosed Paleozoic rocks. East of 69°W, trend changes do occur. South of the Apure fault there is a magnetic high with an east-southeasterly trend, and north of this there are northeasterly highs and lows (stippled area, Fig. 3).

In the El Baúl Uplift exposures of the Guacamayos Group have been studied paleomagnetically by MacDonald and Opdyke (1974). These are rhyolites, rhyodacites and related rocks for which they report K-Ar whole-rock ages of about 195 m.y. (Jurassic). NRM intensities are high and vectors are down and to the north. In field position, without demagnetization or bedding corrections, these directions are in the general direction of the present Earth's field and would reinforce the induced magnetization. Although the outcrop area of the Guacamayos Group is outside the area flown, the anomalies just to the south and west of El Baúl could well be related to the presence of volcanics similar to those

at El Baúl. In addition, Jurassic igneous rocks are also a possible cause of the special character of the series of anomalies in the zone 50 to 100 km south of El Baúl (stippled area, Fig. 3), which straddles the proposed Apure fault. Younger Jurassic basalts (162 m.y.) are known from a well in the Espino Graben 300 km to the east (Feo-Codecido and others, 1984), so that potential post-Paleozoic sources of magnetic anomalies are present over a wide area east of the Andes.

VII. Río Meta Province.

South of the zone cited above as the boundary of the Barinas and Río Meta provinces (Fig. 3), the general character of the magnetic anomalies changes to that of greater amplitude and shorter wavelength, characteristic of patterns over the Guayana Shield to the east (from maps provided to the author by E. Herrero, MEM). Part of this is no doubt caused by the shallowing of the basement approaching the shield, but the rather abrupt change in character, best seen on the map (Fig. 1), suggests a difference in the character of basement or an abrupt shallowing of basement. It is tempting to relate this to an extension of the Altamira fault of Feo-Codecido and others (1984). Their location would have to be moved about 100 km northward to coincide with the province boundary (located on Fig. 3). The fault they proposed is a normal fault, down to the north. It is shown by them as a possible extension of the southern boundary fault of the Espino Graben, in which Jurassic redbeds and basalt have been found in a well. Thus, it seems reasonable that the Barinas-Río Meta province boundary (Fig. 3) could be the location of the southern boundary fault of a graben or graben complex similar to the Espino Graben. The aeromagnetics support prediction of a fault, down to the north, at the location proposed here. North of the fault lower Paleozoic and possibly Jurassic rocks are preserved on Precambrian basement, whereas south of the fault in the Río Meta province, Guayana Shield Precambrian rocks lie under thin, more recent sedimentary cover.

DISCUSSION AND CONCLUSIONS

Magnetic provinces, which have been delineated in western Venezuela based on trends and the character of magnetic anomalies, identify areas of geologic similarity. The Guajira and Paraguaná peninsulas, Sierra de Perijá, and the basement of western Maracaibo Basin have been proposed as accretionary terranes (e.g., Klitgord and Schouten, 1982; Sykes and others, 1982; Maze, 1984) versus the eastern Maracaibo Basin and the Venezuelan Andes. Where induced magnetization is dominant, the aeromagnetic anomalies reflect the current geometric distribution of magnetic rocks and do not supply critical data in identifying accretionary terranes. However, if remanence dominates and its vector is in a different direction from that of the present Earth's field, the magnetic anomalies produced will reflect this different direction. Thus, analyses of these NRM-dominated field magnetic anomalies could be a useful tool in identification of these terranes.

There is no evidence in Figure 1 unequivocally pointing to an accretionary terrane.

Although there are too many source parameters to make positive identifications or confident models at this stage, some observations and suggestions can be made.

1. Caribbean-South American Plate Boundary

The east-west magnetic trends of the Guajira-Paraguaná magnetic province are terminated abruptly by northeasterly Perijá province trends in the western Gulf of Venezuela south of the Guajira Peninsula coast, and less abruptly by northwesterly trends of the Coro province on the southern Paraguaná Peninsula. This line of demarcation (Fig. 3) appears to be a fault zone—the Guajira-Paraguaná fault zone—separating a northern zone or block from the northern Perijá and Coro provinces to the south.

The proposed fault zone is under Cenozoic cover and is nowhere exposed. On the Paraguaná Peninsula it coincides with a fault shown on several maps (e.g., Bellizzia and others, 1976), and on the Guajira Peninsula it is projected to be south of the pre-Tertiary exposures. The proposed fault zone is parallel to subparallel to the Oca fault to the south and to the Cuiza fault, which is 30 km to the north and possibly a branch of it. These and possibly other east-west zones to the north may separate the pre-Tertiary blocks on the Paraguaná and Guajira peninsulas, possibly causing rotation or tilting in the process. This is one explanation for the differential displacement between units on these peninsulas as indicated by paleomagnetic measurements.

The paleomagnetic declination data suggest that there has been a rotation of the Guajira block by 90° relative to that of Paraguaná (MacDonald and Opdyke, 1972; Skerlec and Hargraves, 1980). The aeromagnetic data do not cover more than a small portion of the Guajira Peninsula and do not cover the area of the MacDonald and Opdyke (1972) study, and thus the east-west anomalies there and eastward in the Gulf of Venezuela may not be characteristic of the rest of the peninsula. However, there is a strong suggestion from the aeromagnetic trends (and thus the inferred geologic trends) that eastern Guajira, Paraguaná, and the intervening Gulf of Venezuela have been a single tectonic province since acquisition of these lineations. The paleomagnetic data suggest that this must be later than the time of emplacement (intrusion or tectonic) of the Santa Ana Complex in Early Cretaceous time, e.g., after the arrival of those crustal segments to their present relative position.

Muessig (1984) has proposed that during the Oligocene and Miocene, the Falcón Basin and adjacent offshore basins developed within a zone of extensional tectonics. He calculated that a minimum of 35 to 40 km of extension occurred in the area of the Paraguaná and Guajira peninsulas and the Venezuelan and Netherlands Antilles. The results of these right-lateral offsets on east-west transcurrent faults, e.g., Oca and Cuiza faults, produced Tertiary differential motion of crustal units between these major zones of dislocation. Within each crustal unit there are fault blocks, east-west faults, some in sets, faults oblique to the east-west shear direction, and probably intrusions into the crust parallel to these features. These features could and probably do give rise to strong magnetic anomalies and trends, of which some are parallel (northern Paraguaná, Oca zone), and some are oblique (Coro and Falcón Basin) to the major east-west shear trend. The magnetic anomalies document a complex zone of Neogene interaction on this margin of the Caribbean-South American plate boundary between the Oca fault and north of the Paraguaná Peninsula.

2. South of the Oca Fault

The northeasterly trends of the Perijá province appear related to the similar geologic and tectonic trends in the Sierra de Perijá, and to the Jurassic La Quinta graben complex, which includes the Central Lake Graben. Major magnetic anomalies probably come from the Jurassic volcanics or are associated with deeper sources related to graben development.

The eastern Maracaibo Basin and Falcón Basin south of the Coro province, show modest, but trendless anomalies. The lack of significant anomalies over the Paleozoic metamorphic terrane of the Andes suggests that similar rocks might also underlie this area.

The Barinas and Río Meta magnetic provinces are contrasted mainly on the basis of character changes in anomalies along a lineation trending N30°E, which is thought to be the southern fault margin of a graben preserving unmetamorphosed lower Paleozoic sedimentary rocks to the north. This is on strike with the Espino Graben 300 km to the northeast and appears to be a major feature of the interior plains (llanos) of the Guayana Shield. Some magnetic anomalies near El Baúl could be anomalies produced by strongly magnetized Jurassic volcanics similar to those exposed nearby in the El Baúl Uplift.

A primary subdivision of magnetic provinces in western Venezuela is two-fold. On the one hand, from the area north of the Paraguaná Peninsula to south of the Oca fault and the Coro area, these magnetic anomalies document a complex zone of interaction related to major east-west shear forces in Cenozoic time along this segment of the Caribbean-South American plate boundary. On the other hand, to the south the magnetic anomalies are related to the distribution of pre-Cretaceous rocks. This is an area which exhibits normal faults, thrusts, and basin depressions, some of which are related to the Caribbean-South American plate boundary interactions (Sierra de Perijá, Andes, Maracaibo and Falcón basins), but others (Apure and Altamira faults) are related to earlier pre-Cretaceous development of the igneous and metamorphic terrane north of the Guayana Shield.

ACKNOWLEDGMENTS

This study would not have been possible without the interest and encouragement of A. Bellizzia of Lagoven, formerly director, Dirección de Geología, and E. Herrero, Jefe, División Física de la Tierra, Dirección de Investigones, Geoanalíticas y Technológicas,

Venezuela Ministerio de Energía y Minas, Caracas. Particularly helpful were B. S. Hastings and J. O'Brien, Gulf Oil Exploration and Production, Houston; M. Blake, P. Bauman, C. Rine, J. Bialkowski, K. Bilous, and J. Wyckoff of Princeton University; and R. Shagam, R. B. Hargraves, W. J. Hinze, and W. J. Morgan who gave helpful criticism of the manuscript. This work was supported by the National Science Foundation under grant EAR 81-15833.

REFERENCES CITED

Bartok, P., Reijers, T.J.A., and Juhasz, I., 1981, Lower Cretaceous Cogollo Group, Maracaibo Basin, Venezuela: Sedimentology, diagenesis, and petrophysics: American Association of Petroleum Geologists Bulletin, v. 65, p. 1110–1134.

Bellizzia, A., Pimentel, N., and Bajo, R., compilers, 1976, Mapa Geologico Estructural de Venezuela: Venezuela Ministerio de Minas e Hidrocarburos, scale 1;500,000.

Bonini, W. E., Garing, J. D., and Kellogg, J. N., 1980, Late Cenozoic uplifts of the Maracaibo-Santa Marta block, slow subduction of the Caribbean plate, and results from a gravity study, *in* Transactions, Caribbean Geological Conference, IX, Santo Domingo, Volume 1: Universidad Católica Madre y Maestra, Santiago de los Caballeros, Republica Dominicana, p. 99–105.

Case, J. E., 1974, Major basins along the continental margin of northern South America, *in* Burke, C. A., and Drake, C. L., eds., The geology of continental margins: Springer-Verlag, New York, p. 733–741.

Case, J. E., and Holcombe, T. L., 1980, Geologic-tectonic map of the Caribbean region: U.S. Geological Survey Miscellaneous Investigations Map I-1100, scale 1:2,500,000.

Case, J. E., Holcombe, T. L., and Martin, R. G., 1984, Map of geologic provinces in the Caribbean region, *in* Bonini, W. E., and others, eds., The Caribbean-South American plate boundary and regional tectonics: Geological Society of America Memoir 162 (this volume).

Case, J. E., and MacDonald, W. D., 1973, Regional gravity anomalies and crustal structure in northern Colombia: Geological Society of America Bulletin, v. 84, 2905–2916.

Feo-Codecido, G., 1972, Breves ideas sobre la estructura de la falla de Oca, Venezuela, *in* Transactions, Caribbean Geological Conference, VI, Margarita, Venezuela, 1971, p. 184–190.

Feo-Codecido, G., Smith, F. D., Jr., Aboud, N., and de Di Giacomo, E., 1984, Basement and Paleozoic rocks of the Venezuelan Llanos basins, *in* Bonini, W. E., and others, eds., The Caribbean-South American plate boundary and regional tectonics: Geological Society of America Memoir 162 (this volume).

Jordan, T., 1975, The present-day motions of the Caribbean plate: Journal of Geophysical Research, v. 80, p. 4433–4439.

Kellogg, J. N., 1981, The Cenozoic basement tectonics of the Sierra de Perijá, Venezuela and Colombia [Ph.D. thesis]: Princeton, New Jersey, Princeton University, 236 p.

—— 1984, Cenozoic tectonic history of the Sierra de Perijá and adjacent basins, *in* Bonini, W. E., and others, eds., The Caribbean-South American plate boundary and regional tectonics: Geological Society of America Memoir 162 (this volume).

Kellogg, J. N., and Bonini, W. E., 1982, Subduction of the Caribbean plate and basement uplifts in the overriding South American plate: Tectonics, v. 1, p. 251–276.

Klitgord, K. D., and Schouten, H., 1982, Early Mesozoic Atlantic reconstructions from sea-floor-spreading data [abstract]: EOS (Transactions of the American Geophysical Union), v. 63, no. 18, p 307.

Lindsley, D. H., Andreasen, G. E., and Balsley, J. R., 1966, Magnetic properties of rocks and minerals, *in* Clark, S. P. Jr., ed., Handbook of Physical Constants: Geological Society of America Memoir 97, p. 543–552.

MacDonald, W. D. and Opdyke, N. D., 1972, Tectonic rotations suggested by paleomagnetic results from northern Colombia, South America: Journal of Geophysical Research, v. 77, p. 5720–5730.

—— 1974, Triassic paleomagnetism of northern South America: American Association of Petroleum Geologists Bulletin, v. 58, p. 208–215.

Malfait, B. T., and Dinkelman, M. G., 1972, Circum-Caribbean tectonic and igneous activity and the evolution of the Caribbean plate: Geological Society of America Bulletin, v. 83, p. 251–272.

Maze, W. B., 1984, Jurassic La Quinta formation in the Sierra de Perijá, northwestern Venezuela: Geology and tectonic environment of the red-beds and volcanics, *in* Bonini, W. E., and others, eds., The Caribbean-South American plate boundary and regional tectonics: Geological Society of America Memoir 162 (this volume).

Maze, W. B., and Hargraves, R. B., 1984, Paleomagnetic results from the Jurassic La Quinta formation in the Perijá range, Venezuela, and their tectonic significance, *in* Bonini, W. E., and others, eds., The Caribbean-South American plate boundary and regional tectonics: Geological Society of America Memoir 162 (this volume).

Miller, J. B., 1962, Tectonic trends in Sierra de Perijá and adjacent parts of Venezuela and Colombia: American Association of Petroleum Geologists Bulletin, v. 46, p. 1565–1595.

Moticska, P., 1977, Generación de magmas y autometamorfismo por combustión subterranea de carbones y limolitas carbonosas de la formación Marcelina, Perijá, *in* Memoria, Congreso Geológico Venezolano, V, Caracas, Volume 2: Venezuela Ministerio de Energía y Minas, p. 663–691.

Muessig, K. W., 1984, Structure and Cenozoic tectonics of the Falcón basin, Venezuela, and adjacent areas, *in* Bonini, W. E., and others eds., The Caribbean-South American plate boundary and regional tectonics: Geological Society of America Memoir 162 (this volume).

Onstott, T. C., and Hargraves, R. B., 1980, Paleomagnetic data and the Proterozoic apparent polar wander curve for the Venezuelan Guayana shield, *in* Transactions, Caribbean Geological Conference, IX, Santo Domingo, Volume 2: Universidad Católica Madre y Maestra, Santiago de los Caballeros, Republica Dominicana, p. 475–508.

Perez, O. J., and Aggarwal, Y. P., 1981, Present-day tectonics of the southeastern Caribbean and northeastern Venezuela: Journal of Geophysical Research, v. 86, p. 10, 791–10, 804.

Peter, G., 1972, Geologic structure offshore north-central Venezuela, *in* Transactions, Caribbean Geological Conference, VI, Margarita, Venezuela, 1971, p. 283–294.

Reford, M. S., 1964, Magnetic anomalies over thin sheets: Geophysics, v. 29, p. 532–536.

Rod, E., 1956, Strike-slip faults of northern Venezuela: American Association of Petroleum Geologists Bulletin, v. 40, p. 457–476.

Silver, E., Case, J. E., and MacGilavry, H. J., 1975, Geophysical study of the Venezuelan borderland: Geological Society of America Bulletin, v. 86, p. 213–226.

Skerlec, G. M., and Hargraves, R. B., 1980, Tectonic significance of paleomagnetic data from northern Venezuela: Journal of Geophysical Research, v. 85, p. 5303–5315.

Slichter, L. B., 1942, Magnetic properties of rocks, *in* Special Papers, Number 36: Geological Society of America, p. 293–298.

Sykes, L. R., McCann, W. R., and Kafka, A. L., 1982, Motion of Caribbean plate during last 7 million years and implications for earlier Cenozoic movements: Journal of Geophysical Research, v. 87, p. 10,656–10,676.

Tschanz, C. M., Marvin, R. F., Cruz, J., Mehnert, H. H., and Cebula, G. T., 1974, Geologic evolution of the Sierra Nevada de Santa Marta, northeastern Colombia: Geological Society of America Bulletin, v. 85, p. 273–284.

Vasquez, E., and Dickey, P., 1972, Major faulting in northwestern Venezuela and its relation to global tectonics, *in* Transactions, Caribbean Geological Conference, VI, Margarita, Venezuela, 1971, p. 191–202.

Vierbuchen, R. C., 1979, A working hypothesis for the post-Miocene tectonics of the southeastern Caribbean region: EOS (Transactions of the American Geophysical Union), v. 60, no. 18, p. 395.

—— 1984, The geology of the El Pilar fault zone and adjacent areas in northeastern Venezuela, *in* Bonini, W. E., and others, eds., The Caribbean-South American plate boundary and regional tectonics: Geological Society of America Memoir 162 (this volume).

Ward, D. E., Goldsmith, R., Cruz, J., and Restrepo, H., 1973, Geología de los cuadrángulos H-12 Bucaramanga y H-13 Pamplona, Departamento de Santander: Colombia Instituto Nacional de Investigaciones Geológico-Mineras, Ministerio de Minas y Petróleos, Bogotá, Boletín Geológico, v. 21, no. 1-3, p. i-132.

MANUSCRIPT ACCEPTED BY THE SOCIETY SEPTEMBER 1, 1983

Geological Society of America
Memoir 162
1984

Basement and Paleozoic rocks of the Venezuelan Llanos basins*

Gustavo Feo-Codecido
Corpoven, S.A.
Apartado 61373
Caracas, Venezuela

Foster D. Smith, Jr.
Servicios Técnicos Mobil C.A.
Apartado 60167
Caracas, Venezuela

Nelson Aboud
Corpoven, S.A.
Apartado 61373
Caracas, Venezuela

Estela de Di Giacomo
Maraven, S.A.
Apartado 829
Caracas, Venezuela

ABSTRACT

The igneous and metamorphic rocks of Venezuela may be classified in three geographic units: a southern Precambrian Shield, an intermediate belt of Paleozoic age and a northern border of Mesozoic to Tertiary age. The shield is exposed in the south; it extends northward under a sedimentary cover to an inferred contact with the metamorphosed Paleozoic or Mesozoic-Tertiary rocks. South of this contact, in the eastern states of Anzoátegui and Monagas, the shield's oldest known sedimentary cover is of Cretaceous age; to the west, it is of Cambrian age.

A graben is identified in the State of Guárico, located approximately along the contact of the Paleozoic basement with the Precambrian Shield. It contains a 2,390 m (7,840 ft) column of sedimentary rocks of Carboniferous and Jurassic age, not previously identified in the Barinas-Apure and Eastern Venezuela basins. Preliminary interpretation of recently acquired aeromagnetic data indicates that this feature forms part of a much longer graben, hitherto not known, which may extend westward from near Barcelona, Anzoátegui, to the Colombian border at approximately 70° 00' West Longitude, 7° 10' North Latitude, a distance of 600 km (375 mi).

*Published by permission of Petróleos de Venezuela, S.A. (PDVSA), Corpoven, S.A., and Maraven, S.A. (Affiliates of PDVSA).

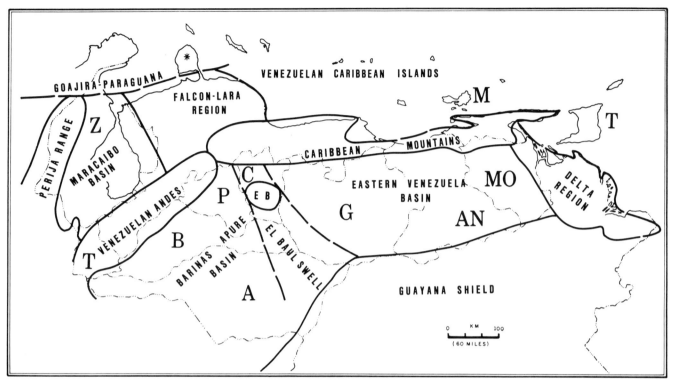

Figure 1. Geomorphological provinces of northern Venezuela and location of States referred to in the text: Z - Zulia, A - Apure, G - Guárico, T - Táchira, P - Portuguesa, AN - Anzoátegui, B - Barinas, C - Cojedes, MO - Monagas, M - Island of Margarita, T - Trinidad, EB - El Baúl Massif, *Paraguaná Peninsula.

INTRODUCTION

The term "basement" is used to denote a complex of igneous and metamorphic rocks, undifferentiated in this report except as to radiometric age. For many years the basement of the Venezuelan Llanos basins, individually known as the Barinas-Apure and Eastern Venezuela basins (Fig. 1), was interpreted to be a relatively uncomplicated extension of the Precambrian Guayana Shield, with inliers of the pre-Cretaceous unmetamorphosed sedimentary rocks of the Carrizal and Hato Viejo formations (Renz and others, 1958, p. 561, 567). This interpretation has undergone changes, but with little formal published documentation.

The publications most pertinent to this paper are the Ministry of Mines and Hydrocarbons (now Ministerio de Energía y Minas) geologic-structural 1:500.000-scale map of Venezuela (Bellizzia and others, 1976) and the treatise on the geology of Venezuela by González de Juana and others (1980). Both show basement depressions in excess of 2,745 m (9,000 ft) subsea in the western part of the Eastern Venezuela basin; González de Juana and others (1980, p. 141) give a section showing some 1,220 m (4,000 ft) of Paleozoic Carrizal and Hato Viejo fill. This paper offers evidence for the following interpretations:

An igneous-metamorphic basement of Paleozoic age is present between the southern Precambrian Shield and the northern complex of allochthonous rocks of Mesozoic to Tertiary age (Fig. 2). It is documented from the states of Zulia, Táchira, and Barinas eastward to central Guárico (Fig. 1). It may be present in the subsurface from central Guárico eastward across northern Venezuela at depths not yet drilled.

The Paleozoic basement from Barinas to central Guárico includes granites, syenites, schists, and other metamorphics which have yielded radiometric age determinations ranging from 433 to 277 m.y. in age. Beyond the southern border of this orogenic belt of Late Ordovician to Permian age, the Paleozoic rocks are represented by the Cambrian sedimentary rocks of the Carrizal Formation. The apparent sharp division between the two units suggests a tectonic contact.

A graben is present in southern Guárico, with Precambrian(?) basement estimated to be at a depth of the order of 6,100 m (20,000 ft) subsea. Petróleos de Venezuela S.A. aeromagnetic data, now being interpreted (March 1983), suggest that this graben extends across Guárico, from Barcelona, Anzoátegui to the Colombian border. In southern Guárico, this "Espino Graben" contains 1,643 m (5, 390 ft) of red beds, provisionally referred to as the La Quinta Formation, intercalated with a 113 m (370 ft) thick basaltic flow dated at 162 m.y. The deepest well, Maraven's NZZ-88X (17, Figs. 4, 5), penetrated below the red beds into an interval with Carboniferous flora at 3,435 m (11,268 ft) subsea followed below by barren sediments (not Carrizal Formation) to a total depth of 4,178 m (13,708 ft) subsea.

In the southern part of the eastern Llanos basins, the Piarra

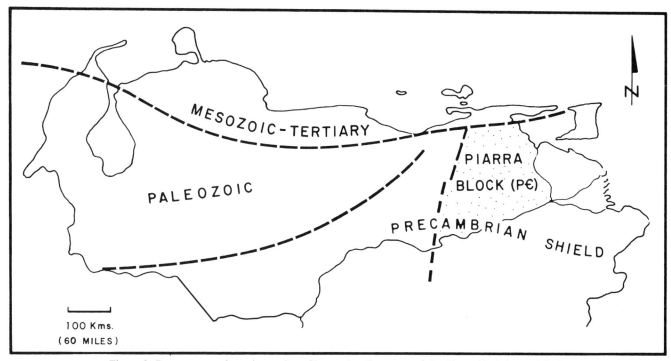

Figure 2. Basement provinces in northern Venezuela. Basement of the Piarra block is overlain by Cretaceous or Tertiary sediments.

block of eastern Anzoátegui and Monagas, Cretaceous and Tertiary sediments lie on basement of Precambrian age. It is not known whether this basement continues north to a contact with the Mesozoic-Tertiary basement of the coast, or to a contact with an interposed belt of Paleozoic basement.

Previous Work

The Guayana Shield, the crystalline rocks of the Andes and Perijá ranges, the Mesozoic-Tertiary basement near the coast and the El Baúl outcrop of the Llanos basins have been described in publications which are summarized by González de Juana and others (1980, p. 31-89, 117-118, 122-124, 142-152, 305-407). We abstract as follows:

The Guayana Shield records four periods of orogenesis (Mendoza, 1977, p. 2237):

Gurian	3,600-2,700 m.y.
Pre-transamazon	2,600-2,100 m.y.
Transamazon	2,000-1,700 m.y.
Orinocan	1,200-800 m.y.

The reported ages of the crystalline rocks of the Andes range from 720 to 225 m.y. Precambrian and Paleozoic rocks are present, as well as the Lower Mesozoic El Carmen granodiorite. The Paleozoic rocks range from 500 to 390 m.y. in age. Two ages, 310 and 370 m.y., are reported from Perijá.

With regard to the allochthonous Mesozoic-Tertiary complex (Bellizzia, 1972) of the Caribbean Mountains, Santamaría and Schubert (1974) place the igneous rocks of the Paraguaná

Peninsula, the islands off the Venezuelan coast and the Caribbean Mountains in two groups: 130 to 114 m.y. and 84 to 30 m.y., plus a 5-m.y. age for dacites on the coast south of the Island of Margarita. González de Juana and others (1980) cite Mesozoic to Tertiary ages for metamorphic units, plus the Paleozoic Sebastopol-El Tinaco Complex of 425 m.y.

The El Baúl massif exposure in Cojedes has been described by Feo-Codecido (1954, 1961, 1963), Martín Bellizzia (1961, 1968), Bellizzia and others (1976), and González de Juana and others (1980). An alkaline granite of 287 ± 10 (Rb/Sr) to 270 ± 10 m.y. (K/Ar) has produced a metamorphic aureole in sediments which include the trilobite-bearing Mireles Formation of Tremadocian age (Cambro-Ordovician) (Rod, 1955; Frederickson, 1958; González de Juana and others, 1980, p. 118–119, 152).

A later thermal event also occurred in the same area of El Baúl, and is represented by the Triassic-Jurassic Guacamayas Group volcanics whose absolute age is reported as 192 ± 3.8 and 195 ± 3.9 m.y. (K/Ar) (MacDonald and Opdyke, 1974, p. 208; González de Juana and others, 1980, p. 174).

This paper adds new age data (Table 1), data on significant wells (Table 2), and interpretations with regard to the basement and Paleozoic sedimentary rocks of the Llanos basins.

DISCUSSION AND INTERPRETATION

The radiometric ages given in Table 1 may represent the age of original crystallization, the age of the last deformation, or possibly a mixed age as a result of partial alteration of an older

TABLE 1. NEW AGE DATA: IGNEOUS AND METAMORPHIC ROCKS

Location	Well	Lithology	Method	Age (m.y.)	Date
34	MCH-12-7X	Leucogranite	K feldspar (K/Ar)	1025 ± 31	1982*
19	SDZ-78X	Salmon colored	Biotite (K/Ar)	1366 ± 41	1982*
19	SDZ-78X	Gneissoid granite	K feldspar (K/Ar)	1110 ± 33	1982*
19	SDZ-68X	Gneissoid granite	? (K/Ar)	869 and 708	1982*
38	SCZ-27X	Gneissoid granite	Biotite (K/Ar)	1772 ± 53	1982*
38	SCZ-75X	Granite diorite	Whole rock (K/Ar)	1397 ± 42	1982*
38	SCZ-82X	Granite	Biotite (K/Ar)	1264 ± 38	1982*
38	SCZ-82X	Granite	Biotite (K/Ar)	1497 ± 45	1982*
22	Hamaca-2	Quartz diorite gneiss	Rb/Sr isochron	1785 ± 15	1976*
24	CN wells	Quartz monzonite	Rb/Sr isochron	1900 ± 10	1980+
4	Agua Linda-1	Granite pegmatite	Feldspar (Rb/Sr)	865 ± 60	1966*
4	Agua Linda-1	Granite pegmatite	Muscovite (Rb/Sr)	463 ± 45	1966*
4	Agua Linda-1	Schist	Muscovite (K/Ar)	411 ± 12	1966*
4	Agua Linda-1	Schist	Muscovite (K/Ar)	406 ± 8	1966*
4	Agua Linda-1		Rb/Sr isochron	433 ± 50	1966*
7	SZW-3	Gneiss or schist	Muscovite (K/Ar)	412 ± 21	1981+
13	GXB-2	Granite	Muscovite (K/Ar)	330	1979*
13	GXB-8	Granite	Muscovite (K/Ar)	321	1979*
14	MCH-2-3X	Pelitic hornfels	Biotite (K/Ar)	347 ± 10	1982*
17	NZZ-88X	Basalt flow	Whole rock (K/Ar)	162 ± 8	1982+
2	QMC-1X	Granite	Whole rock (K/Ar)	138.4 ± 6.9	1979+

Note: Numbers in location column refer to well locations on Figures 4 and 5.
*Analysis by a U. S. oil company laboratory.
+Analysis by a U. S. commercial laboratory.

TABLE 2. WELL DATA

Location	Well	Lat (N)	Long (W)	Remarks
2	QMC-1	10° 58'35"	70° 59'14"	4259 m (13974 ft), granite, 138 m.y.
3	SLC-1-2X	9° 30'39"	71° 31'01"	5646 m (18524 ft), schist.
3	SLC-1-5X	9° 28'44"	71° 32'55"	5801 m (19031 ft), schist.
4	Agua Linda-1	7° 37'25"	70° 54'31"	3537 m (11603 ft), pegmatite and schist, 433 m.y. isochron.
8	Apure-1	7° 39'36"	70° 01'34"	N.P. at 2683 m (8804 ft), top Carrizal Fm. at 2504 m (8215 ft).
5	Apure-2	7° 14'09"	70° 29'11"	N.P. at 2290 m (7514 ft), top Carrizal Fm. at 2271 m (7452 ft).
12	Aprue-3	7° 53'47"	69° 22'14"	N.P. at 1896 m (6221 ft), top Carrizal Fm. at 1851 m (6073 ft).
26	Nutrias-1	8° 04'55"	69° 19'01"	N.P. at 2093 m (6868 ft), top Carrizal Fm. at 2050 m (6727 ft).
6	La Heliera-1	6° 15'-	70° 59'-	N.P. at 2597 m (8519 ft).
7	SZW-3	8° 08'48"	70° 09'35"	2760 m (9055 ft), gneiss or schist, 412 m.y.
9	Ticoporo-1	7° 53'17"	69° 58'04"	2632 m (8635 ft), schist.
10	23-M-2X	11° 40'17"	69° 31'20"	2399 m (7879 ft), gneiss, 114 m.y.
11	15'GU-501	9° 08'01"	69° 16'56"	1035 m (3396 ft), granite.
13	GXB-2	9° 18'07"	66° 56'22"	1552 m (5093 ft), granite, 330 m.y.
13	GXB-8	9° 22'05"	66° 56'22"	1719 m (5639 ft), granite, 321 m.y.
29	Gorrin-1	8° 40'59"	66° 53'52"	855 m (2805 ft), hornfels.
15	Machete-2X	8° 33'40"	66° 27'30"	806 m (2643 ft), hornfels.
14	MCH-2-3X	8° 50'05"	66° 40'00"	1105 m (3624 ft), hornfels, 347 m.y.
30	MCH-2-4X	8° 57'05"	66° 41'00"	1434 m (4706 ft), syenite.
28	MCH-2-5X	8° 51'05"	66° 52'00"	1014 m (3325 ft), syenite.
31	MCH-8-5X	8° 45'17"	65° 59'46"	N.P. at 3750 m (12303 ft), top Carrizal Fm. at 2553 m (8375 ft).
34	MCH-12-7X	8° 09'25"	65° 41'15"	251 m (824 ft), granite, 1025 m.y.
27	MCH-13-1X	7° 59'15"	67° 26'25"	912 m (2991 ft), hornfels.
38	SCZ-27X	8° 21'58"	64° 36'41"	508 m (1666 ft), gneissoid granite, 1772 m.y.
38	SCZ-75X	8° 27'32"	64° 38'36"	709 m (2325 ft), granite diorite, 1397 m.y.
38	SCZ-82X	8° 26'10"	64° 37'00"	661 m (2170 ft), granite, 1497 m.y.
*	SDZ-11X	8° 26'00"	64° 48'40:	N.P. at 669 m (2196 ft), top Carrizal Fm. at 638 m (2094 ft).
37	SDZ-16X	8° 32'45"	64° 49'10"	N.P. at 979 m (3213 ft), top Carrizal Fm. at 917 m (3073 ft).
19	SDZ-43X	8° 17'04"	64° 54'02"	N.P. at 640 m (2099 ft), top Carrizal Fm. at 428 m (1405 ft).
19	SDZ-68X	8° 14'37"	64° 51'46"	361 m (1185 ft), gneissoid granite, 869 m.y.
19	SDZ-78X	8° 13'48"	64° 58'18"	351 m (1150 ft), gneissoid granite, 1366 m.y.
+	NZZ-12X	8° 39'55"	65° 21'00"	N.P. at 2238 m (7341 ft), top Carrizal Fm at 1354 m (4441 ft).
33	NZZ-45X	8° 25'13"	65° 37'15"	N.P. at 631 m (2071 ft), top Carrizal Fm. at 589 m (1931 ft).
17	NZZ-88X	8° 46'59"	65° 23'25"	N.P. at 4178 m (13708 ft), basalt flow at 1838 m (6030 ft), 162 m.y.
18	Carrizal-1	8° 33'01"	65° 04'09"	N.P. at 1472 m (4830 ft), top Carrizal Fm. at 985 m (3232 ft).
18	Carrizal-2X	8° 35'01"	65° 03'19"	N.P. at 2912 m (9553 ft), top Carrizal Fm. at 1084 m (3558 ft).
16	Santa Rita-1	8° 03'40"	·66° 14'12"	455 m (1492 ft), granite.
36	Suata-1	8° 15'22"	65° 12'37"	N.P. at 758 m (2487 ft), top Carrizal Fm. at 334 m (1096 ft).
36	Zuata Este-1X	8° 16'12"	65° 12'40"	1155 m (3788 ft), "basalt."
35	Tres Matas-1	8° 27'54"	65° 23'43"	N.P. at 784 m (2571 ft), top Carrizal Fm. at 771 m (2529 ft).
32	Socorro-1	8° 57'19"	65° 43'00"	N.P. at 1901 m (6236 ft), top Carrizal Fm. at 1738 m (5702 ft).
38	Hato Viejo-1	8° 24'09"	64° 42'31"	710 m (2330 ft), chlorite schist and granite.

TABLE 2. WELL DATA (continued)

Location	Well	Lat (N)	Long (W)	Remarks
38	Maco-1	8°25'50"	64°39'45"	799 m (2622 ft), "igneous-metamorphic."
38	Santa Clara-1	8°26'01"	64°38'34"	670 m (2197 ft), schist.
22	Hamaca-2X	8°35'52"	69°19'19"	1201 m (3940 ft), quartz diorite gneiss, 1785 m.y. isochron.
39	SG-139-X	9°05'28"	64°20'48"	N.P. at 3634 m (11921 ft), top Carrizal Fm. at 3420 m (11219 ft).
20	OG-338-S	8°52'41"	64°24'13"	N.P. at 2922 m (9587 ft), top Carrizal Fm. at 2217 m .(7273 ft).
21	WGDT-1	8°42'31"	63°15'03"	2654 m (8706 ft), gneissoid granite, 5 ft of greenstone.
23	Soledad-1(CVP)	9°17'46"	62°54'27"	4267 m (13998 ft), "granite."
25	Guarao-1	9°13'15"	60°18'45"	N.P. at 5273 m (17300 ft).
1	Mara field	9°02'-	73°43'-	Approx 3050 m (10000 ft), granite, 304 m.y.
24	CN wells	8°42'-	63°15'-	Approx. 1060 m (3500 ft), quartz monzonite, 1900 m.y. isochron.

Note: Numbers in location column refer to well locations on Figures 4 and 5. Approximate locations are also given in geographic coordinates. Remarks include depth to basement, lithology, absolute age, top Carrizal Formation. N.P. = basement not penetrated. All depths in meters and feet subsea. Wells with location numbers 26 to 30 are shown on Figure 5 only.
  *Well SDZ-11X is approximately 13 km south of SDZ-16X, location 37.
  +Well NZZ-12X is approximately 14 km south-southeast of NZZ-88X, location 17.

rock by a much younger deformation. The Paleozoic ages given in Table 1 may be the result of later metamorphism of Precambrian rocks of the Guayana Shield and also metamorphism of earlier Paleozoic sedimentary rocks, including the Cambrian Carrizal Formation. Our interpretation of these data is as follows:

Precambrian rocks

The northern limit of the Guayana Shield may be at the boundary with the Paleozoic or Mesozoic-Tertiary basement (Fig. 2), or it may be that those younger igneous and metamorphic rocks are thrust over the Precambrian Shield which itself continues northward at depth. The Precambrian Iglesias Group of the Andes (González de Juana and others, 1980, p. 93-94) and the 865 m.y. pegmatite feldspar in well Agua Linda-1 (4, Figs. 4, 5) suggest that the shield formerly extended northward beyond its presently ascertainable northern limit. If the feldspar was a Rb-poor feldspar, the 865-m.y. age could be anomalously old. This could be caused by resetting of the whole-rock system at about 400 m.y. or younger, such that Sr was accepted into the feldspar, possibly from muscovite.

A large area of uplifted Precambrian basement, the Piarra block (Fig. 2), occupies eastern Anzoátegui and Monagas. More than 50 wells confirm that Cretaceous or younger sediments lie directly on Precambrian basement as far north as well Soledad-1 (23, Figs. 4, 5) of the Corporación Venezolana del Petróleo.

The western boundary of the block is placed along the trend of the Trico fault (Fig. 4). To the west of the boundary, in the Cambrian basin (Fig. 3), some 100 wells have penetrated the Tertiary and Cretaceous sedimentary sequence and entered the unmetamorphosed Carrizal Formation of Cambrian age. Our

data confirm a Precambrian age of the basement at Hamaca-2X (22, Figs. 4, 5) and the CN wells (24, Figs. 4, 5), but the northern boundary is unknown. The greenstone in WGDT-1 (21, Figs. 4, 5) might indicate a Paleozoic thermal event. In our maps the northern boundary is shown as being a contact with the Mesozoic-Tertiary basement under the eastern Caribbean Mountains, with no intervening Paleozoic basement. The eastern boundary may be along the Bohordal fault.

Munro and Smith (1984) show the basement of the Piarra block to be cut by the Urica fault of Late Tertiary age. Along that fault, in the Soledad-1 area (23, Figs. 4, 5), the basement of the northern block is depressed 2500 m (8200 ft) relative to the southern block. The map (Fig. 4) suggests a possible 100 km (60 mi) right-lateral offset along the Urica fault. Movement of this magnitude is *not* considered to have occurred.

Paleozoic rocks

Basement of acid igneous and metamorphic rocks of Paleozoic age is present in the area from the Perijá Range to western Barinas-Apure basin and thence east to Las Mercedes in central Guárico. A gravity high near Zaraza, Guárico, may mark its easternmost limit (Bellizzia and others, 1976; Bonini and others, 1977). The boundary with the northern Mesozoic-Tertiary basement is placed along the Oca fault zone (Fig. 4) where granites of the Mara field wells (1, Fig. 4) have been dated at 304-200 m.y. (González de Juana and others, 1980, p. 151), and to the south of well QMC-1 (2, Fig. 4) where a slightly foliated granite has been dated at 138 m.y. (Table 1), possibly a mixed age as a result of partial alteration of a much older granite (Paleozoic?) by a much younger deformation (Eocene?). This granite underlies in appar-

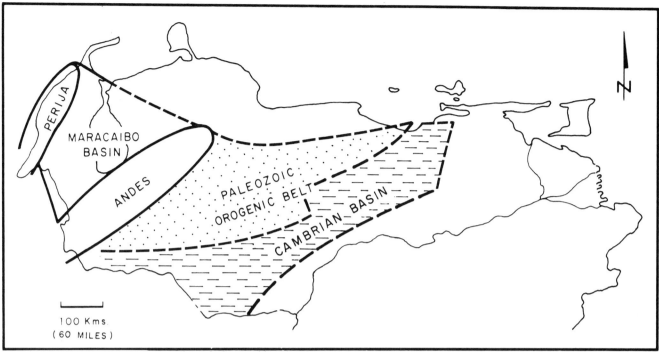

Figure 3. Geographic units of Paleozoic rocks in northern Venezuela.

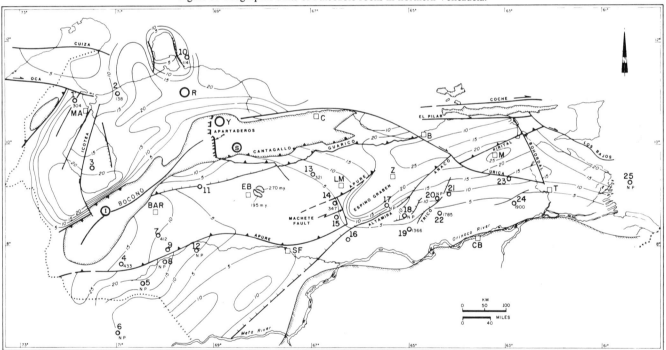

Figure 4. Depth to basement in thousands of feet in northern Venezuela, contour intervals at 5000 ft (1524 m) subsea. Maximum altitude above sea level 5007 m (16426 ft) west of Barinas and 2765 m (9071 ft) near Caracas. Barbs on downthrown side of normal faults. Triangles point in direction of fault plane dip on thrust faults. Towns indicated by squares: MA = Maracaibo, BAR = Barinas, C = Caracas, EB = El Baúl, SF = San Fernando, LM = Las Mercedes, Z = Zaraza, B = Barcelona, CB = Ciudad Bolívar, M = Maturín, T = Tucupita. Approximate location: I = Iglesias Group, S = Sebastopol-El Tinaco Complex, Y = Yaritagua complex, R = Xenoliths, (Muessig, 1978). Hachured outline = approximate area of outcrop of igneous and metamorphic rocks. Small circles and large numbers indicate well locations and location numbers; see Tables 1 and 2 for well data. Small numbers beside the well location indicate radiometric age of basement in millions of years. "N.P." indicates that the well did not enter basement.

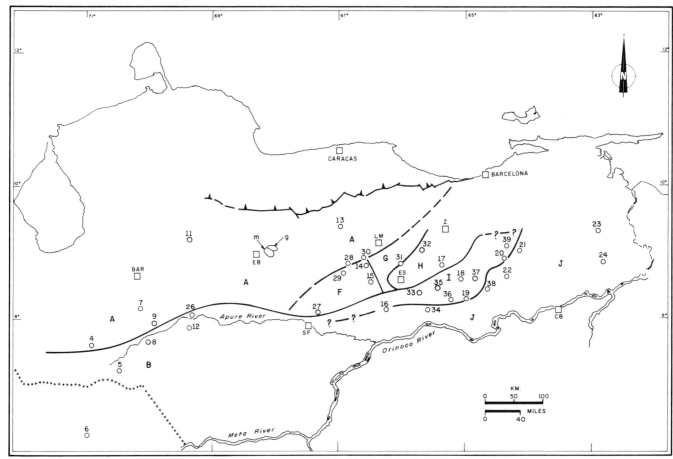

Figure 5. Generalized pre-Cretaceous geology of the Llanos basins. Squares = towns. ES = Espino, see Figure 4 caption for other town names. Letters A through J = Geologic areas. A = Paleozoic igneous or metamorphic rocks at well locations 4, 7, and 13 (Figs. 4, 5); El Barbasco Group metasediments (m, Fig. 5) and El Baúl granite (g, Fig. 5). B = Unfossiliferous Cambrian(?) Carrizal Formation, inferred to lie on Precambrian basement, fossiliferous lowermost Ordovician sediments at location 6 (Figs. 4, 5). F = Hornfels metasediments, of Devonian age at location 14 (Figs. 4, 5). G = Slightly metamorphosed, unfossiliferous Manapire facies of the Cambrian Carrizal Formation, inferred to lie on Precambrian basement. H = Espino Graben area. Carrizal Formation verified at locations 31 and 32 (Fig. 5), inferred to lie on Precambrian basement. Sedimentary rocks of Carboniferous and Jurassic age at location 17 (Figs. 4, 5). I = Fossiliferous Cambrian Formation, underlain by or interfingered with the Hato Viejo Formation along the eastern border and overlying inferred Precambrian basement. Carrizal Formation on Precambrian basement verified at location 19 (Figs. 4, 5). J = Precambrian basement, with absolute ages determined at well locations 34, 19, 38, 22, and 24 (Fig. 5). Small circles and numbers indicate well locations; see Tables 1 and 2 for well data.

ent erosional unconformity limestones of Cretaceous (probably Aptian) age. In the Falcón-Lara region this boundary must pass to the south of well 23-M-2X (10, Fig. 4) where the age of the gneiss has been reported as 114 m.y. (Cigego 1974, unpublished report), and thence east to the Apartaderos fault. This fault (apparently identified for the first time in this report) constitutes the abrupt western limit of the exposed Mesozoic-Tertiary basement of the Caribbean Mountains. The southernmost exposure of this basement is at the Cantagallo thrust zone where Mesozoic crystalline rocks are in overthrust contact with the Paleocene-lower Eocene flysch of the Piedmont Nappe which terminates at the Guárico thrust zone (Bellizzia, 1972; Stephan, 1977).

South of the Cantagallo thrust zone (Fig. 4), the nearest basement with an absolute age date determination is the El Baúl granite of Permian age (Feo-Codecido, 1963), so the boundary between the Paleozoic and the Mesozoic-Tertiary basement must be to the north of the El Baúl massif and the GXB-2 and GXB-8 wells (Table 1; 13, Figs. 4, 5). We infer that the Cantagallo thrust zone itself forms the southern limit of the crystalline rocks of Mesozoic age.

The westernmost outcrop of the Mesozoic-Tertiary crystalline rocks of the Caribbean Mountains is along the Apartaderos fault and its possible northern extension which exposes the metamorphics of the Yaritagua Complex (Y, Fig. 4) and related units

(González de Juana and others, 1980, p. 323-330). Foliated calc-silicate gneiss, granite, and schist xenoliths in trachybasalts (R, Fig. 4), reported by Muessig (1978, p. 264), suggest that, in northern Falcón, these rocks may extend westward in the subsurface to the longitude of the Paraguaná Peninsula.

The massive granite orthogneisses and schists of the Caribbean Mountains are not exposed west of the Apartaderos fault. In their place, klippen of a pumpellyite-prehnite facies or nonmetamorphosed Cretaceous sedimentary rocks ride on the Piedmont Nappe of Paleocene-lower Eocene flysch (Stephan, 1977). The boundary between the Paleozoic and the Mesozoic-Tertiary basement must be placed between the Andes and the Apartaderos fault. We infer it to coincide with the Apartaderos fault for the following reasons:

1. Along the Cantagallo thrust zone, the Villa de Cura Group metavolcanics which are assigned to the Mesozoic-Tertiary basement are thrust over Paleocene-lower Eocene flysch. This may have been caused by long distance overthrusting or by rapid uplift followed by southward sliding of the crystalline rocks over the flysch. It suggests that the relationship of the Mesozoic-Tertiary basement to the flysch along the Cantagallo thrust zone may continue to hold true along the Apartaderos fault. In other words, the Mesozoic-Tertiary basement may not extend under the flysch to the west of the Apartaderos fault.

2. Preliminary analysis of the 1981-82 Petróleos de Venezuela, S.A., aeromagnetic survey data suggests that, immediately west of the Apartaderos fault, the nature of the basement rocks undergoes a marked change, or else the basement is deeper than we believe it to be. We infer that the Mesozoic-Tertiary crystalline complex of the western Caribbean Mountains comes to an abrupt end at the Apartaderos fault.

3. Bell (1972, Fig. 5) presents a diagrammatic map showing an inferred north-south offset, during the Cretaceous, of the Benioff zone and its related area of metamorphism. This offset corresponds approximately with the Apartaderos fault, now exposed. The inferred offset area of metamorphism supports the interpretation of a western limit of the Mesozoic-Tertiary basement along the Apartaderos fault.

The 265 m.y. (U/Pb) granite of the Paraguaná Peninsula and the 425 m.y. (Rb/Sr) igneous rocks of the Sebastopol-El Tinaco Complex (Feo-Codecido and others, 1974; González de Juana and others, 1980, p. 309, 812) indicate a possible former northern extension of the belt of Paleozoic basement in the region now occupied by the Mesozoic-Tertiary basement. Little is known about the basement in the Maracaibo basin. Wells SLC-1-2X (3, Fig. 4) and nearby SLC-5-1X passed from Cretaceous limestones into chlorite schists of unknown age.

In the Andes and Perijá regions, marine sedimentation continued throughout the Paleozoic (with some important local interruptions) as evidenced by the fossiliferous formations of Ordovician, Silurian, Devonian, and Permo-Carboniferous age (González de Juana and others, 1980, p. 120-121, 124-126, 136, 140; Pierce and others, 1961). The Paleozoic of the Andes is also represented by igneous rocks emplaced throughout the era and by

Paleozoic sedimentary rocks, now metamorphosed (Cerro Azul, Mucuchachí, and equivalent formations). While outside the scope of this report, it may be mentioned that the relationship of those Paleozoic sedimentary rocks to the Paleozoic metamorphic and igneous rocks may be the result of displacement of microcratonic plates in pre-Cretaceous tectonism and of displacement during orogeny in the Tertiary.

The Paleozoic Orogenic Belt (Fig. 3) comprises granites, schists, and other metamorphic rocks of Paleozoic age which are overlain by Cretaceous or Tertiary sedimentary rocks in the states of Barinas, Portuguesa, Cojedes, and Guárico, and which are exposed in the El Baúl area. The nature of the Apure fault zone (Fig. 4) suggests that the belt is allochthonous, at least along its southern border. Its northern boundary is not clearly defined in the Andes region; farther east, its northern border is obscured by the Mesozoic-Tertiary allochthon of the western Caribbean Mountains.

The unmetamorphosed Cambrian basin is represented by sedimentary rocks of the Carrizal and Hato Viejo formations which occupy the position of a foreland trough to the south of the Paleozoic Orogenic Belt (Fig. 3), with only a part of its former sedimentary prism still preserved.

The Apure fault zone

A tectonic contact is inferred along the Apure fault zone (Fig. 4) between the Paleozoic Orogenic Belt and the older Paleozoic sedimentary rocks or the Precambrian Guayana Shield.

In the west, the Cretaceous sedimentary sequence of wells Apure 1, 2, 3 and Nutrias-1 (8, 5, 12, 26; Figs. 4, 5) lies on sedimentary rocks which are identified as the Carrizal Formation on the basis of lithology and electric-log characteristics and for which a Cambrian age is inferred. Twenty to forty kilometers to the north (12-25 mi) in the Barinas wells, Cretaceous sedimentary rocks lie on granites and schists which, in Agua Linda-1 and SZW-3 (4, 7, Figs. 4, 5) are of Paleozoic age. We infer that the Carrizal Formation of the Apure and Nutrias wells lies on Precambrian basement, because the Carrizal Formation has not been involved in the thermal event which produced the Paleozoic granites and schists; hence we show the Apure fault zone to be located between Ticoporo-1, (9, Fig. 4, 5) and the Apure and Nutrias wells, with the Paleozoic Orogenic Belt overthrusting the unmetamorphosed Carrizal Formation.

In southern Guárico we recognize five areas, each with its characteristic sequence of basement and Paleozoic rocks:

1. Las Mercedes, with a Paleozoic basement of granite, syenite, and diorite overlain by Cretaceous sedimentary rocks. The southernmost wells, MCH-2-4X (30, Fig. 5) and MCH-2-5X (28, Fig. 5) penetrated a quartzose alkaline syenite (Area A, Fig. 5, in the Las Mercedes area).

2. Gorrín. Highly metamorphosed, fine-grained sedimentary rocks, pelitic hornfels with abundant biotite in MCH-2-3X (347 m.y.; 14, Figs. 4, 5) and with coarser grained hornfels in Gorrín-1 (29, Fig. 5), and also metamorphosed tuffaceous sandstones in

Machete-2X (15, Figs. 4, 5), all overlain by Cretaceous sedimentary rocks (Area F, Fig. 5).

3. Manapire. Unfossiliferous, slightly metamorphosed, red to maroon arkosic lutite interpreted as Carrizal Formation in MCH-8-5X (31, Fig. 5) and several other wells (Area G, Fig. 5).

4. Espino Graben. Contains Jurassic and, in NZZ-88X (17, Figs. 4, 5), Carboniferous sediments not present elsewhere in the area. One well, MCH-8-5X (31, Fig. 5), penetrated maroon Manapire facies of the Carrizal Formation under the Jurassic "La Quinta" Formation (Area H, Fig. 5).

5) Carrizal. The typical greenish-gray Carrizal Formation described below, found in the subsurface as far east as well SG-139-X (39, Fig. 5) (Area I, Fig. 5).

6. Shield overlain by Cretaceous or Tertiary sedimentary rocks (Area J, Fig. 5).

We map the Apure fault zone along the southernmost-known limit of the Gorrín metamorphics (Area F, Fig. 5), to the contact with the Manapire Carrizal Formation along the Machete fault. This interpretation assumes that the Gorrín metamorphics do not underlie the older Manapire Carrizal Formation to the east, and hence that the Apure fault zone is offset to the north along the Machete fault. Its subsurface trace is inferred to extend between the Las Mercedes Paleozoic basement and the inferred Precambrian basement below the Manapire facies of the Carrizal Formation. The subsurface trace of the Apure fault zone is inferred between Nutrias-1 (26, Fig. 5) and the Gorrín metamorphics in well MCH-13-1X (27, Fig. 5) near San Fernando, Apure.

The Espino Graben

Present data are somewhat conflicting with regard to this structural element. Recent (1981-1982) aeromagnetic data indicate a rather rectilinear depression from some 30 km (18 mi) south of Barcelona, Anzoátegui, through Zaraza, Guárico, and San Fernando, Apure, and thence to the Colombian border at approximately 70° 00′ West Longitude, 7° 10′ North Latitude (Fig. 4). Gravity interpretations (see Bellizzia and others, 1976) show depressions in the Barcelona and Espino areas (Fig. 4), but do not indicate the magnitude of the trend as shown by the aeromagnetic data. Much of the reflection seismic data was not recorded to sufficient depth to be of use. The one line interpreted to date in the graben area of Apure is inconclusive. It is only in the Espino Graben proper (Fig. 4) of southern Guárico that all methods are in agreement. Basement drops from some 610 m (2,000 ft) south of the Altamira fault to some 6,100 m (20,000 ft) north of the fault, rising again to 1,525 m (5,000 ft) in the Las Mercedes area to the north. The graben there contains Cambrian, Carboniferous, and Jurassic sedimentary rocks, the latter documented by a 162-m.y. basalt, interpreted to be a subaerial flow on the basis of petrographic study. Although further study is required, we can at least report the probability of major rifting in the Early to Mid-Mesozoic.

SYNTHESIS OF PALEOZOIC STRATIGRAPHY OF THE LLANOS BASINS

The Paleozoic sedimentary rocks of the Venezuelan Llanos basins represent a fine-grained clastic sequence of Early Cambrian to possibly Early Ordovician age. There is no evidence of Cambrian to Early Ordovician sediments anywhere in the rest of Venezuela; the Caparo Formation of the southwestern Venezuelan Andes is of Late Ordovician age (González de Juana and others, 1980, p. 97, 120). The granite and schists of the Paleozoic Orogenic Belt are probably in part the metamorphic equivalent of the Cambrian to Ordovician sediments. In general, both the Paleozoic sedimentary and the igneous and metamorphic rocks of the Llanos basins are overlain by Cretaceous or (in updip areas) Tertiary sedimentary rocks. In the Espino Graben an unfossiliferous red bed section, with an intercalated basalt flow of Jurassic age, is preserved above a fossiliferous section of Carboniferous age.

We list the Paleozoic stratigraphic units and show their geographic areas and stratigraphic relationships in Figures 5 and 6:

The Hato Viejo Formation (Hedberg, 150, p. 1184; see also González de Juana and others, 1980, and this paper): Medium to coarse-grained, variably arkosic and quartzitic sandstone which grades upward into the Carrizal Formation. The lower part of the unit carries a simple zircon-leucoxene suite of detrital heavy minerals which, however, changes upward into a garnet-biotite suite like that of the overlying Carrizal Formation. Some streaks of green argillite in the Hato Viejo Formation are similar to the fine-grained clastic rocks of the Carrizal Formation. No fossils have been found in this unit.

The Carrizal Formation (original published description in Hedberg, 1950, p. 1183–1185; see also González de Juana and others, 1980, and this paper): A very fine-grained sandstone and siltstone sequence found in the subsurface in the states of Guárico and Anzoátegui where we assign it an Early Cambrian age (see below). Rocks of similar lithology are present in wells Apure 1, 2, 3 and Nutrias -1 (8, 5, 12, 26; Figs. 4, 5) in the states of Apure and Barinas.

El Barbasco Group (Martín Bellizzia, 1961, 1968; Feo-Codecido, 1963; González de Juana and others, 1980, p. 118–120): Very fine- to medium-grained metasedimentary rocks at El Baúl, Cojedes, interpreted to be of Early Paleozoic age by González de Juana and others (1980, p. 119). In an ascending order, the group consists of the following three formations:

The Mireles Formation (Rod, 1955; Martín Bellizzia, 1961; González de Juana and others, 1980): A phyllitic sequence which contains trilobites assigned by Frederickson (1958) to the lowermost Ordovician (Tremadocian).

The Cerrajón and Cañaote formations (Martín Bellizzia, 1961; González de Juana and others, 1980): Metamorphosed fine-grained sandstones and micaceous quartzites.

The Güéjar(?) series of the Colombian well La Heliera-1 (6, Figs. 4, 5): Medium to dark-gray, hard, micaceous shale, and

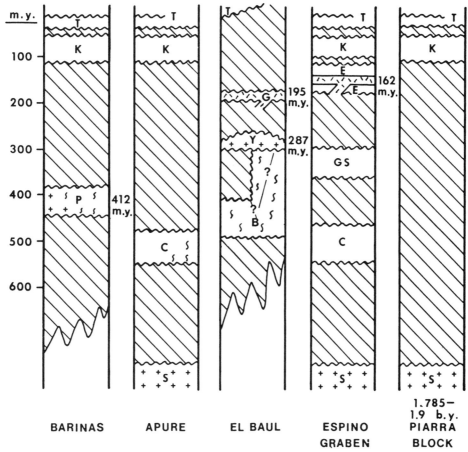

Figure 6. Stratigraphic columns in the Llanos basins. S = Precambrian basement, P = Paleozoic basement, C = Carrizal Formation, Y = El Baúl granite, B = El Barbasco Group metamorphics, G = Guacamayas Group volcanics, GS = Carboniferous sandstones, E = "La Quinta" Formation with an intraformational flow, K = Cretaceous sediments, T = Tertiary sediments. Approximate thickness where known: C = more than 1,830 m (6,000 ft) maximum. B = more than 425 m (1,400 ft). G = approximately 350 m (1,150 ft). GS = possibly 744 m (2,440 ft). E = 1,645 m (5,400 ft). K + T = Barinas = 2,100 m (6,890 ft) to 3,700 m (12,140 ft). Las Mercedes area (Figs. 4 and 5) = 1,675 m (5,500 ft) to 1,920 m (6,300 ft). WGDT-1, location 21 (Figs. 4 and 5) = 2,804 m (9,200 ft). Soledad-1 (CVP), location 23 (Figs 4 and 5) = 4,313 m (14,150 ft).

interbedded white to gray, very fine-grained quartzitic sandstone with calcareous cement. The sandstone occurs in thinly laminated zones within the shales. Some of the shales are slightly metamorphosed with highly contorted bedding. On the basis of macrofossil photographs from La Heliera-1 core, S.A. Root (1980, unpublished report) identified trilobite species *Jujuyaspis keideli* and *Parabolinella* cf *argentinensis*, a dendroid graptolite *Dictyonema* cf *flabelliforme*. He assigns an Early Ordovician (Tremodocian) age to the assemblage.

Granites, schists and phyllites and other metasedimentary rocks of the subsurface of Barinas, Portuguesa, Cojedes, and Guárico. The youngest radiometric age determination of these rocks is 412 m.y. in Barinas and 321 m.y. in Guárico (Table 1).

The El Baúl granite (287-270 m.y.).

Unnamed sandstones in well NZZ-88X (17, Figs. 4, 5) with one interval at 3,435 m (11,268 ft) subsea which contained a

flora representative of the Carboniferous period, barren below to a total depth of 4,178 m (13,708 ft) subsea.

Additional data on the Carrizal Formation

On the basis of areal distribution and thickness, the Carrizal Formation is the most important Paleozoic unit of the Venezuelan Llanos basins. It was first described by Hedberg (1942, p. 200; 1950, p. 1184) from wells Carrizal-1 (18, Figs. 4, 5) and Suata-1 (36, Fig. 5). In 1975, Carrizal-2X (18, Figs. 4, 5), 5 km (3 mi) from Carrizal-1, penetrated 1,827 m (5,995 ft) of the Carrizal Formation between 1,085 m (3,558 ft) and 2,912 m (9,533 ft) subsea and recovered four cores between 1, 247 m (4,090 ft) and 1,626 m (5,335 ft) subsea. This section, the thickest to be penetrated, did not reach the base of the formation.

These cores represent a monotonous sequence of greenish

gray, well-lithified, conspicuously burrowed and bioturbated, micaceous, glauconitic siltstone and claystone with minor interbedded fine-grained sandstone. Cuttings represent the same lithologies, with common color changes to brick red below 2,164 m (7,100 ft) subsea. In the core of NZZ-12X (Table 2) this color change is abrupt and unrelated to bedding. In Carrizal-2X, the kerogen obtained from the cores was of medium yellow color indicative of little thermal alteration, while that from drill cuttings below 2,164 m was of medium brown color.

The following petrographic descriptions are typical:

Sandstone: Quartzose, clayey, glauconitic, very micaceous, silty, bioturbated, very fine grained (average grain size is about 0.1 mm, although grains range from 0.02 to 0.13 mm, medium silt to fine sand). Roundness varies from angular to subrounded, with most grains being subangular.

Quartz grains consist of two types: 1) plutonic igneous and 2) metamorphosed with composite strained grains, both subequant and elongated. Other detrital constituents are abundant mica (muscovite, biotite, and chlorite), sparse plagioclase feldspar, and rare zircon. Authigenic minerals include glauconite and pyrite.

Cement consists of silica, which not only fills pore spaces producing a well-lithified rock, but has also replaced some of the detrital grains. Minor calcite cement occurs in scattered patches; such cement has replaced some of the detrital grains.

Siltstone: Quartzose, very micaceous, clayey, glauconitic, pyritic, sandy (very fine sand), bioturbated. Minerals are the same as described for sandstone. Rock is well indurated and has been cemented by silica.

Claystone: Very silty, pyritic, micaceous, glauconitic, strongly burrowed and bioturbated. Contains few burrows that are filled with silica-cemented, very fine to fine quartz sand.

The Carrizal Formation is usually overlain by Cretaceous sandstones of the Temblador Group or, in the south, by sandstones of Tertiary age. In the Espino Graben (Fig. 4) it underlies the Jurassic "La Quinta" Formation in MCH-8-5X (31, Fig. 5) and other wells. It is present in Carrizal-2X (18, Figs. 4, 5), Zuata Este-1X (36, Fig. 5), and MCH-8-5X and Socorro-1 (32, Fig. 5). It is therefore believed to be present under the Carboniferous sediments in the geographically central well NZZ-88X (17, Figs. 4, 5).

In the east, the formation rests on, or interfingers with the Hato Viejo Formation which has been reported in wells Hato Viejo-1 (38, Fig. 5), Santa Clara-1 and Maco-1 (38, Fig. 5), OG-338-S (20, Fig. 4, 5), Suata-1 and Zuata Este-1X (36, Fig. 5), SDZ-43X (19, Figs. 4, 5), and NZZ-45X (33, Fig. 5). A complete section of 238 m (782 ft) was penetrated in Zuata Este-1X.

It is possible that remnants of the Precambrian Roraima Group are also present below the Carrizal-Hato Viejo formations as suggested by information from wells Santa Clara-1 (38, Fig. 5) and OG-338-S (20, Figs. 4, 5). Santa Clara-1 cored through the unconformity at the base of the Cretaceous Temblador Group and 10 m (33 ft) into silicified, fine to coarse-grained, well-rounded, well-sorted quartz sandstone followed by 3 m (9 ft) of extremely hard, poorly sorted quartzite with well-rounded grains. The quartzite became too hard to core (year 1939) and an additional 15 m (50 ft) were drilled. The descriptions of the Roraima Group are summarized by González de Juana and others (1980, p. 77-86).

On the other hand, after penetrating 637 m (2,090 ft) of Carrizal Formation, well OG-338-S drilled 63 m (206 ft) of uniformly well-cemented, fine-grained calcareous, glauconitic, micaceous sandstone with conglomerate of quartzite and quartz. This could represent the basal sandstone of the Carrizal Formation, that is, the Hato Viejo Formation.

The age of the Carrizal Formation: Our work on acritarch assemblages from the Carrizal Formation in wells Carrizal-2X (18, Figs. 4, 5), SDZ-11X (see Table 2), and SDZ-16X (37, Fig. 5) supports an Early Cambrian age for that part of the formation which has been studied to date. A more complete investigation is being undertaken for later publication. Robertson Research (1981) suggests a Middle Cambrian to Early Ordovician age for acritarchs from other wells in the Carrizal Formation. Stover's (1967) designation of a Late Devonian-Early Mississippian age for acritarchs from wells Carrizal-1 (18, Figs. 4, 5), Tres Matas-1 (35, Fig. 5), Socorro-1 (32, Fig. 5), Hato Viejo-1 (38, Fig. 5), and Suata-1 (36, Fig. 5) is herein revised on the basis of the new samples and in the light of investigations that have been carried out during the years following his work.

Our samples contained undescribed species of small, thick-walled acritarchs and the following identified species: *Baltisphaeridium ciliosum* Volkova 1969, *Michystridium lanatum* Volkova, cf *Estriata minima* Volkova, *Cymatiosphaera sp.* (*Plicatosphaera voluminosa/elementaria* Potter 1974), and the rather common presence of the distinctive *Archaeodiscina umbonulata* Volkova 1968, a species restricted in literature to the Lower Cambrian (Zone 5 of Downie, 1974; Zone O of Vanguestaine, 1974).

SUMMARY AND CONCLUSIONS

At the end of the Paleozoic, the western Llanos basin was represented by a southern Precambrian Shield, in part covered by unmetamorphosed Cambrian, Ordovician(?) and Carboniferous sediments, and a possibly allochthonous northern belt of igneous and metamorphic rocks of Ordovician to Permian age. In the Andes marine sedimentation continued from the Late Ordovician to the Permian, accompanied in places by Paleozoic granitization at depth. In the eastern Llanos basin the Paleozoic sediments were either never deposited or were removed prior to Cretaceous sedimentation.

Rifting occurred in the Early Mesozoic, possibly from northern Anzoátegui west-southwest to the Colombian border, documented by a 2,390 m (7,840 ft) column of Carboniferous and Jurassic sedimentary rocks in the Espino Graben of southern Guárico. The Llanos basins were thereafter subjected to erosion until the Cretaceous transgression.

Across the states of Barinas, Portuguesa, Cojedes, and Guárico, a belt of Paleozoic igneous and metamorphic rocks occupies a median area between the Precambrian Guayana Shield and the Mesozoic to Lower Tertiary allochthonous basement near the Venezuelan coast. This belt may have been a part of the shield, covered by Lower Paleozoic sediments, both of which were later involved in Paleozoic orogenies, or it may have been accreted to the Precambrian Guayana Shield during the orogenies. To the south of the orogenic belt, unmetamorphosed Cambrian sedimentary rocks cover part of the shield in the subsurface of the states of Apure, Guárico, and Anzoátegui. In the State of Guárico, there is a graben with a depth to basement of 6,100 m (20,000 ft), in which are preserved sedimentary rocks of Carboniferous and Jurassic age as yet unknown in the rest of the Llanos basins. Preliminary interpretation (March 1983) of aeromagnetic data indicates that this graben may extend from northern Anzoátegui to the Colombian border, a distance of 600 km (375 mi).

ACKNOWLEDGMENTS

The writers express their gratitude to Corpoven, S.A., and Maraven, S.A., for all facilities given during the preparation of this paper. E. de Di Giacomo acknowledges the assistance of palynologists in the Koninklijke Shell Exploratie en Produktie Laboratorium, Rijswijk, Holland.

REFERENCES CITED

Bell, J. S., 1972, Geotectonic evolution of the southern Caribbean area, *in* Studies in earth and space sciences: The Harry H. Hess Volume, R. Shagam and others, eds., Geological Society of America Memoir 132, p. 369–386.

Bellizzia, G., A., 1972, Is the entire Caribbean Mountain belt of northern Venezuela allochthonous?, *in* Studies in earth and space sciences: The Harry H. Hess Volume, R. Shagam and others, eds., Geological Society of America Memoir 132, p. 363–368.

Bellizzia G., A., Pimentel M., N., and Bajo, O., R., compilers, 1976, Mapa Geológico Estructural de Venezuela: Venezuela Ministerio de Minas e Hidrocarburos, scale 1:500,000.

Bonini, W. E., Pimstein de Gaete, C., and Graterol, V., compilers, 1977, Mapa de anomalías de Bouguer de la parte norte de Venezuela y áreas vecinas: Venezuela Ministerio de Energía y Minas, scale: 1:1,000,000.

Downie, C., 1974, Acritarchs from near the Precambrian-Cambrian boundary—A preliminary account: Review of Palaeobotany and Palinology, v. 18, p. 57–60.

Feo-Codecido, G., 1954, Notas petrológicas sobre formaciones que afloran en la región de El Baúl, Estado Cojedes: Boletín de Geología, Caracas, v. 3, p. 109–121.

——1961, Observaciones radimétricas preliminares en la región de El Baúl, Estado Cojedes, *in* Memoria, Congreso Geológico Venezolano, III, Caracas, 1959: Venezuela Ministerio de Minas e Hidrocarburos, Boletín de Geología, Publicación Especial No. 3, v. 4, p. 1681–1697.

——1963, Notes to accompany the Venezuelan contribution to the edition of a world geological map, scale 1:5,000,000: Asociación Venezolana de Geología, Minería, y Petróleo, Boletín Informativo, v. 6, p. 291–307.

Feo-Codecido, G., and Martín B., C., and Bartok, P., 1974, Guía de la excursión a la Península de Paraguaná: Asociación Venezolana de Geología, Minería y Petróleo, Guía de Excursión, 23 p.

Frederickson, E. A., 1958, Lower Tremadocian trilobite from Venezuala: Journal of Paleontology, v. 32, p. 451–543.

González de Juana, C., Iturralde de Arozena, J. M., and Picard, C., X., 1980, Geología de Venezuela y de sus cuencas petrolíferas: Caracas, Foninves, 1031 p.

Hedberg, H. D., 1942, Mesozoic stratigraphy of northern South America: Eighth American Scientific Congress, U.S.A., 1940, Proceeding IV, p. 195–227.

——1950, Geology of the Eastern Venezuela basin (Anzoátegui-Monagas-Sucre-eastern Guárico portion): Geological Society of America Bulletin, v. 61, p. 1173–1216.

MacDonald, W. D., and Opdyke, N. D., 1974, Triassic paleomagnetism in northern South America: Geological Society of America Bulletin, v. 58, p. 208–215.

Martín Bellizzia, C., 1961, Geología del macizo de El Baúl, Estado Cojedes, Congreso Geológico Venezolano, III, Caracas, 1959: Venezuela Ministerio de Minas e Hidrocarburos, Boletín de Geología, Publicación Especial No. 3, v. 4, p. 1453–1530.

——1968, Edades isotópicas de rocas Venezolanas: Boletín de Geología, Caracas, v. 9, p. 356–380.

Mendoza, V., 1977, Evolución tectónica del Escudo de Guayana, *in* Memoria, Congreso Latinoamericano Geológico, II, Caracas, 1973: Venezuela Ministerio de Minas e Hidrocarburos, Boletín de Geología, Publicación Especial No. 7, v. 3, p. 2237–2270.

Muessig, K. W., 1978, The Central Falcón igneous suite, Venezuela: Alkaline basaltic intrusions of Oligocene-Miocene age: Geologie en Mijnbouw, v. 57 (2), p. 261–266.

Munro, S. E., and Smith, F. D., Jr., 1984, The Urica fault zone, northeastern Venezuela, *in* Bonini, W. E., and others, eds., The Caribbean-South American plate boundary and regional tectonics: Geological Society of America Memoir 162 (this volume).

Pierce, G. R., Jefferson, C. C., Jr., and Smith, W. R., 1961, Fossiliferous Paleozoic localities in the Mérida Andes, Venezuela: American Association of Petroleum Geologists Bulletin, v. 45, p. 342–375.

Renz, H. H., Alberding, H., Dallmus, K., Patterson, J. M., Robie, R. H., Weisbord, N. E., and MasVall, J., 1958, The Eastern Venezuela basin, *in* Weeks, L. G., ed., The habitat of oil: American Association of Petroleum Geologists Symposium Volume, p. 551–600.

Robertson Research (U.S.) Inc., 1981, Palynostratrigraphy of well sections from the Orinoco heavy oil belt, Eastern Venezuela, unpublished report No. 541.

Rod, E., 1955, Trilobites in "metamorphic" rocks of El Baúl, Venezuela: American Association of Petroleum Geologists, v. 39, p. 1865–1869.

Santamaría, F., and Schubert, C., 1974, Geochemistry and geochronology of the Southern Caribbean-Northern Venezuela plate boundary: Geological Society of America Bulletin, v. 85, p. 1085–1098.

Stephan, J. F., 1977, El contacto Cadena Caribe-Andes Merideños entre Carora y El Tocuyo (Edo. Lara): Observaciones sobre el estilo y la edad de las deformaciones Cenozoicas en el Occidente Venezolano, *in* Memoria, Congreso Geológico Venezolano, V, Caracas, Volume 2: Venezuela Ministerio de Energía y Minas, p. 789–816.

Stover, L. E., 1967, Palynological dating of the Carrizal Formation of eastern Venezuela: Asociación Venezolana de Geología, Minería y Petróleo, Boletín Informativo, v. 10, p. 288–302.

Vanguestaine, M., 1974, Espèces zonales d'acritarches du Cambro-Tremodocien de Belgique et de l'Ardenne française: Review of Palaeobotany and Palynology, v. 18, p. 63–82.

MANUSCRIPT ACCEPTED BY THE SOCIETY SEPTEMBER 1, 1983

Printed in U.S.A.

Geological Society of America
Memoir 162
1984

The geology of the El Pilar fault zone and adjacent areas in northeastern Venezuela

Richard C. Vierbuchen
Exxon Production Research
Box 2189
Houston, Texas 77001

ABSTRACT

The right-lateral El Pilar fault system extends from the Gulf of Cariaco in northeastern Venezuela to the east coast of Trinidad. The fault is seismically active, deforms Quaternary strata, and is the boundary between two very different geological provinces. North of the fault is the eastern Cordillera de la Costa, which is composed of Lower Cretaceous metasediments and igneous rocks that accumulated in a tectonically and volcanically active environment, probably a fore-arc setting. These rocks were metamorphosed to greenschist facies during the Late Cretaceous, and deformed by imbricate fold and thrust faults during the Cretaceous and Tertiary. South of the fault is the Serrania del Interior, a fold and thrust belt composed of Cretaceous and Paleogene sediments that were deposited, at least during the Early Cretaceous, in a comparatively stable tectonic environment, probably a passive continental margin, and deformed during post-Middle Eocene time.

In northeastern Venezuela the El Pilar fault consists of two major branches, one of which was not recognized in previous studies. Geologic mapping demonstrates a cumulative dextral displacement that must exceed 20 km. A steep gravity gradient across the fault system suggests that the fault plane is a nearly vertical density discontinuity to a depth of at least 5 to 10 km. Gravity models imply that a relatively dense mafic crust is present north of the fault, and regional geology demonstrates that a similar crust does not exist south of the fault anywhere east of the Gulf of Cariaco. It follows that a total right-lateral displacement of 150 to 300 km may be necessary to account for the steep gravity gradient at the fault.

INTRODUCTION

The El Pilar fault system is an active structural feature of regional importance that is believed to extend at least 400 km from a point east of Trinidad westward to the Cariaco Trench (Fig. 1). This fault system is part of a network of major faults that crosses northern South America and includes the San Sebastian, Oca, Bocono, and Santa Marta faults as well as many smaller ones. The magnitudes, directions, and timing of displacements on the major faults are not well known, nor is the importance of these faults to the present-day and previous boundaries between the Caribbean and South American plates understood.

The El Pilar fault system is west of, and adjacent to, the southern end of the Lesser Antilles island arc. It is the only known structure that could terminate the subduction zone in front of the island arc and accommodate the roughly east-west motion between the Caribbean and South American plates. Therefore, this tectonic setting suggests right-lateral displacement on the El Pilar, or on some undiscovered faults parallel to it. This suggestion is substantiated by focal mechanisms, indicative of right-slip, of earthquakes along the fault (Molnar and Sykes, 1969; Vierbuchen, 1979). Volcanic rocks cropping out in the southern Lesser Antilles are as old as Eocene. If present rates of convergence (about 1 cm/yr: Jordan, 1976) have existed continuously since the Eocene, then as much as 500 km of relative motion could have occurred. One of the major tectonic problems pertaining to

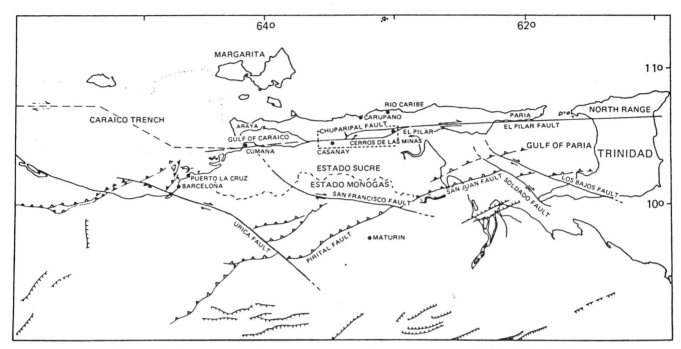

Figure 1. Geologic and geographic reference map of northeastern Venezuela and Trinidad. The dashed
box outlines the map area of Figure 2.

the southern Caribbean plate boundary is the apparent lack of geological evidence for such large displacements.

Throughout its length the El Pilar fault has dramatic topographic or bathymetric expression. At its western end, the fault cuts the eastern side of the Cariaco Trench along an abrupt bathymetric escarpment, forms the sheer coastline at the southern edge of the Gulf of Cariaco, and causes the linear valley through central Estado Sucre, Venezuela. Continuing east, the fault forms the impressive, steeply elevated, linear coast on the southern side of the Paria Peninsula (Fig. 1), crosses the Gulf of Paria, and continues through Trinidad, where it lies at the foot of the North Range.

The geologic expression of the fault is equally dramatic. Along its entire length the El Pilar fault system forms the boundary between two geologic provinces: the Jurassic and Cretaceous metasedimentary, metavolcanic, and ultramafic rocks of the eastern Cordillera de la Costa (Venezuela) and North Range (Trinidad) to the north, and the Cretaceous and Cenozoic sedimentary rocks of the Serrania del Interior to the south. In the eastern Cordillera de la Costa, volcanic rocks, abundant coarse metaclastic deposits, together with the mafic and ultramafic basement that is at least locally unconformably beneath metasedimentary units (Vierbuchen, 1978), all suggest deposition in a tectonically active setting. In contrast, most sediments of the Serrania del Interior record no such tectonism and were apparently deposited in a more quiescent setting.

These geological, seismic, and physiographic manifestations of the El Pilar fault led Molnar and Sykes (1969), Vasquez and Dickey (1972), and Silver and others (1975), to suggest that the El Pilar fault is part of the present-day boundary between the

Caribbean and South American plates. However, these suggestions have not been supported by field studies (Maxwell and Dengo, 1951; Christensen, 1961; Metz, 1964; and Vignali, 1977), which concluded that there is probably less than 15 km of right lateral displacement on the fault.

The present paper reports the results of field work conducted between 1973 and 1977 in northeastern Venezuela. This investigation, which consisted of detailed structural analysis, mapping of 600 square kilometers at a scale of 1:25,000 (Fig. 2, in pocket inside back cover) and gravity modelling, helps to resolve some aspects of the conflict between field studies and regional plate tectonic syntheses. The new results modify previously suggested geological constraints on the magnitude of strike-slip displacement, and further, can be interpreted to imply that a displacement as large as 300 km may have occurred.

PREVIOUS WORK

The history of geological investigations of the El Pilar fault has been well summarized (Metz, 1964; and Schubert, 1979) so only a brief description is presented here. The first systematic geological study was done by Liddle (1928, 1946). The regional tectonic studies of Bucher (1950) and Hess and Maxwell (1953) stimulated a widespread interest in the strike-slip faults of northern South America and inspired subsequent studies of the El Pilar by Rod (1956), Alberding (1957), and Metz (1964). The fault was formally named by Rod (1956), who also gave several qualitative arguments suggesting a right-lateral displacement of 100 km or more.

Alberding (1957) proposed a tentative correlation of metamorphic rocks in the eastern Cordillera de la Costa with those in the main part of the mountain belt near Caracas, implying 574 km of right-lateral displacement on the El Pilar fault. This correlation was apparently contradicted by the first detailed field study of the fault (Metz, 1964), which suggested that several sedimentary units could be matched across the fault and that these correlations limited the maximum possible dextral displacement to 15 km. More recent field work (Vignali, 1977) has also supported a smaller displacement.

Seismic reflection profiles in the Gulf of Cariaco show that the fault extends westward into the Cariaco Trench where it is offset to the north (Lidz and others, 1968). The offset in the fault has been interpreted to imply that the fault predates Cenozoic thrusting in eastern Venezuela (Lidz and others, 1968). Alternatively, Silver and others (1975) proposed that the Cariaco Trench is a rhombochasm between the parallel and offset San Sebastian and El Pilar faults.

REGIONAL GEOLOGIC SETTING

North of the El Pilar fault is an east-west trending mountain range called the eastern Cordillera de la Costa in Venezuela and the North Range in Trinidad. These mountains are underlain by a deformed, low- to medium-grade metamorphic terrane of Mesozoic age. Metamorphic rocks also crop out offshore on Isla Margarita and on smaller islands to the east and west. The most easterly exposures of metamorphic rocks are on Tobago.

The geological history of the eastern Cordillera and North Range is controversial. This controversy stems primarily from the fact that a regional stratigraphy has not been devised for the mountain belt in Venezuela. In the central part of the eastern Cordillera, Seijas (1972) recognized a series of schists and phyllites metamorphosed to the lower greenschist facies and apparently overlain by metaigneous rocks near the coastline. Somewhat different columns have been described in the western and eastern extremes of the mountain range (Schubert, 1971; Gonzalez de Juana and others, 1965). A major task facing future workers is correlation of the different stratigraphies in the eastern Cordillera.

All available information indicates that the metasediments in the eastern Cordillera de la Costa were deposited during the Mesozoic, probably between Late Jurassic and middle Cretaceous time. In the central part of the eastern Cordillera only two formations have yielded fossils. Macsotay (1976, oral communication, and 1968) reports that the older faunas date from Barremian to Aptian (Guinimita Formation), and that the younger faunas are Aptian to Albian (Tunapui Formation). Farther east, on the Paria Peninsula, Gonzales de Juana and others (1974) have suggested a somewhat older age for the metamorphic sequence. Based on a K/Ar date of 128 ± 11 m.y. for granitic rocks believed to intrude metasediments there (Macuro Formation), they propose that some of the sediments were deposited in Late Jurassic time. Alternatively, Kugler (1972), has suggested that the

granitic bodies in Paria may be olistoliths and therefore predate the time of deposition.

Metamorphism of at least part of the eastern Cordillera must be post-Albian because of the age of fossil assemblages preserved in the metamorphosed sediments. Gonzalez de Juana and Vignali (1972) postulate that regional metamorphism occurred during the Late Cretaceous (Campanian to Maestrichtian) on the basis of an apparently post-metamorphic pluton on Margarita with a K/Ar age of 72 ± 6 Ma (Santamaria, 1972).

In the central and eastern parts of the eastern Cordillera de la Costa, the highest grade of metamorphism is characterized by the mineralogical association albite-chlorite-actinolite-epidote-calcite (Copey Formation; Seijas, 1972). However, at the western end of the eastern Cordillera, and on Isla Margarita, there are higher levels of metamorphism. Rocks of amphibolite facies occur on the western tip of Araya (quartz-albite-epidote-almandine subfacies, kyanite has been observed; Schubert, 1971), and eclogites and glaucophone schists on Margarita indicate still higher pressures of metamorphism (Maresch, 1971, 1972).

The deformation of the eastern Cordillera involved thrusting from north to south and folding about east-west trending axes (Christiansen, 1961; Gonzalez de Juana and others, 1965; Sifontes, 1969; Schubert, 1971; and Bladier, 1977). Most formational contacts are now tectonic, a fact that accounts for confusion about the stratigraphy. Folding and thrusting were followed by vertical and strike-slip faulting that continues to the present time. The age of the original compressive deformation is unknown except that it was in part post-metamorphic (Late Cretaceous).

South of the El Pilar fault system, the mountains of the Serrania del Interior are composed of Cretaceous and Tertiary sedimentary rocks that were deposited on the northern flank of the Eastern Venezuela Basin. Subsidence of the Eastern Venezuela Basin began with the deposition of Lower Cretaceous marine and brackish water sediments. On the thinner southern flank of this asymmetric basin, onlapping strata lie unconformably on Devonian to Carboniferous strata (Stover, 1967). The Paleozoic strata unconformably overlie Precambrian rocks and are known only from well cores, but are presumed to be analogous to rocks exposed in the El Baul uplift and the Venezuelan Andes. On the deformed northern limb of the basin, the section is very thick and no rocks older than Barremian have been encountered in outcrop or sursurface.

Generally transgressive marine sedimentation continued from Early Cretaceous to Paleocene time in eastern Venezuela. The Upper Cretaceous and Paleocene sedimentary rocks on the north limb of the basin record a deep marine environment (Hedberg, 1950; Metz, 1964). In most of northeastern Venezuela, deposition of these rocks was followed by uplift and deformation as evidenced by a regional unconformity of Middle Eocene age. After the Middle Eocene, a second depositional sequence began in the Late Eocene; and during Oligocene and Miocene time some 10,000 meters of sediment, mainly derived from the north (the borderland, Paria), were deposited at the axis of the basin

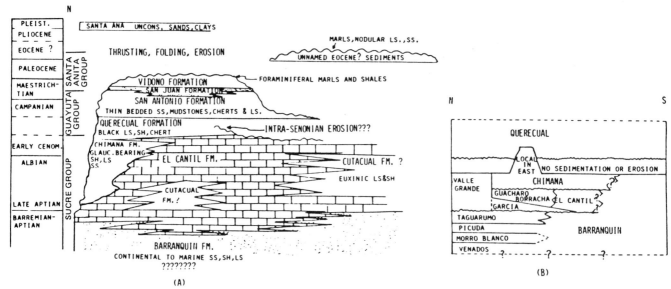

Figure 3. Stratigraphic columns for the northern Serrania del Interior. (a) Modified after Metz (1964). (b) The pre-Querecual stratigraphy of Guillaume and others (1972), which shows the subdivision of the Barranquin Formation.

(Hedberg, 1950). During the Miocene much of northeastern Venezuela was again uplifted and deformed producing another regional unconformity. Above this unconformity lie brackish and fresh water Lower and middle Pliocene deposits. Uplift followed in Plio-Pleistocene times. Total sediment accumulation in the Eastern Venezuela Basin exceeds 13,000 meters and continues today in the Orinoco Delta and Gulf of Paria.

The deformational history of the Serrania del Interior was first studied thoroughly by geologists of the petroleum industry. Early work indicated that folding and faulting of strata in this area began after the Early Eocene. Deformation of the Serrania has produced east-northeast/west-southwest trending fold axes. The southern limbs of the folds are often steeper than the northern limbs, and folds overturned towards the south are not uncommon. Thrust faults transporting material from the north-northwest towards the south-southwest are abundant. The folding and thrusting in the Serrania deforms rocks that are as young as Miocene (Vierbuchen, 1978).

SEDIMENTARY ROCKS NEAR THE EL PILAR FAULT

The map area includes an incomplete section of Lower Cretaceous through Lower Paleocene strata (Fig. 3). Overlying this sequence are scattered exposures of mid-Tertiary strata and, near the El Pilar fault, Pleistocene to Recent rocks.

Sucre Group

The Sucre Group was defined by Hedberg (1950) to include three Lower Cretaceous formations that he supposed to represent a shallow water, transgressive, continental to marine sequence. Rocks of this group are the oldest exposed strata in northeastern Venezuela. The group's cumulative thickness increases toward the north but exceeds several kilometers in the map area (Fig. 2). The group contains no evidence of syndepositional volcanic activity or tectonism in contrast to the abundant volcanic and coarse clastic rocks in the coeval metasediments of the Cordillera. The stratigraphic thickness, gradually transgressive nonmarine to shallow marine depositional environments, and lack of tectonic activity recorded in the Sucre Group imply that it probably accumulated on a passive continental margin.

The basal contact of the Sucre Group is not exposed, but faunas from the lowermost member, the Barranquin Formation, are Barremian. The upper contact is generally abrupt between the youngest member of the Sucre Group (Chimana Formation) and the overlying Guayuta Group, which was deposited in deeper waters. Faunas from the uppermost Chimana Formation yield late Cenomanian ages (Lexico Estratigrafico de Venezuela, 1970).

Barranquin Formation. The Barranquin Formation (Liddle, 1928; Hedberg, 1950; Von der Osten, 1957; Guillaume and others, 1972) was deposited during the Barremian and Aptian and consists of quartzitic sandstone interbedded with arenaceous shale and lesser amounts of limestone. The total exposed thickness is variable, ranging from sections of 2460 meters (Guillaume and others, 1972) to only 200 meters (Rod and Maync, 1954). The formation is conformably overlain by the El Cantil Formation. This contact may be transitional or abrupt. The base of the Barranquin has not been observed (Fig. 3).

South of the Serrania, where Barranquin rocks were not deposited, the Temblador Group constitutes the base of Cretaceous sedimentation. The lower member of this group (Canoa

Formation) is younger than the upper Barranquin but is nonmarine as is the lower Barranquin. Where the Temblador Group has been penetrated by wells, it is seen to overlie, in great discordance, rocks of the Hato Viejo Formation which is considered to be Devonian to Carboniferous in age (Stover, 1967). Thus by analogy, the Barranquin Formation may also have been deposited on top of deformed Paleozoic sediments.

A complete section of the Barranquin Formation is not present in the area adjacent to the El Pilar fault. However, several different stratigraphic levels of this very thick formation are exposed (Fig. 2). These rocks have been displaced southwards by thrust faults and now structurally overlie younger formations. North of the El Pilar fault, most of the Cerros de las Minas is underlain by this formation. Immediately south of the El Pilar fault and west of the Cerros de las Minas are several exposures of the Barranquin Formation surrounded by Pleistocene sediments. Also, south of the El Pilar fault and west of La Pica is another belt of Barranquin Formation. All these areas are mapped as undifferentiated Barranquin Formation. The quality of exposure in the map area was not adequate for unequivocal identification of members (Fig. 3).

The exposures of Barranquin rocks north of the El Pilar fault in the Cerros de las Minas resemble the Venados Member, the lowermost part of the formation. These rocks are best exposed in the valleys of Rio del Medio and Rio Santa Tecla on the north slope of the Cerros. Exposures are commonly of clear, hard, sometimes slightly micaceous, light brown, quartzitic sandstone cropping out in beds 20 to 80 cm thick. Cross-bedding is present but not abundant. Light-colored, micaceous shales are also exposed, but less commonly than the sandstones. Metz (1964) examined these shales for fauna and found them barren. In some areas, such as along the road heading north from Mundo Nuevo, shale layers have been the locus of shearing and minor thrust faulting. Intraformational thrust faults with large displacements were not observed in the Cerros de las Minas, but exposure is not adequate to rule them out. Minor amounts of limestone, in some places shaley and less than one meter thick, and in other places more massive, similar to El Cantil limestones, are also present within this portion of the Barranquin Formation. Metz has described an Early Cretaceous faunal assemblage from one Barranquin limestone bed north of the El Pilar fault. Limestones are far more abundant on the south flank of the Cerros than on the north.

Barranquin rocks exposed south of the El Pilar fault, and west of the Cerros de las Minas, consist of sandstones identical to those north of the fault, blue-grey, bio-clastic limestones in beds .5 to 1 meter thick, and light-colored micaceous shales. Although sandstones are the most abundant rock type, limestones are also common. The significant presence of limestones and the light-colored shales suggest that these rocks may correlate to the Morro Blanco Member of the Barranquin Formation (Fig. 3); however, at the type section, the Morro Blanco Member is less arenaceous. This belt of Barranquin rocks is isolated from the rest of the formation, being cut off by the El Pilar fault on the north and younger rocks on the south. Therefore, its stratigraphic position within the formation cannot be well determined.

In the southwestern part of the map area, Barranquin rocks crop out west of La Pica. In this region the formation consists of light to dark, brown, red, and grey shales interbedded with quartzitic sandstones that are generally more friable than those to the north, and minor amounts of thin-bedded grey limestones. This lithology suggests a correlation to the upper part of the formation (Taguarumo Formation and Picuda Member; von der Osten, 1957). Along the road west of La Pica there are many excellent exposures of these rocks.

The lower contact of the Barranquin exposed within the map area is always a thrust fault. Also, the abundant shale horizons within the formation have been the loci of intense shearing. This probably is true throughout the Serrania and may explain the widely varying estimates of thickness.

El Cantil Formation. Within the map area El Cantil rocks crop out north and south of the El Pilar fault. South of the fault a wide belt of El Cantil rocks trends east-northeast across the map. The massively bedded, blue-grey limestones in this zone form a rugged topography. The entire belt is allochthonous and thrust southwards over younger strata. Thus the normal basal contact with the Barranquin Formation is not present. At the upper contact the younger Chimana and Querecual formations are absent (Fig. 3). Whether this results from lateral facies variation or pre-Querecual erosion is uncertain. At many localities the El Cantil Formation is directly in contact with the San Antonio Formation. Metz (1964) interpreted this juxtaposition as a pre-San Antonio unconformity. However, a major finding of the present study is the recognition of thrust faults that generally account for the El Cantil-San Antonia contact. Therefore, in most localities, it no longer seems necessary to postulate a pre-San Antonio unconformity, which indeed is not present elsewhere in eastern Venezuela.

Excellent exposures of El Cantil rocks, south of the El Pilar fault, are along the highway 2 km south of Casanay (Cerro Gallina). The exposures consist of south-dipping beds of blue-grey limestones averaging 1 m in thickness, separated by thin shale partings. In some places the rocks are highly fractured so that bedding is obscured. Metz (1964, p. 27) described these rocks as "generall well rounded, winnowed, coarse allochem limestones with sparry calcite cement. Allochems generally consists of micrite intraclasts, worm caprinids and fragmented shell material." Limestones in the lower part of this sequence are slightly more argillaceous than those in the upper part.

Limestones north of the El Pilar fault and east of Nueva Colombia are tentatively assigned to the El Cantil Formation. These limestones underlie the Barranquin Formation implying that a thrust fault is present. A similar thrust contact of Barranquin over El Cantil is thought to exist south of the El Pilar fault, to the west. The presence of limestones resembling El Cantil rocks in both the Barranquin and Guinimita formations and the limited outcrop explains the difficulty of a positive identification. Justification for assigning these rocks to the El Cantil Formation is that: 1) they contain a fauna (*Nummuloculina heimi*; identified by Metz, 1964) that is common in El Cantil rocks south of the El Pilar fault but has not been found in the metamorphic rocks or in the Barranquin Formation. The fauna implies that the limestones were deposited between the Albian and Cenomanian and are therefore younger than the surrounding rocks (Barremian-Aptian); 2) the rocks are unmetamorphosed; recrystallization and abundant calcite veins, typical of Guinimita marbles are absent.

Cutacual Formation. The Cutacual Formation was defined by Metz (1964) as "thin, medium-light brown, finely bioclastic limestones, alternating with finely laminated, dark grey, foraminiferal limestones, shales and marls." Thickness of the formation is variable but may exceed 300 meters. Metz suggested that the Cutacual was deposited simultaneously with El Cantil limestones, but in stagnating water behind the reefs. The formation is not regionally extensive, probably because of its environment of deposition. Metz (1964) found that this distinctive and restricted unit crops out both north and south of the El Pilar fault, between the towns of El Pilar and Casanay. For this reason, he concluded that the strike-slip displacement on the fault could not exceed 15 km.

The validity of the original definition of the Cutacual Formation is controversial. Furthermore, recent mapping (Vignali, 1977; Vierbuchen, 1979) indicates that these rocks probably do not crop out north of the El Pilar fault. Vignali (1977) has proposed that the Cutacual Formation actually consists of rocks from the Guayuta Group (Fig. 3) that were misidentified by Metz. However, faunas described by Metz from the type section of the Cutacual are no younger than Early Cenomanian. The formation includes faunal elements as old as late Aptian and is therefore in part considerably older than the Guayuta Group. Moreover, the type section of the Cutacual is not lithologically similar to the Querecual Formation. For these reasons I have retained the Cutacual Formation as it was originally defined. However, no rocks of this formation were found north of the El Pilar fault.

Guayuta Group

The Guayuta Group is a thick, Upper Cretaceous sequence of black carbonaceous limestones and black calcareous shales that are interbedded in the upper part of the sequence with light-grey calcareous sandstones and cherts. Hedberg (1937) defined two formations, the Querecual (not present in the map area of this study), and the San Antonio, using the lithologic distinction between the lower and upper sections. In general the group represents an abrupt change from the shallow, oxygenated seas of Sucre Group deposition to deeper, more stagnant conditions.

San Antonio Formation. San Antonio rocks of the map area are very different from those exposed 150 km to the west at the type section (Hedberg, 1937). Within the map area the dark-colored, carbonaceous limestones, shales, and cherts, typical of the type section, are replaced by light-colored rocks. Metz (1964) has described these rocks in detail. The most common San Antonio facies within the map area is a sandstone-chert-calcareous mudstone association. These rocks vary in color from white to grey or light brown. In the map area, similar light-colored rocks are exposed north and south of the El Pilar fault, but are not present to the west. Metz (1964) considered this relationship as evidence against large strike-slip displacement on the El Pilar fault.

Santa Anita Group

The Santa Anita Group was defined by Hedberg (1950). At the type section on Rio Querecual, the group is 1000 meters thick and has been subdivided into three formations: San Juan, Vidono, and Caratas (Fig. 3). The group crops out along the southern border of the Serrania del Interior. However, farther north, along the El Pilar fault, the uppermost formation (Caratas) is absent by erosion and the lower two formations are sparingly preserved.

Hedberg (1950) described the lithology of the group as "several thousand feet of quartzitic sandstones, dolomitic and calcareous sandstones and siltstones, gray shales, and glauconitic rocks." The lowermost strata are sandstones of Maestrichtian age that are transitional with the underlying Guayuta Group. The upper contact is usually a major erosional unconformity dating from the Middle Eocene. The unconformity marks the first important interruption of the cycle of sedimentation that started with deposition of the Lower Cretaceous Sucre Group. Following erosion and deformation, a new cycle of sedimentation began when the Upper Eocene Mercure Group was deposited.

Unnamed Post-Middle Eocene Formation

Metz (1964) described an unnamed thin sedimentary unit, of unknown age, above the Barranquin Formation in exposures along the Carupano-Maturin road (Fig. 2). The unit first appears in road cuts 1 km north of La Pica, where north-dipping beds of El Cantil limestone overlie grey shales, grey-green, nodular limestones in beds .5 to 1 meter thick, and boulder beds containing blocks of limestone identical to El Cantil types. The contact between the El Cantil and the underlying formation is interpreted as a thrust fault. Faunas of various ages are present in the beds beneath the fault. Metz reports numerous autochthonous Barremian to Aptian forms as well as several reworked Paleocene species (*Globorotalia velascoensis, Trochammina globigeriniformis?, Miliammina sp.*). Metz deduced that the beds containing these fossils must overlie a major unconformity because their lower contact is with the Barranquin Formation. He concludes, on the basis of the reworked Paleocene fauna, that the unconformity is of post-Middle Eocene age.

Study of these outcrops leads me to propose that the mixture of faunal assemblages is due to the incorporation of large blocks of the El Cantil Formation into the younger sediment. The blocks were probably derived from the El Zorro thrust sheet (see below) as it overrode the basin where the younger strata were accumulating. Exposures in the road cuts clearly show blocks up to 4 m in diameter deposited together with beds of coarse conglomerate of the same El Cantil rock type. The autochthonous Barremian-Aptian faunas described by Metz are no doubt either from the fossiliferous El Cantil rubble or from erosional windows into the underlying Barranquin Formation. The reworked Paleocene forms must have been derived from erosion of strata overlying the El Cantil Formation.

This unnamed sedimentary unit vividly records the deformational history of the northern Serrania del Interior. Deposition of the unnamed unit was preceded by uplift and erosion of Paleocene and older strata. Shallow depositional basins formed on the erosional surface. While deposition proceeded in these basins, thrust sheets that were in part subareal, overrode the basins. Debris shed from the thrust sheets was deposited and preserved beneath them. The unnamed deposits near La Pica are the remains of one such basin.

Quaternary Deposits

Quaternary sediments accumulated in and near the El Pilar fault valley. These deposits increase in thickness to the east and west of a high between the towns of Nueva Colombia and Mundo Nuevo (Fig. 2). The Quaternary is more than one kilometer thick in the Gulfs of Cariaco and Paria, respectively west and east of the map area. Gravity models (see below) suggest that Quaternary sediments in the map area are less than several hundred meters thick.

South and east of the town of El Pilar, Quaternary rocks consist of interbedded light-grey to buff-colored fossiliferous limestones, conglomerates with clasts of metamorphic rocks, brown sandstones, and shales of fresh water and marine origin. Metz (1964) referred to these as the Santa Ana Formation and tentatively assigned them to the Pliocene. More recently Macsotay (1976, oral communication) has identified Early Pleistocene faunas in these rocks. The metamorphic clasts in conglomerates from the Santa Ana Formation are the earliest indications of the presence of a metamorphic terrane in eastern Venezuela.

South and west of the vicinity of Mundo Neuvo, Quaternary sediments are composed primarily of well-bedded conglomerates and sandstones that probably originated as alluvial fan deposits. Clasts within the conglomerates range from pebble to boulder size and both metamorphic and sedimentary clasts are present. In the section south of Casanay, relatively finer grained conglomerates with sedimentary clasts overlie coarser grained conglomerates containing metamorphic clasts. In addition, large metamorphic blocks, up to a kilometer in length, are present within or beneath the Quaternary sediments. All of the large blocks are from the Guinimita Formation.

The fan deposits have no doubt undergone a complex history of uplift and erosion since their deposition. Clast size and composition indicate that the source area was initially the metamorphic terrain north of the El Pilar fault. This area was probably uplifted rapidly at the time of fan deposition. The lack of large clasts in the upper part of the fan sequence, and the sudden

dominance of clasts derived from a more distal source of Cretaceous sediments cropping out to the south, suggest slowing of uplift to the north. Even the youngest fan deposits have been tilted and folded. This deformation may continue to the present.

STRATIGRAPHY OF THE METAMORPHIC ROCKS NORTH OF THE EL PILAR FAULT

North of the El Pilar fault metamorphic rocks are exposed from the western end of the Araya Peninsula to the eastern tip of the Paria Peninsula. In Trinidad metamorphic rocks underlie the North Range. Presently, four different stratigraphic columns have been proposed for the Venezuelan portion of this mountain belt. In Venezuela, Schubert (1972) has described the stratigraphy of Araya, Seijas (1971) and Bladier (1977) the central part of the Venezuelan Cordillera, and Gonzalez de Juana and others (1965) the Paria Peninsula. Saunders (1974) summarized the stratigraphy of the North Range.

The map area described here includes part of the central segment of the eastern Cordillera de la Costa of Venezuela. This metamorphic terrane consists of greenschist facies metasedimentary and metaigneous rocks. The highest grade rocks are characterized by the mineralogical association albite-chlorite-actinolite-epidote-calcite (Copey Formation; Seijas, 1972). The metasedimentary rocks include phyllites, metaconglomerates, and marbles. Fossils are sparsely present in the marbles. Metaigneous rocks include metatuffs, spilitized pillow lavas and serpentinite. The entire metamorphic sequence has been folded and thrust toward the south. In addition, the central part of the eastern Cordillera has been intruded by several small igneous bodies of dacitic composition that give K/Ar dates of 5 Ma (Sifontes and Santa Maria, 1972).

Seijas (1972) and Gonzalez de Juana and others (1965) worked out a stratigraphy for the central part of the eastern Cordillera, the basis of which is a simple homoclinal structural interpretation. The early structural models have since been modified by the discovery of major thrust faults in the Cordillera (Bladier, 1977). The few fossils recovered from the metasedimentary rocks are also incompatible with the original stratigraphic order. In light of these findings, and others to be described below, a new stratigraphy is proposed for the central part of the eastern Cordillera (Fig. 4). The major revisions involve reordering of the stratigraphic column and redefinition of some formations.

Copey Formation

Rock types within the formation include metalavas with well-developed pillow structures, metatuffs, serpentinite, quartz-chlorite-epidote schists and metaconglomerates with clasts of basalt, limestone, and serpentinite. The type section is exposed on the highway between Carupano and Playa Copey (Fig. 5). Because the upper contact of the formation is north of the coastline and the lower contact is a thrust fault (Bladier, 1977), only a minimum thickness (700 m) can be estimated.

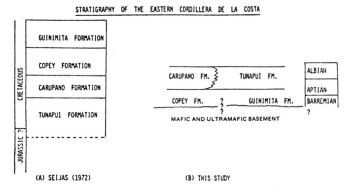

Figure 4. Stratigraphic columns for the central part of the eastern Cordillera de la Costa; (a) is from Seijas (1972) and (b) is that suggested in this study. The revisions shown in (b) are based primarily on new faunal ages (Macsotay, 1968, and personal communication, 1976) and structural interpretations of faulted contacts.

The type section of the Copey Formation consists of some 200 m of sericitic schists and phyllites containing beds of metatuffs. The schists are in turn overlain by a sequence of conglomerates with associated sericite and epidote schists. Clasts in the conglomerates are primarily mafic metavolcanics and less than 1 cm in diameter, but blocks of limestone up to 2 m in diameter are also present. Overlying the conglomerates are at least 300 m of metatuffs and epidote-rich schists, containing lenses of chert and some conglomerates with clasts of serpentinite. Finally, the upper 200 m of the formation are pillow basalts and serpentinized peridotite.

Several aspects of the Copey stratigraphy are puzzling and imply that the present sequence of rocks is not a primary stratigraphy but one that has been reordered by deformation:

1. The serpentinite in the upper part of the formation implies tectonic disruption.

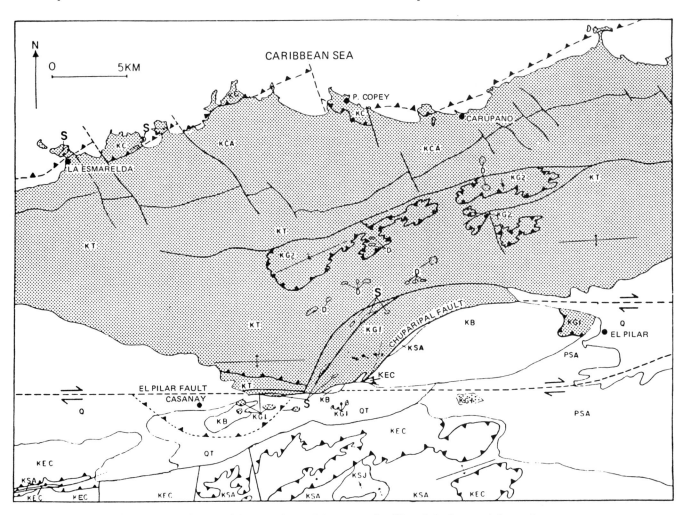

Figure 5. Geologic map of the central part of the eastern Cordillera de la Costa and the northernmost Serrania del Interior. The northern half of the map is modified after Bladier (1977); the southern half is from Vierbuchen (1978). Metamorphic rocks are shaded. Ages are abbreviated as K - Cretaceous, P - Paleocene, Q - Quaternary. Formations are KC - Copey, KG2 - Guinimita (distal facies), KG1 - Guinimita (proximal facies), KCA - Carupano, KT - Tunapui, KB - Barranquin, KEC - El Cantil, KSA - San Antonio, KSJ - San Juan, PSA - Santa Ana, QT - Quaternary terraces and alluvial fan deposits, Q - Quaternary alluvium, S - Serpentinite, D - Dacite (K/Ar age of 5 Ma).

2. The presence of some disruption is confirmed by numerous minor, intraformational, pre- and post-metamorphic thrust faults west of Carupano. For example, near La Esmeralda the major serpentinite body in the central Cordillera is well exposed (Fig. 5) and is clearly thrust southwards over adjacent metatuffs. The contact is visible in sea-cliffs north of the town. Serpentinite in the upper part of the cliffs overlies the tuff. The two rock types are separated by a breccia zone of sheared serpentinite fragments that dips gently northward.

3. At two levels in the formation there are conglomerates bearing clasts of overlying and supposedly younger metaigneous rocks.

These observations indicate that the Copey Formation was deposited in a tectonically and volcanically active setting. The conglomerates in the formation suggest nearby exposures of deformed oceanic crust and marine sediments. The structural style is indicative of compressive deformation. In summary, the evidence suggests deposition near a convergent plate boundary. Rock types and structural features are consistent with those known to occur in fore-arc basins and accretionary wedges. A convergent margin setting is further supported by high-pressure/low-temperature metamorphism along strike in the Araya Peninsula (Schubert, 1971) and to the north on Isla Margarita (Maresch, 1971).

Rocks similar to the Copey Formation, exposed in the Cordillera de la Costa, have undergone a clockwise rotation of 90° since acquiring remnant magnetization and are thought to be obducted remnants of a proto-Lesser Antilles island arc that migrated eastwards during the Late Cretaceous and Tertiary (Skerlec and Hargraves, 1981). Similar relationships may pertain to the Copey Formation.

Guinimita Formation

The Guinimita Formation consists of metaconglomerates, meta-arenites, metalimestones, and quartz-sericite phyllites (Gonzalez de Juana and others, 1965). Fossils from the limestones indicate a probable Barremian age, but may be as young as Aptian (the assemblage comprises the following: Rudisto-*Amphitus Amphitus coelus* faunas as well as *Trigonia tocaimeans?, Exogyra latissima, Nerinea sp., Nerinella sp.,* identified by Macsotay, 1976, oral communication). Because an Albian fauna has now been tentatively identified in the Tunapui Formation (see below), the relative age relationships assumed by previous workers may be incorrect.

Within the central Cordillera the Guinimita crops out in two separate belts (Fig. 5). One of these belts is just north of the El Pilar fault, roughly between the towns of Rio Casanay and El Pilar. There the Guinimita forms an arc-shaped belt that separates the Tunapui Formation from unmetamorphosed Cretaceous rocks to the south. Farther north is a second belt of Guinimita that is surrounded by the Tunapui Formation.

The southern belt is about 300 meters thick (Bladier, 1977). It consists of light-grey limestones, only slightly recrystallized and locally quite fossiliferous, meta-arenites, metaconglomerates containing clasts of both dark- and light-colored quartz, and lesser amounts of grey to black phyllites. Chert is locally present. The northern belt of Guinimita was first recognized by Bladier (1977). He estimated a stratigraphic thickness of about 500 meters. Its lithology consists primarily of thin-bedded, black, recrystallized limestones, quartzo-feldspathic and sericitic schists, grey marbles containing few fossils, small amounts of metaconglomerates with clasts of smokey quartz, and microconglomerates.

The northern belt of Guinimita rocks probably represents a more distal facies than the southern belt, because the clastic rocks are finer grained and the limestone facies suggests deposition in relatively deeper water. The fauna from the northern belt is slightly younger (not Barremian) than the fauna from the southern belt (Victor Campos, 1976, oral communication). Considering these facts, there seems little justification for including the two belts in a single formation. However, until more detailed stratigraphic studies have been completed, the northern belt will be considered part of the Guinimita Formation.

Serpentinite crops out at several places within the southern belt of the Guinimita Formation (Fig. 2). At a quarry near the town of La Pena de Zulia, the contact between sedimentary rocks and a serpentinite body is exposed. Clasts of serpentinite have been found in the overlying Guinimita rocks (V. Campos, 1976, oral communication), implying that the contact was originally sedimentary. Thus it appears that both the Guinimita and Copey formations were, at least in part, deposited unconformably above serpentinite.

Carupano Formation

At the type locality on the coastal highway east of Carupano, the Carupano Formation consists of calcareous phyllites interbedded with dark-colored recrystallized limestone in beds up to 50 cm thick, graphitic-calcareous quartz-mica schists, and rarely, microconglomerates of dark quartz.

The Carupano Formation crops out in an east-west trending belt in the northern part of the Cordillera (Fig. 5). The upper and lower contacts of the formation are not well exposed and may be either sedimentary (Seijas, 1972) or tectonic (Bladier, 1977). The thickness of the formation is about 750 meters, but it is strongly folded, preventing precise measurement.

Tunapui Formation

This formation comprises a thick sequence of quartz-mica schist, metaconglomerate, and minor amounts of quartzite. Tectonic repetition makes it difficult to measure the thickness of the Tunapui Formation, but it is estimated to be 2.5 km (Seijas, 1972).

Tunapui rocks crop out at the northern limit of the map area (Fig. 2). They are strongly folded and in fault contact with the Guinimita Formation. Near the contact of the two formations an Albian fauna (uncoiled ammonites: *Hamites sp., Idiohamites sp., Pseudhelicoceras*) has been identified (Macsotay, 1976, personal communication). The age of the fauna suggests that the Tunapui Formation may be younger than the Guinimita.

SUMMARY OF THE STRATIGRAPHY AND HISTORY OF THE EASTERN CORDILLERA DE LA COSTA

Figure 4 compares my view of the stratigraphy with that of Seijas (1972). Justification for suggesting this new stratigraphic column includes:
1. An improved understanding of the structure of the eastern Cordillera that contradicts the older column especially:
   a. The recognition, by Bladier (1977), of the klippen of Guinimita thrust over the Tunapui Formation. (The northern belt of Guinimita outcrops.)
   b. The new structural interpretation of the southern belt of Guinimita outcrops (see below).
   c. The recognition of tectonic disruption of the Copey Formation.
2. Fossils recovered from the Guinimita and Tunapui formations implying that the former is older.
3. Field relationships and gravity anomalies (see below) indicate that a mafic to ultramafic basement underlies the metasedimentary rocks in the central part of the eastern Cordillera.

The new stratigraphic column suggests a complex history for the development of the eastern Cordillera. The first event in the history is the tectonization, exposure, and erosion of oceanic crust that is implied by the presence of conglomerates bearing serpentinite clasts in the Copey and Guinimita formations, the serpentinite inliers in the Guinimita Formation, and the serpentinite bodies of the Copey Formation. This event was followed by (and likely in part synchronous with) deposition of the Copey and Guinimita formations.

The Copey Formation contains no fossils and is therefore of uncertain age. The Guinimita Formation is, at least locally, in depositional contact with serpentinite but there is no evidence to prove a similar contact in the Copey. Imbricate thrusting of metasedimentary rocks and mafic and ultramafic rocks in the Copey suggests that it was part of an accretionary wedge. If so, rocks in this formation could be of diverse ages and might be both older and younger than the Guinimita, which may have accumulated in the adjacent fore-arc basin.

Deposition of Copey and Guinimita rocks was followed by deposition of the Carupano and Tunapui formations. The Tunapui Formation contains an Aptian-Albian fauna and is therefore at least partially younger than the Guinimita and is assumed to overlie that formation. The Carupano Formation is devoid of fossils. It is more calcareous and contains fewer and finer grained clastic rocks than the Tunapui Formation, and is proposed to be a more distal, lateral facies equivalent of the Tunapui. Sometime after the deposition of these two formations, the sedimentary rocks that now compose the eastern Cordillera were metamorphosed and thrust southwards.

STRUCTURAL GEOLOGY

Deformation of the Sedimentary Terrane

Previous maps of this area (Metz, 1964; Vignali, 1977) show a structure that is largely autochthonous, characterized by broad open folds, minor or no thrust faults, and many vertical faults. My interpretation is different. Detailed studies have shown that thrust faults are pervasive and control the structural style.

Major Thrust Faults

Unequivocal evidence for a low-angle thrust fault was found in the field and led in turn to the postulation of at least two additional overthrusts, one structurally above and the other structurally below the observed fault (Fig. 2).

The best exposed thrust occurs near the base of the El Cantil Formation and has displaced the overlying rocks towards the south. Most of the central portion of the map area (Fig. 2) is underlain by this allochthonous belt of El Cantil limestones. The fault is well exposed on the road between Juan Sanchez and Zorro. This road crosses steep hills underlain by north-dipping limestones. One kilometer from Zorro, at the edge of a broad valley, exposures suddenly change to tightly folded cherts, limestones, and sandstones of the San Antonio Formation. The thrust contact is visible in a road cut and small quarry (Fig. 6). There massive grey El Cantil beds, dipping 25°N, are underlain by a breccia zone 2 to 3 meters wide that contains fragments of El Cantil and San Antonio lithologies. The breccia zone passes into San Antonio rocks, which are deformed into nearly isoclinal folds overturned toward the south. The San Antonio Formation underlies the broad valley south of the contact. This valley was termed the Cangrejal Syncline by Metz because it is flanked on the north and south by hills of the older El Cantil Formation. However, at both sides of the valley the El Cantil actually overlies the San Antonio, proving that the valley is not a syncline but a window eroded through a major thrust sheet. The same relationship occurs between Cerro El Pato and Rio Cumacatal (Fig. 2). The thrust fault is here named the El Zorro Thrust, because it is best exposed near the town of Zorro.

The total length and displacement of the El Zorro Thrust are not known. The fault extends more than 30 km across the map, and the minimum displacement, determined by measuring across strike the distance between the window farthest north and the klippe farthest south, is 5 km. The fault moved simultaneously with deposition of the unnamed post-Middle Eocene formation. The toe of the fault must have moved across an erosional surface (see section above on the unnamed post-Middle Eocene formation).

North of the El Zorro Thrust, and south of the El Pilar fault, a second thrust fault is inferred to place the Barranquin Formation over the El Cantil. Two kilometers southeast of Casanay, Barranquin rocks crop out as an inlier in the Pleistocene basin. The exposed Barranquin generally strikes east-west and dips to-

Figure 6. Exposure of the El Zorro thrust on the road between Cangrejal and Zorro (Fig. 2), approximately 1 km west of Zorro. Massive beds on the left are El Cantil limestones (Aptian-Albian). The underlying thin-bedded unit on the right is the San Antonio Formation (Campanian-Maestrichtian). Note the nearly isoclinal folds in the San Antonio Formation at the right side of the photograph. Axes of these folds are plotted in Figure 7.

wards the north. Less than two kilometers south of these exposures, El Cantil rocks outcrop and also dip to the north. Although the contact is not visible, the inverted stratigraphic relationship requires that faulting has juxtaposed the two formations. Otherwise, given the thickness of the units, a balanced cross section could not be constructed.

Two observations imply that Barranquin rocks north of the El Pilar fault in the Cerros de las Minas are also allochthonous. East of Nueva Colombia, limestones identical to those in the El Cantil Formation appear to underlie the Barranquin Formation (see discussion in the section on the El Cantil Formation). Also, to the north, a sliver of chert crops out between Barranquin sandstones and shales and the metamorphosed Guinimita Formation. Metz (1964) reported that Maestrichtian faunas are present in the cherts. I interpret the cherts as a fault wedge derived from the San Antonio Formation (see cross section B, Fig. 2). The implication is that younger rocks maybe present beneath the Barranquin Formation. For these two reasons, it is proposed that the Cerros de las Minas, between Nueva Colombia and El Pilar, are allochthonous and thrust above younger rocks.

The thrust faults underlying the Barranquin Formation in the Cerros de las Minas and south of Casanay have been trun-

cated by the El Pilar fault. This demonstrates that movement on the El Pilar fault postdates, at least in part, the compressive deformation recorded in the Mesozoic sediments. The truncation could have been accomplished by either vertical movement (southern block upthrown) or strike-slip movement. Upwards movement of the southern block is not consistent with the earlier history recorded in Quaternary strata (see above). Moreover, the great length and straightness of the El Pilar fault strongly imply that it is dominantly a strike-slip fault.

If the thrust sheet in the Cerros de las Minas was broken by striking-slip displacement, it potentially provides a marker for determining the magnitude of that displacement. The "toe" of the sheet might be present on the south side of the fault. I propose that the Barranquin rocks immediately south of the El Pilar fault, between Casanay and Rio Casanay, are part of the displaced "toe" of the Cerros de las Minas thrust plate. As already explained, Barranquin rocks south of the fault are probably also allochthonous and thrust over the El Cantil Formation. Further, they appear to be from the same stratigraphic level (Morro Blanco Member) of the Barranquin Formation that is exposed in the southern part of the Cerros de las Minas. Realigning the thrust plate south of the El Pilar fault with the Cerros de las Minas

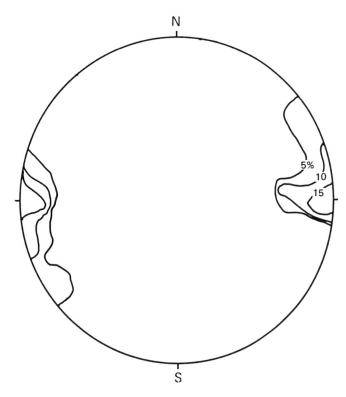

Figure 7. Fold axes from exposures of the San Antonio Formation that are beneath the El Zorro thrust between Cangrejal and Zorro. The contours are based on 125 measurements.

requires a dextral displacement of 10 to 15 km. The displacement must have occurred after thrusting and is therefore of post-Eocene age. The magnitude of movement agrees with that proposed by Metz (1964).

Folding

The axes of major and minor folds in the map area trend east-west to east-northeast. Major folds are generally broad, symmetric structures, some of which have axial lengths over 15 km. These folds are interpreted as rootless, fault-ramp folds. Minor folds are common only within the thin-bedded San Antonio Formation and parts of the Barranquin Formation.

Close association in space of minor folds in the San Antonio Formation with thrust faults, for example, where the El Zorro Thrust is exposed between Juan Sanchez and El Zorro (Figs. 2, 6), indicates that the minor folding was caused by drag during fault movement. Thrust faults must have moved in a direction approximately perpendicular to the fold axes (Fig. 7), that is, from north to south.

Steeply Dipping Faults South of the El Pilar Fault

Steeply dipping faults striking east-northeast and north-south cut sedimentary strata south of the El Pilar fault. The relative age of the two directions of faulting is unknown, but both

sets appear to displace thrust faults. The east-northeast trending Rio Grande Fault is interpreted as a steeply north-dipping reverse fault. In the only exposure of the fault zone, fracture cleavage, minor faults, and tectonic pods all dip vertically. Nevertheless, the intersection of the trace of the fault with topography reflects a steep northern dip. At the eastern end of its exposure, offset on this fault is probably less than 200 m. The north-south striking, steeply dipping faults have a dominately vertical direction of displacement. Where these fault zones are exposed, they commonly contain slickensided surfaces with lineations in both strike and dip directions. Where the separation can be estimated, it is normal and between 50 and 100 m.

Deformation of Quaternary Strata Near the El Pilar Fault

The Quaternary sediments near the El Pilar fault are so poorly exposed that it was not possible to develop a clear picture of their deformational history. However, several scattered exposures provide important clues that the El Pilar fault has been active during Quaternary time. For example, west of El Pilar, in two locations near the fault, young gravel beds have been tilted to vertical dips (Fig. 2). West of Nueva Colombia, the Quaternary is less deformed but broad folds are apparently present. If better exposures existed, some of these areas would undoubtedly reveal intense Quaternary deformation caused by strike-slip faulting.

Summary

Mesozoic and Paleogene deposits in the map area have been deformed by folding and thrusting. Contractional deformation probably occurred during or after Middle to Late Eocene time and before the Quaternary, and was caused by compressive stresses oriented in a roughly N-S direction. During or after thrusting, and probably before the Pleistocene, steeply dipping reverse and normal(?) faults were also active. This later deformation was of considerably less magnitude and, within the map area, limited to minor displacements on east-northeast and north-south trending faults. There is no evidence for more than one period of regional folding. If the various pre-Eocene unconformities postulated by previous workers exist, they are not associated with detectable deformation. The El Pilar fault has been an active dextral strike-slip fault during the Quaternary and may have been active in the pre-Quaternary.

Deformation of the Metamorphic Terrane

The eastern Cordillera de la Costa is underlain by a poly-deformed, low-grade metamorphic terrane. Deformation involved three and locally four periods of folding, pre- and post-metamorphic thrusting, and vertical faulting. Bladier (1977) has identified, in the central part of the eastern Cordillera, a first generation of isoclinal folding, contemporaneous with the earliest stages of metamorphism. These folds were refolded by the second, most intense period of folding, which included development of a pervasive axial plane cleavage. F_2 axial plane cleavage is folded into the broad F_3 macrofolds visible on the geologic maps of the area. All three generations of folds have east-west trending axes.

Bladier envisioned two periods of thrusting, one prior to metamorphism and one after. The two fault-bounded blocks of the Guinimita Formation (Fig. 5) were interpreted by Bladier as having been emplaced as a single sheet during the first phase of thrusting. The second phase of thrust faulting produced the Copey-Carupano contact and numerous intraformational thrusts that truncate foliation.

Vertical faults in two directions postdate thrust faults in the eastern Cordillera (Seijas, 1972; Bladier, 1977). Long, east-west trending faults of minor vertical displacement are common. Shorter north-south to north-northwest trending faults are also present. These two families of faults are approximately parallel to the vertical faults south of the El Pilar, and hence may have a common origin.

Small intrusive bodies of dacitic composition are exposed in the central part of the eastern Cordillera de la Costa. These rocks give K/Ar dates of 5 m.y. Sifontes and Santa Maria, 1972). The intrusions deform metamorphic foliation at their borders. All the known bodies lie in a northeast trending belt that has a large associated negative gravity anomaly. Gravity models indicate that a major intrusion may be present at depth (see below).

The Southern Guinimita Belt and the El Pilar Fault

Bladier (1977) proposed that the two belt of the Guinimata Formation (see section on stratigraphy of metamorphic rocks) were once part of a large thrust sheet that was displaced south-westwards and later breached by erosion. Several observations argue against connecting the two belts of the formation in this way. Serpentinite bodies are concentrated along the decollement surface in the southern belt (see below). However, there is no serpentinite known in the northern belt. Contacts between the northern belt and the underlying Tunapui Formation are well exposed and consist of thick, gently dipping zones of brecciated carbonate. Thus, the basal contacts of the two belts of Guinimita are very different.

The deformation associated with the contacts of the southern belt is also quite different from that at the contacts of the northern belt. The contact at the western edge of the southern belt does not have a shallow dip. It forms a straight line across rugged topography, which implies a near-vertical attitude. To the south this contact is terminated by the El Pilar fault. In most localities, beds adjacent to the contact are vertical or steeply inclined towards the south. For example, on the northwest side of the southern belt (Fig. 2), folded Tunapui rocks are exposed. Farther east, Tunapui rocks turn abruptly to the vertical. The next outcrops are the Guinimita Formation with beds dipping at 70° or more towards the southeast. The Guinimita beds are sheared and minor folds are generally overturned towards the northwest. Still farther west, Guinimita beds flatten to gentle dips.

The east side of the southern belt of Guinimita is in fault contact with the Barranquin Formation. To the south this fault is terminated by the El Pilar fault. The contact with the Barranquin Formation was named the Chuparipal thrust fault by Metz (1964). It has been interpreted as a remnant of a thrust fault that transported the metamorphic terrane from north to south and over the sedimentary rocks (Christensen, 1961; Metz, 1964; and Bladier, 1977). Such an interpretation is consistent with the locally shallow dip of the contact as well as with the east-west striking fold axes, formational contacts, and minor thrusts exposed throughout the Cordillera. Nevertheless, my mapping in the vicinity of the contact has shown that the deformation there is very different from that exposed in the rest of the Cordillera, and therefore an explanation other than simple southward thrusting is required.

Within several kilometers of the contact, east-west trending, horizontal F_2 fold axes, pervasive in the metamorphic terrane, are steeply refolded by north-south striking folds that are not present outside of this zone (Figs. 2, 8, and 9). Also unique to this locality are numerous minor faults that dip gently to the east and west and offset the metamorphic foliation (Fig. 10). Slickensides measured on the fault surfaces almost invariably trend east-west (Fig. 11). In addition, many east-west striking vertical tension gashes are present.

The conclusions to be drawn from these observations are

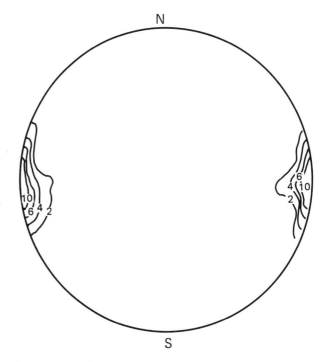

Figure 8. Axes of F-2 folds in the Guinimita Formation (from Bladier, 1977), approximately 5 km north-northwest of the fold axes measured in the same formation that are plotted in Figure 9. The east-west trending fold axes at this locality are typical of the direction of F-2 fold axes throughout the central part of the eastern Cordillera. The only significant exception occurs near the Chuparipal fault where F-2 axes are steeply refolded (see Fig. 9). The contours are based on 46 measurements.

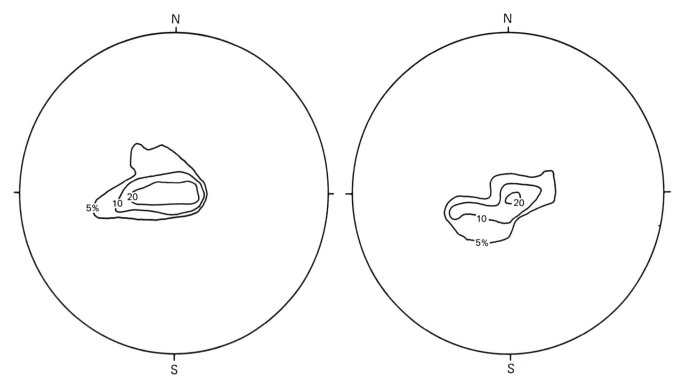

Figure 9. Axes of F-2 folds in the Guinimita Formation measured at outcrops 1 to 2 km northeast of Nueva Colombia (see Fig. 2). In this area, F-2 folds, which normally trend east-west (Fig. 8), have been steeply refolded about a north-south axis. This refolding is known to occur only along the Chuparipal fault, and parallel faults, in the southern belt of the Guinimita Formation. Contours are based on 110 measurements.

Figure 10. Poles to minor faults in the Guinimita Formation, 1 to 2 km northeast of Nueva Colombia (see Fig. 2), near the Chuparipal fault. These faults cut the metamorphic foliation, and F-2 folds and are only known to occur near the Chuparipal fault. Contours are based on 75 measurements.

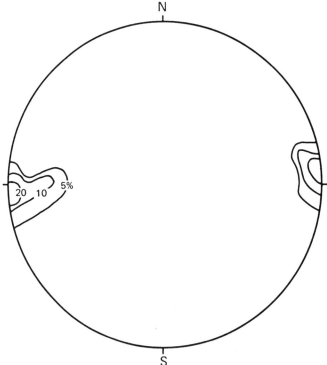

Figure 11. The orientation of slickensides on the minor faults plotted in Figure 10. Contours are based on 75 measurements.

two-fold. An additional phase of deformation is recorded near the (Chuparipal) fault contact between sedimentary and metamorphic rocks north of the El Pilar fault. This deformation is only present close to the fault and is therefore assumed to have been caused by the fault. Minor structures described above show that the deformation resulted from compressive stresses that were approximately horizontal and oriented in an east-west direction. The unique association between minor structures indicative of east-west contraction and the low-angle Chuparipal fault strongly suggests that the fault results from thrusting in an east-west direction, during an isolated and late phase of deformation.

It is proposed that this fault, the Chuparipal thrust fault, is a branch of the El Pilar fault system. Minor structures show that east-west contractional deformation occurred along this curved branch. Such contraction would be predicted to occur along a transpressive strike-slip fault (Sylvester and Smith, 1976). It is further postulated that the curved fault segment was eventually truncated by a younger southern branch (the El Pilar fault of Metz and previous workers) that is more nearly parallel to the direction of displacement. Analogous truncation of curved strike-slip faults is seen in photo-elastic models (Freund, 1974).

Although the Chuparipal thrust has a shallow dip at the surface, it probably steepens at depth and the displacement likely includes a significant amount of strike-slip. Therefore, to eliminate confusion about the origin of this structure, it is proposed that the name "Chuparipal thrust fault" be shortened to Chuparipal fault.

The total strike-slip displacement on the fault is unknown, but some pertinent evidence exists immediately west of the town of El Pilar, where the Guinimita Formation overlies Barranquin sandstones (Fig. 2). This block of Guinimita was probably transported by a combination of dip-slip and strike-slip along the Chuparipal fault. If so, then the minimum dextral displacement is 10 km.

Both sides of the Guinimita arc are interpreted here as upthrown relative to adjacent rocks. If so, the gross structure of the arc is a horst. The internal structure is complicated and involves major and minor vertical and horizontal faults. It is suggested that uplift of the horst was caused by compressive stresses across the curving northern branch of the El Pilar fault. This structural interpretation is shown in cross section B, Figure 2. The structure of the Guinimita arc is remarkably similar to those produced in sandbox and clay cake models of convergent strike-slip faults (Emmons, 1969; Wilcox and others, 1973). Similar structures, called flower structures, have also been documented in numerous localities on seismic profiles across strike-slip faults (Harding and Lowell, 1981).

The dip-slip on the Chuparipal fault must have been at least two kilometers to produce the klippe west of El Pilar at La Pica. The total amount of overthrusting on the Chuparipal cannot greatly exceed the amount of uplift of the Guinimita arc. The magnitude of uplift can, of course, be guessed from the thickness of strata removed from the top of the arc. Considering the thicknesses proposed for the Tunapui and Guinimita formations by

Seijas (1972) and Bladier (1977), uplift of the arc is probably less than 3 to 4 km.

The recognition of this branch in the El Pilar fault systems modifies field constraints on the maximum amount of strike-slip movement. Total dextral displacement across the system must now be considered equal to that proposed above for the southern branch plus an unknown additional amount, greater than 10 km, on the northern branch. Thus, the minimum total displacement is 20 km, and the maximum displacement is not constrained by field observations. Such an interpretation obviously has important implications for models of the tectonic history of the southern Caribbean.

The El Pilar fault system can be redefined as consisting of at least two branches, that in northeastern Venezuela are the boundary between sedimentary and metamorphic terranes. The two terranes have had very different histories, but are now juxtaposed. This implies major displacement by faulting, either thrusting from north to south, dextral strike-slip movement, or, most likely a combination of the two.

GRAVITY STUDIES IN NORTHEASTERN VENEZUELA

The Regional Gravity Field

Several prominent features are obvious on the regional Bouguer anomaly map of the southeastern Caribbean (see map of Bowin, 1976). Two deep lows distort the gravity field: one parallels the Atlantic side of the Lesser Antilles Island Arc and is caused by the great thickness of sediment in front of the arc. The other low is over the Eastern Venezuela Basin. Bonini (1978) has shown that the anomaly over eastern Venezuela is caused by a combination of crustal downwarping and sediment accumulation. The third major feature of the gravity field is a sharp high over the islands of the Lesser Antilles that extends southwestward to Margarita. Bowin (1976) has suggested that the high over Margarita is caused by uplift of metamorphic and igneous basement rocks following the cessation during the Eocene of subduction near Margarita.

The Bouguer gravity over the Eastern Venezuela Basin may be the lowest sea-level value in the world. The Bouguer anomaly map of northern South America (Bonini and others, 1977) shows a minimum of -200 mGal on the coast of the Gulf of Paria (Fig. 12). Total sediment accumulation in the basin approaches 15 km. Moreover, the observed gravity anomalies over the basin suggest that the crust-mantle interface beneath the basin must be depressed at least 10 km (Bonini, 1978).

North of the basin, Bouguer gravity anomalies increase rapidly to +20 mg at the coastline, about 100 km north of the axis of the low in Venezuela. The gradient to the coast is fairly uniform but generally steepens near the El Pilar fault. North of the coast the gradient flattens although gravity values continue to increase to the axis of the Margarita-Lesser Antilles high.

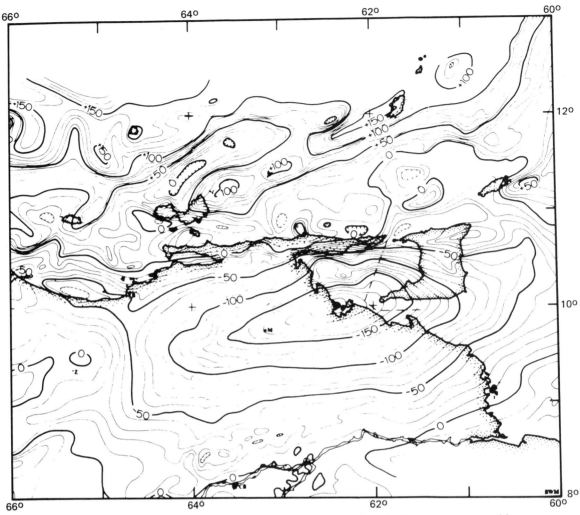

Figure 12. Simple Bouguer anomaly map of northeastern Venezuela and adjacent areas, prepared by
Bonini and others, 1977. Contour interval is 10 mGals.

DETAILED GRAVITY STUDIES NEAR THE EL PILAR FAULT

Figure 13 shows a complete Bouguer anomaly map covering the El Pilar fault zone and the adjacent Caribbean Mountains in Estado Sucre (gravity stations incorporated into the map, in addition to my own field data, were occupied by groups from the Ministerio de Energia y Minas and Cartographia Nacional of Venezuela and from Princeton University). The main features of the map are the steep increase in the gravity field from south to north, and a low south of the city of Carupano (Carupano Low) that is superposed upon this regional increase. Also, at the eastern edge of the map, there is a second low that is caused by thick Quaternary sediments.

Between the western tip of Araya and the eastern shore of the Gulf of Paria, the gravity gradient across the El Pilar fault is uniform except where it is perturbed by the Carupano Low (Fig. 12). The form of the Carupano Low is clearly shown on a residual gravity map (Fig. 14; residual anomalies were calculated

by subtracting a regional gradient obtained from the 1:500,000 maps used to make the maps of Bonini and others, 1977). Thus, the Carupano Low represents a local departure from an apparently consistent density structure across the El Pilar fault in the 150 km long segment between Araya and the Gulf of Paria.

Two, two-dimensional gravity models have been constructed to investigate crustal structure across the El Pilar fault. One model represents a cross section located west of the Carupano Low; the other is through the center of the low (Fig. 13). The model to the west is based on my interpretation of the geology exposed in the eastern Cordillera de la Costa and Serrania del Interior. This model was used as a standard section to study the local perturbation of the Carupano Low.

Obviously, the choice of density values is critical to the gravity interpretation. No density estimates were available from wells or seismic studies. However, the densities of rocks exposed at the surface provide some control for shallow depths. The results of density measurements from samples collected in the map area are summarized in Table 1. An additional constraint on

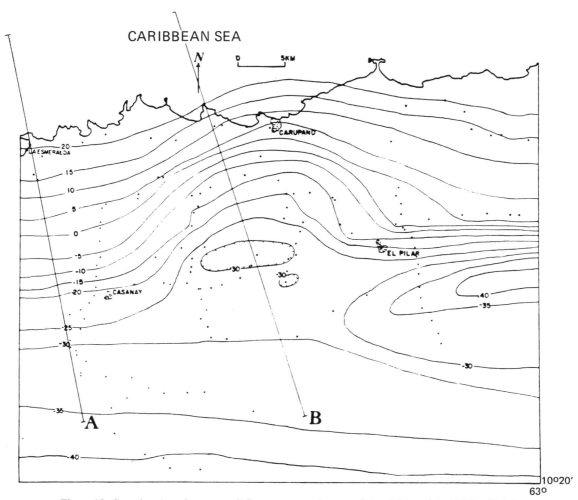

Figure 13. Complete (terrain corrected) Bouguer anomaly map of the vicinity of the El Pilar fault in central Estado Sucre. Lines A and B mark the locations of gravity models of Figure 15 and 16. Black dots are the gravity stations used to make the map.

density values comes from published measurements of density vs. depth of burial for shales in the Eastern Venezuelan Basin (Renz and others, 1958; Hedberg, 1936).

Gravity Model A: This model extends from 10 km north of the coast to 13 km south of the El Pilar fault. The total length of the section is 40 km (Fig. 15). The segment south of the El Pilar fault corresponds to geologic section A-A′, Figure 2.

It was apparent from density measurements and field mapping that the density contrast exposed at ground level is not adequate to account for the steep gravity increase from south to north in the map area. The steepest portion of the gradient associated with Model A is over the El Pilar fault; yet the contrast in bedrock densities across the fault is only about .05 gm/cc. It follows that the cause of the gradient is buried.

The depth to Moho beneath eastern Venezuela is not well known. In Models A and B, the crustal thickness assumed to exist at the coastline is in agreement with the standard sea level crust of 33 ± 2 km calculated by Woollard (1959). The crust was assumed to thin toward the sea and thicken toward the continent.

Seismic refraction data from Edgar (1968) suggest that the depth to Moho beneath the continental shelf of northeastern Venezuela is probably 25 to 30 km. Bonini (1978) has modelled the Moho at a depth of at least 40 km beneath the Eastern Venezuela Basin.

The density structure of the lower crust and upper mantle is also a major unknown. Seismic studies from different parts of the world indicate considerable variation in the density of the lower crust, the upper mantle, and the density contrast across the Moho (see, for example: Talwani and others, 1959; Birch, 1961; Woollard, 1959; Bucher and Smith, 1971; Grow, 1973; Worzel, 1974, 1976). Calculated density contrasts across the Moho are generally between .25 gm/cc and .55 gm/cc. The values used in Models A and B of 2.95 gm/cc for the lower crust and 3.3 gm/cc for the upper mantle agree with the average values calculated by Woollard, 1959.

The effects of the separate parts of Gravity Model A have been isolated and are shown in Figure 15 along with the total calculated anomaly. The crustal thinning, which is assumed to occur toward the sea, accounts for only about one-third of the

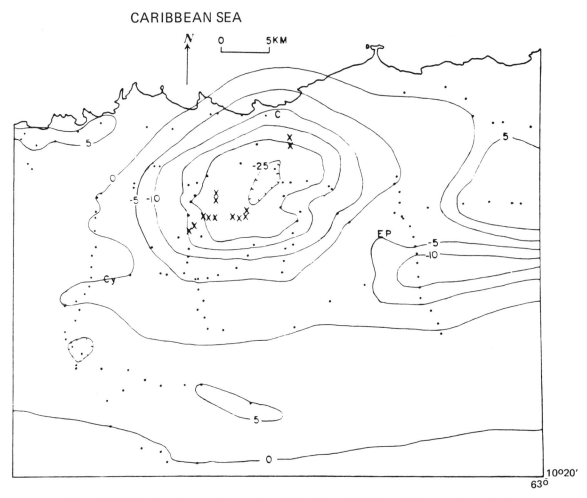

Figure 14. Residual gravity anomaly map derived from Figure 13. X's mark the approximate locations of small dacitic intrusions (K/Ar age of 5 Ma). The 25 mGal residual low (Carupano Low) is thought to be caused by a large, igneous intrusion that underlies the small dacite bodies. C is Carupano, Cy - Casanay, EP - El Pilar.

gravity increase in that direction. Hence, density contrasts in the crust must produce about two-thirds of the gravity gradient; yet as mentioned above, rock densities at the surface are fairly constant. These considerations imply that a strong density contrast may exist at shallow to intermediate depth. Such a density contrast is consistent with the stratigraphic and structural evidence discussed above. This evidence suggests that the metasedimentary rocks of the Cordillera de la Costa are underlain by deformed mafic and ultramafic rocks that were previously part of an oceanic crust. Model A was constructed according to this hypothesis. A single high-density layer in Model A, representing mafic rocks, produces half the gravity increase caused by rocks of the upper crust.

TABLE 1. ROCK DENSITIES (g/cc)

Formation	Lithology	Number of samples	Range	Average
Tunapui	Quartzite	6	2.52-2.58	2.55
Tunapui	Phyllite	27	2.57-2.66	2.61
Carupano	Phyllite	7	2.60-2.64	2.62
Guinamita	Marble	4	2.68-2.73	2.71
El Cantil	Limestone	9	2.59-2.70	2.66
Barranquin	Shale	12	2.51-2.62	2.55

Gravity Model A portrays the El Pilar fault as a vertical discontinuity to a depth of 5 km. Beneath 5 km, crustal densities are assumed to be continuous across the fault. In contrast, the two previous studies of gravity anomalies across the fault (Bonini, 1978; Folinsbee, 1972) interpreted the fault as a thrust contact dipping gently to the north.

In Folinsbee's model, most of the gravity gradient is from a steeply inclined Moho and a lower crust-upper mantle density contrast of .6 gm/cc. This structure required a crustal thickness of 25 km at the coastline. The resulting model still did not produce a sufficiently steep gradient. Bonini assumed densities of 2.7 gm/cc for the metamorphic rocks north of the fault and 2.3 gm/cc for the sediments to the south. My measurements show that the actual values are closer to 2.6 gm/cc and 2.5 to 2.6 gm/cc, respectively. Despite the great density contrast, Bonini's model did not equal the observed gradient. It is in fact very difficult to fit the steep gravity gradient in this section of the Cordillera without assuming that the El Pilar fault has caused a vertical density discontinuity to depths greater than several kilometers.

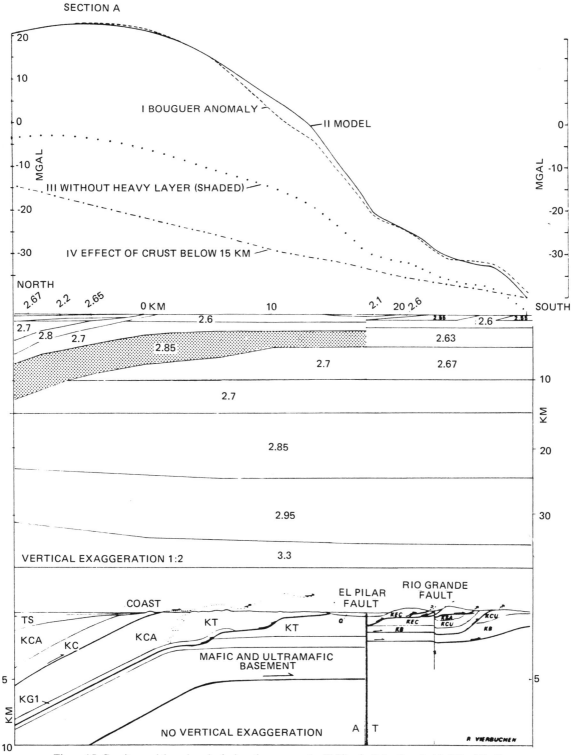

Figure 15. Gravity models and geological sections across the El Pilar fault (section A-A' of Fig. 13). *Top*: observed gravity anomalies and the anomalies calculated from the gravity model. *Middle*: the gravity model used to calculate the anomalies. *Bottom*: the geological section used to constrain the upper 10 km of the gravity model. Formation symbols are defined in Figure 5. Shallow densities in the gravity model were determined by measurements of surface samples (see Table 1). The dotted curve shows the calculated anomaly if the heavy shaded layer in the model (density = 2.85 gm/cc) is replaced by a density of 2.7 gm/cc. This heavy layer represents the basement of the eastern Cordillera, which is believed to be deformed oceanic crust.

The density (or lithologic) discontinuity at the El Pilar fault has important implications for the history of the fault. The contact between the two rock types (i.e., the fault plane) must dip steeply and extend to a depth of at least 5 to 10 km. Because the metamorphic rocks of the Cordillera cannot root to the south (sedimentary rocks of the same age are present immediately south of the fault; see above), it follows that the density discontinuity could not be caused by upward movement of the north side of the fault. On the other hand, it is possible that the discontinuity was caused by southward thrusting of the metamorphic terrane across the El Pilar fault, followed by upward movement of the south block to remove the "toe" of the thrust sheet. However, my study of the Quaternary sediments south of the El Pilar fault does not support the theory that the south block has been uplifted 5 to 10 km; rather, through much of Pleistocene time, the north block apparently moved up relative to the south. The remaining possibility is that large strike-slip displacement has transported the heavy rocks north of the fault into contact with the lighter rocks south of the fault. If so, then the magnitude of this displacement probably exceeds 150 km, because sedimentary rocks of the Serrania extend for at least that distance to the west of the gravity model.

Gravity Model B: Model B extends from 10 km north of the coast to 10 km south of the El Pilar fault. The total length of the section is 40 km (Fig. 16). The segment south of the Chuparipal fault corresponds to geologic section BB', Figure 2. Model B passes through the center of the Carupano Low, a depression in the regional northward increase of Bouguer anomalies. The low has been previously studied by Graterol and Wong (1978), who suggested that it is caused by a buried sedimentary basin. An alternative explanation is presented here.

It is proposed that the Carupano Low is caused by a large, low-density igneous intrusion. Evidence for this hypothesis is shown in Figure 14, where the positions of a number of small dacitic intrusions have been plotted on the residual Bouguer anomaly map. These small intrusions are located close to the axis of the low. The small intrusions have a K/Ar age of 5 m.y. (Sifontes and Santa Maria, 1972). No other igneous rocks of this age are known to exist in the eastern Cordillera. Similarly, there are no other residual Bouguer anomaly lows in the eastern Cordillera analogous to the Carupano Low. Therefore, it is supposed that the two features, the small dacite bodies and the Carupano Low, are related in some way. The relationship suggested here is that the small dacite bodies overlie a large, low-density igneous intrusion that has replaced heavier rocks and thus causes the gravity low.

The hypothesis that the low is caused by a sedimentary basin is rejected for two reasons. First, it does not explain the correlation between the gravity low and the dacite bodies. Second, if a sedimentary basin is present beneath a thin sheet of metamorphic rocks, then it is extremely difficult to account for the regional gravity expression of the El Pilar fault. Model A, which is only several kilometers west of the Carupano Low, shows that the fault causes a significant density contrast to depths of 5 to 10 km.

It follows that the metamorphic terrane of the Cordillera is at least 5 to 10 km thick.

Gravity Model B was derived from Model A with minor adjustments to account for changes in the surface geology and with the addition of lighter material, representing an igneous intrusion and altered country rock, at the axis of the Carupano Low. The residual Bouguer anomaly map shows that the low is not a two-dimensional anomaly. End correlations were therefore made to account for this finite shape. Figure 16 shows that Model B is consistent with the observed Bouguer anomalies.

SUMMARY OF THE CONCLUSIONS DRAWN FROM THE GRAVITY MODELS

Model A shows that the structural interpretation of the eastern Cordillera given above is consistent with the observed Bouguer anomalies. This interpretation involves the presence of a dense layer, presumed to be a tectonized remnant of oceanic crust that is 2 to 3 km thick, beneath the Cordillera. The steep gravity gradient across the El Pilar fault implies that major strike-slip displacement has probably occurred. Rocks equivalent to those in the dense layer on the north side of the fault are not known south of the El Pilar fault in eastern Venezuela. The nearest point at which such rocks could be present is 150 km west, in the Gulf of Cariaco. It follows that the implied minimum right-lateral displacement is 150 km. If a similar argument can be applied as far east as the North Range of Trinidad, then the minimum displacement on the El Pilar fault could be as large as 300 km.

Model B demonstrates that the Carupano Low may be successfully interpreted as a large, low-density intrusion related to the small dacite bodies exposed along the axis of the low. Economically significant mineral deposits in the country rock (primarily silver), which are restricted to the vicinity of the gravity low, are presumed to be related to the intrusion.

CONCLUSIONS

Structural analysis and detailed mapping of the El Pilar fault zone in northeastern Venezuela has modified previously proposed geological constraints on the displacement history of the fault. These constraints were based on correlations between the Cretaceous strata that were believed to crop out on both sides of the fault system in central Estado Sucre, Venezuela. The mapping done in this study has identified a second fault strand (Chuparipal fault) in the El Pilar system. Recognition of this new strand places all Cretaceous sedimentary rocks south of the fault system and all Cretaceous metamorphic rocks north of the fault system. Therefore, the El Pilar can be defined as the fault system that juxtaposed the metamorphic rocks of the eastern Cordillera de la Costa (Venezuela) and North Range (Trinidad) on the north against the very different, unmetamorphosed, coeval sedimentary rocks on the south. It follows that no lithologic correlations can be made between the rocks north and south of the fault system and that the total magnitude of fault displacement is unknown.

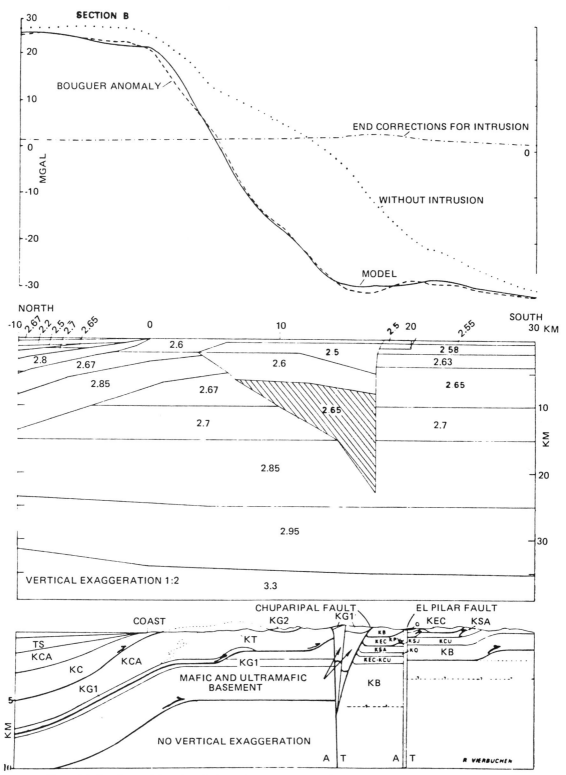

Figure 16. Gravity models and geologic section across the El Pilar fault. See explanation for Figure 15. Section location is B-B' on Figure 13. The shaded area in the model represents an igneous intrusion that is postulated as the cause of the Carupano Low (see Fig. 14). Light rocks above the intrusion result from alteration of the country rock by hydrothermal fluids and by the small dacite bodies that extend upward from the intrusion and crop out at the surface (Fig. 14). The geological section (bottom) is drawn without the intrusion. The gravity model calculated for this section, without the intrusion, is shown by the dotted curve at the top (labeled "without intrusion").

The northern strand of the El Pilar fault system was identified through analysis of minor structures adjacent to a bend in the fault strand. Minor structures demonstrate east-west compressive stresses and dip-slip displacement. Stresses and displacements in this direction are unique in the central Cordillera de la Costa and result from compression across a northerly bend in the east-west trending, right-lateral, El Pilar fault system.

Previously, the Chuparipal fault was thought to be a thrust fault that had transported metamorphic rocks from north to south. However, structural analysis shows that the Chuparipal fault has undergone a combination of dip-slip and strike-slip displacement. Field relationships imply that strike-slip movement has been more than 10 km and that dip-slip has been less than 2 to 4 km. In cross section, the Chuparipal fault is a branch of a pop-up or "flower structure." Such features are common in zones of compressive strike-slip or transpression (Sylvester and Smith, 1976).

Rocks north and south of the El Pilar fault overlap in age but originated in fundamentally different tectonic settings. Barremian to Albian rocks in the north were deposited on deformed oceanic crust, in a volcanically and tectonically active environment, probably a fore-arc setting. Lower Cretaceous rocks to the south were deposited on the craton of South America in a quiescent, shallow-water environment. This striking contrast in history strongly suggests that the rocks across the fault have been juxtaposed by large displacement: either strike-slip, dip-slip, or, most likely, a combination of both.

The steep gravity gradient across the El Pilar fault requires that the fault plane dips steeply to depths of at least 5 to 10 km. This fact, combined with the long, linear trace of the fault and the observed strike-slip focal mechanisms (Molnar and Sykes, 1969; Vierbuchen, 1978), implies that the El Pilar fault is primarily a strike-slip fault.

The high-density crust north of the El Pilar fault in northeastern Venezuela consists of Cretaceous metasedimentary rocks, and mafic and ultramafic igneous rocks. Similar crust cannot exist south of the fault anywhere east of the Gulf of Cariaco because unmetamorphosed, Lower Cretaceous, shallow-marine to non-marine strata are continuously exposed. This fact implies either a right-lateral displacement of at least 150 km or a vertical displacement of at least 5 to 10 km. Because the El Pilar fault does not have the map or cross section characteristics of a dip-slip fault, and because the Pliocene and Quaternary deposits in northeastern Venezuela do not support large vertical uplift south of the El Pilar fault, strike-slip displacement is the best explanation of the field relationships. If this argument is extended to include the

metamorphic rocks north of the fault that occur as far east as the North Range of Trinidad, then the minimum right-lateral displacement on the fault could be as much as 300 km.

The Cretaceous and Paleogene sediments exposed south of the El Pilar fault in the Serrania del Interior have been folded and transported towards the south by thrust faults. In the study area, thrusting was probably post-Middle Eocene, and synchronous with the subsidence of a shallow basin that overlies an Eocene(?) erosional surface. The finding of major thrust faults helped to resolve problems of Cretaceous stratigraphy noted by previous workers.

The history of the metamorphic rocks north of the El Pilar fault in central Estado Sucre, Venezuela, is poorly known. Few fossils are preserved in this region and deformation has disrupted the original stratigraphy. Preliminary identification of some fossils (Macsotay, 1968) and the recognition of imbricate thrust faults in the metamorphic rocks have been used as the basis of a revised stratigraphy for the central part of the eastern Cordillera de la Costa.

Gravity modelling has shown that small, dacitic, Pliocene (Sifontes and Santa Maria, 1972) intrusives exposed in the central part of the eastern Cordillera de la Costa overlie a large intrusive body. The borders of the intrusion are well defined by a residual gravity low of 25 mGal. Economically significant concentrations of silver and other minerals also lie within the gravity low and are probably localized by the intrusion.

ACKNOWLEDGMENTS

Much of the work described in this paper was done while I was a graduate student at Princeton University. Bill Bonini, John Suppe, Jason Morgan, Al Fischer, and Hollis Hedberg, all of the Princeton faculty, provided advice and many valuable suggestions that improved this paper.

My field work in Venezuela was supported by the Ministerio de Energia y Minas. Alirio Bellizzia, Victor Campos, Oliver Macsotay, and Ivon Bladier, all employees of the Ministerio, reviewed the preliminary results of my field mapping along the El Pilar fault. Oliver Macsotay also provided several faunal ages for the metasedimentary units of the eastern Cordillera.

John Maxwell, Jim Case, Reginald Shagam, Bob Hargraves, and Bill Bonini reviewed this paper. Their suggestions resulted in numerous improvements. Partial support for my work in Venezuela was received from NSF grant GA-31877 and from the Princeton Caribbean research fund.

REFERENCES CITED

Alberding, H., 1957, Application of principles of wrench-fault tectonics of Moody and Hill to northern South America, Geol. Soc. Amer. Bull., v. 68, n. 6, p. 785–790.

Birch, F. J., 1961, Composition of the earth's mantle: Geophys. J., v. 4, p. 295–311.

Bladier, I., 1977, Rocas Verdes de la region de Carupano, Venezuela, plano de Despeque de Corrimientos: *in* Boletin de Geodinamica, Comision Internacional de Geodinamica, Grupo 2, Caras, p. 35–49.

Bonini, W. E., 1978, Anomalous crust in the eastern Venezuela Basin and the Bouguer gravity anomaly field of northern Venezuela and the Caribbean borderland, Geologie en Mijnbouw, June, v. 57, p. 117–122.

Bonini, W. E., Pimstein de Gaeta, C., and Graterol, V., 1977, Mapa de anomalias gravimetricas de Bouguer de la parte norte de Venezuela y areas vecinas, Escale 1:1,000,000: Ministerio de Energie y Minas, Direccion de Geologia, Caracas.

Bowin, C., 1976, The Caribbean: gravity field and plate tectonics: Geol. Soc. Amer. Special Paper 169, p. 39.

Bucher, R., and Smith, R., 1971, Crustal structure of the eastern Basin and Range province and the northern Colorado Plateau from Phase velocities of Rayleigh waves: *in* The structure and physical properties of the Earth's crust, Am. Geophys. Union, Geophys. Monogr., no. 14, p. 59–70.

Bucher, W., 1952, Geologic structure and orogenic history of Venezuela, Mem. no. 49, Geol. Soc. Amer., 113 p.

Christensen, R. M., 1961, Geology of the Paria-Araya Peninsula, northeastern Venezuela [Ph.D. thesis]: Univ. of Nebraska, 112 p.

Edgar, N. T., 1968, Seismic refraction and reflection in the Caribbean Sea [Ph.D. thesis]: Columbia University, 43 p.

Emmons, R. C., 1969, Strike-slip rupture patterns in sand models: Tectonophysics, v. 7, no. 1, p. 71–87.

Folinsbee, R. A., 1971, The gravity field and plate boundaries in Venezuela [Ph.D. thesis]: Mass. Inst. Technology and Woods H ole Oceanographic INst., 159 p.

Freund, R., 1974, Kinematics of transform and transcurrent faults: Tectonophysics, v. 21, no. 1-2, p. 93–134.

Gonzalez de Juana, C., Munoz, N. G., and Vignali, M., 1965, Reconocimiento geologico de la parte oriental de la Peninsula de Paria: Asoc. Venezolana Geol. Min. y Petroleo, Bol. Inf., v. 8, p. 255–279.

Gonzalez de Juana, C., and Vignali, C., 1972, Rocas metamorficas y igneas en la Peninsula de Macanao, Margarita, Venezuela: *in* Memoirs of the Sixth Caribbean Geological Congress, p. 63–68.

Gonzalez de Juana, C., Santamaria, F., and Navarro, F., 1974, A few considerations on the age origin and relations of the Dragon Gneiss, Paria peninsula, Venezuela: *in* Contributions to the geology and paleobiology of the Caribbean and adjacent areas, Naturforsch. Ges. Basel, Verh., v. 84, no. 1, p. 153–163.

Graterol, V., and Wong, Jaime, 1978, Anomalia de bouguer residual en la region nor oriental de Venezuela: paper presented at the 8th Caribbean Geological Congress, Geologie en Mijnbouw, v. 57, p. 377.

Grow, J. A., 1973, Crustal and upper mantle structure of the central Aleutian arc: Geol. Soc. Amer. Bull., v. 84, p. 2169–2192.

Guillaume, H. A., Bolli, H. M., and Beckman, J. P., 1972, Estratigrafia del Cretaceo inferior en la Serrania del Interior, oriente de Venezuela: *in* Congresso Geologico Venezolano, 4th, Memoria, v. 3, Venez., Dir. Geol., Bol. Geol., Publ. Esp., no. 5, p. 1619–1655.

Harding, T. P., and Lowell, J. D., 1979, Structural styles, their plate-tectonic habitats, and hydrocarbon traps in petroleum provinces: Am. Assoc. of Petroleum Geol. Bull., v. 63, no. 7, p. 1016–1058.

Hedberg, H. D., 1936, Gravitational contraction of clays and shales: Am. Jour. Sci., v. 31, no. 184, p. 241–287.

—— 1937, Stratigraphy of the Rio Querecual section of northeastern Venezuela: Geol. Soc. Amer. Bull., v. 48, no. 12, p. 1971–2024.

—— 1950, Geology of the Eastern Venezuelan Basin (Anzoategui-Monagas-Sucre-Eastern Guarico Portion): Geol. Soc. Amer. Bull., v. 61, no. 11, p. 1173–1216.

Hess, H. H., and Maxwell, J. C., 1953, Caribbean research project: Geol. Soc. Amer. Bull., v. 64, p. 1–6.

Jordan, T., 1975, The present-day motions of the Caribbean plate: J. Geophys. Res., v. 80, p. 4433–4439.

Kugler, H. G., 1972, The Dragon Gneiss of Paria peninsula (eastern Venezuela), Memoirs of the Sixth Caribbean Geological Congress, p. 113–116.

Liddle, R. A., 1928, The geology of Venezuela and Trinidad, Fort Worth, TexasL: J. P. McGowan, p. 1–552.

—— 1946, The geology of Venezuela and Trinidad, 2nd ed.: Ithaca, New York, Paleon. Research Inst., p. 1–890.

Lidz, L., Ball, M. M., and Charm, W., 1968, Geophysical measurements bearing on the problem of the El Pilar Fault in the northern Venezuelan offshore: Bull. Marine Sci., Institute of Marine Sciences, Miami Univ., v. 18, no. 3, p. 545–560.

Macsotay, O., 1968, Internal report to the Ministerio de Energia y Minas, unpublished.

Maresch, W., 1971, The metamorphism and structure of northeastern Margarita island Venezuela [Ph.D. thesis]: Princeton Univ., p. 7.

—— 1974, Plate tectonics origin of the Caribbean Mountain System of northern South America: Discussion and proposal: Geol. Soc. Amer. Bull., v. 85, p. 669–682.

Mauk, F., 1978, Seismicity and focal mechanisms of the Trinidad inclined seismic zone: Trans. of the Am. Geophys. Union, Abstracts, p. 404.

Maxwell, J. C., and Dengo, G., 1951, The Carupano area and its relation to the tectonics of northeastern Venezuela: Trans. of the Am. Geophys. Union, v. 32, no. 2, p. 259–267.

Metz, H. L., 1964, Geology of the El Pilar Fault zone, State of Sucre, Venezuela [Ph.D. thesis]: Princeton Univ., 102 p.

Ministerio de Minas y Hidrocarburos, 1970, Lexico Estratigrafico de Venezuela: Bol. Geol. (Venezuela), Publ. Esp., no. 4, 756 p.

Molnar, P., and Sykes, L. R., Tectonics of the Caribbean and Middle America regions from focal mechanisms and seismicity: Geol. Soc. Amer. Bull., v. 80, p. 1639–1684.

Perez, O. J., and Aggarwall, Y. P., 1981, Present-day tectonics of the southeastern Caribbean and northeastern Venezuela, J. Geophys. Res., v. 86, p. 10791–10804.

Potter, H. C., 1972, Comments on the geology of northwestern Trinidad in relation to the geology of Paria, Venezuela: Memoirs of the Sixth Caribbean Geological Congress, p. 178–181.

Renz, H. H., 1962, Stratigraphy and paleontology of the type section of the Santa Anita Group and overlying Mercure Group, Rio Querecual, state of Anzoategui, northeastern Venezuela: Asoc. Venezolana Geol. Min. y Petroleo, Bol. In., v. 5, no. 4, p. 89–108.

Renz, H. H., Alberding, H., Dallmus, K. F., Patterson, J. M., Robie, R. H., Weisbord, N. E., and MacVail, J., 1958, The Eastern Venezuelan Basin: Am. Assoc. of Petroleum Geol., Symposium: "Habitat of Oil," p. 551–600.

Rod, E., 1956, Strike-slip faults of northern Venezuela: Am. Assoc. of Petroleum Geol. Bull., v. 40, no. 3, p. 457–476.

Rod, E., and Maync, W., 1954, Revision of Lower Cretaceous stratigraphy of Venezuela: Am. Assoc. of Petroleum Geol. Bull., v. 38, no. 2, p. 193–283.

Santamaria, F. J., 1972, Geochemistry and geochronology of the igneous rocks of the Venezuelan Coast Ranges and southern Caribbean islands and their relation to tectonic evolution [Ph.D. thesis]: Houston, Rice Uni., 112 p.

Saunders, J. B., 1974, Trinidad: in Cenozoic and Mesozoic Orogenic Belts, Spencer, A. M., ed., London, p. 671–682.

Schubert, C., 1971, Metamorphic rocks of the Araya Peninsula, eastern Venezuela: Geologische Rundschau, v. 60, no. 4, p. 1571–1599.

—— 1979, El Pilar fault zone, northeastern Venezuela: brief review: *in* Recent crustal movements, 1977 (Whitten, C. A., eds., et al.), Tectonophysics, v. 52, no. 1-4, p. 447–455.

Seijas, I., and Schubert, C., 1979, Francisco, J., 1972, Geologia de la region de Carupano: *in* Congresso Geologico Venezolano, 4th, Memoria, v. 3, Venez., Dir. Geol., Bol. Geol., Publ. Esp., no. 5, p. 1877–1923.

Sifontes, G.R.S., 1969, Estudio geoeconomica de la region de Carupano: Direccion de Geologia, M.M.H., Caracas, unpublished report.

Sifontes, G.R.S., and Santamaria, F., 1972, Rocas intrusivas jovenes en la region de Carupano: Memoirs of the Sixth Caribbean Geological Congress, p. 121–125.

Silver, E., Case, J., and Macgilavry, H. J., 1975, Geophysical study of the Venezuelan borderland: Geol. Soc. Amer. Bull., v. 86, p. 213–226.

Skerlec, G. M., and Hargraves, R. B., 1980, Tectonic significance of paleomagnetic data from northern Venezuela: v. 85, no. 10, p. 5303–5315.

Stover, L. E., 1967, Palynological dating of the Carrizal Formation of eastern Venezuela: Asoc. Venezoland Geol. Min. y Petroleo, Bol. Inf., v. 10, no. 10, p. 288–290.

Sylvester, A. G., and Smith, R. R., 1976, Tectonic transpression and basement-controlled deformation in San Andreas fault zone, Salton trough, California: Am. Assoc. of Petroleum Geol. Bull., v. 60, p. 1081–2102.

Talwani, M., Sutton, G., and Worzel, J., 1959, A crustal section across the Puerto Rico trench: J. Geophys. Res., v. 64, p. 1545–1555.

Vasquez, E., and Dickey, P., 1972, Major faulting in northwestern Venezuela and its relation to global tectonics: in Memoirs of the Sixth Caribbean Geological Congress, p. 191–202.

Vierbuchen, R. C., 1978, The tectonics of northeastern Venezuela and the southeastern Caribbean Sea [Ph.D. thesis]: Princeton Univ., 175 p.

——1979, A working hypothesis for the post-Miocene tectonics of the southeastern Caribbean region: Trans. of the Amer. Geophys. Union, v. 60, no. 18, p. 395.

Vignali, M., 1977, Geology between Casanay and El Pilar, (El Pilar Fault zone) Edo. Sucre, Venezuela, *in* Abstracts of the 8th Caribbean Geological Congress, p. 5.

von der Osten, E., 1957, Lower Cretaceous Barranquin Formation of northeastern Venezuela: Am. Assoc. of Petroleum Geol. Bull., v. 41, no. 4, p. 679–708.

Wilcox, R. E., Harding, T. P., and Seely, D. R., 1973, Basic wrench tectonics: Am. Assoc. of Petroleum Geol. Bull., v. 57, no. 1, p. 74–96.

Woollard, G. P., 1959, Crustal structure from gravity and seismic measurements: J. Geophys. Res., v. 64, no. 10, p. 1521–1544.

Worzel, J., 1974, Standard oceanic and continental structure: *in* the geology of continental margins, Burk, C. A., and Drake, C., ed., Springer-Verlag, New York, p. 59–66.

——1976, Gravity investigations of the subduction zone: *in* the geophysics of the Pacific Ocean basin and its margins (Woollard Volume), Am. Geophys. Union, Geophys. Monogr., no. 19, p. 1–17.

MANUSCRIPT ACCEPTED BY THE SOCIETY SEPTEMBER 1, 1983

Geological Society of America
Memoir 162
1984

The Urica fault zone, northeastern Venezuela

Stuart E. Munro*
Foster D. Smith, Jr.
Gerencia de Geologia, Corpoven, S.A.
Apartado 61373, Caracas
Venezuela

ABSTRACT

The Urica fault zone is viewed as a major transpressive geosuture in the tectonic evolution of northeastern Venezuela. It can be traced for 350 km from the Coche fault zone on the northwest and, with less certainty, to the area of Tucupita on the southeast, right-lateral displacement being about 35 km. Interpretation of new seismic reflection data shows it as a broad zone of dislocation, 10 km or more in width, the "flower" pattern geometry being characteristic of transpressive strike-slip faults. It separates the Maturin sub-basin into a zone of compression, characterized by imbricate thrusting to the north, from one of tension, distinguished by normal fault sets to the south.

INTRODUCTION

Over the past two decades increasingly detailed geological and geophysical studies, both offshore and onshore, have embraced the Caribbean–South American plate boundary in the region of northeastern Venezuela (Fig. 1) (for an outline of previous work, see Vierbuchen, 1978, 1984). In that extensive area attention has been devoted to the nature and significance of major east-west-trending fractures, such as the El Pilar fault (Metz, 1968; Vierbuchen, 1978), and to the nature of the boundary between the Maturin sub-basin and a zone of imbricate thrusting to the north (De Sisto, 1964). Another prominent tectonic element is a fault set trending northwest-southeast, and this contribution focuses on the largest of these, the Urica fault zone, and adjacent areas (Fig. 1).

NEW SEISMIC DATA AND THEIR PRINCIPAL IMPLICATIONS

First-order implications of new seismic data (Fig. 2 - in pocket inside back cover) are:

1. Whereas surface geological mapping and the geologic map of Venezuela (Salvador and Stainforth, 1968; Bellizzia and others, 1976) suggest that the Urica fault terminated just north of Aguasay (Fig. 1), it has now been traced in subcrop along a more

easterly trend to the area of well Soledad-1 (well S-1, Fig. 1) in the delta of the Rio Orinoco. No modern seismic evidence is available to the east between Soledad-1 and the town of Tucupita (Fig. 1).

2. The Urica fault zone is perceived as a broad zone of dislocation 10 km or more in width, the "flower" pattern geometry (Fig. 2) being characteristic of transpressive strike-slip faults as described by Harding and Lowell (1979).

3. The fault zone separates a zone of compression, characterized by imbricate thrusting to the north, from one of tension, distinguished by normal fault sets to the south (Fig. 2).

4. The zone of imbricate thrusting appears to be far more extensive than formerly believed, continuing well south of the Pirital-Frontal thrust formerly considered to define the southern limit of thrusting. A corollary conclusion is that the Santa Barbara fault (shown diagrammatically on Fig. 2), viewed as a normal fault defining the northern limit of the Maturin sub-basin by De Sisto (1964), is better interpreted as another of the imbricate thrust set.

COMMENTARY

As shown in Figure 1, the Urica fault zone is now defined over a length of some 350 km. At its northwestern extremity it becomes asymptotic with the offshore Coche fault zone. The nature and location of the southeastern termination are less certain. SLAR images show distinct changes in stream courses, tex-

---

*This work was carried out while S.E.M. was on a staff exchange between Britoil plc, 150 St. Vincent Street, Glasgow, Scotland, and Petróleos de Venezuela, S. A., Caracas.

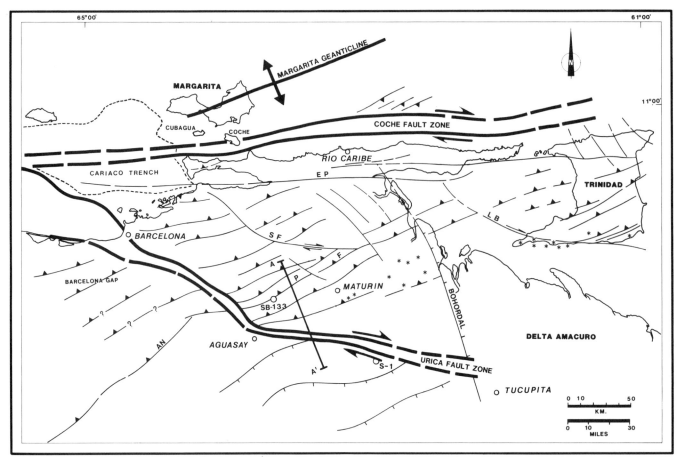

Figure 1. Tectonic map of Paria Block, the area between the Coche and Urica fault zones; the eastern limit is undetermined. Triangles indicate direction of dip of thrust plane. Hachures on downthrown block of normal faults. AN = Anaco fault, EP = El Pilar fault zone, SF = San Francisco fault; P, F = Pirital-Frontal fault, LB = Los Bajos fault, S-1 = well Soledad-1 (CVP), *mud volcano, A–A' = Seismic section, Figure 2.

ture, and topography continuing east of Soledad toward the area of Tucupita, but cannot be traced farther east. There are four possible interpretations for this. (1) Active delta processes may be obscuring the fault trace. (2) The fault may be active at depth but has yet to propagate to the surface. (3) With progressive lateral propagation eastward, the fault will in due course extend beyond the Tucupita area. (4) The fault may terminate as shown on Figure 1 because of dissipation of stresses at the older Bohordal lineament.

The significance of the apparent separation of compressional and tensional terranes by the fault zone requires clarification. At question is the relative timing of the thrusting and movement on the Urica fault. The apparent offset of the imbricate thrust belt by the fault in the area south of Barcelona (Fig. 1) suggests that the Urica faulting occurred after the thrusting. On the other hand, relatively long-term continuation of one regional stress field, with the implication that thrusting and strike-slip faulting have been partly contemporaneous, is suggested by the evident rotation indicated by the pronounced sigmoidal trace of the Urica fault zone.

On the basis of geological mapping, Salvador and Stainforth (1968) inferred 5 to 6 mi (~8–10 km) of right-lateral displacement on the Urica fault and concluded that the principal movement occurred in late Miocene to Pliocene time. A similar sense of displacement was inferred by Wong and others (1982) who found that the map of the second gravity residual shows offset of a positive gravity anomaly. This map suggests a basement offset of 35 km. Indirect support for the displacement estimate of Salvador and Stainforth (1968) is provided by Wilson (1968) who measured 34,400 ft (~10.5 km) of right-lateral offset on the Los Bajos fault in Trinidad (Fig. 1), another of the northwest-southeast fault set. By contrast, Vierbuchen (1978), accepting the age range estimate of Salvador and Stainforth, attempted to relate the origin of the Cariaco Trench (north of Barcelona; Fig. 1) to movements on the El Pilar and Urica faults, and he showed that 50 km of dextral slip along the Urica fault would have been required.

We think that the ultimate cause of the regional stress field responsible for the northwest-southeast fault set, exemplified by the Urica fault zone, is likely to be related to oblique convergence of the Caribbean and South American plates. Vierbuchen (1978, Fig. 27) postulated the formation of a small lithospheric plate (the Paria block) defined by the El Pilar fault, by a subduction zone east of Trinidad, and by a southeast-trending "zone of detachment" that corresponds to the Urica fault zone. We are in broad agreement therewith but suggest that the northern boundary of this small plate is better defined by the Coche fault zone and that in the current youthful stage of development there is no present need to invoke a subduction zone east of Trinidad. In this overview the Urica fault zone can clearly be viewed as a major geosuture in the tectonic evolution of northeastern Venezuela.

REFERENCES CITED

Bellizzia, A., Pimentel, N., and Bajo, R., 1976, Mapa geologico estructural de Venezuela, scale 1:500,000. Ediciones FONINVES, Caracas.

De Sisto, J., 1964, The Santa Barbara fault of northern Monagas: Boletin de la Asociacion Venezolana de Geologia, Mineria y Petroleo, v. 7, p. 99–109.

Harding, T. P., and Lowell, J. D., 1979, Structural styles, their plate-tectonic habitats, and hydrocarbon traps in petroleum provinces: American Association of Petroleum Geologists Bulletin, v. 63, p. 1016–1058.

Metz, H. L., 1968, Stratigraphic and geologic history of the extreme northeastern Serrania del Interior, State of Sucre, Venezuela: Transactions, Caribbean Geological Conference, IV, Trinidad, 1965, p. 275–292.

Salvador, A., and Stainforth, R. M., 1968, Clues in Venezuela to the geology of Trinidad, and vice versa: Transactions, Caribbean Geological Conference, IV, Trinidad, 1965, p. 31–40.

Vierbuchen, R. C., Jr., 1978, The tectonics of northeastern Venezuela and the southeastern Caribbean Sea [Ph.D. thesis]: Princeton University, University Microfilms International, Ann Arbor, Michigan, 48106.

——, 1984, Geology of the El Pilar fault zone and adjacent areas in northeastern Venezuela, *in* Bonini, W. E., and others, eds., The Caribbean-South American Plate Boundary and Regional Tectonics: Geological Society of America Memoir 162 (this volume).

Wilson, C. C., 1968, The Los Bajos fault: Transactions, Caribbean Geological Conference, IV, Trinidad, 1965, p. 87–89.

Wong, J., Passelacqua, H., Graterol, V., and Flores, J., 1982, Interpretacion gravimetrica de la region Guanipa Sur: Report presented at the Primer Congreso Venezolano de Geofisica, 21–25 November, 1982, Caracas.

MANUSCRIPT ACCEPTED BY THE SOCIETY SEPTEMBER 1, 1983

Printed in U.S.A.

Geological Society of America
Memoir 162
1984

Structure and Cenozoic tectonics of the Falcón Basin, Venezuela, and adjacent areas

Karl W. Muessig
Exxon Minerals Company
P.O. Box 2189
Houston, Texas 77001

ABSTRACT

The Falcón Basin in northwestern Venezuela and adjacent offshore basins developed within a zone of extensional tectonics during Oligocene and Miocene times. Extension resulted from right-lateral motion along offset, east-west-trending, transcurrent faults, including the Oca fault in western Venezuela, the Cuiza fault in northern Colombia, and the San Sebastián fault along the coastal areas of central Venezuela. On both local and regional scales, transcurrent and normal faults were active during the early evolution of the basins. These faults define rhomb-shaped pullapart basins in map plan. Extension occurred in a northeast direction causing normal faulting along northwest trends. Basin subsidence was accompanied by crustal thinning and injection of basaltic magmas.

Evidence of Oligocene magmatic activity is found in the central part of the Falcón Basin where volcanic rocks and hypabyssal intrusions are exposed. These rocks are similar to other suites of continental igneous rocks typically associated with rifting environments. Basaltic rocks of both alkalic and subalkalic affinities are present. Xenoliths of the underlying crust and mantle are abundant. Felsic igneous rocks are relatively rare.

Other structures within the Falcón and Bonaire basins formed during later stages of basin development. These include folds and reverse faults of northeast trend and conjugate sets of small transcurrent faults. These structures were amplified by greater compressional stresses during late Miocene and Pliocene time.

A similar Tertiary tectonic regime is postulated for the larger area of the Bonaire Crustal Block, a block that includes the Falcón and Bonaire basins. This block is a broad region of extensional pullapart structures which developed during the Oligocene to Miocene, right-lateral motion between the Caribbean and South American crustal plates. Minimum extension within the northern part of the block, along the Venezuelan and Netherlands Antilles and the Paraguaná and Guajira peninsulas, is estimated at 35 to 45 km in an east-west direction. The extension within this nonrigid block should be considered when determining Tertiary movements between the Caribbean and South American plates.

INTRODUCTION

The Cenozoic tectonic boundary between the Caribbean and South American areas is frequently interpreted as a right-lateral transform boundary through or adjacent to the Falcón Basin in northwestern Venezuela (Stainforth, 1969; Molnar and Sykes, 1969; Malfait and Dinkelman, 1972; Silver and others, 1975). However, the origin of the basin has never been considered in detail. A transform plate boundary through the Falcón region might be expressed as a series of right-lateral transcurrent faults along a line connecting the Oca fault to the San Sebastián fault (Fig. 1). However, the only documented evidence suggesting a buried structure is the presence of *en echelon* folds and block uplifts attributed to right-lateral wrenching (Vásquez and Dickey, 1972). Elsewhere in northern South America, east-west directed, right-lateral, transcurrent faulting apparently began about mid-Eocene (Bell, 1972). This initiation coincides with the cessation of submarine thrusting and flysch deposition in the Siquisique-Barquisimeto area south of the Falcón Basin (Stephan, 1977). Thus it is possible that the Falcón Basin developed during a period of transcurrent faulting that postdates compressional deformation and thrusting of Eocene and earlier times. Indeed, the basin may have developed along transcurrent faults.

Case (1974) suggested that many Neogene basins along the northern South American borderland, including the Falcón Basin, may have originated by north- to northwest-directed extension. Silver and others (1975) briefly examined the basin in relation to its offshore continuation, the Bonaire Basin. They concluded that both developed in a zone of Oligocene to Miocene rifting between the Venezuelan mainland and the Venezuelan and Netherlands Antilles.

Detailed studies of the Falcón Basin geology were undertaken to clarify the history of the basin and to constrain the Tertiary tectonic evolution of the southern Caribbean borderland. This paper summarizes the stratigraphic, structural, geophysical, and petrologic data obtained. Based on these data a tectonic model for the basin is postulated, which reconciles the observed regional extensional regime with the right-lateral, transcurrent faulting. The model is extended to adjacent basins, and its implications for Caribbean–South American plate movements are assessed.

FALCON BASIN STRATIGRAPHY

During Oligocene to Miocene time a large part of northwestern Venezuela was occupied by the marine Falcón Basin (Fig. 1). It was elongate in an east-west direction and bordered by topographic highs to the south, west, and north. The older Paleocene to Eocene flysch basin, which contains allochthonous ophiolites in the Siquisique region (Bellizzia and others, 1972), was deformed, uplifted and was being eroded at this time. It formed the southern high. To the west, the basin shallowed onto the Maracaibo Platform. The Dabajuro high bordered the basin to the northwest. Exploration wells on this high bottom in Upper

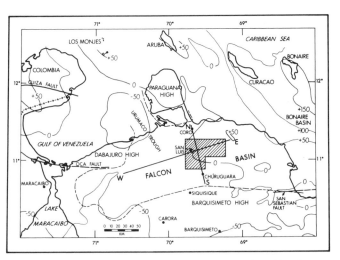

Figure 1. Location map of the Falcón Basin, Venezuela, with paleogeographic features and surrounding structures. Broken line (— – —) indicates the approximate extent of the basin (Wheeler, 1963; Ferrell and others, 1969). Stratigraphic cross section along lines N-S and W-E are shown in Figure 2. Ruled area shows central Falcón Basin study area of Figure 3. Bouguer gravity anomaly contours are included from Bonini and others (1977).

Cretaceous black shales, and an underlying granitic basement is suspected (unpublished report of the Texas Petroleum Company; Ferrell and others, 1969). To the north, the Paraguaná high restricted access to the Falcón Basin to the Caribbean Sea. This high is a basement massif of igneous and metamorphic rocks (MacDonald, 1968; Martín-Bellizzia and Arozena, 1972) ranging in age from Paleozoic through Paleocene (Santamaría and Schubert, 1974). The narrow Urumaco Trough connected the Falcón Basin with the Gulf of Venezuela and separated the Dabajuro and Paraguaná highs. The eastern Falcón Basin opened into the Bonaire Basin (Fig. 1).

The earliest Falcón Basin deposits are considered to be upper Eocene (Wheeler, 1963). Deposition continued intermittently through Oligocene and Miocene time. Sedimentary thicknesses exceeding 9,000 m accumulated and are still preserved in the northern and eastern parts of the basin (unpublished report of the Texas Petroleum Company; Ferrell and others, 1969). In the central basin area, deposition ceased after early Miocene time. During Oligocene to lower Miocene time, approximately 3.5 km of sediments were deposited in the central basin at rapid subsidence rates of 500 to 700 m/m.y. (Díaz de Gamero, 1977).

The Oligocene and Miocene stratigraphy is complicated by many lateral and vertical facies changes (Fig. 2). The stratigraphic section is divided into the lower Oligocene El Paraíso deltaic sands and shales, which by middle Oligocene grade upward into the Pecaya marine shales. Igneous intrusives and extrusives are found within the upper El Paraíso and lower Pecaya formations (Fig. 3). The El Paraíso Formation is a clastic sequence 500 to 1,000 m thick, composed of fine-grained, quartz sandstones and siltstones intercalated with mudstones, shales, and thin lignites. The sandstones are well indurated with siliceous cement and are

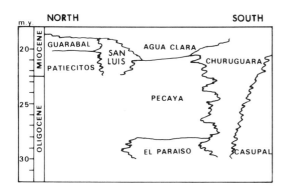

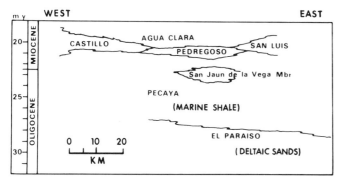

Figure 2. Stratigraphic relationships of the Oligocene and lower Miocene formations in the Falcón Basin. Modified from Díaz de Gamero (1977). Section locations as indicated on Figure 1. Although quite variable, the thickness of the illustrated stratigraphic column is approximately 4,000 m.

frequently cross-bedded, ripple-marked, or heavily bioturbated with tracks, trails, and burrows. Bedding thicknesses are variable. The lower boundary of the El Paraíso Formation is not exposed in the study area, whereas the upper boundary with the Pecaya Formation is gradational. The El Paraíso Formation is an eastward-prograding deltaic complex (Díaz de Gamero, 1977). Thus, to the east, sandstone units and coals thin and disappear, whereas shales become more abundant.

The Pecaya Formation is composed of a monotonously thick (>2,500 m) section of dark gray, pyritic, calcareous shales. Bioclastic limestone lenses and small reefal limestone bodies are present in the shales. These appear to be fore-reef debris shed from the partially synochronous San Luis Formation limestones to the north (Fig. 3). Studies of benthic foraminiferal assemblages in the shales indicate water depths of 1,000 m or more (Díaz de Gamero, 1977). These depths argue against development of the limestone in its present position. Locally anomalous clastic units (San Juan de la Vega Member) are also found within the Pecaya Formation.

To the north and west of the area shown on Figure 3, the upper Pecaya Formation is overlain by, and is in part equivalent to, the Pedregoso calcareous turbidites and shales. This lower Miocene formation was deposited around the southern margin of the San Luis Formation carbonate reefs. North of the San Luis reefs, back-reefal facies are represented by the Patiecitos and Guarabal formations.

To the west, shallow-water clastics of the Castillo Formation replace the Pedregoso and part of the Pecaya formations (Wheeler, 1963). To the south, another complicated carbonate bank is represented by the Churuguara Formation. South of Churuguara the Oligocene to Miocene strata thin rapidly. The Churuguara Formation gives way southward to basin margin conglomerates, sandstones, shales, and coals of the Casupal Formation (Wheeler, 1963).

In the Siquisique area south of the Falcón Basin (Fig. 1), Oligocene to Miocene strata lie in angular unconformity over the Paleocene and Eocene Matatere Formation (Compañía Shell de Venezuela, 1965; Wheeler, 1963). The Matatere is a thick (3,000 m) sequence of turbiditic shales and intercalated graywacke sandstones (Bellizzia and Rodríguez, 1967). It records deposition in a flysch basin that extended south and west to Barquisimeto and Carora and probably to the north under the younger Falcón Basin. Locally within the flysch, olistostrome (Renz and others, 1955) and melange zones are present. Near Siquisique a melange occurs with exotic blocks of basalt, spilite, gabbro, peridotite, chert, limestone, shale, and phyllite (Bellizzia and others, 1972). These exotic block terranes were generated during the Paleogene emplacement of nappes southward into the Matatere flysch basin (Stephan, 1977). Paleomagnetic studies indicate that these blocks, along with many similar allochthonous terranes in northern Venezuela, were tectonically rotated prior to or during their accretion and emplacement (Skerlec and Hargraves, 1980).

By late Eocene both flysch deposition and allochthonous block emplacement had ceased. The flysch was deformed prior to deposition of the first upper Eocene Falcón Basin sediments. The younger Falcón Basin was a more restricted basin with a well defined central axis lying about 45 km north of Siquisique. Paleocene to Eocene events preceding development of the Falcón Basin were essentially synchronous with similar compressional events to the east described by Bell (1971) and Beck (1978) for central Venezuelan areas south of Caracas.

BASIN STRUCTURE

The Falcón Basin is a nearly east-west-trending structural as well as sedimentary feature (Fig. 1). Throughout its history the basin averaged deposition rates of 350 m/m.y. and subsidence rates of 400 m/m.y. By the end of early Miocene time, deposition ceased in the central basin but was continuing in the eastern and northern parts (Wheeler, 1963). During late Miocene and Pliocene times the basin was uplifted and deformed into an east-northeast-trending anticlinorium (Wheeler, 1963; Zambrano and others, 1971). Structural features within the anticlinorium are of the following types: (1) east-west, right-lateral, transcurrent faults, (2) east-west normal faults, (3) northwest to north normal faults, (4) folds with east-northeast axes and east-northeast reverse faults, (5) west-northwest to northwest transcurrent faults, and (6) north to north-northeast transcurrent faults.

These six structural types (inset, Fig. 3) are deformational features predicted by Wilcox and others, (1973) for a terrane

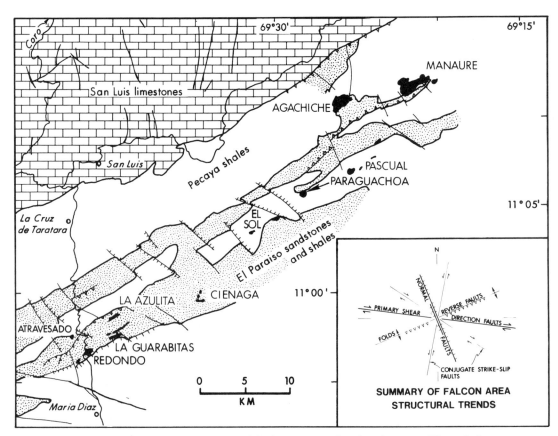

Figure 3. Geologic map of the central Falcón igneous area, location shown on Figure 1. Igneous intrusive and extrusive rocks are solid black. The inset is a summary of structural types and trends found in the Falcón Basin as a whole from published geologic maps by Bellizzia and others, (1976), Bonini and others (1977), and Case and Holcombe (1980).

undergoing right-lateral wrenching or simple shear. However, previous structural syntheses have selectively emphasized different combinations of these structural features in deciphering the evolutionary history of the Falcón Basin. Ferrell and others, (1969; unpublished report of the Texas Petroleum Company) ascribed Oligocene to Miocene subsidence and uplift to simultaneous movement on normal and transcurrent faults, and they attributed Pliocene through Holocene folding to a north-south compression.

Vásquez and Dickey (1972) related the Falcón folds and transcurrent faults to the deformation of the incompetent Oligocene and Miocene sediments by right-lateral movement along buried basement structures represented by the Oca and related faults. Their analysis postulates a rather continuous right-lateral movement from post-Cretaceous through Pliocene times.

As a basis for the tectonic model presented later, the structural features listed above are discussed both on a regional scale and within the area of study.

East-West Transcurrent Faults

East-west trending, right-lateral, transcurrent faults are per-

haps the most widely known faults in northwestern Venezuela and include the Oca, Cuiza, and San Sebastián faults (Fig. 4). The Oca fault is the best known of these. Estimates on its right-lateral displacement range from 15 to 20 km post-Eocene (Feo-Codecido, 1972) to 195 km post-Cretaceous to pre-Pleistocene (Vásquez and Dickey, 1972) to 65 km post-Mesozoic (Tschanz and others, 1974). The Cuiza fault on the Guajira Peninsula is a similar structure. Alvarez (1967) has estimated 15 to 25 km of right-lateral movement on this fault.

The San Sebastián fault (Fig. 4), located just offshore central Venezuela, is frequently cited as the continuation of both Oca and El Pilar fault trends (Rod, 1956; Alberding, 1957; Vásquez and Dickey, 1972; Silver and others, 1975). Evidence of its existence comes from seismic activity, the coastal Venezuelan escarpment, and seismic reflection studies. Vierbuchen (1978) estimated less than 140 km of post-Miocene, right-lateral displacement for this structure.

The displacements on other east-west faults are poorly controlled. East-west faults near Coro have frequently been depicted as transcurrent faults (Smith, 1962; Bellizzia, 1972), but supporting evidence of strike-slip displacement is lacking. A fault explains the rapid thickening of sediments toward the Gulf of Coro

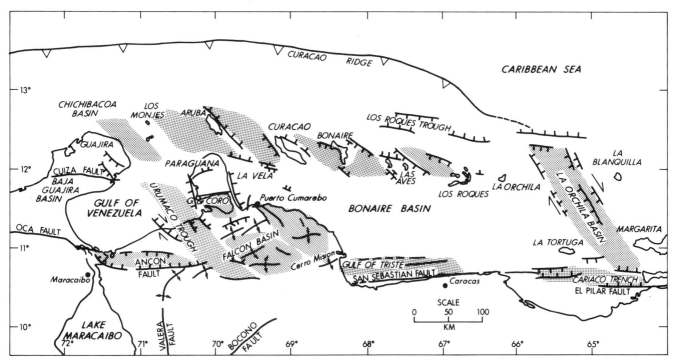

Figure 4. Structural map of northwestern Venezuela and adjacent areas. Shaded areas indicate zones of relative subsidence during the Oligocene and Miocene. The Bonaire and Baja Guajira basins (unshaded) were also areas of subsidence. Structures from Bellizzia and others (1976), Bonini and others (1977), Ferrell and others (1969), Silver and others (1975), and Case and Holcombe (1980).

(unpublished report of the Texas Petroleum Company; Ferrell and others, 1969), but the movement on the fault could well be normal.

Evidence for east-west transcurrent faults in the central basin area is rather slim. Compañía Shell de Venezuela geologists have mapped systems of discontinuous transcurrent faults with this orientation in western and west central Falcón (unpublished report of Compañía Shell de Venezuela; Jaeckli and Erdmann, 1952).

East-West Normal Faults

Several east-west trending normal faults are recognized in the Falcón region, and although the data do not require it, motion on them may also include a strike-slip component. The longest is the Ancon fault (Fig. 4) located south of and parallel to the extension of the Oca fault. Downthrow on this fault is to the north. A normal fault parallel to the southern coast of Paraguaná has a downthrown southern wall, whereas normal faults in northern Paraguaná have downthrown northern walls. Although supported by field evidence, (Martín-Bellizzia and Arozena, 1972; MacDonald, 1968), these faults are primarily delineated by geophysical data (unpublished report of Corporación Venezolana del Petróleo-Institut Français du Pétrole, 1967; Zambrano and others, 1971). Vertical east-west faults (Fig. 5) also border the Siquisique ophiolite blocks and cut unconformably overlying Oligocene sediments (Compañía Shell de Venezuela, 1965).

Northwest to North Normal Faults

Northwest- to north-trending, normal faults of Oligocene to Miocene age are found throughout the Falcón Basin and northern offshore areas. In the central Falcón Basin (Fig. 3) normal fault planes dip both east and west, but the former are more abundant. Where well exposed in road-cuts, small-scale normal faults of similar trend exhibit the characteristics of growth faults. Changes in stratigraphic thickness indicate fault activity or growth during sediment deposition. Changes in sandstone-unit thickness across faults up to 1 m have been noted in Oligocene rocks. Similar features are impossible to detect in the monotonous, Miocene Pecaya shales. The isolated distribution of the San Juan de La Vega Member clastics within the Pecaya Formation (Fig. 2) is probably controlled by a northwest-trending, normal fault bounded graben.

The Paraguaná Peninsula is a horst bordered to the east and west by north-trending normal faults (Fig. 4). East-west cross-sections across the peninsula (unpublished report of Corporación Venezolana del Petróleo-Institut Français du Pétrole, 1967; Zambrano and others, 1971) show these as normal faults active during Oligocene and Miocene sedimentation. Stratigraphic units rapidly increase in thickness across faults. In La Vela Bay east of Paraguaná, test wells penetrate thin red-bed units of probable Oligocene age (Vásquez, 1975), which thicken eastward across normal faults into the Bonaire basin (unpublished report of Corporación Venezolana del Petróleo-Institut Français du Pétrole,

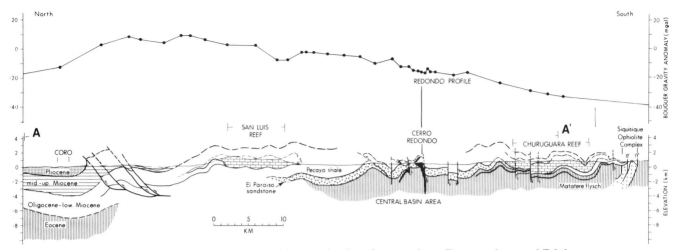

Figure 5. Geologic cross section (A–A′) and regional gravity anomaly profile across the central Falcón Basin. Section location is indicated on Figure 6. Falcón igneous intrusions are solid black.

1967). Similar normal faults of northwest trend were active during sedimentation in the Puerto Cumarebo area in northeastern Falcón (Payne, 1951).

Structures between the islands of the Netherland and Venezuelan Antilles are northwest-trending normal faults, but may have some strike-slip components. These faults form the borders of basins that accumulated about 2 km of Neogene sediments (Silver and others, 1975; Case, 1974; Feo-Codecido, 1971). The Bonaire Basin, which contains from 2 to 5 km of Tertiary sediments (Galavis and Louder, 1971; Case, 1974; Silver and others, 1975), also contains north- to northwest-trending normal faults, as does La Orchila Basin (Los Roques Canyon of Peter, 1972).

The widespread evidence for normal faulting contemporaneous with sedimentation suggests the existence of an extensional tectonic regime during Oligocene and Miocene time. Although these north to northwest faults are rather prominent, their significance has largely been ignored in tectonic models of the Falcón Basin. This neglect is partly a result of recent overprinting by transcurrent motion on the same faults. In this regard, basement structures in the Urumaco Trough area indicate normal faulting (Bellizzia and others, 1976), but surface exposures of younger formations show evidence of transcurrent displacement. Younger transcurrent faulting followed the zones of weakness of earlier normal faults.

Folds and Reverse Faults

Folds are the most conspicuous structural feature of the Falcón anticlinorium. Their axes trend N80°E in the central and eastern basin areas and N50°E in the western part. Fold trends in the offshore and Bonaire Basin areas to the east have easterly axes (Silver and others, 1975). Asymmetric and overturned folds are well developed south of Coro (Fig. 5) where Pliocene formations are overturned to the north and folds are broken by southerly dipping thrust faults. Reverse faults of similar trend are found throughout the basin and within the study area. Anticlines in the

central basin area are slightly asymmetric with steeper northern limbs, axial planes dipping steeply south, and limb dips of 65° to 45°.

West-Northwest to Northwest Transcurrent Faults

Small transcurrent faults of west-northwest to northwest trend are found throughout the basin (Fig. 4) and are particularly abundant in the Puerto Cumarebo and Urumaco Trough areas. As noted previously, transcurrent faulting along northwest trends has overprinted older normal faults in the Urumaco Trough.

In the central basin area, transcurrent faults of northwest to west-northwest trend offset and terminate folds, reverse faults, and topographic ridges of resistant sandstone. Fault traces through resistant sandstones are frequently followed by streams. The majority of these faults exhibit right-lateral displacement. In many cases a significant component of dip-slip as well as strike-slip motion is required to produce observed map patterns.

North to North-Northeast Transcurrent Faults

Compared to other structural features in the basin, north to north-northeast transcurrent faults are rare. Many have both normal and strike-slip movement. Concentrations of these faults are found south of Coro, and in the southwest part of the Falcón Basin where fold trends are more northeasterly. These faults and the west-northwest to northwest transcurrent faults make up a conjugate pair and are hereafter called conjugate strike-slip faults.

In summary, the structural trends in the Falcón system suggest that a primary direction of right-lateral shear was directed slightly south of east (Fig. 3). Early deformation in the Falcón area was confined to east-west transcurrent faults and northwest normal faults. These structures were active during the initiation of the basin. They controlled the rates and patterns of basin sedimentation and subsidence.

Folds and reverse faults and the conjugate strike-slip faults

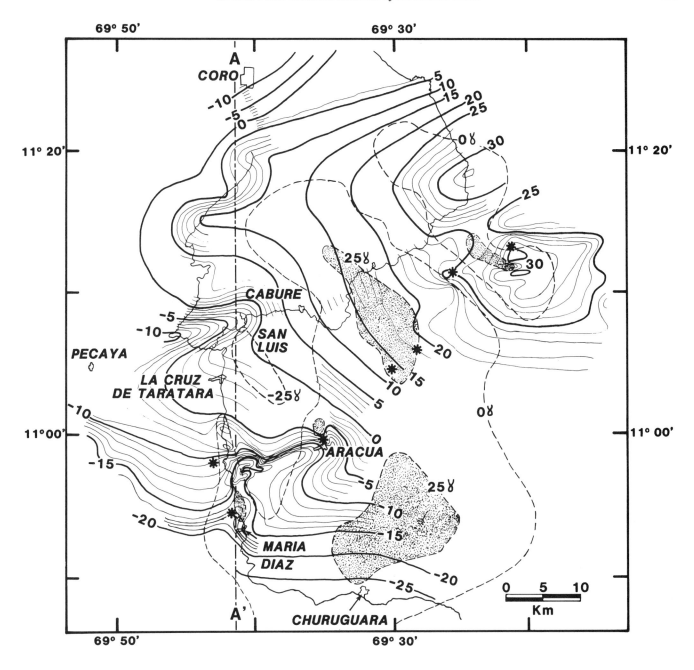

Figure 6. Map of magnetic anomalies and Bouguer gravity anomalies in the central Falcón area. Magnetic anomalies are contoured in 25-gamma intervals (dashed lines); the +25-gamma closures are shaded. Bouguer gravity anomalies (solid lines) are contoured in 5-mgal increments supplemented by 1-mgal contours where data permits. Locations of major igneous exposures (*) are indicated.

are later deformational features imposed during late Miocene to Pliocene time.

GRAVITY AND MAGNETIC ANOMALIES

The regional gravity field of the Falcón and Bonaire basins and adjacent area (Bowin, 1976; Bonini and others, 1977; Bo-

nini, 1978) exhibits several noteworthy features. For most of the southern Caribbean borderland, the zero Bouguer anomaly contour approximates the Caribbean shoreline. The Falcón Basin is the largest regional inland deflection of the zero anomaly in Venezuela, anomalies reaching +50 mgal (Fig. 1). The central basin area is characterized by positive Bouguer anomalies up to 30 mgal (Fig. 6).

In the Bonaire Basin, where Oligocene and Miocene sedimentary thicknesses are similar to the Falcón Basin, Silver and others (1975) used gravity to model a dense, thinned continental crust with 6 km of sediments and a crust-mantle boundary elevated to a depth of about 18 km. In the Falcón Basin a similar model would satisfy the observed gravity. In both regions the positive Bouguer anomalies indicate a density excess in the basins when compared to adjacent continental areas. Injection of dense basaltic material into the basin crust and sediments occurred in the central Falcón Basin. Similar intrusive activity or general crustal thinning may account for the positive anomalies elsewhere in the region.

The regional magnetic field in the central Falcón Basin is rather featureless. Locally, however, the field is influenced by the presence of igneous intrusions and by the basin structure. Anomalies up to 225 gammas are directly associated with igneous intrusions. Larger regions of regional northwest-trending magnetic anomalies up to 25 gammas (Fig. 6) are also loosely associated with outcropping igneous rocks. These elongate trends suggest that the exposed shallow-level sills, flows, and plugs may be fed by northwest-trending dikes at depth. Thus the present northeast trend of igneous outcrops is an artifact of the younger fold trends. The northwest magnetic anomalies may also reflect the normal faults of similar trend discussed previously.

FALCON IGNEOUS ROCK PETROLOGY

Within the central Falcón Basin, basaltic igneous bodies crop out in a northeast-trending 50-km by 10-km belt (unpublished report of Compañía Shell de Venezuela; Brueren, 1949; Coronel, 1970). Outcrop patterns, structural orientations, and the occurrence of border-zones with chilled igneous margins, contact metamorphism, local sedimentary rock inclusions, and breccias were documented by Muessig (1979). These bodies intrude the Oligocene sedimentary rocks as shallow-level sills and plugs or occur as extrusive flows. Potassium-argon isotopic analyses of a trachybasaltic sill yield an age of 22.9 ± 0.9 m.y. (Muessig, 1978). This age is 5 m.y. younger than reported paleontologic ages of surrounding sedimentary rocks. A span of igneous activity from 28 to 23 m.y. can be inferred from stratigraphic constraints. The Falcón igneous rocks are younger than and distinctive in morphology, composition, and origin from both the Siquisique ophiolite blocks (35 km south) and the Paraguaná igneous-metamorphic massif (100 km north).

The Falcón igneous rocks are similar in petrology to other continental igneous rock suites associated with extensional tectonic environments. Basaltic rocks compose about 98% of the igneous suite, whereas felsic rocks are of only minor abundance. Chemical and mineralogic analyses reveal two parental rock types: (1) an alkaline type and (2) a less alkaline type transitional to subalkaline (Muessig, 1979).

Rocks of the first type are alkali olivine basalts and diabases, hawaiites, trachybasaltic diabases, and trachytes. Phenocrysts of olivine and titanaugite, together with sporadic plagioclase, mag-

TABLE 1. REPRESENTATIVE WHOLE-ROCK ANALYSES

	Alkali olivine basalt	Trachybasalt	Subalkaline basalt
SiO_2	45.45	44.52	50.37
Al_2O_3	15.69	15.64	15.18
Fe_2O_3	9.57	10.99	8.44
MgO	5.17	5.33	9.20
CaO	6.29	6.31	8.36
Na_2O	5.14	4.39	2.99
K_2O	1.79	2.59	0.88
TiO_2	2.31	2.40	1.72
MnO	0.17	0.23	0.14
LOI	8.25	7.08	4.56
TOTAL	99.83	99.48	101.84

Trace elements, ppm		
Cr	100	40
Ni	36	45
Cu	25	24
Zr	270	300
Y	30	30
Rb	22.8	33
Ba	600	650
Sr	1027	1005
$^{87}Sr/^{86}Sr$	0.7036	

Note: Alkali olivine basalt from Cerro Atravesado, trachybasalt from La Guarabitas, and subalkaline basalt from Cerro Manaure. For analytical uncertainties see Muessig (1979).

netite, and kaersutite, are set in a groundmass of the above phases with the addition of late stage biotite and analcite. These rocks are nepheline normative, enriched in alkalis (Table 1) and plot in the field of "within plate" basalts as defined by Pearce and Cann (1973) on a Ti, Zr, and Y triangular diagram. Approximately 40% of the exposed igneous rocks have these alkaline affinities.

The transitional to subalkaline types are extensively altered diabase intrusions with relict augite phenocrysts, plagioclase, ilmenite, and pseudomorphs of magnetite and olivine. Compared to the alkaline basalts, these rocks are lower in iron, titanium, and alkalis but higher in silica. Alteration has obscured their nature by introducing silica and leaching alkalis.

Assorted xenoliths and amphibole megacrysts are abundant in alkali basalt lavas near vents at Atravesado, La Azulita, El Sol, and Paraguachoa (Fig. 3). Calc-silicate gneiss is the most common xenolith type (95%), followed by spinel peridotite (4%), gabbro (1%), and rare schist and granite xenoliths. Most of the xenoliths represent a sampling of the underlying crust and mantle through which the basaltic magmas passed. The megacrysts and some of the gabbros may be samples of deeper level magma chambers that were the source of the basalts.

TECTONIC MODEL

The development of a diffuse pullapart basin between paral-

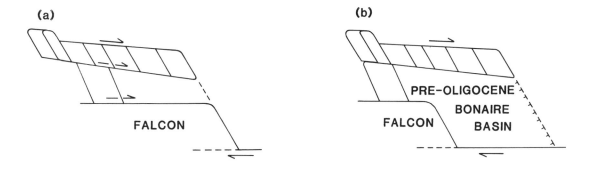

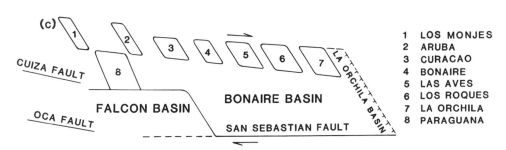

Figure 7. The evolution of the Falcón and Bonaire pullapart basins. Options (a) and (b) represent two configurations of the area prior to extension in the late Eocene. To evolve to the present configuration (c), option (a) requires a greater amount of Oligocene to Miocene extension and right-lateral transcurrent displacement. In option (b) a pre-Oligocene Bonaire basin has a thinner, more mafic crust and requires less extension to evolve to (c).

lel but offset right-lateral transcurrent faults is postulated for the Falcón region. This extensional tectonic model accounts for (1) a complex Oligocene-Miocene stratigraphy and rapid basin subsidence history, (2) synsedimentary east-west transcurrent and east-west and northwest normal faulting, (3) positive gravity anomalies indicative of a relatively thin or dense crust, and (4) alkaline basaltic volcanism and intrusion along northwest trends.

A pullapart basin is a zone of crustal extension and subsidence that develops at a bend in a transcurrent fault system if the motion at the bend is divergent (Crowell, 1974). Subsidence occurs on normal faults that connect parallel sections of the transcurrent fault system. Significant subsidence may also occur along the transcurrent faults. Due to the distribution of stresses within this zone of divergence, the bordering normal and transcurrent faults define a rhomb-shaped basin in map view. In cross-section the structure of the basin is grabenlike. If the pullapart basin is intruded extensively by mafic magma and develops a new simatic crust, then it is a rhombochasm as defined by Carey (1958).

Within a large system of transcurrent faults, pullapart structures may be rather complicated. A classic example has been documented along the San Andreas fault system by Elders and others (1972), Wilcox and others (1973), and Crowell (1974). In

the Salton Trough, for example, 6 km of Pliocene to Holocene sediments were deposited in a pullapart structure while adjacent areas underwent rapid erosion. A rhomb-shaped trough subsided at 600 m/m.y. due to right-lateral movement along offset faults in the San Andreas system. Basaltic and felsic igneous activity is concentrated along the normal fault trends that connect the offset transcurrent faults (Hill, 1977).

The development of pullapart structures in the Falcón Basin reflects the Tertiary tectonic evolution of a larger crustal block termed the Bonaire Block by Silver and others (1975). This block was also subject to right-lateral shearing between the Caribbean and South American plates (Fig. 7).

Falcón Basin Pullapart

In the Falcón pullapart system, extension was dispersed over a wide zone resulting in zones of relative stability and subsidence (Fig. 4). The Paraguaná, Dabajuro, Guajira, and southern mainland (Siquisique) areas remained as highs shedding sediments into grabenlike, pullapart zones. The Falcón Basin, Urumaco Trough, La Vela Bay, and Bonaire Basin were major zones of subsidence.

On a small scale, rhombic grabens are well defined in the Falcón region. A rhombic pullapart basin with distinct structural boundaries and thick sediment accumulations is located north of the Ancon fault (Fig. 4) in the western Falcón Basin. It developed where movement on the Oca fault was transferred southward to the Ancon fault. The Urumaco Trough is also a distinct pullapart structure connecting the Cuiza and Oca fault trends.

The central Falcón Basin is a more diffuse zone of subsidence. A northern boundary fault at the southern margin of the San Luis limestone platform (Fig. 3) is postulated due to the relative stability and the shallow depositional environment of the limestone reefs, in contrast to the deep water, rapidly subsiding environment of the synchronous Pecaya shales. The southern border fault follows east-west scarps in the Siquisique area, which actively shed coarse Miocene sediments northward.

The eastern Falcón Basin and La Vela Bay also exhibit widespread subsidence, which increases in degree to the north and east. The southern border is along fault trends south of the Cerro Misión uplift (Fig. 4). The uplift represents a horst that remained high during subsidence of the surrounding area since early Oligocene time (Hunter, 1972). Northwesterly normal faults with downthrown eastern blocks in the Gulf of Triste indicate that the gulf developed as a pullapart basin to the north of the San Sebastián transcurrent fault.

In addition to pullapart zones of divergence, irregular transcurrent fault systems inevitably develop zones of convergence with associated compressional deformation (Wilcox and others, 1973; Crowell, 1974). The southern border of the Falcón pullapart developed zones of convergence particularly in the western basin. Convergence caused distortion of fold axes in this area to N20°E orientations and development of north- to northeast-trending transcurrent faults.

Bonaire Basin Pullapart

Stratigraphic and structural characteristics of the Falcón Basin appear to extend into the Bonaire Basin (Silver and others, 1975). Northwest-trending normal and east-west-trending normal and transcurrent faults controlled sediment distribution and subsidence in the Bonaire Basin. The San Sebastián transcurrent fault borders the southern margin of the basin (Fig. 4). East-west faults south of Aruba and Curaçao and along the northern basin margin were interpreted by Silver and others (1975) from seismic data as normal faults, but transcurrent motion on these faults is also possible. In conjunction the northwest and east-west fault trends outline the rhombic shape of the Bonaire Basin pullapart structure.

The structures adjacent to the Netherlands and Venezuelan Antilles are also consistent with a pullapart origin. The sedimentary basin between Aruba and Curaçao, for example, exhibits the typical rhombic form with surrounding normal faults (Fig. 4).

Bouguer gravity anomalies for most of the Bonaire Basin are positive (Bonini, 1978; Bonini and others, 1977) despite sedimentary thicknesses of 2 to 6 km. As discussed previously this indicates a thin or relatively dense crust. Modeling of free-air

anomalies across the basin indicates at least 5 km of crustal thinning (Silver and others, 1975). Lacking definitive information on the nature of the basement underlying the Bonaire Basin, the following hypotheses are feasible: (1) the basement and the overlying sediments of the basin have been injected with large amounts of mafic material, or (2) the crust was originally thinner or more mafic prior to basin development. In the second option a preexisting Bonaire Basin (Fig. 7b) would be extended about 150 km during the Oligocene to Miocene pullapart event. The first option requires about 300 km of east-west extension to produce the Bonaire Basin and would imply a major east-west transcurrent fault between the basin and the island chain (Fig. 7a). The transcurrent fault allows extension and subsidence in the basin to exceed that along the island chain.

Additional evidence comes from the Eocene or older Soebi Blanco conglomerates of Bonaire, which contain gneiss, schist, and marble cobbles indicative of a mainland source (Pijpers, 1933; Beets and others, 1977). Silver and others (1975) inferred that the Bonaire Basin formed later than deposition of this conglomerate. Removing the Oligocene to Miocene extension between the Netherlands Antilles and Paraguaná, as proposed by the pullapart model, would place Bonaire and/or Curaçao adjacent to Paraguaná (Fig. 7). Thus, Jurassic basement rocks on Paraguaná could be the source rocks for the conglomerates.

Contrary to the suggestion of Silver and others (1975), the Bonaire Basin may already have existed as a basin (Fig. 7b) during Eocene conglomerate deposition as long as the island chain was continuous from Paraguaná to Bonaire. Extension and breakup of the island chain, however, must have been post-Eocene.

Bonaire Block Pullapart

The Bonaire Block, bounded by the Curaçao Ridge on the north and the Oca–San Sebastián faults on the south, is a small crustal block between the Caribbean and South American plates (Fig. 8). Silver and others (1975) concluded that nonrigid deformation of this block occurred during its clockwise rotation between the major crustal plates.

However, the nature of many of the Neogene basins within this block suggests that a pullapart model may explain their origins. Case (1974) concluded that the northwest- and east-west-trending Colombian basins (Baja Guajira, Chichibacoa; Fig. 4), in addition to the Venezuelan basins between the offshore islands, originated from north to northwest extension within the Bonaire Block. The model presented here redefines the Bonaire Block as a pullapart zone that experienced east to northeast extension between the transcurrent faults that border the Caribbean and South American plates (Fig. 8). Extension occurred mainly during Oligocene and Miocene times.

Geologic and seismic evidence suggest that the Oca, although active, and Cuiza transcurrent faults are not presently zones of major strike-slip displacement (Molnar and Sykes, 1969). Instead, Caribbean–South American plate motion from

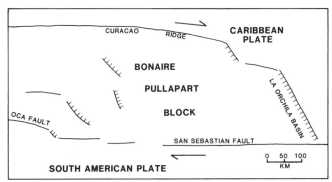

Figure 8. The Bonaire Block pullapart structure and its relationship to the Caribbean and South American plates. The block is identical to that outlined by Silver and others (1975).

late Miocene to Holocene was transferred from the El Pilar–San Sebastián, right-lateral fault systems to a broad zone of compressional deformation and transcurrent movement within the Bonaire Block, and along the Boconó fault and the Curaçao Ridge (Silver and others, 1975; Jordan, 1975; Pennington, 1981). These events are responsible for the overprinting of earlier Oligocene and Miocene northwest-trending normal faults in the Falcón area by transcurrent motion, folding, and thrusting.

Extension within the Bonaire Block

An estimate on the minimum amount of extension within the Bonaire Block pullapart has been calculated assuming that Eocene "basement" sediments (high velocity) attained their present depths by subsidence along normal faults. A similar type of analysis was presented by Illies (1970) to estimate extension in the Rhine Graben. Planar normal faults with dips of 60° are assumed and are consistent with seismic profiles (Silver and others, 1975). The calculation is made for the northern part of the pullapart along the Antillean chain primarily because depth to "basement" is controlled by a variety of geophysical data. The Falcón and Bonaire basins were avoided because of the asymmetric, half-graben faulting and later erosional effects. The basins included are the La Orchila Basin, the basins between all the Antillean Islands, the basin between Aruba and Paraguaná, and the Chichibacoa Basin (Fig. 7). Estimated post-Eocene, east-west extension from the Guajira Peninsula to the La Orchila Basin is about 35 km (Table 2). The value is increased to 45 km by using the Urumaco Trough instead of the Chichibacoa Basin and by adding an extension estimate for the Baja Guajira Basin.

By this argument, at least 35 to 45 km of extension occurred within the Bonaire Block during its Oligocene and Miocene deformation between the Caribbean and South American plates. This is a minimum extension for the following reasons: (1) Major faults along the northern border of the Bonaire Basin may allow greater amounts of east-west extension to the south. (2) Extension within stable blocks (Paraguaná, Guajira) was not considered. (3) Estimates are based on a brittle fracture model that does not

consider crustal stretching. (4) Estimates from basement depths do not account for extension by magma injection. (5) Normal faults may shallow at depth allowing greater extension. (6) Estimates of sediment thickness are probably minimum estimates.

TECTONIC SUMMARY

Prior to initiation of the Falcón Basin, Paleocene to Eocene rocks at the southern border of the basin recorded the culmination of a compressional orogen involving flysch deposition, thrusting, ophiolite obduction, tectonic rotations, and olistostrome development (Skerlec and Hargraves, 1980; Beets and others, 1984). The Antillean islands formed a contiguous unit connected to the Paraguaná Peninsula and the Falcón area (Fig. 7 and 9A).

A major change in the deformation style occurred during the Eocene in northern Venezuela. The change is marked by the cessation of southward thrusting and flysch deposition in both western and central Venezuela by the end of middle Eocene time (Bell, 1972; Maresch, 1974; Stephan, 1977; Beck, 1978). East-west, right-lateral, transcurrent faulting began in northern South America by late Eocene time. Andesitic volcanism and subduction in the Lesser Antilles also started during the Eocene (Bunce and others, 1971; Arculus, 1976). This implies that the Caribbean plate had begun to move eastward with respect to the South American plate regardless of its pre-Eocene displacement (Bell, 1972; Beets and others, 1984).

The Falcón Basin originated by extension during late Eocene to Oligocene time within a pullapart zone caused by transcurrent motion between the Caribbean and South American plates (Fig. 9B). Indeed, the Falcón Basin preserves a record of the sedimentation and deformation that occurred within this zone

TABLE 2. EXTENSION ESTIMATES WITHIN THE NORTHERN PART OF THE BONAIRE BLOCK

Basin	Sediment thickness (km)	Total subsidence (km)	East-west extension (km)
Chichibacoa	4.0	4.2	
Los Monjes-Aruba	1.5	1.7	
Aruba-Paraguaná	1.5	1.7	
Curacao-Aruba	2.0	3.1	
Bonaire-Curacao	2.0	3.5	
Las Aves-Bonaire	1.0	2.7	
Los Roques-Las Aves	0.5	2.5	
La Orchila-Los Roques	0.8	1.8	
La Orchila Basin	1.2	5.1	
TOTAL		26.3	35.1
Urumaco Trough	8.3	8.5	
Baja Guajira	3.0	3.0	
TOTAL (Minus Chichibacoa*)		33.6	44.8

Note: Subsidence equals sediment thickness plus the depth of water in the basin. A basin name with a hyphen (e.g., Curacoa-Aruba) indicates the basin between the two areas. Sediment thickness data compiled from Edgar and others (1971), Zambrano and others (1971), Silver and others (1975), Peter (1972), Case (1974), and Corporación Venezolana del Petróleo-Institut Francais du Pétrole (unpublished report, 1967).
*Alternate estimate of extension not including the Chichibacoa Basin is explained in the text.

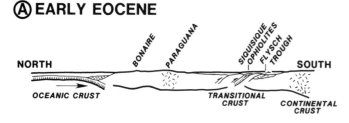

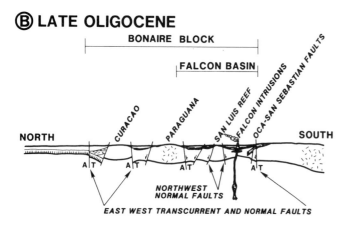

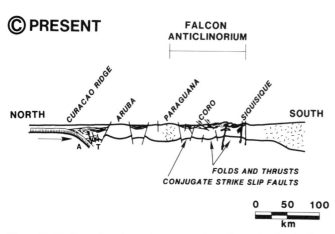

Figure 9. North-south, schematic cross sections illustrating the development of the Falcón Basin and the Bonaire Block (discussed in the text) from late Eocene to present. Extension in the basin occurs in an east-west direction perpendicular to the section. Scale is approximate with no vertical exaggeration. On transcurrent faults, A indicates the block moving away from the reader and T is toward the reader.

of transcurrent faulting. Right-lateral motion along offset, transcurrent faults caused extension in a northeast direction by normal faulting along northwesterly trends, subsidence, crustal thinning, and injection of basaltic alkaline magmas (Fig. 9B).

On a larger scale the Bonaire Block of Silver and others (1975) originated as a diffuse zone of pullapart structures during the Oligocene right-lateral motion between the Caribbean and

South American plates (Fig. 9B). Minimum extension within the northern part of the block was at least 35 to 45 km in an east-west direction. The extension within this nonrigid block must be considered when estimating Caribbean–South American plate movements.

Pullapart structures also developed elsewhere along the Caribbean–South American plate boundary at this time. The La Orchila Basin and the Cariaco Trench are excellent examples. Similar structural basins formed along the northern margin of the Caribbean plate; however, here the motion was in a left-lateral sense with respect to the North American plate (Burke and others, 1980).

In the Bonaire Block, folds, reverse faults, and conjugate sets of secondary transcurrent faults formed later during the development of the pullapart and were amplified by greater compressional stresses during the late Miocene to Holocene tectonic evolution (Fig. 9C). Shallow underthrusting of the block by the Caribbean plate formed the Curaçao Ridge accretionary prism (Silver and others, 1975; Jordan, 1975; Ladd and Watkins, 1979). The Falcón anticlinorium was formed and uplifted by the addition of compressional stress across this generally transform-type plate boundary zone between the Caribbean and South America.

ACKNOWLEDGMENTS

This research was supported by the Princeton University Department of Geological and Geophysical Sciences and the Venezuela Ministerio de Energía y Minas. I am indebted to the many geologists from these organizations who provided field assistance, suggestions, criticism, and above all—stimulating discussions. Among these I am particularly grateful to A. Bellizzia, J. Rios, R. Hargraves, W. Bonini, J. Suppe, R. Vierbuchen, and M. Díaz de Gamero. I thank T. Harding and M. Schuepbach of Exxon Production Research Company for discussing their ideas on wrench faulting and pullapart basins. I also thank H. A. Jordi of Shell Internationale Petroleum and the Venezuela Ministerio de Energía y Minas for providing copies of unpublished geologic studies of the Falcón region.

REFERENCES CITED

Alberding, H., 1957, Application of principles of wrench-fault tectonics of Moody and Hill to northern South America: Geological Society of America Bulletin, v. 68, p. 785–790.

Alvarez, W., 1967, Geology of the Simurúa and Carpintero areas, Guajira Peninsula, Colombia [Ph.D. thesis]: Princeton University, 168 p.

Arculus, R. J., 1976, Geology and geochemistry of the alkali basalt-andesite association of Grenada, Lesser Antilles island arc: Geological Society of America Bulletin, v. 87, p. 612–624.

Beck, C. M., 1978, Polyphase Tertiary tectonics of the interior range in the central part of the western Caribbean chain, Guárico State, northern Venezuela: Geologie Mijnbouw, v. 57, p. 99–104.

Beets, D. J., MacGillavry, H. J., and Klaver, G., 1977, Geology of the Cretaceous and early Tertiary of Bonaire, *in* Guide to the field excursions on Curaçao, Bonaire, and Aruba, Netherlands Antilles: Caribbean Geological Conference, VIII, Curaçao Netherlands Antilles, p. 18–28.

Beets, D. J., Maresch, W. V., Klaver, G. Th., Mottana, A., Bocchio, R., and Beunk, F. F., 1984, Magmatic rock series and high-pressure metamorphism as constraints on the tectonic history of the southern Caribbean, *in* Bonini, W. E., and others, eds., The Caribbean–South American plate boundary and regional tectonics: Geological Society of America Memoir 162 (this volume).

Bell, J. S., 1971, Tectonic evolution of the central part of the Venezuelan coast ranges, *in* Donnelly, T. W., ed., Caribbean geophysical, tectonic, and petrologic studies: Geological Society of America Memoir 130, p. 107–118.

Bell, J. S., 1972, Global tectonics in the southern Caribbean area, *in* Shagam, R., and others, eds., Studies in earth and space sciences: Geological Society of America Memoir 132, p. 369–386.

Bellizzia, A., 1972, Is the entire Caribbean Mountain belt of northern Venezuela allochthonous? *in* Shagam, R., and others, eds., Studies in earth and space sciences: Geological Society of America Memoir 132, p. 363–368.

Bellizzia, A., and Rodríguez, D., 1967, Guía de la excursión a la región de Duaca-Barquisimeto-Bobare: Venezuela Boletín Geología, v. 8, p. 289–309.

Bellizzia, A., Rodríguez, D., and Graterol, M., 1972, Ofiolitas de Siquisique y Río Tocuyo y sus relaciónes con la fallas de Oca: Transactions, Caribbean Geological Conference, VI, Margarita, Venezuela, 1971, p. 182.

Bellizzia, A., Pimentel, N., and Bajo, R., compilers, 1976, Mapa geológico estructural de Venezuela: Venezuela Ministerio de Minas e Hydrocarburos, escala 1:500,000.

Bonini, W. E., 1978, Anomalous crust in the Eastern Venezuela Basin and the Bouguer gravity anomaly field of Northern Venezuela and the Caribbean borderland: Geologie Mijnbouw, v. 57, p. 117–122.

Bonini, W. E., Pimstein de Gaeta, C., and Graterol, V., compilers, 1977, Mapa de anomalías gravimétricas de Bouguer de la parte norte de Venezuela y areas vecinas: Venezuela Ministerio de Energía y Minas, escala 1:1,000,000.

Bowin, C., 1976, The Caribbean: Gravity field and plate tectonics: Geological Society of America Special Paper 169, 79 p.

Bunce, E. T., Phillips, J. D., Chase, R. L., and Bowin, C. O., 1971, The Lesser Antilles and the eastern margin of the Caribbean Sea, *in* Maxwell, A. E., ed., The sea, Volume 4, Part II: New York, John Wiley & Sons, p. 359–385.

Burke, K., Grippi, J., and Sengor, A.M.C., 1980, Neogene structures in Jamaica and tectonic style of the northern Caribbean plate boundary zone: Journal of Geology, v. 88, p. 375–386.

Carey, S. W., 1958, A tectonic approach to continental drift: University of Tasmania Department of Geology Symposium, p. 177–355.

Case, J. E., 1974, Major basins along the continental margin of northern South America, *in* Burk, C., and Drake, C., eds., The geology of continental margins: New York, Springer-Verlag, p. 733–741.

Case, J. E., and Holcombe, T. L., 1980, Geologic-tectonic map of the Caribbean region: U.S. Geological Survey Miscellaneous Investigations Map I-1100, scale 1:2,500,000.

Compañía Shell de Venezuela, 1965, Igneous rocks of the Siquisique region, State of Lara: Asociación Venezolana de Geología Minería y Petróleo Boletín Informativo, v. 8, p. 286–305.

Coronel, G., 1970, Igneous rocks of central Falcón: Associación Venezolana de Geología Minería y Petróleo Boletín Informativo, v. 13, p. 155–162.

Crowell, J. C., 1974, Sedimentation along the San Andreas Fault, California, *in* Dott, R. H., and Shaver, R., eds., Modern and ancient geosynclinal sedimentation: Society of Economic Paleontologists and Mineralogists Special Publication 19, p. 292–303.

Díaz de Gamero, M. L., 1977, Estratigrafía y micropaleontología del Oligocene y Miocene inferior del centro de la Cuenca de Falcón, Venezuela: Escuela de Geología y Minas, Universidad Central de Venezuela, Caracas, GEOS no. 22, p. 3–60.

Edgar, N. T., Ewing, J. I., and Hennion, J., 1971, Seismic refraction and reflection in Caribbean Sea: American Association of Petroleum Geologists Bulletin, v. 55, p. 833–870.

Elders, W. A., Rex, R. W., Meidav, T., Robinson, P. T., and Biehler, S., 1972, Crustal spreading in southern California: Science, v. 178, p. 15–24.

Feo-Codecido, G., 1971, Geología y recursos naturales de la Península de Paraguaná, *in* Symposium on Investigations and Resources of the Caribbean Sea and Adjacent Regions: UNESCO, Paris, France, p. 231–240.

Feo-Codecido, G., 1972, Breves ideas sobre la estructura de la Falla de Oca, Venezuala: Transactions, Caribbean Geological Conference, VI, Margarita, Venezuela, 1971, p. 184–190.

Galavis, J. A., and Louder, L. W., 1971, Preliminary studies on geomorphology, geology and geophysics on the continental shelf and slope of northern South America: Eighth World Petroleum Congress, Caracas, Venezuela, Proceedings, v. 2, p. 107–120.

Hill, D. P., 1977, A model for earthquake swarms: Journal of Geophysical Research, v. 82, p. 1347–1352.

Hunter, V. F., 1972, A middle Eocene flysch from east Falcón, Venezuela: Transactions, Caribbean Geological Conference, VI, Margarita, Venezuela, 1971, p. 128–130.

Illies, J. H., 1970, Graben tectonics in relation to crust mantle interaction, *in* Illies, J. H., and Muller, S., eds., Graben problems: Stuttgart, Schweizerbart, p. 4–27.

Jordan, T., 1975, The present-day motions of the Caribbean plate: Journal of Geophysical Research, v. 80, p. 4433–4439.

Ladd, J. W., and Watkins, J. S., 1979, Tectonic development of trench-arc complexes on the northern and southern margins of the Venezuela Basin, *in* Watkins, J. S., Montadert, L., and Dickerson, P. W., eds., Geological and geophysical investigations of continental margins: American Association of Petroleum Geologists Memoir 29, p. 363–371.

MacDonald, W. D., 1968, Estratigrafía, estructura y metamorfismo de las rocas del Jurásico Superior, Península de Parguaná, Venezuela: Venezuela Boletín Geología, v. 9, p. 441–458.

Malfait, B. T., and Dinkelman, M. G., 1972, Circum-Caribbean tectonic and igneous activity and the evolution of the Caribbean plate: Geological Society of America Bulletin, v. 83, p. 251–272.

Maresch, W. V., 1974, Plate tectonics origin of the Caribbean Mountain system of northern South America: Discussion and proposal: Geological Society of America Bulletin, v. 85, p. 669–682.

Martín-Bellizzia, C., and Arozena, J.M.I., 1972, Complejo ultramáfico zonado de Tausabana-El Rodeo, gabro zonado de Siraba-Capuana y complejo subvolcánico estratificado de Santa Ana, Paraguaná, Estado Falcón: Transactions, Caribbean Geological Conference, VI, Margarita, Venezuela, 1971, p. 337–356.

Molnar, P., and Sykes, L. R., 1969, Tectonics of Caribbean and Middle America regions from focal mechanisms and seismicity: Geological Society of America Bulletin, v. 80, p. 1639–1684.

Muessig, K. W., 1978, The central Falcón igneous suite, Venezuela: Alkaline basaltic intrusions of Oligocene-Miocene age: Geologie Mijnbouw, v. 57, p. 261–266.

——1979, The central Falcón igneous rocks, northwestern Venezuela: Their origin, petrology and tectonic significance [Ph.D. thesis]: Princeton University, 252 p.

Payne, A. L., 1951, Cumarebo oil field, Falcón, Venezuela: American Association of Petroleum Geologists Bulletin, v. 35, p. 1850–1878.

Pearce, J. A., and Cann, J. R., 1973, Tectonic setting of basic volcanic rocks determined using trace element analyses: Earth and Planetary Science Letters, v. 19, p. 290–300.

Pennington, W. D., 1981, Subduction of the eastern Panama Basin and seismotectonics of northwestern South America: Journal of Geophysical Research, v. 86, p. 10753–10770.

Peter, G., 1972, Geologic structure offshore north-central Venezuela: Transactions, Caribbean Geological Conference, VI, Margarita, Venezuela, 1971, p. 283–294.

Pijpers, P. J., 1933, Geology and paleontology of Bonaire (Dutch West Indies) [Ph.D. thesis]: University of Utrecht, Netherlands, 103 p.

Renz, O., Lakeman, R., and Van der Muelen, E., 1955, Submarine sliding in western Venezuela: American Association of Petroleum Geologists Bulletin, v. 39, p. 2053–2067.

Rod, E., 1956, Strike-slip faults of northern Venezuela: American Association of Petroleum Geologists Bulletin, v. 40, p. 457–476.

Santamaría, F. J., and Schubert, C., 1974, Geochemistry and geochronology of the southern Caribbean–northern Venezuela plate boundary: Geological Society of America Bulletin, v. 85, p. 1085–1098.

Silver, E. A., Case, J. E., and MacGillavry, H. J., 1975, Geophysical study of the Venezuelan borderland: Geological Society of America Bulletin, v. 86, p. 213–226.

Skerlec, G. M., and Hargraves, R. B., 1980, Tectonic significance of paleomagnetic data from northern Venezuela: Journal of Geophysical Research, v. 85, p. 5303–5315.

Smith, F. D., Jr., 1962, Mapa geológico-tectónico del Norte de Venezuela: Primer Congreso Venezolana Petroleo, escalo 1:1,000,000.

Stainforth, R. M., 1969, The concept of seafloor-spreading applied to Venezuela: Asociación Venezolana de Geología, Minería y Petróleo Boletín Informativo, v. 12, p. 257–274.

Stephan, J. F., 1977, Una Interpretación de los complejos con bloques asociados a los flysch Paleocene-Eoceno de la cadena Caribe Venezolana: el emplazamiento submarino de la napa de Lara: Caribbean Geological Conference, VIII, Curaçao Netherlands Antilles, Abstracts, p. 199–200.

Tschanz, C. M., Marvin, R. F., Cruz, B., Mehnert, H. H., and Cebula, G. T., 1974, Geologic evolution of the Sierra Nevada de Santa Marta, northeastern Colombia: Geological Society of America Bulletin, v. 85, p. 273–284.

Vásquez, E., 1975, Results of exploration in La Vela Bay: Ninth World Petroleum Congress, Tokyo, Japan, Proceedings, v. 3, p. 195–197.

Vásquez, E., and Dickey, P., 1972, Major faulting in northwestern Venezuela and its relation to global tectonics: Transactions, Caribbean Geological Conference, VI, Margarita, Venezuela, 1971, p. 191–202.

Vierbuchen, R. C., Jr., 1978, The tectonics of northeastern Venezuela and the southeastern Caribbean Sea [Ph.D. thesis]: Princeton University, 175 p.

Wheeler, C. B., 1963, Oligocene and lower Miocene stratigraphy of western and northeastern Falcón Basin, Venezuela: American Association of Petroleum Geologists Bulletin, v. 47, p. 35–68.

Wilcox, R. E., Harding, T. P., and Seely, D. R., 1973, Basic wrench tectonics: American Association of Petroleum Geologists Bulletin, v. 57, p. 74–96.

Zambrano, E., Vásquez, E., Duval, B., Latreille, M., and Coffinieres, B., 1971, Sintesis paleogeográfica y petrolera del occidente de Venezuela: Memoria, Congreso Geológico Venezolano, IV, Caracas, 1969, Venezuela Ministerio de Minas e Hidrocarburos, Boletín de Geología, Publicación Especial no. 5, v. 1, p. 481–552.

MANUSCRIPT ACCEPTED BY THE SOCIETY SEPTEMBER 1, 1983

Geological Society of America
Memoir 162
1984

Paleomagnetic data on the basic igneous intrusions of the central Falcón Basin, Venezuela

Karl W. Muessig
Exxon Minerals Company
P.O. Box 2189
Houston, Texas 77001

ABSTRACT

Paleomagnetic data for basaltic intrusions in the central Falcón Basin document a secondary magnetization associated with alteration during Miocene time. The rocks are extensively altered; however, most of this alteration probably accompanied cooling soon after intrusion. Comparing these secondary magnetic signatures to those of other Cenozoic sites in South America, we conclude that magnetization occurred before the late Miocene folding in the basin and soon after intrusion.

A later stable partial overprinting of this secondary magnetization occurred in which the present geomagnetic field was superimposed on the stable Miocene remanence. At many sites it is difficult to remove this overprinting during demagnetization. The resultant magnetic signature has a steep positive inclination, which in the past has been difficult to interpret.

INTRODUCTION

The Falcón Basin is an extensional pullapart rift of Neogene age which formed within a complex right-lateral transcurrent fault system along the Venezuelan continental margin. The stratigraphic, structural, and tectonic history of the basin is documented in this volume (Muessig, 1984).

The main phase of basin subsidence and filling occurred during Oligocene and Miocene times. Alkaline to subalkaline basaltic sills and dikes intrude the Oligocene-Miocene sediments. One of these sills (La Guarabitas, Fig. 1) was dated at 22.9 ± 0.9 m.y. by K-Ar methods (Muessig, 1978). This age is the youngest basaltic activity reported in Venezuela. Only dacites from the Carúpano area in eastern Venezuela are younger (5 ± 0.5 m.y.; Santamaría and Schubert, 1974).

The Falcón Basin was later inverted by compressional events such that the central basin is now the deformed core of the Falcón Anticlinorium. However, there is little direct evidence for the timing of this deformation in the central basin due to total erosion of younger formations. Unconformities along the northern margin of the basin document some late Miocene folding (Zambrano and others, 1971).

A reconnaissance paleomagnetic study of two Falcón igneous sites was included in the work of Skerlec and Hargraves (1980). They found remanent magnetic vectors with southerly declinations and steep positive inclinations. Surrounding geologic structures do not support a post-intrusion folding event as a total explanation for the inclinations. Instead, a superimposition of two vectors was proposed: one primary reversed prefolding and a stable secondary normal postfolding (Skerlec and Hargraves, 1980).

This report details a more extensive paleomagnetic study of the intrusive rocks in the central basin. It was undertaken to further constrain the timing of magnetization with respect to the intrusive event and to deformation of the basin.

SAMPLING AND ANALYSIS

Samples for paleomagnetic analyses from a particular locality were collected from single continuous outcrops with distances between samples of 1 to 10 m. Cores (2.5 cm diameter) were obtained in the field with a portable drill or drilled from oriented samples in the laboratory. Field orientations were measured with both sun and magnetic compasses, and no significant directional differences were noted between these techniques. The remanent magnetism of cylindrical specimens (2.3 cm long) was measured

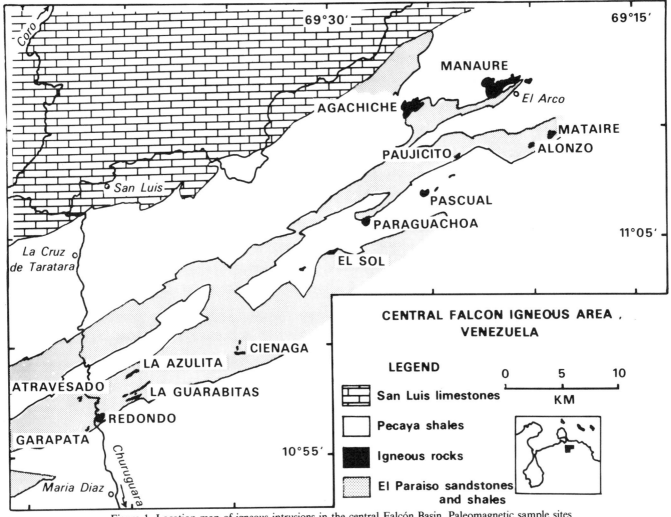

Figure 1. Location map of igneous intrusions in the central Falcón Basin. Paleomagnetic sample sites listed in Table 1 are identified by the name of the particular intrusion.

on a Princeton Applied Research SM-1 spinner magnetometer. The natural remanent magnetism (NRM) and the remanence after 50, 100, 150, 200, 300, and 500 Oe alternating field demagnetization were measured for each sample.

SAMPLE MINERALOGY

In all samples, with the exception of some from La Guarabitas, the primary basaltic mineralogy is extensively altered. Feldspars are replaced by saussurite, chlorite, albite, or sericite; biotite and amphibole when present are partially to completely chloritized; and iddingsite or chlorite are pseudomorphs after olivine. Fresh clinopyroxene is present in most samples; however, in other samples (Atravesado, Paraguachoa) it is marginally to completely altered to chlorite, carbonate, and sphene.

Oxide assemblages and textures indicate an advanced state of oxidation (Haggerty, 1976) and alteration. Pseudomorphic textures preserve oxidation-exsolution lamellae of ilmenite along {111} planes of original titanomagnetites. In many samples such

textures are the only evidence of the prior existence of magnetite. Excepting Atravesado and La Guarabitas, all magnetite has been replaced by sphene or totally oxidized to assemblages of titanohematite, ilmenite, and rutile. Although oxy-exsolved ilmenite lamellae and primary ilmenite are more resistant to later alteration, similar oxidation and sphenitization products have evolved. Chlorite pseudomorphs of olivine and biotite occasionally contain fine-grained hematite and rutile+hematite, respectively. Chrome spinel xenocrysts are surrounded by titanomagnetite or its oxidized products.

The Atravesado samples contain a few large (up to 20-mm diameter) homogeneous titanomagnetites. These are rounded xenocrysts with rims of fine-grained, equant titanomagnetite. Some are composite grains containing magnetite, pyrrhotite, and pyrite. Fine ilmenite lenses have been oxy-exsolved from the magnetite margins. Aside from the large xenocrysts, all other Atravesado oxides are altered to hematite, rutile, and sphene.

The Guarabitas samples contain abundant, fine-grained, partially oxidized magnetite with a spongy texture. Ilmenite, an oxi-

TABLE 1. PALEOMAGNETIC DATA FOR THE CENTRAL FALCON IGNEOUS ROCKS

Site	Location, deg		n	NRM							Demagnetized data				
	Lat,N	Long,W		D,deg	I,deg	k	α_{95} deg	J_N,emu/cm^3	X	Q	ODF Oe	D,deg	I,deg	k	α_{95} deg
Atravesado	10.96	69.71	7	313	30	9	24	6.9×10^{-6}	3.0×10^{-5}	0.24	200	27	56	1	90
Garapata	10.94	69.70	8	25	55	4	34	4.3×10^{-6}	3.7×10^{-5}	0.12	200	358	40	27	11
Redondo*	10.95	69.69	10	27	58	3	37	9.2×10^{-5}			300	282	26	4	26
Guarabitas	10.97	69.66	11	105	17	2	42	9.1×10^{-3}	1.2×10^{-3}	7.6	200	126	11	2	45
Cienaga 1*	10.99	69.58	6	353	81	3	45	2.6×10^{-6}			300	313	87	9	24
Cienaga 2	10.99	69.58	8	325	49	14	16	3.6×10^{-6}	4.7×10^{-5}	0.08	200	312	72	4	30
Paraguachoa	11.09	69.47	6	9	65	2	62	6.6×10^{-7}	2.4×10^{-5}	0.03	200	13	30	4	40
Agachiche	11.18	69.43	6	249	-51	2	74	1.8×10^{-6}	3.5×10^{-5}	0.05	200	192	-40	18	16
Manaure	11.19	69.37	11	345	28	8	17	3.0×10^{-6}	3.6×10^{-5}	0.08	200	1	16	54	6

Note: n is the number of samples; D is the eastward declination from north; I is the inclination, downward positive; k is Fisher's precision parameter; α_{95} is the semiangle of cone of 95% confidence; J_N is the intensity of remanent magnetization; X is the magnetic susceptibility; Q is the Koenigsberger ratio ($J_N/X \cdot H$); ODF is the optimum demagnetizing field. Underlined sites meet the statistical criteria $\alpha_{95} \pm 20.0^\circ$ after demagnetization. Cerros Atravesado, Redondo, and Cienaga are intrusive plugs; Cerros Garapata, Guarabitas, and Manaure are sills; and Paraguachoa is a tuffaceous agglomerate pile.

* Data from Skerlec, and Hargraves (1980).

dation product, and pyrite are present in minor amounts. Other oxide alteration products include a mineral identified as maghemite, which exhibits magnetic properties but has a high, hematite-like reflectivity; and an unidentified Fe, Ti silicate detected by electron microprobe analysis.

DATA

The very low intensities of remanent magnetism and low susceptibilities (Table 1) are consistent with the rather highly altered mineralogy. The NRM intensity is typically less than 10^{-5} emu/cm^3; Cerro Redondo ($<10^{-4}$ emu/cm^3) and La Guarabitas ($<10^{-2}$ emu/cm^3) are exceptions. The Q (Koenigsberger) ratios are also generally low (see Table 1).

At the 200 Oe demagnetization level, intensities have dropped to between the 10^{-6} to 10^{-7} emu/cm^3 level (except Guarabitas), but remanence directions of individual samples show little change from the 100- and 150-Oe levels. Demagnetization of most samples above 200 Oe causes an unstable and viscous remanent magnetism. Repeated demagnetization and measurement of individual specimens at the 500-Oe level frequently yields diametrically opposed directions. For most samples data for the 500-Oe demagnetization level are not reported, since they are generally unreliable. Even at the 200-Oe level, the individual sample directions for many sites are not well grouped.

For the site mean directions, an arbitrary statistical cut off of Fisherian $\alpha_{95} \leq 20^\circ$ has been chosen as a minimum criterion of reliability (after Hicken and others, 1972). Only Cerros Manaure, Garapata, and Agachiche meet this restriction; Figures 2A, 2B, and 2C plot their individual sample NRMs and remanent directions after 200-Oe demagnetization. These figures indicate that alternating field demagnetization greatly improves the clustering of data. The site means and α_{95} cones of confidence have been plotted with the site data.

Cerro Manaure exhibits the best data cluster of all collected sites. Remanent magnetism is stable at the 200-Oe as well as the 500-Oe levels. Fisherian statistics yield a northerly declination, a shallow positive inclination, and an α_{95} of 6°.

Cerro Garapata exhibits a northerly declination, a moderate positive inclination, and an α_{95} of 11°. Cerros Manaure and Garapata were magnetized during a normal polarity interval. Although Cerros Atravesado and Paraguachoa have α_{95}s, which do not meet the reliability criteria, they too have northerly declinations and positive inclinations.

Cerro Agachiche, on the other hand, displays a southerly declination and a moderate negative inclination with an α_{95} of 16°. This implies magnetization during a reversed polarity interval.

INTERPRETATION

From the previously described oxide assemblages and silicate oxidation-alteration phenomena, it is suggested that the measured remanence is held in secondary mineral products.

Magnetite-ilmenite and ilmenite-hematite-rutile assemblages suggest initial oxidation at rather high temperatures (600 °C–700 °C; Haggerty, 1976). This was accompanied by oxidation and deuteric alteration of olivine. High temperature olivine oxidation is suggested by a few olivine pseudomorphs with associated secondary hematite (Haggerty and Baker, 1967). A later lower temperature alteration (200 °C–400 °C; Baker and Haggerty, 1967) may have completely overprinted the high temperature olivine alteration. It seems that the introduction of connate waters from the host rocks altered the feldspars and redistributed Ca into either calcite or prehnite and zeolites, depending on local conditions. Sphenitization of much of the magnetite probably occurred at this time. Some of the Fe was redistributed into chlorites. The minor amount of secondary oxides (hematite) produced from silicate oxidation, coupled with the destruction of most of the magnetite by oxidation and sphenitization, accounts for the present low magnetic intensities.

In light of the above evidence, the measured remanent directions were probably induced during oxidation and alteration of the Falcón igneous bodies soon after intrusion. If intrusion, cooling, and initial alteration all occurred within a short time interval, then the primary and proposed secondary magnetization would be virtually identical in direction.

Considering that sites give approximately reversed northerly and southerly directions, it is fortuitous to assume that intrusion, cooling, and alteration occurred elsewhere with later allochthonous emplacement *without rotation* into the central Falcón area. Indeed, the directions of remanent magnetism in the Falcón rocks are distinctly different from the easterly directions observed for rocks of inferred allochthonous nature in northern Venezuela and the offshore islands (Skerlec and Hargraves, 1980). Paleomagnetic data for the Falcón rocks are consistent with an autochthonous intrusive origin.

Constraints on the timing of post-intrusion magnetization can be obtained from a study of the relationship of magnetization to folding in the Falcón Basin. In the central Falcón area the

igneous intrusions are situated in strata that are folded around east- to east-northeast-trending axes. The Garapata, Agachiche, and Manaure sites are all conformable sills. If magnetization occurred prior to folding, then the remanent directions should be corrected to bring the folded strata back to a horizontal position. Because of the orientations of fold axes, the structural corrections (Table 2) have little effect on declination but substantially decrease inclination (Fig. 2D).

Because the bedding attitude is similar for the three sites, Fisher's k changes little with structural corrections (k=32 in situ versus k=28 structurally corrected; Table 2).

The validity of the structural corrections relies upon comparison of the mean VGP from the Falcón sites with other published Tertiary and Late Cretaceous poles for South America (Table 3). Although rather sparse, the available data for this time period all plot within a narrow, elongate swath (Fig. 3). The structurally corrected, mean VGP for Falcón plots within this swath, whereas the uncorrected mean VGP does not. In addition, the corrected Falcón VGP fits nicely between the group of Eocene and older VGP and the group of Pliocene and younger VGP. This would date the magnetization at Oligocene-Miocene, and thus it is consistent with the age of intrusion and alteration.

TABLE 2. VIRTUAL GEOMAGNETIC POLE (VGP) FOR FALCON SITES WITH $\alpha 95 \leq 20^\circ$, at 200-Oe DEMAGNETIZATION BEFORE AND AFTER STRUCTURAL CORRECTIONS

	200 Oe		VGP		Structurally corrected				VGP	
	D	I	Lat,N	Long,E	DIR	DIP	D	I	Lat,N	Long,E
Garapata	358	40	78	282	0	15	358	25	87	255
Agachiche	192	-40	73	334	350	41	186	-1	77	78
Manaure	1	16	87	90	355	15	1	1	80	106
Mean	5	33	(k=32,α95=22)				3	8	(k=28,α95=24)	
Mean VGP for 3 sites:			82	317					83	93
			(K=57,A95=17)						(K=82,A95=13)	

Note: D is the eastward declination from north; I is the inclination downward. For the applied structural correction DIR is the direction of dip eastward from north; DIP is the amount of dip from the horizontal. VGP for Agachiche has been reversed in calculating Mean VGP. All data are in units of degrees.

TABLE 3. CENOZOIC AND LATE CRETACEOUS PALEOMAGNETIC DATA FOR SOUTH AMERICA

Site	Location, deg Lat,N	Long,W	Age m.y.	D,deg	I,deg	VGP,deg Lat,N	Long,E	Reference
Cerro del Fraile, Arg.	-50.0	72.7	1.0-2.5	352	-65	83	226	Fleck et al., 1972
Basalts, W. Arg.	-37.0	70.0	0.2-0.7	358	-60	85	131	Career and Valencio, 1970
Basalts, W. Arg.	-37.0	70.0	0.2-8.6	359	-59	87	132	"
Basalts, W. Arg.	-37.0	70.0	1.8-2.6	4	-60	85	75	"
Basalts, W. Arg.	-37.0	70.0	5.4-8.6	356	-57	86	194	"
Falcón, Venezuela*	11.1	69.5	23-28			83	93	This study
Abrolho Is., Braz.	-18.0	39.0	46-52	213	31	59	54	Pacca & Valencio, 1972
Rio de los Molino, Arg.	-31.8	64.5	63-68	165	53	77	198	Linares & Valencio, 1975
Pocos de Caldas, Braz.	-21.9	46.6	75	190	41	81	53	Opdyke & MacDonald, 1973

Note: For site locations, N latitude is positive, whereas S is negative.
*The listed age for the Falcón site is the possible range of intrusion ages (Muessig, 1978). The age of magnetization may be younger, but probably no younger than lower Miocene. The VGP is for the structurally corrected Falcón data.

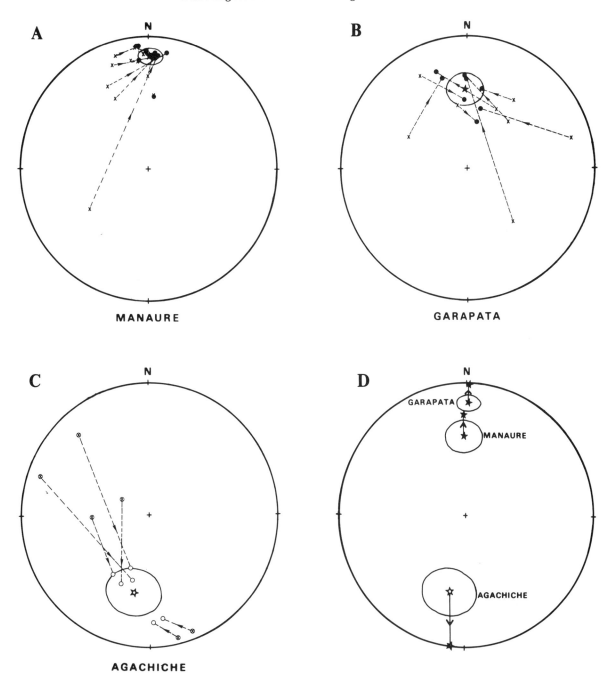

Figure 2. Equal area stereographic plots of directions of magnetization for Falcón paleomagnetic sites with $\alpha_{95} < 20$. For NRM: x is a positive inclination; circled x is negative. For 200 Oe demagnetization: filled circle is a positive inclination; open circle is negative. For the site means of demagnetized data: black star is a positive inclination; open star is negative. Elipses indicate α_{95} confidences for site means of demagnetized data. For A, B, and C, arrows point toward demagnetized data. D is a plot of the directions of site means with and without structural corrections; arrows point toward structurally corrected data.

LATER REMAGNETIZATION

There is good evidence that some of the intensity weakness and scatter in the data are due to an even later remanence overprinting the proposed stable secondary vector.

The paleomagnetic data for Cienaga 1, Redondo (Skerlec and Hargraves, 1980), and for Cienaga 2 have $\alpha_{95} > 20°$ at 200-Oe demagnetization (Table 1). For most samples the NRMs are northerly with steep positive inclinations. During demagnetization, remanent directions are moved progressively southward, while a northerly vector with about 403° inclination is erased (Fig. 4). The resultant site mean has a southerly declination with

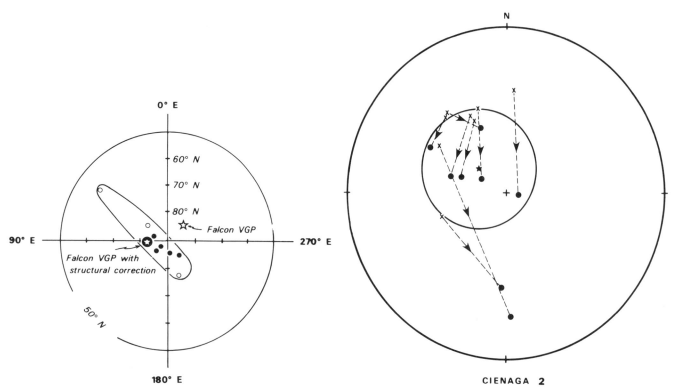

Figure 3. Equal area stereographic plot of Oligocene to Miocene virtual geomagnetic poles (VGP) for Falcón, Venezuela, and Cenozoic and Late Cretaceous poles for South America. Note that the structurally corrected VGP for Falcón falls within the field (elongate ellipse) of South American poles (pre-Oligocene open circle and post-Oligocene filled circle).

Figure 4. Equal area stereographic projection of directions of magnetization for Cerro Cienaga 2, Falcón. For NRM, x is a positive inclination. For 200 Oe demagnetization, filled circle is a positive inclination; black star is the site mean, and the ellipse is the α_{95} cone of confidence. Arrows connect the NRM with the 200 Oe data for each sample.

a steep positive inclination and a rather large scatter ($\alpha_{95} > 20°$). Not only is the steep inclination anomalous, but the positive nature would imply magnetization far south of the present position. A more feasible explanation depicts the remanent direction as a resultant of an original southern remanence with a nonparallel northerly overprint. The original remanence could be very similar to the structurally corrected Agachiche vector (D = 186.7 I = -1.0). The directions of the vectors erased during cleaning are consistent with the overprint being the present geomagnetic field.

Demagnetization to the 500-Oe level of samples from Redondo and Cienaga 1 (Skerlec and Hargraves, 1980) improves the statistics, moves the mean direction southward, but also fails to change the steep positive inclination. The data presented above support the proposal of Skerlec and Hargraves (1980) that the steep inclinations for some Falcón samples might be due to superimposition of secondary and tertiary magnetic vectors.

CONCLUSIONS

The above data suggest that the Falcón intrusions were probably magnetized after intrusion (22.9 m.y. ago) but prior to folding. Since folding in the Falcón Basin is thought to have occurred during late Miocene and Pliocene times (Zambrano and others, 1971), the age of magnetization is constrained to the early or middle Miocene. This stable remanence was most likely acquired soon after intrusion during the cooling and alteration of the rocks.

If so, then the consistent directions obtained from three sites indicate that the Falcón rocks are autochthonous intrusions. The steep inclinations of the remanent directions reported by previous investigators are best explained as a superposition of a very stable present geomagnetic field component upon the Miocene remanent vector.

REFERENCES CITED

Baker, I., and Haggerty, S. E., 1967, The alteration of olivine in basaltic and associated lavas. Part II: Intermediate and low temperature alteration: Contributions to Mineralogy and Petrology, v. 16, p. 258–273.

Creer, K. M., and Valencio, D. A., 1970, Paleomagnetism and rock magnetic studies on the Cenozoic basalts from western Argentina: Geophysical Journal of the Royal Astronomical Society, v. 19, p. 113–146.

Fleck, R. J., Mercer, J. H., Nairn, A.E.M., and Peterson, D. N., 1972, Chronology of late Pliocene and early Pleistocene glacial and magnetic events in southern Argentina: Earth and Planetary Science Letters, v. 16, p. 15–22.

Haggerty, S. E., 1976, Oxidation of opaque mineral oxides in basalts, *in* Rumble, D., ed., Oxide minerals: Mineralogical Society of America Short Course Notes, v. 3, p. Hg1–Hg100.

Haggerty, S. E., and Baker, I., 1967, The alteration of olivine in basaltic and associated lavas. Part I: High temperature alteration: Contributions to Mineralogy and Petrology, v. 16, p. 233–257.

Hicken, A., Irving, E., Law, L. K., and Hastie, J., 1972, Catalogue of paleomagnetic directions and poles: Canadian Department of Energy, Mines and Resources, v. 45, p. 1–135.

Linares, E., and Valencio, D. A., 1975, Paleomagnetism and K-Ar ages of some trachybasaltic dikes from Rio de Los Molino, Province of Cordobo, Republic of Argentina: Journal of Geophysical Research, v. 80, p. 3315–3321.

Muessig, K. W., 1978, The central Falcón igneous suite, Venezuela: Alkaline basaltic intrusions of Oligocene-Miocene age: Geologie Mijnbouw, v. 57, p. 261–266.

Muessig, K. W., 1984, Structure and Cenozoic tectonics of the Falcón Basin, Venezuela, and adjacent areas, *in* Bonini, W. E., and others, eds., The Caribbean–South American plate boundary and regional tectonics: Geological Society of America Memoir 162 (this volume).

Opdyke, N. D., and MacDonald, W. D., 1973, Paleomagnetism of Pocos de Caldas Alkaline Complex, southern Brazil: Earth and Planetary Science Letters, v. 18, p. 37–44.

Pacca, I. G., and Valencio, D. A., 1972, Preliminary paleomagnetic study of igneous rocks from Abrolhos Islands, Brazil: Nature Physical Science, v. 240, p. 163–164.

Santamaría, F. J., and Schubert, C., 1974, Geochemistry and geochronology of the southern Caribbean–northern Venezuela plate boundary: Geological Society of America Bulletin, v. 85, p. 1085–1098.

Skerlec, G. M., and Hargraves, R. B., 1980, Tectonic significance of paleomagnetic data from northern Venezuela: Journal of Geophysical Research, v. 85, p. 5303–5315.

Zambrano, E., Vásquez, E., Duval, B., Latreille, M., and Coffinieres, B., 1971, Sintesis paleogeográfica y petrolera del occidente de Venezuela, *in* Memoria, Congreso Geológico Venezolano, IV, Caracas, 1969: Venezuela Ministerio de Minas e Hidrocarburos, Boletín de Geología, Publicación Especial no. 5, v. 1, p. 481–552.

MANUSCRIPT ACCEPTED BY THE SOCIETY SEPTEMBER

Geological Society of America
Memoir 162
1984

Cenozoic tectonic history of the Sierra de Perijá, Venezuela-Colombia, and adjacent basins

James N. Kellogg
Hawaii Institute of Geophysics
University of Hawaii
Honolulu, Hawaii 96822

ABSTRACT

The four major Cenozoic tectonic phases in the Sierra de Perijá and adjacent basins are the early Eocene tectonic phase, the middle Eocene Caribbean orogeny, the late Oligocene phase, and the late Miocene to present Andean orogeny. Ages of unconformities associated with particularly rapid regional uplift during these phases are early Eocene (53 m.y.), middle Eocene (45 m.y.), late Oligocene (25 m.y.), and Pliocene (3 m.y.). Northwest-southeast compression may have commenced in the Perijá and the Maracaibo Basin as early as the early Eocene. By the middle Eocene the Macoa-Totumo arch had begun to form during intense alpine-type folding and thrusting to the east in Falcón and Lara. During the late Oligocene phase, the Palmar area was uplifted and the most important structural features for hydrocarbon accumulation in the Maracaibo Basin developed. The late Oligocene phase initiated a basement block tectonic style that culminated during the Pliocene in the northwest thrusting of the Santa Marta massif, Sierra de Perijá, and Venezuelan Andes over the adjacent basins. The main uplift of the Sierra de Perijá occurred during the late Miocene-Pliocene Andean orogeny. Right-lateral oblique-slip movement of 90 to 100 km on the Oca fault and left-lateral oblique-slip movement of 100 km on the Santa Marta fault were caused by late Tertiary overthrusting in the Sierra de Perijá and Santa Marta massif. The northwest-southeast shortening that uplifted the Santa Marta massif, Sierra de Perijá, and Venezuelan Andes is related to Caribbean–North Andean convergence along the South Caribbean marginal fault.

During the Pliocene the Panama volcanic arc collided with South America. The North Andean block became detached from the South American plate and is being wedged slowly to the north between the rapidly converging Nazca, Caribbean, and South American plates. The convergence of the three plates has produced rapid subduction at the Colombia trench (6.4 ± 0.7 cm/yr; $088° \pm 7°$), slow subduction at the South Caribbean marginal fault (1.7 ± 0.7 cm/yr; $128° \pm 24°$), and right-lateral shear (1.0 ± 0.2 cm/yr; $235° \pm 5°$) on the Boconó and East Andean fault systems.

INTRODUCTION

Along its entire length of over 200 km, the crest of the Sierra de Perijá is the International Boundary between Venezuela and Colombia and the divide between the Maracaibo Basin of Venezuela and the César Valley of Colombia. The crest of the range is capped by mesas and benches formed of Cretaceous limestones and conglomerates dipping gently to the southeast (6°–8°). A peak 8 km south of Cerro Pintado has the highest elevation (3,650 m) in the range (Figs. 1, 2). At the mountain front on the east side of the Sierra, steeply dipping limestones form prominent flatirons. Well-developed boulder terraces on both sides of the range, but especially near Manaure, attest to rapid uplift and erosion.

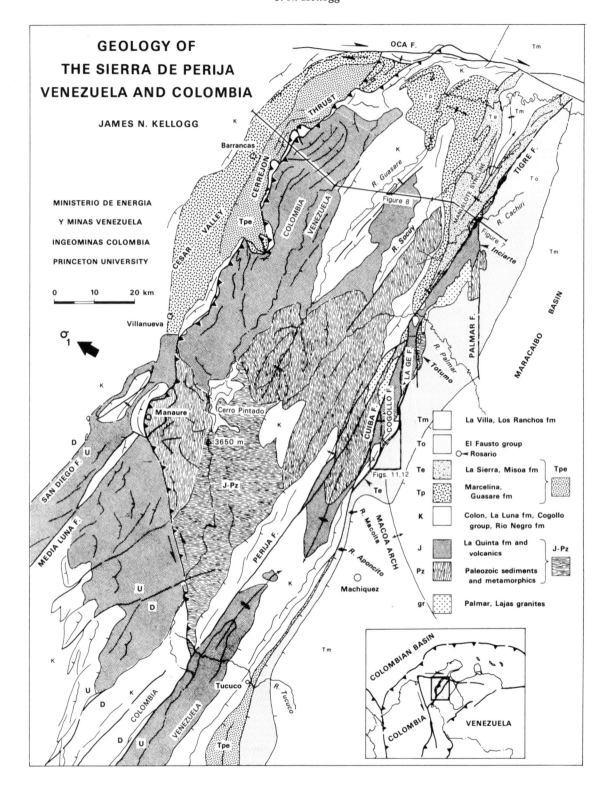

Figure 1. Geologic location map of the Sierra de Perijá. Map symbols on facing page. A radar mosaic (Fig. 2) was used as the base.

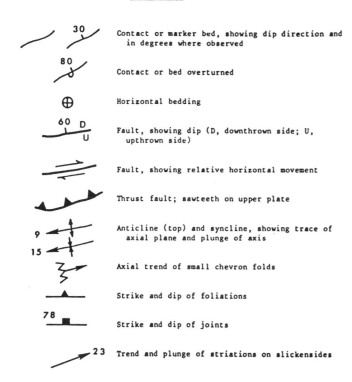

Map Symbols

30	Contact or marker bed, showing dip direction and in degrees where observed
80	Contact or bed overturned
⊕	Horizontal bedding
60 D U	Fault, showing dip (D, downthrown side; U, upthrown side)
	Fault, showing relative horizontal movement
	Thrust fault; sawteeth on upper plate
9 15	Anticline (top) and syncline, showing trace of axial plane and plunge of axis
	Axial trend of small chevron folds
	Strike and dip of foliations
78	Strike and dip of joints
23	Trend and plunge of striations on slickensides

The tectonic history of the Sierra de Perijá was discussed by Miller in 1962. In the present paper a tectonic history of the Sierra is presented that incorporates the author's geologic field mapping and paleontological, radiometric, and stratigraphic data published by others in the two decades after Miller's work.

Before the beginning of Cenozoic tectonic history, at least four major tectonic episodes can be identified in the Sierra de Perijá (Kellogg, 1981): Silurian–Early Devonian; Late Devonian; Late Permian–Triassic; and Jurassic–Early Cretaceous rifting and volcanism. The four episodes all produced unconformities in the stratigraphic column (Fig. 3). All four are also associated with thermal events, unlike the upper Cenozoic block thrusting that produced the present mountain range. In this Memoir, Perijá fission-track ages are discussed by Shagam and Kohn (1984).

The two most important of the pre-Cenozoic tectonic episodes were the Silurian–Early Devonian orogeny and the Jurassic–Early Cretaceous rifting and volcanism. The only regional metamorphism in the Sierra de Perijá occurred prior to deposition of unmetamorphosed Early Devonian (390 m.y.) shales (Caño Grande Formation). Silurian–Early Devonian folding, uplift, and plutonic granitic activity may have been associated with the regional metamorphism.

During Jurassic time, thick arkosic red beds, basaltic andesite flows, ash layers, and welded tuffs were deposited in rift valleys (Maze, 1980, 1984). Geophysical evidence for a Jurassic graben 300 km wide and 5 km deep is found in seismic line PV-D in west Lake Maracaibo 30 to 40 km east of Guamo (Pumpin, 1979). This northeast-southwest-trending graben in west Lake

Maracaibo and one in the Totumo area east of the Tigre-Perijá fault are also indicated by wells that have penetrated the La Quinta volcanics (Pumpin, 1979). In the upper Río Cachirí area, crinoidal limestones of the Permian Palmarito Formation are in direct contact with Lower Cretaceous Río Negro conglomerates (discordance of dip = 30°), suggesting a Jurassic–Early Cretaceous horst west of the Tigre-Perijá fault. On the northwest flank of the Sierra near Cerrejón, more than 5 km of red beds were deposited in a deep basin. The angular unconformity at the top of the La Quinta near Cerrejón is about 30°.

By Cenozoic time, igneous activity in the Sierra de Perijá had ceased, and low-angle block thrusting culminated in the Pliocene uplift of the present mountain range.

EARLY EOCENE UNCONFORMITY

The first significant Cenozoic angular unconformity in the Sierra de Perijá is in the early Eocene (53 m.y.). No significant tectonic activity marked the Cretaceous-Tertiary boundary or the Paleocene. Jurassic–Early Cretaceous rift valleys, red beds, and intermediate volcanics (Pumpin, 1979; Maze, 1980) had been covered by Early Cretaceous conglomerates, sandstones, and massive limestones. By Late Cretaceous time, limestones and shales (La Luna and Colón formations) were being uniformly deposited in a stable marine environment. Southwest of Lake Maracaibo on the eastern flank of the Sierra de Perijá, the Colón Shale is conformably overlain by Cretaceous marine to brackish water shales and thin sandstones and limestones (Río de Oro limestones) of the Mito Juan Formation, passing conformably upward into sandstones, shales, and coals of the Orocue Group of uppermost Cretaceous-Paleocene age. Farther north along the east flank of the Sierra, the Colón and Mito Juan formations are succeeded upward by the Guasare Formation of Paleocene limestones and calcareous sandstones. The Guasare Limestone is well developed in the oil fields of Lake Maracaibo (Hedberg and Sass, 1937).

Rather uniform deposition of shallow marine limestones throughout the Paleocene in the Perijá and the western part of the Maracaibo Basin suggests continued tectonic stability in these areas. The gradients in the Paleocene isopachs shown in Figure 4 are interpreted by Zambrano and others (1971) to be the result of post-Paleocene erosion. In the early Eocene record, however, are the first indications of the coming orogenies. An unconformity is found in the oil fields of Lake Maracaibo, and in the Boscan field northwest of the lake (Zambrano and others, 1971). The unconformity is located between the Guasare Formation and the Misoa Formation Unit "C." The age of the Guasare Formation is well established by molluscs, foraminifera, and pollen as Paleocene (Léxico Estratigráfico de Venezuela, 1970). The Misoa Formation is early to middle Eocene, on the basis of orbitoidforam and pollen evidence (Léxico Estratigráfico de Venezuela, 1970). Post-Paleocene, pre-middle Eocene erosion occurred on the crests of all of the present anticlines in the western Maracaibo Basin (Young and others, 1956). As the axial trend of these anticlines is

Figure 2. A 1:250,000 radar (SLAR) mosaic (Aero-Service Corporation and Goodyear Aerospace, 1971) of the Sierra de Perijá.

northeast-southwest, northwest-southeast compression might have commenced in early Eocene time.

About 250 km northwest of the Sierra de Perijá, the paleo-trench on the west side of the Romeral fault had become active by Paleocene time (Fig. 4). The Romeral fault initially bordered a deep trench filled with abyssal Late Cretaceous and early Tertiary pelagic sediments and turbidites (Duque-Caro, 1979). These sediments are said to be completely absent on the continental platform immediately to the east of the fault. K-Ar ages from the Santa Marta metamorphic belt (MacDonald and others, 1971; Tschanz and others, 1974) indicate Paleocene metamorphism and intrusion of the continental margin. In the Venezuelan Coast

Ranges to the east, Paleocene flysch-molasse deposition marked the initiation of the overthrusting that culminated in the middle Eocene alpine-type orogenesis of the Villa de Cura and Lara nappes.

MIDDLE EOCENE CARIBBEAN OROGENY (45 M.Y.)

A pronounced unconformity and a truncation of the Marcelina, Guasare, and Cretaceous rocks occur across the Macoa-Totumo arch (Fig. 5). This truncation beneath the La Sierra Formation is seen in the subsurface on the Macoa arch (V. D. Winkler and C Key, S. A. Lagoven, Caracas, Venezuela, personal

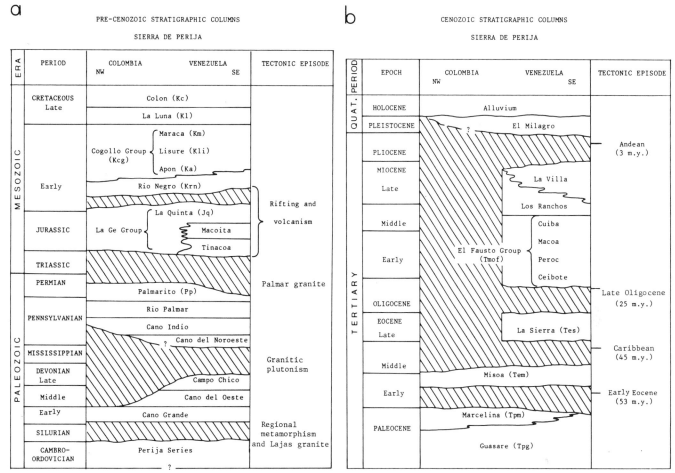

Figure 3. Pre-Cenozoic and Cenozoic stratigraphic columns for the Sierra de Perijá.

commun., 1980) and in outcrop sections southward to the Río Yasa (Miller, 1962). The La Sierra Formation contains the molluscs *Calorhadia (Litorhadia)* sp. and *Carolia.* From these and unspecified fossil evidence Kuyl and others (1955) gave the formation a late Eocene age. The La Sierra Formation can be correlated with the Carbonera Formation in the southern Sierra de Perijá, which contains diagnostic late Eocene molluscs, including *Hannatoma emendorferi, Crommium palmerae, Turritella* sp. of the groups *chira* and *samanensis* and *Raetomya* sp. (Léxico Estratigráfico de Venezuela, 1970). The truncated Marcelina Formation contains no diagnostic fossils, but from its wide association with the Guasare Formation it is assumed to be Paleocene or early Eocene. In the southwestern part of the Maracaibo Basin a pollen zone is missing near the top of the Mirador Formation, indicating a possible early middle Eocene unconformity prior to deposition of the overlying Carbonera Formation (Léxico Estratigráfico de Venezuela, 1970). In Falcón a major unconformity separates the middle Eocene Paují and the late Eocene Santa Rita formations (Zambrano and others, 1971).

In the Sierra de Perijá, the most obvious result of the middle Eocene deformation was the formation of the Macoa-Totumo arch or nose (Fig. 5). A number of faults have approximately the

same north-south trend as the Macoa arch where it meets the Perijá mountain front. These include the Cogollo, La Gé, and Totumo faults (Fig. 1). Prior to late Eocene La Sierra sedimentation, the west side of each fault was downdropped slightly (contrary to the Neogene east side down displacement) (Miller, 1962). The middle Eocene displacements on the Cogollo, La Gé, and Totumo faults may have been related to the uplift of the Macoa arch. These faults are roughly orthogonal to the major middle Eocene thrust faults to the east (075°) but parallel to the thrust faults to the west in Colombia (Fig. 5).

During Eocene time, the Sierra de Perijá north of Machiques (Fig. 1) was in a sandy platform depositional environment (Zambrano and others, 1971). Deltas were forming in the southwest part of the Maracaibo Basin. Northeast of the present lake, 7 km of flysch and shales were deposited in a marginal basin (Fig. 6). The long axis of the basin is aligned northwest-southeast (325-145°). Farther east the motion of the Lara nappe was southeast (140°-170°). The basin may have formed along a marginal tear fault in response to crustal loading by the Lara nappe. In Lara a flysch and molasse with exotic blocks (Matatere Formation) containing early Paleocene to middle Eocene foraminifera was deposited in front of the southeastward thrusting Lara nappe.

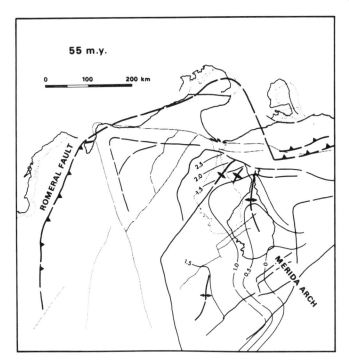

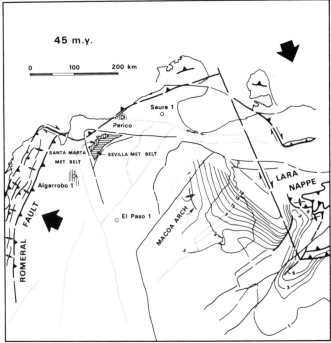

Figure 4. Tectonic reconstruction of the Maracaibo–Santa Marta block in the early Eocene (55 m.y.). Units for the restored Paleocene isopachs are thousands of feet; the contour interval is 500 ft (152 m) (after Zambrano and others, 1971). The present-day coastline is given for reference.

Figure 5. Tectonic reconstruction of the Maracaibo–Santa Marta block in the middle Eocene (45 m.y.). Units for the restored Eocene isopachs are thousands of feet; contour interval is 1,000 or 2,000 ft (305 or 610 m) (after Zambrano and others, 1971). The large arrows indicate the directions of maximum compressive stress for Romeral fault and the Lara nappe. The present-day coastline is shown for reference.

Fold axes that formed during overthrusting were northeast-southwest (050°–080°) (Stephan, 1977b). South-southeastward vergence is indicated by isoclinal microfolds. The nappe fronts were displaced about 250 km. Farther east in the Interior Range, the Piemontine Zone was overthrust by the Villa de Cura nappe in the early middle Eocene (Beck, 1977). The orogeny was characterized by subisoclinal folding (axial trend; 065°–075°), axial plane fracture cleavage, and thrust faults. In northeastern Venezuela and Trinidad, the middle Eocene folding and overthrusting to the south were relatively rapid (Bell, 1972; Vierbuchen, 1978).

In northern Colombia during the middle Eocene orogeny, lateral northwest-southeast compression was at a peak. Caribbean crust was being subducted beneath South America on the newly formed Sinú trench (Duque-Caro, 1979). In the San Jacinto belt to the east of the Sinú trench, steep northeast-trending (015°–025°) fold limbs, tight anticlines and synclines, and thrust faults predominate. The San Jacinto belt and the Western Cordillera were also uplifted about 5 km vertically. This estimate is based on calcium carbonate compensation depth calculations (Duque-Caro, 1972). Tonalitic (quartz diorite) plutonism occurred along the western edge of the platform near the Romeral fault and in the Santa Marta metamorphic belt. K-Ar ages from Santa Marta schists and tonalites collected in six localities (including the Santa Marta batholith) range from 43 to 48 m.y. (Tschanz and others, 1974).

Eocene convergence was occurring on the northern margin of the Caribbean plate as well as on the southern margin. Volca-

nism in the Greater Antilles accompanied thrusting to the northeast in Cuba and the Puerto Rico trench (Monroe, 1968; Malfait and Dinkelman, 1972; Mattson, 1973).

In summary, the Caribbean orogeny was characterized by southeast-verging alpine-type thrust faulting and tight folding in northern Colombia and northern Venezuela and northeast-verging thrust faulting in the Greater Antilles. Nappes were thrust up to 250 km south-southeastward in Lara and in the Interior Range of Venezuela. The Sierra de Perijá area was only peripherally affected by the middle Eocene folding and thrust faulting occurring to the west and east. In the Sierra the most obvious result of the middle Eocene Caribbean orogeny was the formation of the Macoa arch as an outer rise on the basin marginal to the Lara nappe.

After the Caribbean orogeny a dramatic change in Caribbean–North American relative motion occurred. By Oligocene time, thrusting to the northeast in Cuba and in the Puerto Rico trench ended (Monroe, 1968; Mattson, 1973). Volcanism ceased in the Greater Antilles and began in the Lesser Antilles on the eastern margin of the Caribbean plate (Malfait and Dinkelman, 1972). Left-lateral strike-slip motion began in the Cayman trough (Heezen and others, 1973; Holcombe and others, 1973).

LATE OLIGOCENE PHASE (25 M.Y.)

In the Sierra de Perijá a major change took place in the

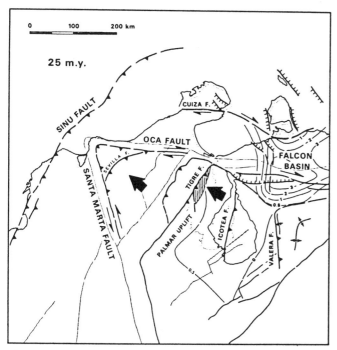

Figure 6. Tectonic reconstruction of the Maracaibo–Santa Marta block in the late Oligocene (25 m.y.). The large arrows indicate the directions of maximum compressive stress for the Santa Marta massif and the Sierra de Perijá. Units for the schematic restored lower Miocene isopachs are thousands of feet; contour interval is 500 or 1,000 ft (152 or 305 m) (after Zambrano and others, 1971). In the hachured area (Palmar uplift) pre-Cretaceous rocks were brought to the surface 25 m.y. ago. The present-day coastline is given for reference.

structural style after the Caribbean orogeny. In the late Oligocene phase, the first basement block uplifts occurred. During the late Oligocene phase the Palmar area was uplifted (Fig. 6), and the most important structural features for hydrocarbon accumulation in the Maracaibo Basin developed.

A major unconformity associated with this phase in Lake Maracaibo is generally assumed to have been caused by Eocene or early Oligocene uplift; however, a careful examination of recent published paleontological evidence as shown below reveals that the unconformity could be as young as late Oligocene. A late Oligocene age for the structural disturbance and uplift is supported by several recent apatite fission-track age determinations from the Sierra de Perijá, which will be discussed in this section.

Southeast of the Tigre fault on the Palmar or Totumo-Inciarte uplift, Eocene, Paleocene, Cretaceous, and pre-Cretaceous crystalline rocks are truncated beneath El Fausto Group shales (Dufour, 1955; Feo-Codecido, 1972). This unconformity over the late Eocene La Sierra Formation is distinctly evident on the Totumo-Inciarte arch (Miller, 1962, Zambrano and others, 1971). Zambrano also traced this unconformity to the top of the late Eocene Carbonera Formation near Alturitas in the southern Sierra. In Lake Maracaibo it overlies the shales of the Pauji Formation, which contain abundant planktonic foraminifera including the uppermost middle Eocene *Porticulosphaera mexi-*

cana zone (Léxico Estratigráfico de Venezuela, 1970). East of the Andes in Barinas the unconformity can be found at the top of the late Eocene Paguey Formation.

The El Fausto Group, which was deposited on the Oligocene unconformity on the Palmar uplift, consists of the Ceibote, Peroc, Macoa, and Cuiba formations. The Ceibote Formation, at the base of the El Fausto Group, has no diagnostic fossils. The *Léxico Estratigráfico* cites an Oligocene palynological age from Kuyl and others (1955), but they give no data for their El Fausto Group age, mention uncertainties in the position of the Oligocene boundaries, and do not refer to the Ceibote Formation specifically. The Peroc Formation has been considered Oligocene to early Miocene on the basis of its relation to units in Lake Maracaibo, but the shallow-water foraminifera found in the formation are of indeterminate age (Léxico Estratigráfico de Venezuela, 1970). The Peroc Formation thins considerably to the east under Lake Maracaibo and passes laterally into the Icotea Formation (Young, 1958). The Icotea Formation, however, contains only redeposited Eocene foraminifera. The overlying La Rosa Formation is rich in fossil foraminifera, including the *Globorotalia fohsi* zone (Dusenbury, 1956) formerly considered late Oligocene but now assigned to middle Miocene. To the south there is a lateral transition from the Peroc to the León Formation (Young, 1958). The León Formation contains molluscs of mid-Miocene age, *Tellina, Chione, Arca, Turritella, Conus,* and *Olivella,* and the early mid-Miocene planktonic foraminifera, *Globigerinatella insueta* (Kuyl and others, 1955; Léxico Estratigráfico de Venezuela, 1970). Thus, the micropaleontological evidence permits a wide range of possible ages for the unconformity, extending from earliest Oligocene through early Miocene.

Recent fission-track age determinations from the Totumo-Inciarte uplift (Kohn, *in* Shagam, 1980) are evidence for a late Oligocene age for the deformation and uplift. The fission-track ages determined from apatite in three samples varied from approximately 22 to 25 m.y. The temperature at which fission tracks in apatite become stable (the blocking temperature) is about 120 °C for rapid uplift or subsidence, but annealing may occur at 50 °C if the temperature is maintained for an extended period of time (Shagam, 1980). As the pre-Cretaceous rocks on the Totumo-Inciarte uplift were significantly eroded at the surface prior to El Fausto deposition, they must have passed through the blocking temperature for apatite at that time. Deposition in the Totumo-Inciarte area was then continuous, and there was no further uplift until Pliocene time. Thus, the only period of uplift that could possibly have been consistent with the late Oligocene (25 m.y.) fission-track ages was the pre–El Fausto phase.

By the end of the Oligocene the pre-Cretaceous crystalline rocks and the overlying Cretaceous, Paleocene, and Eocene sediments east of the Tigre fault were raised 3 to 4 km along the fault and tilted 7° to the southeast, forming the Palmar or Totumo-Inciarte uplift (Fig. 7). Gravity data suggest that near the surface the original dip on the Tigre fault was about 20° ± 10° to the southeast (Kellogg, 1981). Cretaceous limestones were folded as crystalline basement rocks were thrust to the northwest. Geomet-

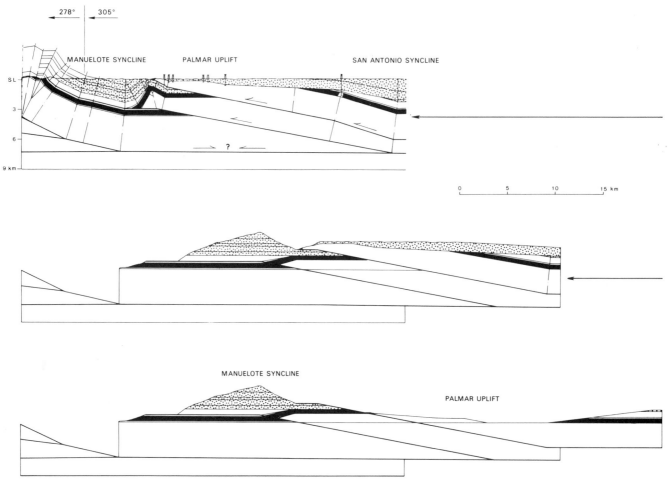

Figure 7. Restored cross sections illustrating the development of the Palmar uplift and Manuelote syncline from the Eocene (bottom) through the Miocene (middle), to the present-day (top). Tertiary rocks are shown in dot pattern and Cogollo Group in black.

ric considerations (Figs. 7, 8) are consistent with up to 11 km of lateral shortening and possible flattening of the fault at a depth of 7 or 8 km.

In Lake Maracaibo the most important structural features for hydrocarbon accumulation were developed during the late Oligocene phase. The period was characterized by folds and high-angle reverse and strike-slip faults trending north-northeast-–south-southwest (Icotea trend) and by intense erosion (Zambrano and others, 1971). As a result of this epeirogenesis the massive Cretaceous limestones of the Cogollo Group and La Luna Formation were fractured enough to constitute high-quality reservoir rocks, especially in the zones most affected structurally, such as the crests of the Lamar, Concepción, and La Paz-Mara anticlines. Petroleum interest is considerably reduced to the south of these oilfields because the massive limestones pass into a shaley facies intercalated between limestone beds, which reduces the fracturing of the Cretaceous section. This is the case in the central and southern Perijá Macoa and Alturitas structures (Salvador and Hotz, 1963). The late Oligocene folds and faults in northern Lake Maracaibo, such as the Icotea or Urdaneta fault, were formed

under west-northwest-verging compression (Zambrano and others, 1971; K. W. Stauffer, Corpoven, Maracaibo, personal commun., 1979). Commercial hydrocarbon accumulations were subsequently trapped on structural highs beneath the late Oligocene unconformity. When the Sierra de Perijá was uplifted in the Pliocene, oils migrated updip to the northwest along the late Oligocene unconformity beneath impermeable El Fausto shales, forming numerous oil and asphalt seeps at the surface (Dufour, 1955).

Strike-slip motion on both the Santa Marta and Oca fault systems probably commenced in the Oligocene. The interpretation presented in this paper of the Palmar uplift as a result of late Oligocene thrusting on the Tigre fault requires about 9 km of right-lateral displacement on the east-west-trending Oca fault system. On the basis of well data north of the Palmar uplift, Feo-Codecido (1972) postulated vertical movement during the Oligocene and 15 to 20 km of right-lateral post-Eocene displacement on the Oca fault. In Falcón, Muessig (1979) ascribed alkaline basaltic intrusions radiometrically dated at 23 to 28 m.y. to tension produced by dextral movement on the Oca fault sys-

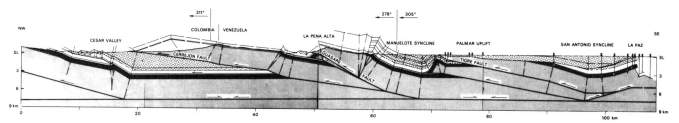

Figure 8. Deep geologic cross section of the northern Sierra de Perijá. The location of the section line is shown in Figure 1. All Tertiary units are shown in the dot pattern; the Cretaceous Cogollo Group is in black; and pre-Cretaceous units are shown in the screen pattern.

tem. Continued right-lateral movement on the east-west-trending fault systems resulted in Miocene subsidence and deposition in the Falcón and Bonaire pull-apart basins (Fig. 6).

Evidence of late Oligocene movement on the north-northwest–south-southeast-trending (343°) Santa Marta Fault system was found in the Algarrobo 1 well (Fig. 5). Drill cuttings from a depth of 8,327 ft (2,538 m) contained Santa Marta phyllite in a pre–lower Miocene boulder conglomerate (T. V. Tolleson of Superior Oil Co. of Colombia, *in* Tschanz and others, 1974). The unconformity is very pronounced in El Paso 1 (Figs. 5, 14) near the Santa Marta fault zone in the César Valley (26 km northeast of Chimichagua) where lower Miocene sediments rest directly on Jurassic La Quinta red beds (Polson and Henao, 1965). West of Santa Marta the margins of the continental platform were being compressed laterally (Duque-Caro, 1979). Upper Oligocene and Miocene pelagic sediments and turbidites were deposited in a trench west of the Sinú fault (Fig. 6) while shallow-water carbonate facies of the same age were being deposited east of the fault.

An apatite fission-track age of approximately 26 m.y. on the Valera granite (Kohn, *in* Shagam, 1980) may indicate late Oligocene uplift in the northern Andes of Venezuela. The uplift may have been the result of displacement on the north-south-trending Valera fault.

There has been no Neogene regional metamorphism or igneous activity in the Santa Marta–Maracaibo block. The late Oligocene phase initiated a basement block tectonic style that culminated in the Pliocene northwest thrusting of the Santa Marta massif, Sierra de Perijá, and Venezuelan Andes over the adjacent basins.

PLIOCENE ANDEAN OROGENY (3 m.y.)

Sierra de Perijá

The main uplift of the Sierra de Perijá occurred during the late Miocene and Pliocene Andean orogeny (Fig. 9). The Sierra was thrust as a block to the northwest over the César Valley on a low-angle thrust fault that extends down to the middle of the crust. The Sierra Nevada de Santa Marta and the Venezuelan Andes were uplifted by similar northwest-verging Pliocene overthrusting. The basement block overthrusts of the Sierra de Perijá, Venezuelan Andes, and the Santa Marta massif are Pliocene-

Pleistocene analogues for Laramide orogenic structures in the middle and southern Rocky Mountains of the United States (Kellogg, 1981). As the Venezuelan Andes overrode the Maracaibo Basin, the southeastern part of the basin subsided and gradually filled with coarse sandstones and conglomerates.

Well logs and seismic data from the Maracaibo Basin indicate the gradual initiation of Andean tectonic activity in the middle Miocene. Maraven's 1978 seismic profile CMO-2 showed an apparent Miocene growth fault on the West Mara anticline. Well-log data suggest that Miocene sediments onlap the northwest flank of the La Paz anticline (Fig. 8). According to Lagoven data (C. E. Key and V. D. Winkler, Lagoven, S. A., Caracas, Venezuela, personal commun., 1979), the Icotea (Urdaneta) fault in Lake Maracaibo began to move in middle Miocene, but the greatest displacement was Pliocene-Pleistocene.

A Pliocene age for the major uplift of the Sierra de Perijá can be inferred from stratigraphic relationships and from apatite

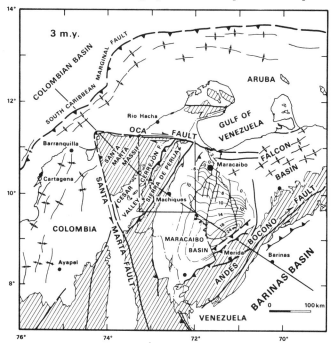

Figure 9. Tectonic reconstruction of the Maracaibo–Santa Marta block during the Pliocene (3 m.y.). Hachured areas are major uplifts with pre-Tertiary rocks exposed. Depths to the late Oligocene unconformity (25 m.y.) are given in thousands of feet; the contour interval is 2,000 ft (610 m) (Zambrano and others, 1971).

Figure 10. Flatirons at the eastern mountain front south of Río Apón (west of Machiques).

fission-track age data. (1) The frontal monocline with its Cretaceous limestone-capped flatirons bordering the Maracaibo Basin is one of the most striking structural features in the mountain range (Figs. 10, 11, and 12). The late Miocene Los Ranchos and La·Villa formations are conformably folded in the frontal monocline with little apparent onlap. One kilometre east of the mountain front near Rio Macoita, La Villa Formation beds dip steeply to the southeast (015°; 55°SE) (Fig. 13), indicating a post–La Villa age for the folding of the frontal monocline. Just east of the monocline, seismic reflection profile PU-10 (Western Geophysical for Corpoven, S.A., 1977) shows displacement of the La Villa Formation on the west-dipping Macoa fault. (2) Deposition on the southeast flank of the Sierra continued without interruption through the middle and upper Miocene. An unconformity then separated the La Villa Formation and the overlying El Milagro Formation (Sutton, 1946; Young, 1958; Léxico Estratigráfico de Venezuela, 1970; Zambrano and others, 1971). The late Miocene age for the La Villa Formation is based on regional correlations. The only fossils found in the El Milagro Formation are silicified wood, but it is probably Pleistocene (Léxico Estratigráfico de Venezuela, 1970), suggesting an approximate Pliocene age for the unconformity. (3) A Pliocene age for the uplift of the Sierra de Perijá is also supported by apatite fission-track age determinations of about 3 m.y. for samples of La

Quinta volcanics (Caño El Tigre) and Río Lajas granite (Kohn, *in* Shagam, 1980).

The Andean uplift of the northern Sierra de Perijá involved northwestward movement on the Cerrejón thrust fault. Field geologic and gravity data (Kellogg, 1981; Kellogg and Bonini, 1982) suggest that near the surface the Cerrejón fault dips 15° ± 10° to the southeast. To account for the total volume of rock uplifted by the Cerrejón thrusting, the fault as shown in Figure 8 must extend to a depth of at least 8 km in the crust. As interpreted in Figure 8, Jurassic La Quinta red beds were displaced 16 to 26 km horizontally and 4.5 km vertically on the thrust. The red beds were emplaced above southeast-dipping shales and sandstones of the Marcelina Formation and the lower part of the Misoa Formation. The thrusting had to at least postdate the early Eocene Misoa Formation. Although no fission-track ages are available for the Colombian side of the Sierra, the high topography (up to 3,650 m) and high Pleistocene terraces (200 to 300 m above river base level south of Manaure) on the Colombian side of the range attest to more rapid Pliocene-Pleistocene uplift there than on the Venezuelan side. In the southern César Valley on the Becerril anticline, Phillips Petroleum Company well and seismic data indicate reverse faults displacing Miocene sediments.

On the northeast flank of the Sierra, the Manuelote syncline and Mostrencos arch also were formed during the Pliocene An-

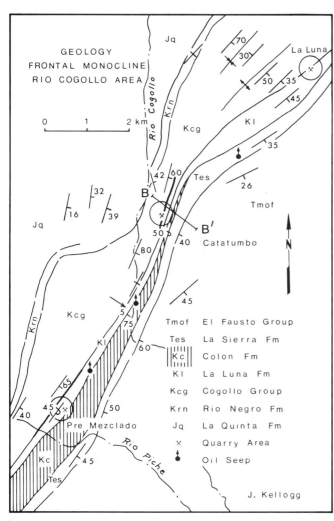

Figure 11. Geologic map of the monocline near Río Cogollo. For location see Figure 1.

dean orogeny. The age of the structures is known by the conformable folding of the late Miocene Los Ranchos Formation within the syncline (Hedberg, 1929). According to the structural interpretation presented in Kellogg (1980, 1981), the Manuelote syncline and the Mostrencos arch were results of a fault bend at a depth of about 3 km. Northwest-southeast compression resulted in 2 to 6 km of shortening and 3 km of vertical displacement on the arch.

Oca Fault Zone

The Sierra de Perijá, the Cerrejón fault, and the Tigre fault are bounded on the north by the east-west (097°) Oca fault system. The model presented in this paper of Perijá orogeny by northwest-southeast shortening (299°) requires oblique right-lateral motion on the Oca fault system. The predicted Pliocene northwest-southeast shortening in the northern Sierra (Cerrejón, Guasare, and Mostrencos arch faults) is about 25 to 35 km. The predicted Oligocene northwest-southeast shortening across the Tigre fault is 7 to 11 km. As the movement is oblique to the Oca

fault system, the apparent displacement will depend on the strike of the displaced structure. Apparent displacement of northeast-southwest-trending structures will be greater than the apparent displacement of northwest-southeast-trending structures. Also, because of crustal shortening in the Sierra de Perijá and Santa Marta massif south of the Oca fault, relative right-lateral displacement will increase to the east.

Estimates of total Tertiary movement on the Oca fault are based on correlation of rock units across the fault. Pre-Cretaceous serpentinite found in the Guajira well Saure 1 (Figs. 5, 14) has been tentatively correlated with serpentinite in the Santa Marta massif 100 km to the west and south of the fault (P. Bartok, F. Jansen, V. Pumpin, Maraven, S. A., Caracas, Venezuela, personal commun., 1979). Serpentinite can be found in a few small ultramafic units in the Sevilla de Santa Marta provinces (Tschanz and others, 1974). Tschanz and others (1974) correlate schist found in the Perico well (Figs. 5, 14) north of the fault with schist from the Santa Marta metamorphic belt south of the fault. If the displacement is northwest-southeast (310°), the true total movement would be about 50 km.

There is no evidence of Paleocene or Eocene activity on the Oca fault, so the Cenozoic movement on the fault was probably associated with the late Oligocene and Andean orogenies in the Sierra de Perijá and Santa Marta massif (Coronel, 1970). Early Andean (middle to late Miocene) activity on the fault is suggested by two apatite fission-track ages of about 12 and 13 m.y. from Toas Island (Kohn, *in* Shagam, 1980).

Toas Island is a small sliver (1.5 by 6 km) in the middle of the Oca fault zone at the entrance to Lake Maracaibo (Fig. 14). The island consists principally of granites, rhyolites, basalts, and Mesozoic sediments (Pimentel, 1977) that are 3 to 4 km higher than immediately adjacent units of the same age. The lack of any significant gravity anomaly suggests that the island has been detached from the basement. The brittle deformational style, small size, and large abrupt vertical displacement are consistent with the flower-structure geometry commonly observed in small slivers along major wrench faults (Harding and Lowell, 1979). Olistostromes formed during uplift on the north side of the island contain blocks of red and white sandstone of probable Miocene age. Post-Miocene fault activity is also supported by strongly deformed Miocene sandstone (125°; 60°NE) found along the Oca fault north of the Sierra de Perijá (2 km east of Las Trojas on Route 1). Along the fault trace in Falcón, Miocene and Pliocene beds of the Coro Formation have been tilted vertically. Maraven's 1978 seismic profiles W-78-7 and W-78-12 (Figs. 15a, 15b), located 80 and 100 km east of the mouth of Lake Maracaibo, also show major vertical movement of late Tertiary sediments in horsts and grabens along the Oca fault zone, demonstrating significant late Tertiary displacement on the fault in Falcón.

Feo-Codecido (1972) estimated only 15 to 20 km of right-lateral post-Eocene displacement on the Oca fault, from the displacement of the Sinamaica depression and San Carlos uplift from the Manuelote syncline and Palmar uplift, respectively. Additional dextral movement may have occurred on the right-lateral

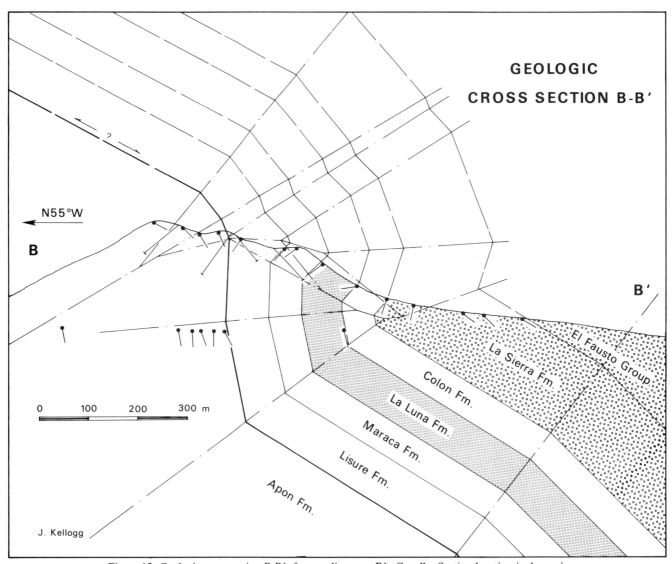

GEOLOGIC

CROSS SECTION B-B'

N55°W

B

B'

0   100   200   300 m

El Fausto Group

La Sierra Fm.

Colon Fm.

La Luna Fm.

Maraca Fm.

Lisure Fm.

Apon Fm.

J. Kellogg

Figure 12. Geologic cross section B-B' of monocline near Río Cogollo. Section location is shown in Figure 11.

strike-slip Matuare fault just north of the Oca, however. In Falcón, Vasquez and Dickey (1972) calculated shortening by folding in the northeast-trending (072°) Falcón anticlinorium as 50 to 60 km in the past 25 m.y. They ascribed the shortening and folding at the surface to dextral shear on the Oca fault zone in the basement. Isopachs for the Pecaya Formation, which was termed Oligocene by Zambrano and others (1971) but which contains primarily early Miocene planktonic foraminifera (Cati and others, 1968), are shown in Figure 14. If the zigzag trend of the present isopachs (Fig. 14) is the result of simple shear distributed across a wide fault zone (Oca) and the linear restored isopachs for 25 m.y. ago shown in Figure 6 are correct, then about 50 km of dextral movement occurred on the Oca fault zone in the past 25 m.y. This apparent 50-km east-west displacement of north-south-trending isopachs would be equivalent to about 60 km of northwest-southeast (298°) displacement.

If the estimated total Tertiary oblique dextral movement on the Oca fault zone in Colombia of 50 km (correlation of Perico well schist and Santa Marta schist by Tschanz and others, 1974) is added to the 40 ± 8 km estimated dextral movement on the Oca fault zone produced by northwest-southeast shortening in the northern Sierra de Perijá (Cerrejon, Guasare, Mostrencos arch, and Tigre faults; Kellogg, 1981), the predicted total Tertiary oblique dextral movement on the Oca fault zone east of the Sierra is 90 ± 8 km. If 60 km of the displacement on the Oca fault zone occurred in the past 25 m.y. (calculated shortening by folding, Vasquez and Dickey, 1972; and displaced isopachs, Fig. 14), then 30 km of oblique-dextral movement are predicted to have taken place on the fault zone in the Oligocene or early Tertiary.

Estimates of the post-middle Eocene sinistral displacement on the Caribbean-North American plate boundary are as high as 1,000 km. These estimates are based on the horizontal displace-

Figure 13. La Villa Formation sandstones dipping steeply to the southeast (015°; 55° SE) 1 km east of the Perijá mountain front near Río Macoita (Fig. 1).

ment of Mesozoic evaporites in Central America (Pinet, 1972) and the opening of the Cayman trough (Holcombe and others, 1973). These estimates of Caribbean–North American movement can be combined with the constraints on relative motion between North America and South America derived by Minster and others (1974) and Ladd (1976) to deduce the post–middle Eocene, Caribbean–South American motion as approximately 600 to 1,000 km (1.5–2.2 cm/yr) (45×10^6 yr) in a west-northwest–east-southeast direction (Jordan, 1975; Minster and Jordan, 1978; K. Burke, Department of Geological Sciences, SUNY, Albany, New York, personal commun., 1980). Post–middle Eocene displacement of this magnitude did not take place on the Oca fault system. The late Tertiary movement on the Oca and Santa Marta faults is the result of shortening in the Sierra de Perijá and Santa Marta massif. This shortening is probably related to upper Tertiary Caribbean–South American and Quaternary Caribbean–North Andean convergence.

North-South Compression

In the northern Sierra de Perijá there is evidence of north-northeast–south-southwest 014° ± 16° (α_{95}) compression from stylolites, folds, and reverse faults (Kellogg, 1981). This compression is perpendicular to the Oca fault. The faulting and folding involves the Misoa Formation, so it is post–middle Eocene. In the Manuelote syncline some of these faults are offset by northwest-ward thrusting (E. Moya, O. Castillo, and Figueroa, Corpozulia, personal commun., 1978). The east-west trend of folds and reverse faults on Toas Island suggests that the uplift of the island may have been caused by north-south compression along the Oca fault (Rod, 1956). A middle to late Miocene age for at least part of the uplift of Toas Island is indicated by two apatite fission-track ages (Kohn *in* Shagam, 1980). The north-south compression has continued into the Pleistocene, because east-west-trending (110°) thrust faults displace the Pliocene-Pleistocene El Milagro Formation (south of La Concepción) near Maracaibo (Bellizzia and others, 1976).

The north-south shortening may be caused by the predicted oblique movement on the Oca fault system. Because of the rigidity of the Maracaibo–Santa Marta block (Dufour, 1955), the deformation near the Oca fault zone is parallel to the fault rather than oblique to it. The shortening is also increased on left-stepping segments of the dextral fault, such as the Montes de Oca at the northern termination of the Sierra de Perijá. Well-log and

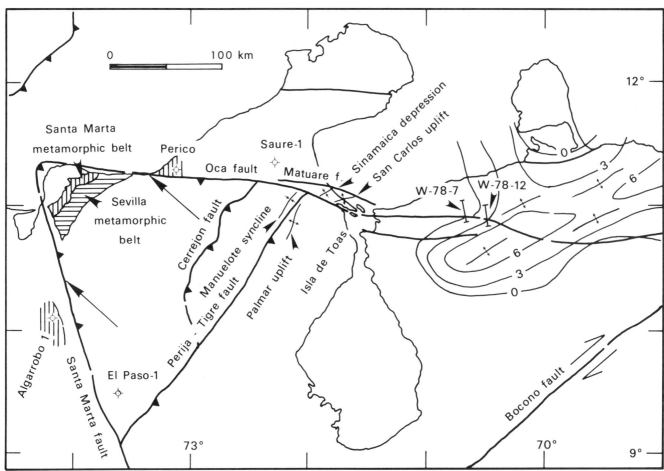

Figure 14. Map of the Oca fault zone. Unrestored lower Miocene isopachs for northwestern Venezuela (Zambrano and others, 1971). Compare with the restored isopachs (Fig. 6) in which the movement on the Oca fault system has been removed.

seismic data from north of the Palmar uplift show that the Oca fault has a reverse profile with the south side up an average of 1 km (Feo-Codecido, 1972).

Santa Marta

The Santa Marta–Bucaramanga fault is a north-northwest-–south-southeast-trending (340°), left-lateral, strike-slip fault that forms the southwest margin of the Santa Marta and Santander massifs. Campbell (1968) claimed that the César and Middle Magdalena basins were contiguous until separated by 110 km of post-Miocene movement, although this is disputed by Polson and Henao (1965). Drill cuttings of phyllitic schist in a pre-Miocene boulder conglomerate from a depth of 8,327 ft (2,538 m) in the Algarrobo 1 well (Fig. 5) are correlated with Santa Marta schists (Tschanz and others, 1974). This correlation demonstrates 100 to 115 km of left-lateral Tertiary separation on the Santa Marta fault system. Subsurface evidence from the El Difícil oil wells 20 km south of Algarrobo 1 also indicates about the same sinistral separation of Precambrian granulites, gneisses, and schists from the Sevilla metamorphic belt. Late Miocene compression

across the fault is suggested by the uplift of the Santander massif. The Guayabo Group sediments that accumulated as the Santander massif began to rise have been identified as late Miocene–early Pliocene (James, 1977). This uplift age is supported by a 6-m.y. apatite fission-track age determination from the massif (Kohn, *in* Shagam, 1980). Pleistocene activity on the Santa Marta–Bucaramanga fault system is shown by deformed terraces and offset stream patterns (Campbell, 1968).

The Santa Marta massif (Sierra Nevada de Santa Marta, Fig. 9) is the highest range in Colombia (about 5,800 m) and one of the highest topographic reliefs (over 9 km) of any coastal mountain in the world. The structural relief is 12 km. The tremendous gravity high (180 mgal relative to the adjacent basins) over the Santa Marta massif (Case and MacDonald, 1973; Bonini and others, 1980; Kellogg and Bonini, 1982) indicates that the crystalline massif is out of local isostatic equilibrium and may have been thrust to the northwest on the Oca and Santa Marta faults (Case and MacDonald, 1973; Bonini and others, 1980). The gravity data are consistent with a low-angle thrust fault extending to the base of the crust. The symmetry of the topography, structure, and gravity anomalies suggests that the thrust-

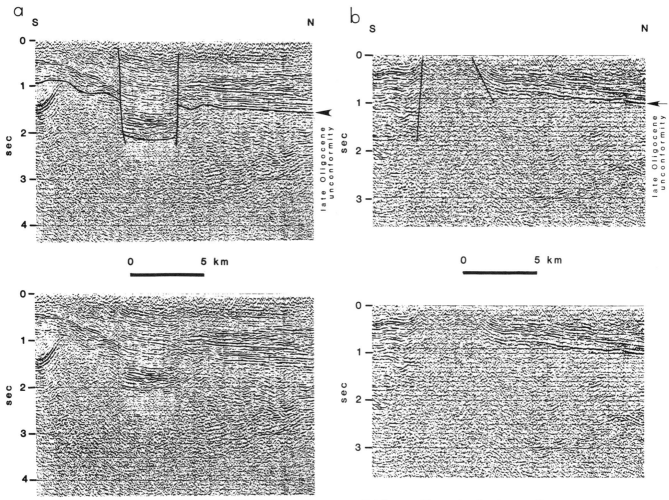

Figure 15. (a) Seismic line W-78-7 from the western Falcón area, Venezuela, showing a pull-apart graben displacing upper Tertiary sediments along the Oca fault (Western Geophysical Co. for Maraven, S.A., Venezuela). Location shown in Figure 14. (b) Seismic line W-78-12 from the western Falcón area, Venezuela, shows a horst displacing upper Tertiary sediments along the Oca fault (Western Geophysical Co. for Maraven, S.A., Venezuela). Location shown in Figure 14.

ing was to the northwest (312°), forming approximately equal angles of 30° to 35° with the Oca and Santa Marta faults. According to this interpretation, the faults accommodated late Tertiary oblique strike-slip (oblique reverse) motion, and their dips are low angle at their junction, increasing along strike to the east and southeast. Fault displacement is not predicted to have continued, at least in the late Tertiary, northwest of their junction. This model also implies that the movements on the Oca and Santa Marta faults have been contemporaneous and of similar magnitudes during the late Tertiary.

To the west of Santa Marta, uplift, folding, and faulting resulted from continued lateral compression during the Pliocene Andean orogeny (Duque-Caro, 1979). Underthrusting and turbidite deposition began on the northwest side of the newly active South Caribbean marginal fault. Mud volcanism and plutonism began in the Sinú trench sediments located west of the Sinú fault.

Venezuelan Andes

Stratigraphic and fission-track data show that the major Cenozoic uplift of the Venezuelan Andes (Fig. 9) was during the Pliocene-Pleistocene orogeny. Thick late Miocene to Pliocene conglomerates of the Betijoque Formation are found on the northwestern flank of the Andes. Rio Yuca Formation conglomerates of the same age were deposited on the southeastern flank (Léxico Estratigráfico de Venezuela, 1970). Late Tertiary oxisols formed near sea level were uplifted at least 2 km during the Pliocene-Pleistocene orogeny (Weingarten, 1977). Rapid uplift is also suggested by the extensive Pleistocene terraces high above the present river base levels (R. Giegengack, Department of Geology, University of Pennsylvania, personal commun., 1979). Further proof of Pliocene uplift of the Venezuelan Andes is supplied by apatite fission-track age determinations on 15 rock samples ranging from 2 to 5 m.y. (Kohn, *in* Shagam, 1980). Using this radiometric age data, Shagam calculated an uplift rate for the Pliocene (5 to 1.8 m.y.) of 0.8 mm/yr.

The gravity low associated with thick deposits of low-density sediments on the northwest flank of the Venezuelan Andes (Fig. 9) can be modeled as northwestward (320°) over-thrusting of the Maracaibo Basin by crystalline rocks of the Andes (Bonini and others, 1980). This model involves a low-angle thrust (22°–25°) extending into the mantle and overriding the Maracaibo Basin by 25 km. This fault is similar to the structural interpretations of the Wind River thrust fault in Wyoming based on gravity data and deep crustal reflection profiling (Smithson and others, 1979). Miocene-Pliocene northwestward folding and thrusting is also apparent in the sedimentological record of the Táchira Depression southwest of the Venezuelan Andes (Macellari, 1984).

In Falcón, late Pliocene northwest-southeast shortening produced northeast-southwest-trending (065°–075°) folds (Bucher, 1952; Wheeler, 1963). This folding may be at least partially related to dextral movement on the Oca fault system. In the Serranía del Interior of eastern Venezuela, mid-Miocene to Pliocene northwest-southeast shortening caused folding (075°) and thrusting (Hedberg, 1950; Salvador and Rosales, 1960; Lamb and Sulek, 1968; Vierbuchen, 1978).

Andean deformation is continuing in the present.

PRESENT-DAY TECTONICS

Introduction

The Caribbean–northwestern South America plate boundary has been controversial and difficult to interpret tectonically. The high seismicity and Holocene displacement on the northeast-trending Boconó fault (Fig. 16) have prompted some to interpret the Boconó fault zone as a Caribbean–South America plate boundary that accommodates east-west convergence (Molnar and Sykes, 1969; Schubert and Sifontes, 1970; Dewey, 1972; Santamaria and Schubert, 1974; Kafka and Weidner, 1981). The geologic, gravity, and seismic evidence for northwest-southeast convergence and the lack of geologic evidence for large displacements on the Boconó fault have led others to interpret the South Caribbean marginal fault (Fig. 16) as part of a diffuse Caribbean–South America plate boundary that accommodates northwest-southeast convergence (Jordan, 1975; Shagam, 1975; Case and Holcombe, 1980; Walper, 1980).

I propose that these features are simultaneous active and accommodate portions of the relative movement between the Caribbean, South America, and Nazca plates.

The Boconó and South Caribbean marginal faults form the eastern and northern boundaries, respectively, of the North Andean block. The North Andean block is bounded on the southeast by the East Andean fault and on the west by the Colombia trench. The North Andean block became detached from South America 3 or 4 m.y. ago and began to be wedged slowly to the

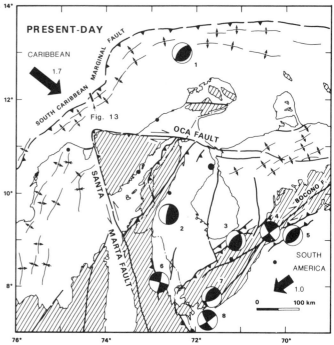

Figure 16. Present-day tectonics of western Venezuela and northeastern Colombia. Bold arrows show present plate motion vectors with respect to the North Andean block derived in this paper. Average velocities are given in centimetres per year. The eight earthquake focal mechanisms are based on P-wave first arrivals. Solution 1 was calculated by Vierbuchen (1978); 2 was by O. Perez, McCann, and A. J. Murphy (Lamont-Doherty Geological Observatory, Palisades, N.Y., unpub. data, 1978); 3, 4, 6, and 8 were by Dewey (1972); and 5 and 7 were by Perez and Aggarwal (1980). All plots are made on equal-area projections of the lower hemisphere of the focal sphere. Dark quadrants are compressional first motions. Source parameters are summarized in Table 1. Hachured areas are major uplifts with pre-Tertiary rocks exposed.

north as the Nazca plate converged rapidly with the Caribbean and South American plates. The interpretation is based on earthquake focal mechanisms, hypocenter locations, seismic reflection profiles, and Quaternary geology.

Uplift Rates

Well-developed Pleistocene boulder terraces suggest rapid present-day uplift of the Sierra de Perijá and Venezuelan Andes (R. Giegengack, Department of Geology, University of Pennsylvania, personal commun., 1979). On the northwest margin of the Sierra de Perijá south of Manaure, terraces are 200 to 300 m above the river base level. On the southeast Perijá mountain front west of Machiques, the Pleistocene terraces contain boulders up to 2 m across (Fig. 17). A crude estimate of the late Pliocene–Pleistocene uplift rate for the Sierra de Perijá can be made from the 2.7-m.y. apatite fission-track ages of Kohn (*in* Shagam, 1980). Assuming a 100 °C closure temperature, an average crustal geothermal gradient of 25 °C/km, and an erosion rate equal to the uplift rate, the uplift rate = 100/(25) (2.7) km/m.y. = 1.5 mm/yr. In view of the low elevation (200 m above sea level) of

TABLE 1. SUMMARY OF SOURCE PARAMETERS FOR EVENTS 1 TO 8 IN FIGURE 16

Event	Date	Origin Time (UT)	m_b*	Depth† (km)	Latitude (deg)	Longitude (deg)	σ_1 (deg)
1	Mar 12, 1968	. .	. .	58	13.2 N	72.3 W	317
2	Nov 17, 1968	001610.1	5.7	175	9.55N	72.60W	280?
3	May 13, 1968	193605.8	4.8	29	9.00N	71.06W	297
				5(ISC)			
4	Jul 19, 1965	041321.2	5.3	20	9.40N	70.28W	085
				31(ISC)			
5	Mar 5, 1975	. .	5.5	shallow	9.04N	69.95W	313
6	Sep 2, 1964	181215.8	4.8	26	8.08N	72.78W	045/075
7	Composite focal mechanism solution for microearthquakes in Uribante-Caparo area						
	1979-1980	. .	. .	8-25	7.8 N	71.7 W	316
8	Dec 21, 1967	113724.5	5.4	29	7.04N	72.02W	087/297
				8(ISC)			

Note: The focal mechanisms in Figure 16 are based on P-wave first arrivals. Solutions 3, 4, 6, and 8 were calculated by Dewey (1972); solution 2 was calculated by O. Perez, McCann, and A. J. Murphy (Lamont-Doherty Geological Observatory, Palisades, NY, unpub. data, 1978); solutions 5 and 7 were by Perez and Aggarwal (1980); and solution 1 was calculated by Vierbuchen (1978). The estimates of the maximum principal stress direction (σ_1) are the azimuths of the axes of maximum pressure (Honda, 1962) for reverse fault solutions and 15 from the axes of maximum pressure for strike-slip fault mechanisms.

*Body wave magnitudes (m_b) are from the Earthquake Data Reports of the United States Coast and Geodetic Survey (USCGS) or the National Oceanic and Atmospheric Administration (NOAA).

†Depths are from the Earthquake Data Reports of USCGS or NOAA and the Bulletin of the International Seismological Center (ISC).

Figure 17. Photograph of one of the Pleistocene terraces on the southeastern Perijá mountain front near Río Negro (west of Machiques). The terraces contain boulders up to 2 m in diameter.

the sample locations, uplift probably did not exceed erosion by more than 200 m. On the other hand, the Pleistocene terraces suggest rapid Pleistocene uplift, so it is unlikely that Pleistocene erosion exceeded uplift by more than 1 km at the sample localities. The resulting range of predicted values for the uplift rate are 1.1 to 1.6 mm/yr. On the Cerrejón thrust fault ($15°$ dip), this uplift rate would be equivalent to a horizontal slip rate of $1.5/\tan 15° = 5.6$ mm/yr. A similar Pleistocene uplift rate of about 1.1 to 1.6 mm/yr can be calculated from the Andean fission-track data. This is a slightly more rapid rate than that of 0.8 mm/yr calculated by Kohn and others (1982) for the Pliocene (3 to 6 m.y.), suggesting an increase in the Andean uplift rate during the Pleistocene.

Northwest-Southeast Compression

The northwest-southeast maximum principal stress direction that has characterized the Maracaibo–Santa Marta block and most of the Caribbean–South American margin throughout the Cenozoic continues to the present. Present-day northwest-southeast convergence across the Caribbean–South American plate boundary has been deduced from North American–South American relative plate motions (Ladd, 1976; Pindell and Dewey, 1982) combined with Caribbean–North American relative motion (Jordan, 1975; Minster and Jordan, 1978) and compressive deformation in the southern Caribbean (Krause, 1971; Shepard, 1973: Case, 1974; Bowin, 1976; Talwani and others, 1977; Ladd and Watkins, 1979; Sinton, 1982). Present-day northwest-southeast compression along the Caribbean–South American margin is also indicated by earthquake focal mechanism solutions that have been determined from P-wave first motions (Dewey, 1972; Vierbuchen, 1978; Perez and Aggarwal, 1980; Kellogg and Bonini, 1982). Five of these focal mechanisms are the thrust fault solutions for events 1, 2, 3, 5, and 7 shown in Figure 16. The source parameters for the earthquakes are listed in Table 1. The events were described in more detail by Kellogg and Bonini (1982).

The maximum principal stress direction (σ_1 in Table 1) can be estimated from the azimuths of the axes of maximum pressure (Honda, 1962) for reverse-fault focal mechanisms. The σ_1 estimates for the shallow reverse faults in Figure 16 are $297°$, $313°$, $316°$, and $317°$. These values correspond well with estimates of maximum compressive stress directions based on structural and gravity data for the Sierra de Perijá ($299°$), the Santa Marta massif ($312°$), and the Venezuelan Andes ($320°$).

There have been few major historic earthquakes along the Oca fault system, but north of Sinamaica the fault does displace a series of Quaternary beach strandlines (Miller, 1962) and a shell horizon with a Carbon-14 age determination of 2,500 yr (Cluff and Hansen, 1969). On May 3, 1849, an earthquake near Maracaibo with a maximum modified Mercalli scale (MM) intensity of VII damaged most large buildings and collapsed others (Cluff and Hansen, 1969). Pleistocene activity on the Santa Marta–Bu-

caramanga fault system is demonstrated by deformation in Pleistocene terraces and offset stream patterns (Campbell, 1968).

Fault plane solutions for earthquakes in eastern Venezuela also indicate present-day, northwest-southeast-trending compression (Rial, 1978; Kafka and Weidner, 1979; Perez and Aggarwal, 1981). The maximum principal stress direction for events in eastern Venezuela can be estimated as $320° \pm 20°$ (Kellogg and Bonini, 1982).

East-West Compression and the Boconó Fault

Present-day east-northeast–west-southwest compression is indicated by the three focal mechanisms determined for events 4, 6, and 8 in Figure 16 (Dewey, 1972; Kafka and Weidner, 1981). It is also suggested by Quaternary displacement and major historic earthquakes on the Boconó fault.

The Boconó fault trends in a northeast-southwest direction ($050°$–$055°$) through the Venezuelan Andes. Measurements of offset stream channels, terrace deposits, alluvial fans, and glacial moraines demonstrate approximately 100 m of Holocene right-lateral displacement (Cluff and Hansen, 1969; Schubert and Sifontes, 1970). The average rate of strike-slip motion on the fault for the past 10,000 yr is approximately $100 \text{ m}/10^4 \text{ yr} = 1$ cm/yr. Up to 800 m of postglacial south-side-up vertical displacement has also been observed on the fault (Giegengack and others, 1976). The largest estimate of right-lateral offset along the Boconó fault is 100 km (Stephan, 1977a) based on an apparent offset of the Caribbean frontal thrust. Rod and others (1958) claimed 30 to 70 km of cumulative right-lateral displacement on the fault, but most estimates are less than 50 km. Therefore, a 1 cm/yr slip rate on the Boconó could only have begun 10 m.y. ago at the earliest and may have begun only 3 or 4 m.y. ago.

Numerous large historic earthquakes have occurred on the Boconó fault system including the great earthquake of March 26, 1812 (MM intensity = XI; magnitude = 8; Cluff and Hansen, 1969). Since the advent of instrumental recordings the Boconó fault zone has been the most seismically active zone in northwestern Venezuela. Events 4, 6, and 8 (Fig. 16) are described in more detail by Kellogg and Bonini (1982).

The $60°$ to $65°$ angle between the Oca and Santa Marta fault systems suggests that Coulomb fracture criteria ($\phi = 30°$) may be applicable to the major oblique wrench faults delineating the Maracaibo–Santa Marta block. If the regional stress regime that results from plate motion causes Navier-Coulomb failure, the maximum compressive stress direction (σ_1) can then be estimated as $30°$ from the fault plane of a strike-slip fault and $15°$ from the axis of maximum pressure in the focal mechanism. The predicted oblique strike-slip (oblique thrust) motion on the Oca and Santa Marta faults implies that the intermediate and least principal stresses are approximately equal in magnitude. As surface wave and geologic information specify the probable fault planes for events 4 and 8, σ_1 can be estimated as east-west ($085°$ and $087°$). Because of a lack of surface wave and geologic information, σ_1 for event 6 may be either $045°$ or $075°$, although $075°$ is more likely in view of the correlation with events 4 and 8. These compressive stress directions are close to the east-west (084°) relative convergence vector for the Nazca and South American plates (Stauder, 1975; Minster and Jordan, 1978). Focal mechanism solutions for earthquakes reveal that a zone of faults with similar dextral strike-slip motion extends along the entire eastern front of Colombia's Eastern Cordillera from Venezuela to Ecuador (Pennington, 1981).

The East Andean and Boconó fault zones mark the present eastern boundary of the tectonically active North Andean convergent zone. If we assume that the North Andean convergent zone is now moving as a semi-rigid block, we can add the South American–North Andean vector (1 ± 0.2 cm/yr; 235° ± 5°) and the Caribbean–South American vector (2.2 ± 0.5 cm/yr; 102° ± 10°) derived by Minster and Jordan (1978) to obtain the Caribbean–North Andean vector shown in Figure 16 (1.7 ± 0.7 cm/yr; 128° ± 24°). The mean direction of the Caribbean–North Andean vector (128° ± 24°) is almost identical to the mean maximum principal stress direction for the Maracaibo–Santa Marta area (130° ± 10°) inferred from earthquake focal mechanism determinations and structural and gravity data (Kellogg and Bonini, 1982).

We can also add the South American–North Andean vector and the Nazca–South American vector (Minster and Jordan, 1978) to obtain the Nazca–North Andean vector (6.4 ± 0.7 cm/yr; 088° ± 7°).

Subduction of the Caribbean beneath South America

Significant Cenozoic underthrusting of Caribbean oceanic lithosphere beneath South American continental lithosphere has been deduced from sedimentological (Duque-Caro, 1979) and gravity data (Case and MacDonald, 1973; Bonini and others, 1980). Present-day subduction along the South Caribbean marginal fault is indicated by folding and thrusting in the deformed belt and earthquake focal mechanisms and hypocenter locations.

The Colombian and Venezuelan basins are separated from the South American coast by a belt of deformed Tertiary and Quaternary sediments. Seismic reflection records from the Colombian Basin (Krause, 1971; Shepard, 1973; Case, 1974; Kellogg and Bonini, 1982) and the Venezuelan Basin (Silver and others, 1975; Talwani and others, 1977; Ladd and Watkins, 1979; Diebold and others, line 119, 1981; Sinton, 1982) show that the belt of deformed sediments is being compressed, folded, and thrust to the north and west over Caribbean oceanic crust. Mud volcanism and diapiric intrusion are presently active near Cartagena and northwest of Santa Marta, as a consequence of lateral compression in the deformed belt (Shepard, 1973; Higgins and Saunders, 1974) and/or methane genesis in the organic-rich Miocene-Pliocene sediments (H. Hedberg, Gulf Oil Exploration and Production Co., Houston, Texas, personal commun., 1983).

In Figure 18, earthquake hypocenters from Sykes and Ewing (1965) and Dewey (1972) have been projected onto a

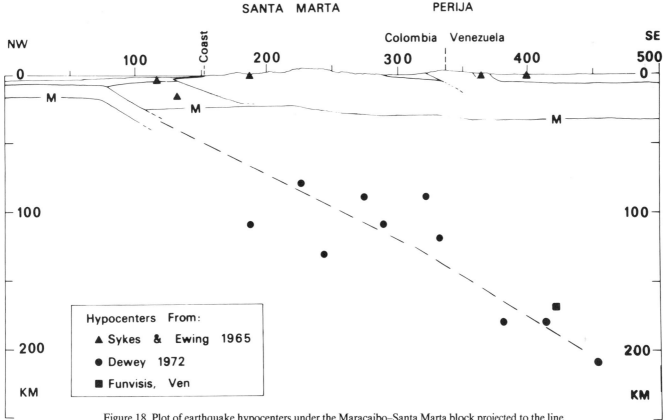

Figure 18. Plot of earthquake hypocenters under the Maracaibo–Santa Marta block projected to the line of section in Figure 9 from the Colombian Basin to the Maracaibo Basin.

northwest-southeast cross section through the Sierra de Perijá and the Santa Marta massif. The earthquakes occurred in a southeast-dipping seismic zone recognized by Dewey (1972) and Pennington (1981). The seismicity terminates 200 km below the Maracaibo Basin. The down-dip length of the Benioff Zone is about 380 to 400 km. Uplift and folding of Miocene-Pliocene pelagic sediments west of the Sinú fault suggest that subduction has moved from the Sinú trench 50 km northwest to the South Caribbean marginal fault in the past 10 m.y. (Duque-Caro, 1979). If one assumes that the thickness of the lithosphere is 100 km, that the Benioff Zone started with an initial length equivalent to the rupture of the lithosphere 100 km/sin 30° = 200 km, and that the equilibration time for a cold slab is 10 m.y. (Isacks and others, 1968), then the horizontal subduction rate is (390–200 km) /10^7 yr = 1.9 ± 0.3 cm/yr. This is similar to the 1.7 ± 0.7 cm/yr Caribbean–North Andean convergence rate derived in this paper.

The P-wave-derived fault plane solutions for events 1 and 2 in Figure 16 are consistent with subduction in a southeast-dipping zone. Event 1 was produced by thrust faulting at a depth of 58 km below the deformed belt southeast of the South Caribbean marginal fault (Vierbuchen, 1978). The nodal plane dipping shallowly to the southeast coincides with the predicted zone of subduction. Event 2 was a deep earthquake (175 km) located southeast of the Sierra de Perijá. The focal mechanism determination (O. Perez, McCann, and A. J. Murphy, Lamont-Doherty Geological Observatory, Palisades, N.Y., unpub. data, 1978) can be explained by east-southeast downdip tension on the downgoing Caribbean lithosphere.

SUMMARY AND CONCLUSIONS

Eight major Phanerozoic tectonic phases have been identified in the Sierra de Perijá. The last four of these occurred in the Cenozoic: the lower Eocene tectonic phase, the middle Eocene Caribbean orogeny, the late Oligocene phase, and the upper Miocene to present Andean orogeny.

During the late Oligocene phase the Palmar area east of the Tigre fault was raised 3 to 4 km along this fault and tilted 7° to the southeast (Fig. 8). The structure was truncated by a major erosional unconformity that is responsible for the most important hydrocarbon trapping in the Lake Maracaibo oil fields. This unconformity had been generally assumed to be late Eocene or early Oligocene, but a reexamination in this paper of recent published paleontological evidence reveals that the unconformity could be as young as late Oligocene. It is also shown that the Oligocene phase was the only period of uplift and erosion that could have been consistent with three late Oligocene (25 m.y.) apatite fission-track ages from the Sierra de Perijá (Kohn, *in* Shagam, 1980).

During the Andean orogeny the main uplift of the Sierra de Perijá occurred, the frontal monocline folded on the southeast flank of the Sierra, and the Manuelote syncline and Mostrencos arch were formed. The predicted Pliocene age is based on strati-graphic relationships and new radiometric data. The apatite fission-track ages of Kohn (*in* Shagam, 1980) were used to estimate the uplift rate for the range: 1.5 mm/yr in the upper Pliocene-Pleistocene.

Right-lateral, oblique-slip displacement of 40 ± 8 km on the Oca fault zone is related to Cenozoic northwest-southeast shortening in the northern Sierra de Perijá. Combining this value with previous estimates of movement on the Oca fault in Colombia, a total Tertiary oblique dextral slip of 90 ± 8 km is predicted for the fault zone east of the Sierra. Published estimates of North American–South American and Caribbean–North American movements have also been used by Kellogg and Bonini (1982) to deduce 600 to 1,000 km of upper Tertiary Caribbean–South American convergence in a west-northwest–east-southeast direction. If these estimates of slip on the Oca fault and Caribbean–South American convergence are correct, the upper Tertiary slip on the Oca fault zone was an order of magnitude too small for the fault to be the Caribbean–South American boundary.

Most of the upper Tertiary Caribbean–South American convergence occurred on the Romeral, Sinú, and South Caribbean marginal faults. The Panama volcanic arc collided with South America 3 m.y. ago, forming the land bridge between North and South America. The Northern Andes were uplifted on deep thrust faults, and the North Andean block became detached from South America and is being wedged slowly to the north as the Nazca plate converges rapidly with the Caribbean and South American plates. Slow present-day subduction of Caribbean oceanic lithosphere beneath North Andean continental lithosphere is indicated by folding and thrusting in the deformed belt and earthquake focal mechanisms and hypocenter locations. The seismic zone dips 30° to the southeast and terminates 200 km below the Maracaibo Basin (Fig. 18).

The Quaternary Caribbean–North Andean convergence (1.7 ± 0.7 cm/yr; 128° (308°) ± 24°) has produced a northwest-southeast maximum principal stress direction (σ_1) in the overriding North Andean plate. The mean σ_1 direction for the Maracaibo-Santa Marta block is 310° ± 10° based on earthquake focal mechanism determinations, structural, and gravity data (Kellogg and Bonini, 1982).

Rapid convergence of the Nazca and South American plates has produced rapid subduction at the Colombia trench on the west side of the North Andean block and right-lateral shear along the Boconó fault on the east side of the North Andean block.

ACKNOWLEDGMENTS

This paper has evolved from my Ph.D. dissertation at Princeton University, which was supervised by Professor W. Bonini. A. Fisher, R. Hargraves, P. Mattson, W. Maze, H. Houghton, A. Espejo, H. Etchart, G. Canelon, N. Pimentel, and J. Xavier shared many helpful ideas with me in the field. R. Shagam also visited the field area and together with B. Kohn kindly made available unpublished fission-track age data for the Sierra de Perijá.

I also benefited from valuable discussion with J. Suppe, W. J. Morgan, H. Hedberg, R. Giegengack, K. Stauffer, H. Duque-Caro, K. Muessig, J. Wittke, and J. Garing. H. Kraus, V. Pumpin, P. Bartok, and F. Jansen of Maraven Exploration and Production in Caracas provided geophysical data for numerous structures in the Maracaibo Basin, including Figure 15. O. Perez of FUNVISIS, Venezuela, and Lamont-Doherty made available several earthquake focal mechanism determinations and microearthquake study data.

The field work was supported by A. Bellizzia of Dirección de Geologia, Ministerio de Energía y Minas, Venezuela, and M. Hermelin of INGEOMINAS, Colombia. The Hawaii Institute of Geophysics supported manuscript preparation. The manuscript was reviewed and considerably improved by changes suggested by H. Hedberg, R. Shagam, R. Vierbuchen, and R. Pujalet. Hawaii Institute of Geophysics contribution no. 1449.

REFERENCES CITED

Beck, C. M., 1977, Tectónica polifásica terciaria de la Faja Piemontina en la parte central de la Serranía del Interior, en Venezuela septentrional: Abstracts, Caribbean Geological Conference, VIII, Willemstad, Curacao, p. 10–12.

Bell, J. S., 1972, Geotectonic evolution of the southern Caribbean area, *in* Shagam, R. and others, eds., Studies in earth and space sciences: Geological Society of America Memoir 132, p. 369–386.

Bellizzia, G. A., Pimentel, M. N. and Bajo, O. R., compilers, 1976, Mapa geológico estructural de Venezuela: Venezuela Ministerio de Minas e Hidrocarburos, scale 1:500,000.

Bonini, W. E., Garing, J. D., and Kellogg, J. N., 1980, Late Cenozoic uplifts of the Maracaibo–Santa Marta block, slow subduction of the Caribbean plate and results from a gravity study: Transactions, Caribbean Geological Conference, IX, Santo Domingo, 1980, Universidad Católica Madre y Maestras, Santiago de los Caballeros, República Dominicana, p. 99–105.

Bowin, C., 1976, Caribbean gravity field and plate tectonics: Geological Society of America Special Paper 169, 79 p.

Bucher, W. H., 1952, Geologic structure and orogenic history of Venezuela: Geological Society of America Memoir 49, 113 p.

Campbell, C. J., 1968, The Santa Marta wrench fault of Colombia and its regional setting: Transactions, Caribbean Geological Conference, IV, Trinidad, 1965, p. 247–260.

Case, J. E., 1974, Major basins along the continental margin of northern South America, *in* Burk, C. A., and Drake, C. L., eds., The geology of continental margins: New York, Springer-Verlag, p. 733–741.

Case, J. E., and Holcombe, T. L., 1980, Geologic-tectonic map of the Caribbean: U.S. Geological Survey Miscellaneous Investigation Series No. I-1100, scale 1:2,500,000, 3 sheets.

Case, J. E., and MacDonald, W. D., 1973, Regional gravity anomalies and crustal structure in northern Colombia: Geological Society of America Bulletin, v. 84, no. 9, p. 2905–2916.

Cati, F., Colalongo, M. L., Crescenti, U., D'Onofrio, S., Follador, U., Pirini Raddrizzani, C., Pomesano Cherchi, A., Salvarorini, G., Sartoni, S., Premoli Silva, I., Wezel, C. F., Bertolino, V., Bizon, G., Bolli, H. M., Borsetti Cati, A. M., Dondi, L., Feinberg, H., Jenkins, D. G., Perconig, E., Sampo, M., and Sprovieri, R., 1968, Biostratigrafia del Neogeno mediterráneo basata sui foraminiferi planctonici: Societa Geologica Italiana, Bollettino, v. 87, p. 491–503.

Cluff, L. S., and Hansen, W. R., 1969, Seismicity and seismic-geology of northwestern Venezuela, Volume I, Evaluation, submitted to Compañía Shell de Venezuela, Caracas: Woodward-Clyde and Associates, 2730 Adelina Street, Oakland, California 94608, various p.

Coronel, G., 1970, Igneous rocks of central Falcón: Asociación Venezolana de Geología, Minería y Petroleo, Boletín Informativo 13, p. 155–162.

Dewey, J. W., 1972, Seismicity and tectonics of western Venezuela: Bulletin of the Seismological Society of America, v. 62, no. 6, p. 1711–1751.

Diebold, J. B., Stoffa, P. L., Buhl, P., and Truchan, M., 1981, Venezuela Basin crustal structure: Journal of Geophysical Research, v. 86, no. B9, p. 7901–7923.

Dufour, J., 1955, Some oil-geological characteristics of Venezuela: Proceedings, World Petroleum Congress, 4th, Rome, Sec. I/A/1, Paper 1, p. 19–35.

Duque-Caro, H., 1972, Ciclos tectónicos y sedimentarios en el norte de Colombia y sus relaciones con la paleoecología: Boletín Geológico: Instituto Nacional de Investigaciones Geológico-Mineras (INGEOMINAS), Bogota, v. 19, no. 3, p. 1–23.

—— 1979, Major structural elements of northwestern Colombia, *in* Watkins, J. S., Montadert, L., and Dickerson, P. W., eds., Geological and geophysical investigations of continental margins: American Association of Petroleum Geologists Memoir 29, p. 329–351.

Dusenbury, A. N., Jr., 1956, Formación La Rosa: Léxico estratigráfico de Venezuela, Ministerio de Minas e Hidrocarburos, Boletín de Geología, Caracas, Publicación Especial.

Feo-Codecido, G., 1972, Breves ideas sobre la estructura de la Falla de Oca, Venezuela: Transactions, Caribbean Geological Conference, VI, Margarita, Venezuela, 1971, p. 184–190.

Giegengack, R., Grauch, R., and Shagam, R., 1976, Geometry of late Cenozoic displacement along the Boconó fault, Venezuelan Andes, *in* Memoria, Congreso Latinoamericano de Geología, II, Caracas, 1973: Venezuela Ministerio de Minas e Hidrocarburos, Boletín de Geología, Publicación Especial no. 7, p. 1201–1226.

Harding, T. P., and Lowell, J. D., 1979, Structural styles, their plate-tectonics habitats, and hydrocarbon traps in petroleum provinces: American Association of Petroleum Geologists Bulletin, v. 63, no. 7, p. 1016–1058.

Hedberg, H. D., 1929, Report on stratigraphy and structure of the Rio Socuy–Rio Guasare area, District of Mara, State of Zulia: Venezuelan Gulf Oil Company, 90 p.

—— 1950, Geology of the eastern Venezuela Basin (Anzoategui-Monagas-Sucre-Eastern Guarico portion): Geological Society of America Bulletin, v. 61, no. 11, p. 1173–1215.

Hedberg, H. D., and Sass, L. C., 1937, Synopsis of the geologic formations of the western part of the Maracaibo Basin, Venezuela: Boletín de Geología y Minería, Caracas, T.1, nos. 2-4, p. 73-112.

Heezen, B. C., Perfit, M. R., Dreyfus, M., and Catalano, R., 1973, The Cayman Ridge: Geological Society of America Abstracts with Programs, v. 5, no. 7, p. 663.

Higgins, G. E., and Saunders, J. B., 1974, Mud volcanoes: Their nature and origin, *in* Contributions to the geology and paleobiology of the Caribbean and adjacent areas: Naturforschende Gesellschaft in Basel, Verhandlungen, Band 84, no. 1, p. 101–152.

Holcombe, T. L., Vogt, P. R., Matthews, J. E., and Murchison, R. R., 1973, Evidence for sea-floor spreading in the Cayman Trough: Earth and Planetary Science Letters, v. 20, p. 357–371.

Honda, H., 1962, Earthquake mechanism and seismic waves: Geophysical Notes, Geophysical Institute, Faculty of Science, Tokyo University, v. 15, supplement, 97 p.

Isacks, B., Oliver, J., and Sykes, L. R., 1968, Seismology and the new global tectonics: Journal of Geophysical Research, v. 73, no. 18, p. 5855–5899.

James, H. E., Jr., 1977, Sedimentology of the iron-oxide-bearing upper Miocene(?) Guayabo Group in the vicinity of Cucúta, Colombia [Ph.D. thesis]: Princeton, New Jersey, Princeton University, 273 p.

Jordan, T. H., 1975, The present-day motions of the Caribbean plate: Journal of

Geophysical Research, v. 80, no. 32, p. 4433–4439.

Kafka, A. L., and Weidner, D. J., 1979, The focal mechanisms and depths of small earthquakes as determined from Rayleigh-wave radiation patterns: Seismological Society of America Bulletin, v. 69, no. 5, p. 1379–1390.

—— 1981, Earthquake focal mechanisms and tectonic processes along the southern boundary of the Caribbean plate: Journal of Geophysical Research, v. 86, no. B4, p. 2877–2888.

Kellogg, J. N., 1980, Cenozoic basement tectonics of the Sierra de Perijá, Venezuela and Colombia, *in* Transactions, Caribbean Geological Conference, IX, Santo Domingo, 1980: Universidad Católica Madre y Maestra, Santiago de los Caballeros, República Dominicana, p. 107–117.

—— 1981, The Cenozoic basement tectonics of the Sierra de Perijá, Venezuela and Colombia [Ph.D. thesis]: Princeton, New Jersey, Princeton University, 246 p.

Kellogg, J. N., and Bonini, W. E., 1982, Subduction of the Caribbean plate and basement uplifts in the overriding South American plate: Tectonics, v. 1, no. 3, p. 251–276.

Kohn, B. P., Shagam, R., and Burkley, L. A., 1982, Mesozoic-Pleistocene fission-track ages on rocks of the Venezuelan Andes and their tectonic implications: Geological Magazine (preprint).

Krause, D. C., 1971, Bathymetry, geomagnetism, and tectonics of the Caribbean Sea north of Colombia, *in* Donnelly, T. W., ed., Caribbean geophysical, tectonic, and petrologic studies: Geological Society of America Memoir 130, p. 35–54.

Kuyl, O. S., Muller, J., and Waterbolk, H. T., 1955, The application of palynology to oil geology, with special reference to western Venezuela: Geologie en Mijnbouw, New Series, v. 17, n. 3, p. 49–75.

Ladd, J. W., 1976, Relative motion of South America with respect to North America and Caribbean tectonics: Geological Society of America Bulletin, v. 87, no. 7, p. 969–976.

Ladd, J. W., and Watkins, J. S., 1979, Tectonic development of trench-arc complexes on the northern and southern margins of the Venezuela Basin, *in* Watkins, J. S., Montadert, L., and Dickerson, P. W., eds., Geological and geophysical investigations of continental margins: American Association of Petroleum Geologists Memoir 29, p. 363–371.

Lamb, J. L., and Sulek, J. A., 1968, Miocene turbidites in the Carapita Formation of eastern Venezuela: Transactions, Caribbean Geological Conference, IV, Trinidad, 1965.

Léxico Estratigráfico de Venezuela, 1970, Dirección de Geología, Ministerio de Minas y Hidrocarburos, Publicación Especial no. 4, 756 p.

MacDonald, W. D., Doolan, B. L., and Cordani, U. G., 1971, Cretaceous–early Tertiary metamorphic K-Ar age values from the south Caribbean: Geological Society of America Bulletin, v. 82, no. 5, 1381–1388.

Macellari, C., 1984, Late Tertiary tectonic history of the Táchira Depression (Southwestern Venezuelan Andes), *in* Bonini, W. E., and Hargraves, R. B., eds., The Caribbean–South American plate boundary and regional tectonics: Geological Society of America Memoir 162 (this volume).

Malfait, B. T., and Dinkelman, M. G., 1972, Circum-Caribbean tectonic and igneous activity and the evolution of the Caribbean plate: Geological Society of America Bulletin, v. 83, no. 2, p. 251–271.

Mattson, P. H., 1973, Middle Cretaceous nappe structure in Puerto Rican ophiolites and their relation to the tectonic history of the Greater Antilles: Geological Society of America Bulletin, v. 84, no. 1, p. 21–37.

Maze, W. B., 1980, Geology and copper mineralization of the Jurassic La Quinta Formation in the Sierra de Perijá, northwestern Venezuela, *in* Transactions, Caribbean Geological Conference IX, Santo Domingo, 1980: Universidad Católica Madre y Maestra, Santiago de los Caballeros, República Dominicana, p. 283–294.

—— 1984, Jurassic La Quinta formation, Sierra de Perijá, northwestern Venezuela: igneous petrology and tectonic environment, *in* Bonini, W. E., and Hargraves, R. B., eds., The Caribbean–South American plate boundary and regional tectonics: Geological Society of America Memoir 162 (this volume).

Miller, J. B., 1962, Tectonic trends in Sierra de Perijá and adjacent parts of

Venezuela and Colombia: American Association of Petroleum Geologists Bulletin, v. 46, no. 9, p. 1565–1595.

Minster, J. B., and Jordan, T. H., 1978, Present-day plate motions: Journal of Geophysical Research, v. 83, no. B11, p. 5331–5354.

Minster, J. B., Jordan, T. H., Molnar, P., and Haines, E., 1974, Numerical modelling of instantaneous plate tectonics: Geophysical Journal (Royal Astronomical Society), v. 36, no. 3, p. 541–576.

Molnar, P., and Sykes, L. R., 1969, Tectonics of the Caribbean and Middle America regions from focal mechanisms and seismicity: Geological Society of America Bulletin, v. 80, no. 9, p. 1639–1684.

Monroe, W. H., 1968, The age of the Puerto Rico Trench: Geological Society of America Bulletin, v. 79, no. 4, p. 487–493.

Muessig, K. W., 1979, The Central Falcón igneous rocks of northwestern Venezuela: Their origin, petrology, and tectonic significance [Ph.D. thesis]: Princeton, New Jersey, Princeton University, 252 p.

Pennington, W. D., 1981, Subduction of the eastern Panama basin and the seismotectonics of northwestern South America: Journal of Geophysical Research, v. 86, no. B11, p. 10753–10770.

Perez, O. J., and Aggarwal, Y. P., 1980, Microseismicity studies in the Uribante-Caparo project, State of Táchira, Venezuela: Fundación Venezolana de Investigaciones Sismológicas, Caracas, Open-File Report no. 021-79, Caracas, 20 p.

—— 1981, Present-day tectonics of the southeastern Caribbean and northeastern Venezuela: Journal of Geophysical Research, v. 86, no. B11, p. 10791–10804.

Pimentel, M.N.R., 1977, Excursion no. 3—(segunda parte): Falla de Oca: Isla de Toas y San Carlos, *in* Memoria, Congreso Latinoamericano de Geología II, Caracas, 1973: Venezuela Ministerio de Minas e Hidrocarburos, Boletín de Geología, Publicación Especial no. 7, v. 4, p. 326–338.

Pindell, J., and Dewey, J. F., 1982, Permo-Triassic reconstruction of western Pangea and the evolution of the Gulf of Mexico/Caribbean region: Tectonics, v. 1, no. 2, p. 179–211.

Pinet, P. R., 1972, Diapirlike features offshore Honduras: Implications regarding tectonic evolution of Cayman Trough and Central America: Geological Society of America Bulletin, v. 83, no. 7, p. 1911–1921.

Polson, I. L., and Henao, D., 1965, The Santa Marta wrench fault: A rebuttal: Transactions, Caribbean Geological Conference, IV, Trinidad, p. 263–266.

Pumpin, V. F., 1979, El marco estructural de Venezuela noroccidental: EPC Report, Maraven, Caracas.

Rial, J. A., 1978, The Caracas, Venezuela earthquake of July 1967: A multiple-source event: Journal of Geophysical Research, v. 83, no. B11, p. 5405–5414.

Rod, E., 1956, Strike-slip faults of northern Venezuela: American Association of Petroleum Geologists Bulletin, v. 40, no. 3, p. 457–476.

Rod, E., Jefferson, C., von der Osten, E., and Miller, R., 1958, Round table discussion—the determination of the Boconó fault: Asociación Venezolana de Geología, Minería y Petroleo, Boletín Informativo, v. 1, p. 69–100.

Salvador, A., and Hotz, E., 1963, Petroleum occurrence in the Cretaceous of Venezuela: Proceedings, World Petroleum Congress VI, Frankfurt, v. 1, p. 115–140.

Salvador, A., and Rosales, H., 1960, Guía de la excursion A-3 Jusepin-Cumana, *in* Memoria, Congreso Geológico Venezolano, III: Venezuela Ministerio de Minas e Hidrocarburos, Boletín de Geología, Publicación Especial no. 3, v. 1, p. 63–74.

Santamaria, F., and Schubert, C., 1974, Geochemistry and geochronology of the southern Caribbean–northern Venezuela plate boundary: Geological Society of America Bulletin, v. 85, no. 7, p. 1085–1098.

Schubert, C., and Sifontes, R. S., 1970, Boconó fault, Venezuelan Andes: Evidence of post-glacial movement: Science, v. 170, no. 3953, p. 66–69.

Shagam, R., 1975, The northern termination of the Andes, *in* Nairn, A.E.M., and Stehli, F. G., eds., The ocean basins and margins: New York, Plenum Press, v. 3, p. 325–420.

—— 1980, Fission track ages on apatite, zircon and sphene from rocks of the Andes, Perijá and Toas island: Venezuela, Dirección de Geología, Ministerio de Energía y Minas, Internal Report no. 9, 35 p.

Shagam, R., and Kohn, B. P., 1984, Apatite, zircon, and sphene fission-track ages of igneous and metamorphic rocks in northern South America, *in* Bonini, W. E., and Hargraves, R. B., eds., The Caribbean–South American plate boundary and regional tectonics: Geological Society of America Memoir 162 (this volume).

Shepard, F. P., 1973, Sea floor off Magdelena delta and Santa Marta area, Colombia: Geological Society of America Bulletin, v. 84, no. 6, p. 1955–1972.

Silver, E. A., Case, J. E., and MacGillavry, H. J., 1975, Geophysical study of the Venezuelan borderland: Geological Society of America Bulletin, v. 86, no. 2, p. 213–226.

Sinton, J. B., 1982, Detailed geophysical studies of two-dimensional structures at active plate margins using seismic refraction and earthquake data [Ph.D. thesis]: Honolulu, University of Hawaii, 275 p.

Smithson, S. B., Brewer, J. A., Kaufman, S., Oliver, J. R., and Hurick, C. A., 1979, Structure of the Laramide Wind River uplift, Wyoming, from COCORP deep reflection data and from gravity data: Journal of Geophysical Research, v. 84, no. B11, p. 5955–5972.

Stauder, W., 1975, Subduction of the Nazca plate under Perú as evidenced by focal mechanisms and by seismicity: Journal of Geophysical Research, v. 80, no. 8, p. 1053–1064.

Stephan, J. F., 1977a, El contacto cadena Caribe-Andes merideños entre Carora y El Tocuyo (Estado Lara). Memoria, 5th Congreso Geologico Venezolano, v. 2, p. 789–816.

Stephan, J. F., 1977b, Una interpretación de los complejos con bloques asociados a los flysch Paleoceno-Eoceno de la Cadena Caribe Venezolana: el emplazamiento submarino de la Napa de Lara: Abstracts, Caribbean Geological Conference, VIII, Willemstad, Curacao, p. 199–200.

Sutton, F. A., 1946, Geology of Maracaibo Basin, Venezuela: American Association of Petroleum Geologists Bulletin, v. 30, no. 10, p. 1621–1741.

Sykes, L. R., and Ewing, M., 1965, The seismicity of the Caribbean region: Journal of Geophysical Research, v. 70, no. 20, p. 5065–5074.

Talwani, M., Windisch, C. C., Stoffa, P. L., Buhl, P., and Houtz, R. E., 1977, Multichannel seismic study in the Venezuelan Basin and the Curacao Ridge, *in* Talwani, M., and Pitman, W. C., III, eds., Island arcs, deep sea trenches and back-arc basins (Maurice Ewing Series): Proceedings of the Symposium, American Geophysical Union, Washington, D.C., no. 1, p. 83–98.

Tschanz, C. M., Marvin, R. F., Cruz, B. J., Mehnert, H. H., and Cebula, G. T., 1974, Geologic evolution of the Sierra Nevada de Santa Marta, northeastern Colombia: Geological Society of America Bulletin, v. 85, no. 2, p. 273–284.

Vasquez, E. E., and Dickey, P. A., 1972, Major faulting in north-western Venezuela and its relation to global tectonics: Transactions, Caribbean Geological Conference VI, Margarita, Venezuela, 1971, p. 191–202.

Vierbuchen, R. C., Jr., 1978, The tectonics of northeastern Venezuela and the southeastern Caribbean Sea [Ph.D. thesis]: Princeton, New Jersey, Princeton University, 175 p.

Walper, J. L., 1980, The tectono-sedimentary history of Caribbean basins and their hydrocarbon potential, *in* Miall, A. D., ed., Facts and principles of world petroleum occurrence: Canadian Society of Petroleum Geologists Memoir 6, p. 887–911.

Weingarten, B., 1977, Tectonic and paleoclimatic significance of a late-Cenozoic red-earth paleosol from the central Andes, Venezuela: Abstracts, Caribbean Geological Conference, VIII, Willemstad, Curacao, p. 221.

Wheeler, C. B., 1963, Oligocene and lower Miocene stratigraphy of western and northeastern Falcón Basin, Venezuela: American Association of Petroleum Geologists Bulletin, v. 47, no. 1, p. 35–68.

Young, G. A., 1958, Correlation of the Oligo-Miocene formations in the districts of Urdaneta and Perijá, State of Zulia: Asociación Venezolana de Geología, Minería y Petroleo, Boletín Informativo, v. 1, no. 4, p. 117–135.

Young, G. A., Bellizzia, A., Renz, H. H., Johnson, F. W., Robie, R. H., and Mas Vall, J., 1956, Geología de las cuencas sedimentarias de Venezuela y de sus campos petrolíferos: International Geological Congress XX, Mexico, Symposium sobre yacimientos de petroleo y gas, tomo 4, p. 161–322.

Zambrano, E., Vásquez, E., Duval, B., Latreille, M., and Coffinieres, B., 1971, Síntesis paleogeográfica y petrolera del occidente de Venezuela, *in* Venezuela Ministerio de Minas e Hidrocarburos, Boletín de Geología, Memoria, Congreso Geológico Venezolano, IV, Caracas, 1969: Publicación Especial no. 5, v. 1, p. 483–552.

MANUSCRIPT ACCEPTED BY THE SOCIETY SEPTEMBER 1, 1983

Printed in U.S.A.

Geological Society of America
Memoir 162
1984

Jurassic La Quinta Formation in the Sierra de Perijá, northwestern Venezuela: Geology and tectonic environment of red beds and volcanic rocks

William B. Maze
Department of Geological and Geophysical Sciences
Princeton University
Princeton, New Jersey 08544

ABSTRACT

The Jurassic age La Quinta Formation in the Sierra de Perijá of northwestern Venezuela consists of red continental sedimentary rocks, interbedded volcanic rocks, and associated hypabyssal intrusives. These strata were deposited in a series of graben that parallel the present Sierra de Perijá and the Venezuelan (Mérida) Andes. Trace element and isotopic data indicate that crustal contamination and/or anatexis were important factors in the petrogenesis of the igneous rocks. Red beds, some with volcanic rocks, of similar type, age, and tectonic setting are found from Mexico to Chile, suggesting that the depositional environment was related to convergent plate margin processes. Tension related to opening of the Caribbean and/or separation of Central America and North America may also have been a factor in Jurassic time. However, paleomagnetic data indicate a differential rotation between the Sierra de Perijá and the Venezuelan Andes that cannot be reconciled with simple models of tension and separation. This suggests that the Sierra de Perijá is an allochthonous terrane emplaced during Jurassic time. Its current setting is the result of a composite of subduction, rifting associated with opening of the Caribbean, and transcurrent motion.

INTRODUCTION

A detailed investigation has been made of the igneous rocks and related copper mineralization in the La Quinta Formation exposed along the Venezuelan flank of the Sierra de Perijá approximately 90 km west of Maracaibo between Río Negro and Río Palmar (Figs. 1 and 2). To that end it was first necessary to establish the stratigraphic and tectonic history of the formation. In addition, reconnaissance field studies of the La Quinta Formation were made in the Venezuelan Andes, and the equivalent Girón Formation was studied near Bucaramanga, Colombia (Santander Massif), and near Manaure, Colombia (east of Valledupar), on the west side of the Sierra de Perijá (Figs. 1 and 2). The stratigraphic and tectonic aspects of the study are presented in this paper.

Initially, the data was interpreted in terms of simple tectonic extension, perhaps related to Jurassic subduction beneath South America. This model appeared to explain adequately the geologic history of the region. As paleomagnetic data (Maze and Hargraves, 1984) became available, however, the possibility of large-scale tectonic rotation and presence of displaced terranes arose. The reader should keep that in mind while reading the following report.

Quebrada (stream) and Cañada (slow-flowing stream) are abbreviated to Q. and C., respectively. The Sierra de Perijá (Perijá Mountains) is referred to as the Perijá, and the Cordillera de los Andes in Venezuela as the Mérida or Venezuelan Andes. The term "front" is used to describe the change in slope at the topographic contact of the mountains and the Maracaibo coastal plain.

The Jurassic–Early Cretaceous (see discussion in Ages section) La Quinta Formation was originally named by Kundig (1938) to include the red sediments of Mesozoic age beneath marine Cretaceous rocks in the Venezuelan Andes. At the type

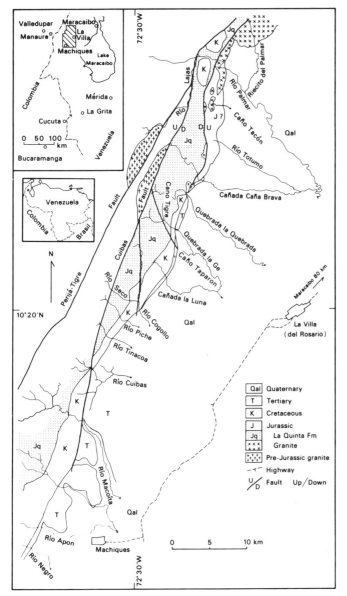

Figure 1. Location map of the field area along the east flank of the Sierra de Perijá. Geography and geology based on Bellizzia and others (1976); 1:100,000 geologic maps by González Padilla and others (1979?); Canelon and others (1979); Creole Petroleum Corp. (1956–1961); and this work. Stipled area represents La Quinta Formation. "J?" represents an area of volcanic material and contact(?) metamorphosed sediment of uncertain age, possibly part of the La Quinta Formation.

locality near La Grita (Fig. 2) the formation is greater than 2,400 m thick, according to the *Léxico Estratigráfico* (1970), but has been measured as 1,610 m by Padron (1978; see also Tarache, 1980) and Schubert and others (1979) and 1,800 m by Hargraves and Shagam (1969). It rests unconformably on Permian-Triassic igneous rocks and Late Paleozoic metamorphic rocks, whereas the top is a conformable gradational contact with basal units of the Río Negro Formation of Cretaceous age. The top is arbitrarily set at the uppermost red bed (*Léxico Estratigráfico*, 1970). The only Jurassic volcanic materials

reported have been "tuffaceous" sandstone lenses and volcanic lithic fragments in the lower 800 m of the type section (Tarache, 1980). The lithic fragments may have been derived from the purple, dacitic tuff unit at the base of the section, dated at 229 ± 15 m.y. (Hargraves and Shagam, 1969; Burkley, 1976).

The name "La Quinta Formation" has since been extended to include the continental red beds and mafic to felsic volcanic rocks that accumulated in a series of grabens (Pumpin, 1979; Bartok and others, 1981) within the region that includes the Venezuelan Andes (Mérida Andes), Maracaibo Basin, and the Sierra de Perijá (Fig. 2). This region will be referred to as the proto–Maracaibo Basin when discussing times other than the present.

The Sierra de Perijá straddles the Venezuela-Colombia border (Fig. 2). The range is 30 to 50 km wide and extends 250 km north-northeast–south-southwest from the Guajira Peninsula in the north to the Santander Massif in the south (Fig. 2). The present topography of the range reflects post-Miocene uplift that produced frontal monoclines with associated passive draping at the margins of blocks. These structures are similar to those in the central and southern Rockies of the United States (Kellogg, 1981, 1984; Miller, 1962). Flatirons capped by thick Cretaceous limestones dominate the mountain front on both sides of the central Perijá. Paleozoic rocks are exposed in some localities, but for the most part the Jurassic-Cretaceous cover has not been stripped off the center of the range.

Regional geology and tectonics have been described by Hedberg and Sass (1937), Liddle and others (1943), Hea and Whitman (1960), Miller (1962), Bowen (1972), and Kellogg (1981, 1984); however, most work has been concentrated on the Cretaceous and younger formations, i.e., those of interest with respect to petroleum exploration. Bowen (1972), Moticska (1973), Moya and Viteri (1977), and Viteri (1978) have described many aspects of the La Quinta Formation within the study area between Río Palmar and Río Negro (Fig. 1).

LA QUINTA FORMATION WITHIN THE STUDY AREA, SIERRA DE PERIJA

The La Quinta Formation within the Río Palmar–Río Negro region (Fig. 1) varies widely in lithology and in environment of deposition from north to south. From Río Palmar almost to Q. La Quebrada the outcrops of the formation rocks are almost entirely contained within a graben that is 2 to 5 km wide (Fig. 1). Farther south the eastern flank of the graben is covered by younger sedimentary deposits. South of Río Cuibas the Jurassic and Cretaceous formations are exposed in a frontal monocline (Fig. 3).

In the northern third of the area, Río Palmar to C. Caña Brava, volcanic rocks dominate the formation. Farther south sedimentary rocks (predominantly red beds) become more abundant, and in Q. La Ge volcanic rocks constitute less than one third of the stratigraphic section. South of Río Cogollo, volcanic rocks are rare to absent. Along Río Macoíta there is volcanic detritus in

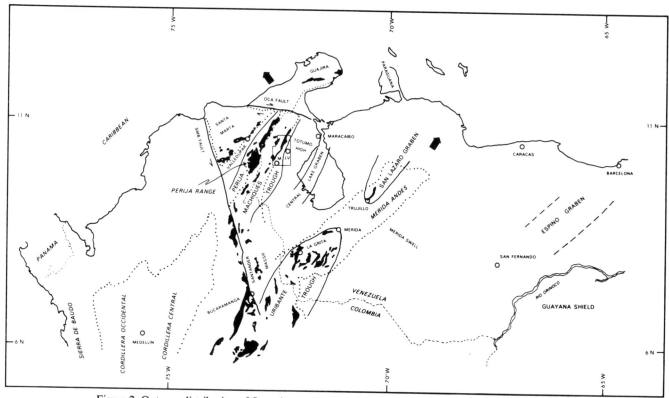

Figure 2. Outcrop distribution of Jurassic strata (solid black area), of which the La Quinta Formation is part, in northern Venezuela and northeastern Colombia. Geology and geography adapted from map by Case and Holcombe (1980) and this work. Jurassic troughs/grabens are outlined and labelled. The Guayana Shield (dotted line) was a high during Jurassic time. The Sierra de Perijá, Mérida Andes (Venezuelan Andes), Santa Marta, and Colombian Andes above approximately 500-m elevation are outlined by heavy dotted lines. Information from Bartok and others (1981), Campbell (1961), Pumpin (1979), and Feo-Codecido and others (1984). M, LV, and SMB Fault represent the towns of Machiques, La Villa (del Rosario), and the Santa Marta–Bucaramanga Fault, respectively. Arrows represent average paleomagnetic declinations as measured in La Quinta Formation rocks (see Maze and Hargraves, 1984). Declinations in the Perijá and Andes are N20°–60°W and N0°–60°E, respectively (Hargraves and Shagam, 1969; Maze and Hargraves, 1984). The rectangular box is the area depicted in Figure 1.

the La Quinta section, but no flows or ash beds have been observed.

Variations in stratigraphic thickness are not well established because of structural complexity and poor exposures. The La Quinta Formation along Río Macoíta is approximately 1,000 m thick (González Padilla, 1973) and according to Hea and Whitman (1960, p. 368) is a minimum of 1,050 m thick in Q. La Ge. I estimate the Q. La Ge section to be as thick as 1,700 m. The total thickness of the volcanic section north of Q. La Ge is unknown, but on the basis of the map distribution is believed to be similar to the Río Macoíta and Q. La Ge sections.

As Q. La Ge is readily accessible and has outcrops of dike, volcanic, and sedimentary rocks, it serves as the basis for the more detailed description of the stratigraphic section given below. Generally, Q. La Ge lithology described in the following section is characteristic of a composite section of northern volcanic and southern sedimentary lithologies.

Stratigraphy of the La Quinta Formation in Quebrada La Ge

In the field the La Quinta Formation was divided into three subsections described below as lower, transitional, and upper (respectively 1, 2, and 3 in Fig. 3). The lower and upper sections are of approximately equal thickness—700 to 800 m, and the transitional section is probably less than 100 m thick. These subsections cannot be precisely defined because abrupt facies changes make it almost impossible to recognize stratigraphic equivalents.

Lower Section (1 in Fig. 3). In general, the lower half of the La Quinta section in the Q. La Ge drainage consists of fine- to medium-grained, dark red-brown arkose with few, if any, volcanic beds. Minor coarse-grained, pale-green, arkosic sandstones typically contain plant fragments and charcoal clasts up to several centimeters in length. Failure to recover pollen from these rocks

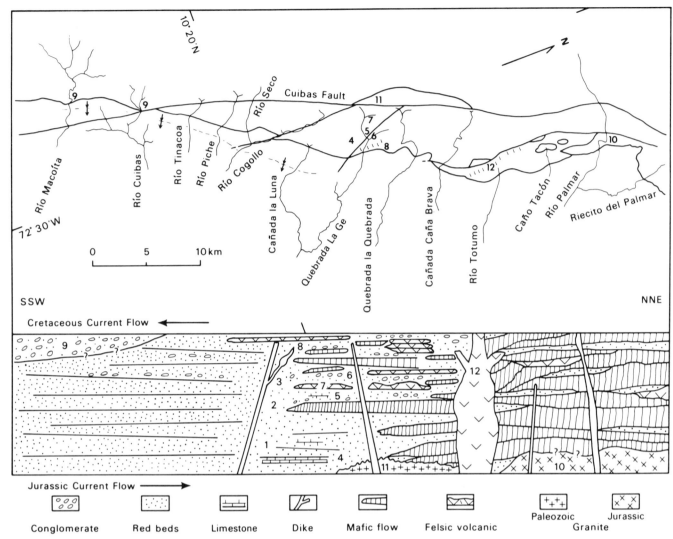

Figure 3. Schematic cross section of the La Quinta Formation and map view of the field area along the mountain front. Lithology reported in the cross section corresponds to that observed in outcrop along the streams, roads, and ridges depicted on the map. Location map and geology are on Figure 1. Numbers from 1 to 12 on both the map and cross section refer to features described in the text. The section is approximately 1,500 m thick. Frontal monocline is represented by the dashed line with double arrows.

suggests a strongly oxidizing environment despite the preservation of many plant fragments (Gil Brenner, SUNY, New Paltz, personal communication on samples supplied by the author). Freshwater carbonate is present in the lower part of the La Quinta sequence in forms varying from small concretions and thin lenses to fossiliferous limestone beds a few meters thick (4 in Fig. 3). Numerous conchostracan fossils (see Paleontology section) are preserved in one of the limestone horizons.

Transition Zone (2 in Fig. 3). The transition zone of the La Ge sequence has the earliest basaltic-andesite flows, ranging in thickness from less than 1 m to greater than 20 m (Fig. 3). In some units the upper parts of the flows are gradational to cobble to boulder conglomerate with fine-grained igneous matrix (called a tuffaceous conglomerate by Moticska and Viteri, personal communication). In Q. La Ge and its tributaries, an extensive

hornblende-andesite flow serves as a stratigraphic marker. This flow, about 20 m thick, is overlain by medium- to coarse-grained, green arkose interbedded with fine-grained, dark red-brown arkose and red to black (rarely green) mudstone. The green beds and the adjacent red beds (5 in Fig. 3) yielded the majority of the fossils (estherids, vertebrate remains, pieces of carbon, plant debris, and several large plant fragments [2–7 cm diameter and 20 cm long]; see section on Paleontology). Although less limestone occurs in this portion than in the lower portion, carbonate concretions are common as spheres and lenses, irregular nodules, and cylinders (possible plant or algal remains). In some places, limey clasts in the red arkose layers appear to have ben mud rip-ups.

Upper Section (3 in Fig. 3). Compared to the lower, the rocks of the upper section are generally coarser-grained, and cobble conglomerates are locally present (6 in Fig. 3). Lenticular conglomerates at several localities have scoured bases and are probably channel deposits (an outstanding example is in lower Caño Tigre). The predominant lithology is dark-red to dark-brown, coarse-grained arkose, bedded on a scale of centimeters to several meters. Trough and tabular(?) cross-bedding with maximum heights of approximately 15 cm were rarely observed. Seventeen measurements were taken, and they indicate that the direction of transport was predominantly northerly (i.e., south to north), but varied from easterly to westerly.

Fine-grained, felsic ash layers (7 in Fig. 3) of centimeter to meter thicknesses are interbedded with the medium- to coarse-grained arkoses at irregular intervals ranging from centimeters to hundreds of meters. At least eight different ash layers varying from 4 cm to 1.5 m thick crop out, irregularly spaced within 200 to 250 m of vertical section. These ash layers are usually reddish-tan to pink, uniformly layered, and conformable with the other sedimentary rocks. Accretionary lapilli and graded bedding characterize some of the felsic ash layers.

The top several hundred meters of the upper section consist of red arkose, with interbedded tuff and minor pebble or cobble conglomerate as well as some layers of light-green sandstone rich in plant remains (8 in Fig. 3). Near the top of the known section several basaltic-andesite flows 1 to 15 m thick are interbedded with red beds (8 in Fig. 3) (confirmed by E. Viteri, personal communication). These flows commonly have vesicular flow tops which are mineralized with native copper in some outcrops. Overlying (or interfingering with?) the mafic flows are felsic to intermediate ash layers, welded tuffs, and flows (confirmed by E. Viteri, personal communication; 8 in Fig. 3). In most places the volcanic rocks are interbedded with sedimentary rocks composed mostly of medium-grained arkose, with some pebble to cobble conglomerate, both matrix and clast-supported. Variations in clast composition from predominantly volcanic to predominantly granitic/metamorphic emphasize the variation in sediment sources (possibly resulting from active tectonism contemporaneous with deposition).

Farther south, the uppermost La Quinta Formation section is cobble conglomerate which contains rounded volcanic rocks of andesitic(?) composition. It is called the Seco Conglomerate member (9 in Fig. 3) (*Léxico Estratigráfico*, 1970) and crops out along Río Cuibas as well as Río Macoíta, where it constitutes the upper several hundred(?) meters of the La Quinta section (González Padilla, 1973).

Igneous Rocks in the Study Area

The volcanic-plutonic rocks of the La Ge–Totumo area (Fig. 1) have been described by Hea and Whitman (1960), Miller (1962), Bowen (1972), Moticska (1973), Moya and Viteri (1977), and Viteri (1978). A wide variety of rock types is present: granite, granodiorite, rhyolite, rhyodacite, dacite, latite, andesite, basaltic-andesite, and basalt. Moticska (personal communication *in* Viteri, 1978, p. 59) reported that north of Río Palmar (outside

the study area) predominantly basic to intermediate rocks crop out, whereas in the Totumo region felsic rocks are anomalously common. Viteri (1978) estimated that in the La Ge–Totumo region (Fig. 1) 44% of the volcanic rocks are basalts and basaltic-andesites, 38% are rhyodacites and dacites, and 18% are rhyolites. The mafic rocks commonly occur as flows, whereas 85% of the rhyolites and rhyodacites occur as tuffs and ashes (Viteri, 1978). The igneous rocks fall into four broad categories: (1) felsic to intermediate plutonic rocks, (2) felsic to intermediate volcaniclastic rocks, (3) intermediate to mafic flow-rocks, and (4) mafic to felsic dikes.

The plutonic rocks are granites and granodiorites of two distinct ages: (1) Jurassic Palmar granite (10 in Fig. 3) and (2) mid-Paleozoic Lajas and upper Q. La Ge granites (11 in Fig. 3; Espejo and others, 1978, 1980; Dasch, 1982; see Fig. 1 and Table 2). The name "Palmar granite" is used to describe granite outcrops along Río Palmar and outcrops of granodiorite in Caño Tacón and along Riecito del Palmar (dated as 167 ± 3 m.y., Dasch, 1982; see Table 1 and discussion in Ages section). In Caño Tacón the granodiorite intruded sedimentary rocks. Although the age of the sedimentary rocks is unknown, it is possible that they are basal units of the La Quinta Formation.

The volcaniclastic rocks occur as tuff (ash-flow tuff, welded tuff, and tuff-breccia) and water-laid(?) volcanic debris. Typical ash beds are reddish to pink and 1 cm to 1 m thick, whereas the tuff-breccias have a wide range of colors and are 1 m to 100+ m thick. Many of the volcaniclastic rocks are high-silica rhyolites, but they range to intermediate compositions.

A sequence of tuff (perhaps as much as 100 m) exists between Q. La Quebrada and C. Caña Brava (Fig. 3). The tuff units are interbedded with red bed and conglomerate layers up to tens of meters thick. The conglomerate clasts are predominantly volcanic material, apparently locally derived. Silicified tree trunks at least 14 by 30 cm in cross section have been found in some of the tuffs north of Q. La Quebrada. Thicknesses of the tuff units and sediment intervals between them vary, and probably were controlled by local topography and proximity to the volcanic center. Even though volcanic rock exposures are absent to the south, numerous volcanic clasts exist in the sedimentary rocks, indicating that there may once have been volcanic activity in that area.

Two distinct types of flows crop out in the field area: (1) basaltic andesite and (2) hornblende andesite. In general, the basaltic-andesite flows have no amphibole, are thinner, less extensive, and have vesicular tops. The hornblende-andesite flows are distinguished by their high percentage (5–10%) of modal amphibole, greater areal extent and thickness of individual flows, and lack of vesicular tops.

Field observations suggest that the entire La Quinta section is cut by dikes, although poor exposures prevent verification of this relationship for the uppermost sedimentary rocks. Viteri (1978) recognized three types of dikes on the basis of field relationships and chemistry: (1) diabase, (2) basaltic andesite, and (3) aplite (granodioritic composition). I have observed these three types, as well as rhyolitic dikes, and found the diabase dikes, which compositionally are alkalic basalts, to be the most abundant.

In some places contact metamorphism affects the country rock for 0.5 m adjacent to a mafic dike several meters wide, and for several meters where irregularly shaped mafic bodies up to 50 m thick occur. The lateral extent of these large mafic dikes is rarely definable, although some lithologies have been traced for 1 km or more. None of the felsic to intermediate dikes observed are as thick or continuous as the mafic dikes, nor do they have such visible metamorphic aureoles.

North of Río Palmar and Riecito del Palmar (Fig. 1), thick sequences (up to several hundred meters) of mafic flows (basalt to andesite; Moticska, 1973; Viteri, 1978) are interbedded with minor amounts of sedimentary rocks (Fig. 3). The thickness and the number of volcanic units decrease to the south (Q. La Ge, Q. La Quebrada, Fig. 3) where they are interbedded with red beds, siltstone, and minor conglomerate. Many of the volcanics found in Q. La Quebrada and Q. La Ge are dacitic to rhyolitic ash layers up to several meters thick or volcaniclastic beds with clast sizes of a few centimeters in diameter. Rhyolitic ashbeds and alkalic dikes crop out along the upper reaches of Río Cogollo. No dikes of any kind were observed to the south of Río Seco.

What is interpreted as a felsic-volcanic neck crops out just south of Río Totumo (12 in Fig. 3). North and south of this neck, a linear arrangement of more felsic eruptive centers(?) parallel to the mountain front suggests that the extent and distribution of felsic volcanism was controlled by faulting that produced the northerly trending graben (Fig. 2).

The concentration of the volcanic rocks in the northern part of the area makes direct comparison of the stratigraphic sections difficult, as the accumulation time for the sedimentary rocks versus the volcanic rocks may have been substantially different. In addition, proximity to a volcanic center meant that the type of detritus accumulating was also different from north to south. However, 17 paleocurrent measurements taken in the La Quinta Formation and 13 in the Río Negro Formation indicate that the overall current flow was from south to north during the deposition of the La Quinta and reversed during Río Negro Formation time (Cretaceous). This current pattern suggests that the development of a volcanic pile in the north did not significantly affect the northward paleoslope, at least on the scale of the field area (Fig. 1).

Geochemistry

Whole-rock X-ray fluorescence (XRF) analyses for major and some trace elements in more than 50 rocks have been made. Sr isotope analyses were made on 15 of the rocks (see Table 2). An AFM diagram with this data (Fig. 4) shows that all the rocks fall within the calc-alkaline field as defined by Irvine and Baragar (1971). The dike rocks, however, can be further classified as alkalic on the basis of CIPW normative values and Na_2O+K_2O versus SiO_2 plots (Irvine and Baragar, 1971). Further plots of geochemical data (Maze, 1983, 1984) point to two fractionation

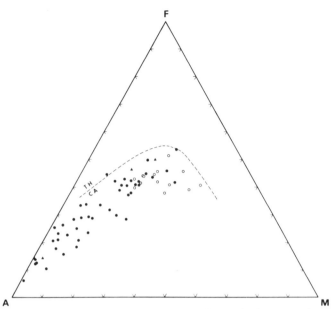

Figure 4. AFM diagram for Venezuelan and Colombian igneous rocks of Jurassic age. Open circles represent dikes, and filled circles represent all other igneous rocks from the Venezuelan field area (Fig. 1). Filled triangles represent rocks from the Colombian side of the Sierra de Perijá near Manaure (Fig. 1). A = wt% Na_2O+K_2O, M=wt% MgO, and F=wt% FeO^*+MnO. Total iron is reported as FeO^*. The dashed line separates the field of tholeiitic rocks (TH) from calc-alkaline rocks (CA), following Irvine and Baragar (1971).

trends: (1) a standard calc-alkaline trend of silica enrichment for the volcanic rocks and (2) apparent control of dike rock composition by pyroxene fractionation.

High Sr isotope initial ratios (>0.706 for all rocks, see Table 2) suggest that anatexis of the lower crust and/or contamination of melts by continental crust have been a significant factor in the petrogenesis of the Perijá igneous rocks.

TECTONIC SETTING OF THE PROTO–MARACAIBO BASIN (DEFINED IN THE INTRODUCTION)

Bartok and others (1981) indicated that grabens filled with red beds extended from northern Venezuela to the Columbia-Ecuador border. By contrast, the regions to the west and east of the Jurassic grabens apparently were not affected by the same extensional event. The Santa Marta Massif to the west was a marine platform with spilites and graywackes, bounded by a Jurassic graben to the southeast and a Jurassic metamorphic belt to the northwest (Tschanz and others, 1974).

A major fault system (of which the Perijá-Tigre Fault is a part) trending approximately N35°E (roughly parallel to the Perijá Range) was active during the Mesozoic and Cenozoic. It occurs in the Perijá and has been detected seismically in the subsurface in the Maracaibo Basin to the east (Miller, 1962, p. 1579; Kellogg, 1981). If the initial topography was of low relief, vertical displacement during Jurassic time along the N35°E faults in the Totumo area (i.e., the area that extends from C. Caña Brava to the north of Riecito del Palmar, Fig. 1) could have amounted to at least the thickness of the La Quinta Formation (1,500 m or more) based on the absence of La Quinta Formation on the upthrown sides.

Normal displacements occurred along these faults during the Jurassic, as evidenced by the grabens in the proto–Maracaibo Basin and to the east (Fig. 2) (Pumpin, 1979; Bartok and others, 1981; Feo-Codecido and others, 1984). The exact number of grabens is not well established, but there were at least four large grabens. The present Sierra de Perijá and Mérida Andes (Fig. 2) each occupy the site of large grabens. The Perijá and Andean grabens existed as depositional troughs from at least as early as middle Jurassic, and perhaps as early as latest Paleozoic, until the Early Cretaceous when marine sedimentation covered the entire region (based on isopach maps by C. V. Campbell, 1961). In the Perijá the Jurassic graben is called the Perijá-Machiques Trough. The Andean graben is actually two grabens, the Uribante Trough in the south and the San Lazaro Graben (Lara Trough) in the north, separated by the Mérida swell (Fig. 2) (Bartok and others, 1981; Stainforth, 1969; Campbell, 1961).

Another graben, smaller than the others, is the Central Lake Graben beneath Lake Maracaibo (Fig. 2) described by Bartok and others (1981) and Pumpin (1979). It runs the length of the west side of Lake Maracaibo in subsurface and can be seen in seismic profiles run across the lake (Pumpin, 1979; Kellogg, 1981, 1984). The boundary faults of the graben do not affect any of the Cretaceous deposits. A positive magnetic anomaly interpreted as indicating the presence of the Central Lake Graben beneath Lake Maracaibo, and extending to the south-southwest, can be seen on a regional magnetic map being compiled by W. E. Bonini of Princeton University.

The Espino Graben has been described as Jurassic age by Feo-Codecido and others (1984) on the basis of magnetics, gravity, coring of red beds and basalt, and a K-Ar whole-rock age on the basalt of 162 m.y. Its relationship to the grabens within the proto-Maracaibo Basin is not known.

The above discussion treats the tectonic setting as one of simple tension, with a series of subparallel grabens opening during the Jurassic. Paleomagnetic data (Maze and Hargraves, 1984) discussed in a later section, however, suggest that the regional tectonic setting may be far more complex.

ENVIRONMENT OF DEPOSITION IN THE STUDY AREA

The graben described here (between Riecito del Palmar and Río Cuibas, Fig. 1) was probably only one of many smaller grabens within the major grabens previously discussed. In the graben studied (Fig. 1) many of the sedimentary rocks contain locally derived volcanic clasts (apparently from earlier beds within the La Quinta Formation) as well as granitic and metamorphic clasts (phyllites, schists). The inclusion of intrabasinal

clasts from the same formation strongly suggests that relatively rapid uplift of the edges of each graben must have occurred, resulting in erosion and redeposition of the early La Quinta Formation rocks. The older metamorphic and plutonic clasts are most likely from the nearby Paleozoic crystalline rocks exposed at the edges of the graben. On the basis of zircon U-Pb ages (Dasch, 1982), there appear to be two distinct ages of granite. The Lajas granite (Fig. 1) is of probable Devonian age (>300 m.y.; Table 2), whereas the Palmar granite has an age of 167 ± 3 m.y. Thus it is more likely that it is only the older Lajas-type granites that are found as clasts in the La Quinta Formation. Variations in petrography make it difficult to identify the parent granite on the basis of clasts.

The paleogeography is envisioned as a narrow graben with central, northward-flowing drainage. Numerous small tributaries transported sediment from the graben walls into the central river. The graben floor was probably alternately dry and wet—either seasonally or because of episodic flooding. Evidence for this comes from the sedimentary facies changes (e.g., conglomerate to mudstone) as well as the presence of Estherida. These small creatures most likely lived in shallow lakes and transient puddles resulting from storms or flooding (Moore and others, 1952). Based on the fossils found, trees and other large flora were evidently quite common, providing habitat for some large animals.

Volcanism, both explosive pyroclastic and flow, caused many rapid changes in the local depositional environment and contributed flow material, ash, and detritus to the La Quinta Formation. Vegetation was buried locally by thick deposits and widespread blankets of ash. Preservation of the thin ash beds suggests that they were deposited in shallow water. Thin limestone beds with Estherida and horizons of limey concretions are indicative of a freshwater lake environment. Most variation in sediment lithology is best explained in terms of meandering streams and diversity of provenance. Faults along the margins of the small graben seem to have been the principal sites of felsic volcanism, particularly in the Totumo area, although minor eruptive centers and dikes are scattered throughout. Mafic volcanism is not clearly related to any fault system.

Accumulation of La Quinta Formation detritus ended with uplift of the Totumo area at the end of Jurassic time. Sediment transport direction (as indicated by measurements of crossbedding) reversed during deposition of the Río Negro Formation, and more sediment came from a source rich in quartz (quartzite and granite) than from Jurassic rocks along the margins of the graben.

POST–LA QUINTA DEPOSITION AND TECTONICS IN THE STUDY AREA

Extension, subsidence, and volcanism associated with deposition, deformation, and redeposition, especially along graben margins, continued throughout "La Quinta time." This was followed by Early Cretaceous subsidence, marine transgression, and sedimentation extending from Peru to eastern Venezuela (for summary, see González de Juana and others, 1980).

Throughout the study area these changes are recorded by white to gray Cretaceous sandstone and conglomerate of the Río Negro Formation or the calcareous basal conglomerate of the Cogollo Group, which overlies the La Quinta Formation. In most cases the uppermost La Quinta sediments and/or the Cretaceous sediments are more quartz-rich than the typical red beds of the La Quinta Formation. The Río Negro/Cogollo Group contact with the La Quinta Formation commonly appears to be unconformable, except along Río Macoíta (Fig. 1) and in a tributary of Q. La Ge; north of this river there is angular unconformity of as much as 30° (e.g., between Q. La Ge and C. La Luna).

Thicknesses ranging from 1,000 to 2,500 m of Cretaceous sediments and 3,000 to 4,500 m of Cenozoic sediments covered the Jurassic section (González de Juana and others, 1980). No volcanic activity is known to have occurred after the La Quinta deposition, although at least three periods of tectonic activity are suggested by unconformities in the Cretaceous/Tertiary succession within the field area (Miller, 1962; Bowen, 1972; Kellogg, 1981, 1984; Maze, 1983). The three periods are (1) post-Paleocene (Kellogg, 1981, 1984, recognizes both an early and a middle Eocene event); (2) post-Eocene; and (3) post-Miocene. All of these events may have involved rejuvenation of the same fault system active in the Jurassic. The post-Miocene event was compressional and resulted in thrusting, block-uplift, and formation of the frontal monocline of the present Sierra de Perijá.

Both left-lateral strike-slip and reverse sense displacements are postulated to have occurred along the N35°E fault system in the basin during Cenozoic time (Rod, 1956). The Perijá-Tigre Fault (Fig. 1), which is part of the N35°E system, has been interpreted by Kellogg (1981) as a low-angle thrust fault. The presence of this fault and numerous fold and fault structures in the subsurface of the Maracaibo Basin (Miller, 1962; Kellogg, 1981) indicates that compression was the dominant force in the proto--Maracaibo Basin during Cenozoic time.

A second series of faults oriented northwest-southeast, apparently conjugate to the N35°E trend (Fig. 1), cuts the frontal structure of the sierra in an en echelon pattern. The faults have apparent normal movement (<1,000 m?) and have thus created a sequence of blocks with uplifted southwestern edges within the frontal structure. The timing of these faults is not certain. They may have been active in Mesozoic time, but no reliable evidence has been found.

AGE RELATIONSHIPS

Within the field area (Fig. 1) two ages of plutonic activity—Devonian and Jurassic—have been identified (Espejo and others, 1978, 1980; Dasch, 1982), whereas the two known ages of volcanic activity are both Jurassic. Felsic tuffs within the pre–La Quinta Tinacoa Formation have been assigned an Early Jurassic age on the basis of fossils (Odreman R. and Benedetto, 1977; Benedetto and Odreman R. 1977), and the La Quinta volcanic

rocks are Middle to Late Jurassic based on paleontology (Odreman R. and Benedetto, 1977) and U-Pb zircon ages (Dasch and Banks, 1981; Dasch, 1982).

SOURCES OF AGE DATA

Seven independent sources of ages exist for the rocks in the Sierra de Perijá (Table 1): (1) U-Pb zircon ages measured by Dasch (1982; see also Dasch and Banks, 1981) in an independent study coordinated with this study, (2) Rb-Sr and K-Ar ages (Etchart and Cordani *in* Espejo and others, 1978), (3) fission-track (FT) ages from apatite, zircon, and sphene (Shagam and others, 1984), (4) a compilation of other, poorly documented radiometric ages—mostly by Shell, S. A.—by Martin B. (1968), (5) K-Ar ages by G. Steinitz on samples supplied by W. Maze and R. Shagam (Steinitz and Maze, 1984), (6) paleontological evidence (Odreman R. and Benedetto, 1977; Benedetto and Odreman R., 1977), and (7) Sr isotope analyses and paleontologic identifications made for this study.

U-Pb Zircon Ages

Samples from four outcrops of granitic rock, two outcrops of volcanic rock, and one outcrop of granitic gneiss were collected by P. O. Banks of Case Western Reserve University and me. Three of the granitic outcrops—Riecito del Palmar (BV-79-7), Río Palmar (BV-79-8), and Caño Tacón (BV-79-9) (Table 1)—are considered by us to represent a single pluton (Palmar granite). Zircon separates from the three sites include some inherited detrital zircon grains, but together define a discordia with a lower intercept of 167 ± 3 m.y. (Dasch, 1982). The upper intercept of 1,400 m.y. has been interpreted by Dasch as "an 'average' age for the inherited Pb component."

The Lajas Granite (BV-79-12) also has a "detrital" component and gives ages of 310 to 385 m.y. for crystallization and 1,050 m.y. for the inherited detrital component (Dasch, 1982).

One volcanic rock sample (PE-5, Table 1; no. 3, Table 2) is from a flow of hornblende andesite in Río Totumo. This gives "nearly concordant results of 163 ± 5 m.y." (Dasch, 1982). No detrital component was reported.

The second volcanic rock (BV-79-5) is a pink, rhyolitic welded tuff-breccia from the upper portion of the La Quinta Formation in a tributary of Q. La Ge (Caño Avispa). U-Pb analysis did not yield an unequivocal age, but reasonable interpretation of the data suggested an age in the range 140 to 160 m.y., and as with the granitic samples, a minor detrital component was reported (Dasch, 1982).

The sample of gneiss is from an outcrop of the Perijá Series located northeast of Riecito del Palmar (outside the study area). This series was originally assigned ages ranging from Devonian to Precambrian (Shagam, 1975). It, however, gave results indicating metamorphism "no earlier than late Devonian but prior to the end of the Paleozoic era" and the presence of a Precambrian detrital component (Dasch, 1982).

TABLE 1. AGES KNOWN FOR ROCKS IN THE SIERRA DE PERIJA*

	Dasch (1982)† Zircon U-Pb	Espejo et al. (1978)	Martin B.§ (1968)	Kohn et al.† (1983) Zircon/Apatite Fission-track		Odreman and Benedetto (1977) Fossils	this paper
				Z	A		
Granites: BV79- Tacón 9 Palmar 8 Riecito 7	167±3	K-Ar Palmar granite bio 216±6 chlor 158±8 chlor 115±6 feld 269±8	Caño Riecito bio 200±40(?) orth 800(?) Mara-13 bio 210 (K-Ar) Caño Emboscado bio 302 42(?)	70±8 98±11 72±9	21.8±2.3 8.1±1 24.9±3 22.1±2.7	. .	. .
BV79-5 La Quinta volcanics PE-5	140-160 163±5	Diorite dike in Palmar granite K-Ar amph 168±5	. .	119±19 114±18 87±9	2.6±0.4	mid- to Late Jurassic	Rb-Sr whole-rock** 156-174 K-Ar whole-rock†† 155±5, 146±7
La Quinta sediments	. .	. .	. .	. .		mid- to Late Jurassic	Fossils Jurassic to Early Cret.§§
Lajas Granite BV79-12	U-Pb 310-385 Pb/Pb 1050	Rb-Sr whole rock 380±25 334±12	orthoclase 370±20(?) 320±20(?)	80±10	2.8±0.6	. .	
Perijá gneiss BV79-10	metamorphism between Late Devonian and Late Permian	. .	Campo Mara muscovite 304±80(?)	126±13	25.1±2.5 24.9±2.2	. .	. .

*Ages in millions of years; (?) indicates the method of dating is unknown; bio = biotite, chlor = chlorite, feld = feldspar, amph = amphibole, orth = orthoclase.

†Sample numbers refer to samples used for U-Pb and Fission-Track dating.

§Lack of almost all field and analytical data makes this information of dubious value, but it is included for completeness.

**Analyzed by M. D. Feigenson and M. So Chyi, Department of Terrestrial Magnetism, Carnegie Institution of Washington, and Princeton University. M.D.F. present address: Department of Geology, Rutgers University, New Brunswick, NJ.

††See Steinitz and Maze (1984). Samples WM-80-3 and SV-73-21 respectively.

§§Paul Tasch, University of Kansas, Wichita; Erling Dorf, Princeton University; Jack Horner, Museum of the Rockies, Bozeman, Montana, formerly of Princeton University.

Rb-Sr Isotope Ages

Whole-rock Sr isotopes determined by M. D. Feigenson and M. So Chyi for 15 volcanic rocks from the Perijá are presented in Table 2. A least-squares fit to the data for the 13 volcanic rocks plus the Caño Tacón granite gives an isochron (R = 0.967) of 174 m.y. The isochron (R = 0.97) determined for the 6 felsic volcanic rocks alone (first 6 samples in Table 2) is 156 m.y. Both of these isochrons are significant at the <1% confidence level. The two ages are in agreement with those determined by Dasch (1982).

Espejo and others (1978) have published Rb-Sr ages and Sr isotope values for the Paleozoic Lajas Granite (see Tables 1 and 2). The U-Pb zircon age of 310 to 385 m.y. measured by Dasch

(1982) is in agreement with the ages of 380 ± 25 m.y. and 334 ± 12 m.y. that Espejo and others (1978) reported on different samples from the same area. I consider sample 278 from the granite at the head of Q. La Ge (Table 2) to be a Paleozoic granite similar to the Lajas Granite (see Figs. 1 and 3) on the basis of location, lithology, and geochemistry.

K-Ar Ages

Martin B. (1968) and Espejo and others (1978) (see Table 1) reported K-Ar ages from a number of rocks in the La Ge–Totumo area which, in general, are in agreement with the U-Pb ages of Dasch (1982). No attempt is made here to reconcile differen-

TABLE 2. Rb-Sr WHOLE-ROCK DATA

Sample number	Location	Rb (ppm)*	Sr (ppm)*	$^{87}Rb/^{86}Sr$[†]	$^{87}Sr/^{86}Sr$[§]	87/86 initial**
White rhyolite tuff breccias						
35	Quebrada La Ge	160 (169)	60 (65)	7.51	0.724168 ±68	0.707
232	Quebrada La Ge	160	150	3.08	0.71712 ±129	0.710
298	Río Seco	145[††](134)	25[††](46)	8.42	0.728574 ±83	0.709
Red ash/glass						
44	Río Totumo volcanic plug	160	360	1.28	0.713625 ±52	0.711
249A	Quebrada La Ge ash	60	440	0.39	0.709012 ±82	0.708
318B	Río Cogollo ash	20[††]	300[††]	0.19	0.70834 ±64	0.708
Hornblende-Andesites						
3	Río Totumo	120[††]	423[††]	0.82	0.70762 ±64	0.706
216	Cañada Caña Brava	80	590	0.39	0.708208 ±233	0.707
233B	Quebrada La Ge	50	760	0.19	0.70624 ±127	0.706
354	Río Cogollo	20	45	1.28	0.706895 ±50	0.704
Alkalic dikes						
302	Río Seco	40 (33)	1160 (1290)	0.074	0.708203 ±157	0.708
304B	Río Seco	80 (31)	600 (171)	0.524	0.7067 ±200	0.706
309	Río Seco	60 (68)	620 (581)	0.338	0.712 ±.005	0.711
Granites						
278	Head of Q. La Ge	220	110	5.78	0.73724 ±81	0.708
BV79-9	Caño Tacón	40	500	0.23	0.70592 ±85	0.705
Espejo and others (1978)	Lajas Granite	294	81.6	10.41	0.767	0.715
		314	39.5	22.97	0.820	0.706

*XRF analyses (10 ppm detection limit for first 15). Numbers in parentheses done by isotope dilution (ID) (analyst M. So Chyi); error is 1%. ID values used when available. Discrepancy between XRF and ID values is apparently due to sample heterogeneity.

[†]2.89(Rb/Sr).

[§]6 analyses by M. D. Feigenson and 9 by M. So Chyi, Department of Terrestrial Magnetism, Carnegie Institution of Washington, and Princeton University. M.D.F. present address: Rutgers University, Department of Geology, New Brunswick, NJ. ± values refer to 4th, 5th, and 6th decimal places unless a decimal point is used.

**Rounded values, assuming 160-m.y. age for all, except 350 m.y. for 278 and two by Espejo and others (1978).

[††]Average of 2 or 3 splits analyzed separately.

ces. In most cases the exact sample locality is unknown, and for some even the analytical technique is unknown.

G. Steinitz (Geological Survey of Israel, Jerusalem) made two whole-rock K-Ar analyses (Table 1; Steinitz and Maze, 1984) of hornblende andesites: (1) sample no. 3, the same as PE-5 analyzed for U-Pb (163 ± 5 m.y., Dasch, 1982), and (2) a sample collected by R. Shagam in Q. La Ge that is equivalent to no. 233B in Table 1 (labelled SV-73-21 in Steinitz and Maze, 1984). On the basis of lithology and field relations, the two samples were inferred to be from roughly the same stratigraphic level, the level of the transition zone in Q. La Ge.

Fission-Track Ages

A complete discussion of fission-track (FT) data from

Perijá rocks is presented by Shagam and others (1984). The ages determined for rocks in the Perijá have been included in Table 1. Sample BV79-5, a tuff close to the top(?) of the volcanic pile, yielded zircon FT ages of 119 ± 19 and 114 ± 14 m.y. that have been interpreted as original crystallization ages (Shagam and others, 1984). The remaining ages in Table 1 could not be simply interpreted. On the other hand, the apatite FT ages correlate well with the Cenozoic tectonic events discussed earlier. At the very least, the FT ages provide control for the minimum age of appreciable thermal disturbance in the field area.

Paleontology

Paleontologic evidence of Jurassic age in the Perijá has been reported mostly by Odreman R. and Benedetto (1977) and

Figure 5. Present location of Late Jurassic red beds—mostly continental—shown in black along the active margin of Latin America. Allochthonous refers to the hachured areas that were accreted or obducted since the end of Jurassic time. Note that areas in central and northern Mexico are of uncertain age, ranging from Triassic to Cretaceous(?). All locations are based on published maps listed in the Bibliography of maps.

Benedetto and Odreman R. (1977). Their work in the Río Palmar area has demonstrated an Early Jurassic age for the Tinacoa Formation previously thought to be Triassic, and Middle to Late Jurassic ages for the overlying Macoíta and La Quinta formations.

Unlike the red-bed sequence of the basal La Quinta Formation in the Mérida Andes, the Macoíta and Tinacoa formations in the Perijá consist of wacke, dark calcareous shale, dark limestone, mudstone, siltstone, pyroclastic debris, and some units that appear to be turbidite sequences (Odreman R. and Benedetto, 1977; Benedetto and Odreman R., 1977). Fish plates (A. G. Fischer, Princeton University, personal communication; Odreman R. and Benedetto, 1977) exist in some of the black shales in the Tinacoa Formation, as well as Estherida, gastropods, and various early Jurassic plant remains. Benedetto and Odreman R. (1977) reported that the Macoíta Formation is entirely continental, whereas the Tinacoa Formation is in part marine. An angular unconformity exists between the two formations in Río Macoíta (Gonzalez Padilla, 1973; Maze, 1983).

During the present study numerous conchostracan fossils (Estherida) were found in the La Quinta Formation of the Sierra de Perijá. Samples of Estherida (identified by Paul Tasch, Wichita State University, Kansas) collected during this study range in age from Late Triassic–Early Jurassic [similar to *Cyzicus*

(Lioestheria) Columbianus (Bock) and *Cyzicus (Euestheria)* sp.] from the lower part of the La Quinta Formation to Late Jurassic–Early Cretaceous [similar to *Pseudoestheria (= Cyzicus (Lioestheria)) pricei* (Cardosa)] from the transition zone. A Late Triassic–Early Jurassic age for Estherida from La Quinta beds suggests that the lower red beds and limestone units in Q. La Ge may be a facies of the Tinacoa and/or Macoíta formations.

The Early Cretaceous age suggested by some of the Estherida is consistent with two other fossils found in Quebrada La Ge. (1) Although poorly preserved, plant fossils collected from a light green sandstone layer may be Cretaceous flora (E. Dorf, Princeton University, personal communication). (2) Vertebrate fossils found in Q. La Ge were tentatively identified by J. Horner (Museum of the Rockies, Bozeman, Montana, formerly of Princeton University) as a mandible of a crocodilian, possibly of Early Cretaceous age.

In the Mérida Andes, the age of the base of the La Quinta Formation is constrained by a U-Pb age on the underlying volcanic tuff of 229 ± 15 m.y. (Burkley, 1976). Plant fossils collected there by Benedetto and Odreman R. (1977) indicate an Early Jurassic age for the lower part of the La Quinta Formation, making it equivalent to the Tinacoa Formation in the Perijá. Evidently Estherida (conchostracan) similar to those from the transition zone in the Sierra de Perijá have been found in the upper La Quinta Formation in the Venezuelan Andes, providing evidence for a Cretaceous age for the top of the Andean La Quinta Formation as well (Padron, cited in Tarache, 1980).

OCCURRENCE OF JURASSIC RED BEDS OF LA QUINTA TYPE, AGE, AND TECTONIC SETTING, FROM MEXICO TO CHILE

An explanation for the La Quinta Formation environment should be compatible with that for rocks of similar type, age, and tectonic setting in the South American and adjacent plates. To this end a paleogeographic map of South America, inspired by maps of Weeks (1947) and Harrington (1962), was constructed using the recent literature and geologic maps (Fig. 5; Table 3), showing the distribution of Late Jurassic continental red beds in Latin America, from Mexico south to Chile. Volcanic rocks occur with some of these red beds. In most areas in South America outcrops of red beds are in the eastern piedmont of the present main Andean chain (sub-Andean; Geyer, 1980) where they originally accumulated in a series of elongate basins and grabens to the east of a volcanic arc and bordering the western margin of the Precambrian shield. Marine sedimentation alternated with and/or followed the continental facies. Table 3 provides a summary of the red beds and references used.

Venezuela and Colombia

As discussed earlier, the La Quinta Formation is known to crop out in much of the Venezuelan Andes and the Sierra de Perijá (*Léxico Estratigráfico,* 1970). In addition, numerous oil

TABLE 3. RED-BED FORMATIONS OF LATE JURASSIC AGE IN LATIN AMERICA

Country	Formation name	Reference*
Mexico	Huizachal Fm (N) Todos Santos Fm (SE) Nazas Fm	López Ramos (1981) Gose and others (1982) Carta Geológica (1976) Blair (1981)
Guatemala	Todos Santos Fm	Nagle and others (1977) Burkart and others (1973) Mapa Geológico (1970)
El Salvador	Todos Santos (Metapán)	Nagle and others (1977) Weyl (1980) Mapa Geológico (1974, 1978)
Honduras	Todos Santos (Metapán, Tegucigalpa?)	Nagle and others (1977) Burkart and others (1973) Weyl (1980) Mapa Geológico (1974) Finch (1981)
Nicaragua	Todos Santos or Metapán	Nagle and others (1977) Weyl (1980) Mapa Geológico Preliminar (1974)
Venezuela	La Quinta Fm Rancho Grande mbr of Cojoro Group	Bellizzia and others (1976) Geyer (1980) Léxico Estr. (1970)
Colombia	Girón Fm in the Perijá, Santander, and Guajira	Cediel (1969) Mapa Geológico (1976)
Ecuador	Chapiza Fm with volcanic Misahualli mbr	Mapa Geológico (1969) Sauer (1971)
Peru	Chapiza Fm Sarayaquillo Fm	Audebaud and others (1973) Mapa Metalogenético (1969) Rivera (1956) Mapa Geológico (1975)
Chile	Tordillo Fm Tordillolitense	Mapa Geológico (1968) Hoffstetter and others (1956)
Argentina	Tordillo Fm	Mapa Geotectónico (1978) Yrigoyen (1972)

*References listed were used to locate red beds on Figure 5. Maps used are
  listed in Bibliography of Maps.
 Metapán not used since Mills and others (1967); Tegucigalpa discarded by
  Carpenter (1954) and in some places Tegucigalpa may be Late Cretaceous or
  Tertiary beds of the Valle de Angeles Formation (Finch, 1981, and written
  communication, 1983).

wells within the Maracaibo Basin—both in the lake and on the coastal plain—have encountered red beds correlated with the La Quinta (Lagoven, personal communication).

As in the type section near La Grita (Fig. 1), discussed earlier, the La Quinta Formation in the rest of the Andes has very little volcanic material (Tarache, 1980; Hargraves and Shagam, 1969; *Léxico Estratigráfico,* 1970). In general, the La Quinta Formation in the Mérida Andes consists primarily of dark red arkoses, lithic arenites, lithic wackes, coarse cobble conglomerates with red to gray arkosic matrix, siltstones, and thin, freshwater limestones (commonly containing Estherida; Tarache, 1980; *Léxico Estratigráfico,* 1970; Houghton, 1977). A relatively rare, dark-green to black facies (fine-grained conglomerate, sandstone, shale, and impure limestone) of the La Quinta Formation occurs near La Mesa (along the San Juan–Jají road) west of Mérida (Fig. 1) (Odreman R. and Ghosh, 1980; Odreman R., personal tour, 1979).

On the Guajira Peninsula (Fig. 3) there are outcrops of Jurassic red beds and volcanic rocks assigned variously to the Rancho Grande member of the Cojoro Group (*Léxico Estratigráfico,* 1970) or to the La Quinta Formation (Macdonald, 1965). The red sediments are interlayered with rhyodacites (Geyer, 1977). Red-bed outcrops similar to the La Quinta of the Perijá are found on Toas Island at the mouth of Lake Maracaibo, associated with basaltic rocks. Red beds on the Paraguaná Peninsula originally reported as Jurassic (Hedberg, 1938a, 1938b, 1942) are now thought to be of Tertiary age (González de Juana and others, 1980).

On the Colombian side of the Perijá the equivalent red beds are assigned to the Girón Formation (type locality near Bucaramanga). In the Santander Massif both the Jordán and overlying Girón Formation are similar to the Venezuelan La Quinta Formation in the Sierra de Perijá, although the thickness of the Girón Formation may exceed 4,600 m (Cediel, 1969).

Although the tectonic setting on the Guajira Peninsula and on Toas Island is not well established, the other red beds in Venezuela and Colombia accumulated in subparallel grabens.

Along the western coast of Colombia and Ecuador, the geologic setting is complicated by the presence of accreted Mesozoic oceanic crust (Goossens and others, 1977; Shepherd and Moberly, 1981; Feininger, 1982, 1983). The Dolores-Guayaquil Megashear (see Shepherd and Moberly, 1981, Fig. 18) is roughly parallel to the western margin of the Jurassic graben depicted by Bartok and others (1981) (Fig. 2, this paper). The terrane to the west of this graben is considered to be allochthonous (Late Cretaceous age) (Goossens and others, 1977; Shepherd and Moberly, 1981). To the east, the geologic setting in Jurassic time is reported to have been a depositional environment in an extensional regime (Bürgl, 1973).

Mexico and Guatemala

There is a nearly continuous zone of continental red beds from northern Mexico south to eastern Guatemala. The age range

of the red beds is somewhat uncertain, and ages assigned vary from Triassic to Cretaceous. In Mexico, these beds are known as the Huizachal Formation in the north and the Todos Santos Formation in the southeast (Viniegra O., 1971; Anderson and others, 1973; Blair, 1981; López Ramos, 1981; Gose and others, 1982). Blair (1981) reported volcanic rocks (porphyritic rhyolite) preceding and interbedded with the Todos Santos Formation in Chiapas, Mexico. The red beds of the Nazas Formation in the state of Durango are assigned a Triassic age on the Carta Geológico (1976), but dated 126 ± 16 m.y. by Halpern and others (1974, cited in Gose and others, 1982).

The type locality of Todos Santos Formation is in Guatemala where the age is more closely constrained to Late Jurassic–Neocomian to Albian (Mapa Geológico, 1970; Anderson and others, 1973; Burkart and others, 1973; Nagle and others, 1977; Weyl, 1980). In Guatemala the Todos Santos is considered by Anderson and others (1973) to have "accumulated as alluvial fan deposits in river valleys more than 1200 m deep." Marine limestone occurs high in the Guatemalan Todos Santos Formation (Anderson and others, 1973). Anderson and others (1973) correlated, at least in part, the Todos Santos of Guatemala with the "continental clastic rocks of the Metapán Series of Sapper (1937) in El Salvador." Deposition in both Mexico and Guatemala was at least partly confined to graben (Weyl, 1980; Blair, 1981; López Ramos, 1981), possibly associated "with rifting and later subsidence of the Gulf of Mexico basin."

El Salvador, Honduras, and Nicaragua

In these three countries the presence of Cenozoic and Mesozoic red beds has resulted in a great confusion in the field and in the literature. Burkart and others (1973) gave a brief review of literature in which the Metapán Formation of El Salvador and the Tegucigalpa Formation of Honduras (see note in Table 3) have been described as possible equivalents of the Todos Santos Formation in Guatemala; however, Burkart (written communication, 1983) has advised me that such correlations are questionable. Further confusion arises from the fact that the Metapán type section is in Guatemala (Mills and others, 1967). Nagle and others (1977) used the name "Metapán" for Jurassic red beds in El Salvador, Honduras, *and* Nicaragua. Finch (1981) used the name "Todos Santos Formation" for a red-bed clastic sequence in Honduras of "Jurassic to Early Cretaceous age, on the basis of its stratigraphic position" and cautioned that the name "Metapán" has not been used by Honduran workers since the publication by Mills and others in 1967 and probably should be dropped (Finch, written communication, 1983; see note in Table 3). Mills and others (1967) described the Honduran Todos Santos as a postorogenic formation up to 2,950 ft thick deposited after folding and block faulting in the Late Jurassic–Neocomian. Weyl (1980) used the name "Todos Santos Formation" for all the above countries and used "Metapán" for another formation. The Mapa Geológico (1978) for El Salvador lists the Todos Santos Formation as Jurassic to Cretaceous—oldest of the red rocks of the "Estratos de Metapán" which range in age from Jurassic to Oligocene(?)—and the Mapa Geológico (1974) of Honduras has the Todos Santos as a Jurassic formation. In Nicaragua the name "Todos Santos" is used for a sequence of limestones, schists, and conglomerates of Late Jurassic–Early Cretaceous age (Mapa Geológico Preliminar, 1974).

Despite the confusion, it is evident that there is a sequence of Late Jurassic–Early Cretaceous red beds in El Salvador, Honduras, and Nicaragua that is best called the Todos Santos Formation. Weyl (1980) summarized the Todos Santos Formation as a "post-orogenic molasse of the Upper Paleozoic (Jaliscoan) tectonic cycle which was deposited in the form of detrital fans in intramontane basins and grabens."

Ecuador and Peru

The red beds of the Chapiza Formation in Ecuador (Hoffstetter, 1956) have been considered to be the facies equivalent of the Girón Formation in Colombia and the Sarayaquillo Formation in eastern Peru (Jenks, 1956). However, the red beds of Peru are also known as the Chapiza Formation (Rivera, 1956). In Peru no volcanic units have been reported, whereas in Ecuador there is a pyroclastic unit known as the Misahualli Member (Sauer, 1971). Deposition of the continental sediments in both Peru and Ecuador occurred in elongate basins parallel to the coast and inland from volcanic activity (e.g., Tschopp, 1953; Ham and Herrera, 1963; Audebaud and others, 1973). In southern Ecuador, Feininger (1980) described the red beds and volcanics of the Chapiza Formation as elements of a "continental volcanic arc." As discussed above in the section on Venezuela and Colombia, the present Pacific margin of Ecuador consists of accreted terrane (Fig. 5), younger than the Chapiza Formation.

Chile and Argentina

The Late Jurassic red beds of Chile and Argentina share the same formation name—"Tordillo" (or "Tordillolitense" in Chile)—and nearly the same geographic location, i.e., along the border between the two countries (Hoffstetter and others, 1956; Yrigoyen, 1972). Volcanic units occur with the red beds in the west and die out to the east. As in Peru and Ecuador, accumulation of these formations took place in narrow basins parallel to the coast and east of the major Jurassic volcanic activity (Hoffstetter and others, 1956; Yrigoyen, 1972). Gust and others (1983) reported bimodal volcanic rocks (Tobifera group equivalents) in narrow grabens formed by Jurassic extension to the east of the Andes between 40°S and Tierra del Fuego. Levi and Aguirre (1981) presented evidence for a "tensional regime" and subsequent subsidence in central Chile during Mesozoic time where "block faulting operated during the development of these basins along regional lineaments."

RIFTING AND THE RED–BED GRABEN ENVIRONMENT

On the basis of the regional geology and tectonic setting for Latin America, it is postulated that the red beds in the above regions accumulated in graben or elongate basins, suggesting an extensional environment, possibly with an oblique-slip component, and inviting consideration of rifting mechanisms. Rifting implies that "the entire thickness of the lithosphere has ruptured under tension" (Burke, 1978). One would expect thin lithosphere to be more easily rifted, but the cause of thin lithosphere and of the tension necessary for rifting may have more than one explanation.

The many treatises written on continental rifts and their associated rocks cover numerous mechanisms in three broad settings: (1) passive margins, (2) midcontinental regions, and (3) active margins. The various mechanisms invoked in the literature have been put forth to explain settings such as those of the late Triassic Newark group, the Oslo Rift, the Rhinegraben, the East African Rift, the North Sea, and the Rio Grande Rift.

1. With reference to passive margins and possibly midcontinental regions, Sleep (1971) and Kinsman (1975) have discussed rifting initiated by doming over a heat source (e.g., mantle plume) with subsequent erosion of the highs and filling of the rifts. McKenzie (1978) proposed simple thermal subsidence of thinned lithosphere to cause tension with no doming. Sawyer and others (1982) and Royden and Keen (1980) described subsidence along the Atlantic margin of the United States using models similar to the model of McKenzie, and got fits close to the observed structure. Likewise, Sclater and Christie (1980) have applied a McKenzie-type thermal subsidence model to the North Sea. Vierbuchen and others (1983) have proposed a thermomechanical model incorporating mechanical heterogeneity to overcome some of the shortcomings of the McKenzie model.

Many of the extensional features along the margins of the Atlantic (passive margins) are classified as aulacogens by Burke (1976). One such feature is postulated for the Newark group that was described by King (1959) as postorogenic detritus filling the isostatically controlled "downfaulted troughs" of the Paleozoic orogeny. Alternatively, the Newark group can be described as filling the graben that formed as the result of rifting and subsidence accompanying opening of the Atlantic. The change in models reflects the increasing use of plate-tectonic processes to explain geologic features, especially along continental margins.

2. Theories to explain midcontinental rifting, as with passive and active margins, have evolved into an immense literature (e.g., Neumann and Ramberg, 1978; Ramberg and Neumann, 1978; Riecker, 1979; Pálmason, 1982). In general the basic concept is that rifting occurs in areas which have high heat flow and a thin lithosphere, but cause and effect are not clearly distinguished. However, once the thin lithosphere exists, the models proposed for passive margins are apparently applicable.

3. Along active margins, extensional basins behind the vol-

canic arc can be floored by oceanic crust (marginal basins) or continental crust (retroarc basins, Dickinson, 1974; Miall, 1981) and filled with continental and/or marine detritus (Tarney and Windley, 1981). Marginal or retroarc basins are apparently active for less than 20 m.y. and are believed in some manner to be related to subduction, which in turn is controlled by the age and type of crust being subducted (e.g., Toksöz and Bird, 1977; Jurdy, 1979; Furlong and others, 1982). Mantle convection above the subducting slab may be responsible for extension of the crust/lithosphere either by heating and thinning (Toksöz and Bird, 1977), mechanical dragging, or a combination of the two. Sudden changes in subduction rates and resultant changes in compression may also be responsible for initiation or termination of backarc spreading (e.g., Furlong and others, 1982).

DISCUSSION OF TECTONIC SETTING

The proto–Maracaibo Basin is difficult to classify as an active or passive margin paleorift (or aulacogen) because of its position on the northwestern "corner" of South America. Depending on the exact time chosen, and the reconstruction used, northwestern Venezuela may have been isolated from the Pacific by Mexico (midcontinental setting; Walper, 1980; Pindell and Dewey, 1982), "exposed" to subduction from the Pacific (Dickinson and Coney, 1980), a passive margin parallel to a spreading center between North and South America (Klitgord and Schouten, 1982; Klitgord and others, 1984), or an allochthonous terrane of unknown regional tectonic setting (for other reconstructions, see Bullard and others, 1965; Van der Voo and others, 1976; Irving, 1977; Buffler and others, 1980; Dickinson and Coney, 1980; Gose and others, 1980; Mooney, 1980; Walper, 1980; Van der Voo, 1981; Pindell and Dewey, 1982; Klitgord and Schouten, 1982; Klitgord and others, 1984). Models invoked for the proto–Maracaibo Basin can thus vary from North Sea–type rift (Sclater and Christie, 1980), to aulacogen (proposed by Burke, 1976), to marginal or retroarc basin, to exotic terrane.

Today the structural trends of the major grabens discussed earlier define a weak fan opening to the north (Fig. 2): Sierra de Perijá and Central Lake oriented N30°–35°E, Mérida Andes oriented N45°–50°E, and Espino trending N50°–60°E. Paleomagnetic data from the La Quinta Formation in the Sierra de Perijá (Maze and Hargraves, 1984) and the Venezuelan Andes (Hargraves and Shagam, 1969), however, suggest a 20° to 120° differential rotation of the Perijá La Quinta with respect to the Andean La Quinta during the Jurassic (see vectors on Fig. 2). The paleomagnetic data from the alkalic dikes in the Sierra de Perijá (Maze and Hargraves, 1984), intruded at the end of La Quinta time, appears to indicate that all rotation had ceased prior to their emplacement.

If it is assumed that the grabens originally formed as parallel structures, the maximum rotation necessary (allowed) to produce the present fan-shaped assembly is 20° to 30°. This rotation does not, however, coincide with that required by the paleomagnetic data (Maze and Hargraves, 1984). Thus it is difficult to reconcile

the paleomagnetic data with simple models of tension and separation to form subparallel grabens. However, the paleomagnetic data may be reconciled with the field evidence by postulating either left-lateral (1 in Fig. 6) or right-lateral (2 in Fig. 6) transcurrent motion along a fault oriented approximately northeast-southwest between the Perijá and Mérida Andes, which moved the Perijá to the south and west from the north side of the Venezuelan Andes or north and east from the south side of Mexico (Fig. 6). Position 1 (Fig. 6) is not considered as likely a possibility because of conflict with relative sense of motion with respect to spreading, space problems in the Gulf of Mexico, and lack of geologic correlation/control. Position 2 is more consistent with the prevailing motion at the time (Klitgord and Schouten, 1982; Klitgord, personal communication) and available paleomagnetic data (Maze and Hargraves, 1984) which suggest northerly motion as well as rotation during the Jurassic.

Criteria for distinguishing between one hypothesis or another include the nature of the igneous rocks, the regional geologic setting, and the paleomagnetic evidence. The calc-alkaline rocks of the La Quinta Formation are not typical of rifts like the Rhinegraben, East African Rift, Newark Basin, or North Sea, which are commonly tholeiitic, yet they are quite similar to those of the midcontinental Rio Grande Rift. They are also similar to the volcanic rocks found at convergent margins such as the present Andean chain in western South America.

The presence of subduction-related rocks of Jurassic age in northwestern Santa Marta (assuming Santa Marta was attached to the Perijá block at the time; Fig. 2; Tschanz and others, 1974) as well as along much of the west coast of South America indicates that northwestern Venezuela was probably "exposed" to subduction in Middle Jurassic time (see also Dickinson and Coney, 1980; Mooney, 1980). The existence of similar Jurassic red beds (some with associated volcanic rocks) in a narrow strip along the west coast from Mexico to Chile (Fig. 5) also suggests an origin related to convergent plate margin processes.

A compilation of paleomagnetic data from Cretaceous and Tertiary rocks in the southern Caribbean (Skerlec and Hargraves, 1980) displays a predominance of east-west directions unlike the data from the Perijá. MacDonald and Opdyke (1972) concluded that the Guajira Peninsula occupied a site to the south of its present position, "possibly off northwestern South America, in Cretaceous times." The data of Maze and Hargraves (1984; see also MacDonald and Opdyke, 1984) suggest that the Perijá graben was rotated in Jurassic time and was unaffected by the later tectonic rotations and movements proposed by MacDonald and Opdyke (1972) and Skerlec and Hargraves (1980). Apparently the allochthonous terranes depicted in Figures 5 and 6 which are accreted to the west coast of Ecuador and Colombia and obducted onto the north coast of Venezuela, "docked" in the Late Cretaceous or early Tertiary, subsequent to events associated with the La Quinta Formation.

In light of the above, it is hypothesized that the La Quinta Formation red-bed–volcanic environment in the Sierra de Perijá was the result of a combination of events. The calc-alkaline igne-

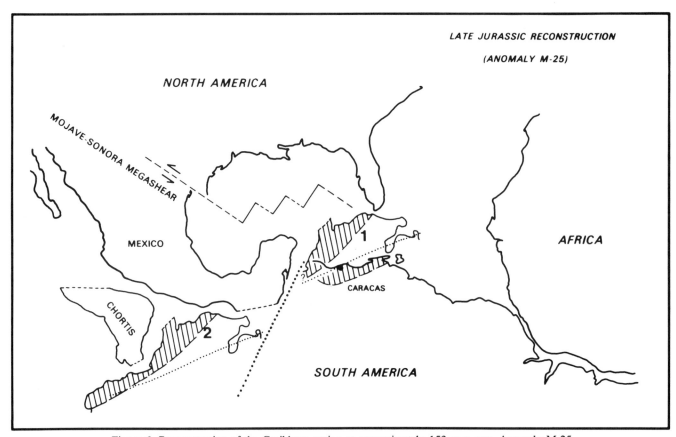

Figure 6. Reconstruction of the Caribbean region at approximately 153 m.y. ago. Anomaly M-25 reconstruction of Klitgord and Schouten (1982) and Klitgord and others (1984) was used for the fit of North America–South America–Africa. The heavy dotted line terminates the northwest coast of South America, starting at the Paraguaná Peninsula in the north and running south to the border between Ecuador and Peru (after Klitgord and Schouten, 1982; Klitgord and others, 1984). Numbers 1 and 2 refer to the two possible locations for the Perijá block using the available paleomagnetic constraints (Maze and Hargraves, 1984). Position 2 is preferred. The light dotted lines correspond to the heavy dotted line and the eastern margin of the Perijá block. The allochthonous terranes of Figure 5 are shown as hachured areas for the ease of orientation, but it is assumed that they were not in existence prior to Late Jurassic–Early Cretaceous time and should not be used as a constraint on the reconstruction.

ous suite suggests a possible association with subduction. Extension and graben formation may have been the result of retroarc tension or tension related to the rifting required to open the Caribbean. Oblique-slip and resultant pull-apart basins may also have been a factor in the formation of La Quinta basins. Paleomagnetic data requires that the terrane be moved in from elsewhere.

CONCLUSIONS

The La Quinta Formation in a small area in the Sierra de Perijá of northwestern Venezuela has been examined in detail and placed in perspective relative to red-bed environments of Late Jurassic age elsewhere in Latin America. At the time of La Quinta deposition there was an extensional regime, resulting in graben formation in what is now northwestern Venezuela. Volcanism was primarily of a calc-alkaline nature. Contamination by Precambrian continental crust is indicated by $^{87}/^{86}$Sr data.

Due to the simultaneous opening of the Caribbean and subduction along the Pacific margin, the setting in Jurassic time was extremely complex; therefore, complete separation of tectonic events may not be possible or appropriate when reconstructing the geologic history of northwestern Venezuela. Paleomagnetic data provide constraints that suggest northwestern South America may not be autochthonous, at least the region west of Lake Maracaibo. Transcurrent movement may have been responsible for both the graben in which the La Quinta Formation was deposited and the confusing paleomagnetic results.

ACKNOWLEDGMENTS

This research is part of my Ph.D. thesis and was supported by the Princeton University Department of Geological and Geophysical Sciences, the National Science Foundation, the Venezuelan Ministerio de Energía y Minas (M.E.M.), and INGEOMINAS in Colombia. I especially thank R. B. Hargraves, A. G.

Fischer, and W. E. Bonini for support, discussion, and criticism, both in the field and at Princeton. Much appreciated encouragement came from R. Shagam of Ben-Gurion University who accompanied me in the field and through countless pages of data, questions, and corrections. B. Kohn of Ben-Gurion University is to be thanked for directing much attention to fission-track ages. I am much obliged to P. O. Banks of Case Western Reserve University for ideas in the field, over the phone, and especially support of L. Dasch's thesis work on Perijá U-Pb ages. G. J. MacPherson, F. J. Spera, S. C. Bergman, V. B. Sisson, K. Kleinspehn, and R. Druhan I thank for their hours spent reading rough drafts and discussing nascent thoughts. M. D. Feigenson and M. So Chyi deserve special mention for doing Sr-isotope analyses. P. Tasch, G. Brenner, E. Dorf, J. Horner, and A. G. Fischer lent an eye to identify fossil remains. T. Anderson and M. Bass critically reviewed and improved my manuscripts. R. C. Finch set me straight on the Mesozoic of Honduras. A. Bellizzia, as director of the Dirección de Geología of M.E.M., made the field work in Venezuela possible. E. Viteri, A. Espejo, V. Campos, N. Pimentel B., C. Figuero, F. Rondon, I. Fierro, O. Odreman, G. Benedetto, R. García Jarpa, R. Carmona, M. Chin-A-Lien, and P. Bustamante were the M.E.M. geologists/paleontologists responsible for logistical support and directions to type localities and key outcrops. G. Canelon provided M.E.M. office space in Maracaibo. H. Etchart and J. N. Kellogg were valuable companions in the office and the field. Estoy muy agradecido a los Sres. Regino de La Villa, Abrahan Carvajal, Adelmo Taborda, y Emiro por sus valiosos servicios en el trabajo de campo.

Bibliography of maps used to locate Jurassic red beds in Latin America

Mexico
Carta Geológica de la República Mexicana, 1976, 1:2,000,000, 4ª edición, compilada por Ing. Ernesto López Ramos.

Guatemala
Mapa Geológico de la República de Guatemala, 1970 primera edición, 1:500,000, 2ª impresión litográfica octubre 1975, Instituto Geográfico Nacional.

El Salvador
Mapa Geológico General de la República de El Salvador, 1974, 1:500,000, Publicado por la Bundesanstalt für Bodenforschung, Hannover, Redacción científica por H. S. Weber, G. Wiesemann, y H. Wittekindt.

Mapa Geológico de la República de El Salvador/América Central, 1978, 1:100,000, 6 sheets in color, Publicado por la Bundesanstalt für Geowissenschaften und Rohstoffe, Hannover. Mapéo Geológico por la Misión Geológico Alemana en El Salvador en colaboración con el Centro Investigaciones Geotécnicas (1967–1971), Redacción científica: H. S. Weber y G. Wiesemann en colaboración con: W. Lorenz y M. Schmidt-Thomé.

Honduras
Mapa Geológico de la República de Honduras, 1974, 1974 primera edición, 1:500,000, compilado por Ing. Reniery Elvir Aceituno, Dirección General de Minas E Hidrocarburos, Ministerio de Recursos Naturales.

Nicaragua
Mapa Geológico Preliminar, 1974, 1:1,000,000, compilado por Ing. M. A. Martinez H., preparado por el Instituto Geográfico Nacional y Servicio Geológico Nacional, República de Nicaragua, Ministerio de Obras Públicas.

Venezuela
Mapa Geológico Estructural de Venezuela, 1976, 1:500,000, compilado e integrado por Bellizzia G., A., Pimentel, M., N., and Bajo, O., R., Ministerio de Minas e Hydrocarburos, Venezuela.

Colombia
Mapa Geológico de Colombia, 1976, 1:500,000, compilación geológica por J. L. Arango Calad, T. Kassen Bustamante, y H. Duque Caro, Ministerio de Minas y Energía.

Ecuador
Mapa Geológico de La República del Ecuador, 1969, 1:1,000,000, Ministerio de Industrias y Comercio, Servicio Nacional de Geología y Minería.

Peru
Mapa Metalogenético, 1969, 1:1,000,000, publicado por Sociedad Nacional de Minería y Petróleo.

Mapa Geológico del Peru, 1975, 1:1,000,000, República del Peru, Ministerio de Energía y Minas, Instituto de Geología y Minería, 4 color sheets and 41-page explanation.

Chile
Mapa Geológico de Chile, 1968, 1:1,000,000, Instituto de Investigaciones Geológicas, director Carlos Ruiz Fuller, Jefe División Geología Regional José Corvalan Diaz.

Argentina
Mapa Geotectónico de la República Argentina, 1978, 1:2,500,000, preparado por el Profesor Doctor Angel V. Borrello, and others, publicado por el Servicio Geológico Nacional.

REFERENCES CITED

Anderson, Thomas H., Burkart, Burke, Clemons, Russell E., Bohnenberger, Otto H., and Blount, Don N., 1973, Geology of the western Altos Cuchumatanes, northwestern Guatemala: Geological Society of America Bulletin, v. 84, p. 805–826.

Audebaud, Etienne, Capdevila, Raymond, Dalmayrac, Bernard, Debelmas, Jacques, Laubacher, Gerard, Lefevre, Christian, Marocco, Rene, Martinez, Claude, Mattauer, Maurice, Megard, Francois, Paredes, Jorge, Tomasi, Pierre, 1973, Les traits geologiques essentiels des Andes Centrales (Perou-Bolivie): Revue de Geographie Physique et de Geologie Dynamique (2), v. XV, fasc. 1-2, p. 73–114.

Bartok, P., Reijers, T.J.A., and Juhasz, I., 1981, Lower Cretaceous Cogollo Group, Maracaibo Basin, Venezuela: Sedimentology, diagenesis, and petrophysics: American Association of Petroleum Geologists Bulletin, p. 1110–1134.

Bellizzia, G., A., Pimentel, M., N., and Bajo, O., R., 1976, Mapa Geológico Estructural de Venezuela, Escala 1:500,000, Ministerio de Minas e Hydrocarburos, Venezuela.

Benedetto, Gianluigi, and Odreman R., Oscar, 1977, Nuevas evidencias paleontológicas en la Formación La Quinta, su edad y correlación con las unidades aflorantes en la Sierra de Perijá y Cordillera Oriental de Colombia: Memoria, V Congreso Geológico Venezolano, tomo 1, Ministerio de Energía y Minas, sociedad Venezolana de Geólogos, Caracas, p. 87–106.

Blair, Terence Cletus, 1981, Alluvial fan deposits of the Todos Santos Formation of Central Chiapas, Mexico [Master's thesis]: University of Texas at Arlington, 134 p.

Bowen, John Myles, 1972, Estratigrafía del Precretáceo en la parte norte de la Sierra de Perijá: Memoria, IV Congreso Geológico Venezolano, tomo II, Ministerio de Minas e Hidrocarburos, Boletín de Geología Publicación Especial no. 5, p. 729–761.

Buffler, Richard T., Watkins, Joel S., Shaub, Jeanne F., and Worzel, J. Lamar, 1980, Structure and early geologic history of the deep central Gulf of Mexico basin, *in* Pilger, Rex H., ed., The origin of the Gulf of Mexico and the early opening of the central North Atlantic Ocean: Baton Rouge, Louisiana State University, p. 3–16.

Bullard, E. C., Everett, J., and Smith, A. G., 1965, The fit of the continents around the Atlantic, *in* Blackett, P.M.S., and Runcorn, S. K., eds., A symposium on continental drift: Royal Society of London Philosophical Transactions, series A, v. 258, p. 47–75.

Bürgl, Hans, 1973, Precambrian to Middle Cretaceous stratigraphy of Colombia: Order from: Charles G. Allen, Consultant Paleontologist, Carrera 10, No. 67-58, Bogutá, Colombia (translated by Charles G. Allen and Norman R. Rowlinson).

Burkart, Burke, Clemons, Russell E., and Crane, David C., 1973, Mesozoic and Cenozoic stratigraphy of southeastern Guatemala: American Association of Petroleum Geologists Bulletin, v. 57, p. 63–73.

Burke, Kevin, 1976, Development of graben associated with the initial ruptures of the Atlantic Ocean: Tectonophysics, v. 36, p. 93–112.

——1978, Evolution of continental rift systems in the light of plate tectonics, *in* Ramberg, I. E., and Neumann, E. -R., eds., Tectonics and geophysics of continental rifts: Proceedings of the NATO Advanced Study Institute Paleo-rift Systems with Emphasis on the Permian Oslo Rift, Oslo, Norway, July 27–August 5, 1977, v. 2, 444 p., p. 1–9.

Burkley, L. A., 1976, Geochronology of the central Venezuelan Andes: [Ph.D. thesis]: Cleveland, Ohio, Case Western Reserve University, 150 p.

Campbell, C. V., 1961, Subsurface maps of the Maracaibo Basin: Jersey Production Research Co., p. 22 (unpublished).

Canelon, G., Etchart, H., and Maze, W., 1979, Mapa geológico de la Hoja Laberinto: Ministerio de Energía y Minas, Venezuela, scale 1:100,000.

Carpenter, R. H., 1954, Geology and ore deposits of the Rosario mining district and the San Juancito Mountains, Honduras, Central America: Geological Society of America Bulletin, v. 65, p. 23–28.

Case, J. E., and Holcombe, T. L., 1980, Geologic-tectonic map of the Caribbean Region: U.S. Geological Survey, Miscellaneous Investigations Series, Map I-1100.

Cediel, Fabio, 1969, Die Girón-Gruppe Eine früh-mesozoische Molasse der Ost-kordillere Kolumbiens: Neues Jahrbuch für Mineralogie und Palaontologie, v. 133, p. 111–162.

Creole Petroleum Corp., 1956–1961, Geologic maps of the Sierra de Perijá, sheets C-2 and D-2, scales 1:100,000 and 1:50,000.

Dasch, Lawrence E., 1982, U-Pb Geochronology of the Sierra de Perijá, Venezuela [Master's thesis]: Cleveland, Ohio, Case Western Reserve University, 164 p.

Dasch, Lawrence E., and Banks, Philip, 1981, Zircon U-Pb ages from the Sierra de Perijá, Venezuela: Geological Society of America 94th Annual Meeting, Abstracts with Programs, p. 436.

Dickinson, William R., 1974, Plate tectonics and sedimentation, *in* Dickinson, W. R., ed., Tectonics and sedimentation: Society of Economic Paleontolo-gists and Mineralogists, Special Publication no. 22, p. 1–27.

Dickinson, William R., and Coney, Peter J., 1980, Plate tectonic constraints on the origin of the Gulf of Mexico, *in* Pilger, Rex H., ed., The origin of the Gulf of Mexico and the early opening of the central North Atlantic Ocean: Baton Rouge, Louisiana State University, p. 27–36.

Espejo C., A., Etchart, H. L., Cordani, U. G., and Kawashita, K., 1978, Geocro-nología de intrusivas ácidas en la Sierra de Perijá Venezuela: II Congreso Geológico Colombiano, 1978, Bogotá, Colombia (tomo II).

——1980, Geocronología de intrusivas ácidas en la Sierra de Perijá, Venezuela: República de Venezuela, Ministerio de Energía y Minas, Boletín de Geo-logía, v. XIV, no. 26, p. 245–254.

Feininger, Tomas, 1980, Eclogite and related high-pressure regional metamorphic rocks from the Andes of Ecuador: Journal of Petrology, v. 21, p. 107–140.

——1982, The metamorphic "basement" of Ecuador: Geological Society of America Bulletin, v. 93, p. 87–92.

——1983, Allochthonous terranes in the Andes of Ecuador and northwestern Peru: Geological Association of Canada, Mineralogical Association of Canada, Canadian Geophysical Union, Victoria 1983, Program with abstracts, p. A22.

Feo-Codecido, Gustavo, Smith, Foster D., Jr., Aboud, Nelson, and DeDiGia-como, Estel, 1984, Basement and Paleozoic rocks of the Venezuelan Llanos Basins, *in* Bonini, W. E., and others, eds., The Caribbean- South American plate boundary and regional tectonics: Geological Society of America Me-moir 162 (this volume).

Finch, R. C., 1981, Mesozoic stratigraphy of central Honduras: American Associ-ation of Petroleum Geologists Bulletin, v. 65, p. 1320–1333.

Furlong, Kevin P., Chapman, David S., and Alfeld, Peter W., 1982, Thermal modeling of the geometry of subduction with implications for the tectonics of the overriding plate: Journal of Geophysical Research, v. 87, no. B3, p. 1786–1802.

Geyer, Otto F., 1977, El Jurásico de las penínsulas de la Guajira (Colombia) y de Paraguaná (Venezuela): 8th Caribbean Geological Conference, Curaçao, 9–24 July, 1977, p. 57–58.

Geyer, Von Otto F., 1980, Die mesozoische Magnafazies-Abfolge in den nordli-chen Andean (Peru, Ekuador, Kolumbien): Geologische Rundschau, v. 69, p. 875–891.

González de Juana, Clemente, Iturralde de Arozena, Juana Ma., Cadillat, Xavier Picard, 1980, Geología de Venezuela y de sus cuencas petrolíferas: Ediciones Foninves, Caracas, tomo I, 407 p.; tomo II, 624 p.

González Padilla, L., 1973, Guia de la excursión no. 2—Sierra de Perijá, Ruta Villa del Rosario–Río Macoíta: II Congreso Latino americano de geología, Caracas, República de Venezuela, Ministerio de Minas e Hidrocarburos, Dirección de Geología, tomo I, p. 312–317.

González Padilla, L., Canelon C., G., and Etchart K., H., 1979(?), Mapa geológico de la Hoja Machiques Río Negro: Ministerio de Energía y Minas, Venezuela, scale 1:100,000.

Goossens, P. J., Rose, W. I., Jr., and Flores, Decio, 1977, Geochemistry of tholeiites of the Basic Igneous Complex of northwestern South America: Geological Society of America Bulletin, v. 88, p. 1711–1720.

Gose, W. A., Scott, G. R., and Swartz, D. K., 1980, The aggregation of Mesoa-merica: Paleomagnetic evidence, *in* Pilger, Rex H., ed., The origin of the Gulf of Mexico and the early opening of the central North Atlantic Ocean: Baton Rouge, Louisiana State University, p. 51–54.

Gose, Wulf A., Belcher, Robert C., and Scott, Gary R., 1982, Paleomagnetic results from northeastern Mexico: Evidence for large Mesozoic rotations: Geology, v. 10, p. 50–54.

Gust, D. A., Biddle, K. T., Phelps, D. W., and Uliana, M. A., 1983, The tectonic setting of Middle Jurassic volcanism in southern South America [abs.]: EOS, Transactios of the American Geophysical Union, v. 64, no. 45, p. 893.

Halpern, M., Guerrero, G. J., and Ruiz, C. M., 1974, Rb-Sr dates of igneous and metamorphic rocks from southeastern and central Mexico: A progress report [abs.]: Reunion Annual Union Geofísica Mexicana, p. 30–32.

Ham, Cornelius K., and Herrera, Leo J., 1963, Role of subandean fault system in tectonics of eastern Peru and Ecuador: American Association of Petroleum Geologists Memoir 2, p.47–61.

Hargraves, R. B., and Shagam, R., 1969, Paleomagnetic study of La Quinta Formation, Venezuela: American Association of Petroleum Geologists Bul-letin, v. 53, p. 537–552.

Harrington, Horacio J., 1962, Paleogeographic development of South America: American Association of Petroleum Geologists Bulletin, v. 46, no. 10, p. 1773–1814.

Hea, J. P., and Whitman, A. B., 1960, Estratigrafía y petrología de los sedimentos precretáceos de la parte norte-central de la Sierra de Perijá, Estado Zulia, Venezuela: Congreso Geológico Venezolano III, Caracas, 1959, Memoria, tomo I, p. 315–376.

Hedberg, H. D., 1938a, Informe del comite de nomenclature estratigráfica al II congreso geológico Venezolana. Apendice II: Boletín Geología y Minería, Caracas, t. 2, no. 2–4, p. 265–269.

——1938b, Report of the committee on stratigraphic nomenclature to the second

Venezuelan geological congress at San Cristobal. Appendix I: Boletín Geología y Minería, Caracas, t. 2, no. 2–4, p. 239–242 (ed. en inglés).

—— 1942, Mesozoic stratigraphy of northern South America: 8th American Scientific Congress Proceedings, Washington, v. 4, p. 195–227.

Hedberg, H. D., and Sass, L. C., 1937, Synopsis of the geologic formations of the western part of the Maracaibo Basin, Venezuela: Boletín Geología y Minería, Caracas, v. 1, no. 2–4, p. 73–112.

Hoffstetter, Robert, 1956, Lexique Stratigraphique International, volume V, Amerique Latine, Fascicule 5a, Ecuador: Centre National de la Recherche Scientifique, Paris, 191 p.

Hoffstetter, Robert, Fuenzalida, Humberto, and Cecioni, Giovanni, 1956, Lexique stratigraphique, international, volume V, Amerique Latine, Fascicule 7, Chile: Centre National de la Recherche Scientifique, Paris, 444 p.

Houghton, Hugh F., 1977, Sedimentology of the Jurassic La Quinta Formation, Sierra de Perijá and Mérida Andes, Venezuela. Phase I: Regional reconnaissance, measurement and description of the type section: Internal report to the Venezuelan Ministerio de Energía y Minas from Princeton University.

Irvine, T. N., and Baragar, W.R.A., 1971, A guide to the chemical classification of the common volcanic rocks: Canadian Journal of Earth Science, v. 8, p. 523–548.

Irving, E. M., 1975, Structural evolution of the northernmost Andes, Colombia: U.S. Geological Survey Professional Paper 846, 47 p.

Jenks, William F., editor, 1956, Handbook of South American geology, an explanation of the Geologic Map of South America: Geological Society of America Memoir 65, 378 p.

Jurdy, D. M., 1979, Relative plate motions and the formation of marginal basins: Journal of Geophysical Research, v. 84, p. 6796–6802.

Kellogg, James Nelson, 1981, The Cenozoic basement tectonics of the Sierra de Perijá, Venezuela and Colombia [Ph.D. thesis]: Princeton, New Jersey, Princeton University.

—— 1984, Cenozoic tectonic history of the Sierra de Perijá and adjacent basins, *in* Bonini, W. E., and others, eds., The Caribbean– South American plate boundary and regional tectonics: Geological Society of America Memoir 162 (this volume).

King, Philip B., 1959, The evolution of North America: Princeton, New Jersey, Princeton University Press, 189 p.

Kinsman, David J. J., 1975, Rift valley basins and sedimentary history of trailing continental margins, *in* Fischer, A. G., and Judson, Sheldon, eds., Petroleum and global tectonics: Princeton, New Jersey, Princeton University Press, 322 p.

Klitgord, K. D., and Schouten, H., 1982, Early Mesozoic Atlantic reconstructions from sea-floor-spreading data [abs.]: EOS, Transactions of the American Geophysical Union, v. 63, no. 18, p. 307.

Klitgord, Kim D., Popenoe, Peter, and Schouten, Hans, 1984, Florida: A Jurassic transform plate boundary: Journal of Geophysical Research (in press).

Kundig, E., 1938, Las rocas pre-cretáceas de los Andes Centrales de Venezuela, con algunas observaciones sobre su tectónica: Boletín Geología y Minería, Venezuela, v. 2, p. 21–43.

Levi, B., and Aguirre, L., 1981, Ensialic spreading-subsidence in the Mesozoic and Paleogene Andes of central Chile: Journal of the Geological Society, v. 138, no. 1, p. 75–81.

Léxico Estratigráfico de Venezuela, 1970, (Segunda Edición): Boletín de Geología, Publicación Especial no. 4, República de Venezuela, Ministerio de Minas E Hidrocarburos, Dirección de Geología, 756 p.

Liddle, R. A., Harris, G. D., and Wells, J. W., 1943, The Río Cachirí section in the Sierra de Perijá, Venezuela: Bulletin of American Paleontology, v. 27, no. 108, p. 273–365.

López Ramos, E., 1981, Geología de México: Tesis Resendiz (Division Comercial), Mexico, tomo III, Segunda Edición, Abril de 1981, 446 p.

MacDonald, William D., 1965, Geology of the Serranía de Macuira area Guajira Peninsula, northeast Columbia: Fourth Caribbean Conference, Trinidad, 1965, p. 267–274.

MacDonald, William D. and Opdyke, Neil D., 1972, Tectonic rotations suggested by paleomagnetic results from northern Colombia, South America:

Journal of Geophysical Research, v. 77, no. 29, p. 5720–5730.

—— 1984, Jurassic paleomagnetic directions from the Santa Marta Massif, Colombia: Implications to paleogeography and tectonics, *in* Bonini, W. E., and others, eds., The Caribbean–South American plate boundary and regional tectonics: Geological Society of America Memoir 162 (this volume).

Martin Bellizzia, C., 1968, Edades isotopicas de rocas Venezolanas: Boletín de Geología X, República de Venezuela, Ministerio de Minas e Hidrocarburos (Caracas) Dirección de Geología, p. 356–379.

Maze, William B., 1983, Jurassic La Quinta Formation in the Sierra de Perijá, northwestern Venezuela: Geology, tectonic environment, paleomagnetic data, and copper mineralization of red beds and volcanics [Ph.D. thesis]: Princeton, New Jersey, Princeton University.

—— 1984, Igneous rocks of the La Quinta Formation in the Sierra de Perijá of northwestern Venezuela: Petrogenesis and geochemistry: Proceedings of the Tenth Caribbean Geological Conference held in Cartagena, Colombia, August, 1983 INGEOMINAS, Bogotá (in press).

Maze, W. B., and Hargraves, R. B., 1984, Paleomagnetic results from the Jurassic La Quinta Formation in the Perijá Range, Venezuela and their tectonic significance, *in* Bonini, W. E., and others, eds., The Caribbean–South American plate boundary and regional tectonics: Geological Society of America Memoir 162 (this volume).

McKenzie, D. P., 1978, Some remarks on the development of sedimentary basins: Earth and Planetary Science Letters, v. 40, p. 25–32.

Miall, Andrew D., 1981, Alluvial sedimentary basins: Tectonic setting and basin architecture, *in* Sedimentation and tectonics in alluvial basins: Miall, A. D., ed., Geological Association of Canada Special Paper 23, p. 1–33.

Miller, J. B., 1962, Tectonic trends in Sierra de Perijá and adjacent parts of Venezuela and Colombia: American Association of Petroleum Geologists Bulletin, v. 46, no. 9, p. 1565–1595.

Mills, R. A., Hugh, K. E., Feray, D. E., and Swolfs, H. C., 1967, Mesozoic stratigraphy of Honduras: American Association of Petroleum Geologists Bulletin, v. 51, p. 1711–1786.

Mooney, Walter D., 1980, An east Pacific–Caribbean ridge during the Jurassic and Cretaceous and the evolution of western Colombia, *in* Pilger, Rex H., ed., The origin of the Gulf of Mexico and the early opening of the central North Atlantic Ocean: Baton Rouge, Louisiana State University, p. 55–73.

Moore, Raymond C., Lalicker, Cecil G., and Fischer, Alfred G., 1952, Invertebrate fossils: McGraw-Hill Book Company, Incorporated, p. 544–545, 766 p.

Moticska, P., 1973, Guia de la excursión No. 2—Sierra de Perijá, Complejo volcánico-plutónico de El Totumo-Inciarte: II Congreso Latino Geológico, Caracas, República de Venezuela, Ministerio de Minas e Hidrocarburos, Dirección de Geología, p. 300–311.

Moya, E., and Viteri, E., 1977, Nuevas evidencias sobre magmatismo y metalogénesis en el complejo volcánico-plutónico de El Totumo-Inciarte, Sierra de Perijá, Estado Zulia: V Congreso Geológico Venezolano, Caracas, República de Venezuela, Ministerio de Minas e Hidrocarburos Dirección de Geología, Memoria II, p. 693–711.

Nagle, Frederick, Rosenfeld, Joshua, and Stipp, Jerry J., 1977, Guatemala, where plates collide: A reconnaissance guide to Guatemalan geology: Coral Gables, Florida, Miami Geological Society, University of Miami, 72 p.

Neumann, E. -R., and Ramberg, I. B., editors, 1978, Petrology and geochemistry of continental rifts: Proceedings of the NATO Advanced Study Institute Paleorift Systems with Emphasis on the Permian Oslo Rift, Oslo, Norway, July 27–August 5, 1977, v. 1, 296 p.

Odreman R., O., and Benedetto, Gianluigi, 1977, Paleontología y edad de la Formación Tinacoa, Sierra de Perijá, Estado Zulia, Venezuela: Memoria V, Congreso Geológico Venezolano, tomo I, p 15–32.

Odreman Rivas, Oscar E., and Ghosh, Santosh, 1980, Estudio paleoambientalpaleontológico de facies de la Formación La Quinta, cerca de Mérida: Boletín de Geología, XIV, no. 26, República de Venezuela, Ministerio de Energía y Minas, p. 89–104.

Padron, V., 1978, Geología de la región al suroeste de La Grita, Estado Táchira [B.Sc. thesis]: Caracas, Universidad Central Venezuela, 178 p., referenced in

Tarache, 1980.

Pálmason, G., editor, 1982, Continental and oceanic rifts, Geodynamic series volume 8: Washington, D.C., and American Geophysical Union, Geological Society of America, 309 p.

Pindell, James, and Dewey, John F., 1982, Permo-Triassic reconstruction of Western Pangea and the evolution of the Gulf of Mexico/Caribbean region: Tectonics, v. 1, p. 179–211.

Pumpin, V. F., 1979, El marco estructural de Venezuela noroccidental: EPC Report, Maraven, Caracas.

Ramberg, I. B., and Neumann, E.-R., editors, 1978, Tectonics and geophysics of continental rifts: Proceedings of the NATO Advanced Study Institute Paleo-rift Systems with Emphasis on the Permian Oslo Rift, Oslo, Norway, July 27–August 5, 1977, v. 2, 444 p.

Riecker, Robert E., editor, 1979, Rio Grande Rift: Tectonics and magmatism: Washington, D.C., American Geophysical Union, 438 p.

Rivera, Rosalvina, 1956, Lexique Stratigraphique International, volume V, Amerique Latine (sous la direction de R. Hoffstetter) Fascicule 5b, Peru: Centre National de la Recherche Scientifique, Paris, 131 p.

Rod, E., 1956, Strike-slip faults of northern Venezuela: American Association of Petroleum Geologists Bulletin, v. 40, p. 457–476.

Royden, L., and Keen, C. E., 1980, Rifting process and thermal evolution of the continental margin of eastern Canada determined from subsidence curves: Earth and Planetary Science Letters, v. 51, p. 343–361.

Sapper, C., 1937, Mittelamerika: Handbuch der regionalen Geologie: 8, 4a, Heidelberg, Steinman und Wilckens, 160 p.

Sauer, Walther, 1971, Geologie von Ecuador: Beitrage Zur Regionalen Geologie der Erde, Gebruner Borntraeger, Berlin, 316 p.

Sawyer, D. S., Swift, B. A., Sclater, J. G., Toksöz, M. N., 1982, Extensional model for the subsidence of the northern United States Atlantic continental margin: Geology, v. 10, p. 134–140.

Schubert, C., Sifontes, R. S., Padrón, V. E., Vélez, J. R., and Loaiza, P. A., 1979, Formación La Quinta (Jurásico) Andes Merideños: Geología de la sección tipo: Acta Científica Venezolana, 30, p. 42–55.

Sclater, John G., and Christie, P.A.F., 1980, Continental stretching: An explanation of the post-mid-Cretaceous subsidence of the central North Sea Basin: Journal of Geophysical Research, v. 85, no. B7, p. 3711–3739.

Shagam, Reginald, 1975, The northern termination of the Andes, Chapter 9, in The ocean basins and margins, Nairn, Alan E. M., and Stehli, Francis G., eds., Volume 3, The Gulf of Mexico and the Caribbean: New York, Plenum Press, p. 325–420.

Shagam, R., Kohn, B., Banks, P., Dasch, L., Vargas, R., Rodriguez, G. I., and Pimentel, N., 1984, Cretaceous-Pliocene fission track ages on apatite, zircon, and sphene from Toas Island and the Perijá Range, western Venezuela, and Santander Massif, eastern Colombia: Regional tectonic implications, in Bonini, W. E., and others, eds., the Caribbean–South American plate boundary and regional tectonics: Geological Society of America Memoir 162 (this volume).

Shepherd, Glenn L., and Moberly, Ralph, 1981, Coastal structure of the continental margin, northwest Peru and southwest Ecuador, in Kulm, L. D., and others, eds., Nazca plate: Crustal formation and Andean convergence: Geological Society of America Memoir 154, p. 351–391.

Skerlec, G. M., and Hargraves, R. B., 1980, Tectonic significance of paleomagnetic data from northern Venezuela: Journal of Geophysical Research, v. 85, no. B10, p. 5303–5315.

Sleep, N. H., 1971, Thermal effects of the formation of Atlantic continental margins by continental break up: Geophysical Journal of the Royal Astronomical Society, v. 24, p. 325–350.

Stainforth, R. M., 1969, The concept of seafloor spreading applied to Venezuela:

Asociación Venazolano de Geología, Minas y Petróleo Boletín Informativo, v. 12, no. 8, p. 256–274.

Steinitz, Gideon, and Maze, William B., 1984, K-Ar ages on hornblende-andesite from the Sierra de Perijá, western Venezuela, in Bonini, W. E., and others, eds., The Caribbean–South American plate boundary and regional tectonics: Geological Society of America Memoir 162 (this volume).

Tarache, Crisalida, 1980, Stratigraphy and uranium potential of the La Quinta Formation of Jurassic age, North-Central Tachira, Venezuela [M.S. thesis]: Golden, Colorado School of Mines, Geology Department, 103 p.

Tarney, J., and Windley, B. F., 1981, Marginal basins through geological time, in The origin and evolution of the Earth's continental crust, a Royal Society discussion organized by S. Moorbath, F.R.S., and B. F. Windley, held on 21 and 22 February, 1980: Philosophical Transactions of the Royal Society of London, A301, p. 217–232, p. 33–47.

Toksöz, M. Nafi, and Bird, Peter, 1977, Formation and evolution of marginal basins and continental plateaus, in Talwani, Manik, and Pitman, Walter C., III, eds., Island arcs deep sea trenches and back-arc basins, Maurice Ewing Series 1: American Geophysical Union, p. 379–393.

Tschanz, C. M., Marvin, R. F., Cruz, B., Mehnert, H. H., and Cebula, G. T., 1974, Geologic evolution of the Sierra Nevada de Santa Marta, northeastern Colombia: Geological Society of America Bulletin 85, p. 273–284.

Tschopp, H. J., 1953, Oil explorations in the oriente of Ecuador: American Association of Petroleum Geologists Bulletin, v. 37, p. 2303–2347.

Van der Voo, R., Mauk, F. J., and French, R. B., 1976, Permian-Triassic continental configurations and the origin of the Gulf of Mexico: Geology, v. 4, p. 177–180.

Van der Voo, Rob, 1981, A paleomagnetic comparison of various reconstructions of the Atlantic-bordering continents which have implications for the origin of the Gulf of Mexico, in Pilger, Rex H., Jr., ed., The origin of the Gulf of Mexico and the early opening of the central North Atlantic Ocean: Proceedings of a symposium February 19–20, 1981, Houston Geological Society, Part II, p. 1–5.

Vierbuchen, R. C., George, R. P., and Vail, P. R., 1983, A thermal-mechanical model of rifting with implications for outer highs on passive continental margins: in Watkins, J. S. and Drake, C. L., eds., Studies in continental margin geology: American Association of Petroleum Geologists Memoir 34, p. 765–778.

Viniegra O., Francisco, 1971, Age and evolution of salt basins of southeastern Mexico: American Association of Petroleum Geologists Bulletin, v. 55, p. 478–494.

Viteri A., Enrique, 1978, Genesis del cobre nativo asociado a rocas volcánicas de la Formación La Quinta en la Sierra de Perijá, Venezuela: Ministerio de Energía y Minas, Venezuela, Boletín de Geología, v. XIII, no. 24, p. 47–82.

Walper, Jack L., 1980, Tectonic evolution of the Gulf of Mexico, in Pilger, Rex H., Jr., ed., The origin of the Gulf of Mexico and the early opening of the central North Atlantic Ocean: Proceedings of a symposium, Baton Rouge, Louisiana State University, p. 3–16.

Weeks, L. G., 1947, Paleogeography of South America: American Association of Petroleum Geologists Bulletin, v. 31, p. 1194–1241.

Weyl, Richard, 1980, Geology of Central America, second edition: Berlin, Gebruder Borntraeger, 371 p.

Yrigoyen, Marcelo R., 1972, Cordillera principal, in Geología Regionál Argentina, Resultados del primer simposio de geología regionál Argentina realizado en Córdoba (11–15 de Septiembre de 1969), Centenario de su Fundación: Academia Nacional de Ciencia, p. 345–364.

MANUSCRIPT ACCEPTED BY THE SOCIETY SEPTEMBER 1, 1983

Geological Society of America
Memoir 162
1984

K-Ar ages on hornblende-andesite from the
Sierra de Perijá, western Venezuela

Gideon Steinitz
Geological Survey of Israel
Jerusalem, Israel

William B. Maze
Department of Geological and Geophysical Sciences
Princeton University
Princeton, New Jersey 08544

ABSTRACT

Whole-rock K-Ar ages of 155 ± 5 m.y. and 146 ± 7 m.y. were obtained from two hornblende-andesite flows associated with the Jurassic La Quinta Formation of northwestern Venezuela and are in agreement with others reported for this formation. These data provide age control on stratigraphic relationships and paleomagnetic measurements.

Abundant volcanic rocks characterize a part of the southeastern Perijá Mountains of western Venezuela (see Maze, 1984, for an account of previous investigations and the current status of studies). Some of the mafic, intermediate, and felsic lavas and tuffs may be of Triassic age, but most are interbedded with the La Quinta Formation of believed Middle to Late Jurassic age, possibly extending into the Early Cretaceous, based mainly on floras identified by Benedetto and Odreman (1977). A set of alkalic mafic dikes intrude the La Quinta Formation and cut many of the lavas and tuffs. Distinctive paleomagnetic vectors (Maze and Hargraves, 1984) of the dike suite compared to the other volcanic rocks and red beds of the La Quinta Formation suggest that the dikes represent a separate, younger, igneous event.

Volcanism ceased prior to deposition of the unconformably overlying marine Cretaceous section. The basal unit of the latter (Río Negro Formation) is poorly fossiliferous, but most workers infer a probable Barremian age (see correlation table 1 of the Geologic Map of Venezuela, Bellizzia and others, 1976), corresponding to a maximum absolute age of about 114 m.y. (Odin and others, 1982). This, in turn, sets minimum ages for the La Quinta Formation and its interbedded volcanic rocks and for the mafic dike suite.

Age relationships are further complicated by the presence of a fault-block of coarse quartzose conglomeratic sandstones and some volcanic rocks which are lithologically distinct from typical La Quinta sedimentary rocks and are of unknown age (see Maze, 1984, for further discussion).

The problem of correlation and absolute age dating might have been a matter of local interest only, but for the results of the paleomagnetic studies (Maze and Hargraves, 1984; Maze, 1984) that further suggest major differential displacement between the La Quinta Formation of the Venezuelan Andes and the La Quinta Formation of the Perijá Range. Inasmuch as the Perijá Range constitutes a significant tectonic element in the borderland of the South American plate adjacent to the boundary with the Caribbean plate, it is now pertinent to plate-tectonic reconstructions to establish ages of stratigraphic boundaries and igneous and other thermal events in greater detail and with finer precision than are currently available.

The K-Ar ages on hornblende-andesite reported here are on a sample (WM-80-3) collected by Maze in 1980 in Río Totumo (Fig. 1) and a sample (V-73-21) collected by R. Shagam in 1973 in Quebrada La Ge. The latter is interpreted as lava intercalated in La Quinta red beds; the former occurs in country rocks of uncertain correlation. Both flows were also sampled at these two locations for a paleomagnetic study (Maze and Hargraves, 1984).

Attempts to separate hornblende for K-Ar dating were unsuccessful, so whole-rock powders were used. The freshest whole-rock samples were selected and weathered rinds removed, but in thin section there is abundant indication of secondary alteration in the form of Fe-oxide as rims about hornblende and pyroxene and as crack fillings. The ages were measured by Steinitz at the Israel Geological Survey, Jerusalem. Analytical data and calculated ages for the two samples are given in Table 1.

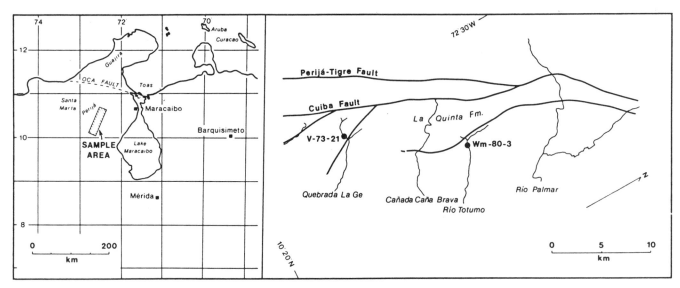

Figure 1. Sample location map. For geologic detail, see Maze (1984). For sites sampled for paleomagnetic studies, see Maze and Hargraves (1984).

TABLE 1. ANALYTICAL DATA AND CALCULATED AGES

Sample	%K	^{40}K ppm	%Ar rad	^{40}Ar rad ppm	Age $\pm$ 1σ m.y.
WM-80-3	2.43	2.90	92.03	2.73×10^{-2}	155 $\pm$ 5
V-73-21	2.11	2.52	66.06	$2.23 \times .0^{-2}$	146 $\pm$ 7

Note: Using IUGS accepted decay constants published by Steiger and Jäger (1977). For analytical procedures see Steinitz and others (1984).

Approximate age equivalence of the two andesite occurrences is indicated with a possibility that the El Totumo rocks may be slightly older. Paleomagnetic data for the two sites (Maze and Hargraves, 1984) support the latter conclusion in that vectors are approximately reversed with respect to each other, both corrected and uncorrected for structural rotation. The ages straddle the boundary between the Middle and Late Jurassic (Odin and others, 1982) establishing a maximum age for the alkalic mafic dike suite and possibly a minimum age for the unnamed rocks in the fault-block of which WM-80-3 is a part. Overlapping precision of a U-Pb zircon age of 163 ± 5 m.y. (Dasch, 1982; sample from the same outcrop as WM-80-3) and the K-Ar age for the Totumo locality suggests that the latter represents the age of cooling on original crystallization or very shortly thereafter. A zircon fission-track age of 87 ± 9 m.y. on a sample from the Totumo locality (Shagam and others, 1984) indicates only a low-temperature thermal history following the setting of the K-Ar clocks.

Additional radiometric dating, especially of the uppermost lavas and tuffs in the La Quinta Formation and the mafic dike suite, is clearly indicated, preferably in parallel with further paleomagnetic studies.

ACKNOWLEDGMENTS

We thank our colleagues at the Geological Survey of Israel, Ben Gurion University of the Negev, Princeton University, and the Dirección de Geología, Ministerio de Energía y Minas, Venezuela.

REFERENCES CITED

Bellizzia, A., Pimentel, N., and Bajo, R., 1976, Mapa Geológico Estructural de Venezuela, Escala 1:500,000: Ministerio de Minas e Hidrocarburos, Venezuela.

Benedetto, G., and Odreman R., O., 1977, Nuevas evidencias paleontológicas en la Formación La Quinta, su edad y correlación con las unidades aflorantes en la Sierra de Perijá y Cordillera Oriental de Colombia, *in* Memoria, Congreso Geológico Venezolano, V., Caracas, Volume 1: Venezuela Ministerio de Energía y Minas, p. 87–106.

Dasch, Lawrence E., 1982, U-Pb Geochronology of the Sierra de Perijá, Venezuela [Master's thesis]: Case Western Reserve University, Cleveland, 164 p.

Maze, William B., 1984, Jurassic La Quinta Formation in the Sierra de Perijá, northwestern Venezuela: Geology and tectonic environment of red beds and volcanic rocks, *in* Bonini, W. E., and others, eds., The Caribbean–South American plate boundary and regional tectonics: Geological Society of America Memoir 162 (this volume).

Maze, W. B., and Hargraves, R. B., 1984, Paleomagnetic results from the Jurassic La Quinta Formation in the Perijá Range, Venezuela, and their tectonic significance, *in* Bonini, W. E., and others, eds., The Caribbean–South American plate boundary and regional tectonics: Geological Society of America Memoir 162 (this volume).

Odin, G. S., Curry, D., Gale, N. H., and Kennedy, W. J., compilers, 1982, The Phanerozoic time scale in 1981, *in* Odin, G. S., ed., Numerical dating in stratigraphy: New York, John Wiley and Sons, p. 957–960.

Shagam, R., Kohn, B., Banks, P., Dasch, L., Vargas, R., Rodriguez, G. I., and Pimentel, N., 1984, Tectonic implications of Cretaceous-Pliocene fission-track ages from rocks of the circum-Maracaibo Basin region of western Venezuela and eastern Columbia *in* Bonini, W. E., and others, eds., The Caribbean–South American plate boundary and regional tectonics: Geological Society of America Memoir 162 (this volume).

Steiger, R. H., and Jager, E., 1977, Subcommission of geochronology: Convention on the use of decay constants in geo and cosmochemistry: Earth and Planetary Science Letters, v. 36, p. 359–362.

Steinitz, G., Lang, B., Mor, D., and Dallal, C., 1984, The K-Ar Laboratory at the Geological Survey of Israel, Current Research, 1982: Geological Survey of Israel, Jerusalem (in press).

MANUSCRIPT ACCEPTED BY THE SOCIETY SEPTEMBER 1, 1983

Geological Society of America
Memoir 162
1984

Paleomagnetic results from the Jurassic La Quinta Formation in the Perijá Range, Venezuela, and their tectonic significance

W. B. Maze
R. B. Hargraves
Department of Geological and Geophysical Sciences
Princeton University
Princeton, New Jersey 08544

ABSTRACT

The paleomagnetism of 150 oriented hand samples (32 sites) of volcanic, red bed, and dike rocks, constituting the La Quinta Formation in the Perijá Range of Venezuela, has been studied. After demagnetization, consistent vectors from volcanic rocks and red beds are bipolar but show considerable scatter both before and after correction for present dip. A northwest-southeast declination predominates, however, in contrast to the northeast-southwest declination previously recorded in La Quinta strata from the Venezuelan Andes. As the declination predicted from data from the South American craton is northerly, differential rotation betwen the Perijá Range and Mérida Andes, causing opening to the north, is suggested. Six dikes give consistent northerly vectors, indicating emplacement after the differential rotations, probably in Early Cretaceous time.

INTRODUCTION

The La Quinta Formation in the Perijá Range of northwestern Venezuela consists of a sequence of continental sedimentary and igneous rocks 1,000 to 1,700 m thick (Maze, 1984). Red arkosic sandstone and siltstones are the dominant sedimentary rocks, with some conglomerate as well as minor occurrences of limestone and shale. The igneous rocks, which in the field area constitute as much as 30% of the La Quinta exposures, may be loosely divided into three categories: (1) flow rocks, (2) volcaniclastic rocks, and (3) dikes. Two types of flow rocks are distinguished: vesicular basaltic-andesite flows 1 to 3 m thick and nonvesicular porphyritic hornblende-andesite flows 5 to 15 m thick. Flow-banding is a common feature of the hornblende-andesite. Volcaniclastic rocks interbedded with the sedimentary and flow rocks are mostly rhyolitic and rhyodacitic and occur as tuffs, welded tuffs, and debris flows from centimeters to hundreds of meters thick. Flattened clasts and flow-banding are indicators of bedding which, however, may not truly reflect a paleohorizontal attitude. Dikes cut all rock types within the La Quinta Formation. All dikes sampled for this study are alkalic basalts with chemistry distinctly different from that of the flow and volcaniclastic rocks.

The paleontologic age of the La Quinta Formation has been established as Middle to Late Jurassic by Benedetto and Odreman (1977). U-Pb, Rb-Sr, and K-Ar isotopic ages on the volcanic rocks range from about 170 to 140 m.y. (Espejo and others, 1980; Dasch, 1982; Steinitz and Maze, 1984; Maze, 1984). Red bed formations of generally similar lithology and age are known also in the Venezuelan Andes (the type locality for the La Quinta Formation; Hargraves and Shagam, 1969), on the Colombian side of the Perijá Range and at Santander (Giron Formation; Cediel, 1969), the Guajira Peninsula (MacDonald and Opdyke, 1972), and Isla de Toas (Pimentel, 1975; Fig. 1).

Oriented samples were collected in the hope that paleomagnetic data might (a) aid in local stratigraphic correlations within the Perijá Range and (b) elucidate the structural history of the Perijá-Maracaibo-Andes block by comparison with results from the La Quinta in the Venezuelan Andes (Hargraves and Shagam, 1969) and rocks of similar age elsewhere in cratonic South Amer-

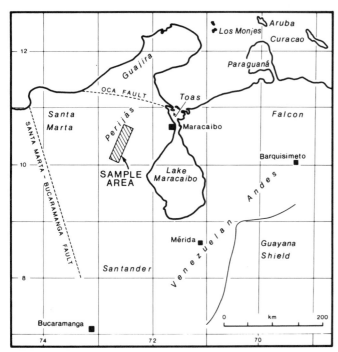

Figure 1. Map showing area sampled in Perijá Range and location of Toas, northwestern South America.

ica. The results, presented below, were of no value with respect to local correlation, but do provide evidence of differential tectonic rotation between the Perijá Range and Andes.

SAMPLING AND MEASUREMENT

One hundred and fifty oriented hand samples were collected from 32 sites as shown in Figures 1 and 2. The samples include red sedimentary rocks as well as mafic to intermediate or felsic intrusive and extrusive rocks (see Table 1). Standard 2.5-cm cores were drilled from these samples in the laboratory.

Paleomagnetic study included measurement of natural remanent magnetism (NRM) (Schonsted DSM 1) and stepwise alternating field (AF) demagnetization of selected specimens to 70 milliteslas (mT) (700 Oe) using a 2-axis tumbler. Thirty-two specimens were thermally demagnetized in a field <20 nanoteslas (nT) ($\leq$20 gamma) to 650 °C. These data were analyzed by vector subtraction and plotting of Zijderveld (1967) diagrams (Fig. 3) whereby single-component vectors could be identified. Site-mean results reported in Table 1, however, are those showing minimum dispersion at a specified optimum demagnetizing field (ODF) or temperature (ODT).

PETROLOGY AND MAGNETIC PROPERTIES

Volcanic Rocks

The basaltic andesites are dark, fine-grained rocks occurring as flows with vesicular tops and brecciated bottoms; the

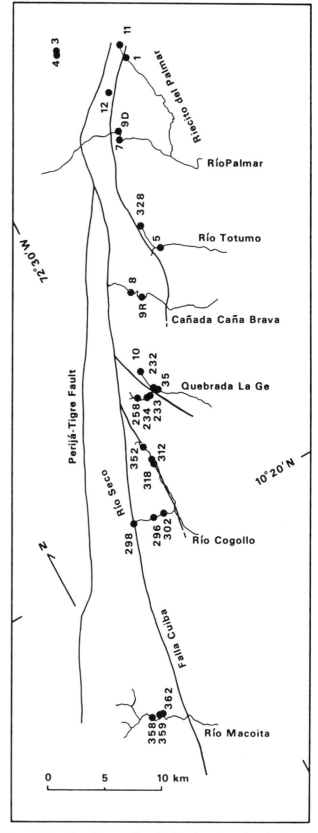

Figure 2. Map showing sampling site locations. Locations for sites 3, 4, 11, and 12 are approximate.

TABLE 1. PALEOMAGNETIC MEASUREMENTS ON PERIJÁ RANGE SAMPLES
UNCORRECTED FOR BEDDING

Site No.	Petrology	N	D_n	I_n	k	$\bar{J}_n$	MDF_{mT}	ODF/T	D	I	k	α_{95}
	Volcanic rocks											
3	agglomerate	4	290	−9	13	1×10^{-5}	40	20	290	−1	9	32
4	volcaniclastic	4	213	−44	2	3×10^{-6}	40	20	179	−41	15	24
5	hornblende andesite	7(4)	347	25	20	5×10^{-4}	20	50	340	−2	148	8
7	dacite porphyry	5(4)	228	67	3	1×10^{-6}	50	50	137	57	44	12
8	basaltic andesite	5	328	39	2	5×10^{-4}	7.5	40	303	−9	14	21
11	pyroclastic	3	257	69	13	1×10^{-5}	5	30	224	50	12	37
12	vesicular lava	5	325	15	6	1×10^{-4}	5	30	334	5	8	29
35	rhyolite tuff	5	123	57	12	4×10^{-7}	30	30	150	38	32	14
232	rhyolite tuff	5	142	−25	105	1×10^{-4}	40	40	143	−26	114	7
233	hornblende andesite	9	87	39	2	1×10^{-4}	40	30	130	−26	9	19
258	basaltic andesite	5	154	25	114	5×10^{-4}	50	30	155	23	121	7
298	rhyolite tuff	5	304	70	9	1×10^{-5}	20	20	293	58	12	23
328	volcaniclastic	4	143	28	140	8×10^{-6}	40	20	146	20	169	7
	Dikes											
1	alkali basalt	5	341	40	2	1×10^{-4}	7.5	20	357	43	11	24
7	alkali basalt	7	349	47	5	4×10^{-4}	20	30	350	27	65	8
9D	alkali basalt	4	12	61	11	5×10^{-4}	20	20	357	26	16	23
302	alkali basalt	4	348	37	8	6×10^{-4}	7.5	30	343	25	10	31
312	alkali basalt	2	354	45	651	2×10^{-4}	6	20	00	49	229	17
352	alkali basalt	4	06	44	23	3×10^{-4}	12.5	30	04	42	12	28
Toas	basalt	4	328	46	6	9×10^{-4}	50	30	350	37	56	17
	Red Beds						$MDT^{\circ}C$					
4		2	06	19	2	8×10^{-6}	300°	400°	343	−1	69	31
9R		4(3)	339	42	10	6×10^{-6}	250°	500°	300	−5	5	62
10		5	330	60	10	2×10^{-5}	200°	555°	261	13	10	26
234		6	00	57	4	1×10^{-5}	100°	305°	27	45	7	29
296		4	01	43	16	9×10^{-5}	150°	600°	75	39	17	23
302		4	347	23	24	2×10^{-5}	450°	600°	334	−9	51	13
312		2	346	59	89	2×10^{-5}	75°	200°	328	41	4	90
318		3	328	25	9	6×10^{-5}	300°	530°	303	6	39	20
352		5	15	53	44	3×10^{-5}	400°	555°	37	58	106	8
358		5	348	54	4	1×10^{-6}	150°	500°	24	14	5	36
359		5	359	36	16	1×10^{-5}	250°	400°	29	23	8	29
362		5	344	51	43	7×10^{-6}	150°	500°	348	67	2	85

Note: Site number and location are indicated on Figure 2; N = number of samples collected and used in analysis. Smaller numbers in parentheses, where indicated, were used in ODF/ODT analyses. D, I = magnetic decliniation and inclination (+ down); subscript n refers to original NRM; k and α_{95} are Fisher's statistical parameters. $\bar{J}_n$ is the average intensity (emu/cm^3) or NRM. MDF(mT) or MDT($^{\circ}$C) are the mean demagnetizing field or temperature, which is the average alternating field (mT) or temperature ($^{\circ}$C) required to reduce NRM to half of its original intensity. ODF/T is the optimal demagnetizing field or temperature, giving the best site-mean statistics. Whole-rock K-Ar ages on hornblende andesite from sites 5 and 233 are 155 ± 5 and 146 ± 7 m.y., respectively (Steinitz and Maze, 1984). U-Pb ages on zircons from the hornblende andesite at site 5 are 163 ± 5 m.y. (Dasch, 1982). These zircons gave a 90 ± 9 m.y. fission-track age (Shagam and others, 1984).

hornblende-andesite flows are coarser grained (porphyritic) and lack vesicular tops. Clinopyroxene, plagioclase, Fe-Ti oxides, and apatite are present in both, but amphibole occurs only in the latter. Although the degree of alteration varies considerably from site to site, the feldspars in the more mafic rocks are somewhat saussuritized, whereas the pyroxenes and amphiboles are usually fresh. Conspicuous hematite-rich coronas occur around horn-blende phenocrysts. Primary opaques are relict titanomagnetites, with both ilmenite lamellae and host spinel variably replaced by hematite. Little evidence of maghemite was seen. The mineralogy

and texture suggest that the alteration is deuteric, occurring at high temperatures during the cooling of these mafic lavas rather than much later at lower temperatures.

The felsic volcanic rocks commonly are pervasively altered and silicified, with a few wispy chloritic relicts remaining after original ferromagnesians. Original clinopyroxene and amphibole can be identified in the least-altered rocks, but there is little evidence of primary oxide phases.

With the exception of some of the more fragmental volcaniclastics and those felsic lavas more intensely altered by hy-

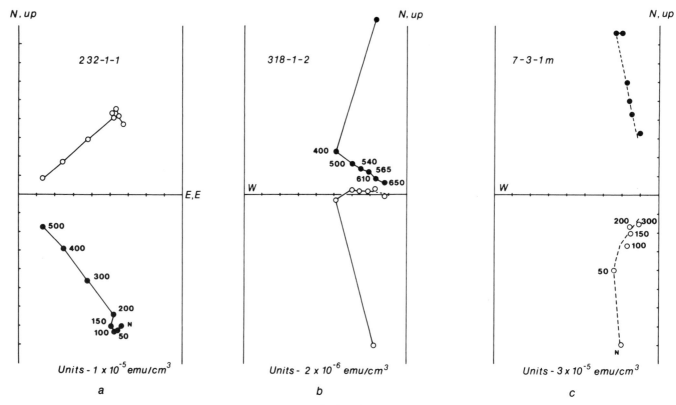

Figure 3. Zijderveld plots illustrating demagnetization of various rock types, in situ coordinates; rhyolite tuff (a); red bed (b); dike rock (c).

drothermal activity, the median demagnetizing field (MDF) of all other volcanic rocks is high (20 to 50 mT, see Table 1), suggesting that single-domain magnetite or hematite is the carrier of the remanent magnetism. This is consistent with the observation of hematite rimming and veining the primary spinels and as an alteration product of hornblende in some samples. Radiometric ages of the hornblende andesite at site 5 and site 233 are given in Table 1.

Sedimentary Rocks

The red beds sampled are medium- to fine-grained, well-bedded red or red-brown to purple arkosic sandstones and siltstones. AF demagnetization of some samples (during which little change in NRM direction occurred) revealed a MDF of between 10 and 20 mT. On thermal demagnetization, most samples showed a steady decline to between 0.5 and 0.2 of the initial moment (J_N) at 500 or 525 °C, followed at higher temperatures by more irregular behavior. Most samples showed a continuing steeper drop in intensity after 500 or 525 °C, but some showed a slight recovery between 560 and 600 °C. The steeper drop in intensity above 500 °C to 530 °C is not accompanied by any significant change in direction, which remained constant in individual samples even after demagnetization to 650 °C; the within-site grouping, however, was only fair (see Table 1). The site mean

result showing minimum dispersion at one of the 500, 530, 560, 600, or 650 °C demagnetization steps is reported in Table 1.

Dikes

The basaltic dikes, varying from 0.3 to 5 m in thickness, are in general more equigranular and appear finer grained than the volcanic rocks. Alteration has affected the plagioclase and oxide grains and caused minor chloritization of the pyroxenes; however, the pyroxene microphenocrysts are usually quite fresh. As in the volcanic rocks, the larger opaque grains in the dikes show evidence of high-temperature oxyexsolution of ilmenite from magnetite followed by intense alteration of the ilmenite lamellae and partial replacement of the relict spinel by hematite.

The remanent magnetism in all seven dikes sampled is constant in direction and of intermediate stability with respect to AF demagnetization (MDF 6 to 20 mT).

SUMMARY AND DISCUSSION OF PALEOMAGNETIC RESULTS

The site means of the best single-component vectors that have been isolated by systematic demagnetization of all samples are listed in Table 1 and illustrated in Figures 4 and 5.

It is apparent that with or without correction for bedding

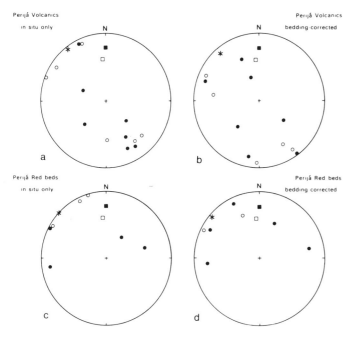

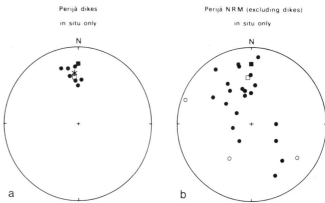

Figure 5. Equal-area projections of demagnetized dike vectors (a) and NRM of all Perijá sites (uncorrected for bedding attitude) (b). Dike mean from Table 2 is plotted as *. Symbols as in Figure 4.

Figure 4. Equal-area projections of demagnetized vectors from volcanic rocks (a, b) and red beds (c, d) before and after structural (bed) correction (when structural correction is possible). Open circles are upper hemisphere, and solid circles are lower hemisphere. Open square represents present Earth's field, and solid square is present theoretical dipole field. Statistics are in Tables 1 and 2. Means from Table 2 plotted as *.

there is considerable scatter in vector directions for both red beds and volcanic rocks. The dikes, on the other hand, are well grouped.

Red Beds and Volcanic Rocks

Both normal and reverse magnetizations are evident in the volcanic rocks and possibly in the red beds (see Fig. 4). Positive and negative inclinations are associated with both polarities, but the average is nearly horizontal both before and after bed correction (Table 1). As the vectors are northwest-southeast, and the prevailing strike is northeasterly, structural corrections primarily affect the inclinations, but in general, where correction shallows one site vector, it steepens another, and the average remains approximately the same.

The volcanic-rock data (Fig. 4a, 4b), which include the magnetizations most likely to have been acquired during initial cooling, are the most consistent in orientation at individual sites. Unfortunately, the foliation interpreted as bedding in these units may not be dependable as an indicator of a paleohorizontal attitude, as discussed in the introduction. The scatter between site-mean vectors, therefore, may in part reflect errors in structural correction, as well as the errors to be expected from secular variation. Some of the volcanic units are intensely altered, and the magnetization in these may be entirely secondary. As tectonism associated with volcanism probably occurred during the deposi-

tion of the La Quinta Formation (and tectonism has certainly occurred since), the secondary magnetizations, even if occurring soon after deposition or eruption, may be of arbitrary orientation.

Bedding in the red beds is quite clear, but the timing of magnetization in such strata is notoriously uncertain (Larson and others, 1982). Given the turbulent geologic history of this area since Jurassic time, the scatter in these data is not surprising, even if it can be assumed that single-component magnetizations have been isolated.

Dikes

The six dikes sampled are comparatively well grouped (Fig. 5a and Table 1), in situ D = 354, I = 36, k = 47. Country-rock bedding attitude is known only for three of the dikes (sites 302, 312, and 352, Fig. 2), and correction for this flattens the inclination somewhat (N = 3, D = 06, I = 27, k = 33), but it remains relatively high. A dike cutting granite on Isla de Toas (Fig. 1) has a similar NRM (Table 1). This mean vector is close to that of the theoretical dipole field for Maracaibo at the present time (D = 360°, I = +20°). The stability of the dike vectors to alternating fields in excess of 20 mT, however, convinces us that the magnetization is primary and not a modern-day overprint.

The absolute age of the dikes is not known. Petrochemically, they are distinctly alkaline (Maze, 1984) and hence unlike any of the La Quinta volcanic rocks which they intrude. They could thus be considerably younger than the La Quinta strata. The absence of any evidence of igneous activity in the Maracaibo area since the beginning of Cretaceous sedimentation, however, suggests that they are no younger than Early Cretaceous, and possibly they are directly related to the closing stages of La Quinta volcanism.

Thermal Overprint

The in situ site mean NRM directions (predemagnetization) of all La Quinta samples measured in this study are projected in

TABLE 2. PALEO-DECLINATION, INCLINATION, AND LATITUDE OF MARACAIBO RELATIVE TO
          PRESENT THEORETICAL DIPOLE FIELD (a); MEAN DECLINATIONS AND
                    INCLINATIONS OF PERIJÁ ROCKS (b)

(a) Time m.y.	Declination°	Inclination°	Latitude
180	13°E	+4	2°N
160	13°E	+10	5°N
140	13°E	+6	3°N
120	7°E	+6	3°N
100	0	+10	5°N
80	10°W	+14	7°N
60	10°W	+14	7°N
40	1°W	+10	5°N
20	3°W	+8	4°N
Present theoretical dipole field	0°	+20	10°30'N
Present field direction	4.5°W	+39	---

(b) Mean Directions after ODF or ODT	D°	I°	Confidence angles	M.L.E.(K1/K2)
Volcanic Rocks				
in situ (12)	323	-2	13.5/28.4	-7.4/-2.3
bedding corrected (10)	319	+9	16.4/35.9	-6.5/-2
Red Beds				
in situ (5)	311	0	6.3/36.1	-65.3/-2.7
bedding corrected (5)	309	+3	15.1/35.7	-12.3/-2.9
Perijá Dikes				
in situ (6)	354	36	3.3/10.4	-155.4/-16.2
bedding corrected (3)	6	27	4.2/16	-192.2/-13.9

Note: (a) As inferred from paleogeographic history of cratonic South America
(Barron and others, 1981). Present field direction from Magnetic Variation Epoch
1975.0 and Magnetic Inclination or Dip Epoch 1975.0 maps prepared and published by
the Defense Mapping Agency Hydrographic Center, Washington, D.C. 20390. (b) From
measurements of La Quinta rocks in the Perijá Range after ODF or ODT demagnetization,
calculated using Bingham statistics (Onstott, 1980). Number of samples used in
calculating the mean given in parentheses. Samples with low Fisher statistics and/or
unreliable geologic field relationships were excluded. Bedding control known for
only 3 Perijá dikes. The two confidence angles are a measure of the ellipticity of
the vector probability distribution (cf. α_{95}) and the MLE is the "maximum
likelihood estimate", analogous to Fisher's k (see Onstott, 1980).

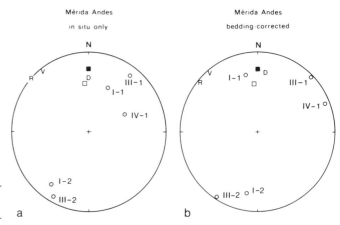

Figure 6. Equal-area projections of La Quinta red-bed normal and reverse group means from three of the four sites studied in detail in the Mérida Andes before correction (a) and after correction (b) for bedding attitude (Hargraves and Shagam, 1969, Fig. 10). Omitted are vectors shown in the original figure from the Permian-Triassic tuffs (III-1) and from locality V, now considered to have undergone intense brecciation (see Giegengack, 1984). Perijá means for red beds, volcanic rocks, and dikes from Table 2 are plotted as R, V, and D, respectively. Symbols as in Figure 4.

Figure 5b. A similarity to the mean-dike vector is apparent, and indeed, the vectors erased during early stages of both AF and thermal demagnetization trend northerly, in general, with moderately steep positive inclination (e.g., Fig. 3b, 3c). This suggests the pervasive regional presence of a viscous overprint parallel to that of the dikes and presumably acquired at the time of their emplacement.

TECTONIC SIGNIFICANCE

The disruption of Pangea, starting in Jurassic time, involved the separation of South America from both Africa and North America. Spreading associated with the formation of the South Atlantic and the Caribbean required subduction on the Pacific margins. The northwestern border of the Guayana shield (Fig. 1) must have borne the brunt of the stresses and deformations associated with these events. The Perijá Range is located in the extreme northwestern segment of South America, much of which was deformed, modified, or possibly accreted during this interval.

As a baseline to which our results should be compared, we have determined the magnetic declination and inclination of Maracaibo, as a function of time, from the best published estimates of the paleogeography of cratonic South America (Table 2; Barron and others, 1981). Although the near-horizontal average inclination of the Perijá La Quinta strata (Table 2) is consistent with that predicted, the northwesterly declination is distinctly

anomalous. Counterclockwise rotation of at least a large part of the Perijá Range is indicated.

The northerly declination of the dikes is consistent with that predicted from the craton, which suggests that the counterclockwise rotation of the Perijá Range was completed by the time of dike emplacement. The mean inclination of the dikes, however, is anomalously steep for times older than 20 m.y. (Table 2), implying a paleolatitude of at least 13°N at the time of intrusion. Alternatively, large blocks of the Perijá Range could have been tilted northward since Cretaceous time, but for this there is no independent evidence. Gose and Swartz (1977) have determined a transient excursion of Honduras 20° north of its present latitude in the Albian, which might be related to the dike inclination anomaly reported here.

The La Quinta paleomagnetic record in the Mérida Andes, reported previously (Hargraves and Shagam, 1969), was likewise highly variable and deemed in large part to be secondary. The only unique feature there was a prevailing northeasterly declination (Fig. 6), for which tectonic rotation was invoked. The rotations implied by the La Quinta paleomagnetic data in the Mérida Andes (~0–60° clockwise) and the Perijá Range (20°–60° counterclockwise) are thus in opposite directions. The implied opening in Late Jurassic or earliest Cretaceous time may have set the stage for the Maracaibo basin structurally, but the thick accumulation of sediment began only in Eocene time. This resulted from erosion of a source area developed to the northeast after collision with an island arc in latest Cretaceous time (Beets and others, 1984). The predominant present and recent tectonic style in the ranges surrounding the Maracaibo basin, however, is overthrusting to the northwest: a Tertiary compressive regime incompatible with the Jurassic tension required by this study. The La Quinta strata themselves were deposited in a series of graben, and perhaps the rotations indicated by the paleomagnetic data are part of this transient episode, heralding the thick accumulation of sedimentary strata in the Maracaibo basin.

The paleomagnetic data reported here, however, are few and of uneven quality; they can at best be considered to suggest the exceedingly complex tectonic history experienced since Jurassic time by all terranes on the northwestern margin of South America.

REFERENCES CITED

Barron, C. J., Harrison, C.G.A., Sloan, J. L., and Hay, W. W., 1981, Paleogeography, 180 million years ago to the present: Eclogae Geologicae Helvetiae, v. 74, p. 433–470.

Beets, Dirk J., Maresch, Walter V., Klaver, Gerard Th., Mottana, Annibal, Bocchio, Rosangela, Beunk, Frank F., and Monen, Hendrik P., 1984, Magmatic rock series and high-pressure metamorphism as constraints on the tectonic history of the southern Caribbean, *in* Bonini, W. E., and others, eds., The Caribbean–South American plate boundary and regional tectonics: Geological Society of America Memoir 162 (this volume).

Benedetto, G., and Odreman, R., O., 1977, Nuevas evidencias paleontológicas en la formación La Quinta, su edad y correlación con las unidades aflorantes en la Sierra de Perijá y Cordillera Oriental de Colombia: Memoria, Congreso Geológico Venezolano, V., Caracas, v. 1, p. 87–106.

Cediel, F., 1969, Die Giron-Gruppe: Eine früh-mesozoische Molasse der Ostkordillere Kolumbiens: Neues Jahrbuch für Geologie und Paläontologie, Abhandlungen, v. 133, 2, p. 111–162.

Dasch, L. E., 1982, U-Pb geochronology of the Sierra de Perijá, Venezuela [Master's thesis]: Case Western Reserve University, Cleveland, 164 p.

Espejo, A., Etchart, H., Cordani, U., Kawashita, K., 1980, Geocronología de intrusives ácidas en la Sierra de Perijá, Venezuela: Venezuela Ministerio de Energía y Minas, Boletín de Geología, v. 14, p. 245–254.

Giegengack, Robert, 1984, Late-Cenozoic environments of the Central Venezuelan Andes, *in* Bonini, W. E., and others, eds., The Caribbean–South American plate boundary and regional tectonics: Geological Society of America Memoir 162 (this volume).

Gose, W. A., and Swartz, D. K., 1977, Paleomagnetic results from Cretaceous sediments in Honduras: Tectonic implications. Geology, v. 5, p. 505–508.

Hargraves, R. B., and Shagam, R., 1969, Paleomagnetic study of La Quinta Formation, Venezuela: American Association of Petroleum Geologists Bulletin, v. 53, p. 537–552.

Larson, E. E., Walker, T. R., Patterson, P. E., Hobblitt, R. P., and Rosenbaum, J., 1982, Paleomagnetism of the Moenkopi Formation, Colorado Plateau: Basis for long-term model of acquisition of chemical remanent magnetism in red beds: Journal of Geophysical Research, v. 87, p. 1081–1106.

MacDonald, W. D., and Opdyke, N. D., 1972, Tectonic rotations suggested by paleomagnetic results from northern Colombia, South America: Journal of Geophysical Research, v. 77, p. 539–546.

Maze, William B., 1984, Jurassic La Quinta Formation in the Sierra de Perijá, northwestern Venezuela: Geology and tectonic environment of red beds and volcanics, *in* Bonini, W. E., and others, eds., The Caribbean–South American plate boundary and regional tectonics: Geological Society of America Memoir 162 (this volume).

Onstott, T. C., 1980, Application of the Bingham distribution function in paleomagnetic studies, Journal of Geophysical Research, v. 85, p. 1500–1510.

Pimentel, N. R., 1975, Falla de Oca: Isla de Toas y San Carlos: (Excursion no. 3 (segunda parte), *in* Memoria, Congreso Latinamericano de Geología, II, Caracas 1973, Volume 1: Venezuela Ministerio de Energía y Minas, p. 326–338.

Shagam, R., Kohn, B., Banks, P., Dasch, L., Vargas, R., Rodriguez, G. I., and Pimentel, N., 1984, Cretaceous-Pliocene fission track ages on apatite, zircon, and sphene from Toas Island and the Perijá Range, western Venezuela, and Santander massif, eastern Colombia: Regional tectonic implications, *in* Bonini, W. E., and others, eds., The Caribbean–South American plate boundary and regional tectonics: Geological Society of America Memoir 162 (this volume).

Steinitz, Gideon and Maze, William B., 1984, K-Ar ages on hornblende-andesite from the Sierra de Perijá, western Venezuela, *in* Bonini, W. E., and others, eds., The Caribbean–South American plate boundary and regional tectonics: Geological Society of America Memoir 162 (this volume).

Zijderveld, J.D.A., 1967, A. C. demagnetization of rocks: Analysis of results, *in* Collinson, D. W., Creer, K. M., and Runcorn, S. K., eds., Methods in paleomagnetism: New York, Elsevier, p. 254–286.

ACKNOWLEDGMENTS

This research was supported by NSF grants EAR 8008207, EAR 8212736, and the Ministerio de Energía y Minas of Venezuela. We are grateful to Joe Wolinski for assistance in sample preparation, and to Naoma Dorety for all magnetic measurements and data reduction. Reviews by W. E. Bonini, W. D. MacDonald, R. Shagam, and R. Van der Voo improved the manuscript and are greatly appreciated.

MANUSCRIPT ACCEPTED BY THE SOCIETY SEPTEMBER 1, 1983

Printed in U.S.A.

Geological Society of America
Memoir 162
1984

Preliminary paleomagnetic results from Jurassic rocks of the Santa Marta Massif, Colombia

William D. MacDonald
Department of Geological Sciences
State University of New York
Binghamton, New York 13901

Neil D. Opdyke
Department of Geology
University of Florida
Gainesville, Florida 32611

ABSTRACT

From the Sierra Nevada of Santa Marta at the northern terminus of the Andes, preliminary paleomagnetic results from ten sites in two formations characteristically have stable directions with northerly declinations and moderate positive inclinations. Most site mean directions are close in attitude to the present Earth's field and are not indicative of large tectonic rotations. An antipodally reversed direction is obtained at one site by thermal demagnetization, suggesting magnetization in an ancient paleofield nearly coaxial with the present field. Few cratonic Jurassic directions have been reported from South America, but at least one such study shows Jurassic paleofield directions close to the present field directions. In defining the Mesozoic polar wander path for South America, distinguishing Jurassic from present field directions will be a continuing challenge.

INTRODUCTION

As part of a paleomagnetic survey of Mesozoic rocks from South America (MacDonald and Opdyke, 1972; Opdyke and MacDonald, 1973; MacDonald and Opdyke, 1974), Jurassic red beds and volcanic rocks were sampled on the southeast flank of the Sierra Nevada of Santa Marta at the north end of the Andes. Preliminary results are summarized here for 75 samples from 11 sites in three Jurassic formations as mapped by Tschanz and others (1969). The red beds and volcanic rocks of this study are approximately coeval with those in the Perija mountains nearby to the east, from which paleomagnetic results are reported by Maze and Hargraves (1984).

GEOLOGIC SETTING

The Santa Marta massif lies at the northern end of the Colombian Andes. The oldest rocks here are 1,300-m.y.-old gneisses which are invaded by numerous batholiths of Jurassic and younger age (MacDonald and Hurley, 1969; Tschanz and others, 1969, 1974). These crystalline rocks are unconformably overlain and partly roofed by Jurassic terrestrial clastic rocks and silicic to basic volcanic rocks. Cretaceous marine strata unconformably overlie the Jurassic section.

This study presents preliminary paleomagnetic results for Jurassic red beds and volcanic rocks along the southeast flank of the massif (Fig. 1). In the region sampled, the prevailing dip is southeasterly, gentle to moderate. Many Mesozoic formations have been mapped in this area (Tschanz and others, 1969). Faulting, facies changes, limited exposure, and scarcity of radiometric age dates complicate the stratigraphic correlation of these strata. Sampling in the present study emphasizes formations for which

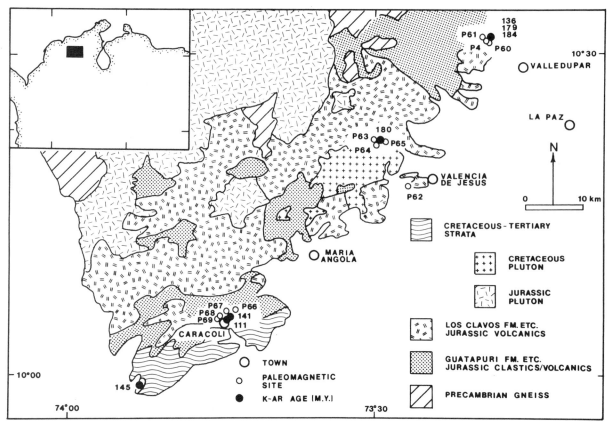

Figure 1. Geologic sketch map of the southern Sierra Nevada of Santa Marta, Colombia, showing locations of paleomagnetic sites and radiometric ages. After Tschanz and others (1969, 1974).

K-Ar radiometric ages are available. Ages noted herein have been recomputed from Tschanz and others (1974) by the method of Dalrymple (1979).

PALEOMAGNETIC SURVEY

Three rock formations were sampled: Los Clavos, Guatapuri, and Caja de Ahorros. Because only one site (P62) was sampled in the silicic volcanic Caja de Ahorros formation, and because the age of this unit is not well controlled, this formation is not discussed further. The distribution of the other units is sketched in Figure 1.

Los Clavos Formation

Rhyolitic to dacitic ignimbrites, flows, dikes, and breccias characterize this unit. Radiometric ages range mainly from 179 to 184 m.y., as recomputed from Tschanz and others (1974). Six sampling sites in two groups lie about 30 km apart near Valledupar and Valencia de Jesus (Fig. 1). The lithologies sampled include rhyolitic ignimbrite (P4, P60, P61), felsic dike (P63), granitic porphyry (P64), and epidotized agglomerate (P65). Bedding attitudes could not be discerned at sites P63, P64, and P65.

Guatapuri Formation

Red beds predominate in this formation, intercalated with silicic to basic flows. Four sites were drilled, in diverse lithologies: tuffs and flows (P66), red beds (P67), keratophyre (P68), and basalt (P69). Near Caracoli, a basalt at P69 yields a recomputed age of 141 m.y. From a nearby keratophyre (P68), we obtained a younger, presumably reset, K-Ar age of 111 ± 4 m.y. About 60 km to the northeast, near Valledupar, Guatapuri red beds rest on older basalts with recomputed ages of 165 and 181 m.y. (Tschanz and others, 1974).

Paleomagnetic Results

Paleomagnetic site mean directions are summarized in Table 1. After serial alternating field (a.f.) demagnetization of pilot samples, the remaining samples at each site were demagnetized at a single step. The a.f. demagnetizing sequence was 0, 2.5, 5, 7.5, 10, 15, and 20 millitesla (mT). Typical samples show a univectorial decay toward the origin after removal of a low coercivity component by fields below 10 mT (Fig. 2). Site P69, a scoriaceous basalt flow, showed poor clustering after a.f. demagnetization at 20 mT, and was thermally demagnetized. Six samples from site

TABLE 1. PALEOMAGNETIC SITE MEANS FOR JURASSIC UNITS,
SIERRA NEVADA DE SANTA MARTA, COLOMBIA

Site	n	Field (mT) or temperature (°C)	In situ inc.	dec.	Tilt-corrected inc.	dec.	κ	α₉₅
Guatapuri Fm (10.1°N, 073.7°W; approx. age 141 m.y.)								
P66	10	25mT	71.9	289.9			3.9	28.3
P67	9	20mT	42.0	346.4	57.6	350.5	290.2	3.0
P68	8	20mT	38.4	012.8	50.8	023.0	208.3	3.9
P69	7	20mT	02.8	224.7			1.4	--
P69	4	600 °C	-26.1	184.6	-31.1	186.8	906.4	3.1
Los Clavos Fm (10.4°N, 073.4°W; approx. age 180 m.y.)								
P60	5	5mT	17.3	337.4	31.6	348.1	5.9	34.5
P61	6	15mT	17.0	342.8	29.0	353.3	19.4	15.6
P63	7	15mT	49.5	016.4			32.9	10.7
P64	6	10mT	50.3	326.9			6.0	29.7
P65	6	15mT	-26.4	046.1			6.7	27.8
Caja de Ahorros Fm (10.3°N, 073.4°W; age possibly 140-180 m.y.)								
P62	5	20mT	-35.8	340.7			1.4	--

Note: n is number of samples per site; a.f. demagnetizing field is in millitesla (mT); thermal demagnetization level is in degrees Centigrade (°C); κ is Fisher's (1953) precision parameter; α₉₅ is semi-apical angle of cone of 95% confidence about site mean direction.

P69 were thermally demagnetized in the sequence 120, 210, 300, 400, 450, 500, 550, 600, 650, and 750 °C. Directions for four of the six samples so treated moved into a tight cluster, for which the Fisher (1953) statistics are reported in Table 1 at the 600 °C demagnetization level. P69 is the only reversed site detected.

Bedding tilt corrections were applied to site-mean directions for sites at which bedding was evident: P60, P61, P67, P68, and P69 (Table 1). Because the regional dip is to the southeast, with interruptions related mainly to faulting, the tilt correction increases the inclination and adds a small clockwise increment to the declination.

Formation means and paleomagnetic poles are reported in Table 2. In calculating the Los Clavos formation mean, site P65 (Table 1) was omitted. That site is an epidotized agglomerate for which the site mean direction diverges widely from other Los Clavos directions. Site P66, excluded from the Guatapuri formation mean, has scattered directions, high coercivities, and an unusual near-vertical inclination (Fig. 3). In retrospect, thermal demagnetization might possibly have been of benefit to this site.

DISCUSSION

An important issue in this preliminary study is, Do the Santa Marta results represent Jurassic paleofield directions or a younger

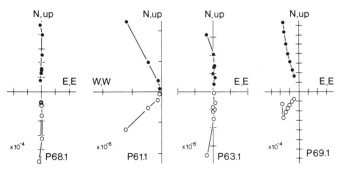

Figure 2. Typical orthogonal projections of a.f. demagnetized vectors for samples from the Jurassic of Santa Marta. Filled circles are projections into the horizontal plane, and open circles are projections into the vertical east-west plane. Units are emu/cm³.

TABLE 2. JURASSIC PALEOMAGNETIC MEANS AND POLES

Formation	Sites	Mean direction inc.	dec.	α₉₅	Paleomagnetic pole lat.(N)	long.(E)	dp	dm
Guatapuri	P67*,P68*,P69*	36.4	002.4	20.8	79.6	-061.1	14.1	24.2
	P67†,P68†,P69§	47.2	007.6	26.7				
Los Clavos	P60*,P61*,P63*,P64*	34.7	344.6	29.1				
	P60†,P61†,P63*,P64*	41.3	351.2	21.1	74.3	-104.5	15.7	25.7

*In situ, a.f. demagnetized
†Tilt-corrected, a.f. demagnetized.
§Tilt-corrected, thermally demagnetized, polarity reversed for computation of mean.

overprint? The reversed site (P69) indicates that a paleofield is represented.

Few paleomagnetic studies of coeval rocks from cratonic South America are available for comparison. A study by Schult and Guerreiro (1979) of Jurassic basalts from the Maranhao basin of Brazil is relevant. The pole for those 158-m.y.-old basalts is at 85 °N, 083 °E, with an alpha-95 of 6.9°. The 95% confidence circle of that pole overlaps the present geographic North Pole. Diabase dikes from the Guayana shield of Venezuela have a pole at 70°N, 066°E, with dp = 2.6, dm = 5.0 (MacDonald and Opdyke, 1974). Although that pole is significantly different from the present pole, its recomputed age is 204 ± 5 m.y., near the Triassic-Jurassic boundary, estimated at 204 ± 4 m.y. (Odin and others, 1982). This diabase dike pole, therefore, probably represents a field 40 to 60 m.y. older than the Santa Marta rocks studied here.

The Santa Marta results do not indicate large tectonic rotations such as have affected the La Teta lava of the Guajira Peninsula nearby to the north (MacDonald and Opdyke, 1972). Parenthetically, we note here that the age estimate of La Teta lava has been revised to 143 ± 7 m.y., placing it near the end of the Jurassic rather than in the Cretaceous.

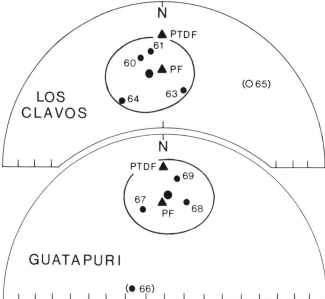

Figure 3. Site-mean directions and formation-mean directions with circles of 95% confidence for Los Clavos and Guatapuri Formations, Santa Marta massif. Site-means in parentheses have been excluded from the calculation of the formation-means. Data from Tables 1 and 2.

Several Jurassic paleomagnetic analyses have been made elsewhere in the mobile orogenic zone peripheral to South America. A complete summary of those results is beyond the scope of this note. It is significant to observe that several research groups that have studied Jurassic rocks from that region of South America have reported paleomagnetic directions nearly parallel to those of the present theoretical dipole field (Creer and others, 1972; Vilas, 1974; Shackleton and others, 1979). Reversals are common in some of the units analyzed.

CONCLUSION

It is anticipated that distinguishing Jurassic paleomagnetic directions from younger overprints in Santa Marta, as elsewhere in South America, will be a continuing challenge. Associated ambiguities in the age of magnetization will undoubtedly complicate the development of the Jurassic polar wander path for South America, as well as the analysis and identification of allochthonous tectonic terranes in the Andes, a major task which has barely begun.

ACKNOWLEDGMENTS

This research was initiated under grant GA-1548 from the National Science Foundation. Conversations with C. M. Tschanz of the U.S. Geological Survey on Santa Marta geology, and with R. B. Hargraves of Princeton University on paleomagnetism of South America have been enlightening. G. E. MacDonald provided helpful advice in computer matters.

REFERENCES CITED

Creer, K. M., Mitchell, J. G., and Abou Deeb, J., 1972, Palaeomagnetism and radiometric age of the Jurassic Chon-Aike Formation from Santa Cruz province, Argentina: Implications for the opening of the South Atlantic: Earth and Planetary Science Letters, v. 14, p. 131–138.

Dalrymple, G. B., 1979, Critical tables for conversion of K-Ar ages from old to new constants: Geology, v. 7, p. 558–560.

Fisher, R. A., 1953, Dispersion on a sphere: Proceedings, Royal Society of London, v. 217A, p. 295–305.

MacDonald, W. D., and Hurley, P. M., 1969, Precambrian gneisses from northern Colombia, South America: Geological Society of America, Bulletin, v. 80, p. 1867–1872.

MacDonald, W. D., and Opdyke, N. D., 1972, Tectonic rotations suggested by paleomagnetic results from northern Colombia, South America: Journal of Geophysical Research, v. 77, p. 539–546.

——1974, Triassic paleomagnetism of northern South America: American Association of Petroleum Geologists Bulletin, v. 58, p. 208–215.

Maze, W. B., and Hargraves, R. B., (1984), Paleomagnetic results from the Jurassic La Quinta Formation in the Perija Range, Venezuela, and their tectonic significance, *in* Bonini, W. E., and others, eds., The Caribbean--South American plate boundary and regional tectonics: Geological Society of America Memoir 162 (this volume).

Odin, G. S., Curry, D., Gale, N. H., and Kennedy, W. J., 1982, The Phanerozoic time scale in 1981, *in* Odin, G.S., ed., Numerical dating in stratigraphy: New York, John Wiley and Sons, p. 957–960.

Opdyke, N. D., and MacDonald, W. D., 1973, Paleomagnetism of Late Cretaceous Poços de Caldas alkaline complex, southern Brazil: Earth and Planetary Science Letters, v. 18, p. 37–44.

Schult, A., and Guerreiro, S.D.C., 1979, Paleomagnetism of Mesozoic igneous rocks from the Maranhao basin, Brazil, and the time of opening of the South Atlantic: Earth and Planetary Science Letters, v. 42, p. 427–436.

Shackleton, R. M., Ries, A. C., Coward, M. P., and Cobbold, P. R., 1979, Structure, metamorphism, and geochronology of the Arequipa massif of coastal Peru: Geological Society (London), Journal, v. 136, part 2, p. 195–214.

Tschanz, C. M., Jimeno, A., and Cruz, J., 1969, Mapa geologico de reconocimiento de la Sierra Nevada de Santa Marta, Colombia: Instituto Nacional de Investigaciones Geologico-Mineras, Bogota, scale 1:200,000.

Tschanz, C. M., Marvin, R. F., Cruz, J., Mehnert, H. H., and Cebula, G. T., 1974, Geologic evolution of the Sierra Nevada de Santa Marta, northeastern Colombia: Geological Society of America Bulletin, v. 85, p. 273–284.

Vilas, J.F.A., 1974, Palaeomagnetism of some igneous rocks of the middle Jurassic Chon-Aike formation from Estancia La Reconquista, province of Santa Cruz, Argentina: Royal Astronomical Society, Geophysical Journal, v. 39, p. 511–522.

MANUSCRIPT ACCEPTED BY THE SOCIETY SEPTEMBER 1, 1983

Geological Society of America
Memoir 162
1984

Paleomagnetic results from rhyolites (Early Cretaceous?) and andesite dikes at two localities in the Ocana area, northern Santander Massif, Colombia

R. B. Hargraves
Department of Geological and Geophysical Sciences
Princeton University
Princeton, New Jersey 08544

R. Shagam
Department of Geology and Mineralogy
Ben-Gurion University of the Negev
Beer Sheva, Israel 84120

R. Vargas
G. I. Rodriguez
Ingeominas
Bogota, Colombia

ABSTRACT

The mean remanent magnetization vector of six samples from andesite dikes at one locality after 15 mT ODF is D = 309, I = +77, k = 13, α_{95} = 20. Five rhyolite samples from each of two sites at a second locality yield: 50 mT ODF, D = 194, I = +30, k = 166, α_{95} = 6; 30 mT ODF D = 158, I = +46, k = 101, α_{95} = 8. These rocks are probably of Mesozoic age and the magnetization appears primary; but the vectors are unlike those for the stable South American craton at any time during the Phanerozoic. These data suggest that complex tectonic rotations have occurred in at least this part of the Santander Massif.

INTRODUCTION

In order to test the feasibility of paleomagnetic studies on rocks of the Santander Massif, suites of samples were collected at three localities in the vicinity of Ocana, Norte de Santander, in the Cordillera Oriental of Colombia. Sample localities are shown in Figure 1 (collections by Shagam, Vargas, and Rodriguez).

Locality I, at water's edge in a narrow gorge of the R. Algodonal, is a fresh outcrop of granite with a pronounced spaced (~0.5 cm) tectonic foliation. The six samples came from three (at least) thin (15–25 cm thick) closely spaced, andesitic dikes that clearly transect the foliation of the granites. Relationships between these dikes are obscured by alluvium.

Locality II consists of two sites alongside the Ocana-Abrego road (Fig. 1). The first (a), a small culvert, exposed fresh, tough,

dark gray rhyolite porphyry with K-feldspar, plagioclase, and quartz phenocrysts. Samples from the second site (b) were collected from a long, high roadcut in rhyolite porphyry near the crest of a hill about 350 m southeast of the culvert. The dark fresh rhyolite produces an off-white kaolinitic regolith in contrast to the yellows of weathered granite in the area. The intimate mixture of those colors in the field underscores the problem of mapping igneous rock facies in detail, and to date most local workers group granite and volcanics in a single granite-rhyolite complex.

A few kilometres north of locality II, granite shows a highly convolute contact with upper Paleozoic (Carboniferous?) carbonates suggestive of intrusive relationships, although there is no overt sign of contact metamorphism. Localities I and II (Fig. 1)

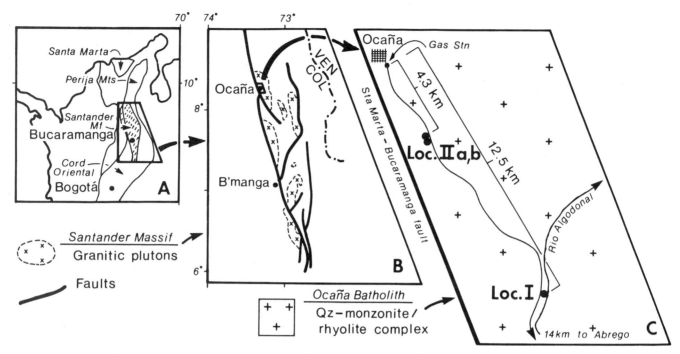

Figure 1. Location map. A = regional setting; B = Santander Massif to show distribution of granitic plutons; C = rough sketch map of part of Ocana batholith to show localities sampled for paleomagnetic study.

fall in a granite-rhyolite complex (the Ocana batholith) assigned an Early Jurassic age on the geologic map of Colombia (Arango and others, 1976). This probably derives from field observations and isotopic age determinations of Goldsmith and others (1971). They did not date the granite but inferred that it was probably of Early Jurassic age on the basis of its petrographic similarity to three other batholiths to the south which gave near-identical K-Ar (biotite) ages in the range of 192 to 196 ± 7 m.y. They did measure a K-Ar (sanidine) age of 127 ± 3 m.y. on the rhyolite near Ocana. Their location map suggests that the sample locality must have been very close to (if not the same as) our locality II roadcut. Noting that dacite porphyry dikes, ~50 to 100 km farther south, cut an Early Cretaceous formation, Goldsmith and others (1971) conceded the possibility that the rhyolite is of Early Cretaceous age. Hence, it may represent a separate phase of igneous activity distinctly younger than the granites.

The northern Santander Massif passes northeastward through a dogleg into the Perija Mountains where there is evidence for Middle Jurassic granites and a wide range of volcanic rocks, including rhyolite; these range in age from the Late Jurassic possibly into earliest Cretaceous time (see data compiled by Maze, 1984). From the foregoing, one can only conclude that the rhyolites of locality II have an age in the range of Permian to earliest Cretaceous. The pronounced tectonic foliation in the granite at locality I is puzzling; it cannot be ascribed to Tertiary shearing on the major Santa Marta–Bucaramanga fault to the west inasmuch as the andesitic dikes clearly postdate it (no Tertiary igneous activity is known on the massif). Although the

regional map suggests that locality I pertains to the same granite-rhyolite complex as locality II, it may constitute an early (Permian?) phase of a lengthy igneous event. The age of the dikes relative to the rhyolites of locality II is not known.

DISCUSSION OF PALEOMAGNETIC DATA

Oriented samples were cored and their paleomagnetism measured in the Rock Magnetism Laboratory, Princeton University. The results obtained are shown in Table 1.

As is indicated by Zijderveld plots (Fig. 2a) and by the similarity between the NRM directions of individual samples both before and after demagnetization (as well as in the site means), the remanence of the andesite dikes, although weak, is approximately single component up to 20 mT, after which it

TABLE 1. MAGNETIC DATA (IN SITU) FOR SAMPLES FROM SANTANDER MASSIF

Locality/ site	Rock type	n	D_{NRM}	I_{NRM}	k	α_{95}	ODF_{mT}	D	I	k	α_{95}
I	andesitic dikes	6	342	74	10	23	15	309	77	13	20
II a	rhyolite porphyry	5	188	35	141	7	50	194	30	166	6
II b	rhyolite porphyry	5	157	49	81	9	30	158	46	101	8
							50	154	54	26	15
(mean of IIa & b)		(2)	175	43	18	62	(30)	176	40	17	66

n = number of samples; D = declination, I = inclination, κ = Fisher's kappa, α_{95} = cone of 95% confidence; ODF = optimal demagnetizing field in millitesla (mT).

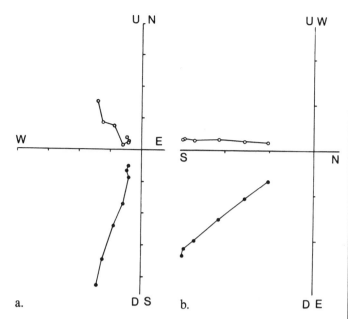

Figure 2. Representative Zijderveld plots illustrating A. F. demagnetization behavior in eight steps, N - 50 mT: a, andesite dike sample 19-2-1, units 1×10^{-3} A/m; b, rhyolite sample 20-3-1, units 1×10^{-2} A/m.

begins to scatter. These dikes are so thin that it is unlikely that secular variation is averaged out, and the scatter between samples from different dikes is attributed to this cause.

After elimination of a small soft component, the NRM of the rhyolites (sites IIa and IIb) is stronger, consistent and extremely stable to 50 mT (see Fig. 2b). The vectors for these two sites, though broadly similar, are yet significantly different; their mean is also given in Table 1.

These magnetizations are stable but clearly very different from the magnetization of the present earth's field; hence, there is no reason to suspect that the magnetizations are other than primary. The NRM directions, however, are quite unlike those for the stable South American craton at any time in the Phanerozoic. Clearly, radical tectonic rotations have occurred, but the sense and magnitude are completely unknown. In this regard it is pertinent to recall the results reported by Scott (1978) on the Payande red beds, reputedly of Triassic age. Taken in conjunction with the data from Jurassic and Cretaceous rocks farther north, as shown in Figure 3, it is clear that the aggregation of northwestern South America was a complex process (Feininger, 1983).

The data presented here demonstrate that rock types suitable for paleomagnetic study are present in the Santander Massif; a thorough study of this tectonic block, together with isotopic age dating, may prove extremely rewarding.

ACKNOWLEDGMENTS

This research was supported by Ingeominas, Colombia and National Science Foundation Grant EAR 8008207. We thank Naoma Dorety who made all the magnetic measurements.

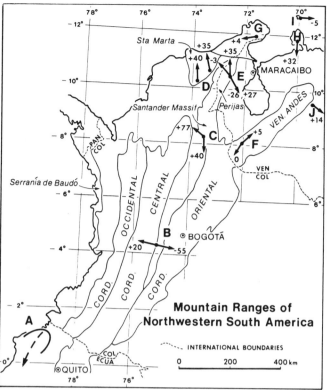

Figure 3. Northwestern South America with some geologic and paleomagnetic evidence for allochthony and major tectonic rotations. A, northern limit of allochthonous terranes in Ecuador that extend south (broken arrow) to northern Peru (Feininger, 1983). Paleomagnetic evidence of rotations and/or translations are displayed by arrows representing various generalized Mesozoic-age vectors at localities: B, Payande red beds, Scott (1978); C, this study; D, MacDonald and Opdyke (1984); E, Maze and Hargraves (1984); F, La Quinta red beds, Hargraves and Shagam (1969); G, La Teta lavas, MacDonald and Opdyke (1972); H, Santa Ana Complex, Skerlec and Hargraves (1980); I, Aruba batholith, Stearns and others (1982); J, Cerro Pelon complex, Skerlec and Hargraves (1980).

REFERENCES CITED

Arango C., J. L., Kassem B., T., and Duque-Caro, H., compilers, 1976, Mapa geologico de Colombia: Colombia Instituto Nacional de Investigaciones Geologico-Mineras, Ministerio de Minas y Energia, Bogota, scale 1:1,500,000.

Feininger, T., 1983, Allochthonous terranes in the Andes of Ecuador and northwestern Peru [abs.]: Geological Association of Canada, Abstract Volume 8, p. A22.

Goldsmith, R., Marvin, R. F., and Mehnert, H. H., 1971, Radiometric ages in the Santander Massif, Eastern cordillera, Colombia Andes: U.S. Geological Survey Professional Paper 750-D, p. 44–49.

Hargraves, R. B., and Shagam, R., 1969, Paleomagnetic study of La Quinta Formation, Venezuela; American Association of Petroleum Geologists, Bulletin, v. 53, p. 537–552.

MacDonald, W. D., and Opdyke, N. D., 1972, Tectonic rotations suggested by paleomagnetic results from Northern Colombia, South America; Journal of Geophysical Research, v. 77, p. 5720–5730.

MacDonald, W. D., and Opdyke, N. D., 1984, Preliminary paleomagnetic results from Jurassic rocks of the Santa Marta massif, Colombia, *in* Bonini, W. E., and others, eds., The Caribbean–South American plate boundary and regional tectonics: Geological Society of America Memoir 162 (this volume).

Maze, W. B., 1984, Jurassic La Quinta Formation in the Sierra de Perija, northwestern Venezuela: geology and tectonic environment of red beds and volcanics, *in* Bonini, W. E., and others, eds., The Caribbean–South American plate boundary and regional tectonics: Geological Society of America Memoir 162 (this volume).

Maze, W. B., and Hargraves, R. B., 1984, Paleomagnetic results from the Jurassic La Quinta Formation in the Perija Range, Venezuela, and their tectonic significance, *in* Bonini, W. E., and others, ed., The Caribbean–South American plate boundary and regional tectonics: Geological Society of America Memoir 162 (this volume).

Scott, G. R., 1978, Translation of accretionary slivers: Triassic results from the Central Cordillera of Colombia [abs.]: EOS (American Geophysical Union Transactions), v. 59, p. 1058–1059.

Skerlec, G. M., and Hargraves, R. B., 1980, Tectonic significance of paleomagnetic data from northern Venezuela; Journal of Geophysical Research, v. 85, p. 5303–5315.

Stearns, C., Mauk, F. J., and Van der Voo, R., 1982, Late Cretaceous–Early Tertiary paleomagnetism of Aruba and Bonaire (Netherlands Leeward Antilles); Journal of Geophysical Research, v. 87, p. 1127–1141.

MANUSCRIPT ACCEPTED BY THE SOCIETY SEPTEMBER 1, 1983

Geological Society of America
Memoir 162
1984

Structural style, diapirism, and accretionary episodes of the Sinú–San Jacinto terrane, southwestern Caribbean borderland

Hermann Duque-Caro
Instituto Nacional de Investigaciones Geológico-Mineras
Ingeominas
Bogotá, Colombia

ABSTRACT

A further review of the principal structural and stratigraphic characteristics of the northwestern Colombia area, in the southern Caribbean borderland has led me to revise previous interpretations, now recognizing it as a new tectonostratigraphic province in which diapirism is the principal stress factor in modeling what here is termed the Sinú–San Jacinto terrane.

Gravity loading of pelagic and hemipelagic sediments by denser turbidites triggered diapirism along the trench margins during the two principal accretionary episodes in the evolution of this terrane. The magnitude of deformation combined with uplift characterizes this diapirism as orogenic. The scales of deformation and uplift also appear directly related to the intensity of turbidite sedimentation. Lateral compressional stresses previously proposed are considered a minor contribution to the modeling of the Sinú–San Jacinto terrane.

The shaping of this terrane took place in two episodes corresponding to the development of the San Jacinto and Sinú accretionary wedges. The main diapiric events of the San Jacinto belt were during Paleocene to middle Eocene and during late Eocene to early Oligocene times, and the main diapiric events of the Sinú belt were during middle Miocene and during late Miocene to early Pliocene times. Each episode of intense diapirism was followed by uplift accompanied by little deformation. The diapirism appears related to deep-sea sedimentation, whereas uplift was accompanied by shallower marine and terrestrial sedimentation. Two major unconformities are characteristic of each major diapiric event: middle Eocene and early Oligocene for the San Jacinto belt and late middle Miocene and pre–late Pliocene for the Sinú belt. The former tectonic-sedimentary stages are revised, and three new names, "Sinuian," "Turbacian," and "Sincelejian," are proposed.

INTRODUCTION

The Sinú–San Jacinto terrane, which occupies the northern extension of the Western Geologic province of Colombia (Figs. 1, 2) is a continental borderland zone striking obliquely to the present coast; to the east and to the west are limited by the Romeral and Colombia structural lineaments, respectively (Duque-Caro, 1979). The exposed sedimentary sequences include highly deformed pelagic and hemipelagic facies, somewhat less deformed and very thick turbidite sequences, and only gently deformed shoaling marine sequences topped in places by lacustrine-alluvial strata. While the sequence of facies is essentially the same in both belts, the ages are notably different: In the eastern (San Jacinto) belt, the deformed rocks are of Cretaceous to early Oligocene age, while in the western (Sinú) belt, they are of Oligocene to early Pliocene age.

The older, deep-water facies emerge in narrow, steep-flanked anticlines, which are separated by broad synclines. In parts of the Sinú belt, mud volcanoes are associated with anticlines (Duque-Caro, 1979). In the subsurface, these units are

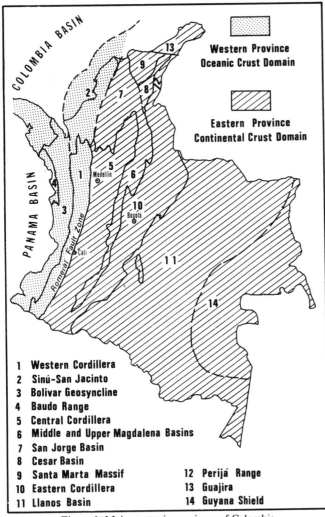

1 Western Cordillera
2 Sinú–San Jacinto
3 Bolivar Geosyncline
4 Baudo Range
5 Central Cordillera
6 Middle and Upper Magdalena Basins
7 San Jorge Basin
8 Cesar Basin
9 Santa Marta Massif
10 Eastern Cordillera
11 Llanos Basin
12 Perijá Range
13 Guajira
14 Guyana Shield

Figure 1. Major tectonic provinces of Colombia.

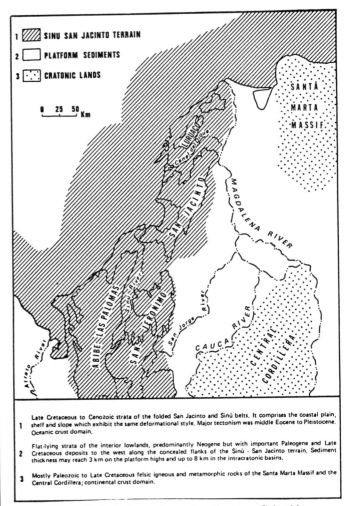

1 Late Cretaceous to Cenozoic strata of the folded San Jacinto and Sinú belts. It comprises the coastal plain, shelf and slope which exhibit the same deformational style. Major tectonism was middle Eocene to Pleistocene. Oceanic crust domain.

2 Flat-lying strata of the interior lowlands, predominantly Neogene but with important Paleogene and Late Cretaceous deposits to the west along the concealed flanks of the Sinú - San Jacinto terrain. Sediment thickness may reach 3 km on the platform highs and up to 8 km in the intracratonic basins.

3 Mostly Paleozoic to Late Cretaceous felsic igneous and metamorphic rocks of the Santa Marta Massif and the Central Cordillera; continental crust domain.

Figure 2. Geologic provinces of northwestern Colombia.

readily seen in seismic profiles in which complex, diapiric anti-clines appear as transparent domes surrounded and outlined by coherent flat or gently dipping reflections.

Previous interpretations (Duque-Caro, 1979) invoke lateral compressional stresses normal to the continental margin asso-ciated with a continuing interaction between the Caribbean oce-anic (Colombia Basin) and northern South America continental crusts as the principal forces operating from Late Cretaceous to Holocene time. This lateral compression resulted in the successive accretion of the San Jacinto and Sinú belts, the two distinct components of the Sinú–San Jacinto terrane. In the same fashion, the illustrated accretion of the San Jacinto belt in particular (e.g., Figs. 9–13, Duque-Caro, 1979) was conventionally adopted from current models of plate-tectonics.

A reevaluation of some other known geologic evidences not interpreted before, such as the structural style, geomorphology, and the distinctive pattern of highly deformed pelagites and he-mipelagites underlying thick turbidite strata, now indicates that while the gross accretion of these terranes is probably the result of

plate convergence, the deformation of the sediments may be only indirectly related to regional stresses and shortening.

The purpose of this study is therefore oriented toward eval-uating and integrating the geologic data of previous and new observations in the context of the revised regional frame for the Sinú–San Jacinto terrane.

STRUCTURAL STYLE

The Sinú–San Jacinto terrane, commonly called the Mon-tañas de María and the Sinú trough, or the unstable region or geosyncline, comprises two accretionary wedges, the San Jacinto and the Sinú belts (Duque-Caro, 1979). Each displays similar structural styles composed of narrow, steep, and elongated anti-clinal structures and broad gentle synclines.

San Jacinto Belt

This accretionary feature (Duque-Caro, 1979), which forms

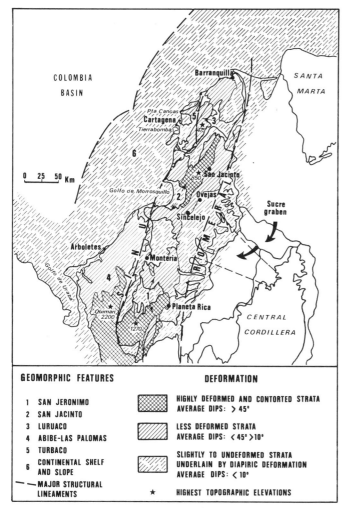

Figure 3. Major geomorphic features and deformational characteristics of the Sinú–San Jacinto terrane.

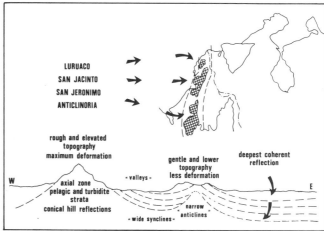

Figure 4. Schematic cross-section of the deformational style of the San Jacinto belt, based on surface and subsurface information.

the eastern part of the Sinú–San Jacinto terrane (Fig. 3), lies immediately adjacent to the platform. It is limited both to the east and to the west by the Romeral and Sinú structural lineaments, respectively (Duque-Caro, 1979; Fig. 3). Two kinds of landforms characterize this tectonostratigraphic feature: (1) Minor structural units, expressed by single conical hills of intensively contorted pelagic and hemipelagic strata that are mostly located toward the western flanks of the belt, for example, in the Golfo de Morrosquillo area, where these features appear as isolated masses emerging from the surrounding flat terrain and reaching altitudes of 20 m (Duque-Caro, 1979). (2) Major complex structural units, separated by low-lying swampy terrain that increases in altitude southward. Conical hills reach elevations of 475 m on the Luruaco (Fig. 3), 850 m on the San Jacinto, and 1,270 m on the San Jerónimo anticlinoria. The axial zone of these features shows rough and elevated topography (Fig. 3) and corresponds to narrow, elongated, and tight anticlinal structures, commonly faulted along the strike. The oldest strata along the cores of these anticlines, cherts and siliceous mudstones of Late Cretaceous age, are

surrounded by turbidites of Paleocene to middle Eocene age (Duque-Caro, 1979). Toward the nucleus zone of the uplift, the cherts and siliceous mudstones and the turbidites are contorted, but commonly the cherts and siliceous mudstones appear more intensely deformed (macrofolded and microfolded) than the overlying turbidite strata. Average dips of more than 45° in the nuclear zones characteristically wane gradually on the flanks to average less than 45° (Fig. 3). Concurrently, the rough terrain gives way to a more gentle and rounded topography as altitude decreases (Figs. 3, 4). Here, both in surface and subsurface, distinctive wide and elongated synclines are less deformed than are the anticlines. Characteristically, the peripheral and elongated anticlinal structures are rounded, and their exposed nuclei are younger than are those of the axial zone, for example, in the San Jacinto area (Figs. 3, 4).

The seismic expression of this structural style is also very distinctive. Conical reflections and domelike transparent zones ("rough topography") in the eastern flanks of the San Jacinto belt are particularly characteristic below the deepest coherent reflection and toward the corresponding areas of the uplifts (Fig. 4). Here, these conical reflections and domelike transparent zones ("rough topography") coincide with and would seem derived from the contorted pelagite, hemipelagite, and turbidite strata previously discussed. This phenomenon has also been observed in other areas of the Caribbean, for example, in the adjacent areas of the Beata Ridge, Colombia Basin side (Duque-Caro, 1979), below Houtz and Ludwig's (1977, Fig. 2) prominent reflector (PR). There, this seismic expression would also be derived from a deformed Horizon B" and associated volcanics (Duque-Caro, 1979, p. 337).

The fundamentally different S-shaped structures of the Sincelejo-Ovejas area (Fig. 3; structures 2 and 3 *in* Duque-Caro, 1979, Fig. 4) are associated with the tectonics of the platform margin and will be discussed below.

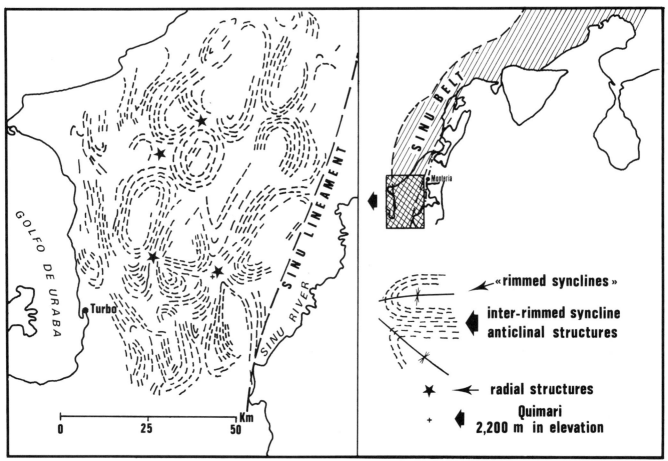

Figure 5. Structural style of the Sinú belt based on photogeologic and radar imagery information.

Sinú Belt

This tectonic feature, the second accretionary wedge of the Sinú–San Jacinto terrane, adjoins the west flanks of the San Jacinto belt, from which it is separated by the Sinú structural lineament (Fig. 3; Duque-Caro, 1979). Like the San Jacinto belt, it comprises two distinctive kinds of landforms: (1) Simple conical masses of mud up to 20 m high emerge isolated mostly along the major fault zones following the main trend of the belt. The Cañaverales and El Totumo mud volcanoes along the Sinú lineament in the Cartagena-Barranquilla area, the Punta Canoas and Arboletes mud volcanoes along the coastal margins (Fig. 3), and hills on the sea floor off the Magdalena delta (Shepard, 1973) are examples. (2) Larger conical hills of deformed sediments, increasing in altitudes southward, are found from the Turbaco to the Abibe–Las Palomas anticlinoria (Fig. 3). The axial zones of the Abibe–Las Palomas anticlinoria exhibit the maximum deformation of the Sinú belt (Fig. 3). There, two conspicuous hills, at the Alto de Quimarí area, emerge as the highest (2,200 m) and most distinctive topographic elevations of the area. Radial patterns composed of narrow, steep, and elongated anticlinal structures, always coinciding with the highest topographic elevations, surrounded by four or more wide and arcuate synclines ("rimmed

synclines," Fig. 5) build the distinctive structural style of the Sinú belt. Here, as in the San Jacinto belt, highly deformed pelagic and hemipelagic beds compose the nuclei of the anticlinal structures and always are overlain by turbidites. In the Canal del Dique and Montería-Arboletes areas (Fig. 3), lower Miocene pelagic carbonates and radiolarian chert beds are overlain by middle Miocene turbidite strata. Geomorphically, radial drainage patterns diverge from the centers of the structures. The corresponding seismic expression appears to be the same displayed in outcrops: narrow, steep, anticlinal structures, mostly characterized by domelike transparent or blank zones of almost no reflection.

To the north in the Turbaco anticlinorium (Fig. 3), topographic elevations decrease and maximum altitudes up to 200 m are characteristic. There, numerous vents at the top of the structure extrude mud, for example, in Loma de los Volcanes. In the Turbaco anticlinorium, the deformed cores remain concealed below less-deformed strata of Pliocene-Pleistocene age; however, seismic sections both onshore and offshore show the characteristic domelike transparent or blank zones surrounded by clear horizontal reflections.

The continental shelf and slope (Fig. 3), the third component of the Sinú belt, have the same characteristic structural style as shown by seismic and exploratory well information. Tierrabomba and all of those islands bordering the present coastal margin of northwestern Colombia are built on the complex anticlinal units characteristic of the Sinú belt.

Discussion

Diapiric structures of mud or shale are a characteristic feature of the Sinú belt. The structural parallels between the two belts make it seem likely that they also played a role in the development of the contorted anticlinal cores of the San Jacinto belt.

Diapirism may be defined as the process in which materials from deeper crustal levels have discordantly intruded into higher levels (O'Brien, 1968). Small-scale clay diapirism of this sort has been investigated in the "mud-lumps" of the Mississippi delta (Morgan, 1961; Morgan and others, 1968). The materials that form the diapirs in the Colombian coastal belt are the pelagic and hemipelagic muds that form the basal part of the sedimentary sequence. The beds they pierce are the subsequent turbidite sequences. Presumably, it was the load of several kilometres of turbidites that produced a gravity inversion and overpressure in the underlying pelagic beds and induced their mobilization. The observation that sandstones and wildflysch conglomerates immediately overlie pelagite strata (cf. Duque-Caro, 1979, p. 344) on the highest topographic elevations of the Sinú–San Jacinto terrane makes it appear probable that the diapirs rose out of the axes of turbidite basins, where the load would have been greatest.

In the same fashion, the effect of diapiric deformation appears to be restricted, both vertically and laterally, to those beds composed of the piercing pelagites and hemipelagites and the overlying turbidites. Multichannel seismic records of the Colombia Basin and the adjacent Sinú belt (Lu and McMillen, 1982) illustrate this phenomenon. Undeformed landward-dipping reflectors pass beneath the diapirically deformed lower slope of the Sinú belt. This would indicate that sedimentation prior to the diapirically deformed reflections was mostly pelagic without much influx of terrigenous sediments and therefore with the non-appropriated conditions for diapirism.

In summary, the Sinú–San Jacinto terrane is characterized by the following: (1) There is a distinctive structural style of narrow, steep, and elongated anticlinal structures separated by broad synclines ("rimmed synclines"). (2) Highly deformed pelagic and hemipelagic strata occur in the nuclei of anticlines and are overlain by thick turbidite strata. (3) Rough and elevated topography coincides with maximum deformation, in contrast with the gradual decrease of deformation and topographic elevations farther from the nucleus zone of the anticlinoria. (4) Mud volcanoes are common in the Sinú terrane, and shale diapirs have been recognized in the subsurface. (5) The construction of the Sinú belt appears to be a historical repetition of the growth of the San Jacinto belt. (6) In each case, deposition of pelagic and hemipelagic sequences, presumably on the deep ocean floor, was succeeded by an episode of very thick turbidite sedimentation, in what may have been a trench. In each case, this was succeeded by

intense deformation of the basal pelagites—at least in the anticlines—and a lesser degree of deformation of the turbidites, decreasing toward the synclines. This deformation has at least a strong diapiric component.

How much of the accretion of these two belts to the continent has resulted from simple sedimentary accumulation? What was the role of diapirism? How much lateral compression was involved, and when did it occur? The answers to these questions are not clear, but in order to answer them even partially we must turn to a more detailed examination of the stratigraphy and depositional-deformational history of this terrane.

The Sinú–San Jacinto terrane resulted from two progressive accretionary episodes, San Jacinto and Sinú, of deformation and emergence during the early Cenozoic pre-Andean and late Cenozoic Andean phases of the Andean orogeny (Duque-Caro, 1979). Each accretionary prism began as deep ocean floor and ended up as a belt of land of shoal water accreted to the continent. Previously, I have proposed three major tectonic and sedimentary stages, mainly from areas located within the San Jacinto belt, to characterize the evolution of northwestern Colombia (Duque-Caro, 1972, 1975): (1) the Late Cretaceous to middle Eocene Cansonian, (2) the late Eocene to middle Miocene Carmenian, and (3) the late Miocene to Pliocene Tubarian. The corresponding sedimentary sequences are separated by regional unconformities, which were recognized by both the foraminiferal paleobathymetric variations from deep to shallow and by facies changes from deep-water turbidites to shallow facies.

On the basis of new knowledge gathered from the recognition of two progressive accretionary episodes and from the diapiric mechanism previously discussed, the terminology applied to the San Jacinto belt is revised and three new stages, the Sincelejian, the Sinuian, and the Turbacian, are proposed.

SAN JACINTO ACCRETIONARY EPISODE

Three major tectonic and sedimentary stages reflect the deformation and emergence history of this accretionary episode: (1) the deep-sea Cansonian, (2) the shoaling marine Carmenian, and (3) the terrestrial Sincelejian. Diapirism during the late Cansonian and early Carmenian was followed, for the most part, by emergence accompanied by little deformation during the late Carmenian and Sincelejian stages.

Cansonian Deep-Sea Stage

This episode was originally described from the Arroyo Alferez section (Duque-Caro, 1972) in the San Jacinto belt. Cansonian rocks crop out mainly along the axial zone of the belt and include the most intensively deformed strata of the Sinú–San Jacinto terrane (Figs. 3, 6): Late Cretaceous to middle Eocene pelagites, hemipelagites, and turbidites of 4,000 m and more thick. To the south, where these deposits are best exposed and where the highest topographic elevations are located, the former Late Cretaceous pelagites and hemipelagites are associated with

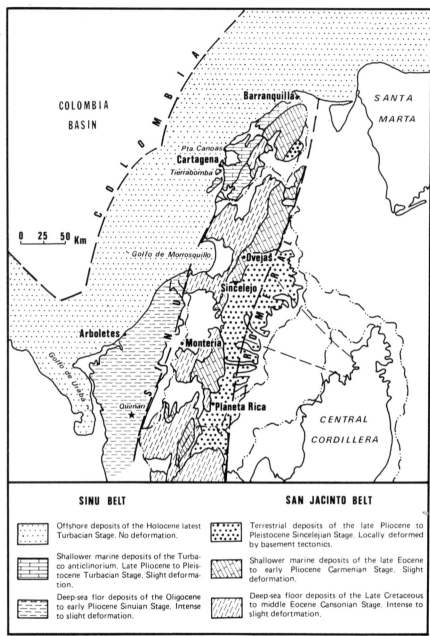

Figure 6. Major sedimentary and tectonic stages of the Sinú–San Jacinto terrane.

mafic and ultramafic intrusives and interbedded basaltic flows. There, they are characteristically underlain by another sequence of Late Cretaceous turbidites, more than 2,000 m in thickness (Duque-Caro, 1979). The chert beds at the top of this pre-Cansonian turbidite sequence (Fig. 7) have been correlated with the Caribbean seismic reflector Horizon B" (Duque-Caro, 1979).

All of the stratigraphic sequences of the Cansonian Stage were deposited in a deep-sea environment, marginal to the Romeral Paleotrench (Duque-Caro, 1979).

Late Cansonian Diapirism. A Paleocene unconformity has been suggested based on both paleobathymetry and lithologic character changes from the mainly Late Cretaceous pelagites and

hemipelagites to the turbidites (Duque-Caro, 1975, 1979). New evidences in outcrops show that the stratigraphic contact between the pelagites and the overlying Paleocene turbidites appears unconformable. However, some of the deformational characteristics earlier discussed, for example, the contorted appearance of the cherts and siliceous mudstones versus the less contorted turbidite strata, may be tectonic as a result of the diapiric emplacement of the pelagites. This diapiric emplacement took place during the turbidite deposition as an effect of gravity loading of the denser turbidites earlier discussed.

This event marked the initial phase of deformation and emergence of the San Jacinto belt and was active until middle

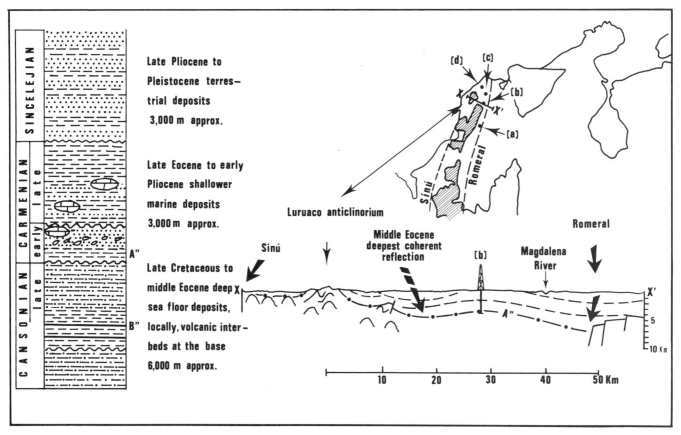

Figure 7. Stratigraphy and cross-section of the San Jacinto belt based on surface, seismic data, and well-log interpretation.

Eocene time. The distinctive early Tertiary basins of northwestern Colombia began to take shape during this time interval.

Middle Eocene Unconformity. An unconformity marks the closing of the former diapiric episode, and with this, the closing of the deep-sea Cansonian Stage and the beginning of the shallowing marine Carmenian Stage (Duque-Caro, 1972, 1973, 1975, 1979). Previous field observations made within the San Jacinto anticlinorium (Duque-Caro, 1972) confirm the existence of an angular unconformity between the deep-water turbidite facies of the Cansonian Stage and the overlying shallower facies of the Carmenian Stage. In the Arroyo Alferez section (Duque-Caro, 1972) there is marked contrast between the dips of the Cansonian strata and those of the overlying Carmenian beds. On the average, the Cansonian Stage shows dips above 45° in contrast with dips of the Carmenian Stage which average less than 45°.

In subsurface, combined biolithostratigraphic well-log information and seismic data on the eastern flanks of the Sinú–San Jacinto terrane show in most cases a hemipelagic calcareous mudstone interval always lying on top of the upper Cansonian turbidites. This mudstone interval contains the middle Eocene *Globorotalia bolivariana* fauna of the basal Carmenian Stage, which has always been found located by the deepest coherent seismic reflection. I use this mudstone interval, therefore, to date the unconformity between the Cansonian and the Carmenian

stages. In the same fashion, I correlate the former deepest coherent seismic reflection (Figs. 7, 8) with the Caribbean seismic Horizon A" (Duque-Caro, 1975, 1979). This correlation is supported by the occurrence of middle Eocene calcareous and siliceous mudstones and cherty beds resting on top of the Cansonian turbidites in the vicinity of the Luruaco (Arroyo Henequen) and Montería areas, among other localities (Fig. 6).

Carmenian Shoaling Stage

I have now included the former Carmen and Tubará paleo-bathymetric cycles (Duque-Caro, 1972) within a single tectonic-sedimentary episode to include all of the marine sediments deposited in those distinctive basins left by the previous Cansonian diapirism. They consist of mostly marine terrigenous clastic sediments, facially variable, and of hemipelagites (Fig. 7). The basal part of the Carmenian sequence is characteristically composed of hemipelagites of middle Eocene age that unconformably overlie the turbidites of the late Cansonian Stage. The uppermost part unconformably underlies the late Pliocene-Pleistocene terrestrial Sincelejian Stage. The lithologic facies of the Carmenian Stage range in age from middle Eocene to early Pliocene and are found mainly on the east flanks of the San Jacinto belt, where they may reach an overall thickness of 3,000 m (Figs. 6, 7). Characteristically, the Carmenian deposits are filling those

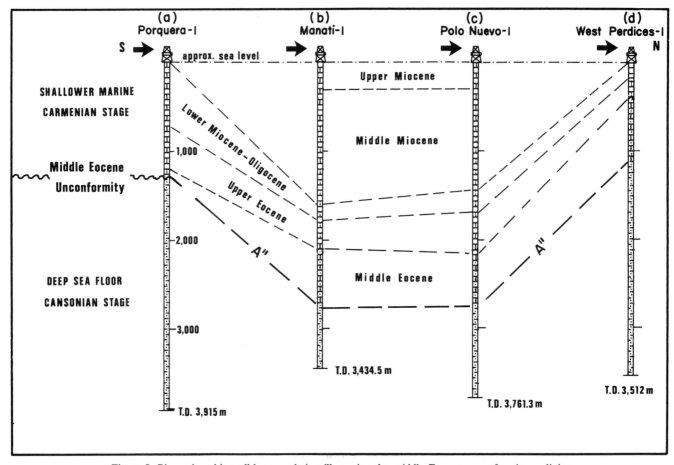

Figure 8. Biostratigraphic well-log correlation illustrating the middle Eocene unconformity outlining structural highs and lows. Refer to Figure 7 for locations of wells.

topographic lows (valleys) left by the Cansonian uplift. The unconformable relation to the previous Cansonian deposits is readily recognized in surface and subsurface sections where the middle Eocene to Pliocene stratigraphic record is more nearly complete and continuous toward the structural lows (Figs. 7, 8). On the flanks of the structures and above the structural highs, the stratigraphic record appears incomplete, particularly the late Eocene-Oligocene interval that occurs in the lows and is commonly missing above the structural highs. The wells Porquera-1 and Perdices-1, which were drilled on shallow structural highs, and the wells Manatí-1 and Polo Nuevo-1, drilled on deeper structural highs (Fig. 8), demonstrate this structural and stratigraphic phenomenon.

The depositional environment of the Carmenian Stage exhibits shallower facies than the older Cansonian Stage, and it is variable both vertically and laterally. Foraminiferal paleobathymetric interpretations (Duque-Caro, 1972, 1973, 1975, 1979) suggest several cycles of deep to shallow episodes. However, comparisons of periods of deformation and emergence from northern Colombia with deepening and shallowing episodes suggested by foraminiferal paleobathymetry indicate that (a) periods of deformation and emergence correspond with apparent bathymetric changes from deep to shallow, but that (b) not all such

inferred bathymetric changes have a corresponding structural counterpart. This would suggest that foraminiferal "paleobathymetric oscillations" do not necessarily indicate depth variations of the seafloor through time. This phenomenon will be dealt with in a separate paper.

Early Carmenian Diapirism. In the same fashion as the late Cansonian diapirism, an angular unconformity exists between the basal hemipelagites and the overlying coarse clastics of the Carmenian Stage. New observations in outcrops, toward the flanks of the Cansonian uplift (cf. Fig. 5), show middle Eocene hemipelagites (cherts and mudstones) of the basal Carmenian Stage, highly deformed and contorted, immediately underlying less-deformed, coarse, clastic facies. Much shearing has been observed at the stratigraphic contact between these two facies, particularly affecting the hemipelagites, for example, in the Luruaco (Arroyo Henequén) and Montería areas, among other localities (Fig. 6). The above factual evidence now indicates that the apparent angular unconformity between the middle Eocene hemipelagites and the immediately overlying coarse clastic deposits may be, instead, tectonic, and the unconformity marks the second major but less intense episode of diapirism within the tectonic evolution of the San Jacinto accretionary wedge. This event, which occurred during the thick, coarse clastic and intense deposition of the early Carmenian Stage, remodeled the older middle Eocene basin relief and ended up with the next hemipelagite deposition during Oligocene-Miocene time.

As earlier discussed, sedimentation appears to be a major factor controlling diapirism in the Sinú–San Jacinto terrane. Therefore, the characteristic distribution of the anticlinal structures of the younger Carmenian uplift is marking the new position of the depositional axis of the basin during the early Carmenian time (cf. Fig. 4) prior to this early Carmenian diapirism.

Early Oligocene Unconformity. Like the middle Eocene unconformity that marked the closing of the late Cansonian diapirism at the end of the turbidite deposition, the early Oligocene unconformity marks the closing of the early Carmenian diapirism at the end of the coarse clastic deposition. In outcrops at the flanks of the highest topographic elevations built by the anticlinal structures of the early Carmenian uplift, there is another hemipelagite sequence (Carmen Shale Formation) of late Oligocene to early Miocene age. These deposits are filling unconformably the lowest and deepest synclinal depressions (valleys) left by the early Carmenian diapirism, for example, in the San Jacinto and Sincelejo areas (Figs. 3, 4, 7).

Late Carmenian Emergence. The late Carmenian, which comprises the late Oligocene to early Pliocene marine deposits, was a period characterized by mostly emergence. In fact, no diapiric deformation has been observed during this time interval along the eastern flanks of the San Jacinto belt where these deposits are exposed. Here, it is important to record and discuss some previously suggested stratigraphic disturbances within this same episode (Duque-Caro, 1972, 1975, 1979). These have been discussed, mostly arguing foraminiferal paleobathymetric variants.

Oligocene-Miocene Diastrophism. Stainforth (1968) invoked paleobathymetric foraminiferal evidence from Oligocene-Miocene deposits to define diastrophic movements involving uplift and erosion in northern South America; however, in northwestern Colombia, for example, in the San Jacinto and Carmen-Zambrano areas, the Oligocene-Miocene stratigraphic record seems to be continuous (Duque-Caro, 1975), even though a paleobathymetric change was suggested by foraminifera within the Oligocene-Miocene Carmen Shale (Duque-Caro, 1975, 1979). No unconformity appears to be associated with this within the San Jacinto belt. As commented earlier, foraminiferal paleobathymetric variants in some instances, but not always, have a corresponding structural counterpart. Why does foraminiferal paleobathymetry coincide in some areas but not in others with periods of deformation and emergence? Is this phenomenon a mere coincidence? Or, instead, is there another factor controlling bathymetric distribution of biotas? (Duque-Caro, in prep.)

Middle Miocene Disturbance. A second major paleobathymetric change occurs between the Carmen and Tubará cycles (Duque-Caro, 1972). No angular unconformity has been recognized within the San Jacinto belt, although major tectonic and orogenic disturbances have been demonstrated in other areas of

Colombia and northern South America (Haffer, 1970; Van Houten, 1976; Kellogg, 1981; and others). Emergence of the Carmenian deposits of the San Jacinto belt at this time, even though they could occur, cannot presently be proven with the known physical evidences.

Pre-Late Pliocene Emergence. A gentle angular unconformity marks the contact of the slightly deformed bed rocks of the expanded Carmenian Stage with the overlying late Pliocene to Pleistocene terrestrial Sincelejian Stage (Fig. 7). This unconformity witnesses the emergence of the Carmenian deposits and, with this, the closing of the Tertiary marine episode within the San Jacinto belt and the adjacent platform after a period of emergence without much deformation.

Sincelejian Terrestrial Stage

This new name, adopted from the town of Sincelejo, is proposed for the fluvial and lacustrine sediments cropping out at the eastern flanks of the Sinú–San Jacinto terrane particularly well exposed in the Sincelejo-Ovejas region (Fig. 6). There, they overlie unconformably the older marine deposits and build the sedimentary cover of the easternmost flanks of the San Jacinto accretionary wedges and the platform. They are of late Pliocene-Pleistocene age and correspond to the last stage of the sedimentary development of the San Jacinto belt (Figs. 6, 7). The sandstones, mudstones, and conglomerates are variable in thickness, ranging from a few hundred to a few thousand metres. In the Sucre tectonic depression (Fig. 3; Duque-Caro, 1979), these deposits thicken abruptly to an estimated 4,000 m at the easternmost flanks of the Sinú–San Jacinto terrane.

Pleistocene Emergence. A last tectonic episode affecting the San Jacinto belt is recorded in the uplifted topography of the San Jerónimo Serranía and the S-shaped structures deforming the late Pliocene–Pleistocene Sincelejo Group (Dueñas and Duque-Caro, 1981) at the eastern margins of the San Jacinto belt. These have resulted from transcurrent faulting, emergence, and subsidence of the platform, mainly controlled by basement tectonics (Duque-Caro, 1979). Only the easternmost flanks of the Sinú–San Jacinto terrane, close to the margin of the platform (Romeral lineament, Fig. 6) where the Sincelejian sediments are located, were affected by this tectonic disturbance

SINU ACCRETIONARY EPISODE

Two new tectonic-sedimentary stages are here proposed to characterize the evolution of the Sinú belt: (1) the Sinuian deep-sea stage and (2) the Turbacian shoaling stage. Diapirism, in the same fashion as in the San Jacinto belt, was characteristic at an initial stage of development, which then was followed for the most part by a period of emergence up to the present.

Sinuian Deep-Sea Stage

I propose this name to characterize the tectonic and sedi-

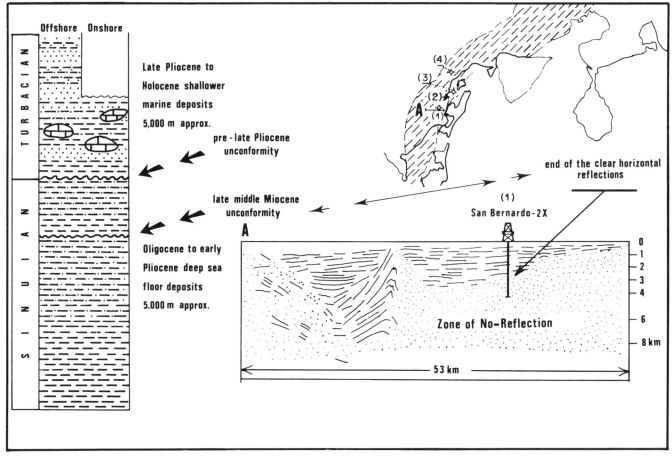

Figure 9. Stratigraphy and offshore cross-section of the Sinú belt based on seismic data and biostratigraphic well-log correlation.

mentary evolution of those deposits that crop out to the south of the Sinú belt, in the complex Abibe–Las Palomas anticlinoria, for example, in the Montería-Arboletes road (Fig. 6). They include pelagites and hemipelagites (siliceous and calcareous mudstones and cherts) and turbidites of early Miocene to early Pliocene age that may reach 5,000 m in thickness (Duque-Caro, 1979). Sediments older than Oligocene age have not been found in outcrops within the Sinú belt.

Early Sinuian Diapirism. A stratigraphic break is suggested by the contrasting structural style between the lower Miocene pelagites and the overlying middle Miocene turbidites. Whereas the pelagites are intensively deformed, for example, in the Canal del Dique and Montería-Arboletes areas, the overlying turbidites appear less deformed. This phenomenon recalls the relationships between the pelagites and the turbidites of the San Jacinto Cansonian Stage. Therefore, this contact, stratigraphically unconformable in appearance, also suggests diapiric emplacement of the pelagites. This diapirism took place during the middle Miocene turbidite deposition as an effect of gravity loading of the denser turbidites discussed earlier. This phenomenon marked the initial phase of deformation and emergence of the Sinú belt.

Late Middle Miocene Unconformity. This feature marks the climax of diapirism of the Sinú belt during the Sinuian Stage.

On the surface, for example, in the Arboletes region, Pliocene-Pleistocene strata at the flanks of characteristic "rimmed synclines" unconformably overlie middle Miocene strata. Commonly, the stratigraphic contact, coinciding with this unconformity, is sheared due to the diapiric nature of the older strata. In the subsurface, I have observed a characteristic biostratigraphic interval composed of arenaceous and reworked foraminiferal microfauna of Oligocene to middle Miocene age in samples from both onshore and offshore exploratory wells. In seismic records, this biostratigraphic interval appears to coincide at depth with the end of the clear horizontal reflections and the beginning of the blank zones (Figs. 9, 10). Therefore, since the middle Miocene was the time when the Sinú belt underwent major diapirism and emergence, particularly in the Abibe–Las Palomas anticlinoria, I think that this biostratigraphic interval is a useful tool to recognize the late middle Miocene unconformity. The resultant "rimmed synclines," which were left submerged after this diapirism, built a characteristic pattern of small rounded basins (diapiric basins) where the shallower deposits of the younger Turbacian Stage began to accumulate (Fig. 9). The end of the middle Miocene time was characterized by great instability in northwestern Colombia (Haffer, 1970; Van Houten, 1976; Duque-Caro, 1979) and in northern South America (Kellogg,

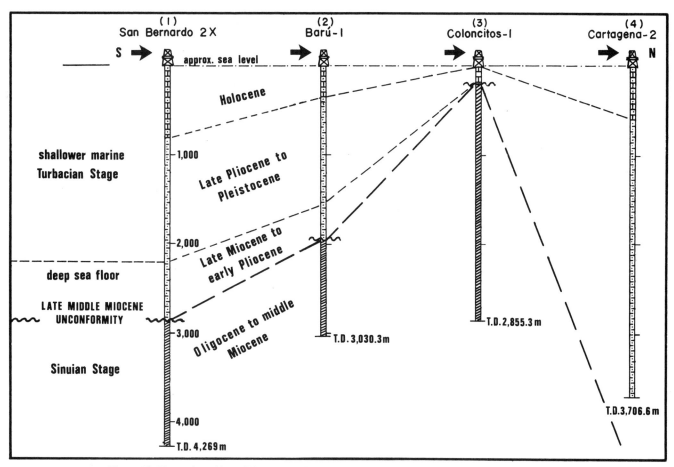

Figure 10. Biostratigraphic well-log correlation illustrating structural highs and lows that are outlined by the late middle Miocene unconformity. Refer to Figure 9 for location of wells.

1981, among others). Pre–late Miocene volcanic sands and silts at the bottom of DSDP sites 154 and 154A, Colombia Basin, record the igneous volcanic activity in Central America at the end of middle Miocene time (Edgar and others, 1973). Late middle Miocene was also the time when water circulation between the Atlantic and the Pacific Oceans first ceased as a result of the initial emergence of the Panama Isthmus (Edgar and others, 1973; Duque-Caro, in prep.).

Pre–Late Pliocene Diapirism. Pre–Late Pliocene uplift is suggested by the absence of post–early Pliocene marine deposits in the southern onshore part of the Sinú belt. In the Punta Canoas area (Fig. 6) of the Turbaco anticlinorium, late Miocene to early Pliocene strata are diapirically deformed, which indicates that diapirism was still active during this time interval. This activity coincides with the pre-Pliocene uplift of the San Jacinto belt which closed the Tertiary marine episode within the eastern region of the Sinú–San Jacinto terrane and the adjacent platform.

Turbacian Shoaling Stage

This new stage, adopted from the town of Turbaco, is proposed to characterize the tectonic and sedimentary evolution of the shallower marine sediments that were deposited within the small rounded basins formed after the Sinuian Stage diapirism.

These deposits, the second sedimentary stage of development of the Sinú accretionary episode, are well exposed to the north in the Cartagena-Barranquilla onshore area and to the south at the coastal margins of the Abibe–Las Palomas uplift (Fig. 6). They consist of clastic terrigenous and hemipelagic sediments of late Pliocene to Pleistocene age, including reefal limestones in the Cartagena-Barranquilla area. The thickness is variable and reaches 4,000 m in offshore sections, whereas onshore it is only 243 m in the well Coloncitos-1 (Fig. 10). These deposits overlie unconformably the older Sinuian Stage and form the tops and the flanks of the Turbaco anticlinorium in the Cartagena area. There, they reach elevations in the order of 170 m (Cerro La Popa) and 220 m (Turbaco region) above sea level. Seismic records show clear horizontal reflections of sediments (Fig. 9) filling the depressions left by the diapiric disturbances of the underlying strata which are highly deformed.

Formerly, these deposits were thought to be part of the Tubará paleobathymetric cycle (Duque-Caro, 1975, 1979). However, based on the present knowledge of the tectonic and sedimentary evolution of northwestern Colombia, these two facies cannot be associated because the Tubará cycle, of late Miocene to early Pliocene age (Duque-Caro, 1972), belongs to the distinctive tectonic and sedimentary province of the San Jacinto belt.

Pleistocene Emergence. Pleistocene disturbances that caused the last emergence of the Sinú belt involved some deformation at the Turbaco anticlinorium and the uplifted westernmost coastal portions of the Abibe–Las Palomas anticlinoria: the late Pliocene–Pleistocene marine deposits were uplifted in the order of 200 m above sea level. At the western flanks of these uplifts, Turbacian strata, tilted and in general dipping west, are resting unconformably on the Sinuian strata. As mentioned earlier, the stratigraphic contact between these two units is sheared due to the diapiric nature of the Sinuian strata. Middle Miocene diapiric anticlinal structures seem to have been continuously emerging after the Sinuian diapirism without much deformation of the overlying Turbacian strata. Those islands bordering the present coastal margin of northwestern Colombia were built on the diapiric anticlinal structures left by the Sinuian diapirism and have been emerging since Pleistocene time. The Turbacian Stage, like the Sincelejian Stage in the San Jacinto belt, was for the most part a period of emergence without diapiric deformation.

Offshore Deposits. We may include in the Turbacian Stage the Pleistocene to Holocene deposits forming the marginal coast beach terraces and the cover of both the islands and the offshore area of northwestern Colombia (Fig. 6). In offshore sections, overlying disconformably the older late Pliocene–Pleistocene deposits, are undercompacted clastic and carbonate deposits that reach a thickness of approximately 1,000 m (Fig. 9). In seismic records, these deposits are expressed by the upper horizontal reflections, and along the coastal margins, they form beach terraces. In some of the bordering islands, for example, in the Tierrabomba island, beach terraces of 20 and 3 m above sea level are present. Porta and Solé de Porta (1960) mapped these terraces and dated the highest as late Pliocene, and the lowest as 2850 B.P. (Porta and others, 1963). The older date would correspond to the time of uplift on the late Pliocene–Pleistocene onshore deposits of the Turbacian Stage, and the second indicates a last emergence during Holocene time. Doubt has been cast on some of these ages by the work of William D. Page (1982, personal commun.) who has obtained several C^{14} radiometric datings from beach terraces to the south of the Golfo de Morrosquillo (Fig. 6). Nearly all of the ages reported are in the order of 2,000 and 5,100 yr ago, even for terraces as high as 36 m above high tide. Ages older than 10,000 yr were assumed and are in the range of 80,000 to 125,000 yr.

SUMMARY AND CONCLUSIONS

Antecedents and Facts

The Sinú–San Jacinto terrane of northwestern Colombia comprises the San Jacinto and the Sinú belts, two accretionary wedges characterized by narrow, steep, and elongated anticlinal structures separated by broad and gentle synclines (Fig. 11).

Highly deformed pelagites and hemipelagites always over-

lain by turbidites are characteristic of the nuclear zone of the anticlinal structures that form the highest topographic elevations.

The rough and elevated topography always coincides with maximum deformation, and both decrease distally from the nucleus zone of the anticlinoria.

Multichannel seismic records of the Colombia Basin and the adjacent Sinú belt show undeformed landward-dipping reflectors passing beneath the deformed lower slope of the western margin of the Sinú–San Jacinto terrane.

The lithologic facies of the Sinú–San Jacinto terrane can be classified according to their depositional setting in (a) deep-sea deposits, composed of pelagites, hemipelagites, and turbidites; (b) shallower marine deposits composed of marine terrigenous clastic sediments and hemipelagites; and (c) terrestrial deposits composed of fluvial and lacustrine sediments.

The stratigraphic record of the San Jacinto belt, as currently known, ranges in age from Late Cretaceous to middle Eocene for the deep-sea deposits, middle Eocene to early Pliocene for the shallower marine deposits, and late Pliocene to Pleistocene for the terrestrial deposits. In the Sinú belt the respective stratigraphic record ranges from Oligocene to middle Miocene for the deep-sea deposits and late Pliocene to Holocene for the shallower marine deposits. The late Miocene to early Pliocene stratigraphic record is poorly exposed in the Sinú belt. Only in the Cartagena area is the late Miocene record represented by radiolarian and foraminiferal mudstones accompanied by younger, shallower marine foraminiferal and molluscan mudstones.

Interpretation

1. The margin of northwestern Colombia grew by successive accretion of two belts, each of which passed through a similar (but not synchronous) history of tectonic and sedimentary events. This history involves (a) deposition of pelagic and hemipelagic sediments in deep-sea environments; (b) deep burial under turbidites (trench deposits?); (c) deformation by diapiric intrusion of the former into the latter, with development of steep, narrow anticlines and broad synclines; (d) further burial under continental margin sediments, accumulated in progressively shoaler settings, partly as a result of sedimentation, but probably in part as a result of uplift; (e) partial emergence, with deposition of fluvial and lacustrine sediments in remaining basins. Deformation during (d) and (e) was gentle but led to formation of some discordances.

2. The San Jacinto belt demonstrates such a history extending from Late Cretaceous to Pleistocene time. Diapirism occurred in the Paleocene–early Oligocene interval. The Sinú belt records Oligocene-Holocene history, diapiric deformation occurring between mid-Miocene and early Pliocene time.

3. The structural deformation seen at the surface and in seismic sections appears largely attributable to diapirism.

4. The shoaling and eventual emergence of the sedimentary pile includes several possible components. One is the simple accumulation of sediments at the surface. The possibility of thickening by lateral compression cannot be wholly dismissed but lacks

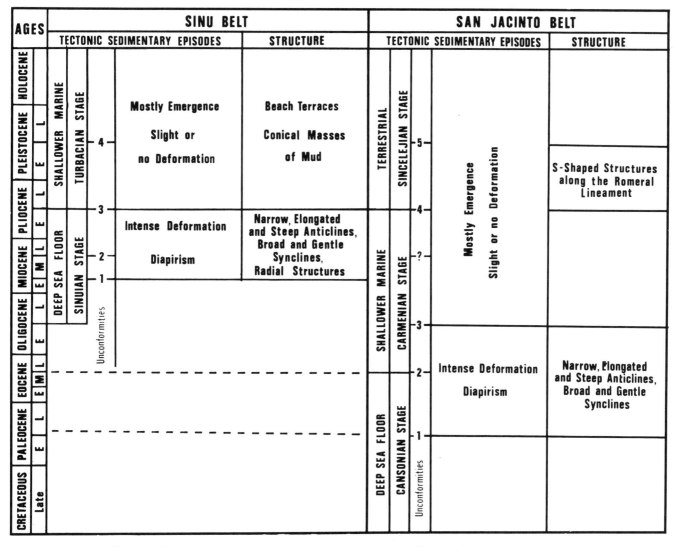

Figure 11. Accretionary episodes and characteristic structures of the Sinú–San Jacinto terrane.

observational support at this time. Yet another possibility is that the sedimentary pile was added to at the bottom by structural accretion of sediments scraped off a subducting crustal slab. If so, then the deepest parts of the sedimentary body must have a very different structure than the parts exposed.

5. Viewed in this light, the tectonic history of this terrane seems to be due to a complex combination of regional forces (plate convergence) and internal gravity-driven diapirism.

6. The great magnitude of uplift and internal deformation qualifies the tectonic-evolutionary processes of the Sinú–San Jacinto terrane as orogenic.

ACKNOWLEDGMENTS

I am deeply grateful to Alfred G. Fischer for his encouraging comments and suggestions during the review process of this paper. F. B. van Houten, R. Shagam, T. A. Konisgmark, and J. W. Ladd also have read and reviewed it and have contributed enormously to its improvement. To all of them, my sincere thanks.

REFERENCES CITED

Dueñas, H., and Duque-Caro, H., 1981, Geología del cuadrángulo F-8: Boletín Geológico, Ingeominas, v. 24, no. 1, p. 1–35.

Duque-Caro, H., 1972, Ciclos tectónicos y sedimentarios en el norte de Colombia y sus relaciones con la paleoecología: Boletín Geológico Ingeominas, v. 19, no. 3, p. 1–23.

——1973, The geology of the Montería area: Colombia Society of Petroleum Geologists and Geophysicists, 14th Annual Field Conference, Guidebook, p. 397–431.

——1975, Los foraminíferos planctónicos y el terciario de Colombia: Revista Española de Micropaleontología, v. 7, no. 3, p. 403–427.

——1979, Major structural elements and evolution of northwestern Colombia, *in* Watkins, J. S., and others, eds., Geological and geophysical investigations of continental margins: American Association of Petroleum Geologists Memoir 29, p. 329–351.

Edgar, N. T., Saunders, J. B., and others, 1973, Site 154, initial reports of the

Deep Sea Drilling Project, Volume 15: Washington D.C., U.S. Government Printing Office, p. 407–471.

Haffer, J., 1970, Geological climatic history and zoogeographic significance of the Urabá region in northwestern Colombia: Caldasia, v. 10, no. 50, p. 603–636.

Houtz, R. E., and Ludwig, W. J., 1977, Structure of Colombia Basin, Caribbean sea, from profiler-sonobuouy measurements: Journal of Geophysical Research, v. 82, no. 30, p. 4861–4867.

Kellogg, J. N., 1981, The Cenozoic basement tectonics of the Sierra de Perijá, Venezuela and Colombia [Ph.D. thesis]: Princeton University, 230 p.

Lu, R. S., and McMillen, K. J., 1982, Multichannel seismic survey of the Colombia Basin and adjacent margins, *in* Watkins, J. S., and Drake, C. L., eds., Studies in continental margin geology: American Association of Petroleum Geologists Memoir 34, p. 395–412.

Morgan, J. P., 1961, Mudlumps at the mouths of the Mississippi River, *in* Genesis and paleontology of the Mississippi River mudlumps: Louisiana Geological Survey Bulletin 35, Part I, p. 55–59.

Morgan, J. P., and others, 1968, Mudlumps: Diapiric structures in Mississippi delta sediments, *in* Braunstein, J., and O'Brien, G. D., eds., Diapirism and

diapirs: American Association of Petroleum Geologists Memoir 8, p. 145–161.

O'Brien, G. D., 1968, Survey of diapirs and diapirism, *in* Braunstein, J., and O'Brien, G. D., eds., Diapirism and diapirs: American Association of Petroleum Geologists Memoir 8, p. 1–9.

Porta, J., and Solé de Porta, N., 1960, El cuaternario marino de Tierrabomba: Boletín Geológico, Universidad Industrial de Santander, no. 4, p. 19–44.

Porta, J., and others, 1963, Nuevas aportaciones al Holoceno de Tierrabomba: Boletín Geológico, Universidad Industrial de Santander, no. 12, p. 35–44.

Shepard, F. P., 1973, Seafloor off Magdalena delta and Santa Marta area, Colombia: Geological Society of America Bulletin, v. 84, p. 1955–1972.

Stainforth, R. M., 1968, Mid-Tertiary diastrophism in northern South America, *in* Transactions, Caribbean Geological Conference, IV, Trinidad, 1965, Caribbean Printers, Arima, Trinidad, p. 159–177.

Van Houten, F. B., 1976, Late Cenozoic volcaniclastic deposits, Andean foredeep, Colombia: Geological Society of America Bulletin, v. 87, p. 481–495.

MANUSCRIPT ACCEPTED BY THE SOCIETY SEPTEMBER 1, 1983

Geological Society of America
Memoir 162
1984

Late Cretaceous condensed sequence, Venezuelan Andes

Santosh K. Ghosh*
Universidad de Los Andes
Mérida, Venezuela

ABSTRACT

The 1- to 3-m-thick Tres Esquinas sequence of Santonian to Campanian age is a foraminiferal pelletal packstone consisting of shells, intraclasts, coated grains, pellets, and minor detrital quartz—all diagenetically replaced to varying degrees by apatite, glauconite, dolomite, pyrite, and silica. The phosphatization and glauconitization of the unit may be related to early diagenesis in a geologic setting in which upwelled waters were combined with low influx of detrital sediments during a marine depositional stillstand coincident with an expanded oxygen-minimum layer. Dolomite formed in the postlithification stage mainly at the expense of calcite cement and glauconite. Finally, pyrite and granular quartz formed diagenetically from precursor opal and black mono-sulfides, respectively. The condensed sequence, probably 8 to 10 m.y. long, is significant in the context of Cretaceous sedimentation in that it separates the Coniacian La Luna Formation representing the culmination of Cretaceous transgression from the overlying Colon Formation of normal marine shelf environment. The regressive trend, first documented in the Colon Formation, reached its peak during the Tertiary Period.

INTRODUCTION

The Tres Esquinas marker horizon, 1 to 3 m thick, has long been recognized informally as the glauconitic zone in the Cretaceous of Maracaibo Basin. Stainforth (1962) assigned a formal name to this unit as the Tres Esquinas Member of the Colon Formation. Widely exposed in the Venezuelan Andes, it always occurs at the contact zone between the La Luna and Colon formations (Ghosh, 1980). This report briefly describes the profile of the Tres Esquinas glauconitic phosphorite, including its mineralogy, diagenetic history, and genesis, based on a preliminary study of some outcrops in the vicinity of the city of Mérida, Mérida State (Fig. 1).

GEOLOGIC SETTING

The Tres Esquinas Member overlies the La Luna Formation, which is a deep-water, basinal, cherty limestone-shale sequence containing an abundant ammonitic and planktonic foraminiferal assemblage (Fig. 2). The shaly sequence of the su-

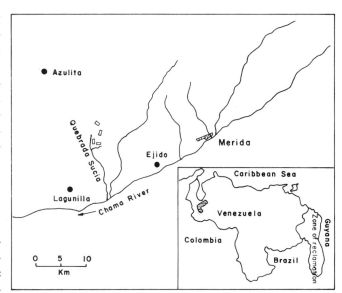

Figure 1. Location map of the study area showing the Quebrada Sucia outcrops. Inset shows location of Mérida State and the study area within it.

*Present address: INTEVEP S. A., Apartado 76343, Caracas 1070 A, Venezuela.

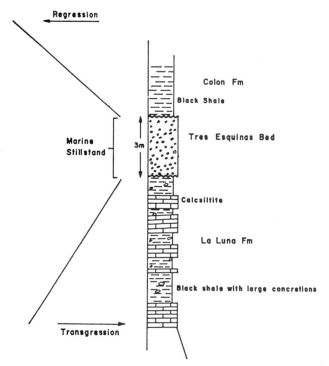

Figure 2. Stratigraphic position of the Tres Esquinas condensed sequence between the La Luna and Colon formations at the contact of a Cretaceous transgressive-regressive pair.

prajacent Colon Formation represents the closing phase of Cretaceous sedimentation in western Venezuela. Locally, in the southern part of the Maracaibo Basin, a fossiliferous unit (the Socuy Limestone) containing a mixture of pelagic and benthic foraminifera overlies the Tres Esquinas bed. The Socuy Lime-

stone is the lower member of the Colon Formation (Sellier de Civrieux, 1952).

CHARACTERISTICS OF THE CONDENSED SEQUENCE

In the Quebrada Sucia occurrence (Fig. 1) the bed is continuously exposed in the river bed for a distance of 400 m. The grayish green color is related to the presence of abundant glauconite grains (Fig. 3). The sequence lacks stratification due to abundant bioturbation. The main framework components are sand-sized foraminiferal shells, comminuted bone and tooth fragments (Fig. 4), pellets and superficially coated grains, all replaced to varying degrees by aphanitic apatite and glauconite. Impregnation by bituminous matter is common, especially within the foraminiferal chambers. Minor detrital quartz is also present. The bed is traversed by many cross-cutting calcite veins in the outcrops. The P_2O_5 content of five samples, measured by colorimetric analysis, from different horizons within the phosphorite sequence is variable and ranges from 4% to 10%. Two other important constituents are pyrite and dolomite, the latter being wholly a replacement product of the original intergranular calcite. All mineral phases have been confirmed by X-ray diffraction analysis.

In thin section the rock is a phosphatized bioclastic pelletal packstone. The dark brown apatite appears to have replaced mainly the framework elements (Fig. 5). The superficially coated grains commonly show a fossil in the core and a thin outer rind of concentric growth ring. Also, there are some homogeneous pellets of apatitic composition. The foraminifera in the nucleii of many coated grains have been completely replaced by apatite. The

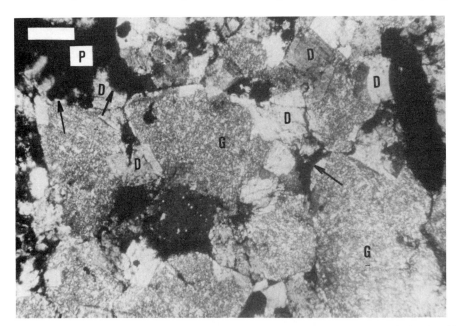

Figure 3. Photomicrograph showing glauconite grains (G) partially replaced at the margin by rhombic dolomite (D). The two arrows at the upper left show pyrite replacing both glauconite and dolomite (D). The arrow on the middle right shows pore-filling cement. Scale bar is 0.5 mm.

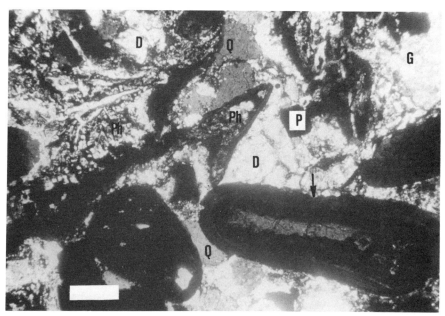

Figure 4. Photomicrograph showing phosphatic bone fragments (Ph), detrital quartz (Q), glauconite (G), pyrite (P), and dolomite (D). Arrow indicates the corroded outermargin of a phosphatic bone fragment at the contact of younger dolomite crystals. Scale bar is 0.1 mm.

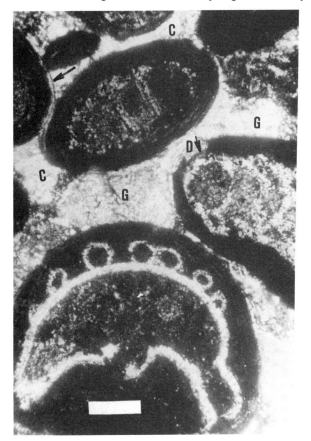

Figure 5. Photomicrograph showing superficially coated foraminifera in the midst of glauconite (G) grains and dolomitized calcite cement (C). Upper left arrow shows clear apatitic cement. Middle right arrow indicates a dolomite grain (D) replacing the dark cryptocrystalline phosphatic coating on a foraminiferal shell. Scale bar is 0.1 mm.

intraclasts, originally of calcium carbonate composition, are in general larger (up to a few millimeters), irregularly shaped and may contain broken shells and fish remains. The presence of intraclasts perhaps suggests a period of erosion, disintegration, and redeposition of calcareous material prior to phosphatization during a period of marine depositional stillstand (Kennedy and Garrison, 1975a, 1975b).

The glauconite grains occur as a framework element mixed with other bioclastic grains. Some infaunal burrows have been filled with a mixture of glauconite and phosphatized grains. Some of the glauconite grains have morphologies similar to those of fecal pellets and may show cracks filled with carbonate cement. Glauconitization of the interstitial area also have been observed in some areas.

Dolomite, rarely intergrown with ankerite, typically occurs as rhombic grains replacing intergranular calcite, glauconite (Fig. 3), and to a lesser extent phosphatized foraminifera. The intergranular calcite cement occurs very haphazardly because of pervasive replacement by dolomite. Occasional phosphatization of the intergranular cement has been observed.

Pyrite occurs in two very different forms: as large nodules (Fig. 6) and as finely disseminated grains (Fig. 4) throughout the rock. The haphazardly oriented nodules could be related to pyritization of preexisting burrow traces. The finely disseminated pyrite notably shows its cubic crystalline outline penetrating into the other minerals, such as apatite, calcite, glauconite, and dolomite, thus attesting to its late origin. Secondary silica in the form of granular quartz and relicts of opaline silica may occur as fossil skeletons, pore-filling cement, and also as an isopachous rim cement around pyrite grains.

No sedimentary structures are present in these beds, possibly

Figure 6. Outcrop view of the condensed bed traversed by calcitic veins. Large pyrite nodules (lighter color) are abundant. Some are circular in outline (arrows), suggesting pyrite mineralization within vertical burrow traces. Scale bar is 15 cm.

due to intense bioturbation. Rare local lag deposits of bones of saurians (Renz, 1980) and shark teeth may occur in this sequence.

DIAGENETIC PROCESSES

Fig. 7 attempts to show the sequence of events leading to the formation of the glauconitic phosphorite. After the deposition of the La Luna Formation, a period of marine stillstand ensued when there was virtually no sedimentation for prolonged periods. At this eogenetic stage (Choquette and Pray, 1970) of diagenesis, apatite and glauconite formed almost simultaneously, that is, succeeding each other rapidly and, perhaps, alternately (Giresse, 1980) on the sea floor at the sediment water interface and also slightly below it. That apatite and glauconite affected only the biogenic fraction and not the other mineral phases corroborates their relatively early origin. Opaline silica (a precursor of quartz) developed slightly later, either replacing the fossil skeletons or as infillings of fossils. In the reducing environment, bacterial action would have induced sulfate reduction, leading to H_2S generation and promoting precipitation of black monosulfides (Scholle, 1971).

Once the phosphatization and glauconitization occurred early in the diagenetic stage, prelithification eodiagenesis was completed by abundant bioturbation and loss of unstable organic matter. Evidence of bioturbation in this unit is plentiful in the form of straight and radiating horizontal (1 cm wide and up to 15 cm long) unidentified animal traces. Also, in some outcrops, reticulate horizontal burrow systems of the type *Thalassinoides* are present. The intergranular calcite cement developed next in the sequence. The uneven distribution of calcite could be due to

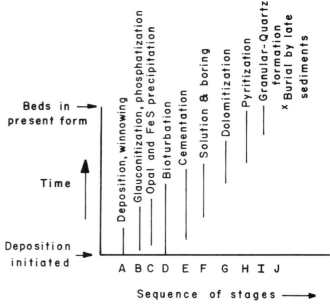

Figure 7. Probably sequence of events leading to the development of the Tres Esquinas condensed bed. Stages A to D represent events in the prelithification phase, whereas stages E to I belong to the postlithification phase.

interference by burrowing organisms. Evidence of occasional phosphatization of the intergranular calcite implies the relatively early origin of at least part of the calcite cement (Fig. 7). Postlithification but preburial diagenesis was mainly brought about by the boring organisms and solutions. Physical abrasion also could have occurred at this stage.

Subsequent events in the postlithification stage involved dolomitization, pyritization, and secondary quartz formation. Dolomite typically occurs as well-developed rhombs and replaces mainly calcite and glauconite but also apatite. Textural evidence clearly indicates that dolomite is entirely a diagenetic mineral and not a primary cement. Pyrite forms next as euhedral grains at the expense of earlier formed FeS and tends to replace all other mineral phases. Secondary quartz, mostly occurring as void-filling and rim cement, would have developed at the expense of earlier formed skeletal and pore-filling silica, concomitant with or slightly postdating pyrite formation. An occasional isopachous rim cement around pyrite crystals corroborates the later origin of quartz.

GENETIC MODEL

A model is proposed here to interpret the observed diagenetic glauconitic phosphorite in terms of paleoenvironment and paleogeography. Various authors (Summerhayes, 1970; Parker, 1975; and Birch, 1979, 1980) have noted similar stratigraphic association of diagenetic glauconitic phosphorite with unconformity surfaces related to eustatic sea-level oscillations. As with many modern and ancient analogs (D'Anglejan, 1967; Summerhayes and others, 1973; Jarvis, 1980), the Tres Esquinas phosphorite would have formed in or adjacent to belts of upwelling waters within the zone of oxygen minimum. The oxygen-minimum layer is best developed at the eastern margin of tropical oceans, where upwelling promotes high fertility and deep waters are relatively old and impoverished in oxygen (Fischer and Arthur, 1977). Several evidences, such as the enrichment in organic matter, reduced iron, and pyrite in the Tres Esquinas Member and in the La Luna Formation, imply that the phosphates formed as a replacement of calcareous material in a low oxygen, organic-rich milieu like that prevailing off South-West Africa and off Peru-Chile (Baturin, 1971; Veeh and others, 1973; Goldhaber, 1978; Giresse, 1980). The apatite mineralization took place by replacement of calcareous material in regions of low rate of deposition in the outer shelf to slope areas remote from sources of detrital influx (Manheim and others, 1975). Upwelling of nutrient-rich waters from basinal areas triggered a high biological productivity in the photic zone and subsequent sedimentation of dead zooplanktons on the marine sea floor (Birch, 1979). Eventually, excessive bloom of phytoplanktons would have poisoned the waters sufficiently to have caused mass mortality of reptilian and fish population leading to further P enrichment of the bottom sediments (Bromley, 1967). Numerous phosphatic bone beds within the subjacent basinal La Luna Formation may lend support to their origin adjacent to belts of upwelling waters. High level warm seas, as was prevailing during and after the peak of Cretaceous transgression, documented in the La Luna Formation, are usually associated with expanded ocean-wide oxygen minimum and phosphogenesis (Piper and Codispoti, 1975; Fischer and Arthur, 1977; Sheldon, 1980).

Dissolution of organic remains and reprecipitation in the form of a poorly crystalline apatite (Suess, 1978) occurred mainly in the foraminiferal shells but also as occasional apatite pellets. Glauconite also would have developed in such a milieu where a relatively high concentration of silica was achieved through the upwelling waters. McCrae (1972) observed that the presence of glauconite may well be an index to moderately high dissolved silica and iron levels and low oxygen content in sea water. A low rate of deposition, a preexisting calcareous substrate, an abundant supply of organic matter, and a moderate to plentiful supply of dissolved iron in the marine basin would have facilitated the formation of glauconite (Manheim and others, 1975; Giresse, 1980; Odin and Letolle, 1980; Odin and Matter, 1981).

As to the common occurrence of glauconite and apatite in the same bed, it is interesting to speculate on their relative time of origin. Textural data being inconclusive, it is thought that the conditions required for phosphatization and glauconitization are different though similar, that is, reducing to intermittently anoxic for apatite as opposed to mildly reducing for glauconite (Bentor, 1980; Jarvis, 1980; Odin and Letolle, 1980). Both facies are not in equilibrium with sea water at the same time. However, both processes can take place rapidly and can succeed each other in time and space in response to even subtle changes in the environment (Giresse, 1980). Presence of abundant burrows indicating aerobic conditions implies that phosphate producing anoxic conditions must have been short-lived and intermittent. In a condensed sequence, such as the Tres Esquinas bed, the probability of a glauconite-apatite association will logically be much higher.

Concomitant with the deposition of phosphate and glauconite in a reducing environment, hydrogen sulfide was probably generated by sulfate reduction leading to precipitation of black monosulfides (FeS). Opaline silica also would have precipitated at about the same time, preferably in the skeletal matter. Soon thereafter, the whole assemblage of phosphatized shell debris, glauconite, monosulfide of iron, and opaline silica cement was buried and subsequently underwent further diagenesis, being completely cut off from then onward from the ambient chemical environment of the sea floor. Dolomite formation is thought to have occurred at this stage. The common association of apatite and dolomite in the Tres Esquinas Member does not represent a true paragenesis, as the two minerals formed in contrasting chemical conditions. Apatite formation requires a Mg^2/Ca^2-ratio below that of sea water, whereas dolomitization requires a ratio above that of sea water (Bentor, 1980). Dolomitization probably occurs here with the aid of carbon dioxide produced by endosedimentary bacteria interacting with primary calcite and depleting the connate water of its Mg (Mattes and Mountjoy, 1980). Presence of organic matter in the sediments could have facilitated dolomitization (Zenger and Dunham, 1980).

In the next stage of diagenesis, pyritization from precursor monosulfides took place in the same reducing anoxic environment. It is not well understood as to why pyrite formation is so late, but surely it is not a change in Eh, because the sediment was thoroughly reducing from the time of burial onward. Granular

TABLE 1.   FAUNAL EVIDENCE FROM THE TRES ESQUINAS BED AND ADJACENT FORMATIONS

Stratigraphic unit		Author	Fossil evidence	Age assigned
Colon Formation	Top	Renz (1977)	Sphenodiscus	Maestrichtian
		Renz (1977)	Baculites, Euhomaloceras	Campanian
Colon Formation (Socuy Member)	Base	Sellier de Civrieux (1952)	Globotruncana fornicata (Plummer), G. cretacea Cushman, G. rosetta (Carsey), Rugoglobigerina rugosa (Plummer), Globigerina cretacea Cushman	late Campanian
Tres Esquinas Member		Renz (1977)	Texanites Texanus	Santonian
		Ford and Houbolt (1963)	Globotruncana calcarata Cushman	late Campanian early Maestrichtian
La Luna Formation	Top	Renz (1980)	Barroiceras, Peroniceras, Gauthiericeras, Prionocycloceras and Prionocyclus	Coniacian
La Luna Formation	Base	Renz (1977)	Mariella (M) cf. worthensis (Adkins and Winton), Acanthoceras	Cenomanian

secondary quartz developed partly at the expense of its opaline precursor, concomitant with or slightly postdating pyrite formation.

AGE OF THE CONDENSED SEQUENCE

The condensed sequence clearly marks a prominent break in faunal as well as lithologic and sedimentological aspects between the La Luna and Colon formations. The discontinuity along the Tres Esquinas sequence is of a paraconformable nature, as the beds above and below do not show any angular relationship. Faunal data from the condensed sequence as well as the overlying and underlying formations (Table 1) tend to indicate a post-Coniacian but pre-Maestrichtian age for the Tres Esquinas bed. Evidently there exist divergent opinions, regarding the age of the hiatus or period of reduced sedimentation represented by the Tres Esquinas bed. Sutton (1946) assigned a Santonian age to the period of reduced sedimentation, whereas Sellier de Civrieux (1952) observed that the break might extend from late Santonian to early Campanian. Cushman and Hedberg (1941) suggested that the period of reduced sedimentation represented by the Tres Esquinas bed might extend to the Middle Campanian. In the light of the above discussion and the Santonian age assigned by Renz

(1977) to this unit, it is tentatively suggested that at least the entire Santonian and a large part of the Campanian are represented by the Tres Esquinas condensed sequence.

SYNTHESIS

The vast areal extent of this thin unit all over western Venezuela and traceable to the Magdalena Valley in Colombia (Ward and Goldsmith, 1973) indicates that a much more detailed regional analysis will be necessary in order to interpret the genesis of the Tres Esquinas bed and the Upper Cretaceous paleogeography prevailing at that time. However, local data from the area around Mérida provide valuable insight into the nature of the sequence and its geologic implications. The condensed sequence documents a relatively long stratigraphic break, probably 8 to 10 m.y. during Santonian-Campanian time, and coincides with a marine stillstand at the top of a transgressive episode. A marine upwelling phenomenon would have contributed to the additional P, Ca, and silica input into the biogenic sediments of the condensed bed occurring within an expanded oxygen-minimum layer. Presence of a coeval glauconitic sequence from the Eastern Cordillera (Tellez, 1978) and the Magdalena Valley in Colombia suggests that such surfaces of nondeposition or very slow deposition may serve as useful tools in correlating stratigraphic breaks within the Andean Cretaceous sequence and perhaps on a wider regional scale. Fischer and Arthur (1977) observed that Coniacian-Santonian euxinic conditions, though not as widespread as in the preceding Aptian-Albian interval, have been documented in western Atlantic (DSDP site 4), carbonaceous chalks in the Caribbean Sea, and black shales in the South Atlantic (DSDP site 40). It is also worthwhile to mention that successful correlation has been made of many Cretaceous disconformities between the Western Interior and the coastal plains of North America and carbonate platform-shelf facies of central and Western Europe on the basis of eustatic stillstand events

(Kauffman, 1981). Further studies in South America leading to recognition of widespread regional stratigraphic breaks, such as those represented by the Tres Esquinas Member, may eventually extend such correlation to the southern continent.

ACKNOWLEDGMENTS

I am thankful to C. Benjamini and R. Shagam of Ben-Gurion University of The Negev, Israel, and A. Fischer of Princeton University for their many helpful comments. I also thank my colleague Oswaldo Gallango for his help in preparing the various illustrations in the text.

REFERENCES CITED

Baturin, G. N., 1971, Stages of phosphorite formation on the ocean floor: Nature Physical Science (London), v. 232, no. 29, p. 61–62.

Bentor, Y. K., 1980, Phosphorites—the unsolved problems, *in* Bentor, Y. K., ed., Marine phosphorites: Geochemistry, occurrence, genesis: Society of Economic Paleontologists and Mineralogists Special Publication 29, p. 3–18.

Birch, G. F., 1979, Phosphatic rocks on the western margin of South Africa: Journal of Sedimentary Petrology, v. 49, p. 93–110.

——1980, A model of penecontemporaneous phosphatization by diagenetic and authigenic mechanisms from the western margin of southern Africa, *in* Bentor, Y. K., ed., Marine phosphorites: Geochemistry, occurrence, genesis: Society of Economic Paleontologists and Mineralogists Special Publication 29, p. 79–100.

Bromley, R. G., 1967, Marine phosphorites as depth indicators: Marine Geology, v. 5, p. 503–510.

Choquette, P. W., and Pray, L. C., 1970, Geologic nomenclature and classification of porosity in sedimentary carbonates: American Association of Petroleum Geologists Bulletin, v. 54, p. 207–250.

Cushman, J. A., and Hedberg, H. D., 1941, Upper Cretaceous foraminifera from Santander del Norte, Colombia, S. A.: Cushman Laboratory of Foraminiferal Research Contribution, v. 17, pt. 4, p. 79–100.

D'Anglejan, B. R., 1967, Origin of marine phosphorites off Baja California, Mexico: Marine Geology, v. 5, p. 15–44.

Fischer, A. G., and Arthur, M. A., 1977, Secular variation in the pelagic realm, *in* Cook, H. E., and Enos, P., eds., Deep water carbonate environments: Society of Economic Paleontologists and Mineralogists Special Publication 25, p. 19–50.

Ford, A., and Houbolt, J. J., 1963, Las microfacies del Cretaceo de Venezuela Occidental: Leiden, E. J. Brill, 59 p.

Ghosh, S. K., 1980, Anatomia de una capa dura marine—un caso de los Andes Venezolanos [abs.]: XXX Convención Anual de ASOVAC, v. 1, p. 26.

Giresse, P., 1980, Phosphorus concentrations in the unconsolidated sediments of the Tropical Atlantic Shelf of Africa south of the Equator—Oceanographic comments, *in* Bentor, Y. K., ed., Marine phosphorites: Geochemistry, occurrence, genesis: Society of Economic Paleontologists and Mineralogists Special Publication no. 29, p. 101–138.

Goldhaber, M., 1978, Euxinic facies, *in* Fairbridge, R. W., and Bourgeois, J., eds., The encyclopedia of sedimentology: Stroudsburg, Pennsylvania, Dowden, Hutchinson and Ross Inc., p. 296–299.

Jarvis, I., 1980, The initiation of phosphatic chalk sedimentation—The Senonian (Cretaceous) of the Anglo-Paris Basin, *in* Bentor, Y. K., ed., Marine phosphorites: Geochemistry, occurrence, genesis: Society of Economic Paleontologists and Mineralogists Special Publication 29, p. 167–192.

Kauffman, E. G., 1981, Regional disconformities and eustatic history: Cretaceous Trans-Atlantic test case [abs.]: American Association of Petroleum Geologists Bulletin, v. 65, p. 944.

Kennedy, W. J., and Garrison, R. R., 1975a, Morphology and genesis of nodular chalks and hardgrounds in the Upper Cretaceous of southern England: Sedimentology, v. 22, p. 311–386.

——1975b, Morphology and genesis of nodular phosphates in the Cenomanian glauconitic marl of south-west England: Lethaia, v. 8, p. 339–360.

Manheim, F. T., Rowe, G., and Jipa, D., 1975, Marine phosphorite formations off Peru: Journal of Sedimentary Petrology, v. 45, p. 243–251.

Mattes, B. W., and Mountjoy, E. J., 1980, Burial dolomitization of the upper Devonian Miette buildup, Jasper National Park, Alberta, *in* Zenger, D. H., and others, eds., Concepts and models of dolomitization: Society of Economic Paleontologists and Mineralogists Special Publication 28, p. 259–297.

McCrae, S. G., 1972, Glauconite: Earth Science Reviews, v. 8, p. 397–440.

Odin, G. S., and Letolle, R., 1980, Glauconitization and phosphatization environments: A tentative correlation, *in* Bentor, Y. K., ed., Marine phosphorites: Geochemistry occurrence, genesis: Society of Economic Paleontologists and Mineralogists Special Publication 29, p. 227–237.

Odin, G. S., and Matter, A., 1981, De glauconiarum origine: Sedimentology, v. 28, p. 611–641.

Parker, R. J., 1975, The petrology and origin of some glauconitic glauco-conglomeratic phosphorites from the South African continental margin: Journal of Sedimentary Petrology, v. 45, p. 230–242.

Piper, D. Z., and Codispoti, L. A., 1975, Marine phosphorite deposits and the nitrogen cycle: Science, v. 188, p. 15–18.

Renz, O., 1977, The lithologic units of Cretaceous in Western Venezuela: Memoria, Congreso Geológico Venezolano, V, Caracas, Volume 1, Venezuela Ministerio de Energía y Minas, p. 45–58.

——1980, Cretaceous ammonites of Venezuela: Caracas, Maraven, a subsidiary of Petroleos de Venezuela, S. A., 132 p.

Scholle, P., 1971, Diagenesis of deep water carbonate turbidites, Upper Cretaceous, Monte Antola Flysch, Northern Apennine, Italy: Journal of Sedimentary Petrology, v. 41, p. 233–250.

Sellier de Civrieux, J. M., 1952, Estudio de la microfauna de la sección tipo del Miembro Socuy de la Fm. Colon, Distrito Mara, Estado Zulia: Boletín Geologia, v. 2, p. 231–330.

Sheldon, R. P., 1980, Episodicity of phosphate deposition and deep ocean circulation—an hypothesis, *in* Bentor, Y. K., ed., Marine phosphorites: Geochemistry, occurrence, genesis: Society of Economic Paleontologists and Mineralogists Special Publication 29, p. 239–248.

Stainforth, R. M., 1962, Some new stratigraphic units in western Venezuela: Asociación Venezolana de Geología Minería y Petróleo, v. 5, p. 279–282.

Suess, E., 1978, Remineralization mechanism of phosphates from sediments: Jerusalem, Tenth International Sedimentological Congress, v. 2, p. 650.

Summerhayes, C. P., 1970, Phosphate deposits on the northwest African continental shelf and slope [Ph.D. thesis]: University of London, 282 p.

Summerhayes, C. P., Birch, G. F., Rogers, J., and Dingle, R. V., 1973, Phosphate in the sediments off south-western Africa: Nature, v. 243, p. 509–511.

Sutton, F. A., 1946, Geology of the Maracaibo Basin, Venezuela: American Association of Petroleum Geologists Bulletin, v. 30, no. 10, p. 1621–1741.

Tellez, N. A., 1978, Geología del area Labateca-Ragonvalia: Boletín Geología, v. 12, p. 5–37.

Veeh, H. H., Burnett, W. C., and Souter, A., 1973, Contemporary phosphates on the continental margin off Peru: Science, v. 181, p. 844–845.

Ward, D. E., and Goldsmith, R., 1973, Geología de los cuadrangulos H-12 Bucaramanga y H-13 Pamplona-Departamento de Santander: v. 21, no. 1–3, p. 1–132.

Zenger, D. H., and Dunham, J. B., 1980, Concepts and models of dolomitization—An introduction, *in* Zenger, D. H., and others, eds., Concepts and models of dolomitization: Society of Economic Paleontologists and Mineralogists Special Publication 28, p. 1–10.

MANUSCRIPT ACCEPTED BY THE SOCIETY SEPTEMBER 1, 1983

Geological Society of America
Memoir 162
1984

Late Cenozoic Guayabo delta complex in southwestern Maracaibo Basin, northeastern Colombia

Franklyn B. Van Houten
Department of Geological and Geophysical Sciences Princeton University
Princeton, New Jersey 08544

Harold E. James
Gulf Research and Development Company
Houston, Texas 77036

ABSTRACT

The Guayabo Group is an oolitic ironstone-bearing delta complex which accumulated in the southwestern part of the Maracaibo Basin in late Miocene–early Pliocene(?) time. The 1,000 to 1,700-m-thick deposit comprises the paralic to fluvial Cúcuta, Cornejo, and Urimaco Formations. These are characterized by combinations of seven different vertical sequences of sediment. Sequence 1 consists of channel sandstones containing small quartz and chert pebbles, and burrow-mottled mudstones, deposited on alluvial plains with braided and low-sinuosity rivers. Sequence 2 is made up of color-mottled mudstones and thick channel sandstones which accumulated on the upper delta plain where channelways were more sinuous. Sequence 3 consists of thin, narrow channel sandbodies deposited in crevasses at high angles to major distributaries, and associated drab mudstone enclosing thin sandy lentils. Sequence 4 includes elongate channel sandstones with complex internal geometry and abundant carbonaceous fragments, interbedded with burrowed drab mudstone, which accumulated along distributaries and bar fingers in very shallow water. Sequence 5 is carbonaceous, muddy sheet and channel sandstones and mudstones, as well as local lignite, which accumulated in interdistributary marshes. Sequence 6 is made up of carbonaceous mudstone grading upward through laminated siltstone and mudstone to prograding sheet sandstones deposited along a microtidal shoreline. Repeated decrease in detrital influx led to local development of thin oolitic ironstones in marginal embayments during stillstand or the initial stage of renewed transgression. Sequence 7 includes drab mudstone and discontinuous sheet sandstones formed as distal splays in embayments where thin oolitic ironstones developed locally during interruptions of detrital influx.

During early and middle Cenozoic time, intermittent deformation and uplift of the Cordillera Central of Colombia produced three thick conglomerates in the Andean foredeep and repeated regional uplift and erosion of the more stable craton in northeastern Colombia and adjacent Venezuela. In late Oligocene and early Miocene time, waning uplift of the Cordillera Central and subdued uplift of basement blocks in northeastern Colombia and adjacent Venezuela were accompanied by fine-grained detrital sedimentation both in the foredeep and on the fractured craton to the northeast. In late Cenozoic time major deformation and uplift produced coarse-grained deposits in the Andean foredeep and in marginal basins around the basement blocks. Detritus shed eastward from the Santander Massif in the Cordillera Oriental accumulated in the low-energy Guayabo delta complex. Initial paralic deposits composed of combinations of sequences

3 to 7 constitute the Cúcuta Formation. The succeeding Cornejo Formation consists of more varied combinations of sequences 2 to 7 in nonmarine to paralic sediments with associated marginal marine oolitic ironstones. Increased uplift of the massif supplied coarser fluvial deposits of the Urimaca Formation (sequences 1 to 6) which prograded eastward across the southwestern corner of the Maracaibo Basin. Late Pliocene and Pleistocene regional uplift led to excavation of the area by northward-flowing drainage systems.

INTRODUCTION

Our paper presents a description and reconstruction of the late Miocene–early Pliocene(?) low-energy Guayabo delta complex in the southwestern corner of the Maracaibo Basin. We relate its development to late Cenozoic deformation of the craton in northeastern Colombia and southwestern Venezuela. Study of this group yields a useful record of fluvial and paralic processes and products of a small delta complex which is correlative with well-known, finer-grained distal facies in the central part of the basin. These deposits also provide unusually good information about the environment in which numerous thin oolitic ironstones developed.

Regional Setting

During early Cenozoic time, a broad, unbroken lowland of aggradation east of the Cordillera Central included the present sites (Fig. 1) of the Magdalena Valley, the Cordillera Oriental, the Cordillera de Merida, and the Maracaibo–Santa Marta block (Van Houten and Travis, 1968; Irving, 1975; Kellogg, 1981). In Miocene–early Pliocene time the Santander Massif (Dallmus and Graves, 1972), the Sierra de Perija (Kellogg, 1981), and the Cordillera de Merida (Zambrano and others, 1971) rose above the fractured craton and supplied detritus preserved in the Guayabo and correlative deposits. Renewed deformation of the Santander Massif in Pliocene time folded and faulted the Guayabo Group, and associated regional uplift terminated sedimentation in the southwestern corner of the Maracaibo Basin.

GENERAL FEATURES

Stratigraphy

In the 500-km² study area in the vicinity of Cúcuta (Fig. 1), the late Miocene–early Pliocene(?) Guayabo Group conformably overlies the Oligocene–early Miocene Leon Formation. Muddy Leon deposits are increasingly silty toward the upper contact which is marked by local scouring and lag deposits containing fragments of wood and reptilian bones and teeth. The coarsening-upward Guayabo Group consists of as much as 1,240 m of nonmarine to paralic mudstone and sandstone (Fig. 2), comprising from bottom to top the Cúcuta, Cornejo, and Urimaco Formations. These names have been proposed and defined by James (1977, p. 5–13) because the essentially equivalent lithofacies, the Isnotu and Betijoque Formations, in southwestern Venezuela

Figure 1. Map of Colombia and adjacent Venezuela, with index map of Cúcuta area. Guayabo delta complex outlined by dots.

(Fig. 5) were derived from the Cordillera de Merida and are not an integral part of the Guayabo delta complex.

The Cúcuta Formation consists of 110 to 150 m of drab mudstone interbedded with thin lenses and sheets of ripple-bedded, fine-grained sandstone. The few channel deposits in the upper part of the formation are composed of medium-grained sandstone. The Cornejo Formation consists of 390 to 610 m of drab and color-mottled mudstone, thin sheet sandstones, thicker sheet and fining-upward channel sandstones, and local oolitic ironstones and lignites about 10 to 15 cm thick. The Urimaco Formation, 450 to 1,085 m thick, is marked by the presence of quartz- and chert-pebble conglomerates in many of its coarse-grained, fining-upward channel sandbodies that are overlain by

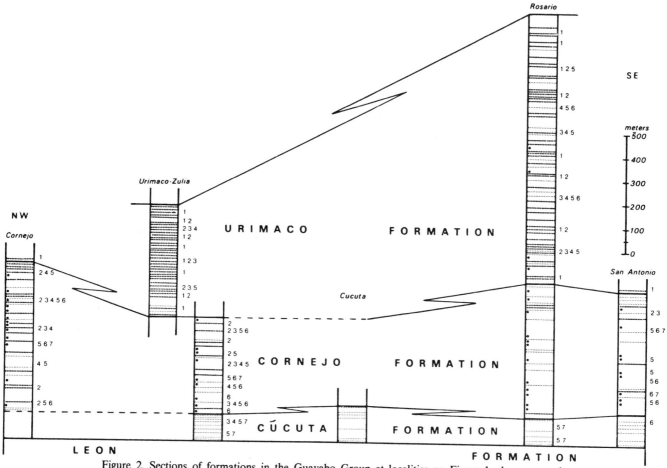

Figure 2. Sections of formations in the Guayabo Group at localities on Figure 1; shows general distribution of major sandbodies (sheet, fine dotted lines; channel, coarse dotted lines), of facies sequences (numbered) identified in Figure 4, and of main intervals of oolitic ironstones (large dots). Group coarsens upward, and each formation coarsens westward toward source area.

color-mottled mudstone. Subordinate drab mudstone, sheet and channel sandstone, as well as rare oolitic ironstone and lignite like those in the Cornejo Formation, are also present.

Fossils

Body fossils scattered through mudstones and muddy sandstones in the Cúcuta and Cornejo Formations consist of fragments of turtle shells and teeth, vertebrae, and fragments of crocodile ribs. These remains suggest an aquatic, paralic environment. Poorly preserved molds of gastropods and pelecypods are of little diagnostic value.

Trace fossils are common throughout the Guayabo Group. *Stuffed burrows* 0.3 to 0.7 cm in diameter, with meniscus fecal filling, are most abundant in muddy sandstones and overlying color-mottled mudstone, as well as in oolitic ironstones commonly overlain by transgressive marine deposits. Sinuous and dendritic branching burrows are also present on the surface of some ferriferous oolites. The stuffed burrows in the color-mottled

deposits imply adequate organic matter in the original sediment. The thorough bioturbation suggests accumulation slow enough to permit the extensive burrowing. In contrast, carbonaceous matter and traces of lignite are preserved in mudstones with few burrows.

Ophiomorpha, the callianassid burrow, is common in crossbedded, well-sorted, fine- to medium-grained sheet sandstones. This association indicates that the crustaceans inhabited migrating sandwaves. Although *Ophiomorpha* is known to be most common in sandy shoreline deposits, it does occur in sediments of a wide range of marine to brackish-water environments. *Thalassinoides*, a network of burrows on a bedding surface, is present in sandy mudstones, sandstones, and oolitic ironstones. It was not found associated with *Ophiomorpha*.

Most of the *plant fossils* in the Guayabo Group are casts of roots and slender stumps in sandy mudstone and muddy sandstone. Presumably, many plant remains were destroyed by burrowers. Commonly, the casts of roots are difficult to differentiate from burrows. Most of the indeterminate structures occur in

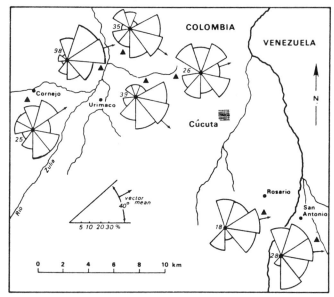

Figure 3. Paleocurrent roses, Guayabo Group, in 40° modal classes. Number of measurements given for each locality.

mudstone beneath transgressive marine deposits. Branching casts that decrease in diameter downward probably are fossil roots, and their association with *Ophiomorpha* suggests they were roots of salt-tolerant trees. Rare, crudely circular vertical structures as much as 17 cm in diameter in sandstone may be casts of stumps.

Dispersal Data

Directional indicators (Fig. 3) were measured in about 250 cross-bedded sandbodies in the Guayabo Group. In addition, local directions of migrating channels were determined for 18 sandbodies in the Urimaco Formation.

Dispersal was generally eastward but varied locally. There is an upward trend toward less variation in the successively coarser-grained formations. Paleocurrent data corroborate the reconstructed pattern of facies, indicating a western source and easterly paleoslope.

Vertical Sequences

In order to facilitate discussion and reconstruction of the Guayabo delta complex we have identified seven recurring vertical sequences of lithofacies and their sedimentary structures and fossils (Fig. 4; Appendix 1). Each repeated sequence summarizes significant features related to processes of sedimentation. Sequence 1 accumulated in channels of braided and other low-sinuosity streams and on their floodplains; sequence 2 in channels and levees of meandering streams on the upper delta plain; sequence 3 in crevasse and proximal splay channels oriented at high angles to larger channels; sequence 4 in lower delta plain distributaries and bar fingers in very shallow water; sequence 5 in interdistributary swamps and marshes with abundant vegetation; sequence 6 in brackish to marine embayments along a coast with

		1 sst	1 mdst	2 sst	2 mdst	3	4	5	6 L mdst	6 sst	6 U mdst	7
sand	pebbles	X										
	coarse	X		X								
	medium	X		X			X	X		x		x
	fine	X		X		X	X	X		X		X
	very fine					X	X			X		X
	mud		X		X	X	X	X	X		X	
grading	fining up	X		X				X				X
	coarsening up									X		
base of sandbodies	sharp	X		X		X	X			X		
	gradational		X		X					X	X	
top of sandbodies	sharp						X			X	X	X
	gradational	X		X			X	X		X		
shape of sandbodies	sheet		X		X		X			X		X
	wide, sinuous	X		X			X					
	elongate				x	X	X	x				
oriented parallel to	slope	X		X			X					
	shore									X		
bedding	parallel	X		X			X			X		X
	large trough	X		X			X	X		X		
	small tabular					X				X		X
	large tabular	X		X								X
ripples	current						X					X
	oscillation									X		
casts	load											X
	groove	X										X
shrinkage cracks			X		X	X	X			X		
plant remains	fragments						X	X	X			X
	shredded									X		X
	roots, stumps			X				X	X			
burrows	large	X		X			X			X	X	X
	small		X		X		X	X		X	X	X
color of mud	drab					X	X	X	X			
	mottled		X		X					X		

Figure 4. Major features in seven facies sequences in Guayabo Group: X, dominant; x, subordinate. Description in Appendix 1.

low tidal range; and sequence 7 in tranquil interdistributary coastal embayments locally intruded by distal crevasse splay.

Within the Guayabo Group the vertical sequences occur in differing combinations in the three generally coarsening-upward formations as well as in the proximal to distal facies of each formation (Fig. 2). Sequence 1 occurs only in the Urimaco Formation. Sequences 2, 3, and 4 are common in the Cornejo and Urimaco Formations; 3 and 4 also occur locally in the upper part of the Cúcuta Formation. Sequence 5 is present in all of the formations. Sequence 6 is widespread in the Cornejo Formation; it also occurs in the uppermost distal Cúcuta Formation and locally in distal Urimaco deposits. Sequence 7 is common in the Cornejo and Cúcuta Formations. Lateral and vertical distribution of these sequences depicts a prograding alluvial plain and low-energy delta complex comparable to that of the shallow-water Holocene Guadelupe Delta on the Texas Coastal Plain (Donaldson and others, 1970).

PETROLOGY

Mudstones

Mudstones constitute 50% to 80% of the Guayabo Group. Drab mudstone is either burrowed (bioturbated) or is carbonaceous and contains thin layers of lignite. Color-mottled varieties are commonly associated with fining-upward fluvial sequences, especially in the Cornejo Formation.

The predominant kind of clay in Guayabo mudstones is mixed-layer illite-montmorillonite together with montmorillonite. Illite and kaolinite are, with few exceptions, minor constituents. Montmorillonite predominates in some mudstones associated with oolitic ironstones. Berthierine is present in some of the ironstones.

The mixed-layer illite-montmorillonite probably was the product of incomplete weathering of igneous and metamorphic rocks of the Santander Massif that was the source of most of the Guayabo detritus. The subordinate kaolinite, in contrast, was eroded from Cretaceous marine strata in the source area..

Sandstones

Grain size, sorting, and kind of bedding vary with the style of sandbody. Ripple- and small-scale cross-bedded, well-sorted fine-grained sandstone predominates in sheet sandbodies as much as 2 m thick. Tabular and trough cross-bedded, medium- to coarse-grained sandstone and small-pebble conglomerate occur in multistory channel sandbodies as thick as 10 m. Bioturbation has rendered many sandstones structureless. The lower part of many of the channel sandstones is marked by patchy ferruginous cement, whereas the top of some sheet sandstones is burrowed and ferruginized.

Most of the Guayabo sandstones are chert litharenites containing as much as 25% chert. A few of the small-pebble conglomerates contain nearly 80% chert. K-feldspar averages 0.5% to 7% and plagioclase less than 3% of the sand grains. Metamorphic rock fragments constitute as much as 10%. Grains of zircon and tourmaline range from well-rounded to euhedral or angular. The assemblage of minerals in the Guayabo sandstones points to derivation from metamorphic and nonvolcanic igneous rocks as well as Mesozoic sedmentary rocks, and supports other evidence that the detritus was eroded from the Santander Massif.

Oolitic Ironstones

As many as 34 thin (less than 15 cm thick), generally unbedded sandy ironstones containing goethite and berthierine ooids occur in the Cornejo and lower Urimaco Formations (James and Van Houten, 1979). They recur in every 15 to 25 m of section in the west and less than 1 m apart locally in the east. Spherical symmetrical ooids (0.25–0.5 mm, rarely to 2 mm, in diameter) are composed of goethite; some of these contain a minor amount of berthierine. In contrast, many of the berthierine ooids are irregularly elongate and distorted as a result of plastic deformation. Collophane nodules enclosing goethite ooids are present in many of the ironstones.

Nuclei of the ooids include structureless goethite or berthierine, quartz grains, ferruginous silty mudstone (fecal pellets), and ooid fragments; more rarely, pieces of shells and fish teeth are in goethitic ooids. Detrital sand grains in the ironstones normally are angular to subangular and are smaller than the iron-rich ooids.

RECONSTRUCTION

The late Miocene–early Pliocene(?) Guayabo Group accumulated on an eastward-sloping alluvial plain and low-energy lobate delta complex that prograded into a seaway with a tidal range of less than 0.5 m. Brackish to marine water filled the interdistributary embayments where thin oolitic ironstones accumulated repeatedly. The delta complex probably covered 5,000 to 6,000 km^2, approximately the size of the Rio Grande Delta and less than one quarter the size of the Mississippi Delta today.

Provenance

Paleocurrent data and facies relations in the Guayabo Group require a western source. Petrographic analyses and regional reconstruction point to derivation from the Santander Massif that had begun to rise in Miocene time. By late Miocene time the massif had been deeply dissected and locally stripped to its crystalline basement. The Mesozoic sedimentary cover yielded part of the detritus shed eastward, and part came from incompletely weathered basement rocks. No significant amount of detritus was supplied by the rising Cordillera de Merida to the east.

Guayabo Delta Complex

The Guayabo Group, like correlative deposits flanking the Sierra de Perija and Cordillera de Merida, is a small delta complex that contributed to the late Cenozoic aggradation of the Maracaibo Basin. Fining-upward fluvial cycles accumulated on the upper delta plain with proximal braided and distal meandering streams, while prograding coarsening-upward detrital sequences were built on the paralic lower delta plain and delta front.

Initial sedimentation of the Guayabo Group succeeded widespread accumulation of Oligocene–early Miocene Leon mudstone. The first Guayabo deposits consisted of drab mud and thin sheets of fine-grained sand that formed in a nearshore, delta-front environment (Cúcuta Formation). Thin, narrow channel sands in the upper part of the formation heralded development of distributaries of a lobate delta. As the delta prograded eastward (Cornejo Formation), mud accumulated on floodplains and natural levees while sand was concentrated in point bars on the lower delta plain. Along the embayed shoreline, levees were smaller and crevasse-filling was more important. Mud and bar-finger sand extended the distributary system into a shallow microtidal marine environment. Between the distributaries burrowed and root-disturbed organic-rich mud and local peat accumulated in marshes. Where distributary abandonment reduced sediment supply, ferruginous ooids developed during still-stands and initial stages of transgression, before influx of detritus was renewed. This repeated pattern is like Fisher's (1964) reconstruction of glauconite development in asymmetric cycles within alternating marine and nonmarine Eocene sequences of the northern Gulf Coast. Continued eastward progradation of the delta spread pebbly sand across the Cúcuta area. This episode of maximum progradation (Urimaco Formation) reflects both increased uplift of the source area and sedimentation in excess of local subsidence of the Maracaibo block. Aggradation of the southwestern corner of the basin ended with onset of erosion of the delta plain and shift of the major drainage in the Cúcuta area from eastward to northward.

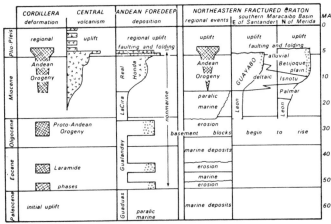

Figure 5. Principal Cenozoic activity in Cordillera Central, Andean foredeep, and southern Maracaibo Basin area. Names of orogenies and phases are those of Campbell (1974).

REGIONAL SYNTHESIS

During early Cenozoic time a broad alluvial plain stretched several hundred kilometres northeastward across the Andean foredeep (Van Houten, 1976) to the more stable craton where several deltas and their shoreline lay in the vicinity of Lake Maracaibo (Bockmeulen and others, 1983). Two pulses of deformation and uplift of the Cordillera Central (Fig. 5) shed coarse detritus across the foredeep while local uplift and erosion of cratonic blocks occurred in northeastern Colombia and southwestern Venezuela. Renewed deformation in Oligocene time (Fig. 5) produced another thick conglomerate along the western part of the foredeep (Anderson, 1972). On the craton to the northeast, mid-Cenozoic deformation initiated strike-slip and basement-block displacement as well as regional erosion (Kellogg, 1981). Subdued uplift of the Santander Massif, Sierra de Perija, and Cordillera de Merida at this time led to partial stripping of their sedimentary mantle.

Waning uplift of the Cordillera Central and the northeastern basement blocks in late Oligocene to early or middle Miocene time (Fig. 5) induced widespread muddy sedimentation. In the west this produced floodplain and lacustrine montmorillonitic mudstone (La Cira and Colorado Formations) that record the beginning of explosive volcanism in the Cordillera Central (Van Houten, 1976). Farther northeast, correlative paralic mudstones (Leon, Icotea) accumulated on a mid-Cenozoic unconformity while uplift and erosion prevailed in the central part of the Maracaibo Basin (Bockmeulen and others, 1983).

Late Miocene deformation, uplift, and increased volcanism along the Cordillera Central (Fig. 5) marked the initial stage of late Cenozoic orogeny (Wellman, 1970). This activity is reflected in significantly increased detrital influx into the proximal part of the foredeep (Real Formation in the north; volcanic-rich Honda Group in the south). It culminated in the accumulation of coarse early Pliocene(?) Neiva Conglomerate (Howe, 1974) and of Mesa volcaniclastic debris across nonorogenic fans (Van Houten,

1976) in the aggrading foredeep. During late Pliocene and Pleistocene regional uplift, sporadic explosive volcanism along the Cordillera Central spread sheets of pyroclastic debris across the dissected Magdalena lowland to the east.

The cratonic blocks in the northeast also rose in late Miocene and early Pliocene time. Prograding nonvolcanic sandy detritus from them (Cúcuta, Cornejo, Isnotu, Los Ranchos Formations) accumulated conformably on older muddy sediments around the margin of the Maracaibo Basin. Relief on the Santander Massif was relatively subdued during most of this episode so that low-energy deltaic sedimentation marked by repeated development of thin oolitic ironstones prevailed in the southwestern corner of the basin. Aggradation there ended after increased uplift of the massif had supplied the coarse alluvial plain deposits of the Urimaco Formation that prograded across the Cúcuta area. To the east, deformation and uplift of the Cordillera de Merida produced the thick, coarse nonmarine Betijoque Formation along the northern flank of the range. This aggradation apparently continued after Guayabo sedimentation had ceased (Fig. 5; Zambrano and others, 1971) and spread northward across the Maracaibo Basin. The final phase of deformation ended with faulting and folding of strata around the basement blocks. Subsequent regional uplift and erosion in late Pliocene and Pleistocene time led to excavation of the basin-fill in northeastern Colombia and southwestern Venezuela by northward-flowing drainage systems.

ACKNOWLEDGMENTS

This review is based mostly on James's (1977) detailed study of the Guayabo Group; some information is from regional reconstructions. Financial support for the research was provided by National Science Foundation Grant GA-14874 and by Princeton University. Logistics for field work were arranged by INGEOMINAS, Colombia.

APPENDIX 1. VERTICAL SEQUENCES IN THE GUAYABO GROUP (JAMES, 1977, p. 52–134).

Sequence 1

Channel sandstones overlain by color-mottled mudstone have erosional bases, gradational tops, intraformational mud clasts, and tabular or elongate shapes. Several sandbodies are fining-upward, as in migrating point-bar deposits. Lateral variation in orientation of some of the paleochannels supports this interpretation. In contrast, multiple levels of erosion in pebbly sandstones, large-scale cross-bedding, and parallel bedding within 1 m of the top of 5-m-thick sandbodies, as well as a local absence of grading, suggest more complex channel development in braided reaches.

Multistoried sandbodies were produced by reoccupation of channels or by intersection of channels of differing orientation as subsidence and sedimentation continued. Many of the intraformational chips are gray, pointing to at least local non-oxidizing conditions in the enclosing sandstone; only a few have been reddened by oxidation after burial.

Muddy sand grading upward into sandy mud accumulated mainly by vertical accretion on natural levees and floodplains. In these aerated sediments, roots and burrowers destroyed primary structures and commonly obscured the grading, and much of the original carbonaceous material was consumed. Color-mottling was produced by soil-forming processes active above the water table.

Sequence 2

Sandbodies have erosional bases, an upward decrease in size of cross-beds and sand grains, and variation in channel orientation characteristic of point bars in sinuous channels. These sandstones are associated with widespread mudstones whose abundant burrowing, stump casts, and color-mottling point to oxidizing conditions on floodplains.

Deposits of sequence 2 are more common than those of sequence 1 down the easterly paleoslope. Along this profile, marked reduction of pebble and sand sizes in less than 30 km is accompanied by an increase in deposits of meandering distributaries on the upper delta plain.

Sequence 3.

Massive drab mudstone, which rarely has distinct burrows, is associated with thin, small-scale, cross-bedded, fine-grained sandstone in narrow channels and sheets. These sandy deposits decrease rapidly in thickness and grain size away from major channels, suggesting deposition in crevasses and proximal splays. The specific relations among the thin sandbodies are difficult to reconstruct, however.

Sequence 4.

Elongate channel sandstones with complex internal structure and abundant carbonaceous fragments are associated with thoroughly burrowed drab mudstone. Paleocurrent directions are predominantly downslope in the commonly cross-bedded distributary channel sandbodies. Truncation and rapid gradation in the upper part of the sandstones suggest development in or near an environment with enough energy to prevent preservation of subaerial levees. Beyond the sinuous distributary channels, poorly bedded mudstone and muddy sandstone indicate currents running at high angles to channel axes in water only a few metres deep. These channel and bar-finger sediments probably formed at the end of the prograding stage of delta construction and were then covered by muddy deposits of the transgression following abandonment.

Sequence 5.

Carbonaceous, muddy, thin-sheet, and thick-channel sandstones and associated drab mudstone containing abundant plant fragments are extensively disrupted by plant roots. Very locally, there are thin, ripple-bedded, fine-grained sandstones. The channel sandbodies accumulated either in delta distributaries or in tidal channels. Localized loading of the sandbodies on a muddy substrate apparently caused rapid dewatering and brecciation of the underlying mudstone. Association of the root-disturbed deposits with *Ophiomorpha*-bearing sandstone points to accumulation in interdistributary swamps near a brackish or normal marine shoreline.

Sequence 6.

Above a persistent parallel-bedded sandstone or an oolitic ironstone 10 to 15 cm thick, the succession coarsens upward from gray carbonaceous marine mudstone, through interbedded mudstone, siltstone, and very fine grained sandstone, to ripple-, cross-, and parallel-bedded sandstone locally containing *Ophiomorpha,* and into burrowed and root-disturbed color-mottled mudstone. Conditions producing these shoaling-upward successions changed from reducing marine in the lower part to oxidizing shoreline in the upper part. Each succession ended with delta lobe abandonment, followed by a stillstand and renewed transgression producing an oolitic ironstone or thin sandstone that locally may have been a lag deposit. Some of the ferruginized mudstones at the top of the sequence probably record development of an iron-rich hardground in very shallow water.

Sequence 7.

In coarsening-upward successions like those in sequence 6, carbonaceous gray mudstone containing well-bedded, fine-grained, sheet sandstones overlies a thin, massive, poorly sorted ferruginized sandstone, or a burrowed goethitic or berthierine-rich oolitic ironstone. The interbedded, fine-grained sandstones are characterized by sharp basal contacts above burrowed mudstone, groove and load casts, small-scale cross-bedding, current ripple-bedding, and syndepositional slumping. These sequences apparently were distal splay deposits that accumulated in embayments where ferruginous oolites developed during interruptions of detrital influx. Their habitat presumably was somewhat more reducing than that part of the embayments where sequence 6 prevailed.

REFERENCES CITED

Anderson, T. A., 1972, Paleogene nonmarine Gualanday Group, Neiva Basin, Colombia, and regional development of the Colombian Andes: Geological Society of America Bulletin, v. 83, p. 2423–2438.

Bockmeulen, H., Barker, C., and Dickey, P. A., 1983, Geology, and geochemistry of crude oils, Bolivar Coastal Fields, Venezuela: American Association of Petroleum Geologists Bulletin, v. 67, p. 242–270.

Campbell, C. J., 1974, Colombian Andes, *in* Spencer, A. M., ed., Mesozoic-Cenozoic orogenic belts: The Geological Society, London, Special Publication no. 4, p. 706–724.

Dallmus, K. F., and Graves, 1972, Tectonic history of Maracaibo and Barinas Basins: Transactions, Caribbean Geological Conference, VI, Margarita, Venezuela, 1971, p. 214–226.

Donaldson, A. C., Martin, R. H., and Kanes, W. H., 1970, Holocene Guadalupe Delta of Texas Gulf Coast: Society of Economic Paleontologists and Mineralogists, Special Publication 15, p. 107–137.

Fisher, W. L., 1964, Sedimentary patterns in Eocene cyclic deposits, northern Gulf Coast region, *in* Merriam, D. F., ed., Symposium on Cyclic Sedimentation: Kansas Geological Survey Bulletin 169, v. 1, p. 151–170.

Howe, M. W., 1974, Nonmarine Neiva Formation (Pliocene?), Upper Magdalena Valley, Colombia: Regional tectonism: Geological Society of America Bulletin, v. 85, p. 1031–1042.

Irving, E. M., 1975, Structural evolution of the northernmost Andes, Colombia: U.S. Geological Survey Professional Paper 846, p. 1–47.

James, H. E., Jr., 1977, Sedimentology of the iron-oxide-bearing Upper Miocene(?) Guayabo Group in the vicinity of Cúcuta, Colombia [Ph.D. thesis]: Princeton University, Princeton, New Jersey, 214 p.

James, H. E., and Van Houten, F. B., 1979, Miocene goethitic and chamositic oolites, northeastern Colombia: Sedimentology, v. 26, p. 125–133.

Kellogg, J. N., 1981, The Cenozoic basement tectonics of the Sierra de Perija, Venezuela and Colombia [Ph.D. thesis]: Princeton University, Princeton, New Jersey, 233 p.

Van Houten, F. B., 1976, Late Cenozoic volcaniclastic deposits, Andean foredeep, Colombia: Geological Society of America Bulletin, v. 87, p. 481–495.

Van Houten, F. B., and Travis, R. B., 1968, Cenozoic deposits, Upper Magdalena Valley, Colombia: American Association of Petroleum Geologists Bulletin, v. 52, p. 675–702.

Wellman, S. S., 1970, Stratigraphy and petrology of the nonmarine Honda Group (Miocene), Upper Magdalena Valley, Colombia: Geological Society of America Bulletin, v. 81, p. 2353–2374.

Zambrano, E., Vasquez, E., Duval, B., Latreille, M., and Coffinieri, B., 1971, Sintesis paleogeografica y petrolera del occidente de Venezuela, *in* Memoria, Congresso Geologica Venezolano, IV, Caracas, Volume 1: Venezuela Ministerio de Energia y Minas, p. 483–552.

MANUSCRIPT ACCEPTED BY THE SOCIETY SEPTEMBER 1,

Geological Society of America
Memoir 162
1984

Late Tertiary tectonic history of the Táchira Depression, southwestern Venezuelan Andes

Carlos Macellari
Institute of Polar Studies and
Department of Geology and Mineralogy
Ohio State University
Columbus, Ohio 43210

ABSTRACT

New field observations in the Táchira Depression indicate the presence of conspicuous thrust faults that were previously mapped as normal faults. This new information, combined with the sedimentological study of a Miocene-Pliocene fluvial sequence (La Copé Formation), indicates a succession of late Cenozoic tectonic events in the area. During late Oligocene and up to middle(?) Pliocene time, regional compressive stress oriented at 120° produced two sets of superimposed folds and caused thrust faulting toward the northwest. During the middle(?) to late Pliocene, compressive stresses changed to 040°, producing low-angle thrust faulting toward the northeast. This latter stress direction was the result of the relative southwest movement of the Venezuelan Andes toward the Cordillera Oriental of Colombia. After thrusting, normal faulting occurred in the area, probably simultaneously with the final uplift of the nearby Andes.

INTRODUCTION

The Táchira Depression is a topographic saddle between the Venezuelan Andes to the northeast and the Cordillera Oriental of Colombia to the southwest (Fig. 1). A Jurassic to Quaternary sedimentary sequence more than 7 km thick has accumulated here. There sediments were affected by a series of tectonic events that produced a structurally complex terrain (Kehrer, 1938; Renz, 1960; Trump and Salvador, 1964; Ramírez and Campos, 1972; Useche, 1975, unpub. report). The oldest rocks are Jurassic red beds, which are unconformably overlain by transgressive marine Lower Cretaceous sandstone, "middle" Cretaceous limestone and chert, and Upper Cretaceous black shale (Fig. 2). The Tertiary rocks represent a regressive sequence of Paleocene to lower Oligocene siltstone, sandstone, and coal of marine to paralic origin. This study concentrated on the La Copé Formation, which comprises Miocene-Pliocene fluvial sediments deposited in angular unconformity above the older rocks. Quaternary sediments unconformably overlie the La Copé Formation. These two unconformities provide valuable information for the understanding of the tectonic history of the Táchira Depression.

LA COPE FORMATION (MIOCENE-PLIOCENE)

The La Copé Formation is composed of poorly cemented conglomerate, diamictite, sandstone, sandy mudstone, and mudstone and is divisible into two lithologically distinct members (Macellari, 1982). The conglomeratic lower member was deposited by a system of braided rivers. This member thins rapidly toward the southwest and is vertically and laterally followed by the upper member. The upper member comprises cycles of coarse-grained sandstone (with intraformational clasts) below siltstone and mudstone. These fining-upward cycles are point-bar deposits, characteristic of high-sinuosity, meandering rivers (Allen, 1965; Moody-Stuart, 1966). Maximum thickness of the La Copé Formation in the Táchira Depression is 1,680 m. Paleocurrent indicators (cross-beddings and tree-trunk orientations), variation in maximum clast size, pebble and sandstone composition, and distribution of sedimentary facies indicate that these sediments had a source located northeast of the depression (Macellari, 1981).

The La Copé Formation is probably late Miocene to early Pliocene in age. The youngest sediments beneath this formation

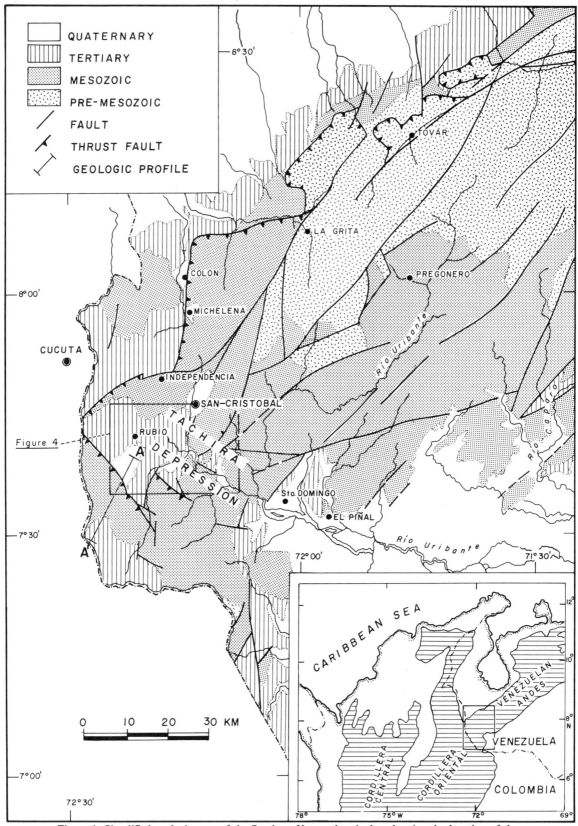

Figure 1. Simplified geologic map of the Southern Venezuelan Andes, showing the location of the Táchira Depression.

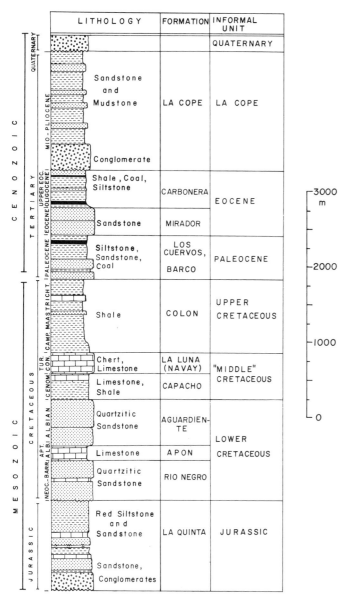

Figure 2. Stratigraphic column of the Táchira Depression.

are the upper Eocene–Oligocene shale and sandstone of the Carbonera Formation. Based on pollen remains, Trump and Salvador (1964) suggested a Miocene-Pliocene age for the Guayabo Group. Judging from the field area covered in that work, the samples analyzed were probably collected from the La Copé Formation in the Táchira Depression. Conglomerates similar to La Copé conglomerates and deposited just prior to the Andean orogeny (Bürgl, 1967) in the Venezuelan Andes and in the Cordillera Oriental of Colombia are latest Miocene to earliest Pliocene in age.

The La Copé Formation unconformably overlies the Lower Cretaceous Aguardiente to the Eocene Carbonera formations (Figs. 2 and 3). The discordance can only be appreciated on a regional basis not within single outcrops. Figure 3 shows the subcrop geology beneath the La Copé Formation. This pattern

suggests that these Miocene-Pliocene sediments were deposited over the nose and flanks of a large, southward-plunging anticline whose axis is roughly parallel to the Venezuelan Andes.

REGIONAL STRUCTURE OF THE TACHIRA DEPRESSION

The entire Táchira Depression is broken by numerous faults that obscure well-developed, regional folds with axes plunging 210°. Anticlinal crests and synclinal troughs are about 4 to 6 km apart. Anticlinal structures occur north of Rubio, in La Petrolea (southwest of Santa Ana), and in El Corozo, 6 km to the south of San Cristóbal (Figs. 4 and 6).

Faults trend in almost every direction. However, many major faults in the northern part of the Táchira Depression trend 15° to 30°, whereas a trend 130° (parallel to the Bramón Fault) is common in the southern part. Previous investigators (Kehrer, 1938; Bucher, 1952; Trump and Salvador, 1964; Useche, 1975) did not recognize large thrust faults in the Táchira Depression, although they mapped the reverse fault located east of San Cristóbal. Field work conducted by the author revealed several additional reverse and thrust faults in the area, which provide new clues for the interpretation of the geology of the southwestern Venezuelan Andes.

Bramón Fault (southwest corner of Fig. 4 and Fig. 6). "Middle" Cretaceous coquina of the Capacho Formation and Lower Cretaceous sandstone (Aguardiente and Río Negro formations) structurally overlie Miocene-Pliocene siltstone and sandstone of the La Copé Formation. The contact can be observed near Quebrada La Blanca (southwestern part of Fig. 4). Cretaceous beds are strongly fractured, folded, and overturned for more than 1 km away from the trace of the fault (Fig. 5). The irregular trace of the fault is evidence of the low-angle character of the thrust.

Unnamed Fault (A) (south of Santa Ana, Fig. 4 and A on Fig. 6). Here shale of the Colón Formation (Upper Cretaceous) rests on Eocene shale and coal beds of the Carbonera Formation. On the upthrown (south) side of the contact, the beds are overturned for 2 km perpendicular to the fault trace, producing a structure similar to that shown in Figure 5 for the Bramón Fault. This fault is interpreted as an extension of the Bramón Fault, which has been offset by younger normal faults.

Unnamed Fault (B) (1.5 km south of San Josecito, Fig. 4 and B on Fig. 6). There is only sparse field evidence of its nature; because it is parallel to the Bramón and A faults, fault B could also be a thrust.

Unnamed Fault (C) (Fig. 4 and C on Fig. 6). Two lines of evidence indicate this fault is a thrust. First is the irregular fault trace. Second, along the trace of fault E (Fig. 6), hot springs and sulfurous emanations occur at Río Zuñiga (point 1, Fig. 6) and Quebrada de Aza (point 2, Fig. 6). Farther to the west, hot springs and sulfurous emanations are found in the Quebrada La Mona (point 3, Fig. 6) and in El Corozo (point 4, Fig. 6). These four places have the only emanations of this kind in the area, and

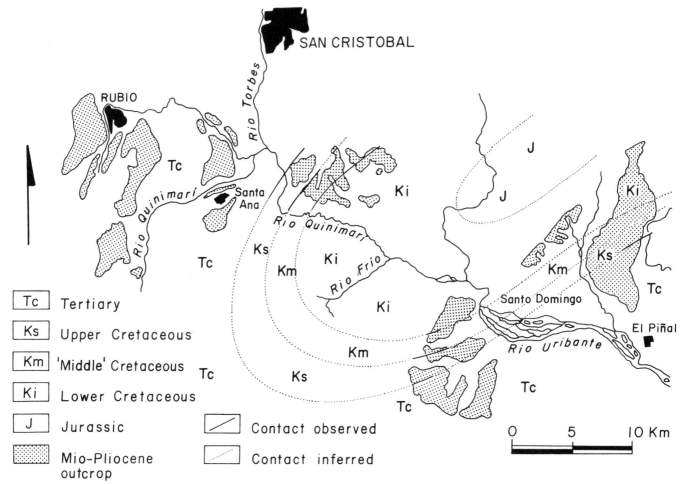

Figure 3. Present distribution of Miocene-Pliocene sediments in the Táchira Depression and inferred paleogeology at the time of deposition of the La Copé Formation. See Figures 1 and 4 for location.

they lie along a straight line coincident with the visible trace of fault E. Consequently, fault E continues, without offset in the subsurface west of point 2 (Fig. 6) where it is covered by rocks thrust southeast along fault C. Fault C places "middle" and Upper Cretaceous rocks over Miocene-Pliocene and Upper Cretaceous strata.

Unnamed Fault (D) (center of Fig. 4 and D on Fig. 6). Here Paleocene and Eocene rocks are faulted northeastward upon Miocene-Pliocene conglomerate and sandstone. Direct evidence of the nature of this fault was not observed in the field, but its sinuous trace indicates that it is a thrust.

San Cristóbal Fault (Fig. 6). East of San Cristóbal this high-angle reverse fault brings Jurassic red beds in contact with Lower Cretaceous sandstone and limestone. This fault terminates toward the south, and only minor displacement is observed when it reaches the Río Torbes (Fig. 4). Rocks near the fault contact are strongly brecciated and locally overturned. The displacement is small because beds on either side of the fault are always from the same formation or from a closely related stratigraphic interval.

Capacho Fault (Fig. 6). The conspicuous Capacho Fault is

located north of the Táchira Depression. Previous workers called the Capacho Fault a reverse fault (Trump and Salvador, 1964), a right-lateral fault (Ramírez and Campos, 1972), and a normal fault (Ramírez and Campos, unpub. revised map, MEM, Venezuela). The locally overturned beds of the upthrown side, the tightly folded anticline close to the fault (point 5, Fig. 6), and the irregular trace of the fault (west of Libertad) suggest that the Capacho is a thrust fault along which the Cretaceous La Luna and Capacho formations moved northwestward over the Eocene Carbonera and Mirador formations. The nature of the fault at its eastern end (point 6 on Fig. 6) is not known. If the Borotá anticline represents the northward continuation of the anticline east of Libertad, then a right-lateral displacement of up to 6 km could be present on the Capacho Fault here.

In summary, the thrust and reverse faults can be grouped in two classes: those with a northeast trend, located to the north of the Táchira Depression and with relatively minor displacements (San Cristóbal and Capacho faults), and those with a northwest trend, located to the south of the depression and with relatively major displacements. The folds in the Lower Cretaceous to Miocene-Pliocene sediments formed first, before other major tec-

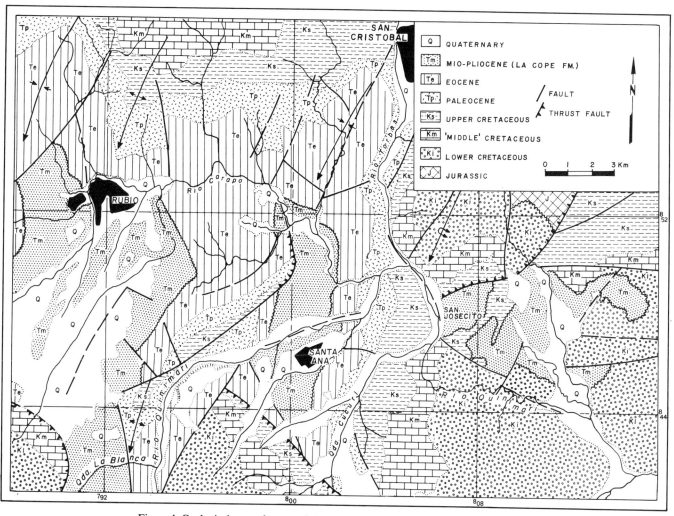

Figure 4. Geological map of a part of the Táchira Depression. See Figure 1 for location.

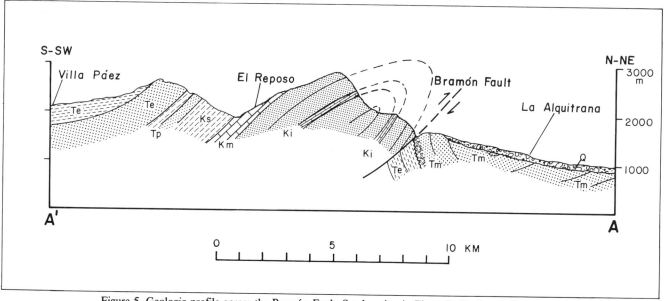

Figure 5. Geologic profile across the Bramón Fault. See location in Figure 1. Ki: Lower Cretaceous;
Km: "middle" Cretaceous; Ks: Upper Cretaceous; Tp: Paleocene; Te: Eocene; Tm: Miocene-Pliocene;
Q: Quaternary.

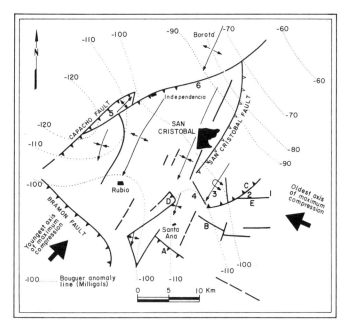

Figure 6. Generalized tectonic map of the Táchira Depression with Bouguer gravity contours superimposed (Bouguer anomaly data from Hospers and Van Wijnen, 1959). Faults with black triangles are interpreted as thrust faults; faults with white triangles are interpreted as high angle reverse faults.

tonic disturbances of the sedimentary sequence. Folds are truncated and displaced by both thrust and normal faults. Normal faulting, which clearly cuts across both folds and thrust faults, came last.

SEQUENCE OF TECTONIC EVENTS IN THE TACHIRA DEPRESSION

The relationship between regional structure and the La Copé Formation indicates four stages in the late Tertiary development of the Táchira Depression:

Stage 1

During late Oligocene–early Miocene times, the area northeast of the Táchira Depression was slowly uplifted. Shale of the León Formation (late Oligocene–early Miocene) of the Maracaibo Basin were not deposited (or alternatively, were stripped off before the Miocene-Pliocene) in the Táchira Depression and in the Santo Domingo area. The uplift was probably related to regional, large-scale folding about an axis oriented 030° and plunging to the south (Fig. 3). Erosion of this uplifted structure shed the Miocene-Pliocene molasselike sediments of the La Copé Formation into the depression.

Stage 2

During the Pliocene, after deposition of La Copé Formation, compression became more intense, producing smaller scale folds

with axes parallel to the large-scale folds of stage 1. Final compression at the end of this stage resulted in localized thrust and reverse faulting (San Cristóbal and Capacho faults).

Stage 3

During the middle(?) to late Pliocene, compressive stresses were reoriented and resulted in thrusting toward the northeast (Bramón and related faults).

Stage 4

After the major compressive phases, normal faulting affected the area, probably as a consequence of the final uplift of the Andes. Normal and strike-slip faulting took place very recently in the Táchira Depression.

DISCUSSION

The tectonic development of the Táchira Depression may be understood in terms of the relative movement of two large, rigid blocks: the Cordillera Oriental of Colombia to the southwest and the Venezuelan Andes to the northeast. The Táchira Depression in a narrow sense is a triangle confined by three major faults: the San Cristóbal Fault to the east, the Capacho Fault to the northeast, and the Bramón Fault to the southwest.

According to Bucher (1952), the most significant features of the area are four en echelon faults that trend northwest-southeast, indicating that maximum elongation was perpendicular to these faults. Field mapping shows that two and possibly three of Bucher's en echelon faults are thrust faults (Bramón, A, and B(?) faults, Fig. 6). Consequently, the axis of maximum compression and not the axis of greatest elongation was perpendicular to these faults.

The Bouguer gravity field (based on data of Hospers and Van Wijnen, 1959) shows strong negative values west of Libertad on the Capacho Fault and northeast of Santa Ana. The negative anomalies west of Libertad result from the presence of the thicker sediments of the Maracaibo Basin. Negative anomalies northeast of Santa Ana are probably due to an unusually thick sequence caused by the thrusting of Cretaceous rocks northeastward over Eocene and Miocene-Pliocene strata along faults A, B(?), and D. If so, the thrust displacement was about 10 km or more.

No single direction of principal stress produced all the features present in the Táchira Depression. A principal compressional stress direction oriented approximately 120° produced the folds that developed before and immediately after deposition of the La Copé Formation. The Bramón and related A and B(?) faults, in turn, resulted from a compressive stress oriented approximately 040°. These compressive forces are consistent with very different conceptual models of the tectonic evolution of the Venezuelan Andes.

If the Capacho Fault is a wrench fault of right-lateral displacement, then the 120° orientation of compression could be a

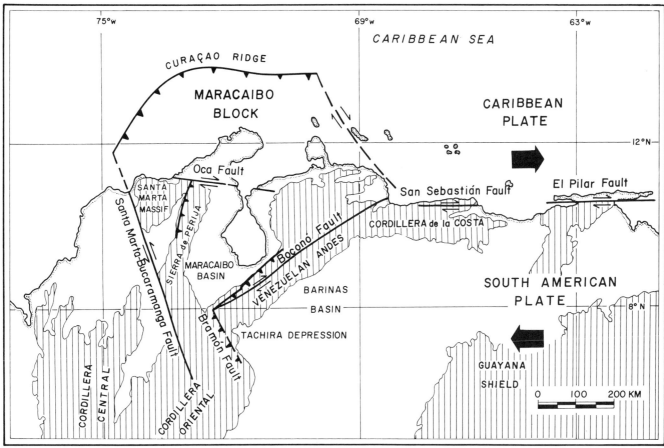

Figure 7. Plate tectonic model of northwestern South America, showing relationships between the Caribbean plate, the South American plate, and the Maracaibo block (modified from Bowin, 1976).

secondary stress produced by this fault (see, for example, Wilcox and others, 1973). In this case, folds of the Táchira Depression would be drag folds. DeRatmiroff (1971) described extensive imbricate thrusting of the northwestern flank of the Andes toward the Maracaibo Basin. He attributed this thrusting to compressive stress from 120° during the Miocene-Pliocene. Similarly, Kellogg (1980), Bonini and others (1980), and Kohn and others (1984) considered this stress orientation as a regional feature that affected not only the Venezuelan Andes but also the Sierra de Perijá and the Santa Marta Massif. Thus, it seems that the compressive stress from 120° was a regional tectonic feature affecting a great part of the Venezuelan Andes and not related to wrench faulting along the Capacho Fault.

The compressive stress oriented 040° would have had another origin. Mencher (1963), Schubert and Sifontes (1970), Dewey (1972), and Santamaría and Schubert (1974) proposed that right-lateral displacement along the Boconó Fault started in the latest Tertiary and continues to the present (Fig. 7). Right-lateral displacement is consistent with the thrust movement of the Bramón Fault (Fig. 7). Consequently, I propose that the thrust faulting along the Bramón Fault is the response to the relative displacement of the Venezuelan Andes toward the Cordillera Oriental of Colombia. This hypothesis is supported by the fre-

quency of shallow seismic activity in this area (the greatest in northwestern Venezuela). From focal mechanism solutions, Dewey (1972) concluded that the area is currently being subjected to compressive forces oriented east-west, but the focal mechanism solution for a recent (January 27, 1970) earthquake inside the depression could also indicate northeast-southwest compression. The direction of movement or amount of displacement along the Boconó Fault remains uncertain. Rod (1956) calculated a right-lateral displacement of 33 km in the Boconó region. Bushman (1965) described a displacement of 45 km in southeastern Lara and up to 65 km in the Andes proper. DeRatmiroff (1971) cited 75 km in the Boconó region of the northeastern Andes, but only 15 km in the southern part between Tovar and La Grita. Schubert and Sifontes (1970) and Giegengack and Grauch (1972) described recent lateral movement along the Boconó Fault of up to 250 m, based on the offset of Pleistocene glacial moraines and other recent depositional features. Shagam (1972, 1975) and Schubert (1968, 1969) thought that vertical displacement predominated along the Boconó Fault and that most of the lateral displacement occurred only very recently. To date, there is no unequivocal evidence against large lateral displacement along the Boconó Fault. According to Stephan (1977), the main problem with large displacements along this fault is accommodating this movement in Táchira State. Northeastward thrusting could (as discussed above) accommodate 10 km or more of displacement by thrusting along the Bramón and related faults. Large vertical movements are a common feature in wrench faults (Reading, 1980). In the case of the Boconó Fault, both vertical and horizontal displacements could have taken place at different times or simultaneously.

Schubert and Sifontes (1970), Dewey (1972), and Santamaría and Schubert (1974) considered the Boconó Fault as one of the present boundaries between the South American and Caribbean plates. Shagam (1972, 1975), Giegengack and Grauch (1976), and Giegengack and others (1976) disagreed with this idea. Clearly the contact between the Caribbean and South American plates is complex, and the boundary may well be a general zone of contact. Bowin (1976) suggested that the trapezoidal "Maracaibo" block separates a part of the South American from the Caribbean plate.

The Maracaibo block (Fig. 7) is delimited by the Boconó Fault on the southeast, the Santa Marta–Bucaramanga Fault on the southwest, the Curaçao Ridge on the north, and a vaguely defined northeast border roughly parallel to the Santa Marta–Bucaramanga Fault. Evidence of thrusting to the north, near the Curaçao Ridge, was presented by Silver and others (1975), and this zone shows shallow seismic activity (Bowin, 1976). The Santa Marta–Bucaramanga Fault has been described as a left-lateral wrench fault, as a normal fault, and as a high-angle reverse fault (Campbell, 1968; Tschanz and others, 1974). Shallow seismic activity along this fault is not very important (Bowin, 1976). The Oca Fault, exposed in the middle of the Maracaibo block, probably had as much as 15 to 20 km of right-lateral movement (Feo-Codecido, 1972), which ended in the Pliocene when the plate boundaries changed and the Oca Fault became inactive (Vasquez and Dickey, 1972).

The presence of a Maracaibo block as a tectonic element between the Caribbean and South American plates is, as outlined below, consistent with the structural history of the Táchira Depression.

During late Oligocene to early Pliocene time, compressive forces were oriented northwest-southeast. Left-lateral movement occurred along the Santa Marta–Bucaramanga Fault; two stages of folding affected the Táchira Depression; the Oca Fault moved right-laterally and thrust faulting occurred in the Curaçao Ridge. Greater movement along the Santa Marta–Bucaramanga Fault (up to 110 km) than on the Oca Fault (as much as 15–20 km) might have resulted in the clockwise rotation of the Venezuelan Andes that has been indicated paleomagnetically (Hargraves and Shagam, 1969; Shagam and Hargraves, 1972).

During late Pliocene time, northeast-southwest compression dominated. Right-lateral movement began on the Boconó Fault, and in response, northeastward thrust faulting occurred in the Táchira Depression along the Bramón and related faults.

CONCLUSIONS

The structure of the Táchira Depression can be explained as the consequence of the relative movements of two rigid blocks (Venezuelan Andes to the northeast and the Cordillera Oriental of Colombia to the southwest), separated by a thick sedimentary pile which absorbed most of the stresses produced by the movement of these two blocks.

Two different directions of maximum compression affected the area during late Cenozoic time. The first was oriented approximately 120°. It was responsible for the folds and probably for the thrust faults with a relative displacement to the west-northwest. Later, the principal compressive axis shifted to 040°, producing the thrust faults with relative movement to the north-northeast. If the La Copé Formation age is correctly interpreted as Miocene-Pliocene, then the principal tectonic phase of the Andean orogeny in the Táchira Depression occurred during Pliocene time.

ACKNOWLEDGMENTS

The Ministerio del Ambiente, Venezuela, supported field work for this study. I am greatly indebted to R. Bustos and D. Elliot for invaluable discussion on several aspects of this study. R Shagam, T. A. Anderson, F. B. Van Houten, and T. DeVries reviewed the manuscript.

REFERENCES CITED

Allen, J.R.L., 1965, Fining-upwards cycles in alluvial successions: Geological Journal, v. 4, p. 229–246.

Bonini, W. E., Garing, J. D., and Kellogg, J. N., 1980, Late Cenozoic uplifts of the Maracaibo-Santa Marta block, slow subduction of the Caribbean plate and results from a gravity study, *in* Transactions, Caribbean Geological Conference, IX, Santo Domingo, 1980: Universidad Católica Madre y Maestra, Santiago de los Caballeros, República Dominicana, p. 99–105.

Bowin, C., 1976, Caribbean gravity field and plate tectonics: Geological Society of America Special Paper 169, 79 p.

Bucher, W. H., 1952, Geological structure and orogenic history of Venezuela: Geological Society of America Memoir 49, 111 p.

Bürgl, H., 1967, The orogenesis in the Andean System of Colombia: Tectonophysics, v. 4, p. 429–443.

Bushman, J. R., 1965, Geología del area de Barquisimeto, Venezuela: Boletín de Geología (Venezuela), Caracas, v. 6, no. 11, p. 3–11.

Campbell, C. J., 1968, The Santa Marta wrench fault of Colombia and its regional setting, *in* Transactions, Caribbean Geological Conference, IV, Port-of-Spain, Trinidad and Tobago, 1965, p. 247–261.

DeRatmiroff, G., 1971, Late Cenozoic imbricate thrusting in Venezuelan Andes: American Association of Petroleum Geologists Bulletin, v. 55, p. 1336–1344.

Dewey, J. W., 1972, Seismicity and tectonics of western Venezuela: Seismological Society of America Bulletin, v. 62, p. 1711–1751.

Feo-Codecido, G., 1972, Breves ideas sobre la estructura de la falla de Oca, Venezuela, *in* Transactions, Caribbean Geological Conference, VI, Margarita, Venezuela, 1971, p. 184–190.

Giegengack, R., and Grauch, R. I., 1972, Boconó Fault, Venezuelan Andes: Science, v. 175, p. 558–560.

—— 1976, Quaternary geology of the central Andes, Venezuela: A preliminary assessment, *in* Memoria, Congreso Latinoamericano de Geología, II, Caracas, 1973: Venezuela Ministerio de Minas e Hidrocarburos, Boletín de Geología, Publicación Especial no. 7, v. 1, p. 241–283.

Giegengack, R., Grauch, R. I., and Shagam, R., 1976, Geometry of late Cenozoic displacement along the Boconó fault, Venezuelan Andes, *in* Memoria, Congreso Latinoamericano de Geología, II, Caracas, 1973: Venezuela Ministerio de Minas e Hidrocarburos, Boletín de Geología, Publicación Especial no. 7, v. 2, p. 1201–1226.

Hargraves, R. B., and Shagam, R., 1969, Paleomagnetic study of La Quinta Formation, Venezuela: American Association of Petroleum Geologists Bulletin, v. 53, p. 537–552.

Hospers, J., and Van Wijnen, J. C., 1959, The gravity field of the Venezuelan Andes and adjacent basins: Verhandelingen der Koninklijke Nederlandse Akademie Van Wetenschappen, Afd. Natuurkunde, Eerste Reeks, v. 23, p. 1–95.

Kehrer, L., 1938, Algunas observaciones sobre la estratigrafía en el Estado Táchira: Boletín de Geología y Minas (Venezuela), Caracas, v. 2, no. 2-4, p. 44–56.

Kellogg, J. N., 1980, Cenozoic basement tectonics of the Sierra de Perija, Venezuela and Colombia, *in* Transactions, Caribbean Geological Conference, IX, Santo Domingo, 1980: Universidad Católica Madre y Maestra, Santiago de los Caballeros, República Dominicana, p. 107–117.

Kohn, B. P., Shagam, R., and Burkley, L. A., 1984, Mesozoic-Pleistocene fission-track ages on rocks of the Venezuelan Andes and their tectonic implications, *in* Bonini W. E., and others, eds., The Caribbean–South American plate boundary and regional tectonics: Geological Society of America Memoir 162 (this volume).

Macellari, C., 1981, Late Cenozoic deposits of the Táchira Depression and the Tectonic evolution of the southern Venezuelan Andes [M.Sc. thesis]: Columbus, Ohio, The Ohio State University, 167 p.

—— 1982, El Mio-Plioceno de la Depresión del Táchira (Andes Venezolanos): Distribución Paleogeográfica e Imlicancias Tectónicas: Geos, Caracas, v. 27, p. 3–14.

Mencher, E., 1963, Tectonic history of Venezuela, *in* Childs, O. E., and Beebe, B. W., eds., Backbone of the Americas, Tectonic history from pole to pole: American Association of Petroleum Geologists Memoir, v. 2, p. 73–87.

Moody-Stuart, M., 1966, High- and low-sinuosity stream deposits, with examples from the Devonian of Spitsbergen: Journal of Sedimentary Petrology, v. 36, p. 1102–1117.

Ramírez, C., and Campos, V. C., 1972, Geología de la región de la Grita-San Cristóbal, Estado Táchira, *in* Memoria, Congreso Geológico Venezolano, IV, Caracas, 1969: Venezuela Ministerio de Minas e Hidrocarburos, Boletín de Geología, Publicación Especial no. 5, v. 2, p. 861–893.

Reading, H. G., 1980, Characteristics and recognition of strike-slip fault systems, *in* Ballance, P. F., and Reading, H. G., eds., Sedimentation in oblique-slip mobile zones: Oxford, London, Blackwell Scientific Publications, p. 7–27.

Renz, O., 1960, Guía de la excursión C-7, Andes Suroccidentales, sección de Santo Domingo a San Antonio (Estado Táchira), *in* Memoria, Congreso Geológico Venezolano, III, Caracas, 1960: Venezuela Dirección de Geología, Publicacioín Especial no. 3, v. 1, p. 87–91.

Rod, O., 1956, Strike-slip faults of northern Venezuela: American Association of Petroleum Geologists Bulletin, v. 40, p. 457–476.

Santamaría, F., and Schubert, C., 1974, Geochemistry and geochronology of the southern Caribbean–northern Venezuelan plate boundary: Geological Society of America Bulletin, v. 85, p. 1085–1097.

Schubert, C., 1968, Geología de la Región de Barinitas-Santo Domingo, Andes Venezolanos Surorientales: Boletín de Geología (Venezuela), Caracas, v. 10, no. 19, p. 183–266.

—— 1969, Geologic structure of a part of the Barinas Mountain Front, Venezuelan Andes: Geological Society of America Bulletin, v. 80, p. 443–458.

Schubert, C., and Sifontes, R. S., 1970, Boconó Fault, Venezuelan Andes: Science, v. 175, p. 560–561.

Shagam, R., 1972, Evolución tectónica de los Andes Venezolanos, *in* Memoria, Congreso Geológico Venezolano, IV, Caracas, 1969: Venezuela Ministerio de Minas e Hidrocarburos, Boletín de Geología, Publicación Especial no. 5, v. 2, p. 1201–1258.

—— 1975, The northern termination of the Andes, *in* Nairn, A. E., and Stehli, F. G., eds., The ocean basins and margins: New York, London, Plenum Press, v. 3, p. 325–415.

Shagam, R., and Hargraves, R. B., 1972, Estudio paleomagnetico de las formaciones Sabaneta y Mérida, Andes Venezolanos, *in* Memoria, Congreso Geológico Venezolano, IV, Caracas, 1969: Venezuela Ministerio de Minas e Hidrocarburos, Boletín de Geología, Publicación Especial no. 5, v. 2, p. 1157–1182.

Silver, E. A., Case, J. E., and MacGillavry, H. J., 1975, Geophysical study of the Venezuelan borderland: Geological Society of America Bulletin, v. 86, p. 213–226.

Stephan, J. F., 1977, El contacto cadena Caribe-Andes Merideños entre Carora y El Tocuyo (Estado Lara), *in* Memoria, Congreso Geológico Venezolano, V, Caracas, Volume 2: Venezuela Ministerio de Energia y Minas, p. 789–816.

Trump, G. W., and Salvador, A., 1964, Guidebook to the geology of western Táchira: Asociación Venezolana de Geología Minería y Petroleo, Caracas, p. 1–25.

Tschanz, C. M., Marvin, R. F., Cruz, J. B., Mehnert, H. H., and Cebula, G. T., 1974, Geologic evolution of the Sierra Nevada de Santa Marta, northeastern Colombia: Geological Society of America Bulletin, v. 85, p. 273–284.

Vasquez, E., and Dickey, P. A., 1972, Major faulting in north-western Venezuela and its relation to global tectonics, *in* Transactions, Caribbean Geological Conference, VI, Margarita, Venezuela, 1971, p. 191–202.

Wilcox, R. E., Harding, T. P., and Seely, D. R., 1973, Basic wrench tectonics: American Association of Petroleum Geologists Bulletin, v. 57, p. 74–96.

MANUSCRIPT ACCEPTED BY THE SOCIETY SEPTEMBER 1, 1983

Printed in U.S.A.

Geological Society of America
Memoir 162
1984

Late Cenozoic tectonic environments of the Central Venezuelan Andes

Robert Giegengack
Department of Geology
The University of Pennsylvania
Philadelphia, Pennsylvania 19174

ABSTRACT

The Andes of western Venezuela have been interpreted as a segment of the plate boundary between South America and the Caribbean. The details of the late Cenozoic history of the Venezuelan Andes do not support the widely held perception that a narrow axial fault zone (the Boconó Fault Zone) has accommodated all of the strike-slip displacement that has occurred between those plates in late Cenozoic time.

Field data not fully considered in previous interpretations include:

1. An extensive breccia zone, closely related (where exposures reveal its base) to a high-angle thrust fault that rises within the Boconó lineament and rolls over to the northwest to carry rocks as old as Precambrian(?) gneiss, early Paleozoic(?) schist, and late Paleozoic phyllite over a terrane of flat-lying sediments as young as early Miocene.

2. A deep red, locally bauxitic paleosol that formed near sea level during a long period of relative tectonic stability *after* emplacement of the thrust sheet, and that has been arched a minimum of 3 km over parts of the Central Venezuelan Andes in late Quaternary time.

3. A complex system of interrelated high-angle faults along which Quaternary displacement has been well documented, amounting to (a) ⩾250 m of right-lateral strike-slip displacement since deposition of morainal landforms during the penultimate glaciation and (b) at least 3 km of vertical displacement adjacent to a series of linear central grabens.

The grabens have been interpreted as pull-apart basins and may represent evidence of substantial right-lateral strike-slip movement along the Boconó Fault in latest Quaternary time. The post–early Miocene pre–late Quaternary history of the cordillera, however, clearly shows that for much of that interval of time the stress field imposed on the mountain range has been more complex and has included a substantial component of compression.

INTRODUCTION

This paper presents a tectonic model for the Central Venezuelan Andes in later Tertiary time that is based primarily on analysis of surficial features and sediments (Giegengack, 1977a). The late Tertiary tectonic pattern that has emerged from this study differs from patterns extrapolated from bedrock evidence from Cretaceous and earlier periods (e.g., Shagam, 1976; González de Juana and others, 1980). This difference results both from the probability that the tectonic setting has changed appre-

ciably in Tertiary time and from the perspective that is brought to bear on a tectonically active terrane by a surficial geologist, which may differ somewhat from the perspective of the reconnaissance structural/stratigraphic geologist. The principal thrust of this paper is to present the data of this continuing analysis and to develop a tectonic model, necessarily still tentative, of the strain history of the Central Venezuelan Andes from mid-Miocene(?) time until the present, as constructed from those data.

TECHNIQUES USED

The techniques applied in this study of late Cenozoic tectonic history of the Andes emphasize the extent to which bedrock structure and late Cenozoic perturbation of Quaternary landforms and sedimentary bodies are illuminated by the configuration of the present landscape, as revealed both by field observation and study of air photographs.

Supplemental stratigraphic-structural investigations were undertaken in those regions where the details of structural configuration did not yield readily to geomorphic analysis, or where that analysis revealed structural relationships inconsistent with the data base of earlier studies, on which other tectonic models have been founded.

This project will eventually yield a detailed Quaternary tectonic map and a conventional map of the surficial geology of the major central valleys of the Central Venezuelan Andes: those products are not ready to be presented with this paper.

The most important consequence of this study is the development of a neotectonic model that bears little resemblance to models derived from either extrapolation of structural relations among rocks of Cretaceous and older ages or from inferences based on analysis of modern seismicity. The latter studies include models based on the limited seismic data collected directly in the last 30 to 50 years (Molnar and Sykes, 1969; Bolt and others, 1972; Dewey, 1972; Kelleher and others, 1973; Larotta, 1975; Kafka and Weidner, 1980, 1981; Pennington, 1981; Pennington and others, 1979) and on inferences drawn from the distribution of modified Mercalli intensities of various historic earthquakes, as reconstructed from historic accounts (Kelleher and others, 1973; Larotta, 1975). Neither the distribution of earthquake foci nor the sense of first motion for the few earthquakes for which such analyses have been completed is consistent with the geomorphic evidence that documents events of the last 10^6 to 10^7 years. This discrepancy should be attributed solely to the differences in time scale and should give pause to those who would extrapolate patterns inferred from a few decades of empiric data to processes that occur over much longer periods of geologic time.

GEOGRAPHIC SETTING

The Central Venezuelan Andes represent the easternmost finger of the 5 distinct mountain ranges that fan out across northern Colombia and western Venezuela as the northward termination of the Andes (Shagam, 1976). The two easternmost ranges, the Perijás and the Venezuelan Andes, differ from the rest of the Cordilleran chain by the total absence of young volcanic rocks. The elevations reached in these ranges (5,002 m at Pico Bolivar in the Venezuelan Andes; 3,840 m at the highest point in the Perijás) are the result of recent tectonic uplift, not accumulation of volcanic rocks.

Specific locations and geographic features mentioned in this paper are located on the map of Figure 1.

The Venezuelan Andes consist of two parallel ranges separated by a deep, linear central valley. The central valley is discontinuous and consists of a series of fault-bounded basins ranging in size from a few kilometres square to 50 by 15 km (the Tabay-Estánquez Graben of Giegengack, 1977b; the La Gonzáles Graben of Renz, 1956, and Schubert, 1980) to 25 by 25 km, in the case of the Táchira Depression (Macellari, 1981, 1984). The central valleys are drained by a moderately well integrated system of major rivers, some of which (e.g., the Rio Chama, the Rio Motatán) carry runoff from chains of individual basins. High-altitude divides along the axis of the central valley separate the major drainage systems within that valley. Deep canyons through which the principal streams escape to the Maracaibo drainage across the northwest range or to the Orinoco drainage to the southeast separate the two ranges into a series of individually named cordillera.

The linear central valley has long been recognized as a chain of intermontane fault-bounded basins, the boundaries of which have severely constrained the development of the several drainage systems along which these basins have been integrated (Rod, 1956; Fig. 2). Cataclasis of bedrock adjacent to the faults bounding the central basins was also described by Rod (1956).

Bedrock structures within the grabens are obscured by the mass of Quaternary sediment that veneers the valley floors and clings precariously as erosion remnants high on the valley walls; incision of this sedimentary fill has provided dramatic mountain-terrace scenery. Although the bodies of Quaternary sediment typically are mapped as a single stratigraphic unit (Qal or Qt, for Quaternary terraces) on the latest geologic maps published by the Ministry of Energy and Mines, they can be referred, on both sedimentologic and morphologic grounds, to a series of complex Quaternary stratigraphic sections that have not yet been successfully correlated from basin to basin.

PREVIOUS WORK

Thoughtful observations of contemporary glacial activity and inferred glacial-age depression of the regional snowline in the Venezuelan Andes, as determined from geomorphic analysis, were made as early as 1888 (Giegengack and Grauch, 1975; Schubert, 1971). The first neotectonic evaluation of the central valleys to gain general attention was contained in the paper by Rod (1956) who described the topographic expression and the structural setting of the Boconó Fault. This fault is a first-order feature that runs the length of the Venezuelan Andes from Puerto Cabello on the Caribbean to the Táchira Depression on the border with Colombia (Figs. 1, 2, 5, and 9). In the Central Venezuelan Andes the Boconó Fault, as Rod defined it, represents one border or the other of each of the linear intermontane basins.

Rod (1956) described the now-classic locality on the crest of the divide between the Rio Chama and the Rio Santo Domingo, where glacial-age moraine landforms are offset in a right-lateral

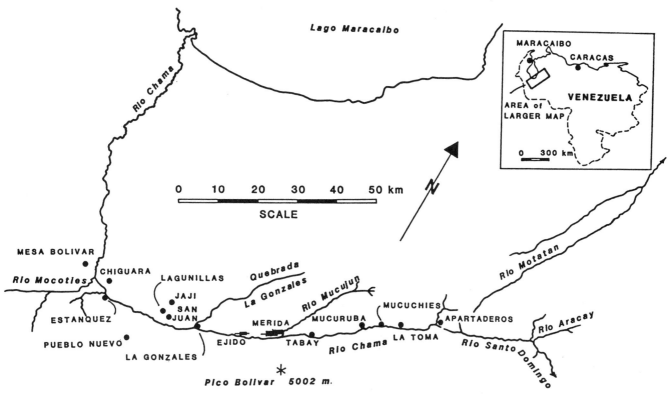

Figure 1. Map of features and localities mentioned in the text.

sense along the Boconó Fault; Rod estimated that displacement as 80 to 100 m. That modest estimate was sufficient to lead many of the present generation of global synthesists to identify the Boconó Fault as the surface expression of a major transform along which the Caribbean Plate was transposed hundreds of kilometres eastward with reference to South America (e.g., Molnar and Sykes, 1969; Dewey, 1972; Kelleher and others, 1973; Santamaria and Schubert, 1974; Jordan, 1975; Bowin, 1976; Vierbuchen, 1978; Kafka and Weidner, 1980; Schubert, 1981; Pennington, 1981).

The combined stratigraphic, structural, and geomorphic evidence presented in this paper demonstrates that the Boconó Fault in late Tertiary time has been the locus of measurable displacement far more complex than simple right-lateral strike-slip.

ROCK UNITS DEFINED IN THIS STUDY

The stratigraphic framework of the Central Venezuelan Andes has been summarized by Shagam (1976) Shagam and others (1984), and González de Juana and others (1980). Recently published maps by Bellízzia and others (1976) and Case and Holcombe (1980) permit this framework to be placed in a larger regional perspective.

Two new rock units, whose significance has emerged from study of the late Tertiary stratigraphy and tectonics of the Central

Venezuelan Andes, are presented here; they are the Chama Valley Breccia and the Pueblo Nuevo Paleo-Oxisol.

The Chama Valley Breccia

The investigation described here was focused early on the bodies of Quaternary sediment that blanket the floors of the fault-bounded basins and are arrayed as flights of stepped terraces against the northwest margins of the basins. Within this complex were encountered masses of rock of an anomalous texture not readily assignable either to the column of Quaternary sediment or to the underlying bedrock, although on existing maps these bodies have been assigned to both. The rock in question consists of a disaggregated mass of fracture-bounded, angular, pebble to cobble-size particles assignable by virtue of lithology to one or another of the sedimentary or metamorphic rock units known from the Central Venezuelan Andes. In many outcrops, the rock is a loose, unconsolidated mass which nonetheless retains considerable competence in an erosional environment by virtue of the tightly interlocking geometry of the contacts among adjacent rock particles. No cement occupies the fractures, and no preferred orientation either of elongate particles or of the fracture pattern suggests a consistent sense of shear. In most exposures no displacement has occurred along or across the fracture surfaces; in other outcrops the fractures are open by as much as a few cen-

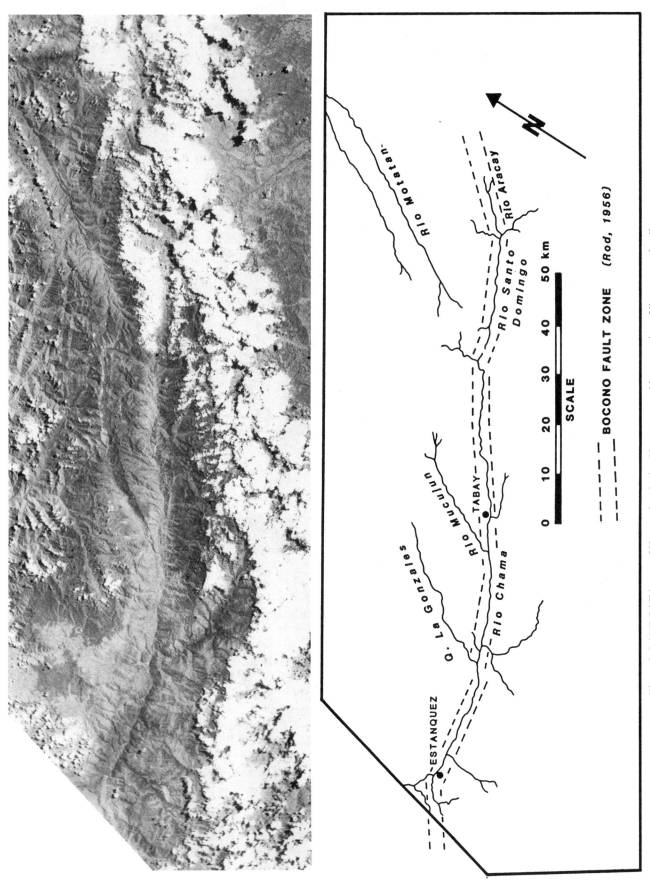

Figure 2. LANDSAT image of Venezuelan Andes. Note topographic expression of linear central-valley system. Compare Figures 1 and 5. Photo courtesy NASA.

timetres, and the space between particles is filled with a fine-grained matrix of a lithology identical to that of the adjacent particles. Throughout this range in texture, an original sedimentary or metamorphic fabric is recognizable on the outcrop; the fracture geometry is clearly a later, superimposed effect.

Where first encountered, under and near the village of Mucuchies (Figs. 1 and 5), this rock was given the field designation of a breccia; as the extent of its regional development became apparent, it came to be known as the Chama Valley Breccia. On Figure 5 it is mapped as a stratigraphic unit; comparison of that map with any map of the bedrock geology of any part of the Central Venezuelan Andes will show that the boundaries of the Chama Valley Breccia cut across map contacts between lithostratigraphically defined formations.

The brecciated texture is developed in amphibolite gneiss of the Precambrian(?) Sierra Nevada Formation, in granite intrusives into that unit, in several of the Paleozoic phyllite- to schist-grade complexes, and in the arkosic sandstones and shales of the Triassic(?)-Jurassic La Quinta Formation.

Distended Textures in Rocks of the Chama Valley Breccia.

In some outcrops within the breccia zone the configuration of original rock structures is still clearly discernible. The rock is cut by many intersecting, closely spaced fracture planes on which shear displacement has not occurred. Original structures such as bedding or folds can be traced across these outcrops. Other rock masses within the breccia zone have clearly been distended or dilated by the drawing apart of adjacent fracture-bounded fragments; these fragments range in size from microscopic particles to particles a few tens of centimetres in longest dimension. Where distension is of the order of a few percent, the principal effect of that distension is the enhancement of rock porosity, with some minor relative rotation of adjacent blocks. Where the fractures that bound the blocks have remained unfilled, the breccia has lost all competence and behaves as (and has been mistaken for) an unconsolidated sediment. In other outcrops the fractures are filled with a silt-size matrix composed of finely comminuted grains of a lithology identical to that of blocks adjacent to the fractures. This matrix appears to have been injected into the fractures as a slurry.

As the extent of distension becomes greater, it becomes progressively more difficult to reconstruct the initial configuration of the bedrock mass. The most distended rock masses have acquired a secondary texture that has altered them to a tight, competent, massive rock of silt-size fragments in which subparallel stringers of widely spaced, larger, angular particles evoke a memory of an original sedimentary or metamorphic configuration. Such macroscopically identifiable particles may represent as little as 1% of the volume of the rock. Figure 3 is a schematic representation of this full range of textural distension.

In rocks with the most distended textures, the superficial resemblance of the subparallel stringers of angular blocks, particularly in low-grade metamorphic terranes, to sedimentary bedding has led to the assignment of many such bodies to the column of Quaternary sediment (i.e., conglomerate stringers in a silt ma-

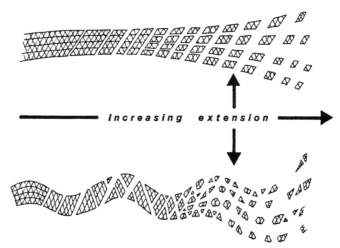

Figure 3. Range of textures encountered in bodies of distended-texture breccia.

trix). The potential for this kind of misidentification is enhanced by the fact that the breccia, typically far more erodible than nearby, unbrecciated rock masses, is easily eroded to the level at which adjacent constructional rock masses are accumulating. Thus, for example, the alluvial fans underlying the towns of Mucuchies, Mucurubá, Ejido, San Juan, and Lagunillas are in fact complex landforms; the upper surfaces are constructional where they are underlain by Quaternary sediment and erosional where they truncate masses of brecciated bedrock. Other masses of brecciated bedrock rise above such alluvial landforms as low, rounded hills.

Even in rocks of the most distended texture, however, the components of each individual stringer invariably are all of the same lithology; typically, that lithology differs somewhat from that of immediately superjacent or subjacent stringers. No known sedimentary process is capable of such fine discrimination, which is repeated over and over in many outcrops and characterizes the most distended textures wherever they are exposed. In some outcrops, tightly folded structures can be reconstructed, outlined by the configuration of subparallel extended stringers of blocks of identical lithology suspended in a finer matrix. Typically, the geometry of the folds is truncated by the surface of the landform under which the folds are developed; the structures thus cannot be attributed to Quaternary deformation by mass-wasting processes of bodies of sediment deposited on the present landscape.

A clear distinction between undisputed sedimentary textures and those preserved in the most distended rocks of the breccia zone can be made where masses of distended-texture breccia have been incorporated as individual giant blocks in accumulating bodies of Quaternary alluvium. Examples are frequently encountered in the mass of Quaternary sediment that covers the floor of the central graben in the region between Ejido and Estánquez. Most such blocks truncate bedding in the alluvium and lie at angles such that organization of individual stringers within a given mass of breccia is clearly discordant with bedding in the alluvium. Some such blocks are larger than the outcrops in which

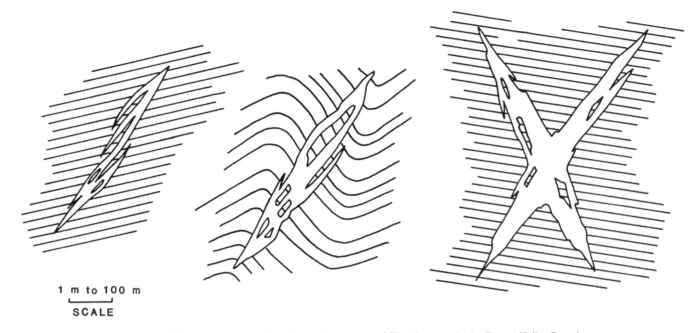

1 m to 100 m
SCALE

Figure 4. Schematic representation of range in geometry of filled fractures in the Chama Valley Breccia (note variation in scale).

their characteristic textures appear, and thus the entire outcrop is distended-texture breccia. Detailed, very local mapping is necessary to demonstrate that these outcrops represent blocks of distended-texture breccia lying in bodies of Quaternary alluvium.

A new road from San Juan to Jají (Fig. 1) was cut through Chama Valley Breccia in 1975–1976; road cuts along this route permit examination of 15 km of more or less continuous exposure through the Chama Valley Breccia in which examples of the full range of distended texture can be observed. The road is cut principally in predominantly red arkosic sandstone, conglomerate, and shale assigned to the La Quinta Formation of Triassic(?)-Jurassic age. The La Quinta Formation here is deformed into broad folds whose axes strike N45°E to N55°E. Bedding is in places overturned on the southeast-facing limbs of anticlines. Bedrock in the road cut is brecciated throughout, but in 90% of exposures examined the textures are not greatly distended. The remaining 10% displays the full range of distension, and provides valuable clues to reconstruction of the mechanism responsible for these textures.

In many areas of outcrop along the road cut the brecciated bedrock is cut by a set of wedge-shaped fractures whose formation clearly postdates development of the brecciated texture; the fractures are filled with a fine-grained rock matrix. The fractures observed range in width from a few centimetres to 5 or 6 m; most become narrower as the road bed is approached, but many continue beneath the level of the road. Figure 4 is a schematic representation of the forms assumed by typical fracture fillings observed in the field. The fractures bear no systematic relationship to rock bedding, but appear to maintain a constant strike parallel to hinges of the major folds on the northwest flank of the graben.

Dips typically range through 30° on either side of vertical; a few fractures show gentler dips. At several localities along the road cut, two or more fractures intersect, typically at angles close to 60° and 120°.

The fractures are filled with a dense, fine-grained mass of rock lithologically attributable to the La Quinta Formation; suspended in that mass are larger, angular blocks of sedimentary rock recognizable as lithologic and textural continuations of sedimentary beds that appear to terminate at the margins of the fractures. It is possible to trace a given bed to the edge of a fracture, follow it from recognizable block to recognizable block across the fracture, and pick up the same bed in the expected position on the other side. Thus, there is no measurable shear displacement along the fracture; apparent net deformation in the process of fracture formation is restricted to the drawing apart of the two sides of a given fracture apparent in the plane of the outcrop. Blocks of rock detached from those sides lie suspended in the fine-grained rock matrix that fills the fractures. The filled fractures thus represent bodies of distended-texture breccia still in place in their loci of formation. These bodies appear much more resistant to erosion than is the fractured, friable adjacent bedrock.

Relationship of Breccia to Thrust Fault. A secondary discovery that led to the model of origin of the distended-texture breccia that appears below came when students from the University of Pennsylvania determined that rocks assigned to the Chama Valley Breccia are associated with a young thrust fault, the northeast extension of a system described by deRatmiroff (1971), Useche (1977), and Garcia J. (1978). This thrust appears at the southwestern limit of the map of Figure 5. The cross section exposed along the southwest wall of the canyon where the Rio

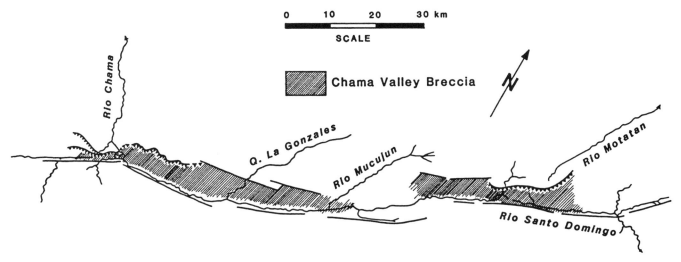

Figure 5. Map of distribution of Chama Valley Breccia and major faults that limit that distribution. Compare Figure 2.

Chama cuts through the northwest range of the Andes shows that thrusting occurred along at least two surfaces, producing a stack of overlapping slivers. The highest sliver exposed in the canyon carries schists and phyllites assigned to the Tostós Formation of Paleozoic age (Garcia J., 1978) over sediments of the Triassic(?)-Jurassic La Quinta Formation. A lower thrust carries the La Quinta Formation over a relatively undeformed sequence of Cretaceous sediments. The thrust surfaces dip steeply to the southeast at the southeast end of the canyon and roll over to become horizontal toward the northwest. The thrust-fault surfaces are truncated against an east-west-trending normal fault with at least 600 m of vertical displacement of the northwest block relative to the southeast block. The thrust surfaces in the northwest block have been removed by erosion. Dramatic drag folds in the Cretaceous Aguardiente Formation confirm this sense of displacement, but the magnitude of that displacement is apparent to anyone who stands on the massive sandstones of the Aguardiente Formation near the river bed and looks up at the same strata at the top of the cuesta through which the river flows a few hundred metres downstream.

The thrust surface has not been identified farther to the northwest, where the Cretaceous and younger sediments are covered by the thick sequence of later Tertiary sediments of the Maracaibo Basin. However, southwest of the canyon of the Rio Chama, near the village of Mesa Bolivar, metamorphic rocks assigned to the Tostós Formation overlie undeformed sediments (the Palmár Formation) of lower Miocene age (Useche, 1977; Garcia J., 1978), indicating that some, at least, of the thrust displacement occurred after early Miocene time.

In the section along the Rio Chama the rocks of the lowest thrust surface are so severely brecciated as to be assigned only with some difficulty to the named formations. The Cretaceous rocks under the lowest sliver, however, show very little evidence of disturbance only a few metres below the thrust plate.

Rocks of the Chama Valley Breccia are associated with a high-angle reverse fault mapped by Kovisars (1969, 1971) and Grauch (1971, 1972) northeast of Mucuchíes (Fig. 5). Here, too, development of breccia and of fractures filled with distended-texture breccia is restricted to the upper plate of the fault. Chama Valley Breccia is well exposed between these two localities (see map of Fig. 5) but has not elsewhere been demonstrated to lie on the upper surface of a thrust fault. In several quebradas along the northwest side of the Rio Chama between the Chama Canyon and Lagunillas, the extreme topographic relief makes it possible to document that the thickness of the breccia exceeds 1.5 km. Distended-texture breccia has been observed in place as fracture fillings only in rocks interpreted to lie over the distal, near-horizontal part of the thrust plate.

Distribution of the Chama Valley Breccia. The map of distribution of brecciated textures in bedrock (Fig. 5) is dramatically supported by the identical pattern of distribution of a characteristic "grain" to the topography on LANDSAT images of the Central Venezuelan Andes (Fig. 2). That grain results from the topographic expression of the development within the incompetent rock of the breccia zone of a series of closely spaced, parallel, normal faults, not all of which are apparent in the severely brecciated outcrops examined.

On the LANDSAT imagery (Fig. 2), the geomorphic expression of the Chama Valley Breccia terminates abruptly to the northwest. At several localities this truncation has been identified in the field as the termination of the Chama Valley Breccia against unbrecciated rock in the upthrown block of a normal fault of large displacement.

Summary of the Chama Valley Breccia. The breccia zone, then, can be characterized as follows:

1. It is a zone of severe textural modification of bedrock ranging in lithology from the amphibolite gneiss of the Sierra Nevada Formation to the arkosic sediments of the La Quinta Formation.

2. No development of a consistent fabric expressed by orien-

tation of fractures within the breccia is apparent, except for the large-scale orientation of later-generation fractures filled with distended-texture breccia.

3. Bodies of distended-texture breccia occur as fracture fillings within the breccia; or as isolated masses whose connection, if any, with the brecciated rock "in place" cannot be established; or as transported masses of a great range in size lying in bodies of younger alluvial sediment. Distended-texture breccia is present along the entire length of the mapped breccia zone.

4. The breccia zone is confined to the upper plate(s) of a thrust fault at both the southwest and northeast limits of its distribution on the map presented here. Between these localities, bedrock relationships are obscured either by the great thickness of the breccia itself or by a cover of Quaternary alluvium.

5. Distended-texture breccia in place as fracture fillings has been observed only in what is interpreted to be the distal, near-horizontal part of the thrust plate.

6. The breccia zone is truncated to the northwest by a series of normal faults that bring unbrecciated rock units into juxtaposition with the breccia.

7. No representative of the Chama Valley Breccia has yet been identified in the sedimentary column northwest of the Andes. If present, and it seems likely that the thrust plate extended far northwest of where it is last observed, it is buried under the later Tertiary section in the Maracaibo Basin.

8. Individual blocks of unbrecciated bedrock of various ages lie on the valley floor in contact with breccia. The largest of these blocks is southwest of Ejido, where a section of Cretaceous sediment overlying schists of the Tostós(?) Formation crops out in a linear hill that extends 6 km parallel to the Rio Chama on its northwest side; this hill is surrounded on its other sides by masses of breccia. Several blocks of unbrecciated Cretaceous and lower Tertiary sediment lie against the northwest wall of the Chama Valley at the confluence of the Rio Mucujún, a principal tributary draining the high country northeast of Mérida.

Many isolated blocks of Mucuchies Formation, a coarse terrestrial basin fill of later Tertiary age (Grauch, 1971; Murphy, 1977) lie in breccia in the graben developed around the town of Mucuchies.

Geomorphic Expression of the Chama Valley Breccia.
The breccia zone is a zone of acute bedrock incompetence. This is clearly shown by the difference in topographic expression between the two sides of the Chama Valley:

Along its southeast margin, the graben terminates against a linear, near-vertical fault of great displacement, along which high-grade metamorphic rock has been carried at least to the elevation of Pico Bolivar, 5,002 m above sea level. For most of its length within the central valley, the Rio Chama flows close to the southeast wall; little Chama Valley Breccia has been observed between the river and the fault margin of the valley. Tributary valleys (quebradas) draining the southeast wall of the valley are short and steep, typically flow directly on bedrock, carry little sediment, and enter the Rio Chama through a narrow belt of small but well-defined alluvial landforms.

The northwest side of the valley is underlain by severely brecciated bedrock from Estánquez to Apartaderos. The breccia is cut by a system of remarkably parallel linear normal faults. The faults all strike within a few degrees of N50°E, and with very few exceptions, the sense of latest movement on them has been down to the southeast. In at least one locality, the major landslip at Chiguará, mass-wasting processes have contributed measurably to this displacement; elsewhere, the linearity and persistence of individual normal-fault traces indicate that the tectonic pattern has not been obscured by later surficial processes. Quebradas draining the northwest wall of the valley are long and slope more gently than those on the southeast. They typically carry large amounts of suspended and bed-load sediment during times of heavy runoff, flow across valley floors underlain by alluvium, at least in their last few kilometres of flow, and enter the Rio Chama through a thick complex of alluvial landforms. The fan under the towns of San Juan and Lagunillas, deposited by the Quebrada La Gonzales in the recent past, has a surface area of 25 km^2 and is at least 200 m thick. Where brecciated bedrock on the valley floor has not been fully buried by alluvium, it projects as low, rounded hills above the depositional surfaces of the alluvium, in marked contrast to the sharply dissected fresh bedrock exposed on the mountain ranges flanking the valley.

These differences between the two sides of the valley can be attributed directly to the difference in the erodibility of the underlying bedrock: the breccia is much more vulnerable to erosion than is the fresh, unaltered crystalline bedrock on the southeast wall of the graben.

Comparison of the Chama Valley Breccia with Other Tectonic Breccias.
The Chama Valley Breccia differs from other regionally developed breccia units in three important respects:

1. No systematic tectonic fabric imposed on the brecciated mass has yet been identified, except for the large-scale pattern of the orientation of the filled fracture wedges superimposed on the brecciated texture.

2. The breccia is uncemented and, except for the bodies of distended-texture breccia, which are well indurated, is highly permeable.

3. Distended-texture fracture fillings like those described here have not been identified elsewhere, so far as a detailed search of the literature has revealed.

From the point of view of late Tertiary tectonics, the breccia appears to be the most instructive mappable rock unit in the Central Venezuelan Andes. Thus, despite the uncertainties described here with reference to the three-dimensional configuration of the brecciated mass, and despite the failure of a detailed literature search to identify an analog that is faithful even to development of the distended-texture fracture fillings, the model that appears below is offered as an explanation, however speculative, of the tectonic process(es) that produced the Chama Valley Breccia.

Origin of the Chama Valley Breccia.
The model assumes that the breccia everywhere bears the same relationship to the reverse fault described here in the canyon of the Rio Chama and

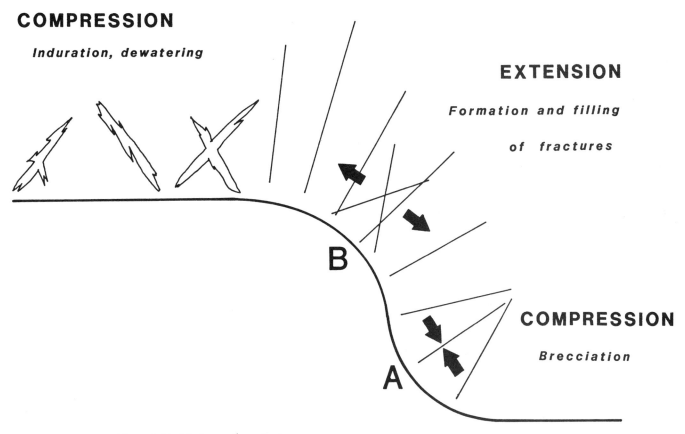

COMPRESSION

Induration, dewatering

EXTENSION

Formation and filling

of fractures

B

COMPRESSION

Brecciation

A

Figure 6. Model of tectonic mechanism proposed to account for the range of textures in rocks overlying the post–early Miocene thrust fault in the Central Andes of Venezuela.

at Apartaderos; that is, the brecciated texture is developed in the rock mass over the lowest thrust surface.

The thrust-fault surface is nearly vertical where it can be examined rising out of the floor of the Chama Valley, near the southwest margin of the present grabens, and becomes horizontal in the presumed direction of transport, to the northwest. If this feature is a typical thrust fault, the vertical portion cannot persist to great depth; accordingly, the model assumes that the fault surface becomes horizontal again to the southeast and that it now terminates at some unknown depth against the normal fault or faults that define the southeast limit of the chain of grabens (Fig. 6). At the time that movement occurred along it, the thrust-fault surface would have had the stepped configuration shown in Figure 6. Steps or ramp regions on thrust faults are described from many localities, although few have been shown to dip as steeply as the ramp postulated here. The near-horizontal attitude of the lower Miocene rocks over which the thrust is emplaced at Mesa Bolivar indicates that the configuration of the thrust surface cannot be attributed to post-faulting deformation.

As the rock on the upper plate passed through the concavity at point A on Figure 6, it was subjected to a severe compressive stress, the magnitude of which is greater with greater distance away from the thrust surface. It was in this stress environment that the intense, fine-textured, pervasive brecciation occurred.

As the rock over the thrust surface moved up the vertical part of that surface, the intensity of the compressive environment was reduced. At the time of fault movement (post–early Miocene or later), the thrust surface was not far below the ground surface. The brecciated mass was a highly permeable, unconfined aquifer saturated to the elevation of the contemporary water table, at that time probably only a few metres beneath the surface of the landscape. The pore-water pressure in the unconfined aquifer was then fully hydrostatic and may have represented 2 to 5 km of hydrostatic head at the thrust surface itself. The pore-water pressure at the base of the upper plate may have been sufficient to reduce normal stress enough to prevent development of a shear fabric in the rock on either side of the thrust surface.

As the rock on the upper surface of the thrust plate passed over the convexity at point B in Figure 6, the stress environment changed as tension replaced compression. As the rock passed over the convexity, the least principal stress, σ_3, assumed an orientation tangent to the thrust surface; σ_2 was horizontal and parallel to the strike of the breccia zone (N50°E); and σ_1, the greatest principal stress, was represented by the component of the gravitational stress perpendicular to the thrust surface. The geometric relationship of the axes of the stress ellipsoid to the thrust surface would have been the same everywhere over the convexity, but the ratio of σ_3 to σ_1 would have been a function of radius of curvature of the convexity. In this stress environment the brecciated rock was fractured along a series of intersecting planes at ~60° to the thrust surface. As these fractures opened, they were instantly filled with a slurry of fine-grained rock flour, produced during the earlier compressive deformation, injected into the developing fracture by the pressure differential between the hydrostatic pore-water pressure and the reduced pressure in the opening fractures. If movement along the fault was sporadic in a series of discrete displacements, each fracture would have been opened and filled essentially instantaneously, thereby suspending the larger blocks observed to "float" in the mass of distended-texture breccia; if displacement along the fault occurred by more or less continuous creep, then individual fractures would have been filled incrementally as they opened, supporting larger blocks at either margin of a given fracture as the width of the fracture slowly grew.

If a mechanism such as that just described did in fact operate, then the ideal body of distended-texture breccia would be a wedge, tapering toward the thrust surface, dipping at 60° to the plane of the thrust, and elongated in the direction of the strike of the thrust fault. A large number of the filled fractures observed in the road cuts along the San Juan–Jají road resemble this ideal form. Some intersecting fractures have even been observed to dip at 60° SE and 60° NW, and to intersect as a lineament oriented at N50°E, the approximate strike of the breccia body, the presumed thrust fault, and the system of normal faults that cut the breccia (Figs. 2 and 5).

As the brecciated and fractured mass moved off the convexity and onto the horizontal portion of the thrust surface, the tensional environment was replaced once again by a compressional stress field, but one less severe than had prevailed as the rock passed through the concavity previously encountered on the thrust surface. In this latest compressional environment the masses of distended-texture breccia were compacted and dewatered, producing in them a denser, more competent texture than that which prevails in adjacent zones of the breccia. Any microscopic fabric that might have developed in the fracture fillings, and so far none has been identified, should reflect this latest episode of compression.

In the subsequent episode of regional arching of the Central Venezuelan Andes, evidence for which will be detailed later, a chain of grabens opened in the incompetent rock of the near-surface breccia zone approximately along the axis of the crest of the arch. Erosion of the breccia exposed along the northwest margins of those grabens delivered great quantities of alluvial sediment to the graben floors; masses of the more resistant distended-texture breccia exposed by that erosion slid as discrete blocks into the accumulating sedimentary piles. Release of seismic energy during development of the system of normal faults along the northwest margins of the grabens may have contributed to the failure of individual masses of distended-texture breccia.

A consequence of this model is the probability that the upper plate over the thrust actually "erupted" somewhere at the ground surface. At the most distal point where it can be observed,

its termination against the graben-limiting border faults to the northwest, the breccia is at least 1.5 km thick. It seems likely that the leading edge of an erupting thrust mass would have been far to the northwest of the present northwest margin of the graben.

The model presented here is speculative and, perhaps, inadequately constrained by field information. If the breccia were a less important component of the rock column in the Venezuelan Andes, and if it were of only minor inferred significance to the late tectonic history of the mountain range, the temptation to force an understanding of the mechanism of its origin would be less acute. But the observations presented here indicate that the Chama Valley Breccia may be the feature of greatest tectonic significance yet described in the Central Venezuelan Andes, and the model outlined here explains its origin in a manner consistent with available field evidence.

Discussion of Tectonic Analogs. The literature does not offer many descriptions of breccia bodies like the Chama Valley Breccia, nor does it contain descriptions of thrust faults that resemble in geometry and depth of occurrence the model described here. Serra (1977) has described several thrust surfaces from the Wyoming-Idaho overthrust belt of geometry similar to the model presented here, but for each of these examples the entire ramp is represented in a single outcrop, hence on a much smaller scale. Furthermore, Serra's examples lack rock as severely brecciated as the rock of the Chama Valley Breccia. The geometry of joints and fractures associated with the thrusts he described is analogous to what would be generated by stresses developed in the model described here for the Chama Valley Breccia.

Another possible analog is the pile of thrust sheets that, according to Harland (1959, 1961, 1969, 1971), Lowell (1972), and Maher and others (1981), was "extruded" in Tertiary time from the shear zone along the major strike-slip fault in Caledonian Spitzbergen. These imbricate tabular rock bodies were thrust out over the ground surface on either side of that fault as a *transpression welt* as the two plates moved past one another with a substantial component of convergence (Fig. 7).

A similar structural-topographic geometry was described by Suggate (1963), Walcott and Cresswell (1979), and Sibson and others (1981) along portions of the Alpine Fault in New Zealand. They described individual superficial nappes of schist that have been thrust as far as 2 km over Quaternary glacial moraines in the central part of the Southern Alps, presumably in the course of net right-lateral transcurrent displacement (Fig. 7). These nappes are confined to the northwest side of the fault, supporting the inference of net uplift on the southeast side.

The position of the Venezuelan Andes on that limb of the Caribbean/South American plate boundary where oblique convergence seems the most probable geometry of plate interaction permits consideration of a transpression mechanism for emplacement of the thrust sheets as a working hypothesis, despite the major differences between the Chama Valley Breccia and the rocks described from Spitzbergen and New Zealand. DeRatmiroff (1971), upon considering the configuration of imbricate thrust sheets along the margins of the Venezuelan Andes, de-

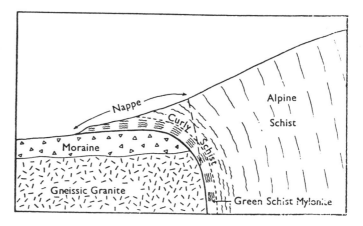

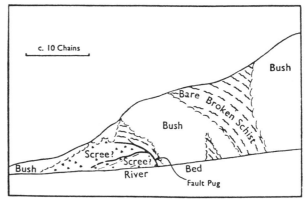

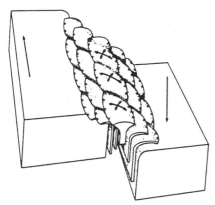

Figure 9. Conceptual diagram of an upthrust-bounded welt created by convergent strike-slip or transform motion. Portions of two plates or blocks are shown, the one moving at a low convergent angle into the other in the same plane, causing each segment to move a small distance successively past the other in a right-slip direction. A space problem is created, and the easiest direction of relief for the crowded material is upward. A welt with downward-tapering wedges and upthrust margins is created. The upthrusts are not necessarily so symmetrically disposed as shown and the deeper faults may tend to coalesce and braid rather than be parallel.

Folding precedes faulting in this type of deformation, which explains the observations by Orvin (1940) that the faults are younger than the folds in the Spitsbergen Tertiary orogenic belt.

FIG. 3.—Superficial overthrusting at the Alpine Fault. Above: Diagrammatic representation by Wellman (1955). Below: Section exposed at Gaunt Creek, Waitangi-taona River, South Westland, from a sketch by the author. (Note that to match the diagram above, the sketch is shown in mirror image of the actual exposure on the south-west bank of Gaunt Creek.

Figure 7. Potential structural analogs of the Chama Valley Breccia. Left: diagram of the *transpression welt* from Spitzbergen (Lowell, 1972, p. 3099); right: superficial overthrusting along the Alpine Fault in New Zealand (Suggate, 1963, p. 109).

scribed the late Tertiary deformation of the Cordillera as a "mushroomlike deformation with marginal imbricate thrusting" (p. 1336), presumably the result of oblique convergence. Although his map (p. 1340–1341) shows the extension to the southwest of the thrust fault associated with the Chama Valley Breccia, he described no anomalous rock textures associated with that thrust.

Other potential textural analogs of the Chama Valley Breccia include the Hoodoo Hills Havoc of Southern Nevada (Guth, 1980, 1981); the Amargosa Chaos from along the eastern margin of Death Valley (Noble, 1941; see also Drewes, 1959, 1960); the Riggs Chaos (Kupfer, 1960; see also Drewes, 1960) from the Silurian Hills, 60 km southeast of exposures of the Amargosa Chaos; other structurally contorted terranes developed over flat thrust sheets in the extreme southern part of the Basin and Range province; and some of the chaotic units enfolded in the Franciscan mélange of the California Coast Ranges (e.g., Phipps and others, 1979; Phipps, 1984). In many of these cases the unusual textures of the rocks have been interpreted to have developed, at least in part, during failure of a rock mass under gravitational

stress. In some cases a mass of brecciated rock is inferred to have slid back down a fault surface along which it had been tectonically driven earlier.

Although blocks of the Chama Valley Breccia have moved down topographic gradients under the influence of gravity, in some cases along the original thrust surface, enough of the rock mass remains in place over the thrust surface to demonstrate that the pervasively brecciated texture of the Chama Valley Breccia predates any downslope movement. Bodies of distended-texture breccia that have moved downslope under gravitational stress have suffered no further deformation in the process, as is attested by the extreme competence of the masses of distended-texture breccia incorporated in bodies of Quaternary sediment on the valley floors.

As a geomorphic feature, the breccia zone is as striking and persistent a component of the Central Andes as is the chain of linear central grabens; if the model presented here of its mechanism of origin is a reasonable approximation, it should be considered the single most important tectonic feature of the Cordillera as well. The condition of the rocks in the breccia zone has clearly

354   R. Giegengack

influenced both the development and the present configuration of the central graben through which the Rio Chama flows.

Independent of the series of investigations reported here, Carlos Macellari (1981, 1984) mapped and described a major thrust fault of middle Tertiary age in the Táchira Depression, ~60 km beyond the southwestern limit of the map of Figure 5. Macellari did not describe brecciation on the scale of the Chama Valley Breccia, nor has he observed bodies of distended-texture breccia associated with the thrust fault, which suggests that the geometry of the thrust fault at depth beneath the Táchira Depression may be quite different from the configuration of the same(?) fault farther to the northeast. It seems likely, however, that the middle Tertiary thrust sheet in Táchira was emplaced in response to the same compressional stress field responsible for the Chama Valley thrust fault and the associated brecciated textures.

The Pueblo Nuevo Paleo-Oxisol

The development of the Andes as a positive topographic feature postdates emplacement of the breccia zone. This is shown most clearly by the mineralogy and distribution of an extensive paleo-oxisol, inferred to have formed near sea level, that is now best preserved on crests of divides between canyons of quebradas tributary to the Rio Chama on the southeast side of the central graben southwest of Pico Bolivar.

Composition of the Paleo-Oxisol. Both the composition of the paleo-oxisol and its stratigraphic position distinguish it clearly from the thin, stony, erosionally unstable soil veneers that are forming today on bedrock surfaces exposed on valley walls and on the surfaces of bodies of alluvium on the valley floors.

Exposures in the walls of valleys that dissect the paleosol-veneered upland surface are abundant and instructive. South of the Rio Chama, B-horizon thicknesses ranging from a few centimetres to 40 m were recorded; thin O-horizons or A-horizons were observed on some exposures. North of the Rio Chama both the thickness of the B-horizon and the total thickness of the soil is less. On some exposures recognizable masses of deeply weathered bedrock can be observed protruding through what is left of the original soil-veneered surface.

The color of the paleosol is variable, ranging from white to strong brown. The dominant colors, those that present a striking contrast with the colors of soils forming today, are reddish brown

(2.5 YR, 5 YR, and 7.5 YR [dry] of the Munsell Color System). Lateral color variation appears inconsistent and is abrupt on some outcrops, but a consistent reduction in both value and chroma with depth in the soil profile is apparent in all exposures examined.

Weingarten (1977) collected and analyzed 124 samples, of which 102 were collected south of the Rio Chama and 22 north of the river. The samples collected are not all representative of the same depth within the paleosol, since it was not always clear how far beneath the original soil surface each sample lay at the time of collection. Composition of the soil was determined by qualitative X-ray diffraction and by limited chemical analysis. The X-ray analysis was performed by Weingarten (1977); selected chemical parameters were determined by the Merkle Laboratory of the Pennyslvania State University (Weingarten, 1977).

Mineralogy of the Paleo-Oxisol. The clay-mineral fraction of paleosol samples taken south of the Rio Chama consists of kaolinite, illite-muscovite, minor amounts of 2:1-type clays, and quartz. Gibbsite is an important component of some of these soils and is absent from others. Paleosol samples from localities north of the Rio Chama, where the soil is developed primarily on brecciated rocks of the La Quinta Formation, are similar but lack gibbsite and yield less-well-defined diffraction peaks for kaolinite, implying a poorer quality of crystallization of that mineral. Iron-oxide minerals and quartz particles larger than fine silt were removed prior to clay-mineral analysis; the percentages of these components were not systematically determined.

Average values for measurements of cation-exchange capacity (C.E.C.), base-saturation, and soil pH are shown in Table 1. On the basis of its composition, Weingarten (1977) classified this soil as an oxisol.

It seems likely that the chemical and mineralogic differences between the 102 samples of paleo-oxisol from south of the river and those of the 22 samples taken from the terrain north of the river, like the physical differences observed on the outcrops, are results of the greater depth within the soil profile from which samples from the northern terrain were collected, a consequence of the deeper erosion that has occurred on terrain underlain by the incompetent rock of the Chama Valley Breccia.

Distribution of the Paleo-Oxisol. Close to the Rio Chama, the typical paleo-oxisol remnant lies across the crest of a narrow divide separating two steep tributary canyons cut after

TABLE 1. AVERAGE CATION-EXCHANGE CAPACITY (C.E.C.), % BASE SATURATION,
AND pH VALUES FOR SAMPLES OF THE PUEBLO NUEVO PALEO-OXISOL

Exchangeable cations	SOUTH				NORTH			
	K	Mg	Ca	TOTAL	K	Mg	Ca	TOTAL
C.E.C. (meq/100 g soil)	0.12	1.1	1.5	9.1	0.16	1.9	1.8	11.3
% Base saturation	1.3	12.0	16.5	29.8	1.4	16.8	15.9	34.1
pH		5.4				5.2		
		(range 4.8-6.3)				(range 4.2-6.6)		

Note: These values are averages for 102 samples from terrain south of the Rio Chama and 22 samples from terrain north of the Rio Chama. Information from Weingarten (1977).

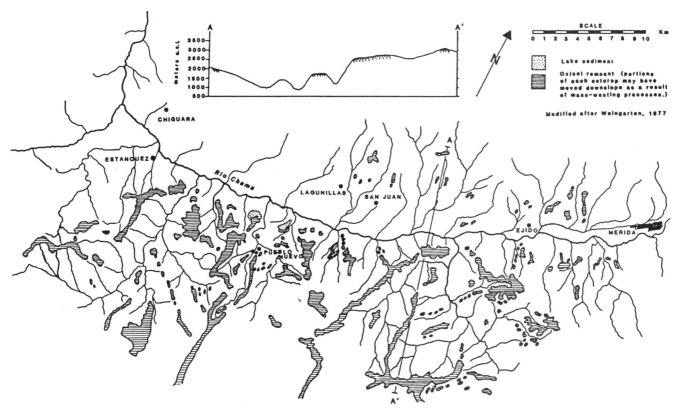

Figure 8. Distribution of oxisol remnants and paleolacustrine sediments adjacent to the valley of the Rio Chama, Central Andes, Venezuela.

collapse of the central graben (Fig. 8). The soil is developed chiefly on phyllites and slates of the Mucuchachí Formation. Locally, the soil lies on white granite of presumed intrusive origin and on an amphibolite gneiss assigned to the Sierra Nevada Formation. There is no detectable difference in the composition of the soil from place to place that can be attributed to variation in bedrock mineralogy.

Remnants of similar paleosols have been identified along the Rio Santo Domingo–Rio Chama divide and in several localities in and adjacent to the valley of the Rio Boconó, ~50 km northeast of the northeast limit of Figures 1 and 5. The detailed investigation summarized here (Weingarten, 1977) has thus far extended only to the exposures southeast of the Rio Chama and to nearby localities on the northwest side of the river (Fig. 8).

The altitude of paleosol-veneered landscape remnants ranges from 1,300 m above sea level at the extreme southwestern end of the region mapped in Figure 8 to 3,100 m above sea level at the northeastern limit of that map. That margin is 8 km southwest of Pico Bolivar, which at 5,002 m is the highest point in the Venezuelan Andes. This increase in altitude is not achieved by means of a systematically rising gentle slope. The land rises in a series of steps, with many minor reversals; individual steps are tilted as much as 10° to 15° from the horizontal. The variation in elevation from one soil remnant to another shows that the surface on which the soil has been developed has either been warped or

cut up by a system of normal faults, along which individual soil-veneered surfaces have been offset with reference to one another as the blocks on which the relict soil is preserved moved along those faults. Along the steep slope that descends into the Rio Chama from the southeast, a belt of paleosol-veneered surfaces dips back to the southeast at angles in excess of 30°. These surfaces represent the tops of toreva blocks, rotated back to the southeast as they moved down northwest-dipping, concave-upward failure surfaces activated during the foundering of the Chama graben.

The paleosol-veneered surface is being rapidly destroyed today as tributary-stream networks grow headward from the Rio Chama. Dissection has reduced a number of paleosol-veneered inter-stream divides to knife-crested ridges, from whose summits all traces of original landscape elements have been lost. Many of these crests now lie below what was the original landscape surface and no longer preserve the uppermost horizons of the oxisol profile.

Near-vertical cuts through the soil profile in the walls of the canyons show details of the horizontal zonation of the soil. It is clear from these exposures, from the topographic contrast between the flat-topped, soil-veneered divide crests and the precipitous valleys in between, and from the absence of a similar well-developed soil on any surface below the ridge crests that the soil is a relict feature, developed on a terrane of low relief that is now being rapidly incised and eliminated. On several larger remnants the details of ground-surface morphology, including a series of small, shallow ponds and associated low-relief drainage networks, are well preserved.

Distribution of Modern Oxisols. There is a consensus among soil scientists (e.g., Buol and others, 1980; Birkeland, 1974) that formation of oxisols requires relatively high mean annual temperatures (~22 °C or higher). These conditions prevail today in the Central Venezuelan Andes only below 800 m above sea level. Formation of oxisols also requires abundant rainfall and a long period of landscape stability. Contemporary oxisols are largely restricted to tropical and subtropical shield areas, where under ideal conditions gibbsite and other aluminum oxides can accumulate to concentrations sufficient to qualify the soil as a bauxite. Oxisols form most readily over basic rocks or over transported materials already intensively weathered in the source area (Buol and others, 1980). The occurrence of this paleo-oxisol *in situ* over bedrock of intermediate to acid composition implies an even longer period of stable conditions for its formation than may be the case for many contemporary oxisols.

Climatic conditions appropriate for the development of oxisols prevail today on the lowlands flanking the Central Venezuelan Andes, but the modern soils forming there are not so described. The topographic relief in the Andes is so great today that all exposed surfaces are either being eroded too rapidly or covered too rapidly by products of erosion to allow an oxisol to develop.

Age of the Paleo-Oxisol. The age and configuration of the paleo-oxisol is important to the reconstruction of the late Tertiary history of the Andes. The paleo-oxisol is not forming today; indeed, it is being rapidly obliterated. Its present distribution on ridge crests and uplands indicates that its formation predates development of the tectonic environment in which uplift and fault displacement of that surface occurred. No trace of the oxisol has been observed on any of the terrane interpreted to lie *under* the thrust fault or its projection. No such soil could develop on a terrane as unstable tectonically as this region must have been during emplacement of the thrust sheet and development of the Chama Valley Breccia. Thus it seems most probable that the soil developed across a stable, low-altitude terrane of low relief on which the geometry of the post–early Miocene thrust fault and associated tectonic rock textures had already developed.

No effective means of dating iron oxide, alumina, or other authigenic component of the paleo-oxisol geochemically or isotopically has yet been developed. The cation-exchange capacity of the oxisol, while lower than that of many soils, is still too high to permit the assumption that any authigenic component of the paleosol has remained a closed chemical system since its origin.

In at least three localities overlooking the Chama valley from the southeast, the paleo-oxisol is overlain by lake sediments, the most instructive of which are those near the village of Pueblo Nuevo (Fig. 8). Dispite the existence of many exposures of sediments of both central and littoral facies of what is interpreted to have been a chain of shallow, fault-bounded lake basins, no fossils have yet been recovered that would permit an age to be assigned to the lake sediments or a minimum age to be imposed on the underlying paleo-oxisol.

Interpretation of the Paleo-Oxisol. The paleo-oxisol-veneered landscape remnants today lie at altitudes incompatible with the climatic requirements thought necessary for formation of soils of that character, and in a tectonic environment that clearly does not provide the long-term landscape stability that is also required. The simplest reconstruction of the environment of formation of the oxisol is that it developed in a warm, humid climate over a low-altitude, low-relief terrane on which breccia associated with a post–early Miocene thrust fault had already been emplaced. The terrane remained close to sea level and tectonically stable during the millions of years necessary for development of the oxisol, and was subsequently arched with a *minimum* amplitude of 2,300 m (at least 4,200 m if the paleo-oxisol-veneered terrane is projected over Pico Bolivar). The soil-veneered terrane was broken up into a series of distinct blocks along normal faults as that terrane was draped over the crest of the Cordillera in the latest (Pleistocene?) episode of Andean uplift. The accelerated erosion and dissection triggered by that uplift is today rapidly reducing the area of preserved paleo-oxisol-veneered terrane. Details of landscape morphology that developed on the soil-veneered terrane are well preserved even on those landscape remnants that have been steeply tilted by movement along normal faults; thus, little time has elapsed since the event that elevated the soil-veneered terrane.

If the period of time since inception of the uplift event is assumed to be 1 m.y., a value consistent with the quality of preservation of relict drainage morphology on the paleo-oxisol-veneered landscape remnants, and if the amplitude of post-soil-development arching is assumed to be 5 km, the rate of that uplift then is of the order of 5 mm/yr, a value consistent with estimates of rates of landscape uplift in other parts of the world where contemporary topographic relief is high. A fission-track study of uplift rates on the northwest range of the Venezuelan Andes (Kohn and others, 1984) has yielded a value of 0.8 km/m.y. or 0.8 mm/yr; this value should be considered a minimum value if significant updoming of isotherms occurred.

DIRECT EVIDENCE OF LATE QUATERNARY TECTONIC EVENTS

Evidence presented here shows that considerable tectonic activity has taken place in the Venezuelan Andes in latest Quaternary time. The most dramatic testimony of the intensity of late Cenozoic tectonic events is, of course, the amplitude of present relief of the Central Venezuelan Andes. The Cordillera has been high enough to have supported mountain glaciation through at least one, and possibly three, glacial cycles (Giegengack and Grauch, 1975, 1976), despite its proximity to the Equator. Offset of glacial landforms produced during the latest of those cycles provides additional testimony of the intensity of late Cenozoic tectonic events.

No radiocarbon determinations unequivocally date or even bracket time(s) of occupation of any terminal moraines in the Venezuelan Andes. The difficulties of relating glacial chronologies in disparate areas of the tropics to each other has been discussed by Giegengack and Grauch (1972a, 1976) and Hastenrath (1981, p. 61–67), among others. Glaciation at lat 14°S in the Andes of Peru culminated between 28,000 and 14,000 radiocarbon years B.P. (Mercer and Palácios, 1977). This time span correlates well with the classical Wisconsinan/Würm of higher northern latitudes. In the absence of radiocarbon ages closely tied to local glacial events, however, the glacial stages described for the Venezuelan Andes should not be correlated with periods of mountain glaciation at other latitudes and/or on other continents.

In an earlier publication. Giegengack and others (1976) summarized the evidence for reconstructing the geometry of displacement along the Boconó Fault Zone as follows:

1. The Bonconó Fault Zone is a wide tectonic zone of severe weakness that has been active for a long time.

2. The sense of most recent displacement along the fault is not consistent: at many localities right-lateral strike-slip displacement of significant magnitude is apparent; at some of these and at other localities significant vertical displacement with north side down was observed. No left-lateral displacement was unequivocally identified anywhere. At several localities (e.g., Boconó, Mucuchies, Mucurubá) surficial sediment that predates the modern drainage configuration lies directly across a continuation of the fault trace and is not displaced, either horizontally or vertically.

3. The late history of drainage changes within the central Andean valleys points to a complex history of interaction of climtic and tectonic events.

4. Surprisingly, at many localities masses of glacial and alluvial sediment within the fault zone behaved as competent rigid blocks during faulting events that must have been of severe seismic magnitude and frequent recurrence.

Extension of that study to a large region and to a longer span of geologic time leads to the following additional conclusions:

5. The best evidence of right-lateral strike-slip displacement is preserved at localities where the Boconó Fault Zone crosses high-altitude divides at the heads of major drainage networks whose linear geometry is controlled by the configuration of the fault zone. Between these restricted localities the fault zone traverses a series of fault-controlled grabens on the floors of which a discontinuous veneer of Quaternary alluvium is cut by many, subparallel faults. Evidence of right-lateral strike-slip displacement along these faults is not apparent, however, either because it did not occur or because it has been accommodated along so many closely spaced fracture surfaces that it cannot be identified on any single fault.

6. The faults limiting the individual grabens along the trace of the Boconó Fault Zone appear to be nearly vertical at the ground surface. Net vertical displacement in the recent past, while rarely directly measurable at any specific locality, has been of the order of 4 km between the flanks and the center of the Tabay-Estanquez Graben; comparable magnitudes of displacement have occurred in other grabens.

Figure 9 is a map of the Central Andes of Venezuela on which are summarized the principal data of fault displacement for which morphologic and/or stratigraphic evidence is available. A similar map that differs significantly in detail has recently been published by Schubert (1981).

Schubert (1980, 1981, 1982) is impressed by the clear record of recent right-lateral strike-slip displacement of Quaternary morphologic features and by the pivotal position assumed by western Venezuela in many contemporary models of the geometry of plate interaction in northwestern South America. In the publications cited above, the assertion offered earlier (Schubert and Sifontes, 1970; Santamaria and Schubert, 1974; Schubert and Henneberg, 1975) that the Boconó Fault represents the contemporary plate boundary between South America and the Caribbean Plate is repeated. Following terminology of Crowell (1974), Schubert has extended the direct evidence to postulate that the string of fault-bounded grabenlike features along the Boconó Fault Zone represents a chain of pull-apart basins at releasing bends of a major strike-slip fault. He postulates at least 7 to 9 km of right-lateral strike-slip displacement to account for the geometry of those basins. This analysis, while provocative, ignores several important considerations:

1. The only unequivocal evidence of right-lateral displacement is provided by the offset morainal loops in the valley of the Rio Aracay and across the Rio Chama–Rio Santo Domingo divide, where Schubert and Sifontes (1970) measured 66 m of right-lateral strike-slip displacement of those moraines. Rod (1956) had earlier estimated 80 to 100 m of right-lateral strike-slip displacement for the same locality. Evidence of 250 m of right-lateral strike-slip displacement since morainal landforms were deposited during the penultimate glaciation has been presented by Giegengack and others (Giegengack and Grauch, 1972b, 1975; Giegengack and others, 1976). However, no other direct evidence of right-lateral displacement from other offset bedrock features or displaced landforms has been described. At several localities along the trace of most recent movement along the fault (Boconó, Mucuchies, Mucurubá, Lagunillas), bodies of Quaternary sediment demonstrably older than the faulted moraines lie across the fault zone and are not displaced in a right-lateral sense.

2. The extreme relief and high absolute altitude of the Cordillera and the evidence of quite recent rapid uplift cannot be directly explained without a strong component of convergence along the fault zone. Evidence detailed here indicates that the demonstrable recent vertical uplift is an order of magnitude greater than lateral offset that has yet been demonstrated for the same period of time. Indeed, the magnitude of lateral displacement thus far demonstrated could easily represent local adjustment to the space problem that must result from several kilometres of compression-induced uplift.

3. A regional appraisal of the structure and morphotectonics of the zone of convergence among the Caribbean Plate, the

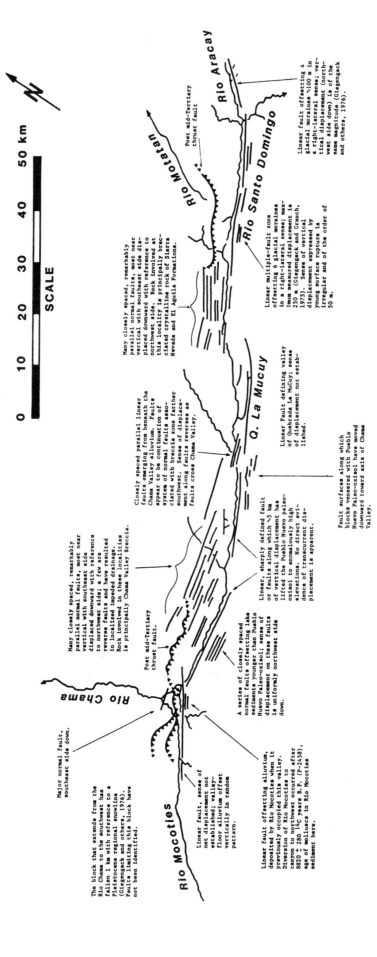

Figure 9. Summary of direct evidence for Quaternary displacement along faults and groups of faults in and adjacent to the major central valleys of the Central Andes of Venezuela.

South American Plate, and the Nazca Plate (e.g., Pindell and Dewey, 1982; Case and Holcombe, 1980; Pennington, 1981; Shagam, 1976) does not necessarily lead to the conclusion that all of the required oblique displacement between South America and the Caribbean had to be accommodated along a single lineament. However broad and complex the Boconó Fault Zone might be, it is a zone far less broad and complex than the region that lies between rock clearly belonging to the South American Plate and rock that can be confidently assigned to the Caribbean or Nazca Plate.

It may yet be demonstrated that tens of kilometres of right-lateral strike-slip displacement have been accommodated along the Boconó Fault Zone. Significant lateral displacement may have occurred within the Chama Valley Breccia Zone, which may yet be shown to be a *transpression welt*. But the evidence provided by the chemistry, attitude, and distribution of the remnants of the paleo-oxisol described here indicates that for a large part of the past 10^7 years the region adjacent to the Boconó Fault Zone was a region of low absolute altitude. Lateral displacement along the Boconó Fault Zone may have occurred during the period of development of the paleo-oxisol without leaving any direct evidence of that displacement, but any significant convergence among the plates in question must have been accommodated elsewhere, very likely by movement among the several lithospheric plates north and west of the Maracaibo Basin.

DISCUSSION

Figure 10 is a schematic representation of the interpretation offered here of the history of the changing tectonic environments in the Central Venezuelan Andes since early Miocene time.

The gravity data presented by Bonini (1976) and Bonini and others (1977, 1984), including the several interpretive cross sections presented with the maps, provide a valuable constraint on reconstruction of the subsurface configuration of basins adjacent to the Venezuelan Andes. The sections show that a column of 9 km of soft sediment, presumably of late Tertiary age, is overridden by denser rock along the northwestern Andean front. Little, if any, expression of the Andes themselves is apparent from the gravity contours.

The two interpretive cross sections presented here as Figure 11 are consistent with all available structural and geophysical

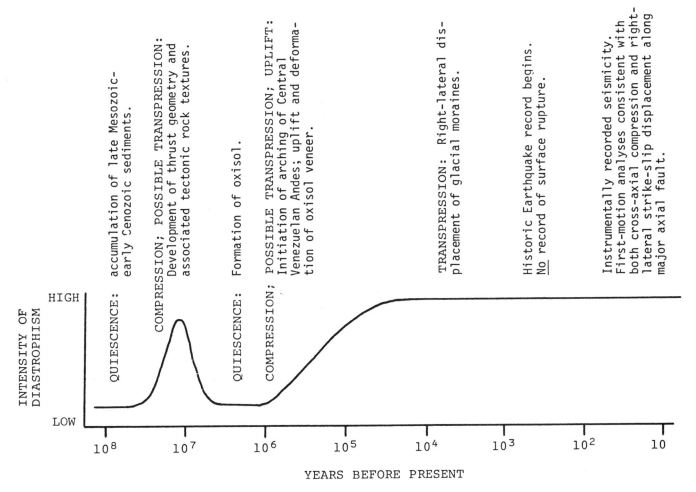

Figure 10. Diagrammatic representation of the history of changing tectonic environments in the Central Venezuelan Andes.

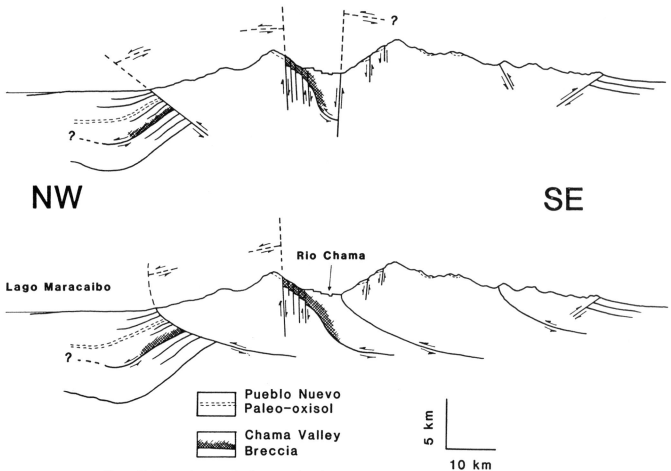

Figure 11. Composite generalized cross-section of topography and structure across the Central Venezuelan Andes: two interpretations.

information, despite their significant differences from one another. The northwest flank of the Venezuelan Andes is best interpreted as a thrust fault that carries crystalline rocks and sediments of Paleozoic, Mesozoic, and Cenozoic age over late Cenozoic sediments of the Maracaibo Basin. Sooner or later this province will attract the attention of petroleum geologists seeking traps in the late Cenozoic sediments, including, perhaps, the distal extension of the Chama Valley Breccia, developed against the underside of the fault surface (for a discussion of recent drilling into similar structural provinces in the Rocky Mountains, see Gries, 1981).

At least some of the oblique displacement between the Caribbean and South America is now being accommodated along the Boconó Fault Zone in the Venezuelan Andes. However, the stratigraphic similarity of terranes on either side of the Boconó Fault Zone (e.g., see Shagam, 1976; Kovisars, 1969, 1971; Grauch, 1971, 1972) is not compatible with many kilometres of right-lateral strike-slip displacement along the fault.

One might rather think of the northwest corner of South America as a cluster of fault-bounded lithospheric plates of subcontinental dimensions jostling one another in response to rela-

tive displacements among the subcrustal blocks on which they lie. An illustrative analogy is provided by the behavior of individual blocks of ice on a frozen river as the spring breakup occurs and the blocks push and grind together in response to the inexorable downslope movement of the fluid on which they ride (for vivid contemporary accounts of the annual ice breakup on various Alaskan rivers, see Brown, 1976, or McPhee, 1976). No observer of a breakup event would contest the assertion that net displacement of the ice floes is in the downstream direction and that the dominant stress field imposed on the ice blocks, at least in the earliest stages of the breakup, is one of convergence along the flow axis of the stream. Yet relative displacement among blocks produces tensional effects, as in the brief appearance of open water as two blocks are pulled apart; shear effects, as blocks grind past one another; and compressional effects, as individual blocks are pushed far up on the river banks, ride over one another, are steeply upended, and even flip over. With appropriate scale corrections in both time and materials, one may liken the northwest corner of South America to the situation that might prevail if five or six large blocks of ice should become trapped in an eddy over the slight depression of a whirlpool (a convergent triple junction?). Even with the best flow data available, few hydraulic engineers would attempt to predict the geometry of interaction along each of the ice to ice interfaces represented.

SUMMARY

Evidence of late Cenozoic tectonic environments in the Central Venezuelan Andes presented here takes three forms:

1. The configuration of a late Cenozoic thrust fault and the distribution and inferred relationship to that thrust fault of a thick, extensive breccia unit displaying a remarkable range of rock textures attributable to a complex tectonic history.

2. The geochemistry, attitude, paleogeomorphology, and distribution of remnants of a paleo-oxisol that is interpreted to have developed close to sea level over much of the terrane that is now the highest part of the Central Venezuelan Andes.

3. Direct determination of the sense and magnitude of displacement of Quaternary landforms and sediments along recently active faults.

Within the constraints imposed by those data, the following conclusions can be offered (refer also to Figs. 10 and 11, and Fig. 12, a series of sketches depicting, in schematic cross section, the events outlined below):

The Central Venezuelan Andes were subject to severe compression at least as late as early to mid-Miocene time, and possibly later, resulting in the formation of the Chama Valley thrust fault and breccia zone. This compressional stress field may have involved a component of oblique shear, direct evidence of which is not apparent. There is no evidence that the terrane where the Venezuelan Andes now lie was high at that time. This tectonic environment may represent the easternmost expression of the net convergence that prevailed among the South American, Caribbean, and Nazca Plates in later Tertiary time (e.g., see Pindell and Dewey, 1982, Figs. 23, 24, and 26).

After emplacement of the thrust sheet, the terrane southeast of the Maracaibo Basin remained close to sea level for the order of 10^7 years while a deep oxisol developed across rocks of a wide range of lithologies. The uppermost horizons of this soil approach the composition of commercial-grade bauxite. At this time the interplate convergence required by most late Tertiary continental reconstructions (e.g., see Pindell and Dewey, 1982, Figs. 24 and 26) must have been accommodated somewhere in the terrane of structural complexity in northern Colombia.

The period of soil formation was interrupted on the order of 10^6 years ago by renewed imposition of compression, in response to which the terrane that is now the Venezuelan Andes was arched and uplifted. At this time crystalline rocks were also thrust over later Tertiary and Quaternary sediments along the northwest front of the Andes, and a string of grabenlike depressions opened along the axis of the Cordillera. Normal faults associated with the development of these depressions cut the sole of the thrust plate described earlier. At this time the previously stable oxisol-veneered surface was broken up into a series of fault-bounded blocks as that surface was draped over the rising core of the

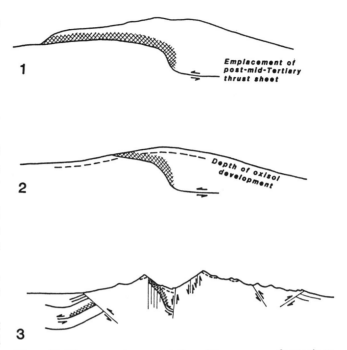

Figure 12. Diagrammatic representation of the sequence of tectonic environments that developed in the Central Andes of Venezuela in late Cenozoic time.

Cordillera. At least 250 m of right-lateral strike-slip displacement occurred along what has since been called the Bocono Fault Zone late during that arching episode; no direct evidence of transcurrent offset of greater magnitude has been documented.

Contemporary seismic evidence can be invoked to support both convergence and right-lateral strike-slip displacement in the Venezuelan Andes at the present time. It seems likely that the rate of uplift is as great now as it has been in the past million years; the paleo-oxisol is being rapidly incised and removed from surviving remnants of that relict surface. For the past million years, then, it seems likely that the Venezuelan Andes participated in accommodating some of the convergent stress imposed on northwest South America by oblique convergence of the Caribbean and South American Plates and, at least for the most recent part of that time, some of the right-lateral strike-slip displacement as well. Some of that complex displacement has been accommodated along the Bocono Fault Zone; some has probably been accommodated along other faults in the Venezuelan Andes, and some has been taken up by displacement along tectonic features far removed from the Venezuelan Andes.

The foregoing analysis is based on field evidence that must still be viewed as preliminary. The model presented here may well be altered as the meager evidence now available is augmented, but it should serve at least temporarily as a late Tertiary tectonic framework to which other data can be referred. The complex structural province of northwestern South America will provide challenging and frustrating opportunities to many generations of field geologists, geophysicists, and regional tectonic synthesists before its potential to instruct us is exhausted.

ACKNOWLEDGMENTS

The Ministério de Energía y Minas (formerly Ministério de Minas e Hidrocarburos) of the Republic of Venezuela fully supported all field phases of this work from 1969 through 1978. Throughout that period, Alírio Bellízzia, in several different capacities in the central Ministry offices, maintained close and supportive contact with this project and repeatedly expressed a well-informed interest in its progress. Other geologists in the Ministry who have offered assistance and constructive criticism include Alfredo Menendez, Alberto Vivas, Enrique Araujo, Enrique Lavíe, Juan Rios, Oliver Macsotay, Anibal Espejo, Luis González Silva, and Alfredo Sabater. Raúl Garcia Jarpa and Victor Campos supported this work, both as administrators in Caracas and as field colleagues in the Andes. The following geologists, assigned at various times to the Andean office in Mérida, participated in the field aspect of this study: Gustavo Canelon, Fernando Rondon, Ignácio Fierro, Santosh Ghosh, Jean-Francois Stephan, Gianluigi Benedetto, Oscar Odreman, Armando Useche, Richard Aarden, Juana de Arozena, Peter Moticska, and Armando Diaz Quintero.

Carlos Ferrer, Lionel Vivas, Rigoberto Andressen, and Jesús Pettit, of the Instituto de Geografia of the Univérsidad de Los Andes in Mérida, offered the perspective of the physical geographer to this analysis of surficial geology.

Reginald Shagam, first as a colleague at the University of Pennsylvania and later as a member of the geology faculty at Ben-Gurion University of the Negev in Israel, has maintained a vital interest in Andean tectonics and contributed much to this work, both in the field and elsewhere.

The following University of Pennsylvania students, who participated in this work at various stages, have made valuable contributions to an understanding of Andean tectonics: Leons Kovisars, Richard I. Grauch, Baruch Weingarten, Cinda Graubard, Donald Murphy, G. Murray Smith, W. Richardson McKinney, Laura Stokking, and Donald Segal. R. Ian Harker, Hermann W. Pfefferkorn, Barry P. Kohn, and Stephen P. Phipps, faculty colleagues at the University of Pennsylvania, have contributed valuable criticism to this study.

Parts of this work have been presented as seminars and invited papers at several meetings and institutions, where useful suggestions were offered by Arthur H. Bloom, William E. Bonini, Hannes K. Brueckner, John F. Casey, Terry Engelder, Lawrence Gilpin, Winthrop Means, Lucian Platt, John E. Sanders, Steve Schamel, Grant Skerlec, Manfred Strecker, Lawrence Teufel, Donald Turcotte, and F. B. Van Houten.

Vicente and Francisco Molinas, António Rondon, and José, Oswaldo, and Lucídio Rivas, as Ministry drivers, mechanics, and field assistants, contributed a lot of hard work and infectious enthusiasm to the success of the field phases of this study.

REFERENCES CITED

Bellízzia G., A., Pimentel M., N., and Bajo O., R., compilers, 1976, Mapa geológico estructural de Venezuela: Venezuela Ministério de Minas e Hdirocarburos, scale 1:500,000.

Birkeland, Peter W., 1974, Pedology, weathering, and geomorphological research: New York, Oxford University Press, 285 p.

Bolt, B. A., Fiedler, Guenther, and Dewey, J. W., 1972, Earthquakes and tectonics in western Venezuela: Proceedings, Upper Mantle Symposium, Buenos Aires, 1970, International Upper Mantle Project, v. 2, p. 119–129.

Bonini, W. E., 1976, Gravimetria, *in* Bellízzia G., A., Pimentel M., N., and Bajo O., R., compilers, Mapa geológico estructural de Venezuela: Venezuela Ministério de Minas e Hidrocarburos, scale 1:500,000.

——1984, Magnetic provinces in western Venezuela, *in* Bonini, W. E., and others, eds., The Caribbean–South American plate boundary and regional tectonics: Geological Society of America Memoir 162 (this volume).

Bonini, W. E., Pimstein de Gaete, C., and Graterol, V., compilers, 1977, Mapa de anomalias de Bouguer de la parte norte de Venezuela y áreas vecinas: Venezuela Ministério de Energia y Minas, scale 1:1,000,000.

Bonini, W. E., Garing, J. D., and Kellogg, J. N., 1980, Late Cenozoic uplifts of the Maracaibo–Santa Marta block, slow subduction of the Caribbean plate and results from a gravity study, *in* Transactions, Caribbean Geological Conference, IX, Santo Domingo, 1980: Universidad Católica Madre y Maestra, Santiago de los Caballeros, República Dominicana, v. 1, p. 99–105.

Bowin, C., 1976, The Caribbean: Gravity field and plate tectonics: Geological Society of America Special Paper 169.

Brown, Dale, 1976, Wild Alaska: New York, Time-Life Books, 184 p.

Buol, S. W., Hole, F. D., and McCracken, R. J., 1980, Soil genesis and classification: Ames, Iowa, Iowa State University Press, 404 p.

Case, J. E., and Holcombe, T. L., 1980, Geologic-tectonic map of the Caribbean region: U.S. Geological Survey Miscellaneous Investigations Map I-1100, scale 1:2,500,000.

Crowell, J. C., 1974, Origin of late Cenozoic basins in southern California, *in* Dickinson, W. R., ed., Tectonics and sedimentation: Society of Economic Paleontology and Mineralogy, Special Publication no. 22, p. 190–204.

deRatmiroff, Gregor N., 1971, Late Cenozoic thrusting in Venezuelan Andes: American Association of Petroleum Geologists, Bulletin, v. 55, p. 1336–1344.

Dewey, J. W., 1972, Seismicity and tectonics of western Venezuela: Bulletin of Seismological Society of America, v. 62, p. 1711–1751.

Drewes, Harald, 1959, Turtleback faults of Death Valley, California: A reinterpretation: Geological Society of America Bulletin, v. 70, p. 1497–1508.

——1960, Origin of the Amargosa Thrust Fault, Death Valley Area, California: A result of strike-slip faulting in Tertiary time: U.S. Geological Survey Professional Paper 400B, p. B268–B270.

Garcia, J., Raúl, 1978, Geology of the Tovar Sheet, Mérida Andes: Ministério de Energia y Minas, República de Venezuela, Caracas (unpublished map).

Giegengack, Robert, 1977a, A neotectonic model of the Boconó Fault Zone, Venezuelan Andes: 2nd Congreso Venezolano de Sismologia e Ingeniéria Sísmica, Mérida, Venezuela, Trabaios, Tomo 1, p. 1.1–1.5.

——1977b, Late Cenozoic tectonics of the Tabay-Estánquez Graben, Venezuelan Andes: 5th Congreso Venezolano de Geologia, Caracas: Ministério de Energia y Minas, Sociédad Venezolana de Geólogos, Memória, Tomo II, p. 721–737.

Giegengack, Robert, and Grauch, R. I., 1972a, Boconó Fault, Venezuelan Andes: Science, v. 175, p. 558–561.

——1972b, Geomorphologic expression of the Boconó Fault, Venezuelan Andes, or geomorphology to a fault: Geological Society of America Abstracts with Programs, (Annual Meeting), p. 719–720.

—— 1975, Quaternary geology of the Central Andes, Venezuela—a preliminary assessment: Excursion No. 1, Cordillera de Los Andes, 2nd Congreso Latinoamericano de Geologia, Caracas, Memória, Tomo I, p. 241–283.

—— 1976, Late-Cenozoic climatic stratigraphy of the Venezuelan Andes: Memória, 2nd Congreso Latinoamericano de Geologia, Ministério de Minas e Hidrocarburos, Republica de Venezuela, Bóletin de Geologia, Publicacion Especiál no. 7, Tomo II, p. 1187–1200.

Giegengack, Robert, Grauch, R. I., and Shagam, Reginald, 1976, Geometry of late-Cenozoic displacement along the Boconó Fault, Venezuelan Andes: Memória, 2nd Congreso Latinoamericano de Geologia, Ministério de Minas e Hidrocarbuors, República de Venezuela, Bóletin de Geologia, Publicacion Especiál no. 7, Tomo II, p. 1202–1226.

González de Juana, C., Iturralde de Arozena, J. M., and Picard C., X., 1980, Geologia de Venezuela y de sus cuencas petrolíferas: Caracas, Foninves, 1031 p.

Grauch, R. I., 1971, Geology of the Sierra Nevada south of Mucuchies, Venezuelan Andes: An aluminum-silicate-bearing terrane [Ph.D. thesis]: Philadelphia, University of Pennsylvania, 180 p.

—— 1972, Preliminary report of a late(?) Paleozoic metamorphic event in the Venezuelan Andes: Geological Society of America Memoir 132, p. 465–473.

Gries, Robbie, 1981, Oil and gas prospecting beneath the Precambrian of foreland thrust plates in the Rocky Mountains: The Mountain Geologist, v. 18, p. 1–18.

Guth, P. L., 1980, Geology of the Sheep Range, Clark County, Nevada [Ph.D. thesis]: Cambridge, Massachusetts Institute of Technology, 189 p.

—— 1981, Tertiary extension north of the Las Vegas Valley shear zone, Sheep and Desert Ranges, Clark County, Nevada: Geological Society of America Bulletin, v. 92, p. 763–771.

Harland, W. B., 1959, The Caledonian sequence in Ny Friesland, Spitsbergen: Quarterly Journal of the Geological Society of London, v. 114, p. 307–342.

—— 1961, An outline structural history of Spitsbergen: *in* Raasch, G. O., ed., Geology of the Arctic: University of Toronto Press, v. 1, p. 68–132.

—— 1969, Contributions of Spitsbergen to understanding of tectonic evolution of North Atlantic region: *in* North Atlantic—Geology and continental drift: American Association of Petroleum Geologists Memoir 12, p. 817–851.

—— 1971, Tectonic transpression in Caledonian Spitsbergen: Geological Magazine, v. 108, p. 27–42.

Hastenrath, Stefan, 1981, The glaciation of the Ecuadorian Andes: Rotterdam, A. A. Balkema, 159 p.

Jordan, T. H., 1975, The present-day motions of the Caribbean Plate: Journal of Geophysical Research, v. 80, p. 4433–4439.

Kafka, A. L., and Weidner, D. J., 1980, Seismotectonics of northern South America and the southern Caribbean: Plate boundary or intraplate deformation? [Abs. S47]: EOS (American Geophysical Union Transactions) v. 61, p. 1031.

—— 1981, Earthquake focal mechanisms and tectonic processes along the southern boundary of the Caribbean Plate: Journal of Geophysical Research, v. 86, p. 2877–2888.

Kelleher, John, Sykes, Lynn, and Oliver, Jack, 1973, Possible criteria for predicting earthquake locations and their application to major plate boundaries of the Pacific and the Caribbean: Journal of Geophysical Research, v. 78, p. 2547–2585.

Kohn, B. P., Shagam, R., Burkley, L. A., and Banks, P. O., 1984, Mesozoic-Pleistocene fission-track ages on rocks of the Venezuelan Andes and their tectonic implications, *in* Bonini, W. E., and others, eds., The Caribbean–South American plate boundary and regional tectonics: Geological Society of America Memoir 162 (this volume).

Kovisars, Leons, 1969, Geology of the eastern flank of the La Culata Massif, Venezuelan Andes [Ph.D. thesis]: Philadelphia, University of Pennsylvania, 211 p.

—— 1971, Geology of a portion of the north-central Venezuelan Andes: Geological Society of America Bulletin, v. 82, p. 3111–3138.

Kupfer, D. H., 1960, Thrust faulting and chaos structure, Silurian Hills, San Bernardino County, California: Geological Society of America Bulletin, v. 71, p. 181–214.

Larotta Sanchez, Jaime, 1975, Riesgo Sísmico de la region de Mérida: Oficina Técnica Especiál del Sismo, Ministério de Obras Públicas, República de Venezuela, 59 p.

Lowell, J. D., 1972, Spitsbergen Tertiary orogenic belt and the Spitsbergen Fracture Zone: Geological Society of America Bulletin, v. 83, p. 3091–3102.

Macellari, C. E., 1981, Late Cenozoic deposits of the Táchira Depression and its bearings to the tectonic evolution of the southern Venezuelan Andes [M.S. thesis]: Columbus, Ohio State University, 98 p.

Macellari, C., 1984, Late Tertiary tectonic history of the Táchira Depression (southwestern Venezuelan Andes), *in* Bonini, W. E., and others, eds., The Caribbean–South American plate boundary and regional tectonics: Geological Society of America Memoir 162 (this volume).

Maher, H., Craddock, C., and Spatz, J., 1981, Tertiary deformation by transpression at Midterhuken, West Spitsbergen: Geological Society of America Abstracts with Programs, Annual Meeting, Cincinnati, Ohio, p. 501.

McPhee, John, 1976, Coming into the country: New York, Farrar, Straus, and Giroux, Inc.

Mercer, J. H., and Palácios M., Oscar, 1977, Radiocarbon dating of the last glaciation in Peru: Geology, v. 5, p. 600–604.

Molnar, Peter, and Sykes, L. R., 1969, Tectonics of the Caribbean and Middle America regions from focal mechanisms and seismicity: Geological Society of America Bulletin, v. 80, p. 1639–1684.

Murphy, D. C., 1977, Stratigraphic and structural significance of the Mucuchies Formation (late Tertiary), Venezuelan Andes: Abstracts, 8th Caribbean Geological Conference, Curaçao, Netherlands Antilles, p. 135–136.

Noble, L. F., 1941, Structural features of the Virgin Spring area, Death Valley, California: Geological Society of America Bulletin, v. 52, p. 941–1000.

Pennington, W. D., 1981, Subduction of the eastern Panama Basin and seismotectonics of northwestern South America: Journal of Geophysical Research, v. 86, p. 10753–10770.

Pennington, W. D., Mooney, W. D., van Hissenhoven, Rene, Meyer, Hansjurgen, and Ramirez, J. E., 1979, Results of a reconnaissance microearthquake survey of Bucaramanga, Colombia: Geophysical Research Letters, v. 6, p. 65–68.

Phipps, S. P., 1984, Cenozoic thrust faulting in the California Coast Ranges [Ph.D. thesis]: Princeton, Princeton University, New Jersey.

Phipps, S. P., Kleinspehn, K. L., and Suppe, J. E., 1979, Ophiolitic olistrostromes in the basal Great Valley sequence, Napa County, California: Geological Society of America Abstracts with Programs, Cordilleran Section, p. 122.

Pindell, James, and Dewey, J. F., 1982, Permo-Triassic reconstruction of western Pangaea and the evolution of the Gulf of Mexico/Caribbean region: Tectonics, v. 1, p. 179–211.

Renz, O., 1956, Cretaceous in western Venezuela and Guajira (Colombia): International Geological Congress, 20th, Mexico, p. 1–13.

Rod, Emile, 1956, Strike-slip faults of northern Venezuela: American Association of Petroleum Geologists Bulletin, v. 40, p. 457–476.

Santamaria, Francisco, and Schubert, Carlos, 1974, Geochemistry and geochronology of the southern Caribbean–northern Venezuela plate boundary: Geological Society of America Bulletin, v. 7, p. 1085–1098.

Schubert, Carlos, 1971, Observaciones geomorfológicas y glaciales en el Area de Pico Bolivar, Sierra Nevada de Mérida, Venezuela: Bóletin Informativo, Asociacion Venezolana de Geologia, Mineria y Petróleo, v. 14, p. 193–216.

—— 1980, Late-Cenozoic pull-apart basins, Boconó Fault Zone, Venezuelan Andes: Journal of Structural Geology, v. 2, p. 463–468.

—— 1981, Are the Venezuelan fault systems part of the southern Caribbean plate boundary?: Geologische Rundschau, Bd. 70, p. 542–551.

—— 1982, Neotectonics of Boconó Fault, western Venezuela: Tectonophysics, v. 85, p. 205–220.

Schubert, C., and Henneberg, H. G., 1975, Geological and geodetic investigations on the movement along the Boconó Fault, Venezuelan Andes: Tectonophysics, v. 29, p. 199–207.

Schubert, Carlos, and Sifontes, R. S., 1970, Boconó Fault, Venezuelan Andes:

Evidence of postglacial movement: Science, v. 170, p. 66–69.

Serra, Sandro, 1977, Styles of deformation in the ramp regions of overthrust faults: 29th Annual Field Conference, Wyoming Geological Association Guidebook, p. 487–498.

Shagam, Reginald, 1976, The northern termination of the Andes: *in* Nairn, A.E.M., and Stehli, F. G., eds., The ocean basins and margins, Volume 3 The Gulf of Mexico and the Caribbean: New York and London, Plenum Press, chap. 9, p. 325–420.

Shagam, R., Kohn, B. P., Banks, P. O., Dasch, L. E., and others, 1984, Tectonic implications of Cretaceous-Pliocene fission-track ages from rocks of the circum-Maracaibo Basin region of western Venezuela and eastern Colombia, *in* Bonini, W. E., and others, eds., The Caribbean–South American plate boundary and regional tectonics: Geological Society of America Memoir 162 (this volume).

Sibson, R. H., White, S. H., and Atkinson, B. K., 1981, Structure and distribution of fault rocks in the Alpine Fault Zone, New Zealand: *in* McClay, K. R., and Price, N. J., eds., Thrust and nappe tectonics: Geological Society of London, p. 197–210.

Suggate, R. P., 1963, The Alpine Fault: Transactions of the Royal Society of New Zealand, Geology, v. 2, p. 105–129.

Useche C., Armando, 1977, Geologia de la region de la Azulita, Estado Mérida: Memória, 2nd Congreso Latinoamericano de Geologia, Ministério de Minas e Hidrocarburos, República de Venezuela, Bóletin de Geologia, Publicacion Especiál no. 7, Tomo III, p. 1773–1786.

Vierbuchen, R. C., Jr., 1978, The tectonics of northeastern Venezuela and the southern Caribbean Sea [Ph.D. thesis]: Princeton, New Jersey, Princeton University, 175 p.

Walcott, R. I., and Cresswell, M. M., 1979, editors, The origin of the Southern Alps: The Royal Society of New Zealand, Bulletin 18, 147 p.

Weingarten, Baruch, 1977, Tectonic and paleoclimatic significance of a late-Cenozoic paleosol from the Central Andes, Venezuela [M.S. thesis]: Philadelphia, University of Pennsylvania, 67 p.

MANUSCRIPT ACCEPTED BY THE SOCIETY SEPTEMBER 1, 1983

Geological Society of America
Memoir 162
1984

Mesozoic-Pleistocene fission-track ages on rocks of the Venezuelan Andes and their tectonic implications

B. P. Kohn
R. Shagam
Department of Geology and Mineralogy
Ben-Gurion University of the Negev
Beer Sheva, Israel 84120

P. O. Banks
L. A. Burkley
Case Western Reserve University
Cleveland, Ohio 44106

ABSTRACT

A total of 45 fission-track ages are reported on 22 apatite, 21 zircon, and 2 sphene concentrates from 23 rock samples from the Venezuelan Andes. Apatite ages range from about 1.4 to 24.0 Ma and are all interpreted as cooling ages related to uplift and erosion of cover rocks. Nine of the apatite ages are on one granitic pluton and adjacent rocks which occur through relief of 2,360 m, and the plot of age versus elevation approximates a straight-line slope of 0.8 km/Ma uplift. Because that rate far exceeds accepted rates of thermal diffusion in rocks, updoming of isotherms is inferred. In that case, elevation-age curves may not give accurate rates of uplift. Moreover, the time span of uplift is likely to be greater than the range of ages actually measured. It is suggested that the problem of updoming of isotherms in interpreting apatite ages may best be resolved by multiple sampling as near as possible to the vertical at two or more sites located at different ranges of elevation.

On the basis of the apatite ages, it is suggested that uplift of the Andes involved first uplift of the leading (northwestern) margin in the Oligocene to Miocene, followed by uplift of the trailing (southeastern) margin in the late Miocene, and in turn by rapid uplift of the central Andes in Pliocene-Pleistocene time. This view differs significantly from previous models of Tertiary Andean uplift.

Zircon ages range from about 60 to 172 Ma with a strong grouping in the range of about 81 to 113 Ma. Ages of 60 and 61 Ma (early Paleocene) are interpreted as uplift ages (and possibly a third age at 72 Ma) and are in conformity with regional stratigraphic evidence. The remaining ages are viewed as mixed ages, that is, ages modified by lengthy residence in the depth zone of partial track annealing. As such they do not have specific geologic significance. Two sphene ages (about 139 and 159 Ma) are also interpreted as mixed ages.

INTRODUCTION

Apart from minor gravels and oxisols (and in one small area some terrestrial Miocene deposits), the stratigraphy of the Venezuelan Andes terminates with the upper Eocene section. The next younger chronostratigraphic level of some regional extent is represented by late Pleistocene moraines. The youngest igneous rocks are of late Permian to early Triassic age; no igneous activity is associated with the Tertiary uplift history, thereby eliminating conventional radiometric methods (e.g., K-Ar) as tools for dating that history. Uplift was accompanied by the deposition of extensive terrestrial piedmont deposits of Oligocene-Quaternary age, but the paucity of fossils has severely limited their usefulness in unraveling the uplift history. Accordingly, the tectonic uplift history has received only cursory attention but for some recent studies by Giegengack (1984).

In recent years fission-track dating has developed into a valuable tool for dating postcrystallization events. Hence, pre-Mesozoic metamorphic and igneous rocks of the Andes are potentially capable of providing information basic to the interpretation of tectonic events during the Oligocene-Pliocene gap in the stratigraphic record. Recognition of that potential stimulated this study.

The fission-track method, as in many other radiometric methods, dates the time at which a mineral cooled through a particular temperature interval. Wagner (1979) described temperature-depth relationships in terms of an intermediate zone where tracks both accumulate and anneal, underlain by a hotter zone in which all tracks are annealed and overlain by a cooler zone in which all tracks are preserved. Clearly, the retention of fission tracks in any given mineral is a function of the cooling history of that mineral. Because the track-retention temperature limit is different for different mineral species, the fission-track method can provide valuable information on the thermal history of a rock containing coexisting datable minerals.

Estimates of the effective track-retention (i.e., blocking or closure) temperature to which a mineral has cooled at the time given by the fission-track age have been discussed extensively in the literature (e.g., Wagner and Reimer, 1972; Haack, 1977; Wagner and others, 1977; Gleadow and Lovering, 1978; Naeser, 1979; Harrison and others, 1979; Gleadow and Brooks, 1979; Naeser, 1981; Gleadow and Duddy, 1981). The effective closure temperature approximates to about the stage at which half of the tracks become stable (in the intermediate zone, above) and varies with the cooling rate (Wagner and Reimer, 1972; Wagner and others, 1977; Haack, 1977). Recent fission-track measurements on apatites from deep boreholes (Naeser, 1979, 1981; Gleadow and Duddy, 1981) indicate that the temperature interval over which geological track fading occurs is much narrower than that predicted from laboratory-derived annealing curves extrapolated into geological time (Wagner, 1968; Naeser and Faul, 1969; Naeser, 1979; Wagner, 1979).

The geological history of the Venezuelan Andes outlined below indicates that the duration of heating of the rocks studied during their last phase of burial was of the order of 10^8 years. This time approximates to an effective closure temperature for apatite of ~ 85 °C and to a temperature of ~ 105 °C for total annealing (Naeser 1979, 1981; Gleadow and Duddy, 1981). Data for the effective closure temperature of other minerals are less well known; following discussions on this subject by Harrison and others (1979) and Gleadow and Brooks (1979), we here assume temperatures of $\sim 200 \pm 50$ °C for zircon and $\sim 250 \pm 50$ °C for sphene. Closure temperatures are not constants for each given mineral but vary through a range of values as a function of cooling rate. The closure temperature in a region of rapid uplift (and hence rapid cooling) will be higher than in more stable areas. Apatite ages, because of their low closure temperature, are uniquely sensitive to low-temperature thermal events and, especially in orogenic belts, to the effects of uplift and erosion (cf. Wagner and others, 1977). In the light of the foregoing it is to be expected that where apatite, zircon, and sphene ages are discordant, apatites will give the youngest ages and sphenes the oldest. Apatite and zircon ages will be younger than those from minerals dated by the K-Ar and Rb-Sr methods whose effective closure temperatures have recently been reviewed by Harrison and others (1979). On the other hand, dates on sphene may be concordant with K-Ar (estimated closure temperature 220 ± 40 °C) and Rb-Sr (300 ± 25 °C) ages on coexisting biotites (Gleadow and Lovering, 1978; Harrison and others, 1979; Gleadow and Brooks, 1979).

GEOLOGICAL HISTORY

The Venezuelan Andes may be simply described in terms of a crystalline core located in the central, most massive, part of the ranges, overlain toward the northeast and southwest by successive skirts of upper Paleozoic, Mesozoic, and Cenozoic formations (Shagam, 1976). There are, in addition, two narrow parallel belts of lower Paleozoic and Precambrian rocks located in the Colorado Horst in the southeastern piedmont of the Andes (Fig. 1). Those aspects of the geological history pertinent to this study are illustrated in Figure 2 from which the following inferences may be drawn for the bulk of the Andes north of the Colorado Horst (Fig. 1):

1. The Rb-Sr and K-Ar ages (Table 1), whether they represent cooling following original magmatic crystallization or following resetting of older metamorphic clocks, cluster in the late Paleozoic to early Mesozoic depending on location, indicating that the zircon and sphene fission-track clocks must have been reset correspondingly. Hence the details of pre–late Paleozoic thermal events would have been erased and may be ignored in this study.

2. Keeping in mind the fact that all the apatite samples come from crystalline rocks below the unconformity under the Jurassic Quinta red beds or marine Cretaceous to Paleogene section, it is clear from the data presented below that the cover of Jurassic-

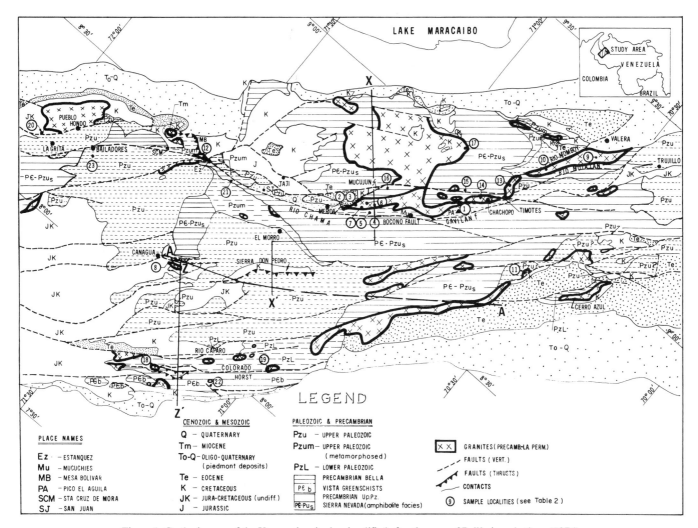

Figure 1. Geologic map of the Venezuelan Andes simplified after the map of Bellízzia and others (1976). Sample numbers and rock types corresponding to the circled map location numbers are given in Table 2. See Figure 4 for a detailed map of the Carmen block. For significance of lines A-A', X-X', and Z-Z', see Figure 11.

Eocene section must have been thick enough to depress the apatites into the zone of total track annealing until post-Eocene "Andean" uplift brought them through the closure temperature, because all apatite fission-track ages are younger than the sedimentary age of the overlying strata.

For the region of the Colorado Horst it is concluded that:

3. The late-Paleozoic orogeny in the bulk of the Andes to the north exerted no thermal effects sufficient to reset the K-Ar clock. Clearly, in this region the initial setting of the fission-track clocks probably occurred at some time(s) in the mid-Paleozoic.

4. After deposition of a far thinner pile of upper Paleozoic sediment compared to the northern Andes, events of uplift, erosion, and subsequent deposition of Jurassic-Eocene section were substantially the same as for the northern Andes. Here too the data for apatite presented below indicate that the corresponding rocks were only brought through the closure temperature for the last time in the course of post-Eocene "Andean" uplift.

On the basis of field work by Shagam (see plate of cross-section in Shagam, 1972a) corroborated by the data of Zambrano and others (1971) and Renz (1977), we estimate the minimum thickness of post-Paleozoic section for each sample as shown in Table 2. In addition, one has to take into account the estimated depth of the sample below the unconformity at the contact with the La Quinta or Cretaceous formations. Where the cover has been removed by erosion, estimates of the former location of the unconformity have been made based on extrapolation from the areas of nearest outcrop taking into account aspects of structural geology such as major faults (see Table 2).

ANALYTICAL METHODS AND RESULTS

Apatite and zircon splits remaining from the U-Pb studies of Burkley (1976) were used in this study. In addition, apatite, zircon, and sphene, in the size range 62.5 to 250 μm, were

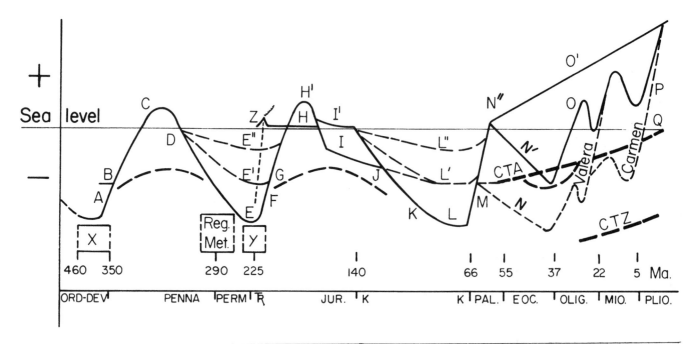

Time	Rock Unit	Geological Event	
		SE Piedmont area (Colorado Horst)	Central and Northern Andes
Devonian	X	Uplifted along curve (A) through CTA (B).	
	X	Cover eroded (C).	
Penna	X	Subsidence accompanied by sedimentation (D). Probably along path E" (evidence for wedging out of section), though E' or E also possible.	Subsidence and sedimentation probably along path E evidenced by 4-7 km of section. Clocks annealed.
End Perm	Y		Orogeny with extensive granitic plutonism.
	Z		Volcanic equivalents of plutonism.
Triassic	Y		Uplift along (F) through CTA (G) with unroofing of cover rocks (H'). Clocks reset.
	X	Uplift probably from above CTA with unroofing of cover rocks (H').	
	Z		Minor erosion compared to X and Y; probably followed path close to sea level (H). Clock set.
Jurassic	X,Y,Z	Subsidence of faulted basins locally in which Quinta sediments accumulated (I); intervening areas underwent peneplanation close to sea level (I').	
Cret.	X,Y,Z	Continued subsidence of whole area with marine transgression, along probable path JKL indicated by consistent Tertiary apatite ages (i.e. depression below CTA).	
End Cret to Eoc.	X,Y,Z	Renewed uplift (M) and erosion (N") contemporaneous with deposition of early Tertiary regressive deltaic section in subsiding graben areas (N'). Uplift not sufficient to bring X,Y,Z, through CTA (N).	
Oligo- Present	X,Y,Z	Tertiary "Andean" uplift. Continuous uplift (O') or in a series of stages (O); latter preferred. In one such event part of Valera pluton uplifted through CTA in Late Oligocene but most bodies uplifted in Late Miocene-Early Pleistocene (P). Updoming of CTA which may presently be close to or above sea level under high Andes (Q).	

Figure 2. Diagrammatic representation of uplift and subsidence history of the Venezuelan Andes pertinent to fission-track measurements. Neither ordinate (elevation) nor abscissa (time) are scaled. Effective closure temperatures shown for apatite and zircon, respectively CTA and CTZ, ignore possible variations related to uplift, cooling rates, and changing geothermal gradients. "Reg. Met" represents last event of regional metamorphism. For detailed regional geological history, see Shagam (1976).

TABLE 1. SUMMARY OF GEOCHRONOLOGICAL STUDIES OF UNITS IN ANDES DATED IN THIS STUDY

Rock unit	Sample number	Previous work			This study (Fission-track age, Ma)		
		Isotopic age (Ma)	Method	Reference	Apatite	Zircon	Sphene
El Carmen Granodiorite		196 ± 10	K-Ar biotite	(1)			
		200 ± 10	K-Ar biotite	(1)			
		200 ± 25	Rb-Sr biotite	(1)			
	LB-71-VE-2	225 ± 25*	U-Pb zircon	(2)	4.7 ± 0.6	92 ± 14	
		200 ± 40	U-Pb apatite	(2)	4.9 ± 0.7	93 ± 10	
	SA-79-1				1.8 ± 0.4		
	SA-79-2				2.5 ± 0.4		139 ± 17
	SA-79-4				3.4 ± 0.6		
	SA-79-5				2.8 ± 0.4		
	SA-79-6				2.7 ± 0.5		
	SA-79-7				2.6 ± 0.5		
Bailadores Rhyolite	SA-79-8					60 ± 9	
Canaguá Granite	LB-71-VE-5	460 ± 20*	U-Pb zircon	(2)	3.3 ± 0.6	119 ± 17	
		470 ± 90	U-Pb apatite	(2)			
Valera Granite		210 ± 5	K-Ar orthoclase	(1)			
		240 ± 5	K-Ar biotite	(1)			
		220 ± 15	K-Ar muscovite	(1)			
		216 ± 25	Rb-Sr orthoclase	(1)			
		245 ± 5	Rb-Sr biotite	(1)			
		373 ± 10	Rb-Sr muscovite	(1)			
		263 ± 30	Rb-Sr whole rock	(1)			
	LB-71-VE-7	595 ± 40*	U-Pb zircon	(2)	21.5 ± 2.1	129 ± 14	
		241 ± 15†	U-Pb zircon	(2)	24.0 ± 2.1		
	SA-79-26				2.5 ± 0.4		
La Soledad Qtz Monzonite		229 ± 2	K-Ar biotite	(3)			
		295 ± 4	K-Ar biotite	(3)			
	LB-71-VE-9	475 ± 65*	U-Pb zircon	(2)	3.0 ± 0.4	124 ± 13	
		400 ± 40	U-Pb apatite	(2)		131 ± 13	
Estánquez Granodiorite	LB-71-VE-27	440 ± 50†	U-Pb zircon	(2)	1.4 ± 0.2	113 ± 16	
						109 ± 15	
Timotes Granodiorite		181 ± 9	K-Ar biotite	(4)			
	LB-71-VE-29	435 ± 30*	U-Pb zircon	(2)	3.8 ± 0.7	86 ± 8	
		230 ± 30	U-Pb apatite	(2)			
Chachopo Granodiorite	LB-71-VE-31	225 ± 25*	U-Pb zircon	(2)	2.8 ± 0.6	72 ± 9	
El Carmen satellite body	LB-71-VE-32				4.9 ± 0.9		
El Valle dike	LB-71-VE-35	225 ± 25*	U-Pb zircon	(2)	4.2 ± 0.7	99 ± 10	
						87 ± 13	
La Culata Adamellite		232 ± 20	Rb-Sr biotite	(1)			
		224 ± 15	K-Ar orthoclase	(1)			
		252 ± 15	K-Ar orthoclase	(1)			
	LB-71-VE-39	225 ± 25*	U-Pb zircon	(2)	3.0 ± 0.5	61 ± 5	
Puente Real Augen Gneiss	LB-71-VE-36	465 ± 15	U-Pb zircon	(2)		107 ± 11	
						99 ± 9	
Tapo-Cambur Granite§	78-4	470 ± 50*	U-Pb zircon	(2)	7.7 ± 2.2	159 ± 27	
Bella Vista Schist§	78-7					172 ± 27	
Rio Quiú Granite§	78-8	465 ± 50*	U-Pb zircon	(2)	5.6 ± 1.5	160 ± 20	159 ± 23
La Quinta Rhyolite		122.5 ± 7.7	K-Ar whole rock	(5)			
		149 ± 10	K-Ar biotite	(5)			
		229 ± 15*	U-Pb zircon	(2)			
	S-87-W				10.8 ± 2.1	89 ± 12	
						80 ± 14	

*Primary age of crystallization.
†Lower concordia intercept.
§Samples from Colorado Horst.

References: (1) Martin Bellízia (1968); (2) Burkley (1976); (3) Schubert (1969); (4) Olmeta (1968); (5) Schubert and others (1979).

TABLE 2.   DETAILS OF SAMPLES, LOCALITIES, AND MEASURED OR ESTIMATED THICKNESS OF SEDIMENTARY COVER OVERLYING SPECIMENS COLLECTED FROM ANDES

Specimen No.	Name and rock type	Location no. on map (Fig. 1)	Current Elevation (±100m)	Thickness of cover beds (m)			
				La Quinta	Cretaceous-Eocene	Depth below unconformity	Total
LB-71-VE-2	El Carmen Granodiorite	1	~3900	0?	~1800	~1000-2000	~2800-3800
SA-79-1	El Carmen Granodiorite	2	1520	"	~1800	>2000	>3800
SA-79-2	El Carmen Granodiorite	3	1840	"	"	"	"
SA-79-4	El Carmen Granodiorite	4	3100	"	"	"	"
SA-79-5	El Carmen contact (bi-q-laminated meta-sst)	5	2500	"	"	"	"
SA-79-6	El Carmen Granodiorite	6	2600	"	"	"	"
SA-79-7	El Carmen Granodiorite	7	2375	"	"	"	"
SA-79-8	Bailadores Rhyolite (Mucuchachí Formation)	23	1920	"	~2500	~4000	~6500
LB-71-VE-5	Canaguá Granite	8	1500	~100	~2200	~200	~2500
LB-71-VE-7	Valera Granite	9	700	0-1000?	~2900	~500?	~3400-4400
SA-79-26	Valera Granite	10	2000	"	"	"	"
LB-71-VE-9	La Soledad Qtz Monzonite	11	1200	0?	~2400	~500-2000?	~2900-4400?
LB-71-VE-27	Estánquez Granodiorite	12	450	~1000	~2100	~300-500?	~3400-3900
LB-71-VE-29	Timotes Grandiorite	13	2300	0?	~1500	?	>1500
LB-71-VE-31	Chachopo Granodiorite	14	3000	"	~1500	?	>1500
LB-71-VE-32	El Carmen Satellite body	15	3800	0	~1500	1000-2000	~2500-3500
LB-71-VE-35	El Valle dyke	16	3200	0	~1800	>2000?	>3800
LB-71-VE-35	Puente Real Augen Gneiss	21	700	~1000	2100	?	>3100
LB-71-VE-39	La Culata Adamellite	17	1800	0?	1400	~2000	~3400
78-4	Tapo-Cambur Granite	18	380	0?	1700	>300	>2000
78-8	Rio Quiú Granite	19	400	0?	1100	>1500	>2600
78-7	Bella Vista Schist	22	400	0?	~1200	>1000	>2200
S-87-W	La Quinta Rhyolite	20	1200	2000	2730	0	4700

Note: Data sources: Zambrano and others (1971); Renz (1977); Creole Petroleum Corp. (Lagoven), internal reports; Socony-Mobil (Maraven), internal reports; Shagam (regional mapping contracted by the Ministerio de Energía y Minas).

obtained from crushed- and ground-rock samples collected by Shagam, using heavy liquids and a Frantz isodynamic magnetic separator.

Both the population method (PM) and the external detector method (EDM) (Naeser, 1976) were used for dating apatites (Table 3), the former when the yield of grains was large and the latter when small. Gleadow (1981) has demonstrated that both methods on the same apatite sample yield concordant ages. In the case of PM, one aliquot of grains was heated to 500 °C for 3 hours to remove fossil fission tracks and irradiated with thermal neutrons, while another was retained for determining fossil fission-track density. Irradiated and nonirradiated apatites were mounted in polyester, polished and etched simultaneously in 7% HNO_3 at 21 °C for 45 to 55 seconds. Apatites dated by EDM were mounted, polished, and etched under identical conditions for PM. These polyester wafers were then irradiated together with

TABLE 3. FISSION-TRACK DATA AND AGES FOR APATITES, ZIRCONS AND SPHENES FROM ANDEAN SAMPLES

Sample No.	Location No. on maps	ρ_s cm^{-2}		ρ_i cm^{-2}		ϕ $\times 10^{15}$ neutrons/cm^2		Number of fields or grains[*]	AGE Ma ($\pm 2\sigma$)
				A P A T I T E					
LB-71-VE-2	1	3.44×10^4	(279)[†]	2.05×10^6	(1384)[†]	4.60	(4274)[†]	300/80	4.7 ± 0.6
		3.55×10^4	(240)	1.11×10^6	(1202)	2.47	(2127)	260/50	4.9 ± 0.7
SA-79-1	2	1.37×10^4	(104)	1.23×10^6	(1664)	2.66	(2251)	280/80	1.8 ± 0.4
SA-79-2	3	2.13×10^4	(240)	1.33×10^6	(1080)	2.58	(3961)	280/60	2.5 ± 0.4
SA-79-4	4	3.08×10^4	(200)	2.00×10^6	(1353)	3.56	(2373)	240/50	3.4 ± 0.6
SA-79-5	5	4.52×10^4	(342)	3.80×10^6	(1542)	3.90	(2111)	280/30	2.8 ± 0.4
SA-79-6	6	2.03×10^4	(132)	1.40×10^6	(1062)	2.98	(1670)	240/100	2.7 ± 0.5
SA-79-7	7	2.43×10^4	(158)	2.45×10^6	(1684)	4.26	(1834)	240/80	2.6 ± 0.5
LB-71-VE-5	8	3.85×10^4	(156)	8.17×10^5	(1326)	1.13	(2437)	150/120	3.3 ± 0.6
LB-71-VE-7	9	3.73×10^5	(1008)	3.57×10^6	(1105)	3.36	(2111)	100/35	21.5 ± 2.1
		3.73×10^5	(1072)	2.34×10^6	(1930)	2.46	(2551)	120/30	24.0 ± 2.1
SA-79-26	10	3.30×10^4	(214)	3.28×10^6	(2220)	4.00	(2274)	240/50	2.5 ± 0.4
LB-71-VE-9	11	9.64×10^4	(339)	1.98×10^6	(1604)	1.00	(2437)	130/60	3.0 ± 0.4
LB-71-VE-27	12	3.72×10^4	(163)	1.53×10^6	(1242)	0.95	(2437)	162/60	1.4 ± 0.2
LB-71-VE-29	13	3.21×10^4	(156)	1.65×10^6	(1114)	3.14	(2772)	180/50	3.8 ± 0.7
LB-71-VE-31	14	2.63×10^4	(86)	1.95×10^6	(1056)	3.41	(2772)	120/50	2.8 ± 0.6
LB-71-VE-32	15	2.90×10^4	(153)	1.61×10^6	(1086)	4.42	(2199)	195/50	4.9 ± 0.9
LB-71-VE-35	16	3.55×10^4	(209)	2.36×10^6	(1117)	4.56	(2199)	218/35	4.2 ± 0.7
LB-71-VE-39	17	5.29×10^4	(143)	1.37×10^6	(1020)	1.25	(2686)	100/55	3.0 ± 0.5
78-4	18	1.34×10^5	(54)	4.09×10^6	(830)	3.85	(4625)	25	7.7 ± 2.2
78-8	19	3.11×10^4	(64)	1.32×10^6	(676)	3.85	(4625)	32	5.6 ± 1.5
S-87-W	20	6.43×10^4	(125)	1.42×10^6	(785)	3.88	(4625)	72/41	10.8 ± 2.1
				Z I R C O N					
LB-71-VE-2	1	3.06×10^6	(942)	1.77×10^6	(246)	0.87	(2251)	9	92 ± 14
		4.65×10^6	(983)	5.96×10^6	(718)	1.95	(1426)	14	93 ± 10
LB-71-VE-5	5	9.75×10^6	(1260)	4.35×10^6	(282)	0.87	(2251)	8	119 ± 17
LB-71-VE-7	9	5.93×10^6	(1296)	6.77×10^6	(740)	2.43	(1712)	10	129 ± 14
LB-71-VE-9	11	5.22×10^6	(829)	6.22×10^6	(989)	2.43	(1712)	6	124 ± 13
		1.47×10^7	(1458)	2.15×10^7	(716)	3.16	(4020)	8	131 ± 13
LB-71-VE-27	12	3.00×10^6	(1090)	1.51×10^6	(274)	0.93	(2437)	10	113 ± 16
		3.78×10^6	(436)	8.65×10^6	(501)	4.10	(2020)	7	109 ± 15
LB-71-VE-29	13	6.40×10^6	(944)	1.43×10^7	(1162)	3.16	(4020)	9	86 ± 8
LB-71-VE-31	14	2.83×10^6	(599)	6.16×10^6	(652)	2.56	(2059)	12	72 ± 9
LB-71-VE-35	16	3.41×10^6	(922)	3.76×10^6	(1015)	1.79	(1712)	12	99 ± 10
		4.88×10^6	(396)	7.84×10^6	(306)	2.30	(2750)	6	87 ± 13
LB-71-VE-36	21	3.90×10^6	(1044)	4.36×10^6	(1168)	1.96	(1712)	10	107 ± 11
		1.26×10^7	(1196)	2.45×10^7	(1164)	3.16	(4020)	6	99 ± 9
LB-71-VE-39	17	3.76×10^6	(1190)	8.43×10^6	(1322)	2.23	(2750)	12	61 ± 5
78-4	18	5.81×10^6	(462)	5.20×10^6	(207)	2.37	(2059)	6	159 ± 27
78-7	22	5.79×10^6	(639)	4.68×10^6	(258)	2.29	(2059)	9	172 ± 27
78-8	19	4.81×10^6	(897)	4.38×10^6	(408)	2.43	(2059)	9	160 ± 20
SA-79-8	23	1.49×10^6	(387)	3.81×10^6	(497)	2.49	(2251)	10	60 ± 9
S-87-W	20	2.52×10^6	(428)	6.10×10^6	(544)	3.55	(1597)	6	89 ± 12
		3.61×10^6	(258)	9.08×10^6	(324)	3.31	(1684)	6	80 ± 14
				S P H E N E					
78-8	19	5.24×10^6	(638)	5.36×10^6	(326)	2.68	(1868)	8	159 ± 23
SA-79-2	3	3.31×10^6	(1073)	2.78×10^6	(451)	1.92	(1426)	8	139 ± 17

Note:

ρ_s = spontaneous track density; ρ_i = induced track density; ϕ = neutron dose.

λ_f = 6.85 × 10^{-17} yr^{-1}, ^{235}U/^{238}U = 7.252 × 10^{-3}, σ^{235} = 580.2 × 10^{-24} cm^2, λ_D = 1.551 × 10^{-10} yr^{-1}.

[*] = Different numbers of fields of view counted in determining the fossil track density (first number) and the induced track density (second number). Single number indicates number of grains used in age determination (by external detector method).

[†] = Numbers in parentheses are tracks counted to determine the reported track density (or the neutron flux).

TABLE 4.   FISSION-TRACK DATA AND AGES FOR ZIRCON FROM FISH CANYON TUFF AGE STANDARD

Sample No.	ρ_s cm^{-2}		ρ_i cm^{-2}		ϕ $\times 10^{15}$ neutrons/cm^2		Number of grains	AGE Ma ($\pm 2\sigma$)
72N8-3	5.48×10^6	(1124)	1.03×10^7	(1055)	0.87	(2251)	11	28.4 ± 2.5
72N8-4	3.05×10^6	(1213)	1.53×10^7	(3048)	2.23	(2750)	12	27.2 ± 2.0
72N8-6	4.52×10^6	(1195)	1.87×10^7	(2472)	1.95	(1426)	11	28.9 ± 2.6
72N8-7	3.94×10^6	(1326)	2.04×10^7	(3435)	2.43	(1712)	12	28.8 ± 2.3
70L-126	3.14×10^6	(646)	8.88×10^6	(914)	1.25	(2686)	13	27.1 ± 3.0
72N-1	5.13×10^6	(2627)	5.23×10^6	(2680)	0.95	(2437)	6	28.8 ± 2.0

Note: Constants and notation used as for Table 3.

external muscovite (Brazil Ruby) detectors which were subsequently etched for 45 minutes in 48% HF at room temperature to reveal induced tracks.

Zircons were dated using mounting and etching techniques described by Gleadow and others (1976). Polished grains were etched for times varying from 10 to 20 hours at a temperature of 215 °C (precision of thermostat reading uncertain) and then irradiated together with external muscovite detectors, which were subsequently etched for up to 45 minutes in 48% HF at room temperature to reveal induced tracks. All spontaneous track counts on zircons were carried out on prismatic faces parallel to the *c* axis. Spontaneous tracks were clearly distinguishable from acicular inclusions which are longer and commonly parallel to one another.

Sphenes were mounted in polyester, polished and etched in 1 HF:2 HNO$_3$: 3 HCl: 6 H$_2$O at 20 °C (Naeser and McKee, 1970). As for zircons, sphenes were dated by the external detector method, induced tracks being counted in muscovite detectors exactly over the same area of the crystal counted for spontaneous tracks. In both cases only grains showing sharp polishing scratches after etching were selected for counting (Gleadow and Lovering, 1977; Gleadow, 1978), and counting results from at least six grains were combined for each age measurement. A factor of 0.5 (Gleadow and Lovering, 1977) was used to correct for different track registration geometries of internal surfaces of crystals and external muscovite detectors. Track counting in all minerals dated was carried out using transmitted light at ×1,250 magnifications under oil immersion.

The thermal neutron dose was determined by track counting in muscovite detectors irradiated in contact with standard NBS glasses 962 and 963 calibrated against NBS Cu foil measurements (Carpenter and Reimer, 1974). In order to monitor lateral variation in neutron flux, three or four glass standards were inserted at intervals along the length of the package sent for irradiation. In cases where flux gradients were detected, the dose received by a given sample was calculated by interpolation between results for the enclosing standards. Zircons and apatites of the Fish Canyon Tuff, Colorado, well dated by K-Ar at 27.9 ± 0.7 Ma (Steven and others, 1967, new constants), were also irradiated together with the NBS standards as an additional check on the neutron dose determinations. Results of fission-track ages on zircon from the Fish Canyon Tuff are given in Table 4.

Table 2 gives details of sample type, locality, and thickness of sedimentary cover. Analytical results for apatites, zircons, and sphenes, and constants used are given in Table 3. The statistical uncertainties on the dates, expressed as two standard deviations throughout, were calculated from the number of spontaneous and induced tracks counted in minerals and induced tracks in external muscovite detectors placed on zircons, sphenes and standards. Errors in the uranium content of the glass standards and in the decay constant were not included but fall well within the precision of the measurement. Table 1 summarizes the fission-track ages for each intrusion or rock unit studied, together with previously determined K-Ar, Rb-Sr, and U-Pb ages.

FISSION-TRACK AGES AND THEIR INTERPRETATION

Because the three minerals studied yield different age ranges (Table 3) and have different track retention properties as a function of temperature, separate consideration of the data for each mineral species is indicated. It is convenient to present the data in reverse chronological order and to discuss in turn the results for apatite, zircon, and sphene.

Apatite Ages

From the fission-track measurements on apatites (Table 3)

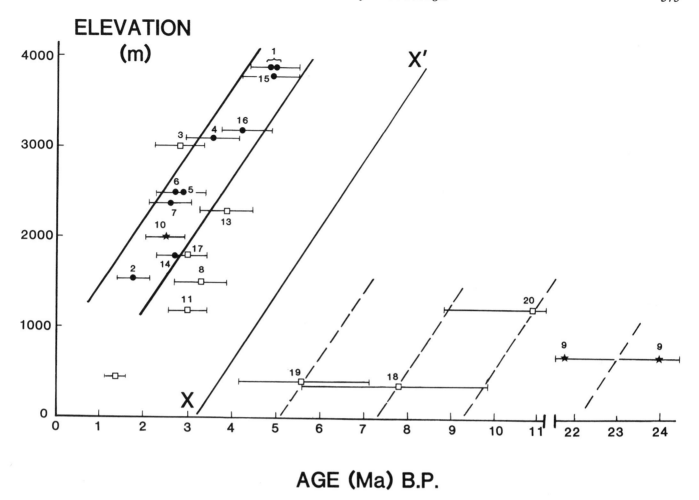

Figure 3. Apatite ages and error bars versus elevation. Numbers identify samples as shown in Table 1; full circles are samples from the Carmen block; stars 9 and 10 are Valera granite sample; open squares are remaining samples. Line X-X' arbitrarily separates young group of ages (left) from older group. Lines bounding Carmen block samples are visual estimate limits of rate of uplift if doming of isotherms is ignored (see discussion in text). Parallel dashed lines through single samples symbolize the fact that those samples probably lie on uplift curves of Carmen type but the slopes shown are arbitrary.

and from the previous discussion, it is clear that the ages do not date times of primary crystallization but rather their cooling history following tectonic uplift.

Shagam (1972b, 1976) suggested that post-Eocene uplift of the Venezuelan Andes crudely culminated in approximately a pyramidal form, canted to the northwest, with most pronounced uplift in the central portion of the ranges and progressively less toward the northeastern and southwestern terminations. During this uplift, as previously explained, the apatite clocks became operative.

Key to the interpretation of the mechanics and chronology of uplift is the elevation versus age plot of the data shown in Figure 3 which gives the apatite ages of nine samples collected from the Carmen tectonic block (Fig. 4). The good approximation to a straight-line slope of the data strongly suggests that elevation is a function of age in that block. The base of uplift must have been rooted below the zone of partial annealing, as shown in

Figure 5, which also shows that the limiting cases of differential uplift can be visualized to occur either by continuous rotation about a horizontally migrating axis or vertically as a consequence of faulting. Clearly, the details of uplift rate will vary depending on the rate of migration of the axis of rotation and its depth relative to the zone of partial annealing. In Figure 5c the axis of rotation is arbitrarily shown to coincide with the closure temperature zone

Evidence concerning the geometry of uplift stems from consideration of a flyschlike sequence of upper Paleozoic section, intruded by the Carmen pluton, which is preserved in outcrop at or close to the contact on the south side. As shown in Figure 6 the metamorphic assemblages and structural characteristics of these rocks change progressively in the prograde direction with decrease in elevation, strongly suggesting that the depth of erosion increases progresively to the southwest. That effect must be related to Tertiary uplift and erosion and cannot be an artefact of

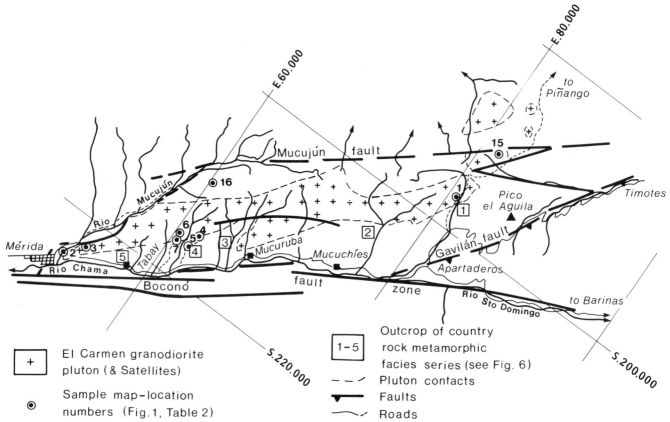

Figure 4. El Carmen block (Fig. 1) expanded to show sampled localities. Boxes (1–5) symbolize regionally metamorphosed country rock with contact metamorphic overprint: 1 = low greenschist facies slate with local andalusite porphyroblasts; 2 = high greenschist facies slate with staurolite porphyroblasts; 3 = laminated (S_0) staurolite-sillimanite schist; 4 = finely and uniformly foliated amphibolite grade pelitic schists, S_0 not recognizable; and 5 = amphibolite grade schists/gneisses with anastomosing foliation. See Figure 6 for significance attached to prograde sequence of metamorphic rocks in outcrop.

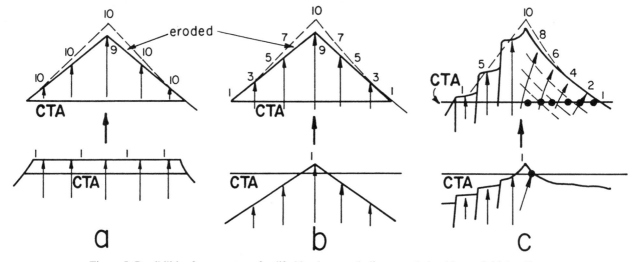

Figure 5. Possibilities for geometry of uplift. Numbers symbolize age relationships. a. Initial uniform uplift through the closure temperature zone for apatite (CTA) followed by differential uplift will yield uniform ages slightly modified by erosion. b. Progressive differential uplift through CTA yields pyramidal block with ages increasing with elevation. Block is continuously modified by erosion. c. Combines vertical uplift of fault blocks on the left with rotational uplift about laterally migrating axes (full-circles) on the right. Internal dashed lines represent isostratigraphic levels. Ages increase with elevation, continuously (right) and discontinuously (left).

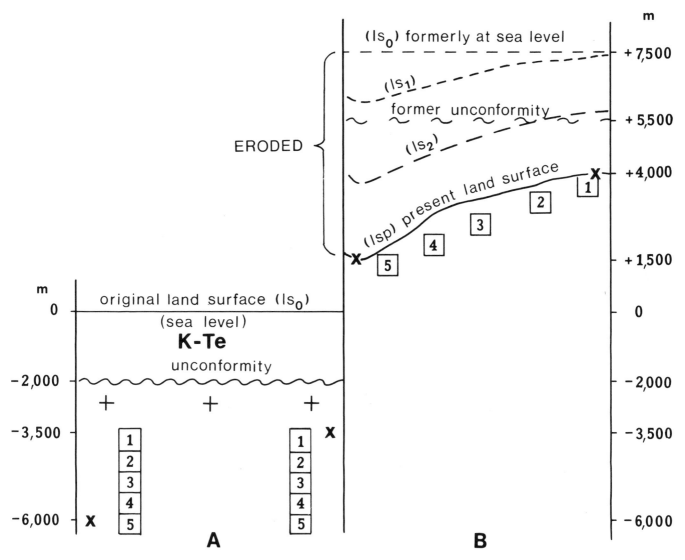

Figure 6. Postulated evolution of topography on the Carmen block implying subvertical uplift history. A. Prior to uplift original landsurface (1s$_0$) approximated sea level. Boxes (1–5) symbolize prograde metamorphic facies series as shown and defined in Figure 4. B. Present landsurface (1s$_P$) and surface exposure sequence of metamorphic country rocks (boxes). Latter imply progressively deeper level of erosion with decrease in altitude and, in turn, that the present topography is the product of differential rates of headward erosion on a block that underwent subvertical uplift. Hypothetical original (1s$_0$) and intermediate (1s$_1$ and 1s$_2$) stages of landform development have been projected above 1s$_P$, but their elevations and ages when formed, especially of 1s$_1$ and 1s$_2$ are not known.

Permian-Triassic orogeny, as indicated in Figure 6. Although it cannot be proven unequivocally, the evidence strongly favors the interpretation of subvertical uplift of the Carmen block in the Tertiary.

Northeastward of the Carmen pluton (Figs. 1 and 4), metamorphic evidence suggests quite a different geometry of uplift. The previously noted flyschlike upper Paleozoic slates in the Pico El Aguila area (~4,300 m) are at low greenschist grade and extend in continuous outcrop to the Valera-Trujillo area (~600 m) at about the *same metamorphic grade*. Although locally there has been some erosion to deeper levels, one may infer approximately the same depth of erosion, to a first approxi-

mation, for that extensive region. We conclude that in that region uplift involved rotation about a laterally migrating axis as depicted in Figure 5c.

All the Carmen block samples come from the region that underwent subvertical uplift through several kilometres which must have involved severe structural dislocation at the boundaries of the block. From the maps of Figures 1 and 4, longitudinal faults such as the Boconó-Gavilán and Mucujún faults are obvious features to define the southeast and northwest boundaries, respectively. The southwestern boundary may be defined either by the same set of cross faults that define the southwest limit of the Culata block to the northwest (Fig. 1) or by the Mucujún fault that may continue under the terrace gravels of Mérida as shown in Figure 4. On the northeast side neither cross faults nor drape folds have been recognized, and the Carmen block wedges out by convergence of the longitudinal faults in the vicinity of Timotes. In the light of the different uplift geometry inferred for that region, it is likely (and is consistent with the field evidence) that the throw on the longitudinal faults decreases toward the northeast. The subvertical uplift for most of the Carmen block implies the corollary of progressive lateral erosion leading to the prsent topography, as illustrated diagrammatically in Figure 6, which may be of considerable importance in the interpretation of uplift rates inferred from elevation-age plots such as Figure 3.

Figure 5c, here chosen as the appropriate geometric model of uplift for the Carmen block and contiguous area to the northeast, is simplified to the extent that the zone of the closure temperature is shown as a horizontal surface fixed in space rather than one progressively updomed as uplift proceeds. Temperature distribution and the possibility or not of thermal doming will depend in large part on the relationships between the rates of uplift and erosion. Clark (1979) inferred updoming of isotherms in the course of uplift of the Alps, whereas Zeitler and others (1982b) concluded that the effect was negligible during the uplift of the Pakistan Himalayas. A computer model of the present isotherms in the Himalayas (their Fig. 8) shows updoming of the 115 °C isotherm by about 300 m at a depth of ~6 km in an area where for the past 15 Ma uplift rates have been in the range of 0.16 to 0.34 mm/yr. They acknowledge that the model is only an order of magnitude approximation. In a subsequent study in the Nanga Parbat region of the Pakistan Himalayas, Zeitler and others (1982a) reported uplift rates for the Pleistocene approaching 1 cm/yr. This rate exceeds by an order of magnitude other values reported from the Himalayas for the recent past and the present.

The plot of Figure 3 suggests a rate of uplift of the Carmen block of 0.8 mm/yr, which is significantly higher than the rate calculated by Zeitler and others (1982b) for one region in the Himalayas, occurring in a shorter, more recent, time span (the past 5 Ma). The rate for the Carmen block is much lower than that determined in another region of the Himalayas (Zeitler and others, 1982a). It appears to us that it is reasonable to conclude, in agreement with Clark (1979), that as uplift proceeded the isotherms were significantly bowed upward. Such updoming if applied to the Nanga Parbat region could aid in reconciling the differences in uplift rates reported.

Wagner and others (1977, p. 9) have also discounted the possibility of bulged isotherms in the Alps because large age differences (by a factor of 7) over relatively small lateral distances (~30 km) would imply unrealistic heat domes. It may be argued, for example, that the narrow width (~10 km) of the Carmen block precludes significant thermal doming. However, ages differ by a factor of only 3, and as outlined in subsequent discussion, this block was uplifted together with its bordering blocks so that thermal doming in the transverse direction may well have exceeded in dimensions the thermal doming in the longitudinal direction.

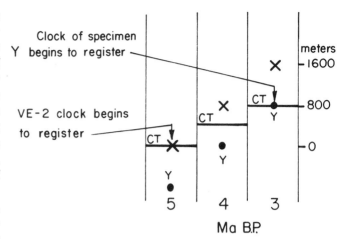

Figure 7. Qualitative representation to show that a specimen (Y) located 800 m *vertically* below VE-2 (X) may give an age suggesting uplift at the rate of 400 m/Ma despite actual uplift of 800 m/Ma because of updoming of the closure temperature zone (CT). Uplift rate of CT is arbitrarily shown as half the tectonic uplift rate.

Thermal doming, to a degree depending on the relationship between rates of uplift and erosion, would be counteracted by the cooling effect of erosion. Clark and Jäger (1969, p. 1157) calculated that in the uplift of the Alps erosion accounts for 30% to 50% of the present heatflow. Thus the doming of the isotherms would tend to lag behind tectonic uplift. The analogy may be drawn to the more subdued profile of the water table compared to land surface at a given time.

Despite the inferred uplift rate of 0.8 km/Ma for the Carmen block (Fig. 3), a sample located 0.8 km vertically below any arbitrary sample, say LB-71-VE-2 (hereafter VE-2), could not be brought through the closure temperature 1 Ma after VE-2 passed through it because of the doming of the isotherms, as shown in Figure 7. Clearly, apatite ages may not register the true rate of uplift.

VE-2 is estimated to have been originally about 1.5 km below the unconformity at the base of the Cretaceous-Eocene cover section of about 2.0 km thickness, the top of which was close to sea level as evidenced by deltaic and coal swamp environments in the Eocene, that is, at a depth of about 3.5 km below sea level. It is now about 4,000 m above sea level. Evidently doming of isotherms has occurred contemporaneously with their overall retreat down stratigraphic section. The uplift history as recorded by the apatite clocks reflects a composite of these two effects as shown in Figure 8, which provides a qualitative framework for reconciling uplift rates based on apatite ages on samples of wide lateral distribution from those based on samples distributed vertically. Even where tectonic uplift and the consequent thermal doming is pronounced, local retreat and embayment of isotherms might be expected where deeply incised valleys formed. This effect would presumably become regional where the rate of erosion exceeds that of uplift. Hence the concept of thermal doming as developed here is not necessarily in contradiction to the findings of Birch (1950) in the Front Range, Colorado. He noted

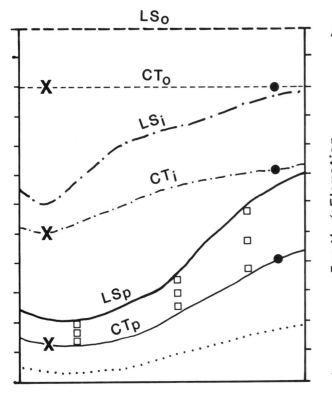

Horizontal distribution of samples

Figure 8. Qualitative model for the evolution of the closure temperature isotherm in response to a combination of rapid tectonic uplift and concomitant erosion. Topographic evolution is reconstructed in the manner of Figure 6. $LS_{o, i, p}$, are original, intermediate, and present land surfaces, respectively; $CT_{o, i, p}$ are original, intermediate, and present closure temperature isotherms, respectively. Dotted line at bottom is additional, higher-temperature, present-day isotherm. Initial doming of isotherms (assuming rapid uplift) is to a degree differentially counteracted by downward cooling related to rate of erosion. Following Birch (1950), isotherms are shown to reflect higher gradients under valleys, lower under ridges. The sense of changing gradients with time may be inferred by reference to the arbitrary depth/elevation units. A suite of specimens (X) collected through time would have yielded a very different rate-of-uplift curve compared to the suite of filled circles. Hence the uplift rate for the Carmen samples (Fig. 3) of wide lateral distribution along LS_P could represent a composite of exaggeratedly fast and slow uplift rates. A better basis for calculating true uplift rate would be to compare ages on samples shown as open boxes where local topography or bore holes permit sampling close to the vertical.

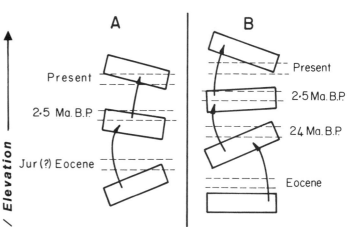

Figure 9. Possible explanations of the two widely differing ages on the apatites of the Valera pluton. A. Age of 24 Ma reflects partial annealing of pre–Late Jurassic age by long-term depression in the zone of partial track annealing (paired dashed lines) followed by rotational uplift in the Pliocene. B. Initial uplift of the northeastern part in the Oligocene followed by rotational uplift of the southwestern part in the Pliocene.

(p. 582) that the vertical gradient of temperature is lower under peaks than under valleys. This is shown qualitatively in Figure 8.

The plot of the data for the Carmen block in Figure 3 shows that age is a function of elevation in this area, that is, uplift occurred at a constant rate. Lacking knowledge of current heat flow, thermal history, and topographic evolution, we are unable to show whether the apparent uplift rate of 0.8 km/Ma is the true rate or to some degree greater or lesser than the true rate. One way to determine the true rate might be to date apatite samples collected as near as possible vertically at two or more sites located at different elevations along the Carmen block. This may be

easier said than done in the light of the results from three pairs of samples which come closest to a vertical distribution among those measured. Map location samples 2 and 3 differ in elevation by about 320 m and in age (ignoring precision of measurement) by 0.7 Ma giving an uplift rate of about 460 m/Ma. Corresponding rate for map samples 4 and 5 (data in Tables 2 and 3) is about 860 m/Ma, and for samples 6 and 7, about 1,250 m/Ma. Clearly, the precision of the measurements is insufficient for calculating uplift rate for samples of relatively narrow vertical separation. Better age-measurement statistics on several samples over wider vertical range may help to overcome this problem.

Until such a program can be undertaken it is concluded that some degree of thermal doming probably occurred in Andean uplift as a consequence of which the time span of uplift was probably greater than that shown by the actual maximum time range measured (cf. Fig. 6).

Some additional insight into the mechanics of uplift comes from consideration of the two apatite ages on the Valera pluton. The elevation-age relationships are reversed with respect to the Carmen plot, and the age range is greater by a factor of about 6. Two possible explanations may be proposed to explain the ages measured (Fig. 9). One possibility (a) is that the Oligocene age may be viewed as a partly annealed age inherited from pre–La Quinta time. Alternatively (b) there was latest Oligocene uplift of the northeast end of the pluton followed by rotation and uplift of the southwestern part in the Late Pliocene. Further discussion must await the measurement of an additional seven samples of Valera granite and contact rocks recently collected for that purpose.

All the remaining rock bodies are represented by single measurements only; useful conclusions as to the conditions of uplift can be drawn from only one (Estánquez, VE-27; 12 on Fig. 1), which at 1.4 Ma was the youngest apatite age measured. This

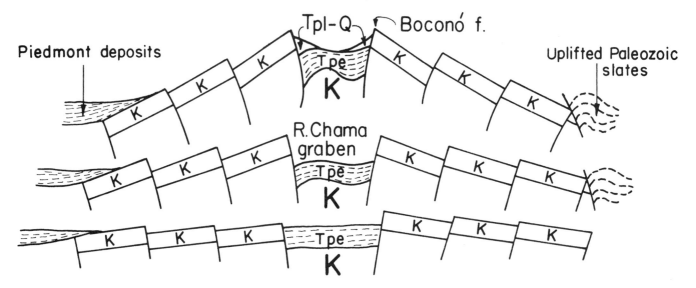

Figure 10. Schematic outline of Tertiary uplift history as originally viewed by Shagam (1972b). Uplift and tilting related to compression results in modest crustal shortening and in mild folding in graben areas separating regions of opposite tilt. Modified after Shagam (1972b, Fig. 5). K = Cretaceous strata; Tpe = Paleocene-Eocene strata; Tpl-Q = Pliocene-Quaternary strata.

sample was collected at an elevation of ~450 m on the pluton that is located in the major longitudinal graben immediately west of the set of faults and drape folds considered to mark the western limit of the Culata block (and possibly the Carmen block too). The Estánquez age does not conform with the slope of the Carmen elevation–age plot. Applying the Carmen uplift curve at face value to the Estánquez mass (i.e., 0.8 km/Ma uplift) implies that the closure temperature isotherm is presently about 2 km below the surface at Estánquez and, in turn, a geothermal gradient of about 40 to 45 °C/km, which borders on the improbable. If, on the other hand, one postulates that the isotherms are not up-domed by uplift the closure isotherm is now closer to about 4 km depth, and uplift of the sample occurred at a rate of about 3 km/Ma, which may be theoretically possible but is difficult to reconcile with the local Pleistocene geology. An explanation of the age on the basis of significant updoming of isotherms combined with heat loss by erosion in the framework of Figure 8 is considered a more rational possibility. This is supported by evidence for about 800 m of vertical movement in the late Pleistocene, inferred by Giegengack and others (1976), in an area immediately adjoining the graben on the north side, about 30 km southwest of the Estánquez pluton. It will take several additional measurements to resolve the uplift history in this region.

On the basis of local geologic relationships, the single ages are inferred to fall on uplift curves of the same sense of slope as the Carmen curve, but no significance attaches to the slope and time span of the dashed lines drawn through those points on Figure 3. They are intended merely to emphasize the probability that such individual sample ages are transient points in an uplift picture rather than the complete event itself. Because of modest topographic relief, it is unlikely that meaningful uplift curves can

be constructed for most of those plutons, with the exception of the Culata batholith (see additional comments below). Nonetheless, the single ages do provide useful information pertinent to the interpretation of the overall uplift history of the Venezuelan Andes as outlined in the subsequent section.

Tectonic Implications of the Apatite Data

Shagam (1972b, 1976) suggested that Tertiary uplift required only modest compression resulting in upward and outward tilting of fault blocks with areas of opposing tilt separated by a subsiding block or graben as shown in Figure 10. Because the Paleogene section is restricted to graben areas, it was suggested that initial uplift began at the close of the Cretaceous (see Fig. 2) so that only the negative graben areas could become sites of Paleogene deposition. In recent years evidence has accumulated suggesting that the structural/tectonic framework of "Andean" uplift was more complex and involved considerably more compression than suggested in Shagam's original models. This evidence may be summarized as follows:

1. A klippe of pre-Mesozoic Tostós schists overlying Miocene section was mapped by DeRatmiróff (1971) and by García and others (1977) in the area extending southwest from Mesa Bolívar. The schists abut the north side of the graben in the Sta Cruz de Mora–Estánquez area.

2. Near Jají, Shagam (1972a, see E–E′ in plate of cross-sections) mapped a major overturned fold in the Upper Cretaceous section. The orientation of the fold suggests a stress field in conformity with that required for the emplacement of the Tostós klippe. The folding, becoming progressively less intense, continues to the northwest where it affects the Eocene section.

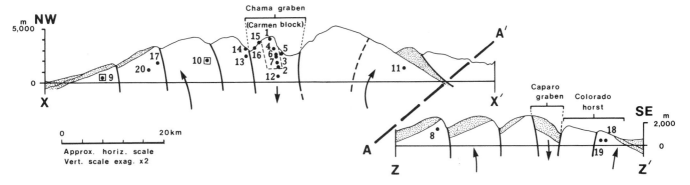

Figure 11. Diagrammatic cross-sections of the central (X-X') and southwestern (Z-Z') parts of Figure 1. Dashed line A-A' (see Fig. 1 for location) represents approximate boundary separating northern (Chama) and southern (Caparo) graben systems, each with its associated tectonic blocks of opposite tilt. Arrows symbolize uplift of blocks and relative subsidence of graben. Apatite samples dated by the fission-track method have been extrapolated into the cross-sections (map location numbers; cf. Figs. 1 and 4, and Table 3). Because the Chama graben and its extensions are oblique to the trend of the ranges, the representation of some samples is more nearly accurate geographically than structurally. For example, the two Valera samples (9, 10) indicated by boxes actually occur in a single block (cf. Fig. 1). Dashed line defines the limits of exposure of El Carmen granodiorite pluton; dotted areas represent cover section (cf. Table 2); blank areas are mainly pre-Mesozoic rocks.

3. A remarkable breccia zone of late Miocene–Pliocene(?) age has been defined by Giegengack (1984) in the graben along the north side of the Rio Chama valley extending at least from Estánquez northeast beyond Mérida. The zone is several kilometres wide and in the area north of San Juan appears to constitute a subhorizontal zone. The implication is that of a major overthrust resembling the Tostós klippe in movement geometry.

4. The southeastern border of the Sierra Don Pedro Cretaceous tilt block (south of El Morro in the central Andes) is defined by a thrust fault mapped by Shagam (1972a, see cross-section F–F').

5. The gravity measurements of Bonini and others (1980) strongly suggest that the mass of the Andes has been moved about 25 km northwest over the southern margin of the Maracaibo Basin section.

The pattern of graben and associated tilt blocks is complex; that shown in the cross-section of Figure 11 is diagrammatic but is a not unreasonable simplification of the relationships in the central part of Figure 1. Locations of the apatite samples with respect to the structural pattern are shown in Figure 11, and plainly there is profound statistical imbalance in the representation of different structural blocks. Tectonic interpretation of the data is correspondingly tentative.

In the original Tertiary uplift model (Fig. 10), the grabens were viewed as originally formed at the close of the Cretaceous and maintained, becoming progressively more appressed, until the present. The premise of original formation at the close of the Cretaceous still appears reasonable; the principal evidence in this regard is not so much the occurrence of Eocene section conformably over the Cretaceous in the graben areas as the *absence* of Eocene section on the Cretaceous in the areas of outward tilting. The obvious deficiency in the original model is that maintenance

of the graben throughout the Tertiary should be reflected in a complete Tertiary fill sequence liberally interspersed with angular unconformities representative of uplift phases. In reality, the present graben fill is composed almost entirely of terrace gravels of probable Pleistocene age. The section intermediate between the Eocene and the Pleistocene deposits has been identified locally only in the Mucuchíes area where Grauch (1971) mapped steeply tilted gravels unconformably overlain by Pleistocene gravels. Pollen analysis of a sample from the tilted gravel (Grauch, 1971) yielded an uncertain age in the Miocene-Pliocene range. It is concluded, therefore, that the original graben sediment trap was filled at the close of the Eocene and that renewal of the graben as a sediment trap did not occur until well into the Neogene. In the vicinity of Mérida the two graben phases were approximately superimposed in space, but Tertiary deposits are so few and scattered that it is not known whether such superimposition was characteristic for the entire Andes. It is suggested in succeeding discussion that a portion of the northern graben complex (e.g., the El Aguila–Valera branch) was only instituted in the younger event.

One significant line of evidence bearing on the younger phase of graben formation stems from the work of Weingarten (1977a, 1977b) who identified thick oxisols at elevations exceeding 3,000 m in areas adjacent to the Rio Chama graben on the south side. The chemistry and mineralogy of these deposits along with geomorphic considerations support the hypothesis that they were formed closer to sea level. The preservation of thick deposits on ridge crests at high elevations suggests relatively recent major uplift. Late rather than early Neogene age for the start of the younger phase of uplift is a reasonable inference.

In the elevation-age plot of Figure 3 the dashed line X–X' arbitrarily separates closely grouped younger uplift age curves from a wider range of older uplift curves. This inference might be challenged on the grounds that brief initial uplift sufficient to trigger the apatite clocks may have been followed by stillstand and, in turn, a major phase of uplift contemporaneous with the younger uplift events. Indeed, Shagam and others (1984) strongly urge such an uplift history for the Santander Massif. In the case of the Venezuelan Andes the same effect of stepped uplift phases can reasonably be inferred, but geomorphic evidence suggests that the major share of the uplift occurred in the initial phase.

A broad correlation of uplift age group vis-à-vis tectonic setting is evident. The young ages are associated with the northern graben system (Estánquez, Carmen, Chachopo, Timotes, and south Valera). Two other young ages are on plutons (Canaguá and Soledad) located along regional strike from each other and, although not in identical structural settings, close to a line separating the two major graben systems and their associated tilt blocks (see Fig. 11). The older ages occur near the leading (Quinta and north Valera) and trailing (Tapo-Cambur and Quiú) margins of the Andes, characterized by outward (respectively, north and south) tilt blocks.

Exceptions to these generalizations are the Culata sample (a young age near the leading margin) and north Valera sample (which can be viewed as an old age in a branch of the graben system). Integration of the fission-track data with the geologic data outlined previously suggests that the sequence of events in the Tertiary uplift history of the Venezuelan Andes might have been as follows (see Fig. 12):

1. Initial uplift at the close of the Cretaceous or early Paleocene that resulted in the formation of principally two longitudinal graben which were filled with Eocene sediment. The uplift was sufficient to make the Andes positive but not enough to bring the crystalline rocks through the closure temperature for apatite.

2. Depending on the significance of the old Valera age (see Fig. 9 and related discussion), overthrusting may be viewed as having begun in the Valera area in the late Oligocene and extended southwest to the Quinta volcanic area by the mid-Miocene; alternatively, the start of the northwest thrusting of the Andes over the Maracaibo Basin is represented by the Quinta volcanic sample only and began in the mid-Miocene. Bonini and others (1980) did not specify the age range of thrusting inferred from gravity data. Kellogg (1981) postulated that phases of thrusting of similar sense in the Sierra de Perijá (northwest of the Andes and the Maracaibo Basin) occurred in pre- and post-Miocene time. There is no overt evidence in the Andes to enable us to distinguish between the possibilities noted above. In this phase of thrusting, graben formation was not renewed or initiated.

At this time the southern part of the Valera granite was not brought through the closure temperature. The Pico El Aguila--Valera branch of the graben system did not then exist. The Culata age of 3 Ma poses a problem in the tectonic framework outlined above. One might have expected, for example, that the Culata curve would fall in some intermediate age bracket between the northern Valera and the Quinta uplift curves or at least resemble the Quinta curve. This is clearly not the case. An explanation based on early uplift followed by subsidence and annealing of the age prior to a second young phase of uplift is not readily consistent with the concept of regional thrusting of the entire Andean mass. It is possible that the sole sample representative of the biggest granitic body in the Andes was fortuitously collected in a local graben and that detailed regional sampling of the batholith will reveal an older uplift curve.

3. Then began a phase of uplift of the trailing margin in the late Miocene. This is interpreted to have been caused by underthrusting from the southeast in the course of northwest tectonic transport when the frontal thrust became locked in place because of frictional resistance or related cause. The precision of the Quiú and Tapo-Cambur ages overlap so that two distinct phases of uplift did not necessarily occur.

4. The locking in place of the leading margin combined with underthrusting of the trailing edge then subjected the intervening mass of the Andes to intense compression that resulted in uplift of the Canagua and Soledad slivers between the anvils of opposed tilt block regimes outlined previously. At the same time the northern graben system underwent major rejuvenation and parts such as the Valera branch may only then have been initiated. If thermal doming occurred (see Fig. 6 and related discussion), it is reasonable to conclude that uplift of the graben system began some time before the oldest Carmen ages measured. Initial formation of that system probably overlapped with the waning stages of underthrusting of the trailing margin. The northern graben system may have begun to exist as a sediment trap in the late Miocene, but as the regional stresses increased, both the graben and its bordering blocks were uplifted. In that uplift the Miocene(?)-Pliocene gravels at Mucuchíes were tilted, and the features indicative of intense compression, previously listed, were produced.

The presence of thin young gravel deposits (probably Pleistocene) in the Rio Caparo graben indicates that the graben was also uplifted, but the topography suggests that the effect was minimal compared to the events affecting the northern graben system.

The uplift of the Carmen block has been interpreted in terms of uniform rate of uplift, which is not unreasonable in view of the close approximation of the data to a straight-line curve. The Estánquez age was interpreted as possibly related to cooling in the erosional component of uplift. It is possible, however, that the late stages of uplift occurred in one or more surges rather than as a uniform-rate process. The principal geologic indication for stepwise uplift is the fact that present fill of the northern graben system is by far dominated by gravels tentatively dated as of Pleistocene age. It will take detailed study of these deposits (in some localities attaining ~400 m thickness) together with further apatite age measurements to resolve the details of the final uplift stage.

The Venezuelan Andes constitute but one of several major tectonic elements in the region extending northwest to the Santa Marta block of northeastern Colombia. The Tertiary tectonic setting of the Andes in that region is discussed by Shagam and

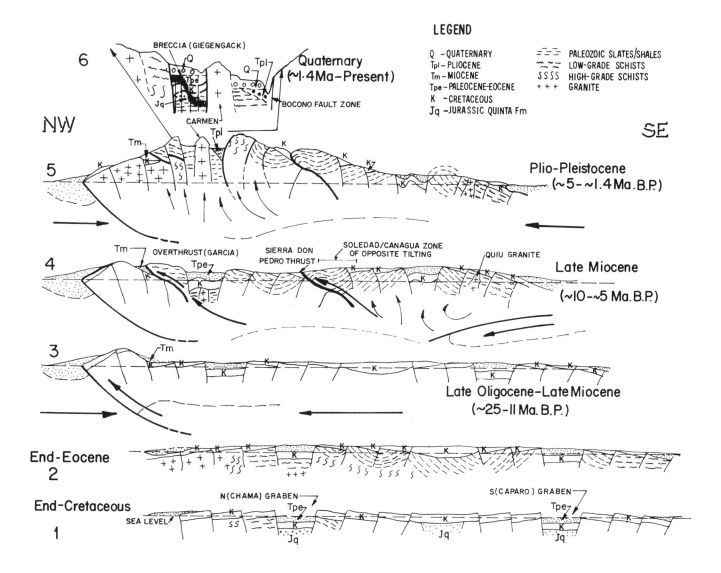

Figure 12. Schematic sections (composites of regional stratigraphic and structural relationships) summarizing the Tertiary uplift history of the Venezuelan Andes. 1. Latest Cretaceous: gentle uplift resulting in tilted blocks and graben formation but insufficient to set basement apatite clocks; Bailadores (23) zircon clock begins to register, and possibly Chachopo (15) and La Culata (17) too. 2. Latest Eocene: smoothing of topography as grabens are filled and minor erosion occurs. 3. Late Oligocene–late Miocene: northeast thrusting of the Andes over the Maracaibo Basin results in frontal uplift which sets the apatite clocks of part of the Valera pluton and the Quinta volcanics. 4. Late Miocene: locking in place of the frontal thrust results, with continued compression, in underthrusting of the trailing (southeastern) margin, which in turn results in the Tostós (García) and Sierra Don Pedro thrusts, and to uplift of the trailing margin where the Quiú and Tapo-Cambur apatite clocks begin to register. 5. Pliocene-Pleistocene: the effects of continued compression are transferred to the north-central region where rapid major uplift occurs with renewal of a graben system and formation of the Giegengack breccia. The Canaguá-Soledad clocks are set followed shortly by those of the Carmen and related bodies in the northern graben system. 6. Quaternary: the graben is filled with coarse gravels and depression of the closure temperature zone because of rapid erosion leads to setting of the Estánquez clock. For simplicity, Quinta red beds (Jq) have been eliminated in sections 2–5.

others (1984) following presentation of the fission-track ages for Toas Island and the Perijá Mountains of Venezuela and for the Santander Massif of the Cordillera Oriental of Colombia.

Zircon Ages and Their Interpretation

Fission-track ages measured on Andean zircons are listed in Table 3. The ages range from the Late Jurassic to the late Paleocene, but one may readily distinguish the three Jurassic ages on granite and schist of the Colorado block from the remaining ages of Cretaceous-Paleocene range distributed along the length of the Andes north of the Rio Caparo. In the latter group, 11 of the 18 ages fall in the range of 81 to 113 Ma.

The zircon ages may be interpreted by one or a combination of the following possibilities:

1. Cooling ages following original crystallization of volcanic or plutonic rocks, or a metamorphic event. In the outline of the geological history (see Fig. 2 and Table 1), it was noted that the last plutonic event in the Colorado block occurred in the early Devonian and in the Andes north of the Rio Caparo at the close of the Permian. Clearly, the zircon ages do not represent cooling ages of the plutonics following original crystallization.

An intriguing possibility is posed by the Quinta volcanic sample (S-87-W; 20 on Fig. 1) on which Burkley (1976) obtained a U-Pb age of about 229 Ma. Schubert and others (1979), who obtained K-Ar ages on the same volcanics of about 149 Ma (biotite) and 123 Ma (whole rock), referred to tuffs in the overlying Quinta red beds, raising the possibility that volcanism (and therefore high temperatures and heat flow locally) continued into latest Jurassic–earliest Cretaceous times. This would explain the late registry of the K-Ar clocks and suggests the possibility that the zircon fission-track clocks were only activated when, in the course of the subsequent marine transgression of the Cretaceous, the closure isotherm overtook the cooling volcanic pile in the course of *subsidence*. Such a concept is in marked contrast to that of cooling by *uplift* and erosion proposed in explanation of the apatite ages. We think this a remote possibility because of the long time interval (about 60 Ma) that elapsed between the registry of the K-Ar and fission-track clocks. Also, Shagam examined La Quinta red beds in the field and thin section and concluded that volcanic components are of reworked sedimentary rather than primary tuffaceous origin. This suggests that heating related to the volcanism had ceased well prior to the time of sedimentation of the red beds. Cooling is more likely to have been a cause of basin subsidence than to have occurred toward the close of the phase of sedimentation.

2. Cooling ages related to a thermal event subsequent to the event of original crystallization and cooling. The fact that most of the ages in the region north of the Rio Caparo are closely grouped in the range of 81 to 113 Ma raises the possibility that there was a mild thermal event at that time insufficient to produce igneous products or metamorphic overprint but sufficient to cause partial annealing of original cooling ages. We do not favor this possibility because the time range corresponds to that of subsidence and

sedimentation of the Cretaceous section and the inference of regional cooling rather than heating.

3. Uplift ages, as in the case of apatite, signifying the time when the sample was raised through the closure temperature for zircon. That temperature (see discussion in the Introduction) corresponds to a depth in the approximate range of 6 to 8 km, depending on the geothermal gradient. Maximum estimates of cover (see Table 2) fall well short of that range with the exception of the Bailadores sample (23) and possibly for the three samples from the Colorado block. Even assuming a high thermal gradient and complete annealing of original ages one would expect resetting of the clocks in the course of uplift and erosion to register Paleocene or Oligocene ages as Andean uplift proceeded in the framework discussed for the apatite ages. Only the Bailadores sample fully conforms with that framework; two other samples Chachopo (15) and La Culata (17) show ages that could be interpreted as dating uplift, although there is considerable uncertainty regarding their depth of burial. All the other ages north of the Rio Caparo, as noted, correspond to the time when the region was clearly still subsiding. In the case of the specimens from the Colorado block it is barely conceivable that the cover consisted mainly of the Precambrian and lower Paleozoic sections themselves and that they underwent intense imbricate thrusting leading to uplift and erosion at and following the close of the Permian, thereby causing the zircon clocks to register at about the time of the start of sedimentation of the Quinta red beds. However, the fact that remnants of upper Paleozoic section are still preserved argues against such a profound event of erosion in Permian-Jurassic time.

4. Mixed ages resulting when a clock set in an original cooling event is depressed into the intermediate zone of partial annealing and subsequently uplifted. The question here is again whether thickness of cover and geothermal gradient were adequate to depress samples into that zone. In the light of the geologic history, the effect is likely to have been at a maximum at the close of the Cretaceous or the Eocene when thickness of cover was at a maximum. The resulting ages would not reflect a thermal event of specific age but merely the degree of partial annealing. This appears to us to be the most rational framework in which to explain the bulk of the zircon ages. The weakness in this hypothesis is the lack of evidence for adequate thickness of cover rocks (see Table 1) to depress the rocks to depths presumably in the range of 6 to 8 km for partial annealing. This problem is further complicated by the inferred low geothermal gradient during the time of basin subsidence and sedimentation. The solution to this problem may reside in the definition of the partial annealing temperature for zircon which has been defined entirely on the basis of laboratory studies. Conceivably the process is kinetic, and the partial annealing temperature is inversely related to residence time so that for periods of time of the order of 10^7 to 10^8 Ma far thinner cover section might suffice to depress samples into the zone of partial annealing, as shown for apatite from drill holes (Naeser, 1979, 1981; Gleadow and Duddy, 1981) compared to laboratory-derived annealing curves.

In summary, we consider that the Bailadores sample (60 Ma) and probably the Culata (61 Ma) and Chachopo (72 Ma) samples represent uplift ages. The three Jurassic ages from the Colorado block may also be uplift ages but are more likely to represent mixed ages as are the remaining samples north of the Rio Caparo. The Bailadores sample comes from a locality south of the northern graben system and the Paleocene age for this sample at least is in conformity with the suggestion that the initial phase of Andean uplift occurred at the close of the Cretaceous in areas outside the graben.

Sphene Ages and Their Interpretation

Only two sphene ages are available (Table 3), one on the Quiú granite in the Colorado block and the other on the Carmen pluton near Mérida. The first is concordant with the zircon age on the same sample, which may be interpreted as a measure of support for the analytical precision although it does not assist the overall problem of interpretation discussed above. Little can be said about the single Carmen sphene age, but its age-elevation relationships compared to other ages on the Carmen pluton (see Tables 1 and 3) suggest that additional zircon and sphene ages along the length of the pluton may provide valuable information concerning the depth to and history of evolution of the intermediate zone of partial annealing.

As in the case for the apatite ages, the regional implications of the zircon and sphene ages presented above are discussed by Shagam and others (1984) following presentation of the corresponding data for the region northwest of the Andes.

CONCLUDING REMARKS

It appears reasonable to claim, in the light of the data presented, that the fission-track dating method provides a valuable tool for the study of the thermal history and tectonism particularly of regions devoid of contemporary igneous/metamorphic events. On the other hand, there is clearly a need to tackle the problems of thermal doming and partial annealing vis-à-vis the apatite and zircon-sphene ages, respectively.

ACKNOWLEDGMENTS

We are deeply indebted to H. Feldstein for carefully supervising the irradiation of our samples at the Nahal Soreq Nuclear Research Center, Israel. The cost of sample collection was borne by the Dirección de Geología, Ministerio de Energía y Minas, Venezuela. Our thanks are extended to A. Bellízzia and R. García, respectively the former and present director, and to their colleagues for logistic assistance rendered.

We have profited from discussions with M. Eyal (Ben-Gurion University), R. F. Giegengack (University of Pennsylvania) and W. E. Bonini, R. B. Hargraves, J. N. Kellogg, and W. Maze (all of Princeton University), and we thank two reviewers (C. W. Naeser and T. M. Lutz) for their notably constructive criticism.

Most laboratory procedures were done in facilities provided by the Department of Geology and Mineralogy, Ben-Gurion University, Israel. This study was partly supported by a National Science Foundation grant (GA10129) to Banks and Shagam at the Department of Geological Sciences, Case Western Reserve University. Shagam further expresses appreciation to S. M. Savin and the administrative staff at Case Western Reserve University for generous hospitality, services, and space provided yearly en route to Israel from Venezuela. We are grateful to C. P. Hughes and his coeditors of *Geological Magazine* who generously permitted the withdrawal of this contribution for publication in this volume.

Facilities provided to Shagam by the Department of Geological and Geophysical Sciences, Princeton University, especially the secretarial assistance of K. Bilous and J. Bialkowski and the re-drafting of some figures, are gratefully acknowledged.

REFERENCES CITED

Bellízzia, G. A., Pimentel, M. N., and Bajo, O. R., compilers, 1976, Mapa geológico estructural de Venezuela: Venezuela Ministerio de Minas e Hidrocarburos, scale 1:500,000.

Birch, F., 1950, Flow of heat in the Front Range, Colorado: Geological Society of America Bulletin, v. 61, p. 567-630.

Bonini, W. E., Garing, J. D., and Kellogg, J. N., 1980, Late Cenozoic uplifts of the Maracaibo–Santa Marta Block, slow subduction of the Caribbean plate and results from a gravity study, in Transactions, Carribean Geological Conference, IX, Santo Domingo, 1980: Universidad Católica Madre y Maestra, Santiago de los Caballeros, Républica Dominicana, v. 1, p. 99–105, (received 1983).

Burkley, L. A., 1976, Geochronology of the central Venezuelan Andes [Ph.D. thesis]: Cleveland, Case Western Reserve University, 150 p.

Carpenter, B. S., and Reimer, G. M., 1974, Calibrated glass standards for fission track use: National Bureau of Standards Publication 260-349, 17 p.

Clark, S. P., Jr., 1979, Thermal models of the central Alps, in Jäger, E., and Hunziker, J. C., eds., Lectures in isotope geology: Berlin, Heidelberg, New York, Springer Verlag, p. 225–230.

Clark, S. P., Jr., and Jäger, E., 1969, Denudation rate in the Alps from geochronologic and heat flow data: American Journal of Science v. 267, p. 1143–1160.

DeRatmiróff, G., 1971, Late Cenozoic imbricate thrusting in Venezuelan Andes: American Association of Petroleum Geologists Bulletin, v. 55, p. 1336–1344.

García, R., Rondón, F., and Canelón, G., 1977, Mapa geológico de la región de Tovar-Guaraque-Mesa Bolívar-Bailadores, Estados Mérida y Táchira: Dirección de Geología, Ministerio de Minas e Hidrocarburos, Caracas, Venezuela.

Giegengack, R. F., 1984, Late-Cenozoic tectonic environments of the Central Andes, in Bonini, W. E., and others, eds., The Caribbean–South American plate boundary and regional tectonics: Geological Society of America Memoir (this volume).

Giegengeck, R. F., Grauch, R. F., and Shagam, R., 1976, Geometry of late Cenozoic displacement along the Boconó fault, Venezuelan Andes, in Memoria, Congreso Latinoamericano de Geología, II, Caracas, 1973: Venezuela Ministerio de Minas e Hidrocarburos, Boletín de Geología, Publicación

Especial no. 7, v. 2, p. 1201–1225.

Gleadow, A.J.W., 1978, Anisotropic and variable track etching characteristics in natural sphenes: Nuclear Track Detection, v. 2, p. 105–117.

——— 1981, Fission track dating methods: What are the real alternatives?: Nuclear Tracks, v. 5, p. 3–14.

Gleadow, A.J.W., and Brooks, C. K., 1979, Fission-track dating, thermal histories and tectonics of igneous intrusions in east Greenland: Contributions to Mineralogy and Petrology, v. 71, p. 45–60.

Gleadow, A.J.W., and Duddy, I. R., 1981, A natural long term experiment for apatite: Nuclear Tracks, v. 5, p. 169–174.

Gleadow, A.J.W., and Lovering, J. F., 1977, Geometry factor for external detectors in fission track dating: Nuclear Track Detection, v. 1, p. 99–106.

——— 1978, Thermal history of granitic rocks from western Victoria: A fission-track dating study: Geological Society of Australia Journal, v. 25, p. 323–340.

Gleadow, A.J.W., Hurford, A. J., and Quaife, R., 1976, Fission track dating of zircons: Improved etching techniques: Earth and Planetary Science Letters, v. 33, p. 273–276.

Grauch, R. I., 1971, Geology of the Sierra Nevada south of Mucuchíes, Venezuelan Andes: An aluminimum-silicate-bearing metamorphic terrain [Ph.D. thesis]: Philadelphia, University of Pennsylvania, 180 p.

Haack, U., 1977, The closing temperature for fission track retention in minerals: American Journal of Science, v. 277, p. 459–464.

Harrison, M. T., Armstrong, R. L., Naeser, C. W., and Harakal, J. E., 1979, Geochronology and thermal history of the coast plutonic complex, near Prince Rupert, British Columbia: Canadian Journal of Earth Science, v. 16, p. 400–410.

Kellogg, J. N., 1981, Cenozoic basement tectonics of the Sierra de Perijá, Venezuela and Colombia, *in* Transactions, Caribbean Geological Conference, IX, Santo Domingo, 1980: Universidad Católica Madre y Maestra, Santiago de los Caballeros, República Dominicana, p. 107–117.

Martín Bellízzia, C., 1968, Edades isotópicas de rocas Venezolanas: Venezuela Ministerio de Minas e Hidrocarburos (Caracas), Boletín de Geología, v. 10, p. 356–379.

Naeser, C. W., 1976, Fission track dating: U.S. Geological Survey, Open-File Report, 76-190.

——— 1979, Fission track dating and geologic annealing of fission tracks, *in* Jäger, E., and Hunziker, J. C., eds., Lectures in isotope geology: Berlin, Heidelberg, New York, Springer-Verlag, p. 154–169.

——— 1981, The fading of fission tracks in the geologic environment: Data from deep drillholes: Nuclear Tracks, v. 5, p. 248–250.

Naeser, C. W., and Faul, H., 1969, Fission track annealing in apatite and sphene: Journal of Geophysical Research, v. 74, p. 705–710.

Naeser, C. W., and McKee, E. H., 1970, Fission-track and K-Ar ages of Tertiary ash-flow tuffs, north central Nevada: Geological Society of America Bulletin, v. 81, p. 3375–3384.

Olmeta, M. A., 1968, Determinación de edades radiométricos en rocas de Venezuela y su procedimiento por el método K-Ar: Venezuela Ministerio de Minas e Hidrocarburos, Boletín de Geología, v. 10, p. 339–344.

Renz, O., 1977, The lithological units of the Cretaceous in western Venezuela, *in* Memoria, Congreso Geológico Venezolano, V, Caracas, v. 1: Venezuela Ministerio de Energía y Minas, p. 45–58.

Schubert, C., 1969, Geologic structure of a part of the Barinas mountain front, Venezuelan Andes: Geological Society of America Bulletin, v. 80, p. 443–458.

Schubert, C., Sifontes, R. S., Padron, V. E., Velez, J. R., and Loaiza, P. A., 1979, Formación La Quinta (Jurásico) Andes Merideños: Geología de la sección tipo: Acta Ciencia Venezolana, v. 30, p. 42–55.

Shagam, R., 1972a, Geología de los Andes centrales de Venezuela, *in* Memoria, Congreso Geológico Venezolano, IV, Caracas, 1969: Venezuela Ministerio de Minas e Hidrocarburos, Boletín de Geología, Publicación Especial no. 5, v. 2, p. 935–938.

——— 1972b, Andean research project, Venezuela: Principal data and tectonic implications: Geological Society of America Memoir 132, p. 449–463.

——— 1976, The northern termination of the Andes, *in* Nairn, A.E.M., and Stehli, F. G., eds., The ocean basins and margins, Volume 3, The Gulf of Mexico and the Carribbean: New York, Plenum Press, p. 325–420.

Shagam, R., Kohn, B. P., Banks, P. O., and Dasch, L. E., and others, 1984, Tectonic implications of Cretaceous-Pliocene fission-track ages from rocks of the circum-Maracaibo Basin region of western Venezuela and eastern Colombia, in Bonini, W. E., and others, eds., The Caribbean–South American plate boundary and regional tectonics: Geological Society of America Memoir 162 (this volume).

Steven, T. A., Mehnert, H. H., and Obradovich, J. D., 1967, Age of volcanic activity in the San Juan Mountains, Colorado: U.S. Geological Survey Professional Paper 575-D, p. D47–D55.

Wagner, G. A., 1968, Fission-track dating of apatites: Earth and Planetary Science Letters, v. 4, p. 411–415.

——— 1979, Correction and interpretation of fission-track ages, *in* Jäger, E., and Hunziker, J. C., eds., Lectures in isotope geology: New York, Berlin, Heidelberg, Springer-Verlag, p. 170–177.

Wagner, G. A., and Reimer, G. M., 1972, Fission track tectonics: The tectonic interpretation of fission track apatite ages: Earth and Planetary Science Letters, v. 14, p. 263–268.

Wagner, G. A., Reimer, G. M., and Jäger, E., 1977, Cooling ages derived by apatite fission-track, mica Rb-Sr and K-Ar dating: The uplift and cooling history of the central Alps: Memorie degli Istituti di Geologia e Mineralogia dell' Università di Padova, v. 30, p. 1–27.

Weingarten, B., 1977a, Tectonic and paleoclimatic significance of a late Cenozoic paleosol from the central Andes, Venezuela [M.Sc. thesis]: Philadelphia, University of Pennsylvania, p. 67.

——— 1977b, Tectonic and paleoclimatic significance of a late Cenozoic red-earth paleosol from the central Andes, Venezuela, *in* Transactions, Carribean Geological Conference, VIII, Curacao, 1977: C.O. Geologisch Instituut, Netherlands, p. 221 (abstracts).

Zambrano, E., Vásquez, E., Duval, B., Latreille, M., and Coffinières, B., 1971, Síntesis paleográfica y petrolera del occidente de Venezuela, *in* Memoria, Congreso Geológico Venezolano, IV, Caracas, 1969: Venezuela Ministerio de Minas e Hidrocarburos, Boletín de Geología, Publicación Especial no. 5, v. 1, p. 483–552.

Zeitler, P. K., Johnson, N. M., Naeser, C. W., and Tahirkheli, Rashid, A. K., 1982a, Fission-track evidence for Quaternary uplift of the Nanga Parbat region, Pakistan: Nature, v. 298, p. 255–257.

Zeitler, P. K., Tahirkheli, Rashid, A. K., Naeser, C. W., and Johnson, N. M., 1982b, Unroofing history of a suture zone in the Himalaya of Pakistan by means of fission-track annealing ages: Earth and Planetary Science Letters, v. 57, p. 227–240.

MANUSCRIPT ACCEPTED BY THE SOCIETY SEPTEMBER 1, 1983

Geological Society of America
Memoir 162
1984

Tectonic implications of Cretaceous-Pliocene fission-track ages from rocks of the circum–Maracaibo Basin region of western Venezuela and eastern Colombia

R. Shagam
B. P. Kohn
Department of Geology and Mineralogy
Ben-Gurion University of the Negev
Beer Sheva, Israel 84120

P. O. Banks
L. E. Dasch
Case Western Reserve University
Cleveland, Ohio 44106

R. Vargas
G. I. Rodríguez
Ingeominas
Bogota, Colombia

N. Pimentel
Direccion de Geologia
Ministerio de Energia y minas
Caracas, Venezuela

ABSTRACT

A total of 56 fission-track ages are reported on 21 apatite, 19 zircon and 2 sphene concentrates from 24 rock samples collected from Toas Island (3) and the Sierra de Perijá (9), Western Venezuela, and from the Santander Massif (12), Eastern Colombia. All apatite ages, set by cooling on uplift, date points on uplift curves. Portions of such uplift curves are late Oligocene (27–22 Ma) in the southeast piedmont of the Sierra de Perijá; early to middle Miocene (19–14 Ma) in the western piedmont and (16–14 Ma) in the central Santander Massif; middle Miocene (13 Ma) in Toas; late Miocene to early Pliocene (7–4 Ma) in the central and northern Santander Massif; and middle Pliocene (3 Ma) in the Sierra de Perijá.

When apatite data for the Venezuelan Andes are added to the above and integrated with regional geologic evidence, it is concluded that Tertiary uplift became progressively greater and occurred at faster rates but through shorter time spans. On an elevation/age plot this should be revealed as a fan-shaped array of uplift curves. Because of inadequate topographic relief at most localities, this cannot be proven but is suggested as a working hypothesis for future verification.

Zircon ages reported here range from about 50 to 126 Ma. When reviewed together with zircon ages from the Venezuelan Andes, it is considered that 8 or 9 ages from the Venezuelan Andes (3), Santander Massif (3), and Perijás (2 or 3) reflect uplift in end-Cretaceous-Paleocene time. Two dates from Toas and the Perijás may give original crystallization ages of felsic volcanics (120–122 Ma). The remaining 23 ages are interpreted as mixed ages related to partial annealing of clocks set in Permian–early Cretaceous time although a pronounced concentration in the range of 85 to 101 Ma raises the possibility of some regional tectono-thermal event at that time. A satisfactory explanation of the latter remains to be found.

The regions referred to above are among the principal tectonic elements of a triangular continental block with apex in Santa Marta, Colombia, and the Oca and Santa Marta-Bucaramanga faults as sides. In this region "Andean" uplift is considered to have been initiated in end-Cretaceous-Paleocene time in response to northwest-southeast compression affecting the Caribbean and South American plates. Progressive compression in the apical direction of the triangle resulted in progressive interlocking of crustal blocks and final major uplift in unison during the Pliocene-Pleistocene(?). The two boundary faults are interpreted as high-angle, oblique-slip type with modest strike-slip component of movement.

INTRODUCTION

Useful data obtained from a fission-track study of rocks from the Venezuelan Andes (Kohn and others, 1984) invited expansion of the study area to those regions described herein. For the purposes of this study, the term "circum–Maracaibo Basin region" is that defined in Figure 1.

The age data for Toas Island, the Perijá Mountains, and the Santander Massif are discussed separately by area, and the localities sampled appear in the maps of Figures 3, 4, and 7, respectively. Two samples (VE-78-1 and -2) from Toas Island were collected by Shagam, and Pimentel in July 1978, and a third (BV-79-2) by Banks, Pimentel, and Shagam in July 1979. All were collected from in situ outcrop. Samples of the Perijá Mountains were collected in July 1979 by Banks, Rogríguez, and Shagam. All came from in situ outcrop, but for a sample of La Quinta Formation tuff (BV-79-5) taken from a boulder almost 1 m in diameter. The original outcrop could not have been more than about 100 m upslope. The Santander Massif was sampled in August 1979 by Banks, Vargas, Rodríguez, and Shagam; a sample of Santa Bárbara granite (BC-79-14) was taken from a large boulder which could not have been displaced from its source by more than a few metres of elevation. All other samples were collected at in situ outcrops.

Mineral separations were done mainly by Dasch at Case Western Reserve University, some by Kohn at Ben-Gurion University. All laboratory processing and measurement of fission-track ages were done by Kohn mainly at Ben-Gurion University, partly at the University of Pennsylvania. Samples were irradiated at the Nahal Soreq Research Institute, Israel.

ANALYTICAL METHODS AND NOMENCLATURE

For a detailed description of the methods used to derive the fission-track data presented in Tables 1-3, the reader is referred to Kohn and others (1984). They also discuss the temperature significance and manner of interpretation of fission-track ages. All ages were determined using the external detector method.

The term "to set" refers to a clock that begins to record time. The terms "to begin operation" and "to trigger" are used synonymously therewith. Erasure of recorded time is here referred to as annealing, following Fleischer and others (1965). The term "to reset," commonly used by geochronologists to signify erasure of recorded time, is not employed herein in order to avoid confusion in the case of a clock initially set, then completely annealed, and finally set anew (i.e., "reset").

REGIONAL GEOLOGIC AND THERMAL HISTORY FRAMEWORK

Inasmuch as the fission-track clocks may be set or annealed at relatively low temperatures, a knowledge of processes such as uplift and erosion or subsidence and burial, in addition to igneous and regional high-temperature metamorphic processes, may be crucial to proper interpretation of the age data (see Kohn and others, 1984). Salient features of the geologic framework pertinent to this study are given in Figure 2.

Ideally, one should also have a sound knowledge of geothermal gradients with time. The available data are very rudimentary. A compilation of the most reliable measurements in deep boreholes (see González de Juana and others, 1980, p. 895) suggests an average current geothermal gradient in Lake Maracaibo of about 33 °C/km (range of 25 to 46 °C/km). Birch (1950) showed that geothermal gradients under elevated areas are less than under adjacent valleys; if one may extrapolate to the scale of the circum–Maracaibo Basin region, one may infer lower gradients in the area of the ranges, but neither heat flow nor deep borehole temperature data are available for confirmation. Vitrinite reflectance data on samples from Maracaibo wells are not available to us. A value of 0.69% R_0 (i.e., high-volatile bituminous, B rank on the USA scale) was obtained on a sample of coal from the Paleocene Marcelina Formation of the Perijás (Shagam, in prep.), but the lack of other samples in vertical distribution voids any attempt to reconstruct the paleogeothermal gradient. Depending on the residence time and thickness of overburden estimated (see subsequent discussion of Perijá stratigraphy), one might infer a lower geothermal gradient during the Cenozoic uplift history of the ranges. In turn, this permits only very crude estimates of the depths of the blocking temperatures for apatite and zircon clocks, namely, about 2 to 4 km and 4 to 8 km, respectively, during Cenozoic time. Additional finer detail of the geologic setting and thermal history pertinent to this study is presented in the discussion of the results by area.

The fission-track ages reported here for the thin, widely spaced eruptive and hypabyssal volcanics show marked discordance with the ages measured by the higher temperature K-Ar, Rb-Sr, and U-Pb clocks. This indicates that all those fission-track ages, with two possible exceptions, must be interpreted in a framework other than that of original cooling on crystallization.

The cooling history of the plutonic rocks is less clear. Most are of batholithic proportions and were intruded at epizonal depths (3–7 km) inferred from the age and nature of adjacent country rock. Relatively "wet" magmas are indicated by the

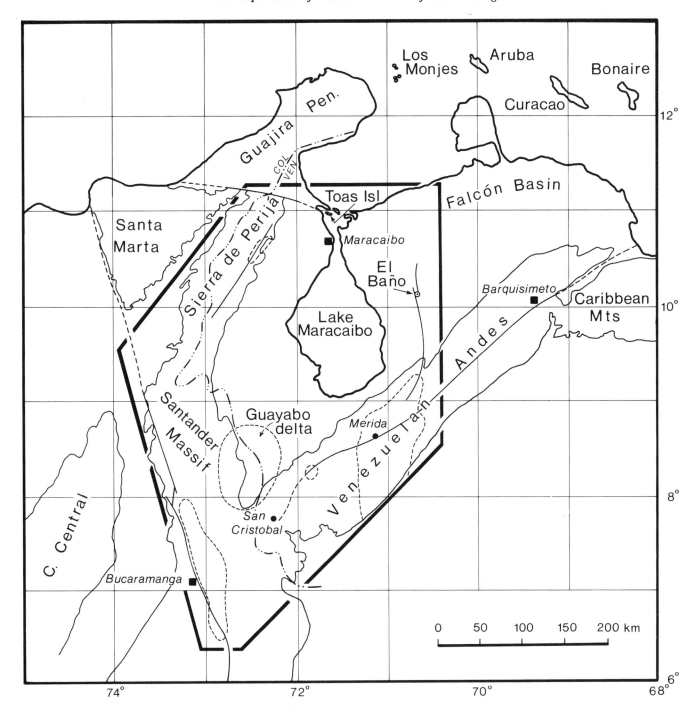

Figure 1. Geographic map of western Venezuela–eastern Colombia to show the circum–Maracaibo Basin region (heavy-lined frame) and the regions sampled for this study: Toas Island and the areas enclosed by dashed lines in the Sierra de Perijá and Santander Massif (commonly abbreviated to "the Perijás" and "the Massif," respectively). The four areas enclosed by dashed lines in the Venezuelan Andes were sampled for the study by Kohn and others (1984).

TABLE 1. DETAILS OF TOAS ISLAND SAMPLES AND ESTIMATED DEPTH OF BURIAL

Specimen No.	Name and rock type	Current elevation (±1 m)	Thickness of cover beds (m)			Depth (m) below unconformity (U_{1a} or U_{1b})	Est. total depth of burial (m)
			La Quinta	Cret.-Eocene	post-Eocene		
S-78-1	rhyolite dike	2	-0 - 50	250, 7000	100 - 1000	20 - 100	5000 - 6000
S-78-2	altered bi (chl) granodiorite	2	-0 - 50	250, 7000	100 - 1000	20 - 100	5000 - 6000
BV-79-2	altered bi (chl) granodiorite	1	-0 - 50	250, 7000	100 - 1000	20 - 100	5000 - 6000

Fission-track data and ages for apatites, zircons, and sphenes from Toas Island

Sample	ρ_s cm^{-2}	ρ_i cm^{-2}	ϕ × 10^{15} neutrons/cm^2	Number of fields or grains [*]	AGE Ma (± 2σ)
			A P A T I T E		
S-78-2	1.23 × 10^5 (320)[†]	1.75 × 10^6 (1298)[†]	3.09 (2059)[†]	96/55	13.3 ± 1.8
BV-79-2	1.52 × 10^5 (329)	2.10 × 10^6 (1422)	2.69 (2834)	80/50	12.0 ± 1.5
	1.53 × 10^5 (460)	5.33 × 10^6 (1331)	7.68 (4971)	110/15	13.5 ± 1.5
			Z I R C O N		
S-78-1	2.67 × 10^6 (925)	3.30 × 10^6 (572)	2.43 (2059)	10	120 ± 14
S-78-2	2.91 × 10^6 (914)	3.74 × 10^6 (588)	1.96 (1712)	7	93 ± 11
	2.86 × 10^6 (696)	4.88 × 10^6 (594)	3.05 (4020)	6	109 ± 13
			S P H E N E		
S-78-2	2.10 × 10^6 (544)	1.93 × 10^6 (250)	2.23 (1712)	8	147 ± 24
	2.42 × 10^6 (1072)	2.01 × 10^6 (456)	1.92 (2952)	8	140 ± 17

Note: See Figure 3 for sample locations.

ρ_s = spontaneous track density; ρ_i = induced track density; ϕ = neutron dose.

λ_f = 6.85 × 10^{-17} yr^{-1}, ^{235}U/^{238}U = 7.252 × 10^{-3}, σ^{235} = 580.2 × 10^{-24} cm^2, λ_D = 1.551 × 10^{-10} yr^{-1}.

[*]Different numbers of fields of view counted in determining the fossil track density (first number) and the induced track density (second number). Single number indicates number of grains used in age determination (by external detector method).

[†]Numbers in parentheses are tracks counted to determine the reported track density (or the neutron dose).

TABLE 2. DETAILS OF SIERRA DE PERIJA SAMPLES AND ESTIMATED DEPTH OF BURIAL

Specimen No.	Name and rock type	Current elevation (±50 m)	Thickness of cover beds (m)			Depth (m) below unconformity (U_{la} or U_{lb})	Est. total depth of burial (m)
			La Quinta	Cret.-Paleoc.	post-Paleoc.		
PE-5	El Totumo porph. (aug,hbl,pl) andesite	190	1000*	2500*	500 - ?*	0 - 200	>4000[†]
BV-79-5	La Quinta red welded tuff/breccia	240	100*	2500*	500 - ?*	0	>3100[†]
BV-79-7	Riecito del Palmar; bi-granodiorite	200	1500	2500*	500 - ?*	0 - 2000*	>4500[†]
BV-79-8	Rio Palmar; bi-granite	110	1500*	2500*	500 - ?*	0 - 2000*	>4500[†]
BV-79-9	Caño Tacón; hbl-bi-granodiorite	320	1500*	2500*	500 - ?*	0 - 2000*	>4500[†]
BV-79-10	Perijá Series; Qz-feld-bi-musc. gneiss	160	0 - 1500*	2500*	1000 - ?*	0 - 3000*	>3500[†]
BV-79-12	Lajas; bi-granite	210	0 - 1000*	2500*	500 - ?*	0 - 1000*	>3000[†]
SP-79-3	La Quinta rhyolite	220	500 - 1000	2500*	500 - ?*	0	>3500[†]
SP-79-4	La Quinta red tuff	220	500 - 1000	2500*	500 - ?*	0	>3500[†]
SP-79-6	Perijá Series gneiss	160	0 - 1500*	2500*	1000 - ?*	0 - 3000*	>3500[†]

Fission-track data and ages for apatites, zircons, and sphene from the Sierra de Perijá

Sample No.	ρ_s cm^{-2}	ρ_i cm^{-2}	ϕ × 10^{15} neutrons/cm^2	Number of fields or grains	AGE Ma (± 2σ)
		A P A T I T E			
BV-79-5	3.70×10^4 (200)	2.18×10^6 (1472)	2.48 (2834)	180/50	2.6 ± 0.4
BV-79-7	1.09×10^5 (414)	7.99×10^5 (1296)	2.94 (3248)	140/120	24.6 ± 3.0
	1.21×10^5 (345)	2.24×10^6 (1529)	8.16 (4625)	35/25	27.0 ± 2.7
BV-79-8	9.57×10^4 (466)	2.49×10^6 (1010)	3.41 (3151)	180/30	8.0 ± 1.0
BV-79-9	1.69×10^5 (550)	1.49×10^6 (1208)	3.10 (4279)	120/60	21.6 ± 2.3
BV-79-10	3.07×10^5 (746)	1.81×10^6 (1225)	2.41 (2834)	90/50	25.1 ± 2.5
	3.02×10^5 (816)	5.75×10^6 (1935)	7.68 (4625)	24/11	24.7 ± 2.2
BV-79-12	2.69×10^4 (100)	1.87×10^6 (1009)	3.11 (4279)	138/40	2.7 ± 0.6
SP-79-4	1.43×10^5 (466)	1.03×10^6 (1538)	3.20 (2302)	120/110	27.2 ± 3.1
		Z I R C O N			
PE-5	2.44×10^6 (776)	6.53×10^6 (1037)	3.80 (2081)	9	87 ± 9
BV-79-5	2.30×10^6 (424)	3.19×10^6 (294)	2.67 (2834)	8	117 ± 19
	2.86×10^6 (384)	4.59×10^6 (308)	3.34 (3151)	6	127 ± 20
BV-79-7	2.57×10^6 (467)	5.91×10^6 (537)	2.68 (2834)	7	71 ± 9
BV-79-8	2.96×10^6 (614)	7.11×10^6 (737)	3.80 (2081)	8	96 ± 11
BV-79-9	2.26×10^6 (633)	4.46×10^6 (624)	2.23 (2750)	9	69 ± 8
BV-79-10	5.53×10^6 (1078)	1.03×10^7 (950)	3.81 (2081)	6	124 ± 13
BV-79-12	9.90×10^5 (412)	1.71×10^6 (712)	2.23 (2750)	7	79 ± 10
		S P H E N E			
SP-79-6	3.00×10^6 (1136)	5.82×10^6 (1102)	3.60 (2426)	11	113 ± 11

Note: Constants and notation used as for Table 1.

*Because of erosion and related uncertainties concerning geologic history indicated figures are based on gross extrapolations from adjacent areas. See discussion in text.

[†]Total depth of burial could have attained ∿8 km, depending largely on thickness estimates for the post-Paleocene formations.

TABLE 3. DETAILS OF SAMPLES FROM THE SANTANDER MASSIF AND ESTIMATED DEPTH OF BURIAL

Map Location No.	Specimen No.	Name and rock type	Current elevation (±75 m)	Thickness of cover beds (m)			Depth (approx.) below unconformity (m) (U_{1a} or U_{1b})	Est. total depth of burial (m) [*]
				Girón/ Jordán	Cret./ Paleoc.	post- Paleoc.		
C-1	C-78-1	Bi-qz-f. orthogneiss	3400	0-1000	2500	0 (?)	∿1000	3500-4500
1	BC-79-1	Pescadero pink bi-granite	650	1000-3000	3000	0-500	∿ 300	4300-6800
3	BC-79-3	Sta. Bárbara qz-monzonite	2100	0-1000	2500	0 (?)	∿1000	3500-4500
4	BC-79-4	B'manga bi-qz-f. gneiss	1950	0-1000	2500	0 (?)	∿1000	3500-4500
5	BC-79-5	Córcova gray qz-monzonite	1500	0-1000	2500	0 (?)	∿1000	3500-4500
6	BC-79-6	Silgará mica schist	625	1000-3000	3000	0-500	∿ 300	4300-6800
7	BC-79-7	C. Pintado leuco-granite	1640	0-1000	2500	0 (?)	∿1000	3500-4500
8	BC-79-8	C. Pintado bi-rich granodiorite	1550	0-1000	2500	0 (?)	∿1000	3500-4500
11	BC-79-11	B'manga bi-qz-f. gneiss	3550	100- 500	2500	0 (?)	∿1000	3600-4000
12	BC-79-12	Bi-qz-f. orthogneiss	3700	100- 500	2500	0 (?)	∿1000	3600-4000
13	BC-79-13	Páramo Rico hbl-bi-tonalite/granodiorite	3680	100- 500	2500	0 (?)	∿1000	3600-4000
14	BC-79-14	Sta. Bárbara qz-monzonite	3460	100- 500	2500	0 (?)	∿1000	3600-4000

Fission-track data and ages for apatites and zircons from the Santander Massif

Sample No.	ρ_s cm^{-2}	ρ_i cm^{-2}	$\phi \times 10^{15}$ neutrons/cm^2	Number of fields or grains	AGE Ma (± 2σ)
A P A T I T E					
BC-79-1	1.35×10^5 (366)	1.75×10^6 (1890)	2.90 (2017)	100/80	13.7 ± 1.7
	1.38×10^5 (298)	4.43×10^6 (1198)	7.20 (2493)	80/20	13.8 ± 1.9
BC-79-3	6.50×10^4 (246)	1.51×10^6 (1632)	2.75 (1544)	140/60	7.3 ± 1.1
	6.48×10^4 (189)	4.02×10^6 (814)	7.20 (2493)	100/20	7.1 ± 1.2
BC-79-4	8.97×10^4 (388)	1.52×10^6 (1642)	3.42 (2178)	160/80	12.4 ± 1.5
BC-79-5	3.70×10^4 (135)	1.22×10^6 (1648)	2.97 (2048)	135/100	5.5 ± 1.0
BC-79-6	1.21×10^5 (458)	1.27×10^6 (1376)	3.23 (2178)	140/80	18.9 ± 2.2
	1.23×10^5 (200)	1.21×10^6 (1312)	3.03 (2048)	60/80	18.9 ± 3.0
BC-79-7	1.64×10^4 (120)	7.62×10^5 (1545)	3.03 (2048)	270/150	4.0 ± 0.8
	1.65×10^4 (103)	1.88×10^6 (976)	7.04 (4971)	72/18	3.8 ± 0.8
BC-79-8	4.96×10^4 (268)	1.27×10^6 (1370)	3.03 (2048)	200/80	7.3 ± 1.0
BC-79-11	4.07×10^4 (374)	1.63×10^6 (1656)	3.23 (2178)	200/75	5.0 ± 0.6
	4.09×10^4 (345)	4.92×10^6 (1023)	8.16 (4565)	53/10	4.2 ± 0.5
BC-79-12	4.41×10^4 (286)	1.27×10^6 (1722)	3.03 (2048)	240/100	6.5 ± 0.9
BC-79-13	2.34×10^5 (190)	2.90×10^6 (1812)	3.23 (2178)	30/42	16.0 ± 2.5
BC-79-14	1.34×10^5 (290)	1.75×10^6 (1779)	2.90 (2017)	80/75	13.6 ± 1.8
C-78-1	5.29×10^4 (458)	2.21×10^6 (2390)	4.15 (4279)	320/40	6.1 ± 0.7
	5.31×10^4 (409)	3.81×10^6 (1686)	7.16 (4971)	80/12	6.1 ± 0.6
Z I R C O N					
BC-79-1	3.90×10^6 (803)	7.61×10^6 (784)	2.97 (2048)	12	93 ± 10
BC-79-3	3.89×10^6 (678)	7.72×10^6 (673)	2.97 (2048)	10	91 ± 11
BC-79-4	3.18×10^6 (986)	5.89×10^6 (978)	2.97 (2048)	12	98 ± 10
BC-79-5	1.73×10^6 (642)	4.79×10^6 (890)	2.97 (2048)	12	66 ± 7
	8.09×10^5 (327)	1.84×10^6 (372)	2.23 (2750)	6	60 ± 9
BC-79-8	3.47×10^6 (535)	6.19×10^6 (503)	2.90 (2017)	12	99 ± 12
BC-79-11	3.25×10^6 (1016)	8.81×10^6 (1323)	2.97 (2017)	12	67 ± 7
BC-79-12	6.88×10^6 (1350)	8.83×10^6 (867)	2.30 (4348)	6	109 ± 10
BC-79-13	4.19×10^6 (1361)	8.08×10^6 (1311)	2.97 (2017)	8	94 ± 8
BC-79-14	3.93×10^6 (1308)	6.74×10^6 (1082)	2.97 (2017)	11	106 ± 10
C-78-1	3.57×10^6 (489)	1.19×10^7 (818)	2.70 (2834)	6	50 ± 6
	7.09×10^6 (668)	1.64×10^7 (771)	2.30 (4348)	6	61 ± 7

Note: See Figure 7 for sample locations. Constants and notation used as for Table 1.

[*] Main uncertainties related to depth below the unconformity (U_{1a}/U_{1b}) and thickness of Girón/Jordán Formations.

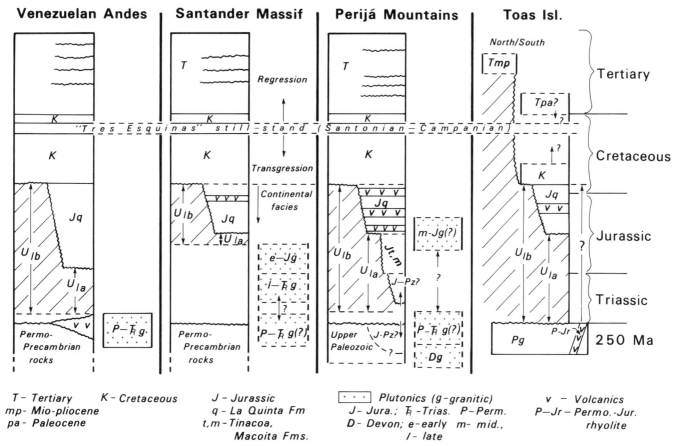

Figure 2. Generalized stratigraphic columns for the circum–Maracaibo Basin region to bring out the principal similarities and differences. Consistent presence of major unconformities separating Late Jurassic (U_{1a}) and Cretaceous (U_{1b}) formations from older rocks, and marked similarities in the stratigraphic succession contrast with differences in the ages and volumes of major magmatic events. Thus intrusion of the Palmar granite in the Perijás postdates U_{1a}. There was extensive Late Jurassic volcanism in the Perijás, less so in Santander and Toas, and probably none in the Venezuelan Andes. Also, a longer Triassic(?)-Jurassic record appears to occur in the Perijás. Wavy lines in the Tertiary section diagrammatically represent uplift events not everywhere coeval. The Tres Esquinas stillstand is a useful regional marker horizon, although it has been eroded away on Toas.

abundance of dike, vein, and other mineralization phenomena. Spera 1980) developed a theoretical model of solidification times for various compositions and conditions of intrusion. His cooling curve for conditions resembling those outlined above predicts that the time to attain the granite solidus $\leqslant 1$ Ma. Because all the fission-track ages are younger than the K-Ar, Rb-Sr, and/or U-Pb ages on the same plutons by 10^7 to 10^8 yr (see Table 4), we conclude that the fission-track clocks were initiated by secondary cooling events not related to cooling on original crystallization.

TOAS ISLAND

Local Geologic Setting and Localities Samples

Located near the mouth of Lake Maracaibo (see Fig. 1), Toas Island has been repeatedly referred to in papers concerned with the nature of the Oca fault. A geologic map and cross-section are shown in Figure 3.

U/Pb isotopic measurements on a granite sample by Dasch (1982) gave a straight line discordia plot with upper intercept at 252 ± 50 Ma, which he viewed as the age of original crystallization, and a lower intercept at 15 ± 60 Ma attributed to increased fluid migration (and resultant leaching of Pb) related to brecciation wrought by movements on the Oca fault. Some basaltic dikes in the granite are best correlated with basalt lavas in the Jurassic La Quinta Formation. On the other hand, as explained below, there is some uncertainty concerning the age of felsic dikes in the granite; they are most commonly considered cogenetic with the granite, but they could be coeval with felsic volcanics associated with the La Quinta Formation in the Perijá Mountains (see Maze, 1984).

As to its structural setting, Toas is viewed as a block within the Oca fault zone. The imaginative cross-section of Rod (1956)

TABLE 4. SUMMARY OF ALL PUBLISHED ISOTOPIC AGES ON ROCK UNITS IN TOAS ISLAND, SIERRA DE PERIJA, AND SANTANDER MASSIF DATED BY THE FISSION-TRACK METHOD IN THIS STUDY

Region & Rock Unit	Sample number	Previous studies			This study (F-t age, Ma)			Footnotes
		Age	Method	Ref.	Apatite	Zircon	Sphene	
TOAS ISLAND								
Granite	BV-79-2	252 ± 25	U-Pb, zircon	(1)	13.3 ± 1.5 / 12.0 ± 1.5			*
	VE-78-2				13.5 ± 1.8	93 ± 11 / 109 ± 13	147 ± 24 / 140 ± 17	
(Granite, Mara)	?	210 ± ?	K-Ar, biotite	(2)				†
Rhyolite dike	VE-78-1					120 ± 14		
SIERRA DE PERIJA								
Granitic plutons								
"Palmar"								§
Riecito	BV-79-7				24.6 ± 3.0 / 27.0 ± 2.7	71 ± 9		
Palmar	BV-79-8	167 ± 3	U-Pb, zircon	(1)	8.0 ± 1.0	96 ± 11		**
Tacón	BV-79-9				21.6 ± 2.3	69 ± 8		
Palmar (≈Riecito?)	P-23A	216 ± 6	K-Ar, biotite	(3)				
	P-23B	158 ± 8	K-Ar, chlorite	(3)				††
	P-30	115 ± 6	K-Ar, chlorite	(3)				††
Palmar (≈Tacón?)	ZV-100	269 ± 8	K-Ar, feldspar	(3)				§§
Lajas	BV-79-12	310 - 385	U-Pb, zircon	(1)	2.7 ± 0.6	79 ± 10		***
Lajas	VE-17B	380 ± 25	Rb-Sr, whole rock	(3)				†††
Lajas	VE-17D	334 ± 12	Rb-Sr, whole rock	(3)				†††
La Quinta Volcanics								
Welded tuff	BV-79-5	140 - 160	U-Pb, zircon	(1)	2.6 ± 0.4	117 ± 19 / 127 ± 20		**
Hbl andesite	PE-5 (=WM-80-3)	163 ± 5	U-Pb, zircon	(1)		87 ± 9		§§§
Hbl andesite tuff	P-25A	168 ± 5	K-Ar, hornblende	(3)				
	SP-79-4				27.2 ± 3.1			
Metamorphic rocks								
Perijá gneiss	BV-79-10	post-Late Dev.-pre end Pz.	U-Pb, zircon	(1)	25.1 ± 2.5 / 24.7 ± 2.2	124 ± 13		****
Perijá gneiss	SP-79-6						113 ± 11	
SANTANDER MASSIF								
Mesozoic plutons								
Riebeckite granite	IMN-12264	160 ± 30	Rb-Sr, whole rock	(4)				††††
Rio Negro	IMN-10894	177 ± 6 / 172 ± 6	K-Ar, biotite	(4)				
Cerro Pintado	BC-79-7				4.0 ± 0.8 / 3.8 ± 0.8			††††
Cerro Pintado	BC-79-8				7.3 ± 1.0	99 ± 12		††††
Pescadero	IMN-11547	193 ± 6	K-Ar, biotite	(4)				
Pescadero	BC-79-1				13.7 ± 1.7 / 13.8 ± 1.8	93 ± 10		††††
Aguablanca	IMN-13201	196 ± 7	K-Ar, biotite	(4)				
Sta. Bárbara	IMN-10924	192 ± 7	K-Ar, biotite	(4)				
Sta. Bárbara	IMN-11045	194 ± 7	K-Ar, biotite	(4)				
Sta. Bárbara	BC-79-3				7.3 ± 1.1 / 7.1 ± 1.2	91 ± 11		††††
Sta. Bárbara	BC-79-14				13.6 ± 1.8	106 ± 10		††††
Páramo Rico	BC-79-13				16.0 ± 2.5	94 ± 8		††††
Córcova	IMN-13197	111 ± 4 / 195 ± 7	K-Ar { biotite / muscovite	(4)				
Córcova	BC-79-5				5.5 ± 1.0	66 ± 7 / 60 ± 9		††††
Precamb-Pz metamorphics								
Silgará phyllite	IMN-12255	198 ± 8	K-Ar, whole rock	(4)				
Silgará phyllite	IMN-12257	221 ± 8	K-Ar, whole rock	(4)				
Silgará schist	BC-79-6				18.9 ± 2.2 / 18.9 ± 3.0			††††
Orthogneiss	IMN-12262	413 ± 30	K-Ar, hornblende	(4)				†††
Orthogneiss	IMN-12256	450 ± 80	Rb-Sr, whole rock	(4)				††††
Orthogneiss	C-78-1				6.1 ± 0.7 / 6.1 ± 0.4	50 ± 6 / 61 ± 7		††††
Orthogneiss	BC-79-12				6.5 ± 0.9	109 ± 10		††††
Bucaramanga Gneiss	BC-79-4				12.4 ± 1.5	98 ± 10		†††† †††
Bucaramanga Gneiss	IMN-13199	680 ± 140 / 198 ± 7 / 189 ± 4	Rb-Sr, whole rock / K-Ar, biotite / K-Ar, biotite	(4)				
Bucaramanga Gneiss	BC-79-11				5.0 ± 0.6 / 4.2 ± 0.5	67 ± 7		††††
Bucaramanga Gneiss	IMN-12263	945 ± 40	K-Ar, hornblende	(4)				

*U-Pb is upper intercept age.
†Drill core; location ∿40 km WSW of Toas.
§Includes all outcrop areas on Figure 4.
**U-Pb is lower intercept age.
††Chlorite after biotite.
§§Feldspar species not given.
***Arcuate array; secondary discordance.

†††Model age.
§§§Concordant U-Pb age.
****U-Pb data cluster indicates complex isotope evolution history; only limits to regional metamorphism can be set.
††††U-Pb age currently being measured at CWRU, Cleveland, OH.

References

(1) Dasch, 1982.
(2) Martín-B., 1968.
(3) Espejo and others, 1980.
(4) Goldsmith and others, 1971.

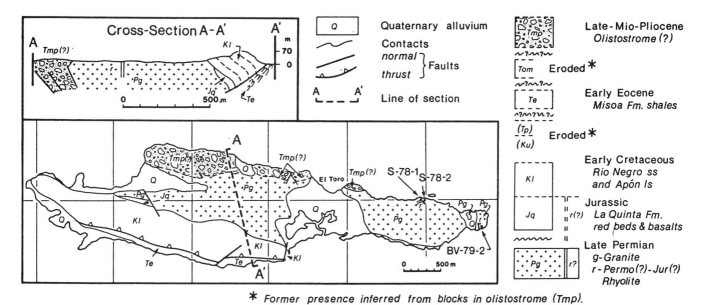

Figure 3. Geologic map (showing sample localities) and cross-section of Toas Island (modified after Pimentel, 1975). See Rod (1956) for additional data.

resembles the flower-structure geometry described by Harding and Lowell (1979) and by Harding and others (1984). There is considerable divergence of views on the nature of the Oca fault (see discussion in the section on regional tectonic interpretation), but it is clear that a component of vertical movement must have been involved in arriving at the current tectonic setting of the island.

The remnant of cover section now exposed does not exceed about 350 m on the south coast and even thinner olistostrome occurs on the north. The possible maximum thickness of cover can only be estimated by extrapolation from the isopach maps of Zambrano and others (1971) from adjacent areas to the south of Toas and could have exceeded 7 km, of which by far the major fraction would have been of Eocene age. Such a thickness with a residence time of about 10^7 yr would have largely annealed the zircon fission tracks, and hence we postulate a maximum in the range of 5 to 6 km (Table 1).

The two sites sampled (see Fig. 3) are a road cut within a few metres of the shore on the north coast at Las Cabeceras about 1 km east of the port of El Toro, and a wave-cut bench on the south coast close to Punta Cabecera (Los Chipitos). Friable, intensely brecciated gray biotite granite is exposed in the road cut at the first locality. The granite is further characterized by abundant slickensides and by patches with a penetrative foliation of variable orientation from outcrop to outcrop. A thin (40 cm) rhyolite dike shows strong curvature like a fold nose (plunge 70°N) and may represent a meso-scale ring dike. The dike shows primary flow lamination and secondary fractures but is virtually unbrecciated or sheared compared to the adjacent granite which was sampled (VE-78-2) about 30 m east of the dike (VE-78-1).

At the second locality the less-weathered, coarse-grained, red porphyritic biotite granite (BV-79-2) is also intensely brec-

ciated and sheared. Here too thin rhyolite dikes were noted, one with pronounced antiformal flow banding. Dasch's (1982) U-Pb age determination was made on zircons from this same sample.

The structural characteristics of the igneous and sedimentary rocks considered in the light of their ages and the fact that the Oca fault was active in the Tertiary raise the possibility that brecciation and shearing of the granite were contemporaneous with original crystallization at the close of the Permian. The effects of shearing during Tertiary activity of the Oca fault may have been concentrated along and in close proximity to spaced shears in the granite, thus fortuitously not affecting the rhyolite dike. On the other hand, the cover section was readily fractured along bedding planes and joints.

Fission-Track Ages and Their Interpretation

It is convenient in this and subsequent sections to discuss the fission-track ages in reverse chronological order. Pertinent geologic and measurement data appear in Table 1.

Apatite Ages. Within the precision of the measurements, the two ages overlap, and we interpret the ages as recording a phase of uplift, erosion, and attendant cooling which was in operation by the middle Miocene (about 13 Ma). From the geologic evidence provided above one may reasonably infer that the apatite clocks of the two granite samples must have been initially set at some time in the range of end-Permian to Late Triassic or Early Jurassic, and were subsequently annealed as a result of burial in Jurassic-Oligocene time. In the middle Miocene the granite was again uplifted and brought through the closure temperature for apatite. Subsequent uplift of the sampled rocks to their present location close to sea level took place in the range of late Miocene to Holocene time, but it cannot be stipulated whether uplift occurred at a constant or variable rate during this interval. The location in the Oca fault zone favors the latter alternative, and the fact that part of the cover section is still preserved on the granite suggests that the final phase(s) of uplift occurred in the recent past.

As noted by Kohn and others (1984) single ages in regions of active tectonism are in all likelihood transient points in an uplift history that may have extended over a considerable period of time. The near–sea level altitudes of all granite exposures on Toas do not permit construction of a relief-derived uplift curve. Depending on the thickness of the cover section as discussed above, the geothermal gradient, and the rate of uplift, it is conceivable that the apatite ages measured represent points on an uplift curve that may have begun significantly earlier than the ages measured and continued to the present. The apatite ages and their interpretation indirectly provide a broad measure of support for Dasch's interpretation of the lower intercept age on discordia.

Zircon Ages. The limits to possible interpretation of the zircon ages may best be appreciated by considering reasonable models of the geologic history. The zircon clock would have begun to operate as a consequence of cooling related to uplift and the cutting of U_{1a} (Fig. 2) in early Mesozoic time. A corresponding age in the range of about 250 to 180 Ma would have been expected, far greater than that measured. Conceivably, in the course of Jurassic-Oligocene burial the granite was depressed into the zone of partial track annealing (see Wagner, 1979, and discussion of Kohn and others, 1984) resulting in a mixed "age" which would reflect the burial process but not its specific age. Subsequent Tertiary uplift would have brought the zircon back into the zone of track stability. An alternative possibility, that the zircon tracks were completely annealed in the course of Jurassic–mid-Cretaceous burial when a pronounced lowering of the geothermal gradient occurred causing the clock to operate again despite further subsidence and burial up to Oligocene time, is rejected as very unlikely. Total Jurassic–mid-Cretaceous section did not exceed about 500 m, far too little to cause annealing of the zircon clock.

The above possibilities hold also for the rhyolite dike sample if the dike is viewed as cogenetic with the granite. If, however, the age of intrusion should prove to be correlative with the felsic volcanics of the La Quinta Formation in the Perijás, the age of original crystallization could fall in the range of about 180 to 130 Ma. Keeping in mind the assigned precision and the fact that a primary cooling age on such a small body would be indistinguishable from a crystallization age, it is barely possible that the zircon age gives the original crystallization age of the dike. A similar age obtained on a Perijá tuff supports this interpretation, but it will take additional radiometric age dating to be sure.

Sphene Age. The comments on the zircon of the granite also hold for the sphene age reported. This is best interpreted as a mixed age resulting from the effect of burial in Jurassic-Oligocene time on a clock originally set in early Mesozoic time. The older sphene "age" compared to zircon is consistent with its higher closure temperature ($\sim 250 \pm 50$ °C).

The significance of the fission-track ages from Toas Island with respect to the tectonic evolution of the circum–Maracaibo Basin region is discussed in the closing section on regional tectonic interpretation.

THE PERIJA MOUNTAINS

Local Geological Setting and Localities Sampled

General inaccessibility and thick vegetative cover have long hindered geologic exploration of the Perijás. Sample locations appear on the geologic map of Figure 4, and modifications to the regional stratigraphic column as exploration progressed are shown in Figure 5. Additional data pertinent to this study of fission-track dating are as follows:

1. Although the data do not lend themselves to simple interpretation, isotopic measurements on Perijá Series gneiss (U-Pb) and on the Lajas granite (U-Pb and Rb-Sr; see Table 4) indicate, in conformity with the field evidence, that those rocks underwent regional metamorphism and primary crystallization, respectively, before the end of the Paleozoic.

2. As shown in Table 4 a wide range of radiometric ages has been obtained on the Palmar granite. We favor the Middle Jurassic (~ 167 Ma) age of original crystallization based on U-Pb isotopic analyses of three samples by Dasch (1982). This preference is supported by similar U-Pb ages on felsic volcanics in the La Quinta Formation and by field evidence for such volcanics at several levels in the formation described by Moya and Viteri (1977) and by Maze (1984). An attempt to verify the details of field-sampling localities and analytical data for the older Palmar ages proved fruitless. Hence, it is difficult to explain the significant discrepancy with Dasch's ages. It is conceivable, for example, that the problem is simply nomenclatural: the Permian ages could well reflect unroofing of the Devonian Lajas granite, formerly grouped with the Palmar granite, (see "El Palmar" in Léxico Estratigráfico de Venezuela, 1970). Dasch's data are in harmony with the views of Moya and Viteri who proposed intimate age relationships of the granite and volcanics in the framework of La Quinta time, although they considered the granite to postdate the volcanism. We found one intrusive granite contact at sample site BV-79-9 (probably an appendage of the Palmar pluton), but the country rocks, coarse quartzose sandstones, and conglomerates are of uncertain age reflected in the map symbol J-Pz assigned by Kellogg (1981). We think that these rocks are most reasonably viewed as a basal member of the La Quinta Formation fortuitously exposed in a horst.

In light of the foregoing it may be said that the Perijá gneiss and Lajas granite samples cooled through the closure temperature for the K-Ar clock by the end of the Paleozoic, the Palmar granite by the Middle Jurassic, and the La Quinta felsic volcanics by the earliest Cretaceous. The proximity of the Perijá gneiss and Lajas granite to the Palmar granite (although this could be the result of later faulting) suggests that the former two rocks may have had their K-Ar clocks reset in the Middle Jurassic. In view of the

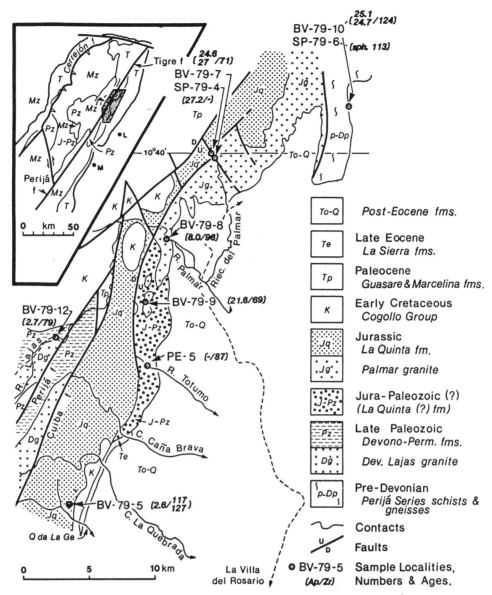

Figure 4. Geologic map of the Palmar uplift and adjacent areas of the southeastern Perijá Mountains showing sample localities. Simplified after Espejo and others, 1979, Canelón and others, 1979, and Kellogg, 1981. Key to inset map: pMz = pre-Mesozoic; J-Pz = Jurassic-Paleozoic(?); Mz = Mesozoic; T = Tertiary; circled dots are L = La Villa del Rosario and M = Machiques. Dashed line is access road leading south to La Villa del Rosario.

relative abundance of basic to felsic volcanics, it is possible that geothermal gradients may have remained high through the end of the Jurassic causing further resetting of the K-Ar clocks of all rocks (cf. Maze, 1984).

3. Interpretation of the zircon ages, especially, is hindered by the absence of any regional pre-Cretaceous isopach data and the almost complete removal by erosion, at most localities, of the cover section which can only be estimated between broad limits (3–8 km; Renz, 1977 and Zambrano and others, 1971 provide estimates of thickness for the Cretaceous only). With respect mainly to the apatite ages, the biggest uncertainty is related to the early to middle Eocene Misoa Formation (Fig. 5). In boreholes in

Lake Maracaibo this formation is found to rest unconformably on Paleocene rocks and attains a thickness of 3,500 to 6,000 m, but to what extent (and thickness) it was deposited over the Perijá ranges is not known. The Misoa is not exposed through a north-south distance of about 70 km between the Manuelote syncline in the north and mountain front exposures west of La Villa del Rosario in the south, a gap that includes the Palmar uplift where most of the samples for dating were collected. Clearly, the presence of the thickness noted of sedimentary section unconformably on the Paleocene section could considerably affect the burial, uplift, and erosion histories of the crystalline rock samples for dating (see below).

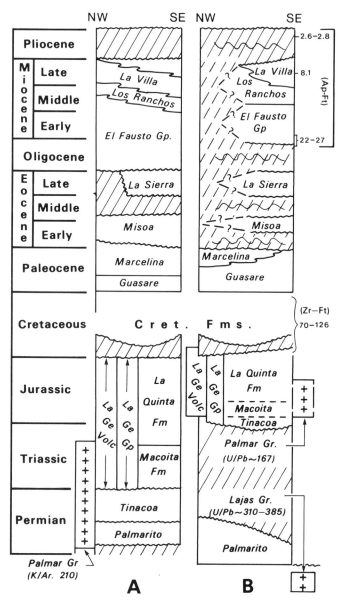

Figure 5. Evolution of views concerning stratigraphic relationships in the Perijás for the late Paleozoic–Pliocene. A. Composite based on Bellízzia and others, 1976, Bowen, 1972, Miller, 1962, and Hedberg and Sass, 1937, summarizing views held until about 1976. B. Present concepts based on composite of Kellogg, 1981, Odreman and Benedetto, 1977, Benedetto and Odreman, 1977, Moya and Viteri, 1977, Maze, 1984, Espejo and others, 1980, and Dasch, 1982. Numbers marked "Ap-Ft" and "Zr-Ft" show corresponding stratigraphic ranges of apatite and zircon fission-track ages, respectively, reported herein. Wavy lines are orogenic phases recognized by Kellogg (1984). Geographic orientations (NW-SE) only relate to Tertiary formations.

A similar problem concerns stratigraphic section of Miocene age (see Fig. 5). Two orthogonal cross-sections by Maze (1980, 1981) show onlap of the Miocene rocks across Lower Cretaceous rocks and Palmar granite suggesting pronounced wedging out toward the north and west. Boreholes in the lake and on the western coastal plain suggest a total thickness of Miocene forma-

tions of almost 4.5 km, but this could have decreased to 1 km over most of the Palmar uplift. The current thickness of those formations decreases to less than 400 m at the mountain front.

4. Kellogg (1981, 1984) defined four tectonic events (see Fig. 5) in the Tertiary uplift history of the Perijá Mountains. Of these he considered the Oligocene and Pliocene events as the most significant. Regrettably the area sampled for dating represents less than one fifth of the physiographic width of the mountain belt in contrast to the near-complete cross-sectional sampling attained in the Venezuelan Andes (cf. Kohn and others, 1984). In part, this reflects problems of accessibility; in part, it is simply related to the restricted distribution of crystalline rocks in outcrop.

Fission-Track Ages and Their Interpretation

The petrography and age data for the Perijá samples are shown in Table 2; samples are fresh to slightly weathered.

Apatite Ages. As in the case of the Toas Island samples, all apatite ages must be interpreted in terms of cooling related to uplift and erosion of cover section. Four of five ages on samples from the Palmar granite outcrop belt correspond approximately to end-Oligocene time and date the period when the samples were uplifted through the closure temperature ($\sim$100 °C) for apatite. No sign of tectonic uplift corresponding to the early and middle Eocene unconformities (see Fig. 5) appears in the apatite ages, but it is conceivable that the clocks were triggered in the early Eocene and annealed as a result of burial under the thick cover of the Misoa Formation to be retriggered at the close of the Oligocene. This possibility can only be conjectured in the light of previous discussion of the paleogeography in Misoa time. Registry and subsequent annealing of the apatite clock in the middle Eocene is unlikely in view of the paltry thickness of the upper Eocene La Sierra Formation. The absence of Oligocene section in the Perijás suggests that although the apatite clocks only became operational toward the close of that epoch, uplift at a slow rate may have begun in latest Eocene–earliest Oligocene time and that it took most of the Oligocene to bring the crystalline basement through the closure temperature for apatite. The subsequent deposition of Miocene piedmont deposits did not result in annealing of the clock, presumably because of the westward wedging out previously noted.

There is no unique explanation for the age of 8.0 ± 1 Ma on sample BV-79-8 collected from the west bank of the Rio Palmar. That age falls in the time of deposition of the La Villa Formation, a piedmont deposit clearly related to shedding of sediment from the mountain belt, but there is no field or subsurface evidence to indicate a significant hiatus at that time (cf. discussion of apatite data for the Santander Massif, Fig. 11). The age might be explained by unusually high local geothermal gradient and/or thickness of cover section requiring protracted uplift for the sample to pass through the closure temperature. Alternatively, and considered less likely, it may be a mixed age resulting from partial annealing of a clock originally set along with the other samples at

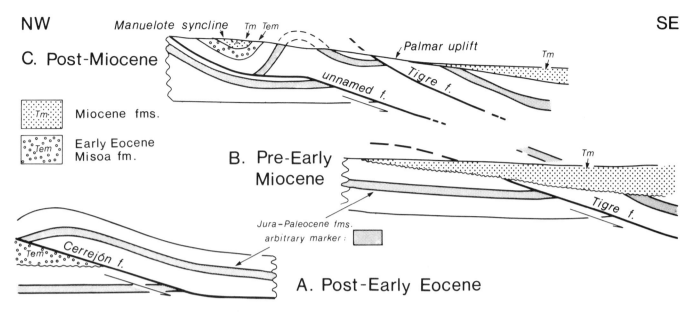

Figure 6. Cartoon to show correlation between the structural evolution of the Perijás as envisioned by Kellogg (1981) and the apatite fission-track ages reported herein. Only a maximum age of mid-Eocene can be set for stage A, and the absence of corresponding ages (about 45 m.y.) suggests either that development of the Cerrejón thrust was coeval with stage B or that sampling for age-dating did not extend sufficiently far to the west. A minimum age for the Palmar uplift is set in B by the overlap of Miocene section and correlates well with the apatite ages of 22 to 27 Ma. Pliocene thrusting (C) is established by the folding of the Miocene in the Manuelote syncline and conforms with apatite ages of 2.6 to 2.8 Ma. Modified after Kellogg's Figure 2-21.

the close of the Oligocene. It would take more detailed sampling and measurement to resolve this problem.

The remaining two samples (BV-79-5 and -12) are external to (west of) the Palmar granite outcrop belt; the former in the La Quinta belt and the latter (the Lajas granite) the only sample west of the Perijá-Tigre fault. Both samples give mid-Pliocene ages, indicating a significant pulse of uplift was occurring at that time. Kellogg (1981, p. 2–26–2-35), on the basis of surface and subsurface structural and stratigraphic data combined with numerous gravity measurements, concluded that the formation of the major Eastern Frontal Monocline of the Perijás occurred in the Pliocene. His view of the sequence of structural events is shown diagrammatically in Figure 6. The two apatite ages are in harmony with that tectonic framework. Here too the possibility of previous setting and annealing of the apatite clock cannot be established for the reasons already given, but we speculate that these two clocks may have begun to operate for the first time in the Pliocene.

Unfortunately, topographic relief is minimal in the area sampled (see elevations given in Table 2) so that uplift curves cannot be constructed. As in the case of the Toas samples, it is inferred that the apatite ages are merely transient points on uplift curves. From regional geological relationships, it is reasonable to infer that the Lajas granite was covered by a minimum of about 3 km of sedimentary section (the top of which was close to sea level) even assuming that there was zero post-Paleocene cover.

For the Lajas granite now to be exposed at an elevation of about 200 m implies uplift at a rate approximating 1 km/Ma.

We considered the possibility that the bimodal apatite ages represent the beginning and ending of one phase of uplift that occurred at a very slow rate. The current occurrence of the corresponding rocks at near identical elevations would be ascribed to the fortuitous effects of Pliocene-Pleistocene faulting and block uplift. This possibility can be rejected as extremely unlikely in view of the relationships of the Miocene formations to the Palmar area characterized by the old apatite ages (cf. Fig. 6). Uplift of that area must have ceased by the earliest Miocene in order for accumulation and preservation of Miocene section to occur. During some unknown subsequent uplift event, the Palmar area was uplifted (possibly as little as 500 m) to its present elevation. This matter is further referred to in discussion of the significance of the apatite ages measured on rocks of the Santander Massif.

The event that was at least partly in progress in the latest Oligocene is reasonably established by the apatite data, but the two geographically widely spaced Pliocene ages and the one age of 8 Ma require further verification. It should be productive also to sample two inliers of Perijá schists northwest of the Perijá-Tigre fault for apatite fission-track dating in order to attain geographically broader coverage of the uplift history.

Zircon Ages. The zircon ages range from Early to Late Cretaceous, overlapping those measured on the Toas samples. However, in light of the differing geologic histories, the interpretation of the Perijá ages is far more complicated and further compounded by uncertainties concerning the thickness of cover section and the thermal history of some samples as noted in the introduction to this section.

With the foregoing in mind the broad possibilities of interpretation of the data are as follows:

La Quinta tuff (BV-79-5). This clock must have begun to operate contemporaneously with deposition at the tail end of La Quinta volcanicity. Stratigraphically it may be close to the base of the Cretaceous section, the oldest unit of which is the largely unfossiliferous Rio Negro Formation which has been tenuously assigned a Barremian age (see Léxico Estratigráfico de Venezuela, 1970). The lower limit of the latter is currently set at 114 ± 2 Ma (Odin and others, 1982). The mean of two separate series of measurements on this sample (122 ± 19 Ma) suggests possible interpretation in terms of the age of original crystallization. Dasch (1982) obtained isotopic data on only two zircon fractions from the tuff, which were insufficient to establish a discordia curve. By comparison with the isotopic results on the Palmar granite, he inferred a slightly younger age and arbitrarily set a minimum age of 140 Ma which he believed was the approximate Jurassic-Cretaceous boundary. With negligible changes in his assumptions, the isotopic data could be interpreted in terms of original crystallization age at about 130 Ma, which is well within the precision of the fission-track measurements. We conclude that the fission-track ages represent original crystallization ages that may have suffered minimal partial annealing, in which case the 3 km of Cretaceous-Paleocene cover was probably augmented by some thickness of Eocene and younger rocks.

Lajas Granite (BV-79-12). Reasonable extrapolation of the base of the Cretaceous section over this granite indicates that it was probably within 1 to 2 km of the surface during the interval represented by U_{1b} and therefore that its clock was operating at that time. This clock may have been completely or partly annealed, or remained unaffected by the thermal effects related to intrusion of the Palmar granite. The measurement (79 ± 10 Ma) could represent a partially annealed clock of no true age significance; alternatively, there may have been sufficient burial for complete annealing of the original (or partly annealed) age, say by mid-Cretaceous time, followed by the uplift of the basin floor during the Tres Esquinas stillstand (see Fig. 2).

Perijá Gneiss (BV-79-10). The zircon clock of this sample must have been triggered during U_{1a}, and the age of 124 ± 13 Ma is amenable to a wide range of interpretations, of which the following are the most reasonable. The age originally set during U_{1a} may have been partly annealed in the course of subsequent burial and/or the heating effects of the Palmar granite. On the other hand, an original age may have been completely annealed by the thermal effects of granitic intrusion and reset during mild uplift associated with U_{1b} (Kellogg reported angular unconformity between the Cretaceous and La Quinta formations in some areas). In that case there would only have been minimal (or no) partial annealing during Cretaceous-Tertiary burial.

Palmar Granite (BV-79-7, -8, -9) and El Totumo Horn-blende Andesite (PE-5). Here too a wide range of possible interpretations of the data exists. Possibly intruded at a depth of 5 to 7 km (probably less for PE-5) in an active volcanic environment, local ambient temperatures may have been high enough to prevent registry of the zircon clocks until regional cooling led to subsidence and the marine transgression of the Cretaceous. In such a framework the zircon clock would be set by cooling in the course of *subsidence* rather than uplift, differences in ages (about 69 to 96 Ma) merely reflecting local differences in cooling rate.

A second possibility is that the clocks were triggered during U_{1b} and then partially annealed during subsequent burial. Interpretation of the data in terms of uplift ages is negated on the one hand by the two results of about 87 and 96 Ma corresponding to a time span known to be one of marine transgression and subsidence. On the other hand, it is favored by the fact that the other two ages (about 69 and 71 Ma) correspond to the early phases of regression following the Tres Esquinas stillstand.

In discussion of the above data the most obvious inference is that unique simple interpretations of the zircon ages are not possible. The similarity of the ages for the La Quinta tuff and Perijá gneiss inclines us to interpret them in terms of original subaerial volcanic and uplift ages, respectively, in the time frame of U_{1b} with possible minimal partial annealing during Cretaceous-Tertiary burial. Interpretation of the age on the Totumo andesite is complicated by the uncertainty as to its habit (extrusive or intrusive) and structural setting. The similarity of the Lajas (79) and two young Palmar granite ages (69, 71) from areas of clearly different tectono-structural setting suggests, in the light of regional stratigraphic relationships, that their interpretation in terms of uplift ages is possible though far from proven. The interpretation of the zircon ages is considered further in the section on regional tectonic interpretation, below.

THE SANTANDER MASSIF

Geologic Setting of the Area Sampled

The location of the Santander Massif with respect to other ranges of the circum–Maracaibo Basin region is shown in Figure 1. Sample localities may be grouped as northern, central, and southern, corresponding to maps A, B, and C, respectively, of Figure 7 (cf. inset map). Sample numbers corresponding to the map location numbers of Figure 7 are given in Table 3. The regional stratigraphy is summarized in Figure 8. Geologic relationships particularly pertinent to this study include the following:

1. Erosion has hindered precise estimates of the thicknesses of formations of the cover section. Those shown in Table 3 are based mainly on data from Julivert and others (1968, especially Figs. 14 and 22, and related text) and Ward and others (1973). It can only be conjectured if (and to what thickness) cover of Eocene-Oligocene age may have existed over the massif. Reasonable estimates suggest that the total thickness of cover section approximated that for the other ranges in the region but was notably thinner than that in the southwest limb of the Cordillera Oriental.

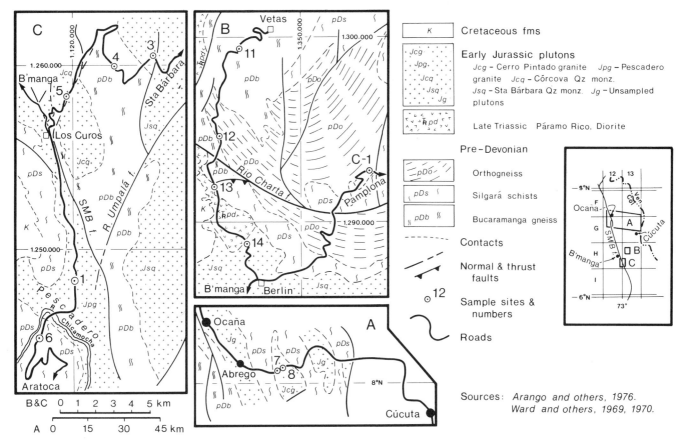

Figure 7. Geologic maps and sample localities of the Santander Massif. Inset and map A modified after Arango C. and others (1976); maps B and C modified after Ward and others (1969, 1970). Sample numbers corresponding to the map location numbers appear in Table 3. Letters F–I and numbers 12 and 13 on inset map are Ingeominas sheet identification code.

2. The most prominent structural feature is the Santa Marta–Bucaramanga (SMB) fault (see Figs. 1, 7, and 15), but there is a striking development of faults oriented northeast-southwest, northwest-southeast, and north-south, transverse or oblique to the north-northwest–south-southeast trend of the topographic range, in marked contrast to the Venezuelan Andes where major faults are parallel to both the range and the regional grain. As a result the Santander Massif exhibits a pronounced mosaic of rhombic blocks. Unfortunately, the structural evolution of the Santander Massif is not known in sufficient detail to test the validity of the uplift events inferred from the apatite ages, in contrast to the situation in the Perijás (Fig. 6).

3. Eastern piedmont deposits (Guayabo delta complex; see Fig. 1 for location) that formed in response to uplift of the Santander Massif do provide a suitable criterion to check the apatite ages. Western piedmont deposits, having derived some of their sediment from the Cordillera Central, are less useful in this regard. The intensity of progradation of the deltaic deposits is described by James (1977) and by Van Houten and James (1984). They were unable, however, to fix the precise age range of the largely unfossiliferous sediments but inferred a maximum age of about 15 Ma on the basis of the age range assigned to the con-

formably underlying, fine grained, paralic León Formation. That the latter extends into the early Miocene is indicated by some sparse faunas (see Léxico Estratigráfico de Venezuela, 1970), but the dating is imprecise.

A younger phase of piedmont deposition is indicated by the presence of thin but coarse boulder deposits of probable Pleistocene-Holocene age sharply prograding beyond the limit of the Guayabo delta. Stratigraphic data are scanty (see "Necesidad Formation" in Léxico Estratigráfico de Venezuela, 1970), but comparable deposits (Milagro Formation) described by Hedberg and Sass (1937) and by Kellogg (1981) from the Perijá foothills point to regional significance of this phase of piedmont deposition and of corresponding uplift in the ranges.

The significance of the piedmont deposits with respect to the apatite ages is discussed with reference to Figure 11 below. Here it is appropriate to note only the following:

• The initial gentle phase(s) of uplift may be represented by the fine muddy sediment of the León Formation rather than by the Guayabo deltaic deposits. Alternatively, the León may be viewed as a lateral facies of the Mugrosa/Colorado Formations (Fig. 8), the sedimentary source having been the Cordillera Central. The former alternative is considered more likely as James

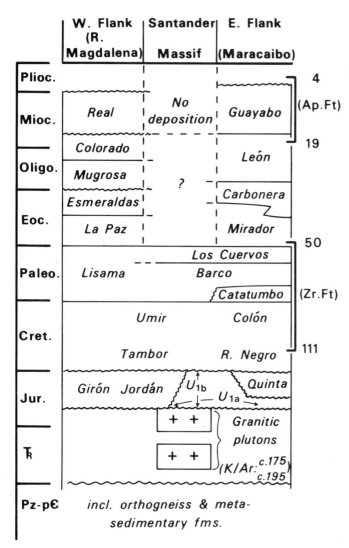

Figure 8. Mesozoic-Cenozoic stratigraphy of the Santander Massif and adjacent areas. Composite of data from Cediel (1968); Julivert and others (1968); Goldsmith and others (1971); Ward and others (1973); Bellízzia and others (1976); and Van Houten and James (1984). Numbers marked "Ap-Ft" and "Zr-Ft" give corresponding stratigraphic ranges of apatite and zircon fission-track ages, respectively, reported herein.

(1977) noted evidence for a crystalline provenance for most of the Guayabo deposits (i.e., there must have been prior removal of much of the cover over the massif).

• There is progressive progradation of increasingly coarser debris up-section with no evidence of angular discordance between or within the Cúcuta, Cornejo, and Urimaco Formations (Fig. 11). These relationships are most reasonably (although not uniquely) interpreted to reflect progressive increase in the rate and amount of uplift of the Santander Massif with time; also, that the bulk of the piedmont deposits accumulated beyond the geographic limit of uplift.

• Renewed piedmont deposition (mainly Pleistocene-Holocene time) of coarse deposits resulted from pronounced rapid uplift. This is supported by the evidence of steep to overturned tilting of Pleistocene terrace deposits near Bucaramanga reported by Julivert (1970).

4. The Santander Massif resembles the Venezuelan Andes in that crystalline rocks dominate the outcrop in the central high mass of the ranges and, as one moves north-northwest and south-southeast along the physiographic length to lower elevations, one finds progressive preservation of late Paleozoic and Mesozoic sedimentary rock units in outcrop. A striking feature of the high crystalline rocks is their deeply weathered friable character in outcrop. In some regions (e.g., near Berlin and in the Ocaña-Abrego area; see Fig. 7) deep regolith, lateritic in places, indicative of intense weathering in a low-level, tropical environment, is preserved. These features suggest that the amount (and rate?) of uplift were notably greater in the outcrop region of the crystalline core and that at least part of the uplift occurred in the recent past.

Fission-Track Ages and Their Interpretation

Analytical data on the Santander samples are given in Table 3.

Apatite Ages. As with the data for the rest of the circum-- Maracaibo Basin area, the apatite ages are best interpreted as cooling ages related to uplift and erosion. In parallel with the samples of the Venezuelan Andes (Kohn and others, 1984), and in sharp contrast to those from the Perijás and Toas, the sampled localities occur through a considerable range of elevation, as shown in Figure 9, which must be taken into account in assessing the uplift history.

Interpretation of the data may be undertaken in two distinct frameworks. The first (**A**) is to view the ages with respect to the fault-bound mosaic of blocks. This approach stems from the results obtained in a tectonic block (El Carmen) in the Venezuelan Andes (Kohn and others, 1984). This approach leads to a model of disjointed block uplifts occurring at different times and rates. The second (**B**) (suggested by reviewer, R. A. Zimmermann) places greater significance on regional grouping of data points largely ignoring the detailed structural mosaic. This view favors a model of continuous uplift though at changing rate marked by break points where uplift curves show a sharp change in slope. Below, the two modes of interpretation (**A** and **B**) are considered in turn and then assessed against the geologic evidence.

A. The Santa Bárbara batholith, represented by some of the southern and central samples, resembles in size and relief the Carmen block in the Andes. We hoped to find a linear uplift curve of Carmen-type and the two granite samples (3S and 14C) do define a zone (a-a' in Fig. 9) of positive slope of uplift rate about 200 m/Ma. However, the validity of the curve is dubious in view of the results on adjacent country rocks of the same tectonic block. Whereas samples 5(S) (Córcova Qz-monzonite) and 13(C) (Páramo Rico tonalite/diorite) are consistent with the two Santa Bárbara granite samples, others, notably 4(S) (Bucaramanga gneiss) and the rest of the central group of samples

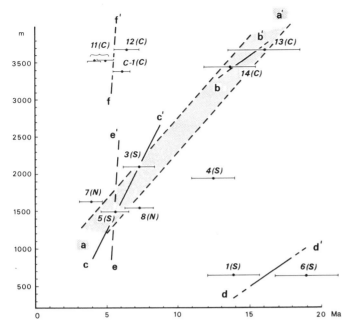

Figure 9. Elevation/age plot of Santander apatite fission-track ages (dots) and precision (bars). Sample C-1 = C-78-1 in Table 3; all other samples have the prefix "BV-79." (N), (C), and (S) signify northern, central, and southern groups of sample localities, respectively. See text for interpretation of possible uplift curves a-a' through f-f' in the framework of separate uplifts of fault-bound blocks (model A). Curve a-a' shown as stippled zone symbolizing measurement precision; such uncertainty applies to all curves shown.

(Vetas block), plot well off a-a'. As explained below the wide spread ages of the central group of samples may possibly be related to the effects of the transverse faulting noted in the introduction, but the detailed map of Ward and others (1970) reveals no suitable fracture to explain the discrepant age of sample 4(S). In the framework specified here of block uplifts one may ascribe the age of sample 4(S) to some local effect (e.g., low thermal gradient) and interpret the remaining sample pairs, 3(S) and 5(S), 13(C) and 14(C), in terms of the poorly defined curve a-a' shown as a stippled zone; alternatively, each sample pair may be viewed as defining the curves (b-b' and c-c') of different slope, c-c' perhaps the more reliable in view of greater topographic separation and precision of measurement. Curves b-b' and c-c' imply successive rotations about a horizontal axis oriented approximately northwest-southeast and located somewhere between the paired outcrop areas. The explanation resembles that proposed by Kohn and others (1984) for the Valera pluton in the Venezuelan Andes (see their Fig. 9).

Samples 1(S) and 6(S) may be distinguished from the remaining (S) samples inasmuch as they occur in a tectonic block (Pescadero) separated from the latter by the major SMB fault. The two samples come from sites at similar elevations about 4 km apart. Within their precision, the two measurements do not overlap, but it is not reasonable to infer separate uplift events for these relatively closely spaced samples especially in view of the absence

of significant faulting in the intervening areas (see map of Ward and others, 1969). A small amount of slow scissorlike rotation in uplift (possibly on the SMB) could explain the earlier triggering of the apatite clock of sample 6(S), or there may have been some local nontectonic cooling effect. On the basis of geologic evidence given below, we think that the data are most reasonably interpreted in terms of a single main pulse of uplift represented by d-d' in Figure 9, of arbitrary slope.

A similar problem to the foregoing arises in the interpretation of the ages on samples 7(N) and 8(N) which are from localities only 1.5 km apart and which probably represent different petrographic facies of one pluton. The ages are interpreted in terms of a single event of uplift (e-e') of unknown rate arbitrarily shown as of intermediate age.

The most serious obstacle to simple interpretation of the data is evident on consideration of the central group of samples which are located in a circle of diameter 11 to 12 km (see Fig. 7) and in a restricted range of elevations (3,400–3,700 m) but which show a wide range of ages (4.2–16.0 Ma). The area is dissected by faults of considerable lateral extension, but the map pattern is not suggestive of especially pronounced displacements. If the samples were located at a depth not far below that of the blocking temperature, the wide range of ages might be explained by relatively small initial displacements on the faults, principally the Rio Charta (Fig. 7), which may also have involved a measure of scissorlike rotation. As previously mentioned the ages on samples 13 and 14 are believed to define an event of uplift such as a-a' or b-b' (Fig. 9). Because of their relatively wide lateral distribution, the presence of intervening faults, and modest differences in elevation, no reliable uplift curves connecting paired samples among 11, 12, and C-1 can be drawn; rather, they would be ascribed to a single event such as f-f' of unknown slope.

The interpretation of the data in the fault-block framework thus leads to four or five separate uplift curves (depending on the choice between a-a' and b-b', c-c') for which the rate can only be approximated for the last three listed.

B. The alternative mode of interpretation of the data is illustrated in Figure 10. The model is largely determined by the ages on the central group of samples. A flat elevation versus age curve of slope ~10 m/Ma extends from about 16 Ma to a break point at about 5 to 8 Ma where it steepens dramatically, corresponding to an uplift rate of 1 to 2 km/Ma. Other similar dog-leg curves can be constructed as indicated in Figure 10. In summary, barely perceptible continuous uplift began in early to middle Miocene continuing almost to the close of the epoch. Relatively sudden and sharp increase in uplift rate then occurred throughout the massif, the resulting erosion fortuitously exposing remnants of the early gentle uplift at different elevations in different areas.

In comparing the apatite ages with the history of development of the Guayabo delta one must expect the former to lag behind the depositional events: a certain amount of uplift and erosion must occur in order to promote downward migration of the 100 °C isotherm and the progressive setting of the apatite clocks. The slower the rate of uplift the greater the time lag; the

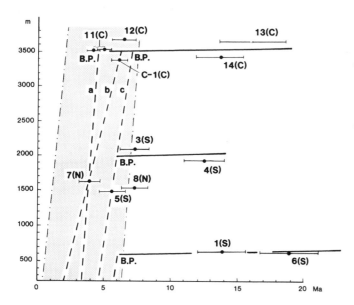

Figure 10. Alternate interpretation (model B) of Santander apatite fission-track ages. Symbols as for Figure 9. This model is characterized by continuous uplift marked by initial very slow rate passing through break points B.P. to subsequent fast uplift rates indicated by steeper curves. During the younger, faster uplift there was differential upward displacement of the remnant older slow phase, which is now encountered at different elevations in different areas. See text for additional commentary. Fast-uplift curves have unknown slopes in stippled area; three possibilities (a, b, c) among many others are shown.

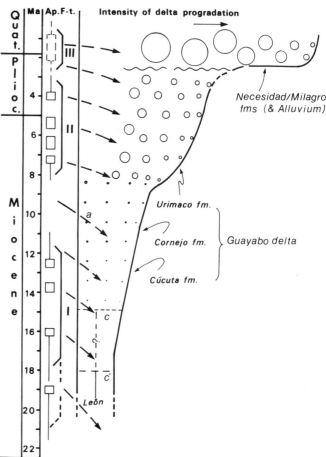

Figure 11. The age range and intensity of progradation of the Guayabo delta complex and overlying Pliocene(?)-Pleistocene deposits vis-à-vis the apatite fission-track ages. Correspondence between the column of apatite ages and that of the piedmont deposits is indicated by the heavy dashed arrows. The slope of the latter is a qualitative representation of the time lag to be expected between deposition of the sediments and setting of the clocks (see text); the faster the uplift the shorter the time lag. For convenience in discussion, the apatite ages are arbitrarily grouped as I, II, and III. The last is shown as a single dashed box to indicate that no corresponding ages have been measured, but their presence below outcrop is inferred from the youngest piedmont deposits (Necesidad and Milagro Formations). Vertical arrows on some age boxes indicate maximum range of precision of measurements for each group. Uplift indicated by the oldest age may be represented by the fine sediments of the underlying León Formation rather than by the Guayabo delta deposits. Horizontal dashed lines C and C' are estimates for the age of the Cúcuta/León contact by Van Houten and James (1984) and the authors, respectively. See text for significance of dashed arrow "a."

latter is partly counterbalanced by a small lag between uplift and the accumulation of *significant* piedmont deposits. The heavy dashed arrows on Figure 11 are intended to symbolize the foregoing, although no quantitative significance can be attached to the slopes indicated. It is seen that there is reasonable consistency between the apatite ages and delta development. The oldest age is represented by the fine sediment of the León Formation and hence inferred to have defined a point on a slow uplift curve. Ages arbitrarily combined as group I (Fig. 11) may signify faster uplift rates as evidenced by the coarser sediments of the Cúcuta and Cornejo Formations. The relationships of Figure 11 suggest to us a slightly older limit for the León/Cúcuta boundary than that roughly estimated by Van Houten and James (1984). James (1977) reported thicknesses up to 155 m and 612 m for the Cúcuta and Cornejo Formations, respectively. Their combined age range can only be roughly estimated at about 7 to 10 Ma. Keeping in mind that piedmont deposits also accumulated west of the massif and that the combined areas of deposition approximate in area that of the massif, one infers about 50 to 100 m/Ma of sediment accumulation requiring a somewhat faster uplift rate than the 10 m/Ma of the flat portion of the uplift curves in Figure 10, although no great reliance can be placed on such estimates with the data available.

The fact that the gap between the group I and II ages is not represented by an obvious hiatus in the deltaic deposits can be

interpreted to signify lowering by erosion of the positive terrane created in uplift phase I (dashed arrow "a" on Fig. 11).

A decided increase in the rate and amount of uplift (group II ages) is reflected in the pronounced progradation and far coarser and thicker deposits (up to 1,085 m measured by James [1977]) of the Urimaco Formation. Yet further progradation of even coarser deposits (Pliocene?-Pleistocene Necesidad/Milagro Formations and overlying alluvium) point to a final distinct phase of uplift, possibly still in progress, at even faster rates of uplift. This is supported by the evidence for patches of deeply weathered remnants of regolith on the high massif. Corresponding rocks have yet to be exposed by erosion, but at some unknown depth below current outcrop a sharp inflection point in the slope of uplift curves can reasonably be predicted. Such inflection points on apatite uplift curves have been described by Wagner and others (1977) for the Alps.

As outlined above, the impression gained is of distinct phases of uplift at progressively increasing rates, which is not consistent with the simple two-phase continuous uplift history of model **B**, above. On the other hand, distinct spurts of uplift do not necessarily support model **A**; the entire massif could have been equally affected by those uplifts. Moreover, it can be argued that relatively minor changes in model **B** (see below) could yield a reasonably close fit to the age and field evidence.

As in many controversies, the final solution may prove to be a compromise between the extremes of models A and B. In the general case, differential uplift of fault-bound blocks is favored by evidence for such effects at several intervals going back to the tectonic environment of deposition of the Jurassic La Quinta and Girón red beds, as described in previous sections. In the specific, Kohn and others (1984) note the very different mechanics and age ranges of uplift of two blocks (El Carmen and Valera) in close proximity one to the other in the Venezuelan Andes. Moreover, the apatite ages in the Perijás (see corresponding section) also argue for distinct phases of uplift affecting different blocks at different times. Additionally, Hedberg and Sass (1937) noted a thickness of about 1,500 m for the El Fausto Group now believed to be of probable early to middle Micoene age. Such a thickness of deposits bespeaks pronounced compensating block uplift in the Perijás rather than gentle epeirogenic uplift. A parallel conclusion is drawn on the basis of the 612 m of Cornejo Formation with respect to uplift of the Santander Massif. Accordingly, we prefer to interpret the older ages of the central group of samples in terms of curves a-a′/b-b′ rather than the flat curve of model **B** (Fig. 10). The indication for progressive increase in the rate of uplift with time favors uplift of the Santa Bárbara block in terms of the two curves b-b′ and c-c′ rather than the single curve a-a′. A degree of compromise between models **A** and **B** can be attained by displacing the flat curve of epeirogenic uplift in model **B** further back in time. The principal evidence here is the fine paralic muds of the León Formation suggesting very gentle uplift through much of the Oligocene and earliest Miocene. In turn, initiation of significant break points in the slopes of uplift curves would begin in early to middle Miocene time rather than latest Miocene, as shown in Figure 10. In further compromise, initial differential uplift of fault-bound blocks may have given way to Pliocene-Pleistocene region-wide uplift because of progressive interlocking of blocks.

The uplift events and characteristics discussed above are summarized in diagrammatic form in Figure 12; the somewhat unorthodox plane of section is unavoidably related to the geographic distribution of the main sampling areas. The principal inferences tentatively drawn are as follows:

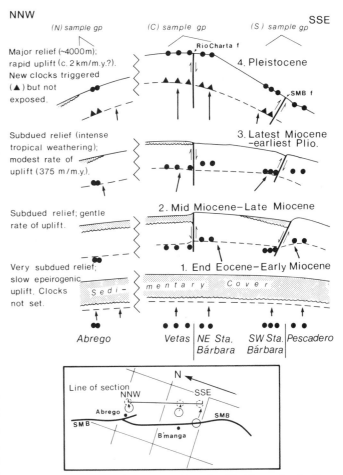

Figure 12. Diagrammatic representation of uplift history for the Santander Massif following model A (Fig. 9). With the possible exception of the oldest age measured (see Fig. 11), apatite ages record stages 2 and 3 only. Stage 1 is predicated on the need for a provenance for the piedmont León Formation and to expose crystalline basement for the provenance of the Miocene Guayabo delta complex. Note that initial uplift of the Pescadero block in stage 2 implies a reverse sense of movement on the SMB fault to that obtaining subsequently. Also, the main uplift component of most samples (especially the central group) occurred in the Pleistocene (stage 4) after initial triggering in stages 2 and 3; hence consideration of the age data on the sole basis of elevation/age plots can be misleading in areas with polyphase uplift history.

1. Whereas apatite fission-track dating suggests only two uplift phases (early to middle Miocene and late Miocene to early Pliocene), additional geologic evidence suggests a four-phase uplift history possibly extending back to the earliest Oligocene and continuing to the present.

2. In broad overview, it appears that initial uplift was of minor amount, occurring at a slow rate through a relatively lengthy time span. This changed progressively to major uplift occurring at faster rates through shorter time spans. On elevation/age plots of apatite ages, some of those characteristics should take the form of a fanlike array of uplift curves becoming progressively steeper toward the present (see Fig. 9). A corollary of the forego-

ing is that the greatest uplift event may not be represented by apatite clocks in current outcrop.

3. The only means of substantiating the postulated earliest phase of uplift would be to date clastic apatites from the León and the basal portion of the Guayabo piedmont deposits.

4. In regions of pronounced subvertical block uplift it is reasonable to expect that uplift curves should show positive slopes on elevation/age plots such as Figure 9. One might thus infer that samples 7(N) and 8(N) fall on an older uplift curve than samples 11(C), 12(C), and C-1(C). However, in view of the lateral separation of the two sampling areas by about 90 km and their location in separate tectonic blocks, the relationships between the two groups of samples could be explained by different amounts and/or rates of uplift during one and the same tectonic event, as shown in Figure 12. The same applies to the relationships between samples 13(C) and 14(C) and 1(S) and 6(S); clearly, from Figure 12 the elevation of the younger b-b' curve (Fig. 9) could be an artefact of the postulated Pleistocene-Holocene event. The important conclusion drawn is that elevation/age curves such as Figure 9 can be misleading unless they are carefully integrated with a wide variety of geologic data.

Because of inadequate sampling, particularly in the transverse (east-west) direction, it is not possible here to suggest an uplift sequence by area other than the fragmentary events shown in Figure 12, in contrast to the sequential stages of uplift of the Venezuelan Andes presented by Kohn and others (1984).

Zircon Ages. The most notable feature of the zircon ages (Table 3) is the bimodal distribution into end-Cretaceous-Paleocene (50–67 Ma) and middle Cretaceous (91–109 Ma) fractions. The zircon clocks must have begun to operate in Middle to Late Jurassic time during the uplift and cooling related to the cutting of the unconformities U_{1a} and U_{1b} (Figs. 2 and 8) below the Mesozoic sedimentary formations. Subsequently, the zircons underwent burial during the deposition of the Late Jurassic and Cretaceous/Paleocene formations. Clearly, one cannot simply interpret those ages to represent earlier phases of a continuous uplift history, the later phases of which are registered by the apatite clocks. To suggest that the three samples that give Paleocene ages represent an event of uplift implies complete annealing of their clocks during burial, and there is doubt whether the thickness of cover section was sufficient for that purpose. The Santander Massif is viewed as constituting a negative but distinct culmination in the Cretaceous separating the areas of more pronounced subsidence of the Bogotá leg of the Cordillera Oriental to the southwest from the northern Santander–Perijá region to the north. If the thickness of Cretaceous deposits was correspondingly less over the massif, it may also signify unusually high geothermal gradients. There is also doubt concerning the thickness of Upper Jurassic Girón Formation over the massif. Thicknesses exceeding 4 km are known west of the massif but may have been considerably less on the massif. In total, there may have been only 3 to 4 km of cover over those samples. The apparent conformable transitional relationships of the Paleocene to the Cretaceous section reported by Ward and others (1973)

are not a serious obstacle to the postulate of an event of Paleocene uplift. In a region of epeirogenic uplift with sparse faunas, distinct hiatuses may easily be missed. Indications that subsidence was drawing to a close in the Maestrichtian are the presence of coal partings. Initial Paleocene uplift of the crystalline floor may have occurred at the same time that subsidence related to compaction was occurring at near-surface depths. We speculate that the three Paleocene ages do not reflect an event of uplift and that relatively shallow burial was compensated by unusually high heat flow. In view of the uncertainties noted above and the regional evidence for phases of post-Cretaceous subsidence prior to the registry of the apatite clocks, it is not considered reliable to interpret these ages with their coexisting apatite ages in terms of one continuous event of slow uplift. The mid-Cretaceous group of ages cannot reasonably be explained on the basis of uplift, nor can they be interpreted on the basis of cooling during subsidence inasmuch as the clocks were operating before the deposition of the Mesozoic formations. It is glib to conclude simply that they represent mixed ages inasmuch as some samples come from localities adjacent to others (e.g., 11 and C-1) that do appear to have been completely annealed. Sharp local differences in geothermal gradient or active faulting during subsidence resulting in significant differences in depth of burial over short lateral distances may have played secondary but important roles in the annealing history.

SYNTHESIS OF FISSION–TRACK AGES

A basis for discussion of the tectonic implications of the fission-track ages may best be established by reviewing all the age data for the circum–Maracaibo Basin region. Apatite ages are summarized in Figure 13. The first uplift phase for which there is reasonable statistical support is that of the Palmar area on the trailing side of the Perijás (A on Fig. 13) which is postulated to have been largely coeval with the initial uplift in the Santander Massif (curve B), both of unknown rate and duration. For the reasons given in the corresponding regional descriptions of the data, curves A and B are shown to imply gentle uplifts at a slow rate of long duration extending possibly from earliest Oligocene through the early Miocene. The oldest age measured in the Venezuelan Andes (about 24 Ma) is tentatively discarded from the viewpoint of uplift history. Although the age has been confirmed by repeated measurement, first results of additional measurements on the same (Valera) pluton currently in progress suggest much younger ages in the vicinity of the old, raising the possibility that it is a partly annealed age originally set in Mesozoic time.

In the middle Miocene, coeval at least with the closing phase of curve B, was the uplift of the northeastern portion of the Santa Bárbara block in Santander. As indicated in the previous section and illustrated in Figure 12, the markedly higher elevation of the curve C samples does not necessarily indicate a far younger uplift curve. To establish whether the faster uplift rate of curve C is real or not requires additional sampling and measurement. In turn, this uplift overlapped partly with that of Toas (D) to be followed by that of the leading (northwest) flank of the Venezuelan Andes

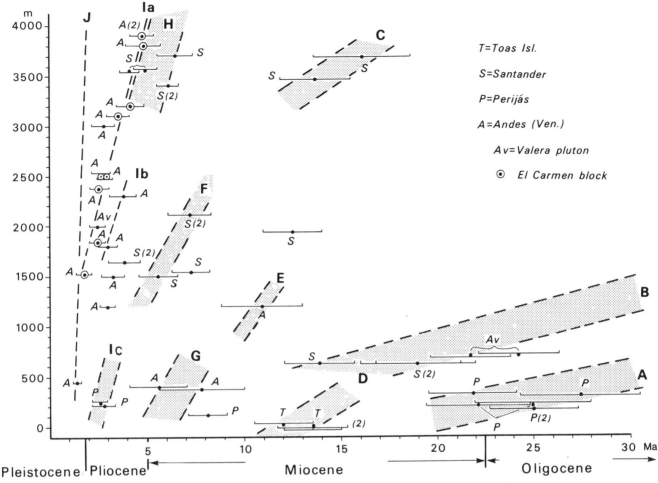

Figure 13. Elevation/age plot of all apatite data from the circum–Maracaibo Basin region. Data for the Venezuelan Andes after Kohn and others (1984). Fan-shaped array of uplift curves is based largely on regional geologic evidence; curve Ia (El Carmen block, Venezuelan Andes) is the only well-established curve based solely on fission-track dating. Apatite ages (dots) and precision (bars) shown; (2) indicates duplicate measurements. Uplift zones are stippled to symbolize precision of measurements.

(E), both of unknown uplift rate and duration. Firmer indication of steeper uplift curves first appears in the late Miocene in the results for the southwestern part of the Santa Bárbara block (F), which was contemporaneous with the uplift of the trailing side of the Venezuelan Andes (G).

What had been initially a counterpoint of block uplifts became increasingly a region-wide harmony of uplifts of all the major blocks during the Pliocene. Uplift of the Vetas block in Santander (H) occurred, closely followed by events in the Venezuelan Andes (Ia, El Carmen block; Ib, axial zone) and Perijás (Ic). Of these only the Carmen uplift is reasonably established as to rate and duration, with the reservations noted by Kohn and others (1984).

The final Pliocene-Pleistocene event (J) is postulated on geologic criteria only, outlined in the previous section, and is believed to have affected all major blocks. A single age of 1.4 Ma in the Venezuelan Andes is the sole fission-track age that conforms with this event, but as Kohn and others pointed out it could

be interpreted in terms of headward erosion only indirectly related to tectonism.

The fanlike array of the uplift curves on Figure 13, implying progressively faster uplift rates of shorter time duration, is based largely on regional stratigraphic evidence outlined in the sections on the Perijás and Santander Massif but is clearly our subjective evaluation. Moreover, Figure 13 is incomplete to the extent that it depicts only those parts of uplift curves for which there are age data. Curve A (Palmar area of the Perijás), for example, defines only an event of uplift that terminated in the earliest Miocene. There must have been one or more subsequent phase(s) of uplift to bring it to its present elevation. Inasmuch as there appears to have been progressive increase in rates of uplift with time, one would expect to find that curve A would pass through one or more inflection points to progressively steeper curves were it possible to obtain and date a series of samples from a deep drill hole. This situation probably obtains for all the curves of Figure 13 so that if and when the total uplift history is defined, we

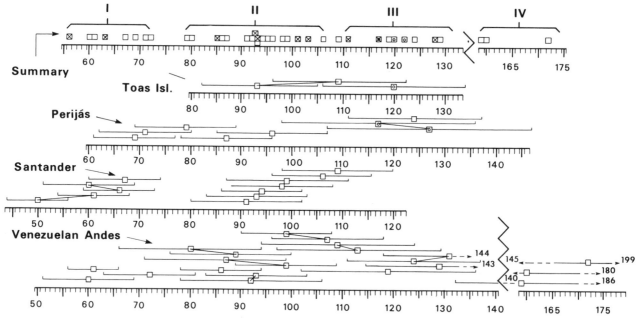

Figure 14. Plot of all zircon fission-track data for the circum–Maracaibo Basin region. Ages (squares) and precision (bars) shown; duplicate measurements indicated by tie-lines between squares. At top is summary of all data (ignoring precision); squares with crosses are average of duplicate measurements. Data for the Venezuelan Andes are from Kohn and others (1984). A geologic basis exists for distinguishing groups I and IV (see text), but the definition of groups II and III is arbitrary and for convenience in discussion only. Ages in group I are viewed as uplift ages; ages in groups II, III, and IV are tentatively interpreted as mixed ages, though squares enclosing circles could represent original crystallization ages.

predict that the fanlike array will show pronounced kinking. A hint of the fan can be seen by comparison of curves Ia, F, and C, but of these only the first can reasonably be considered as well established. Nonetheless, we are emboldened to postulate the progressive change in the nature of the uplift history as a working hypothesis for future evaluation.

A varied and increasingly complex nomenclature for the uplift history of the northern Andes has developed as workers have begun to unravel the sequence of tectonic events (e.g., Giegengack, 1984). Commonly, "Andean" is reserved for Pliocene-Pleistocene uplift events (e.g., Campbell, 1968; Duque-Caro, 1979; Kellogg, 1981, 1984). Thus, for the overall uplift history of the Perijás and environs, Kellogg (1981) suggested an initial event in the early Eocene (about 50 Ma), that is, preceding curve A in Figure 12, and three variously named subsequent phases terminating in Miocene-Pliocene Andean [sic] orogeny. We prefer to view all these events in the framework of "Andean" uplift in view of the connecting threads of uniform tectonic style and overlapping in time of the uplift history. The question then is when the process began. Evidence for an initial phase at the close of the Cretaceous appears in some of the zircon fission-track ages summarized in Figure 14.

Four age groups (I–IV) are shown in Figure 14. Group I corresponds approximately, disregarding precision, to the range of Maestrichtian-Paleocene time, and the ages (3 each from the Venezuelan Andes and Santander and 2 or 3 from the Perijás) have been interpreted as uplift ages. Kohn and others (1984)

inferred that Paleogene section was deposited conformably on Cretaceous rocks only in graben areas and not in adjacent tilt block areas, suggesting an uplift phase at the close of the Cretaceous. Hedberg and Sass (1937, p. 109) inferred a general Caribbean-wide emergence at that time, which is reflected in a pronounced increase in sandy over shaly facies as one passes from Late Cretaceous into Paleocene section in northwestern Venezuela, and by the presence of heavy mineral suites indicative of a crystalline basement provenance. The map and correlation Table I of Bellízzia and others (1976) show pronounced facies changes in short lateral distances and indications of hiatuses along the northern flank of the Venezuelan Andes (see also descriptions of the relevant formations in the Léxico Estratigráfico de Venezuela, 1970). The geologic record clearly points to significant uplift(s) in latest Cretaceous-Paleocene time throughout the circum–Maracaibo Basin region and beyond. As indicated in the discussion of the Santander zircon ages, it would be reading more into the significance of the data than one may reasonably do to view group I ages as representing earlier phases of a continuous uplift history whose terminal phases are represented by the coexisting apatites.

Group II ages fall in the range of 79 to 106 Ma (mainly concentrated in the range 85–101 Ma). This invites speculation of a tectonic pulse at that time (mid-Albian-Cenomanian) which, however, is known beyond doubt to have been one of subsidence and marine transgression regionally, an unpromising framework in which to postulate uplift. For a variety of reasons noted in the regional discussions of the data, it was suggested that group II ages may be partially annealed ages of no true time significance, for lack of viable alternative. In the framework of graben and adjacent uplifted tilt blocks in the Andes (see previous paragraph), Kohn and others reported a group II age of about 93 Ma on sample VE-2 from a graben area (hence more deeply buried to the extent of the thickness of Paleogene section) compared to a group I age of about 60 Ma on sample SA-79-8 collected from the adjacent uplifted block. The interpretation resembles that of Naeser (1979) on apatite ages from Mount Evans, Colorado (his Fig. 12). Our unease over the general "mixed-age" interpretation stems from the fact that all regions (including Toas) are represented in the group, implying a mechanism that would lead to the same remnant age from four widely spaced regions of disparate thermal and tectonic histories. Attempts to relate annealing of the zircon clocks (most if not all of the group II clocks must have been operating in Late Jurassic–earliest Cretaceous time) to processes other than the effects of simple burial have not been rewarding. Reconstruction of hotspot tracks by Morgan (1983) reveals such a feature (Trinidade) suitably located in space but not in time. Extensive basaltic volcanism in the offshore islands north of Venezuela in the Early Cretaceous has been related to southward subduction of the Caribbean by Beets and others (1984), but it is not likely that that thermal regime could have affected the shallowest crustal levels in the Venezuelan Andes. Interpretation of the Maracaibo Basin in terms of a failed spreading center (Burke, 1976) also does not provide a satisfactory basis to explain the group II ages for similar reasons.

The suspicion that the group II ages may have time significance stems from consideration of the concept of thermally controlled subsidence which proceeds at a linear rate as the square root of time for about 70 to 80 Ma. The obvious setting for such a process would be in Late Jurassic–Santonian/Campanian time, with the beginning of cooling occurring as La Quinta plutonism and volcanism began to wane, and the end of the linear rate of subsidence indicated by the stillstand of the Tres Esquinas hardground (see Fig. 2). The group II ages correspond to the final 20 Ma of the postulated cooling event. However, until there is a satisfactory explanation for the annealing of the clocks prior to the initiation of cooling and subsidence, no rational interpretation of the group II results in terms of true ages is possible. This problem merits further study.

Two of the group III ages (one each from Toas and the Perijás) may be original crystallization ages, but the remainder can only be viewed as mixed ages; the problem of explaining them lessened only to the degree that annealing was less. Concerning the three samples of group IV from the southeast piedmont of the Venezuelan Andes, Kohn and others (1984) conceded the possibility that they could represent uplift ages but preferred to view them too as mixed ages. No inference of tectonic significance may be drawn from such a modest sample population.

In summary, it may be said that the fission-track ages and other geologic evidence suggest initiation of Andean block uplift approximately in Paleocene time with further local phases of uplift in the early Eocene, end-Eocene to early Miocene, mid- and late Miocene, terminating with region-wide uplift of all blocks in the Pliocene-Pleistocene, and possibly continuing to the present. The image formed is one of uncoupled crustal blocks being uplifted in response to more local stresses until regional compression led to interlocking of the blocks and their uplift in virtual unison during the Pliocene-Pleistocene.

REGIONAL TECTONIC INTERPRETATION

A rational explanation for the late stages of uplift is immediately apparent on consideration of Figure 15 which shows the circum–Maracaibo Basin ranges to be major tectonic components of a triangular continental block with apex in the Santa Marta region of Colombia and sides defined by the Oca and SMB faults. The postulated regional stress field with principal stress oriented approximately northwest-southeast through at least late Tertiary time is in conformity with the suggestion of Case and MacDonald (1973) of underthrusting of the continent by the Caribbean. Clearly, increasing interlocking of blocks and resulting increase in amount and rate of uplift would be expected as they were compressed toward the confined apical region of the triangle. An elaboration of the concept in terms of slow subduction along a gently dipping plane, the surface trace of which is the southern Caribbean marginal fault (Fig. 15), has been presented by Kellogg and Bonini (1982). Here no distinction is drawn between "underthrusting" and "subduction," though the two terms may imply very different thermal regimes and mechanisms of genesis.

Suitable explanations for the earlier phases of the uplift history are less clear, but alternative possibilities are suggested by two lines of evidence.

1. The region northeast of the Venezuelan Andes is an exotic terrane believed to have been emplaced in latest Cretaceous–Eocene time. The concept was originally developed by Hess, and extensive literature sources may be found in Maresch (1974), Stephan (1977), and Beets and others (1984). The western limit of the allochthonous terrane, according to Stephan, is shown on Figure 15. A slightly more westerly limit indicated by us is intended merely to allow for the effects of erosion and is not based on any specific evidence. The possibility that the allochthon once extended completely over the circum–Maracaibo Basin region is rejected on the grounds of absence of evidence for burial metamorphism in the Cretaceous rocks of the supposed autochthon. Illite crystallinities (Shagam, in prep.) on Cretaceous shales in the region consistently show values well in the range of early diagenesis. The emplacement of such an enormous pile of rocks (Bellízzia, 1972, suggested that the entire Caribbean Mountains including "basement" are allochthonous) through Paleocene-Eocene time may have invited complementary uplift of the circum–Maracaibo Basin region in the framework of a mass-balance effect (although there could have been depression immediately adjacent to the allochthon). The extended period of emplacement of the exotic terrane may have resulted in a corre-

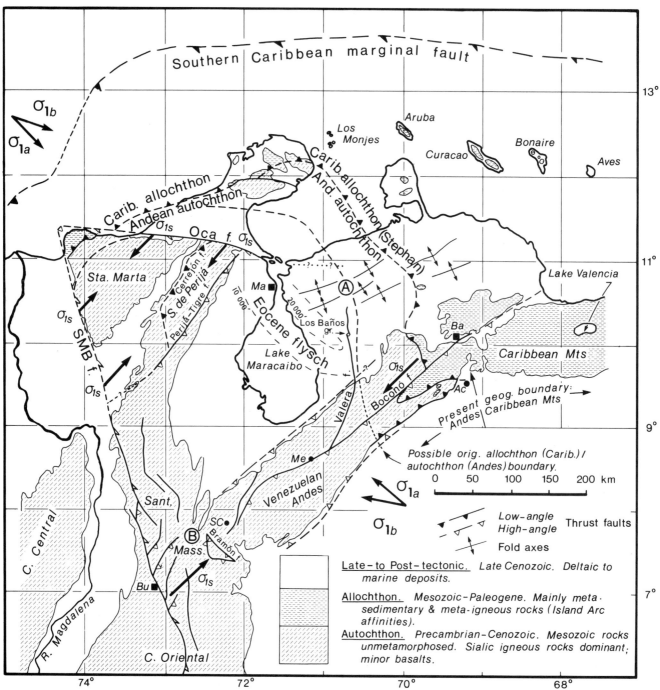

Figure 15. Regional stress field and major tectonic elements. Allochthon was emplaced in end-Cretaceous–Eocene time (Bell, 1972; Stephan, 1977). Principal stress σ_{1a} is that resolved from regional convergence vector σ_{1b} calculated by Jordan (1975) and has operated at least since earliest Oligocene time. Relative northwest thrusting of the triangular crustal plate defined by the Oca and SMB faults involved thrusting of the Venezuelan Andes to the northwest (fault obscured by piedmont deposits) and skin-type thrusts in the Perijás, and required compensating faulting and folding in Falcón (area A) and tear-faulting in the southwest (area B) along the Bramón and other faults (as yet unspecified). Progressive spatial confinement toward the apex of the triangle resulted in progressive interlocking of blocks, harmony of uplift events, and development of subsidiary stresses (σ_{1s} causing longitudinal compression and the pyramidal form of the ranges. The Oca and SMB faults are viewed as oblique-slip type with sense of motion indicated by orientation of teeth. Western margin of the allochthon was affected by uplift events displacing the topographic boundary of the Andes/Caribbean Mountains to the northeast. Map modified after Case and Holcombe (1980).

spondingly lengthy series of oscillating vertical movements to the west, which is in conformity with the evidence for several subtle hiatuses in the Paleocene-Eocene section of the circum–Maracaibo Basin region.

2. An alternative possibility is suggested by the occurrence of the thick Eocene flysch and the El Baño granite inlier east of Lake Maracaibo oriented as shown on Figure 15. Possibly plate interactions at that time resulted in a different regional stress field (oriented more easterly?) than that described by Case and MacDonald (1973) and resulted in a block uplift subparallel to the Santander Massif, of which the El Baño granite is a remnant. However, the sense of emplacement of the exotic terrane in the early Tertiary is consistent with the proposed stress field of Case and MacDonald, and the first alternative posed above is considered the better basis on which to explain the initial uplift phases. In such case, the provenance for the Eocene flysch would have been the transiting allochthon, and its basin of deposition formed in the oscillating mass-balance framework described above.

At some time vertical uplift controlled by the mass-balance mechanism became transformed into relative northwest thrust uplift and a corresponding change from obduction to underthrusting (subduction?), as previously mentioned. The change to the new regime first appears in the latest Oligocene apatite ages in the Perijá piedmont, an uplift event which, for the reasons previously given, may have begun in latest Eocene time. The fact that piedmont deposits became particularly prominent in the region during the Miocene is in harmony with the postulated change in overall tectonic framework. It is logical also to infer that development (or rejuvenation?) of the Oca and SMB faults began at that time.

Muessig (1984) has assembled an impressive mass of evidence and cogently argues for a tensional environment in Oligocene-Miocene time, in contrast to the northwest compression inferred above. He ascribes to right-lateral shear, between the Caribbean and South American plates, the formation of pull-apart basins in an extensive area between the line of the Oca fault (and others farther east) and the line corresponding approximately to the southern Caribbean marginal fault of Figure 15 (see his Figs. 4 and 8). With gentle regional transpression such a shear couple could well have resulted in mild compression oriented approximately northwest-southeast in the region south of the Oca fault.

Relative thrusting to the northwest of the circum–Maracaibo Basin region suggests oblique-slip movement on the two boundary faults as indicated by the teeth on Figure 15. This is in accord with the conclusions of Ward and others (1973) and of Julivert (1970) that there had been a major component of high-angle, dip-slip movement on the SMB but that locally there were indications of left-lateral displacement. In the course of thrusting, structural relief in the range 10 to 15 km developed across those fractures. The sense and amount of movement on the faults imply a corresponding strike-slip component of about 15 to 20 km, in harmony with the estimates of MacDonald and others (1971) and Feo-Codecido (1972) and in contrast to estimates for the Oca of 60 km by Tschanz and others (1974) and about 100 km

by Kellogg (1981) and Kellogg and Bonini, (1982), and for the SMB of about 100 km by Campbell (1968) and by Tschanz and others (1974). The lower estimates proposed here accord with the fact that the geology of the "autochthon" in the northeast Guajira Peninsula (Fig. 15) is akin to that of the Perijás (Maze, 1984). The larger estimates imply a far greater degree of crustal shortening in the direction of the bisector of the apex of the triangle. Bonini and others (1980) suggested that this exceeded 100 km, of which about 25 km was associated with the northwest thrusting of the Venezuelan Andes. The latter estimate is in conformity with the regional geology of those Andes, but the amount of shortening associated with the thrusting of the Perijás is not well established; major imbricate overthrusts mapped by Kellogg (1981, 1984) could be skin-type detachments involving little or no crustal shortening.

A puzzling feature of the Oca fault in view of its supposed structural role is the fact that it appears to die out near the mouth of Lake Maracaibo; one would intuitively have expected its continuation to intersect the northeasternmost extension of the Venezuelan Andes. Muessig (1984; see his Fig. 4 and related text) shows the Oca fault passing eastward into a graben, the southern boundary of which is the Ancon fault. The sense of displacement of the graben is the opposite of that inferred here for the Oca fault and suggests pronounced scissorlike rotation on the fault about a north-south axis located in the mouth of Lake Maracaibo. An alternative suggestion by Vásquez and Dickey (1972), that in eastward continuation, shear on the Oca at depth was transformed into shallow skin-type folding of late Tertiary formations, is ingenious but not mandatory. We take the evidence at face value and conclude that only the thrusting from Toas westward was taken up along the Oca. Keeping in mind the reservations concerning the interpretation of the latest Oligocene apatite age from the Venezuelan Andes, it now appears that northwestward thrusting there occurred in the late Miocene (curve E in Fig. 13). Presumably by that time volume restrictions and resulting frictional resistance hindered further movement on the Oca, and crustal shortening was taken up by folding in Falcón (area A in Fig. 15) and by northwest translation along such faults as the Bramón which affects Miocene-Pliocene(?) fluvial deposits in the southwestern part of the triangle (area B in Fig. 15). The northwest-trending faults in the Santander Massif are viewed in a similar light; left-hand tear motion may have been distributed along several en echelon fractures, each showing no more than a few kilometres of displacement. Macellari (1984) has shown convincingly that the Bramón has a gentle dip to the southwest, and he interprets it as a thrust, but it may also be the near-surface portion of a fracture system with overall flower-structure geometry (Harding and Lowell, 1979) and be vertical at depth. Some of the northeast-trending high-angle reverse faults in the massif could also have formed during the northwest translation of the Venezuelan Andes, their relative youth in the structural history attested to by the fact that some offset the SMB (see maps of Ward and others, 1969, 1970).

An overview of the triangular block of Figure 15 as presented here suggests the following characteristics:

1. It is located entirely within the upper (continental) plate.

2. The three-dimensional shape of the Oca-SMB surface approximates that of a ship's hull with strongly flaring prow in the apical region and bottom imbricate to the descending Caribbean plate boundary.

3. In terms of the uplift history, the definition of the block began in the latest Eocene-earliest Miocene.

4. At least for the circum–Maracaibo Basin portion, uplift was in progress in the area of the Perijás by latest Oligocene, in the Santander Massif by early–mid Miocene, and in the Venezuelan Andes by the late Miocene.

5. In the Pliocene-Pleistocene, compression resulted in a considerable volume problem which was solved by increasing contemporaneity of major uplifts occurring at a fast rate through shorter time spans. Tight packing toward the apex of the triangle would have favored rotation of the boundary faults (clockwise for the SMB; counterclockwise for the Oca) resulting in the broad, gentle, curved traces now seen on the regional scale. Intense normal stresses across those faults at this time would have given rise to subsidiary stresses oriented along the length of the major blocks (see Fig. 15) and would explain the current pyramidal topographic form of the Venezuelan Andes, Perijás, and Santa Marta blocks. This same subsidiary stress field is the likely cause of the minor strike-slip component of motion along such faults as the Boconó and Perijá-Tigre, representing inconsequential shallow spatial adjustments among minor blocks.

Uplift of the Venezuelan Andes affected the western margin of the allochthonous terrane, thereby displacing the geographic boundary with the Caribbean Mountains to the northeast (Fig. 15).

Jaramillo (1978) reported fission-track ages of ~10 Ma on apatite (some erroneously given as zircon?) and 58 ± 6 and 62.4 ± 3.6 Ma on zircon from rocks of the Cordillera Central, Colombia. Apparently Cretaceous-Neogene tectonic effects of regional stresses extended west of the triangular block of Figure 15.

PROPOSED FUTURE STUDIES

The data presented herein and by Kohn and others (1984) provide no more than a reconnaissance survey of the region. As indicated in the discussion of individual areas, there is a need to extend areal coverage and to improve the measurement and statistics and interpretation of the data in regions already sampled. For better understanding of the regional tectonics, the sampling will be extended to the Santa Marta block.

A sample collected from the El Baño granite east of Lake Maracaibo (see Fig. 1) will shortly be processed for fission-track dating. However incomplete, there will then be a thermal history framework to the oil-rich Maracaibo Basin. It is proposed to extend the fission-track dating to core samples recovered from deep boreholes in the basin in parallel with vitrinite reflectance and illite crystallinity studies in order to evaluate the thermal history of the basin. Also, in view of the current high heat flow,

the dating program may provide useful information on the kinetics of annealing of zircon, which would be invaluable to the interpretation of mixed ages.

ACKNOWLEDGMENTS

Only the generous logistical support of the Instituto Nacional de Investigaciones Geológico-Mineras (INGEOMINAS), Bogotá, Colombia, and the Direccion de Geológia, Ministerio de Minas y Energía, Caracas, Venezuela, enabled us to collect the samples dated in this study.

Shagam extends his profound appreciation to S. Judson, former chairman, Department of Geology and Geophysics, for the invitation to spend a sabbatical leave (1982–83) at Princeton University during which this manuscript was written. The current chairman, R. A. Phinney, and all his colleagues proffered such innumerable courtesies, including significant financial support, that perhaps only a poet could find adequate words of appreciation. The invitation of W. E. Bonini and R. B. Hargraves to Shagam to join the editing of the volume in which this study appears was only one of countless acts of generosity that went far beyond normal collegial bounds.

We are also deeply indebted to H. Feldstein for her continued conscientious and efficient supervision in irradiating our samples at the Nahal Soreq Nuclear Research Center, Israel. Some of the fission-track measurements were carried out by Kohn while on sabbatical leave at the Department of Geology, University of Pennsylvania, Philadelphia, Pennsylvania. His grateful appreciation is extended to the chairman, R. F. Giegengack, and all his colleagues for their hospitality and for easy access to the excellent research facilities of the department.

S. Savin, former chairman, Department of Geological Sciences, Case Western Reserve University, Cleveland, kindly arranged for Shagam to use mineral separation and other departmental facilities.

Colleagues and other associates too numerous to list individually provided stimulating discussion of our data, but we alone are responsible for the views expressed herein. The writers much appreciate two extremely useful reviews of the original manuscript by K. W. Muessig and R. A. Zimmermann.

In this, no less than in other cases where manuscripts appear in print, and probably more, the unsung toilers are the technical, library, and administrative staff who kept the wheels of research turning. We express our deepest gratitude to those at Ben-Gurion, Princeton, Pennsylvania, and Case Western Reserve Universities who gave unstintingly of their assistance in those fields. This study was partly supported by a National Science Foundation grant (EAR-7823671) to Banks and Shagam at Case Western Reserve University.

REFERENCES CITED

Arango C., J. L., Kassem B., T., and Duque-Caro, H., compilers, 1976, Mapa geológico de Colombia: Colombia Instituto Nacional de Investigaciones Geológico-Mineras, Ministerio de Minas y Energía, Bogotá, scale 1:1,500,000.

Beets, D. J., Maresch, W. V., Klaver, G. Th., Mottana, A., Bocchio, R., Beunk, F. F., and Monen, H. P., 1984, Magmatic rocks series and high pressure metamorphism as constraints on the tectonic history of the southern Caribbean, *in* Bonini, W. E., and others, eds., The Caribbean–South American plate boundary and regional tectonics: Geological Society of America Memoir 162 (this volume).

Bell, J. S., 1972, Geotectonic evolution of the southern Caribbean area, *in* Shagam, R., and others, eds., Studies in Earth and space sciences: Geological Society of America Memoir 132, p. 369–386.

Bellízzia G., A., 1972, Is the entire Caribbean Mountain belt of northern Venezuela allochthonous?, *in* Shagam, R., and other eds., Studies in Earth and space sciences: Geological Society of America Memoir 132, p. 363–368.

Bellízzia, G. A., Pimentel M., N. R., and Bajo O., R., compilers, 1976, Mapa geológico estructural de Venezuela: Venezuela Ministerio de Minas e Hidrocarburos, scale 1:500,000.

Benedetto, G., and Odreman R., O., 1977, Nuevas evidencias paleontológicas en la Formación La Quinta, su edad y correlación con las unidades aflorantes en la Sierra de Perijá y Cordillera Oriental de Colombia, *in* Memoria, Congreso Geológico Venezolano, V, Caracas, Volume 1: Venezuela Ministerio de Energía y Minas, p. 87–106.

Birch, F., 1950, Flow of heat in the Front Range Colorado: Geological Society of America Bulletin, v. 61, p. 567–630.

Bonini, W. E., Garing, J. D., and Kellogg, J. N., 1980, Late Cenozoic uplifts of the Maracaibo–Santa Marta block, slow subduction of the Caribbean plate and results from a gravity study, *in* Transactions, Caribbean Geological Conference, IX, Santo Domingo, 1980; Universidad Católica Madre y Maestra, Santiago de los Caballeros, República Dominicana, v. 1, p. 99–105.

Bowen, J. M., 1972, Estratigrafia del Precretáceo en la parte norte de la Sierra de Perijá, *in* Memoria, Congreso Geológico Venezolano, IV, Caracas, 1969: Venezuela Ministerio de Minas e Hidrocarburos, Boletín de Geología, Publicación Especial no. 5, v. 2, p. 729–761.

Burke, K., 1976, Development of grabens associated with the initial ruptures of the Atlantic Ocean: Tectonophysics, v. 36, p. 93–112.

Campbell, C. J., 1968, The Santa Marta wrench fault of Colombia and its regional setting, *in* Saunders, J. B., ed., Transactions, Caribbean Geological Conference, IV, Port-of-Spain, Trinidad and Tobago, 1965: Caribbean Printers, Arima, Trinidad and Tobago, p. 247–261.

Canelón, G., Etchart, H., and Maze, W., 1979, Mapa geológico de la Hoja Laberinto: Ministerio de Energía y Minas, Venezuela, scale 1:100,000.

Case, J. E., and Holcombe, T. L., 1980, Geologic-tectonic map of the Caribbean region: U.S. Geological Survey Miscellaneous Investigation Map I-1100, scale 1:2,500,000.

Case, J. E., and MacDonald, W. D., 1973, Regional gravity anomalies and crustal structure in northern Colombia: Geological Society of America Bulletin, v. 84, p. 2905–2916.

Cediel, F., 1968, El Grupo Girón; una molasa Mesozoica de la Cordillera Oriental: Colombia Instituto Nacional de Investigaciones Geológico-Mineras, Ministerio de Minas y Petróleos, Bogotá, v. 16, no. 1-3, p. 5–96.

Dasch, L. E., 1982, U-Pb geochronology of the Sierra de Perijá, Venezuela [M.S. thesis]: Cleveland, Ohio, Case Western Reserve University, 164 p.

Duque-Caro, 1979, Major structural elements and evolution of northwestern Colombia, *in* Watkins, J., and Montadert, L. eds., Geological and geophysical investigations of continental margins: American Association of Petroleum Geologists Memoir 29, p. 329–351.

Espejo, A. C., Canelón, G., and Etchart, H. L., 1979, Mapa geológico de la hoja Tule: Ministerio de Energía y Minas, Venezuela, scale 1:100,000.

Espejo, A. C., Etchart, H. L., Cordani, U. G., and Kawashita, K., 1980, Geo-cronología de intrusivas acidas en la Sierra de Perijá Venezuela: Républica de Venezuela, Ministerio de Energía y Minas, Bulletín de Geología, v. 14, p. 245–254.

Fleischer, R. L., Price, R. B., and Walker, R. M., 1965, Effects of temperature, pressure, and ionization on the formation stability of fission tracks in minerals and glasses: Journal of Geophysical Research, v. 70, p. 1497–1502.

Feo-Codecido, G., 1972, Breves ideas sobre la estructura de la falla de Oca, Venezuela, *in* Cecily Petzall, ed., Transactions, Caribbean Geological Conference, VI, Margarita, Venezuela, 1971: Impreso por CROMOTIP, Caracas, Venezuela, p. 184–190.

Giegengack, R. F., 1984, Late Cenozoic tectonic environments of the central Andes, Venezuela, *in* Bonini, W. E., and others, eds., The Caribbean–South American plate boundary and regional tectonics: Geological Society of America Memoir 162 (this volume).

Goldsmith, R., Marvin, R. F., and Mehnert, H. H., 1971, Radiometric ages in the Santander Massif, eastern Cordillera, Colombian Andes: U.S. Geological Survey Professional Paper 750-D, p. D44–D49.

González de Juana, C., Iturralde de Arozena, J. M., and Picard, C., X., 1980, Geología de Venezuela y de sus cuencas petrolíferas: Caracas, Foninves, 1031 p.

Harding, T. P., and Lowell, J. D., 1979, Structural styles, their plate-tectonic habitats, and hydrocarbon traps in petroleum provinces: American Association of Petroleum Geologists Bulletin, v. 63, p. 1016–1058.

Harding, T. P., Gregory, R. F., and Stephans, L. H., (1984), Convergent wrench fault and positive flower structure, Ardmore Basin, Oklahoma, *in* Bally, A. W., ed., Atlas of seismic and structural styles: American Association of Petroleum Geologists Bulletin (in press).

Hedberg, H. D., and Sass, L. C., 1937, Synopsis of the geological formations of the western part of the Maracaibo Basin, Venezeula: Venezuela Ministerio de Fomento, Boletín de Geología y Minería, English edition, v. 1, nos. 2-4, p. 73–112.

James, H. E., Jr., 1977, Sedimentology of the iron-oxide-bearing upper Miocene(?) Guayabo Group in the vicinity of Cúcuta, Colombia [Ph.D. thesis]: Princeton, New Jersey, Princeton University, 214 p.

Jaramillo M., J. M., 1978, Determinación de las edades de algunas rocas de la Cordillera Central de Colombia por el método de huellas de fisión, *in* Resumenes, Congreso Colombiano de Geología II, Bogotá: Colombia Instituto Nacional de Investigaciones Geológico-Mineras, p. 19–20.

Jordan, T., 1955, The present-day motions of the Caribbean plate: Journal of Geophysical Research, v. 80, p. 4433–4439.

Julivert, M., 1970, Cover and basement tectonics of the Cordillera Oriental of Colombia, South America, and a comparison with some other folded chains: Geological Society of America Bulletin, v. 81, p. 3623–3646.

Julivert, M., and others, 1968, Colombia (Précambrien, Paléozoique, Mésozoique et intrusions d'âge Mésozoique-Tertiaire): Léxique Stratigraphique International, v. 5, 651 p.

Kellogg, J. N., 1981, The Cenozoic basement tectonics of the Sierra de Perijá, Venezuela and Colombia [Ph.D. thesis]: Princeton, New Jersey, Princeton University, 247 p.

———1984, Cenozoic tectonic history of the Sierra de Perijá and adjacent basins, *in* Bonini, W. E., and others, eds., The Caribbean–South American plate boundary and regional tectonics: Geological Society of America Memoir 162 (this volume).

Kellogg, J. N., and Bonini, W. E., 1982, Subduction of the Caribbean plate and basement uplifts in the overriding South American plate: Tectonics, v. 1, p. 251–276.

Kohn, B. P., Shagam, R., Burkley, L. A., and Banks, P. O., 1984, Mesozoic-Pleistocene fission-track ages on rocks of the Venezuelan Andes and their tectonic implications, *in* Bonini, W. E., and others, eds., The Caribbean–South American plate boundary and regional tectonics: Geological Society of America Memoir 162 (this volume).

Léxico Estratigráfico de Venezuela (Segunda Edición), 1970: Venezuela Ministerio de Minas e Hidrocarburos, Direccion de Geología, Boletín de Geología Publicación Especial no. 4; Editorial Sucre, Caracas, Venezuela, 756 p.

MacDonald, W. D., Doolan, B. L., and Cordani, U. G., 1971, Cretaceous–early Tertiary metamorphic K-Ar age values from the south Caribbean: Geological Society of America Bulletin, v. 82, p. 1381–1388.

Macellari, C., 1984, Late Tertiary tectonic history of the Táchira Depression (southwestern Venezuelan Andes), *in* Bonini, W. E., and others, eds., The Caribbean–South American plate boundary and regional tectonics: Geological Society of America Memoir 162 (this volume).

Maresch, W. V., 1974, Plate tectonics origin of the Caribbean Mountain system of northern South America: Discussion and proposal: Geological Society of America Bulletin, v. 85, p. 669–682.

Martín-Bellízzia, C., compiler, 1968, Edades isotópicas de rocas Venezolanos: Venezuela Ministerio de Minas e Hidrocarburos, Boletín de Geología, v. 10, no. 19, p. 356–380.

Maze, W. B., 1980, Geology and copper mineralization of the Jurassic La Quinta Formation in the Sierra de Perijá, northwestern Venezuela, *in* Transactions, Caribbean Geological Conference, IX, Santo Domingo, 1980: Universidad Católica Madre y Maestra, Santiago de los Caballeros, República Dominicana, v. 1, p. 283–294.

——1981, Geology, structure and copper mineralization in the La Quinta Formation between Rio Cogollo and Riecito del Palmar, Sierra de Perijá, Venezuela: Princeton University, Final Project Report to Ministerio de Energía y Minas, Venezuela.

——1984, Jurassic La Quinta Formation in the Sierra de Perijá, northwestern Venezuela: Geology and tectonic environment of red beds and volcanics, *in* Bonini, W. E., and others, eds., The Caribbean–South American plate boundary and regional tectonics; Geological Society of America Memoir 162 (this volume).

Miller, J. B., 1962, Tectonic trends in Sierra de Perijá and adjacent parts of Venezuela and Colombia: American Association of Petroleum Geologists Bulletin, v. 46, p. 1565–1595.

Morgan, W. Jason, 1983, Hotspot tracks and the early rifting of the Atlantic: Tectonophysics, v. 94, p. 123–140.

Moya, E., and Viteri A., E., 1977, Nuevas evidencias sobre magmatismo y metalogénesis en el complejo volcánico-plutónico de El Totumo-Inciarte, Sierra de Perijá, Estado Zulia, *in* Memoria, Congresso Geológico Venezolano, V, Caracas, Volume 2: Venezuela Ministerio de Energía y Minas, p. 693–711.

Muessig, K. W., 1984, Structure and Cenozoic tectonics of the Falcón Basin, Venezuela and adjacent areas, *in* Bonini, W. E., and others, eds., The Caribbean–South American plate boundary and regional tectonics: Geological Society of America Memoir 162 (this volume).

Naeser, C. W., 1979, Fission track dating and geologic annealing of fission tracks, *in* Jager, E., and Hunziker, J. C., eds., Lectures in isotope geology: Berlin, Heidelberg, New York, Springer-Verlag, p. 154–169.

Odin, G. S., Curry, D., Gale, N. H., and Kennedy, W. J., 1982, The Phanerozoic time scale in 1981, *in* Odin, G. S., ed., Numerical dating in stratigraphy: New York, John Wiley & Sons, Ch. 37, p. 957–960.

Odreman R., O., and Benedetto, G., 1977, Paleontología y edad de la Formación Tinacoa, Sierra de Perijá, Estado Zulia, Venezuela, *in* Memoria, Congreso Geológico Venezolano, V., Caracas, Volume 1: Venezuela Ministerio de Energía y Minas, p. 15–32.

Pimentel M., N. R., 1975, Falla de Oca: Isla de Toas y San Carlos (Guide to Excursion no. 3, 2nd part), *in* Memoria, Congreso Latinoamericano de Geología, II, Caracas, 1973: Venezuela Ministerio de Minas e Hidrocarburos, Boletín de Geología, Publicación Especial no. 7, v. 1, p. 326–338.

Renz, O., 1977, The lithologic units of the Cretaceous in western Venezuela, *in* Memoria, Congreso Geológico Venezolano, V, Caracas, Volume 1: Venezuela Ministerio de Energía y Minas, p. 45–58.

Rod, E., 1956, Strike-slip faults of Northern Venezuela: American Association of Petroleum Geologists Bulletin, v. 40, p. 457–476.

Spera, F., 1980, Thermal evolution of plutons: A parameterized approach: Science, v. 207, p. 299–301.

Stephan, J. F., 1977, El contacto cadena Caribe-Andes Merideños entre Carora y El Tocuyo (Estado Lara), *in* Memoria, Congreso Geológico Venezolano, V, Caracas, Volume 2: Venezuela Ministerio de Energía y Minas, p. 789–815.

Tschanz, C. M., Marvin, R. F., Cruz, B., J., Mehnert, H. H., and Cebula, G. T., 1974, Geologic evolution of the Sierra Nevada de Santa Marta, northeastern Colombia: Geological Society of America Bulletin, v. 85, p. 273–284.

Van Houten, F. B., and James, H. E., 1984, Late Cenozoic Guayabo delta complex in southwestern Maracaibo basin, northeastern Colombia, *in* Bonini, W. E., and others, eds., The Caribbean–South American plate boundary and regional tectonics: Geological Society of America Memoir 162 (this volume).

Vásquez, E., and Dickey, P., 1972, Major faulting in northwestern Venezuela and its relation to global tectonics, *in* Cecily Petzall, ed., Transactions, Caribbean Geological Conference, VI, Margarita, Venezuela, 1971: Impreso por CROMOTIP, Caracas, Venezuela, p. 191–202.

Wagner, G. A., 1979, Correction and interpretation of fission-track ages, *in* Jäger, E., and Hunziker, J. C., eds., Lectures in isotope geology: Berlin, Heidelberg, New York, Springer-Verlag, p. 170–177.

Wagner, G. A., Reimer, G. M., and Jäger, E., 1977, Cooling ages derived by fission track, mica Rb-Sr and K-Ar dating: The uplift and cooling history of the central Alps: Memorie degli Istituti di Geologia e Mineralogia dell' Università di Padova, v. 30, p. 1–27.

Ward, D. E., Goldsmith, R., Jimeno V., A., Cruz B., J., Restrepo, H., and Gomez R., E., 1969, Mapa geológico del cuadrángulo H-12 (Bucaramanga): Colombia Insituto Nacional de Investigaciones Geológico-Mineras, Ministerio de Minas y Petróleos, Bogotá, scale 1:100,000.

Ward, D. E., Goldsmith, R., Cruz B., J., Jaramillo C., L., and Vargas I., R., 1970, Mapa geológico del cuadrángulo H-13 (Pamplona): Colombia Instituto Nacional de Investigaciones Geológico-Mineras, Ministerio de Minas y Petróleos, Bogotá, scale 1:100,000.

Ward, D. E., Goldsmith, R., Cruz, B., and Restrepo A., H., 1973, Geología de los cuadrángulos H-12 Bucaramanga y H-13 Pamplona, Departamento de Santander: Colombia Instituto Nacional de Investigaciones Geológico-Mineras, Ministerio de Minas y Petróleos, Bogotá, Boletín Geológico, v. 21, no. 1-3, p. i–132.

Zambrano, E., Vásquez, E., Duval, B., Latreille, M., and Coffinières, B., 1971, Síntesis paleográfica y petrolera del occidente de Venezuela, *in* Memoria, Congreso Geológico Venezolano, IV, Caracas, 1969: Venezuela Ministerio de Minas e Hidrocarburos, Publicación Especial no. 5, v. 1, p. 483–552.

Manuscript Accepted by the Society September 1, 1983

Geological Society of America
Memoir 162
1984

K-Ar hornblende ages from the El Chacao Complex, north-central Venezuela

E. H. Hebeda
E. A. Th. Verdurmen
H.N.A. Priem
Z.W.O. Laboratorium voor Isotopen-Geologie
De Boelelaan 1085
1081 HV Amsterdam
The Netherlands

ABSTRACT

Three hornblendes from the unmetamorphosed El Chacao ultramafic complex intrusive into the Villa de Cura Group yield K-Ar ages between about 98 and 107 Ma, which sets a minimum age to the intrusion.

The El Chacao Complex is a fresh, unmetamorphosed Alaskan-type, zoned, ultramafic complex intruding metavolcanic and related rocks of the Villa de Cura Group in the Serrania del Interior of Venezuela (Murray, 1972, 1973). On the basis of the age of approximately 90 Ma reported by Piburn (1967) for the overlying Tiara volcanics, the Villa de Cura Group has been presumed to be of Early Cretaceous–Jurassic(?) age and to constitute an allochthonous slab emplaced on the foreland of South America. Paleomagnetic study of the El Chacao Complex and related intrusions elsewhere in Venezuela (Skerlec and Hargraves, 1980) revealed that the entire belt had been rotated approximately 90°. A Paleocene reef complex (Morros of San Juan) overlying the Villa de Cura Group shows no evidence of rotation (Hargraves and Skerlec, 1980), which constrains the timing of the tectonic emplacement: the Villa de Cura belt, at least 90 Ma in age, was formed, metamorphosed to blue schists, and then intruded by the El Chacao Complex before the tectonic emplacement, which was completed by Paleocene time.

In order to constrain precisely the timing of these events, K-Ar hornblende dating has been undertaken on three samples from the El Chacao Complex. The samples, from the thesis collection of C. G. Murray and the paleomagnetic sampling collection of G. M. Skerlec and R. B. Hargraves, are a quartz gabbro (ANT 182, Princeton Nr. 273-1), a hornblendite (ANT 183, Princeton Nr. GV-1269), and a biotite-plagioclase hornblendite (ANT 184, Princeton Nr. EL 7-2). The K contents were determined by flame photometry after an ion-exchange procedure to

separate K from the bulk of the other cations, with a Li internal standard and Cs-Al buffer. Argon was extracted in a bakeable glass vacuum apparatus and determined by isotope dilution under static conditions in a Varian GD-150 mass spectrometer. All analyses were made in duplicate. The analytical accuracy is thought to be within 1.5% for K and 2% for radiogenic Ar; these estimated overall limits of relative error are the sum of the known sources of possible systematic error and the precision of the total analytical procedures. The analytical data and calculated K-Ar ages are shown in Table 1.

In view of the rather high closure temperature of hornblende, approximately 550 to 490 °C (Hart and others, 1968; Andriessen, 1978), these ages can be taken as closely approximating the time of intrusion of the El Chacao Complex. Thus, the complex is mid-Cretaceous in age, further constraining the age of the Villa de Cura Group, which it intrudes.

TABLE 1. K-Ar DATA

	K (% Wt)	Radiogenic ^{40}Ar (ppb)	Atmospheric ^{40}Ar (% total ^{40}Ar)	Calculated age (Ma)
ANT 182	0.327	2.48	43	
	0.327	2.50	43	106.6 $\pm$ 3.5
ANT 183	0.343	2.42	44	
	0.343	2.43	46	99.2 $\pm$ 3.5
ANT 184	0.382	2.66	42	
	0.382	2.64	38	97.2 $\pm$ 3.5

ACKNOWLEDGMENTS

We are indebted to R. B. Hargraves, Princeton University, for providing the samples and the geological setting, and to P.A.M. Andriessen, N.A.I.M. Boelrijk, and R. H. Verschure for indispensable help in the laboratory. This work forms part of the research program of the "Stichting voor Isotopen-Geologisch Onderzoek," supported by the Netherlands Organization for the Advancement of Pure Research (Z.W.O.).

REFERENCES CITED

Andriessen, P.A.M., 1978, Isotopic age relations within the polymetamorphic complex of the island of Naxos (Cyclades, Greece): Verh. Nr. 3, Z.W.O. Laboratorium voor Isotopen-Geologie, Amsterdam (Ph.D. thesis, University of Utrecht), 60 p.

Hargraves, R. B., and Skerlec, G. M., 1980, Paleomagnetism of some Cretaceous-Tertiary igneous rocks on Venezuelan offshore islands, Netherlands Antilles, Trinidad and Tobago, *in* Transactions, Caribbean Geological Conference, IX, Santo Domingo, Volume 2: Universidad Catolica Madre y Maestra, Santiago de los Caballeros, Republica Dominicana, p. 509–517.

Hart, S. R., Davis, G. L., Steiger, R. H., and Tilton, G. R., 1968, A comparison of the isotopic mineral age variations and petrological changes induced by contact metamorphism, *in* Hamilton, E. I., and Farquhar, R. M., eds., Radiometric dating for geologists: Interscience Publications, New York, p. 73–110.

Murray, C. G., 1972, Petrologic studies of zoned ultramafic complexes in Venezuela and Alaska [Ph.D. thesis]: Princeton University, Princeton, New Jersey, 188 p.

Murray, C. G., 1973, Zoned ultramafic complexes of the Alaskan type: Feeder pipes of andesitic volcanoes: Geological Society of America Memoir 132, p. 313–335.

Piburn, M. D., 1967, Metamorphism and structure of the Villa de Cura Group, northern Venezuela, [Ph.D. thesis]: Princeton University, Princeton, New Jersey, 148 p.

Skerlec, G. M., and Hargraves, R. B., 1980, Tectonic significance of paleomagnetic data from northern Venezuela: Journal of Geophysical Research, v. 85, p. 5303–5315.

Manuscript Accepted by the Society September 1, 1983

Geological Society of America
Memoir 162
1984

Results and preliminary implications of sixteen fission-track ages from rocks of the western Caribbean Mountains, Venezuela

B. P. Kohn
R. Shagam
Ben-Gurion University of the Negev
Beer Sheva, Israel

T. Subieta
FUNVISIS
Apartado 1892
Caracas 1001, Venezuela

ABSTRACT

Two rocks of the Tinaco Complex (one here so postulated on lithologic grounds) give zircon ages of 49 ± 6 m.y. and 42.7 ± 5.3 m.y. (mean of two ages). Eleven samples of Peña de Mora gneiss (Caracas Group) and Choroní-Tovar "granite" (gneiss) give zircon ages that cluster closely about a mean of 19.7 ± 2.1 m.y. The bimodal zircon ages are explained by a speculative model involving obduction of the Tinaco Complex over the Caracas Group followed by isostatic rebound after southward passage of the Villa de Cura klippe, which set the older ages. In turn, further block uplift during eastward translation of the Caribbean relative to the South American plate led to the setting of the younger group of zircon ages.

A single sphene age of 126 ± 15 m.y. on Choroní "granite" (gneiss) is best viewed as a partially annealed remnant of a Paleozoic(?) clock in Sebastopol gneiss component of the "granite." One apatite separate from Tinaco Complex trondhjemite yields an age of 6.1 ± 1.3 m.y., reflecting a phase of uplift also prominent in the circum-Maracaibo Basin ranges to the west.

INTRODUCTION

A trial survey of fission-track ages in the Caribbean Mountains of north-central Venezuela was indicated when similar studies by Kohn and others (1984) and Shagam and others (1984) on rocks from the circum-Maracaibo Basin ranges provided data useful for tectonic interpretation. The samples were collected by two of the writers (Subieta and Shagam) in October and November 1980 (samples prefixed "SVC" in Table 1) and October 1981 (samples "SV-81"). Sample localities are shown in the map of Figure 1 and the corresponding stratigraphic column in Figure 2. Mineral separations were done at Ben-Gurion University, Israel, and measurement of ages was done by Kohn,

partly there and partly at the University of Pennsylvania, Philadelphia.

Classical views of the orogenic evolution of the Caribbean Mountains in the framework of total autochthony have given way with virtual unanimity to that of allochthony related to collision between the South American plate and an ancestral "Caribbean" plate, resulting in obduction of klippen derived from the latter onto the craton. Still, there is wide divergence of opinion as to the nature, sequence, and timing of obduction. Two extreme views, among others, are those of Bell (1972), who viewed the Villa de Cura Group and related volcanic rocks

415

TABLE 1.   FISSION-TRACK DATA AND AGES FOR APATITE, ZIRCON AND SPHENE FROM THE WESTERN CARIBBEAN MOUNTAINS

Sample No.	Map Location No.	Rock Unit	Elevation (m)	ρ_s cm^{-2}	ρ_i cm^{-2}	ϕ x 10^{15} n·cm^2	* Number of fields or grains	AGE (m.y.) ($\pm$ 2σ)
				A P A T I T E				
SVC-56	1	Tinaco trondhjemite	~450	1.65 x 10^4 (116)[†]	3.42 x 10^5 (564)[†]	2.06 (1999)[†]	260/122	6.1 $\pm$ 1.3
				Z I R C O N				
SV-81-4	2	Peña de Mora Gneiss	750	2.42 x 10^6 (800)	7.17 x 10^6 (1185)	0.912 (2416)	6	18.9 $\pm$ 1.9
SV-81-6	3	Peña de Mora Gneiss	950	3.12 x 10^6 (1706)	7.72 x 10^6 (2112)	0.965 (2416)	6	23.9 $\pm$ 1.7
SVC-7	4	Peña de Mora Gneiss	~20	1.45 x 10^6 (354)	1.18 x 10^7 (1437)	2.30 (4348)	6	17.4 $\pm$ 2.1
SVC-8	5	Peña de Mora Gneiss	~20	1.10 x 10^6 (610)	3.28 x 10^6 (914)	0.913 (2416)	8	18.8 $\pm$ 2.1
SVC-11	6	Peña de Mora Gneiss	1300	9.82 x 10^5 (584)	7.29 x 10^6 (2168)	2.23 (2750)	15	18.4 $\pm$ 1.9
SVC-13	7	Peña de Mora Gneiss	2130	3.24 x 10^6 (810)	1.10 x 10^7 (1374)	0.969 (2416)	6	17.5 $\pm$ 1.7
SVC-15	8	Choroní Granite/gneiss	1600	1.67 x 10^6 (356)	9.77 x 10^6 (1044)	2.30 (4348)	6	24.1 $\pm$ 3.0
SVC-17	9	Choroní Granite/gneiss	1500	2.26 x 10^6 (804)	5.81 x 10^6 (1033)	0.936 (2416)	10	22.3 $\pm$ 2.3
SVC-19	10	Choroní Granite/gneiss	1200	1.79 x 10^6 (731)	1.44 x 10^7 (2937)	2.30 (4348)	9	17.5 $\pm$ 1.9
SVC-20	11	Choroní Granite/gneiss	1100	2.47 x 10^6 (464)	6.35 x 10^6 (548)	0.918 (2416)	6	21.9 $\pm$ 2.9
SVC-24	12	Tovar Granite/gneiss	1900	1.46 x 10^6 (434)	5.08 x 10^6 (756)	0.928 (2416)	6	16.4 $\pm$ 2.1
SVC-27	13	Tinaco gneiss	~30	2.69 x 10^6 (541)	7.73 x 10^6 (719)	2.30 (4348)	8	49.0 $\pm$ 5.8
SVC-48	14	Tinaco Diorite	600	1.26 x 10^6 (550)	3.33 x 10^6 (730)	1.81 (4270)	9	41.9 $\pm$ 4.9
		Tinaco Diorite	600	1.75 x 10^6 (418)	5.68 x 10^6 (679)	2.30 (4348)	7	43.4 $\pm$ 5.6
				S P H E N E				
SVC-17	9	Choroní Granite/gneiss	1500	2.60 x 10^6 (618)	2.28 x 10^6 (555)	1.81 (4270)	5	126 $\pm$ 15

* Different numbers of fields of view counted in determining the fossil track density (first number) and the induced track density (second number). Single number indicates number of grains used in age determination (by external detector method).

† Numbers in parentheses are tracks counted to determine the reported track density (or the neutron flux).

(Fig. 2) as the principal allochthonous element overriding the Caracas Group of the autochthon, and Bellízzia (1972), who suggested that the entire Caribbean Mountain System might be allochthonous. Far more elaboration of the obduction processes has been provided by recent workers who include Maresch (1974), Stephan (1977), Beck (1978), Stephan and others (1980), Beets and others (1984), and Subieta (1983). For all the disagreements in detail, certain elements of orogenic evolution are common to most interpretations. The last (only?) cycle of orogeny is perceived to have begun as a result of plate collision in Late Cretaceous time and to have continued through various phases into Neogene time. The structural concordance of the Villa de Cura Group on the Maestrichtian Paracotas Formation (Fig. 2) has been interpreted to signify that the obduction process was in operation by the close of the Cretaceous. Imbricate southward thrusting affecting Paleocene–early Eocene flysch and related sediments in a foredeep on the southern flanks of the ranges is viewed in the framework of continued southward movements of klippen (though whether or not by gravity sliding is argued). Post-Eocene tectonism involved uplift of the ranges with complementary subsidence of graben (pullaparts?), for example, the Tuy basin (Fig. 1) which was filled with terrestrial mid-Miocene–Pliocene sediments; some further minor thrusting southward over the foreland occurred in late Miocene-Pliocene time. Most workers (e.g., Muessig, 1984) relate the changing tectonic style in late- and post-Eocene time to subparallel eastward translation of the Caribbean plate with respect to the South American plate.

Inevitably small discrepancies must be expected where complex terranes are represented on small-scale maps such as that of Bellízia and others (1976). In two cases unavoidable simplifi- cations of the map have serious consequences for the interpreta-

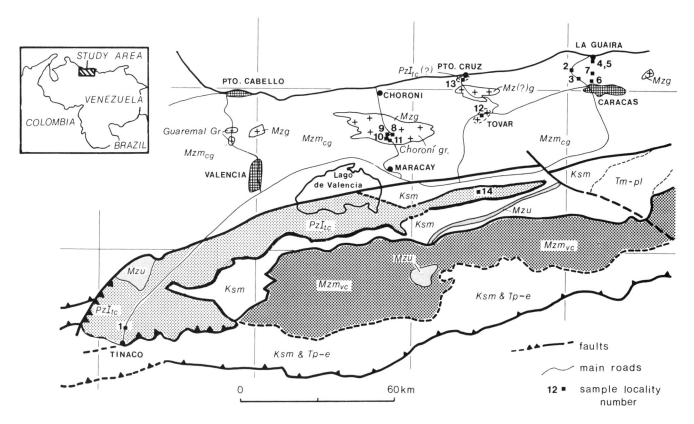

Figure 1. Geologic map of western Caribbean Mountains (simplified after Bellízzia and others, 1976) showing localities sampled for dating by the fission-track method. Key to rock units appears in Figure 2. Specimen numbers corresponding to the map location numbers shown here are given in Table 1.

tion of the fission-track data. One concerns sample SVC-27 (map location 13) which was collected in a river bed near the coast at Puerto Cruz, in an area correlated with the Las Brisas Formation on the map, but which appeared to Subieta and Shagam to be lithologically indistinguishable from typical gneisses of the Tinaco Complex. The regional extent of these outcrops is not known, and they are arbitrarily shown in Figure 1 as constituting a local fault-bound sliver of the Tinaco Complex. Another simplification concerns the map representation of Late Cretaceous granitic plutons ("intrusivas ácidas" on the map legend). Morgan (1967, p. 57) noted that the map representation of such plutons had been exaggerated by failure to exclude some basement gneisses (Sebastopol Gneiss), and further, that the contact between granite and gneiss is obscured by wide areas of migmatite and coarse porphyroblastic microcline gneisses. All the Choroní and Tovar "granite" localities sampled by the writers are moderately to pronouncedly gneissic, and there is hence doubt concerning their proper correlation. The significance thereof is explained below. Even where sample lithologies permit correlation with the formations shown in Figure 2, gross uncertainties remain concerning the correct stratigraphic order and relationships of major rock units because they may have been shuffled in the course of obduction.

FISSION-TRACK AND OTHER PERTINENT ISOTOPIC AGES

Kohn and others (1984) describe the methods used to obtain the fission-track ages, their temperature significance, and possibilities of interpretation. All ages, except for apatite, were determined using the external detector method; the measurement data and calculated ages appear in Table 1.

Because apatite and zircon have closure temperatures that are lower (~100 °C and 200 ± 50 °C, respectively) than those for minerals dated by the K-Ar and Rb-Sr methods, the last two can provide useful data on the minimum age of last annealing of those two fission-track clocks and also provide a cross-check of the fission-track ages in that they should give older ages. Sphene, however, with a closure temperature of ~250 ± 50 °C, may give ages concordant with K-Ar and Rb-Sr on coexisting biotites.

Santamaría and Schubert (1974) reported two K-Ar (biotite) ages, both 30 ± 2 m.y. on the Choroní granite (Fig. 1) and three ages (all biotite) of 30, 31, and 32 m.y. each ± 2 m.y. on the Guaremal granite some 50 km farther west. Morgan (1967) referred to a K-Ar (biotite) age of 33 ± 3 m.y. on the Guaremal granite and a Rb-Sr age, on the same biotite, of 79 ± 5 m.y. González de Juana and others (1980, p. 309) referred to a Rb-Sr

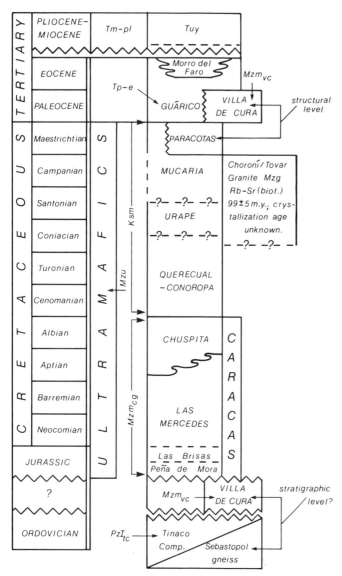

Figure 2. Simplified stratigraphic column of the Caribbean Mountains (after Correlation Chart I, on geologic map of Bellízzia and others, 1976, with minor modifications). Uncertainties in age ranges of formations and even the stratigraphic order of major units stem from complex allochthonous emplacement of large slabs.

model age of 425 m.y. reported by Hess on the Sebastopol gneiss. No isotopic data are available on the pyroxene diorite of the Tinaco Complex, but most workers would probably agree that its minimum age of original crystallization is Mesozoic. Martín B. (1968) compiled K-Ar ages of 112 ± 3 m.y. and 117 ± 3 m.y. on the Tinaco Complex gneiss of the main outcrop belt (Fig. 1), but no analytical details are known. No isotopic ages are available for the Peña de Mora gneisses which attained the epidote amphibolite facies of regional metamorphism, but most workers are agreed that the only (or last?) regional metamorphism had occurred by latest Cretaceous time. Beets and others (1984) suggest specifically the range of Coniacian-Campanian time.

The age data outlined above indicate that apatite and zircon

of the unequivocal magmatic fraction of the Choroní "granite," and presumably of the immediately adjacent Sebastopol gneiss and Caracas Group rocks, were completely annealed ~30 to 33 m.y. ago. The gneissic (Sebastopol) component of the Choroní "granite" that may have been unaffected by the thermal event of magmatic intrusion could have experienced considerably older ages of last annealing of those clocks. Rough extrapolation of the K-Ar ages suggests that the clocks of the pyroxene diorite may have been annealed ~112 m.y. ago. Peña de Mora gneiss samples must have had their fission-track clocks completely annealed in Coniacian-Campanian time, but when the minimum age of complete annealing occurred cannot be specified. Beets and others (1984) outline the complexities of determining paleogeothermal history in a zone of plate collision (see their Fig. 6 and related text); it is most logical to infer that at least the apatite and zircon clocks were least annealed about the same time as the Choroní-Guaremal granites (~30–33 m.y. ago).

In view of the large age discordance ($\sim10^7$–10^8 m.y.) between the fission-track ages (Table 1) and ages of original magmatic crystallization or high grade regional metamorphism, it is likely that the fission-track ages do not reflect points on a simple straight-line cooling curve. They are more reasonably interpreted in terms of cooling related to uplift and erosion of cover rocks. No modern heat-flow data are available for the area of the ranges, and in view of the evidence for allochthonous emplacement of deformed metamorphosed rock units, the nature and thickness of cover section over the samples area for a given time span during orogeny cannot be specified.

DISCUSSION

Apatite and Sphene Ages

Tectonic interpretation of single ages in such a complex terrane can have little value, but indirect implications of those ages are worth brief mention. The one apatite age (6.1 ± 1.3 m.y. on sample SVC-56, map location 1) is especially of interest in that it was the only one of 32 samples collected that yielded apatite in amount and/or U-content adequate for measurement. This is in stark contrast to the abundant apatite suitable for dating obtained from small hand specimens of crystalline rocks of the Venezuelan Andes. There is here a subtle indication that rocks of the Tinaco Complex differ from typical autochthonous craton and may hence also constitute an obducted klippe. This is supported by the trondhjemitic composition of some plutonic rocks (Menéndez, 1967) and the paleomagnetic evidence for large rotations presented by Skerlec and Hargraves (1980). The specific apatite age, corresponding approximately to the Miocene-Pliocene boundary, conforms with abundant evidence for uplift at that time in the circum-Maracaibo Basin ranges to the west (see Kohn and others, 1984; Shagam and others, 1984).

The indirect implications of the sphene age are no less significant in light of the uncertainty concerning Guaremal-Choroní granite/gneiss relationships. If the sphene comes from granitic component, it implies a minimum earliest Cretaceous age of original crystallization almost certainly requiring a major unconformity in the Caracas Group for which there is no evidence, alternatively implying that the Caracas Group is significantly older than currently believed (Fig. 2). A more likely interpretation in terms of Sebastopol gneiss parentage suggests that the sphene age could represent a partially annealed ("mixed") age, of a clock set originally in pre-Mesozoic time and hence of no true age significance. A third possibility is of interest in view of differences of opinion about the metamorphic history of the Caribbean ranges. Whereas Morgan (1970), Maresch (1974), and Beets and others (1984) postulated a single Late Cretaceous phase of regional metamorphism, Stephan and others (1980) suggested that there were two such phases: one of Late Jurassic–Early Cretaceous age and another of Late Cretaceous age. If the sphene age represents a memory of the earlier event postulated by Stephan and others (1980), one would need to explain why it was not annealed in the later event. Unfortunately, the only higher-temperature isotopic ages on the Choroní granite/gneiss are the K-Ar (biotite) ages of 30 m.y. reported by Santamaría and Schubert (1974), and hence there is no means of checking the validity of the sphene age. The most reasonable interpretation of the scanty data is that Morgan's (1967) Rb-Sr age of ~79 m.y. was on true Guaremal granite component, whereas the sphene age reported here is on Sebastopol gneiss component.

An additional five sphene ages on rocks from the circum-Maracaibo Basin ranges are listed by Kohn and others (1984) and by Shagam and others (1984). They range from ~114 to 159 m.y.; one of the samples is gneiss of the Perijá Series which was probably metamorphosed in pre-Devonian time. The precision of the measurement (114 ± 11 m.y.) overlaps that of the Choroní granite/gneiss sample.

Zircon Ages

It is evident from Table 1 that the principal contribution of this study resides in the zircon ages. One group of 11 ages clusters closely about a mean of 19.7 ± 2.1 m.y. (early Miocene); three ages on two samples of the Tinaco Complex are distinctly older (early to middle Eocene). At first glance the two age groups appear to conform neatly with two elements of orogeny previously outlined. One might explain the Eocene ages on the two samples of the Tinaco Complex in terms of a phase of uplift, erosion, and cooling related to isostatic rebound following the passage southward of the main obducted klippe in the Paleocene–early Eocene. The group of 11 early Miocene ages could be related to regional uplift that occurred in response to the phase of lateral translation between the Caribbean and South American plates. On more careful scrutiny, particularly in view of the fact that the two older (Tinaco Complex) samples geographically bracket the 11 younger ages, the need for a more sophisticated interpretation is patent in view of their supposed stratigraphic relationships (Fig. 2). The writers cannot here provide a unique definitive interpretation of the data, but one example (among many possibilities) may help active researchers in the region to use the data in a more rigorous fashion: In this model the Tinaco Complex is viewed as being tectonically emplaced over the Caracas Group (which includes the Peña de Mora gneiss). Subsequently, the two clocks were set by the isostatic rebound mechanism described above. The thickness of the Tinaco Complex was such that uplift was not sufficient for enough erosion to allow cooling and setting of the zircon clocks of the underlying Peña de Mora gneiss. In the Oligocene-Miocene phase of tectonism, a horstlike block of the Coast Ranges was uplifted, the covering Tinaco Complex was eroded away, and the Peña de Mora clocks were set.

An appealing aspect of the model is its contribution to the solution of long-standing metamorphic problems. Explanation of the blueschist facies metamorphism of the Caracas Group in the Coast Group in the Coast Ranges in terms of a subduction mechanism has been rejected by Maresch (1974) and by Beets and others (1984) in favor of burial by obducted slabs. Piburn (1967) was forced to postulate exaggerated thicknesses and recumbent folding of the Villa de Cura klippe in order to attain the necessary depths of burial indicated by the studies of Morgan (1967, 1970). And if such a klippe did exist, why it did not overprint the greenschist-almandine amphibolite facies metamorphism (Menéndez, 1967) of the Tinaco Complex is left unanswered. The additional presence of significant thickness of the Tinaco Complex helps to explain the blueschist grade of the Caracas Group; in turn, the burial of the Tinaco Complex under a far thinner (4–6 km?) Villa de Cura klippe explains the lack of an obvious metamorphic overprint on the gneisses.

Current workers familiar with the regional geology and the evolution of ideas concerning orogeny in the region may easily pick holes in the model. For example, there may be insufficient Oligocene-Miocene Piedmont deposits (e.g., in the Tuy basin, Fig. 1) to match the gross erosion of the Tinaco Complex proposed in the model. The writers' intention, however, is to exemplify the general rather than the specific kind of analysis that may be undertaken.

Additional implications, whether or not one favors the model, may also be of interest. Presumably, isostatic rebound would progress southward in wave form behind the transiting allochthon. The two Tinaco samples have the correct sense of younging, differ in age by about 6 m.y., and are about 45 km apart, suggesting that the klippe advanced at the rate of ~7 km/m.y. A detailed search for intervening remnants of Tinaco Complex outcrop for dating could be useful in order to check that inference. From Table 1 it is seen that there is no reliable correlation between sample elevation and zircon age. Evidently current relief relationships are the product of a younger phase of uplift most likely to have occurred in latest Miocene-Pliocene (Holocene?) time, of which the single apatite age is the only indirect testimony.

A serious obstacle to the useful interpretation of the fission-track data is the paucity of information concerning the thermal history of the orogen. Because of the effects of erosion superim-

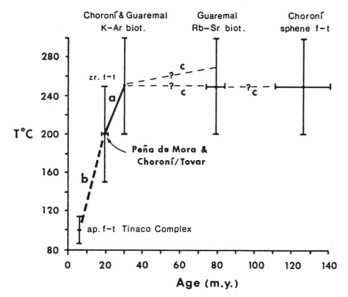

Figure 3. Temperature (*T*)-age plot of available isotopic data to show possible elements of cooling history. Full line a is most reliable curve (narrow time interval; higher and lower temperature data on same Choroní granite complex). Continued cooling to ~100 °C could not have followed a curve of slope significantly different from that of b but lesser accuracy attributed to wide geographic spacing of corresponding samples (see Fig. 1). Lightly dashed curves c implying little or no cooling for 50 to 100 m.y. encompassing the major phase of the orogenic cycle cannot currently be explained.

posed on a complex allochthonous history, as previously noted, it may prove impossible in the foreseeable future to reconstruct paleogeothermal evolution for a given area. However, if the attempt is to be made, the best approach would likely be to date each sample, in a carefully selected network, by as many isotopic clocks as possible. Fragmentary indication of how this might prove useful appears in Figure 3.

The significance of the zircon fission-track ages in the larger regional context is brought out in Fig. 4. The age ranges clearly serve to distinguish the allochthonous from the autochthonous terranes. On the other hand, the small time gap between the oldest ages of the former compared to the youngest of the latter raises for consideration the possibility of some connecting tectonic thread in the evolution of the two regions.

CONCLUDING COMMENTS

In the attempt to define a time reference for the earlier phases of orogeny, numerous efforts (largely futile) have been made to date original crystallization ages of meta-igneous rocks in the Caribbean Mountain system. It is remarkable that there has been no systematic attempt to date para-schists and gneisses, many of which contain minerals amenable to dating by the K-Ar and Rb-Sr methods. The results reported here on the zircons indicate that the corresponding rocks should also be amenable to U-Pb dating. An extensive areal net of such dates used in conjunction with petrologic, paleomagnetic, and structural studies should provide valuable information to resolve major tectonic problems in this complex terrane.

ACKNOWLEDGMENTS

We are extremely grateful to the Dirección de Geología, Ministerio de Energía y Minas, Venezuela, for full logistical support in the sampling program; to C. W. Naeser for kindly supervising the irradiation of our samples at the U.S. Geological Survey, Denver; to the Department of Geological and Geophysical Sciences, Princeton University, for generous supply of facilities and the secretarial and drafting services of J. Bialkowski and C. Rine, to R. Shagam during the writing of the manuscript; and

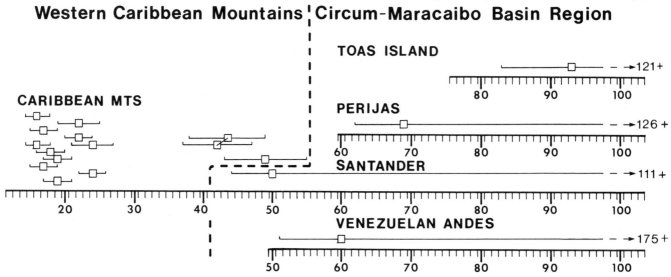

Figure 4. Plot of zircon fission-track ages for the Caribbean Mountains compared to those for the circum-Maracaibo Basin ranges to the west. For the latter region, squares show ages of youngest measurements only; right-hand error bar and numbers show maximum ages measured in each region (for further details see Shagam and others, 1984).

to the Department of Geology, University of Pennsylvania, for the excellent research facilities used by B. P. Kohn for the measurement of most of the fission-track ages.

REFERENCES CITED

Beets, D. J., Maresch, W. V., Klaver, G. Th., Mottana, A., Bocchio, R., Beunk, F. F., and Monen, H. P., 1984, Magmatic rock series and high-pressure metamorphism as constraints on the tectonic history of the southern Caribbean: *in* Bonini, W. E., and others, eds., The Caribbean-South American plate boundary and regional tectonics: Geological Society of America Memoir 162 (this volume).

Beck, C. M., 1978, Polyphase Tertiary tectonics of the Interior Range in the central part of the western Caribbean Chain, Guárico State, northern Venezuela: Geologie en Mijnbouw, v. 57, p. 99-104.

Bell, J. S., 1972, Geotectonic evolution of the southern Caribbean area, *in* Shagam, R., and others, eds., Studies in Earth and space sciences: Geological Society of America Memoir 132, p. 367-386.

Bellízzia, G. A., 1972, Is the entire Caribbean Mountain Belt of northern Venezuela allochthonous?, *in* Shagam, R., and others, eds., Studies in Earth and space sciences: Geological Society of America Memoir 132, p. 363-368.

Bellízzia, G. A., Pimentel, M.N.R., and Bajo, O. R., compilers, 1976, Mapa geológico estructural de Venezuela: Venezuela Ministerio de Minas e Hidrocarburos, scale 1:500,000.

González de Juana, C., Iturralde de Arozena J., M., and Picard, C., X., 1980, Geología de Venezuela y de sus cuencas petrolíferas: Caracas, Foninves, 1031 p.

Kohn, B. P., Shagam, R., Burkley, L. A., and Banks, P. O., 1984, Mesozoic-Pleistocene fission-track ages on rocks of the Venezuelan Andes and their tectonic implications: *in* Bonini, W. E., and others, eds., The Caribbean-South American plate boundary and regional tectonics: Geological Society of America Memoir 162 (this volume).

Maresch, W. V., 1974, Plate tectonics origin of the Caribbean Mountain System of northern South America: Discussion and proposal: Geological Society of America Bulletin, v. 85, p. 669-682.

Martín B., C., compiler, 1968, Edades isotópicas de rocas Venezolanos: Venezuela Ministerio de Minas e Hidrocarburos, Boletín de Geología, v. 10, no. 19, p. 356-380.

Menéndez, A., 1967, Tectonics of the central part of western Caribbean Mountains, Venezuela, *in* Studies in tropical oceanography no. 5: University of Miami, Institute of Marine Science, p. 103-130.

Morgan, B. A., 1967, Geology of the Valencia area, Carabobo, Venezuela [Ph.D. thesis]: Princeton, New Jersey, Princeton University, 220 p.

—— 1970, Petrology and mineralogy of eclogite and garnet amphibolite from Puerto Cabello, Venezuela: Journal of Petrology, v. 11, p. 101-145.

Muessig, K. W., 1984, Structure and Cenozoic tectonics of the Falcón Basin, Venezuela, and adjacent areas: *in* Bonini, W. E., and others, eds., The Caribbean-South American plate boundary and regional tectonics: Geological Society of America Memoir 162 (this volume).

Piburn, M. D., 1967, Metamorphism and structure of the Villa de Cura Group, northern Venezuela [Ph.D. thesis]: Princeton, New Jersey, Princeton University, 148 p.

Santamaría, F., and Schubert, C., 1974, Geochemistry and geochronology of the southern Caribbean-northern Venezuela plate boundary: Geological Society of America Bulletin, v. 85, p. 1085-1098.

Shagam, R., Kohn, B. P., Banks, P. O., Dasch, L. E., Vargas, R., Rodríguez, G. I., and Pimentel, N., 1984, Tectonic implications of Cretaceous-Pliocene fission-track ages from rocks from the circum-Maracaibo Basin region of western Venezuela and eastern Colombia, *in* Bonini, W. E., and others, eds., The Caribbean-South American plate boundary and regional tectonics: Geological Society of America Memoir 162 (this volume).

Skerlec, G. M., and Hargraves, R. B., 1980, Tectonic significance of paleomagnetic data from northern Venezuela: Journal of Geophysical Research, v. 85, p. 5303-5315.

Stephan, J. F., 1977, El contacto cadena Caribe-Andes Merideños entre Carora y el Tocuyo (Estado Lara), *in* Memoria, Congreso Geológico Venezolano, V, Caracas, v. 2: Venezuela Ministerio de Energía y Minas, p. 789-815.

Stephen, J. W., Beck, C., Bellízia, A., and Blanchet, R., 1980, La chaîne Caraibe du Pacifique à l'Atlantique: Bureau de Recherches Géologiques et Minières Mémoire, no. 115, p. 38-59.

Subieta, T., 1983, Estudio tectónica y microtectónico de la Cordillera de la Costa, parte norte-central de Venezuela: Program and Abstracts, 10th, Caribbean Geological Conference, Cartagena, Colombia, August 1983, p. 68.

MANUSCRIPT ACCEPTED BY THE SOCIETY SEPTEMBER 1,

Typeset by WESType Publishing Services, Inc., Boulder, Colorado
Printed in U.S.A. by Malloy Lithographing, Inc., Ann Arbor, Michigan